The Opus Majus of Roger Bacon

(1928)

ALCHYMIA
(From Thurneysser's *Quinta Essentia*, 1570)

Robert Belle Burke

Printed in the United States of America

ISBN 978-1-5128-0999-2

THE
OPUS MAJUS
OF
ROGER BACON

A Translation by

ROBERT BELLE BURKE

PROFESSOR OF LATIN AND DEAN OF THE COLLEGE
UNIVERSITY OF PENNSYLVANIA

VOLUME I

PHILADELPHIA
UNIVERSITY OF PENNSYLVANIA PRESS
LONDON: HUMPHREY MILFORD
OXFORD UNIVERSITY PRESS
MCMXXVIII

FOREWORD

This translation of the Opus Majus *is based on the corrected text of Bridges' edition, London, 1900. Messrs. Williams and Norgate, publishers of Bridges' edition, have kindly granted permission to reproduce the drawings which were copied by Dr. Bridges from the original manuscripts of the* Opus Majus.

I have tried to keep in mind the needs of the reader who desires to follow the Latin text with the aid of a translation, and for this reason I have not attempted as a rule to paraphrase or break up Bacon's sentences.

I gladly take this opportunity to express my deep obligation for aid generously given me in this work by my colleagues, the late Professor W. R. Newbold, Professor Isaac Husik, Professor R. G. Kent, Professor G. H. Hallett, and Professor E. A. Singer, Jr.

ROGER BACON

The Hope-Pinker Statue in the Oxford University Museum

CONTENTS

VOLUME I

ILLUSTRATIONS

VOLUME I

INTRODUCTION

COMPARATIVELY little is known of the life of Roger Bacon. Even the exact year and place of his birth are in doubt. From a statement made by Bacon in the *Opus Tertium* we infer that he was born between 1210 and 1215. From the same source we learn that his family was one of considerable wealth, but had been reduced to utter poverty by its adherence to Henry in his dispute with his barons. He received his collegiate training at Oxford and, as he states in the *Opus Tertium,* devoted more than twenty years to the study of languages and science. "I sought," he writes, "the friendship of all wise men among the Latins; and I caused young men to be trained in languages, in geometrical figures, in numbers, in the construction of tables, in the use of instruments, and in many other necessary things. . . . During this time I spent more than two thousand pounds in those things and in the purchase of books and instruments."

About 1240 Bacon left Oxford and went to the University of Paris, where he received the degree of doctor of theology. It was during his sojourn in Paris that he entered the Franciscan Order about 1247. He subsequently returned to England, but incurring the suspicion of his superiors in the Franciscan Order he was exiled to Paris and placed under restraint in the Paris house.

Guy Fulcodi, who practiced law in Paris for a number of years with great distinction and also served as private secretary to Louis IX, entered the Church after the death of his wife and was elevated to the Papacy in 1265 as Clement IV. While in Paris he had every opportunity to hear of Bacon and his work, and was evidently so favorably impressed that after his elevation to the Papacy he wrote a letter in 1266 to Bacon directing him to transmit copies of all his writings without delay.

As a matter of fact up to this time Bacon had written but little with the exception of a formal treatise *De Multiplicatione Specierum.* Bacon began work at once and composed rapidly the *Opus Majus, Opus Minus,* and *Opus Tertium* and dispatched them together with a copy of the *De Multiplicatione Specierum* to the Pope by the hand of a poor lad named John, whom he had been training for several years. We have no means of determining whether the Pope, who died a few months later, ever received Bacon's works.

In 1271 Bacon published his *Compendium Studii Philosophiae,* in which he vigorously assailed the vice and corruption of his day and in particular the pedantry and false conceit of knowledge rife in the schools. Because of the antagonism thus aroused and the daring novelty of much of his teaching he was finally brought to trial in 1278, condemned, and thrown into prison, where he remained for fourteen years. He was released in 1292, but died shortly afterwards and was buried in the Franciscan Church at Oxford.

The *Opus Majus,* on which Bacon's fame chiefly rests, was written with the purpose of correlating the learning of the thirteenth century and making it available to the Church in its work of elevating and saving mankind. Whewell, in his *Philosophy of the Inductive Sciences,* says: "The *Opus Majus* is a work equally wonderful with regard to its general scheme and to the special treatises with which the outlines of the plan are filled up." Dr. Bridges, in his admirable introduction to the *Opus Majus,* remarks: "The *Opus Majus* remains the one work in which the central thought of Bacon is dominant from first to last: the unity of science and its subordination to the highest ethical purpose conceivable by man."

After several years of study of the *Opus Majus* the translator ventured in a recent article to express his opinion of Bacon's great work in the following words: "In its unity of purpose, in its encyclopaedic range of subjects, in its clarity of statement, in its orderly arrangement of material, in its prophetic scientific vision, in its profound moral earnestness the *Opus Majus* must ever remain one of the few truly great works of human genius." It is certainly within the limits of truth to say that no other great scientific work has ever been inspired by so lofty a purpose as that which ennobled the *Opus Majus* of Roger Bacon.

The *Opus Majus* is divided into seven parts. Part I considers the causes of human error and classifies them as four—Undue regard to Authority, Habit, Popular Prejudice, and False Conceit of Knowledge. Part II shows the close relationship between Philosophy and Theology. Part III emphasizes the importance and utility of the Study of Foreign Languages. Part IV urges the necessity of greater mathematical knowledge, since this science is the key to all other sciences. Its great utility is shown in Astronomy, Optics, Theology, Chronology, Astrology, and the Correction of the Calendar. This section closes with a Treatise on Geography. Part V treats of Optics. It contains a thorough discussion of the General Principles of Vision, Direct Vision, Reflection, and Refraction. Part VI is a plea for Experimental Science and an exposition of the correct method of scientific approach in solving natural problems. Part VII treats of Moral Philosophy, the final and supreme science. It discusses man's relation and duty to God, to his fellows, and to himself, makes the first comparative Study of Religions, and proves the Superiority of the Christian Faith.

Manuscript copies of Bacon's works in more or less mutilated condition are preserved in different parts of Europe, notably in the Mazarin Library in Paris, in the Vatican Library, in the British Museum, in the Bodleian and University College Libraries at Oxford, and at the University of Dublin. The first printed edition of the *Opus Majus* appeared in 1733, edited by Jebb. Bridges' edition with an Introduction was published at the Clarendon Press in 1897. A supplementary volume containing many corrections to the text of the edition of 1897 was published by Dr. Bridges in 1900. The most important contributions in modern times on Bacon and his work are Dr. Bridges' Introduction to the *Opus Majus* and his Analysis in English of its contents, and the masterly treatise of E. Charles, *Roger Bacon, sa Vie, ses Ouvrages, ses Doctrines d'après des textes inédits,* 1861.

Great interest throughout the world was excited some years ago by the announcement of the late Professor W. R. Newbold of the University of Pennsylvania that he had succeeded in deciphering the code in which an unpublished manuscript of Bacon, known as the Voynich Manuscript, is written. At the time of his death Professor Newbold was completing a volume descriptive of his work on the cipher and giving his conclusions. Scholars interested in the work of Bacon will be pleased to learn that this volume will soon appear, edited by Professor Roland G. Kent of the University of Pennsylvania and published by the University of Pennsylvania Press.

THE OPUS MAJUS OF ROGER BACON

PART ONE

Chapter I

A THOROUGH consideration of knowledge consists of two things, perception of what is necessary to obtain it and then of the method of applying it to all matters that they may be directed by its means in the proper way. For by the light of knowledge the Church of God is governed, the commonwealth of the faithful is regulated, the conversion of unbelievers is secured, and those who persist in their malice can be held in check by the excellence of knowledge, so that they may be driven off from the borders of the Church in a better way than by the shedding of Christian blood. Now all matters requiring the guidance of knowledge are reduced to these four heads and no more. Therefore, I shall now try to present to your Holiness the subject of the attainment of this knowledge, not only relatively but absolutely, according to the tenor of my former letter, as best I can at the present time, in the form of a plea that will win your support until my fuller and more definite statement is completed. Since, moreover, the subjects in question are weighty and unusual, they stand in need of the grace and favor accorded to human frailty. For according to the Philosopher in the seventh book of the Metaphysics, those things which in themselves are susceptible of the most perfect cognition are for us objects of but imperfect apprehension. For truth veiled lies hidden in the deep and is placed in the abyss, as Seneca says in the seventh book of his De Beneficiis, and in the fourth of the Quaestiones Naturales; and Marcus Tullius says in the Hortensius that our entire intellect is obstructed by many difficulties, since it is related to those things which are most manifest in their own nature as is the eye of the night-owl or of the bat to the light of the sun, as the Philosopher declares in the second book of the Metaphysics; and as one deaf from his birth is related to the delight of harmony, as Avicenna says in the ninth book of the Metaphysics, wherefore in the investigation of truth the feebleness of our own intellect suffices for us, that we may to the best of our ability put extraneous causes and occasions of error at a farther remove from our weak power of sense perception.

Now there are four chief obstacles in grasping truth, which hinder every man, however learned, and scarcely allow any one to win a clear title to learning, namely, submission to faulty and unworthy authority, influence of custom, popular prejudice, and concealment of our own ignorance accompanied by an ostentatious display of our knowledge. Every man is entangled in these difficulties, every rank is beset. For people without distinction draw the same conclusion from three arguments, than which none could be worse, namely, for this the authority of our predecessors is adduced, this is the custom, this is the common belief; hence correct. But an opposite conclusion and a far better one should be drawn from the premises, as I shall abundantly show by authority, experience, and reason. Should, however, these three errors be refuted by the convincing force of reason, the fourth is always ready and on every one's lips for the excuse of his own ignorance, and although he has no knowledge worthy of the name, he may yet shamelessly magnify it, so that at least to the wretched satisfaction of his own folly he suppresses and evades the truth. Moreover, from these deadly banes come all the evils of the human race; for the most useful, the greatest, and most beautiful lessons of knowledge, as well as the secrets of all science and art, are unknown. But, still worse, men blinded in the fog of these four errors do not perceive their own ignorance, but with every precaution cloak and defend it so as not to find a remedy; and worst of all, although they are in the densest shadows of error, they think that they are in the full light of truth. For these reasons they reckon that truths most firmly established are at the extreme limits of falsehood, that our greatest blessings are of no moment, and our chief interests possess neither weight nor value. On the contrary, they proclaim what is most false, praise what is worst, extol what is most vile, blind to every gleam of wisdom and scorning what they can obtain with great ease. In the excess of their folly they expend their utmost efforts, consume much time, pour out large expenditures on matters of little or no use and of no merit in the judgment of a wise man. Hence it is necessary that the violence and banefulness of these four causes of all evils should be recognized in the beginning and rebuked and banished far from the consideration of science. For where these three bear sway, no reason influences, no right decides, no law binds, religion has no place, nature's mandate fails, the complexion of things is changed, their order is confounded, vice prevails, virtue is extinguished, falsehood reigns, truth is hissed off the scene. Therefore nothing is more necessary of consideration than the positive condemnation of those four errors through the chosen arguments of wise men which shall prove irrefutable. Inasmuch as the wise unite the first three together and condemn them, and since the fourth, owing to its exceptional folly, needs special treatment, I shall first attempt to show the banefulness of the three. But although authority be one of those, I am in no way speaking of that solid and sure authority, which either by God's judgment has been bestowed upon his Church, or which springs from the merit and dignity of an individual among the Saints, the perfect philosophers, and other men of science, who up to the limit of human utility are expert in the pursuit of science; but I am speaking of that authority, which without divine consent many in this world have unlawfully seized, not from the merit of their wisdom but from their presumption and desire of fame—an authority which the ignorant throng concedes to many to its own destruction by the just judgment of God. For according to Scripture "owing to the sins of the people frequently the hypocrite rules"; for I am speaking of the sophistical authorities of the irrational multitude, men who are authorities in an equivocal sense, even as the eye carved in stone or painted on canvas has the name but not the quality of an eye.

Chapter II

THESE three errors sacred Scripture reproves, sainted doctors condemn, canon law forbids, philosophy rebukes; but for reasons previously touched upon in regard to adducing philosophical principles, and since the judgments of philosophers in regard to these three are less widely known, I shall in the first instance adduce those judgments. Seneca indeed condemns all these three banes at once in the book of his Second Epistles near the end in a single statement. He says, "Among the reasons for our evils is the fact that we live according to examples, and are not regulated by reason but influenced by custom. That which if done by few we should not care to imitate, when many begin to do it, we do it also, influenced by numbers more than by higher motives, and an error when it has become general takes for us the place of truth." The Philosopher, in fact, through the whole course of his philosophy attacks unworthy authority and asserts in the second book of the Metaphysics that the chief sources of human error are custom and the influence of the masses. Seneca again in his book on the Happy Life says, "No man errs for himself alone, but he is the cause and author of another's error, and error transmitted from one to another tosses and drives us headlong, and we come to grief by the examples of other men." In his second book on Anger he says because of the evil of custom, "With difficulty are those vices lopped off which have grown up with us." Moreover in his book on Happiness of Life, in opposition to the common opinion he says, "Nothing involves us in greater evils than the fact that we regulate our lives by mere report, reckoning best that which has been so accepted by general consent, and we do not live according to reason but in accordance with our desire to copy others. Hence comes that heaping together of man tumbling over man. For in a great massacre of men, when the mob crowds upon itself, no one falls in such a way as not to drag his neighbor after him, and the first cause those that follow them to stumble into destruction. You may see this happening in every life." He likewise says in the same book, "The mob pits itself against reason in defence of its own bane"; and he adds, "Human affairs are not so well ordered that better counsels please the majority of us," and he continues, "A crowd is the worst of arguments." Marcus Tullius, also, in the third book of the Tusculan Disputations says, "When we have been handed over to school masters, we are not only imbued with divers errors, but truth yields to vanity and firmly established nature herself gives way to mere opinion." Also he says in the Lucullus, "Some complying with a friend or captivated by the mere discourse of some one to whom they have listened, form judgments about matters unknown to them, and driven as it were by stress of weather to some branch of study, no matter what, they cling to it as though to a rock. Many have chosen to remain in error and to defend the opinion to which they are sentimentally attached, rather than without obstinate prepossession to scrutinize that which they assert with the utmost positiveness." Because of the viciousness of custom he asks in the first book on the Divine Nature, "Is not the investigator of nature ashamed to seek testimony to the truth from minds steeped in custom?" Moreover in opposition to the notions of the mob he says in the introduction to the second book of Disputations, "Philosophy is content with few judges, shunning purposely the multitude to which it is an object of suspicion and hatred." In the same second book he says, "I think all things are the more praiseworthy if they occur without the knowledge of the public." But others attack these three errors separately. For in Adalardus' book on Questions about Nature the following query is put about weak authority: "What else is authority of this kind than a halter? Just as brute beasts are led with any kind of halter whithersoever one wishes, and do not perceive whither they are being led nor why; so not a few, captive and bound by beastlike credulity, mere authority leads off into danger." Moreover, in the book on the Eternity of the World he says, "The man who chooses one side of a question because of his love of custom cannot rightly discern the correct opinion." Averroës also at the end of his second book on the Physics says, "Custom is the chief cause hindering us from grasping many clear truths. Just as certain actions though harmful will become easy to the man accustomed to them, and for this reason he comes to believe that they are useful; similarly when one has become accustomed to believe false statements from childhood, the habit so formed will cause him to deny the truth, even as some men have become so used to eating poison that it has become to them a food." Averroës likewise maintains in his commentary on the second book of the Metaphysics that the contraries of the principles, provided they be of general repute, are more gladly received by the multitude and by those who follow the testimony of the majority, than are the fundamental principles themselves." And also Jerome in the introduction to the fifth book of his commentary on Jeremiah asserts that truth is contented with few supporters and is not dismayed by a host of foes. John Chrysostom in his commentary on Matthew says that those who have armed themselves with the multitude, have confessed themselves to be unprotected by truth.

CHAPTER III

The experience of the individual decides any matter which has been proved by authorities with still greater certainty.

FOR we experience in ourselves and in others that these three, involving as they often do evils, are still more frequently connected with the false. But if at any time they are united with the good and the true, the latter are almost always imperfect, and attain only a low rank in science. As a rule the daughter follows the example of her mother, the son the father's, the slave the master's, the porter the king's, the subordinate the superior's, the pupil the master's. Because it is a familiar trait of the sons of Adam to claim for themselves authority and to disseminate in the light their own standards. For all men, according to Aristotle in the fourth book of his Ethics, love their own works, as for example parents their children, poets their measures, and so on through the list. In fact, many have indulged in too great freedom in their writings, and have not hesitated even to insinuate to vicious and brutelike men the thought, Why do you not fill up your pages? Why do you not write even on the back of the sheet? These men are like a lame and blind shepherd with many sheep, which wandering through the by-paths of untruth they have neither the power nor the knowledge to recall to the more healthful pastures of science, and they are like birds desiring to fly without wings, assuming the office of a master before they have acquired the status of a good pupil. These men of necessity fall into so many errors, because lazy people, by comparing themselves with one another, come to reckon themselves lucky; just as when many competitors run a race, he whom despair of winning does not suffer to enter, no matter how precious the prize of victory appears in his eyes, nevertheless reckons himself lucky in comparison with the contestant who as he runs falls into an unseen pitfall. And thus we see with our very eyes that for one example of truth as well in science as in life there are more than a thousand examples of falsehood. For the world is full of examples of this kind, and one example of true perfection easily discovers ten thousand imperfect ones. In numbers nature has formed for us her own illustration of perfection and imperfection. A number is called perfect, the sum of whose aliquot parts exactly equals the number, but there is only one below ten, namely six; and one between ten and one hundred, namely 28; and one between one hundred and a thousand, namely 496; and one between a thousand and ten thousand, namely 8128; and so on. Would that this were the case among men, and it would be sufficient for mankind. But never has this been a fact either in life or science, nor will it be until the final destruction of sin, since not only is there a scarcity of those who are perfect in every virtue and science, but of those who have reached perfection in a single virtue or science. The first are and will be and have always been very few. For they are in fact perfect; since out of ten thousand men one is not found thus perfect either in his way of life or in the matter of wisdom. Would that of the second class of perfection there might be one in ten, and so on with the other classes, so that the perfection of numbers might be maintained in the case of human beings. But it is not so; nay, it is found to be quite the contrary. Similarly concerning custom we prove by experience in our own acts what has just been said with reference to types of individuals. Let any man you please consider his own life from his infancy, and he will find that in the majority of his acts it has been easier for him to reduce to habit what is false and evil. For, in what is good and true, repetition, such is our human frailty, is the mother of satiety, and the wretched man delights himself in the variety of things useful, according to the judgment of the authorities which I mentioned in the beginning; but he is, on the contrary, delighting in things evil and false and harmful to himself and to others. For in most of our actions, unless special grace and divine privilege intervene in the case of some perfect individuals, human corruption diligently persists in what is opposed to truth and salvation; the individual neither feels persistence in sin to be irksome, nor is prone to find vain things contemptible. But if a man from his youth should devote himself to truth in life and science, he in most of his activities persists in his imperfection and takes pleasure therein: perfection, in fact, more frequently saddens him, for it delights very few of us, and especially is this true in the rich fulness of the virtues and sciences; and for this reason youth seldom guards against error, and old age with the greatest difficulty climbs to perfection in anything. Of the mob the same statement holds good. For the great mass of mankind has always erred in regard to God's truth, and only the small body of Christians has received it; moreover, we know that most Christians are imperfect, a fact shown by the small number of saints. The same is true concerning philosophical doctrine; for the mass of mankind has always lacked the wisdom of philosophy. This is shown by the small number of philosophers. In fact, the rank and file of those devoting themselves to philosophy have always remained imperfect. For of famous philosophers Aristotle alone with his followers is stamped with the approval of all wise men, since he organized the branches of philosophy as far as was possible in his age; but nevertheless he did not reach the limit of wisdom, as will be clearly shown below.

Chapter IV

However little authority may be depended upon, it possesses, nevertheless, a name of honor, and habit is more strongly inducive to error than authority; but popular prejudice is more forceful than either of them. For authority merely entices, habit binds, popular opinion makes men obstinate and confirms them in their obstinacy. Habit is a second nature, as the Philosopher says in his book on Memory and Recollection, and in his book of Problems; and for this reason habit brings a greater force to bear than authority. Hence the Philosopher in the tenth book of his Ethics agrees with the sentiment of Jeremiah in regard to the Ethiopian's skin, saying that it is impossible or difficult for him who through habit has grown hardened in evil, to change for the better. Sallust also in his Jugurtha expresses the thought of Solomon when he says, "They spend their old age where they passed their youth." Popular sentiment in fact is far worse than other sentiments. For as Seneca says in the third book of his Declamations, "The multitude once stirred cannot preserve a due moderation." On this point John Chrysostom says in his commentary on Matthew, "They assembled to conquer by numbers him whom they were unable to overcome by their reason." We must consider carefully that the unenlightened throng is not only more energetic in its invitation to evil than the other two sources of error, but is more foolish and further removed from the pale of wisdom. For an example of perfection is drawn upon by an individual to form habit, but it suffices the throng not to sin. For in no rank of the Church is it required that the body at large should possess the perfection of the Church. For even among the religious a small number is fixed at the center of perfection, and the rest wander about on the circumference. Such is the case, as we clearly see, both in the ranks of the secular clergy and of the laity. For as with Moses so with Christ the common throng does not ascend the mountain. Nor in the transfiguration of Christ was the whole body of disciples taken with him, but only the three specially chosen; and when the multitude followed for two years Christ the teacher of perfection as he preached, it afterwards dismissed him, and finally cried out, Crucify him. For the common throng cannot continue anything that is perfect; would that it were avoiding error in life and in pursuit. We see that such is the case among the professors of philosophy as well as in the truth of our faith. For the wise have always been divided from the multitude, and they have veiled the secrets of wisdom not only from the world at large but also from the rank and file of those devoting themselves to philosophy. For this reason wise men of Greece meeting in the watches of the night devoted their leisure, apart from the multitude, to scientific collaboration, of which A. Gellius writes in his book of Attic Nights, the title indicating efforts of this kind made by learned men of Attica, that is Athenians, during the watches of the night in order to avoid the multitude. In this book he says that it is foolish to feed an ass lettuces when thistles suffice him. He is speaking of the multitude for whom rude, cheap, imperfect food of science is sufficient. Nor ought we to cast pearls before swine; for he lessens the majesty of nature who publishes broadcast her mysteries; nor do those matters remain secret of which the crowd is witness, as is shown in the book of Gems. Aristotle also says in his book of Secrets that he would break the celestial seal if he made public the secrets of nature. For this reason the wise although giving in their writings the roots of the mysteries of science have not given the branches, flowers, and fruits to the rank and file of philosophers. For they have either omitted these topics from their writings, or have veiled them in figurative language or in other ways, of which I need not speak at present. Hence according to the view of Aristotle in his book of Secrets, and of his master, Socrates, the secrets of the sciences are not written on the skins of goats and sheep so that they may be discovered by the multitude. For the wisest and most expert frequently find very great difficulty in the books of the ancients. When the Philosopher logically divides probability in his first book of Topics, he separates the multitude from the wise; for he says that that is probable which all men think, or the majority of men, or the wise; for under the term *all men* are included the multitude and the wise alike, therefore by the majority of men is meant the multitude; wherefore the insensate multitude has no partnership with the wise. This results not only from its own folly, but from the fact that in most cases it has a dull, weak head, prone to errors and productive of imperfection, at whose nod it is led in every circumstance, as I have noted in the preceding letter. Therefore, the ignorant multitude is never able to rise to the perfection of wisdom, for it is ignorant of the use of the things of greatest worth; and if by any chance it at length attains them, it turns all into evil, and accordingly by God's just judgment the paths of perfection are denied it; and it attains its greatest good when it is kept from sin. Its own name indicates all that has been said about it; for in all authors it is called the ignorant and insensate multitude. Now ignorance consists in error and imperfection, and therefore error and imperfection are native to the multitude, whose error is more frequent than its wish to perceive, however imperfectly, the truth. For many have been called but few chosen for the reception of the divine truth and of the philosophic as well. For the Philosopher says in the second book of Topics that we must think as the few, although we ought to talk as the many, considering the place and time; because to simulate at times the folly of the multitude is regarded as the highest wisdom, especially when it is in its mad mood. Hence from all these sources are inferred the evil and folly of these three causes of error, and their infinite harm to the human race. For this reason they are under suspicion in every case and not to be trusted.

In an especial degree must popular prejudice be disregarded for the special reasons given, not but that at times these causes of error may hit upon the truth, but because as a rule they are involved in what is false; and very seldom do example and habit contain perfection; which moreover the multitude never attains, as I have explained before.

Chapter V

We can have no protection and defense against these evils, unless we follow the commands and precepts of God and his Scripture and of canon law, of saints and philosophers, and of all the ancient sages. If we hold fast to these commands and precepts, we cannot err nor ought we to be condemned in any particular. The philosophical quotations already given were adduced, accordingly, in the first instance to show the viciousness and folly of these evils, on account of which we might be able and bound to avoid them. But for the reasons given in regard to adducing the testimony of philosophers I have as a rule kept silent in other instances; moreover, it amounts to the same thing as if those testimonials had already been given which must now be added. I am able now, however, to give an intimation of the counsels and testimonies in which counsel or command is more clearly given in regard to avoiding those causes of error. Against popular prejudice, then, let us hold fast the command of Exodus, "Thou shalt not follow the multitude to do evil, nor yield in judgment to the opinion of the many, so that thou mayst depart from the truth." Let us follow also the advice of Tully at the end of his book on Questions, "Although thou art in the eyes of the multitude, do not rest upon its judgment nor agree with it in its estimate of what is most beautiful." Seneca also in his work on Aphorisms invites us and consoles us on our departure from the pathway of the multitude, saying, "Not yet art thou fortunate if the throng has not yet derided thee." Moreover against custom let us follow the advice of Cyprian; "Custom without truth is error of long standing, wherefore leaving error let us follow truth." Augustine teaches that if the truth is manifest, custom should yield to truth, because truth and reason always exclude custom. Isidore also says, "Let custom yield to authority; let law and reason be superior to base custom." For this reason Tully, in his book on the Immortality of the Soul, praising and extolling those who shun custom, says, "It is the mark of a great mind to withdraw its thought from custom." The name of authority, however, is in favor. Therefore our ancestors must be respected whether they possess real authority or apparent authority, which is the authority of the leaders of the multitude. Against the use of apparent authority not only are there proper counsels and commands in accordance with the pathways of God and his saints and philosophers and all men of science, but whatever can be said against the vices of human frailty in real authorities, is applicable to those who abuse assumed authority. Hence if we have counsels and commands against the failures of real authorities, much more so against those abusers of authority. But since truthful authorities, such as saints and leading philosophers, have aided us in the consideration of truth, we must be thankful to them, just as Aristotle in the first book of his Metaphysics is thankful to his predecessors, and at the end of his Elenchi desires to express his personal thanks for their discoveries. Since they have in fact laid the first foundations, we must not only thank them, but we must ascribe to them with a certain amount of reverence credit for the whole structure. Accordingly Seneca maintains in the third book of his Natural Questions, "Whatever has been discovered by men of a later day must none the less be referred to the ancients, because it was the work of a great mind to dispel for the first time the darkness involving nature, and he contributed most to the discovery who hoped that the discovery could be made; and although owing to human frailty they failed in many instances, they must nevertheless be excused." For as Seneca says in the book mentioned, "The ancients must be listened to indulgently."

Chapter VI

But since owing to original sin and the particular sins of the individual parts of the image have been damaged, for reason is blind, memory weak, and the will depraved, and truth and goodness are single, while fallacy opposed to any particular truth and evil hostile to any particular good are of infinite variety; because as the Philosopher, in the second book of his Ethics, explains, it is easy to miss a given mark owing to the manifold possibility of error, while truth and virtue are exact; also truths and virtues are infinite, and there are innumerable gradations in each truth and virtue; it is clear that the human mind is unable to give what is necessary in all, nor can it avoid fallacy and evil in particular ones. For this reason additions can fitly be made to the statements of real authorities, and corrections applied in many cases. This Seneca shows excellently in his book Quaestiones Naturales, declaring in the third book, "Ancient views are too inaccurate and crude; as yet thinkers were groping their way round truth. All things were new to the first investigators, later those same matters came under the file. Nothing is perfect at its commencement." In the fourth book he says, "The time will come when careful study through long ages will bring to light the secrets of nature. A single lifetime is not sufficient for the investigation of such weighty matters. The people of a future age will know much that is unknown to us, and the day will come when posterity will be amazed at our ignorance of things so clear to them." For this reason Priscan says in the introduction to his larger volume that there is no perfection in human discoveries, and adds, "The younger the investigators the more acute," because the younger, that is those of a later age, in the progress of time possess the labors of their predecessors. Since my contention has been proved by the nature of the case and by authority, I shall now establish it in a third way by the result. For thinkers of a later generation have always added to the work of their predecessors, correcting much and changing more, as is shown in the case of Aristotle especially, who discussed critically all the philosophical propositions of his predecessors. Avicenna also and Averroës corrected many of Aristotle's statements. Aristotle is criticized even yet for his statement in regard to the eternity of the world, which he has left too indeterminate; nor is it surprising, since he himself confesses that he did not know everything. For he confesses that he was ignorant in regard to the squaring of the circle, a problem clearly understood in these days; and his ignorance on this point indicates still further ignorance of more important matters. And Avicenna, the chief authority in philosophy after him, as the commentator states on the chapter on the rainbow in Aristotle's book Meteorologics, and as the works arranged by him from Aristotle into a complete system of philosophy make this clear,—Avicenna, I say, declared he was ignorant of the nature of the rainbow, even as the commentator mentioned confesses. In the third book of the Physics he does not hesitate to say that he is ignorant of one of the ten categories, namely, that of habit [*habitus*]. Without doubt errors and false statements are contained in his book on popular philosophy, as in the ninth book of the Metaphysics an error is set forth in regard to the creation of the world, where it is said that God on account of his infinite unity and the fear of attributing to him a diversity of natures can create but one being, that is the first angel, who created the second with the first heaven, and the second created the third with the second heaven, and so on. Also when he maintains in the tenth book that every sin has its final purification in another life, and that sinful souls in every case return to glory he is clearly in error, and so in many instances. Just as Averroës, the greatest after these, refuted Avicenna in many particulars, so also our men of science correct him in more instances, and rightly so, since without doubt he erred in many places, although he spoke very well in others. Moreover, if those greater masters fell into error, much more so have others, their inferiors, done so. Since, however, they have fallen into errors, they have failed much more seriously in the essentials, by heaping up a mass of superfluous and useless matter, and by spreading abroad doubtful, obscure, and perplexing statements; and all these faults in their books are manifest to any one, and are proved by their effect on us. For we are oppressed by such great difficulty in seeing the truth and stagger along, since as a rule one philosopher contradicts another, so that it is a difficult matter for them to agree on a single question of least import, or on a sophism of the most trivial kind, or process of science, as in medicine and surgery and other scientific fields. But not the philosophers only, but even the sacred authors have been subject to some human infirmity in this respect. For they have retracted very many of their own statements. Hence it is that Augustine, who is considered a master in the investigation of hidden truths, composed a book of retractions of his own faulty statements. Jerome also in regard to Isaiah, and elsewhere frequently, does not hesitate to retract his expressed opinion. For in the hurry of dictation he confesses he has frequently been mistaken in his translation, and in many other ways; and so with all the other doctors. Even the sacred authors have mutually corrected their own positions, and in turn vigorously withstood one another. Even Paul withstood Peter, as he himself confesses. Augustine finds fault with the views of Jerome; and Jerome contradicts Augustine in many places. This is clear from the examples of these men, and later writers have corrected the statements of their predecessors. For Origen, generally conceded to be the greatest of the doctors, is disapproved of in many particulars by his successors, because among other reasons he maintained the error of Avicenna in regard to sinful souls, namely, that not one of them will finally be damned. After many sacred and famous doctors have explained Israel to mean the man beholding God, Jerome comes in his commentary on Genesis and proves this exposition false by irrefutable reasons, as will be shown below. Hence he states, although they are men of authority, and we sink under the shadow of those who have interpreted Israel as the man beholding God, yet I prefer to agree with God or the angel who gave the name rather than with the author of some uninspired statement. Catholic doctors also grounded in sacred studies have changed at the present time in public utterances many statements made by the saints, expounding the sacred authors reverently according to their ability with due regard to the truth.

Chapter VII

Since then the facts are as stated, we should not give adherence to all that we have heard and read, but we should examine most carefully the philosophical statements of older writers, that we may make up their deficiencies and correct their errors, with all modesty, however, and consideration. We can reach this pitch of temerity not only because of the necessity to avoid failure or error, but through the examples and patterns of these men, so that in no way are we censurable for our presumption. For Plato says, "My master Socrates is my friend, but truth is still more so"; and Aristotle declares that he would rather be in harmony with truth, than have the approval of Plato's friendship, our very dear teacher. These statements are clear from the life of Aristotle and from the first book of the Ethics and from the book of the Secrets. Seneca says in his book on the Four Cardinal Virtues, "Let not the authority of the speaker influence you. Do not consider who the speaker is but what he says." Boetius in his book De Disciplina Scholarium* says, "It is foolish to put full confidence in the statements of a teacher, for in the first place credence must be given until it becomes evident what he has in mind; later it must be assumed that he made an error in his statement if perchance the pupil is able to find a flaw in the teacher's zealous handling of the subject." Augustine says in answer to Jerome that he is willing to believe that the authors of the sacred Scripture were the only ones who wrote without error, but that in the writings of others, however great their sanctity and learning, he is unwilling to consider there is truth unless they are able to prove their statements by the canon, and by other authors, or by adequate reasons. And in reply to Vicentius he says, "I cannot deny nor ought I that, as in the case of the older writers themselves, so in many of my own works there are many things that may be called in question fairly and without presumption." In the prologue to the third book on the Trinity he says, "In my letters unless you perceive the certainty of a matter, do not retain it as a fact." Likewise to Fortunatus, "Nor ought we to regard the disputations of any, although they be catholic and praiseworthy men, in the same light as canonical Scriptures, so that we may not, while retaining the respect due those men, call in question and reject anything in their writings, if perchance we find that their views have been at variance with the truth as understood with God's help either by others or by us. My attitude toward the writings of others is the same as I desire my readers to assume toward mine." If therefore owing to the necessity of avoiding falsehood and attaining a more perfect state of wisdom we are able and obliged and counseled by sacred writers and worthy philosophers to change their statements, as time and place demand, and to add to their precepts, we have much greater free-

* Not the work of Boetius but of a monk of Brabant of the thirteenth century, by name Thomas (Brabantinus Cantipratanus).

dom in this respect and are urged thereto in matters pertaining to the ignorant throng, and also to its leaders, especially since the popular leaders of this period are not to be ranked in influence with the sacred writers and great philosophers nor with ancient sages, some of whom we have seen in our own time.

Chapter VIII

There is no remedy against these three evils unless with all courage we prefer strong authorities to weak ones, reason to custom, the opinions of the wise to popular prejudice; and let us not trust in the threefold argument, this has a precedent, or is the custom, or the general practice, therefore should be retained. For it is evident from what has been said before in accordance with the opinions of sacred writers and all men of science that the opposite conclusion far more logically follows from the same premises. Although the whole world be influenced by these causes of error, yet let us willingly consider what is opposed to common custom. For it is a great remedy against these evils, according to the statement of Averroës at the end of the second book of his Physics, that evil custom can be removed by the habit of considering what is opposed to it. For he bestows much labor on the opinion which is formed from habitually considering extraneous matters, confirming his conclusion by the result, and saying that for this reason popular opinion is held more firmly than the faith of those philosophizing, because the multitude is not accustomed to consider the opposite side of a question, while those philosophizing take many things into consideration. Therefore let not your Wisdom be surprised, nor your Authority consider it improper if I labor against popular custom and common precedents. For this is the only way of arriving at a consideration of truth and perfection.

Chapter IX

Not only indeed are there these general causes of evil in our vocation and in our life, but there is a fourth one worse than these three likewise common in every rank, and visibly dominating every individual. I have joined together the three causes already mentioned because the wise usually do so, and I have separated this fourth cause from the former ones on account of its especial harmfulness. For this is an extraordinary wild beast, devouring and destroying all reason, namely, the desire to appear wise, by which every man is influenced. For however little and worthless our knowledge we nevertheless extol it; we publish abroad much of which we are ignorant, where we can hide our ignorance, making a clever display that we may glory over nothing. Matters of which we are ignorant, where we cannot make a display of our knowledge, we slight, find fault with, abuse, and bring to naught, that we may not seem ignorant of any matter, glossing over our ignorance like a woman with her finery and meretricious coloring, a foul remedy. Hence we banish by this route from ourselves and from others what is most useful, important, full of every grace, and stable in its nature. This bane, moreover, in addition to its inherent harmfulness reaches the crowning point in its own baseness from the fact that it is the beginning and the source of the three causes of error already mentioned. For owing to excessive zeal in regard to our own feeling and the excusing of our ignorance there arises at once the presumption of weak authority, relying on which we extol what is ours and censure what is another's. Then since every man loves his own labors, we willingly form ours into habit. And since no man errs for himself alone, but delights in spreading abroad his madness among his neighbors, as Seneca says in the book of the Second Epistles, by our inventions we take possession of other men and give our inventions all the publicity possible. It is necessary, however, that these universal causes should be considered first, that error may be avoided and truth shine clear. For in spiritual disease the process is the same as in bodily disease; physicians learn the special and particular causes of a disease through symptoms; but the latter as well the former a knowledge of universal causes precedes, which the physician has to acquire from the processes of nature; since the Philosopher says in his book on Sense and the Sensible that where the principles of natural philosophy end the principles of medicine begin. Likewise accordingly in the cure of ignorance and error, that sane truth may be reached, before taking up the main point at issue the symptoms and particular causes must be shown; but before all universal causes are required without which neither symptoms nor particular causes show anything. For the pathway of learning from the universal to the particular is formed by nature for us, as the Philosopher says in the beginning of the Physics. For if we are ignorant of the universals we are ignorant of what follows the universals.

This fourth cause, moreover, exercised much evil influence of old, just as it is now found to be doing, a fact I make clear in the case of the theologians as well as in philosophy by experience and examples. For Moses, a most sincere man, received the wisdom of the law from God, against which Pharaoh and the Egyptians and the Hebrew people and all the nations murmured to such a degree that the chosen people of God was scarcely willing to receive this wisdom; and nevertheless the law prevailed against adversaries who neglected and hindered the wisdom they did not learn. In like manner Christ the Lord, proceeding with all guilelessness and without a shadow of deceit, and the Apostles, most sincere men, brought wisdom into the world, who met much opposition solely on account of ignorance of so great a novelty, and at length, although with the utmost difficulty, sacred truth was received. Then when sainted doctors wished to give fluent expositions of the divine law, and with the great force of the streams of wisdom to water the Church, they were reckoned for a long time heretics and inventors of falsehoods. For just as the prologues of the blessed Jerome to the Bible and his other works prove, he himself was called a corrupter of Scripture and a forger, and a sower of heresies, and in his own day was overwhelmed and unable to publish his works; but at length after his death the truth of his translation became manifest as well as his exposition, and were so widely circulated among all the churches, that no vestige of the ancient translation, that of the Seventy Interpreters, which the Church used formerly, can be found. So long as the most blessed Pope Gregory exercised his authority, his books were unassailed; but after his death famous men in the Church tried to have them burned, but they were preserved by a most gracious miracle of God and their wisdom has become manifest to the world along with their sweet truthfulness and devotion in the fullest measure. In like manner all the doctors of sacred Scripture encounter hindrance to the truth. For in refreshing men's zeal they have always met contradiction and hindrances, and yet truth gains strength and will do so until the day of Antichrist.

The same is true of philosophy. For Aristotle wished to complete the writings of earlier philosophers and to give new meaning to many things; and although a very wise man, he seemed to meet repulse and his wisdom to be under a cloud until almost our own times. For Avicenna was the first to recall to the full light the philosophy of Aristotle among the Arabs. For the rank and file of those engaged in philosophy were ignorant of Aristotle; for few attained even a slight knowledge of the philosophy of Aristotle before the time of Avicenna, a recent philosopher long after the time of Mahomet. Avicenna, moreover, the especial expounder and chief imitator of Aristotle, encountered many revolts on the part of others. For Averroës, a still greater man after these, as well as others, condemned Avicenna beyond measure; but in these times whatever Averroës says has received the approval of the wise, yet he, too, for a long time was neglected and rejected, and called in question by philosophers famous in their vocation, until gradually his wisdom has become clear and generally approved, although on some matters he has spoken less convincingly. For we know that in our own times objection had long been raised in Paris to the natural philosophy and metaphysic of Aristotle as set forth by Avicenna and Averroës, and through dense ignorance their books and those using them were excommunicated for quite long periods. Since the facts are as stated, and we today approve of the men we have mentioned as philosophers and sacred writers; and we know that every addition and increase in wisdom they have made are worthy of all favor, although in many other matters they have suffered a lessening of their authority, and in many matters they are superfluous, and in certain need correction, and in some explanation, it is clear to us that those who during their individual lives have hindered the evidences of truth and usefulness offered to them by the men mentioned above, have erred too far, and have been very harmful in this respect. But they have done this to extol their own wisdom and to palliate their ignorance. Therefore we ought to apply the same argument to our own case, so that when we reject and revile matters of which we are ignorant, we proclaim our action as a defense of our ignorance, and an exaltation of our scanty knowledge. Therefore let us permit works to be introduced rejoicing in the truth, since without doubt although with difficulty truth will prevail among wise men, until Antichrist and his precursors appear. For the goodness of God has ever been ready to increase the gift of wisdom through a succession of men and to change for the better their philosophic statements in the succession.

Chapter X

But at this point two things are to be discussed, the display of presumed knowledge, and the senseless excusing of ignorance. In regard to the first we ought carefully to consider that since the truths relating to God and his creatures are infinite, and in each there are innumerable gradations, of necessity few facts are known by any one, and for this reason no one should boast of the many things he knows. Since, moreover, our intellect is related to things which are most important as is the eye of the bat to the light of the sun, according to my previous quotation from the Philosopher, there are few things that we really know; for assuredly a subject easily grasped by our intellect is indeed small and of little value, and the difficulty with which the intellect grasps a subject is the measure of the nobility of the subject which is being acquired. But nevertheless all that which our intellect is able to understand and know is small in comparison with those things which in the beginning in its weakness it is bound to believe, such as the divine verities and many secrets of nature and of art completing nature, concerning which no human reason can be given in the beginning; but one must get understanding from God through the experience of the inner light, I mean in the sacred verities of grace and glory, and stirred through the experience of his senses in the secrets of nature and art he must discover their reason.

Moreover, what we believe is far less than that of which we are ignorant; such, for example, are the secrets of God and the hidden things of life eternal which, however, the apostle beheld when caught up to the third heaven, ignorant whether he was in the body or out of the body, matters so great that a man may not speak of them. Likewise in created things; for on account of the great difficulty besetting our intellect it is certain that never, before God is seen face to face, shall a man know anything with final certainty. Therefore if he lived through infinite ages in this mortal state, he would never arrive with certainty at the perfection of wisdom in the multitude of things knowable. For no one is so learned in nature that he knows how to be certain in regard to all the truths involved in the nature and properties of a single fly, nor does he know how to give the particular reasons for its color and explain why it has so many feet, neither more nor less, nor can he give a reason for its members and properties. It is impossible therefore for man to attain perfect knowledge in this life, and it is exceedingly difficult for him to attain imperfect truth and he is very prone and disposed toward whatever is false and empty. Wherefore man ought not to boast of his knowledge nor ought any one to magnify and extol what he knows. For his knowledge is small and of little value in comparison with what he does not understand but believes, and still smaller in comparison with that of which he is ignorant and does not know either by faith or knowledge. And since in comparison with what a man knows those things of which he is ignorant are infinite, and without any comparison greater and better and more beautiful, he is out of his mind who extols himself in regard to his own knowledge. He especially is bereft of reason who makes a display of his knowledge and tries to publish it abroad as something marvelous. Who, moreover, ventures to boast of his knowledge when the whole pith of it acquired by a man, however studious, through thirty or forty years with very heavy expense and labor, may be shown adequately by written and oral instruction to a teachable boy in the course of one year or in less time? For I have proved this in the case of the boy present, who in the midst of great poverty and with little instruction by devoting scarcely a year to increasing his knowledge has so widened his field that all are surprised who know him. For I say fearlessly that although some may know more about philosophy and languages, and many may excel him in various ways, yet there are none among the Latins,* who surpass him in every respect, and he is a

* Adherents of the Western Church.

match for all of them in some things; in some points he excels them. There is no one among the Latins but may listen with profit to this boy. No one so learned, that this boy may not be indispensable in many ways. For although he has learned all that he knows by my counsel, direction, and help, and I have taught him much by written and spoken word, nevertheless he surpasses me, old man though I am, in many ways, because he has been given better roots than I, from which he may expect flowers and wholesome fruits which I shall never attain. Why then shall I boast of my knowledge? Nor should any one do so, however old he shall grow in study, when this boy will more quickly speed to the high points of knowledge, if he continue to be directed by good counsel just as he has been up to the present time. I do not deny that trained scientists will be able through the excellence that is in them to draw forth more easily and quickly many secrets of knowledge than this boy unaided, because he has not made trial of his own powers, nor does he perceive how much he knows, nor is he able to do anything beyond the fundamentals given him. But just as he has better roots than others, as has been said, so indeed if he should be directed by sane and effective counsel beyond the fullness of the source which he now possesses, no older man could overtake him in the full flood of the streams of knowledge. Since, moreover, wise men feel that they are more lacking than fools are willing to acknowledge in regard to themselves, for this reason we see that the wiser men are, the more humbly are they disposed to receive the instruction of another, nor do they disdain the simplicity of the teacher, but behave humbly toward peasants, old women, and children; since simple folk and the ignorant so reckoned know frequently great facts which escape the learned, as Aristotle shows in his second book on Sleep and Wakefulness. For God's conversation is with simple folk according to the Scripture, and experience renders us certain in this respect, since more secrets of knowledge have always been discovered by plain and neglected men than by men of popular fame, because the latter are busied in popular matters. These matters cannot be the great truths of science, as has been shown in my former statements; and I have learned more useful and excellent things without comparison from very plain people unknown to fame in letters, than from all my famous teachers. I have proposed therefore to your Wisdom this example, and sent a person not only for the two reasons noted above, but as a convincing proof that no man should glory concerning his own knowledge, nor despise plain men who know how to set forth matters which God has not granted to men famous in science, and to uncover and bring to light many very important secrets of science not yet grasped by scientists.

Chapter XI

A SECOND fault discoverable here is the fact that ignorance obtains a persuasive position. But when truth is assailed, there is the curse and corruption of evil; then the foulness of ignorance increases the more and is revealed more clearly. It increases because it strives to exclude knowledge from itself and from other things; it is revealed more clearly, since it becomes really known in God's sight and in the sight of all holy men, and according to the judgment of every learned man, although ignorance languishes among the ignorant and foolish, they avoid cloaking their own confusion.* In the third place, since a judge is considered to have knowledge of a case, an ignorant man does not possess the authority of deciding matters of which he is ignorant, and for this reason should he affirm or deny, his decision ought not to stand, nay, it should be the more vigorously opposed, because his judgment, whatever it be, proceeds from ignorance which has no authority. Hence if he should speak the truth, he would not seem to be doing so, and his ignorance would mar his statement. For as the book on the Cardinal Virtues says, a statement has no weight in a case where he condemns who ought to be condemned. Therefore whether a wise man in the public estimation or one really so, or a good man or a holy one should affirm or reject that of which he is ignorant, and especially should he do so to excuse his own ignorance or display his supposed knowledge, he should not win assent in this matter, but disdain and contradiction, although he might be deserving of high praise in other respects.

* As Bridges says in his note on this passage, the sense is not clear.

Causes of Error

CHAPTER XII

I HAVE introduced this cause of all our evils along with the other three particularly for this reason, that we may know that now as in times past many things which are very useful and quite necessary to learning considered absolutely and relatively to the four ways touched upon previously, are denied, neglected, and rejected from ignorance alone; and instead of the vast number of examples to be set forth more at length in the separate divisions of knowledge, I desire now to bring forward certain more important ones. For although the knowledge of tongues and of mathematics is especially necessary to the culture of the Latins, as I have indicated above and shall set forth in the proper place, a knowledge of especial service to the sacred writers and to all the sages of antiquity, we men of today neglect, bring to naught, and reject it, because we are ignorant of those subjects and of their usefulness. Moreover, if certain learned men and sacred writers, either overcome by human frailty or from some rational cause, have neglected certain branches of learning, we men of today obstinately and pertinaciously neglect and reject them, fortifying our neglect with the excuse that learned men and sacred writers neglected them, unwilling meanwhile to consider that in every man there is much imperfection of knowledge both in the case of the sacred writers and of learned men, as has previously been proved clearly and abundantly not only by their examples and authorities, but also by reason and the confirmation of experience. We are not only unwilling, moreover, to consider the human frailty, but not even the cogent reasons many sacred writers and learned men had for avoiding in view of the time and circumstance many useful branches because of men's abuse of them in turning them into obstacles to greater usefulness and safety. Lest, therefore, we should be the cause of our own error, and lest an obstacle to great learning should arise from the fact that we do not understand as we should the ways of the sacred writers and of learned men, we can on the authority of sacred writers and of learned men of old previously given in detail, consider with pious mind and reverent spirit because of the dignity of truth preceding all else,—we can consider, I say, whether the sacred writers and other learned men displayed certain evidences of human imperfection, in which whether affirmed or denied we should not imitate them without consideration.

We know, in fact, that not only have they given us counsel and freedom in this respect, but we see that they themselves have stated many things with great authority which they later retracted with greater humility; therefore there lay concealed in them in earlier days a great imperfection. But if they had lived up to the present time, they would have corrected and changed many more things; and it is a sign of this as well as an argument for the principle in question, that later doctors have changed very many of the statements of the sacred writers, and have piously and reverently interpreted them in a sense not indicated by the letter. Moreover sacred writers contended vigorously with one another and sharply criticized and rejected their respective positions, so that we become disgusted at beholding them, and are astonished above measure. This is evident in the letters of the blessed Augustine and Jerome, and in many other ways. For when Jerome compared himself to a tired ox, which plants its foot more confidently because he had grown old in sacred study, Augustine, his junior though a bishop, replied, the pontiff to the monk, that the aged ox plants its foot more confidently, not because of the vigor of its mind, but because of the old age of its body. And when Augustine asked Jerome many questions, the latter said, "You name questions on different subjects, but I perceive in them constant criticisms of my works. I pass by the formal greetings with which you caress my head; I am silent about the blandishments with which you try to mitigate my reproof, that I may come to the real matters at issue; observance of ceremonies, look you, cannot be a matter of indifference, but is either good or bad; you assert it good, I bad; while you avoid one thing, you fall into another; for while you fear the blaspheming Porphyry, you run into the snare of the Ebionite heresy." Many examples of this kind are collected from the books of the sacred writers and of the histories, which not only in the matters involving dispute, but also in the manner of the reproof show much human frailty in sacred writers, where they affirmed what they should not. But it is clear that they so acted from a lack of definite knowledge; therefore their activities in this respect resulted from a knowledge merely apparent and reckoned as such.

CHAPTER XIII

BUT not alone from the imperfection of this mortality of ours was much rejected by men to whom we should not obstinately cleave, nay, rather we should piously and reverently interpret them to their own honor according to the laws of truth; many very important matters they neglected for definite reasons. One reason is that they had not been translated into the Latin tongue nor arranged by any one of the Latins, and hence it was not surprising that these men did not consider their value. For all the doctors used the books of Plato, because they had been translated, or because they came upon them in Greek; but the books of Aristotle had not been translated at that time. For Augustine was the first translator and expositor of Aristotle, but he expended his effort on the least important and the first of his works, the Categories; nor was the philosophy of Aristotle at that time known to the Greek philosophers, nor to the Arabic, as I explained before. For this reason the sacred writers as well as others neglected the philosophy of Aristotle, of whom they had no experience, yet they praised Plato. Because, moreover, they perceived that Aristotle opposed the views of Plato, they reject Aristotle in many particulars, and say that he assembled together many heresies; just as Augustine says in his book on the City of God, Aristotle while his own teacher Plato was still alive gathered many into his false system. But, nevertheless, on the testimony of all who philosophize, Plato is known to have been altogether inferior to Aristotle. If, therefore, the sacred writers had seen his philosophy they would certainly have used it, and would have extolled it more highly, because they would not have denied a manifest truth, nor would they have rejected the very greatest for the very least. But from the book of the Categories it is clear how much the sacred writers would have praised Aristotle's great efforts, after extolling highly that little book, which does not possess the value of a single straw in comparison with his knowledge scattered in a thousand treatises. For Augustine himself translated it from Greek into Latin for his son, and expounded it carefully, bestowing more praise on Aristotle for this mere nothing, than we would for a great part of his knowledge. For in the beginning of the Categories he says, "Although all science and learning are treated of only through speech, yet no able writer, my son, has been found in any race willing to treat of the origin and beginning of speech itself. Therefore the diligence of the philosopher Aristotle is to be admired, who eager to discuss all things began with an examination of that which he knew had been left undone by all and was necessary to all." And at the end he says, "This, dearest son, is what we have prosecuted with joint labor and turned from Greek into Latin to your profit, that from it also you may receive the excellent fruit of a study perfected by us." Moreover Alcuin, one of the expositors of sacred Scripture and a teacher of Charles the Great, praised highly Augustine's translation of the Categories, and in a metrical preface graced it in these terms: "This little book contains the ten words of nature—words which in an astonishing way belong to everything that can enter our perception. Let him who reads praise the wonderful genius of the men of old, and let him be eager to exercise his own in labor like theirs. Now that which it pleased our master Augustine to transfer from the stored wealth of the ancient Greeks with a Latin key I am sending to you to read, great king, disciple and lover of wisdom, delighting in such a gift."

Boetius, long after the sacred doctors, was the first to begin translating several of Aristotle's works. He translated into Latin some works on logic, a few on natural philosophy, and something from the Metaphysics. We do not now possess either the half or better part. For Aristotle was long unknown even to those engaged in philosophy, not to speak of others, as well as of the rank and file of the Latins. But the sacred writers exalt greatly the works on grammar, logic, rhetoric, and mathematics, and use them freely in the sacred writings. Hence Augustine in the second, third, and fourth books on the Teaching of Christ shows that those subjects have application to divine matters, and elsewhere he similarly touches on the same; and the other sacred writers maintain the same view. But concerning the other works they speak rarely and too little, and more rarely, to be sure, show them much neglect, and sometimes teach that they should be neglected, as we see in Ambrose on the Epistle to the Colossians, and in Jerome on that to Titus, and in Rabinus [*de pressuris ecclesiasticis,**] and so in many other places. But it is clear to all engaged in philosophy and to theologians that these branches are of no value in comparison with the others, nor of any dignity. Therefore if the sacred writers had been practiced in the great sciences of philosophy, they never would have extolled to such a degree the ashes of philosophy, and they would have turned them to sacred uses: for the better and greater sciences are, the more suited are they for divine things. But since only the books on grammar, logic, rhetoric, and mathematics reached their hands, they aided themselves with these according to the grace given them: whatever they could draw from these in a praiseworthy way, they turned more abundantly to the law of God, as is clear in their expositions and respective treatises; and this will be set forth more plainly in its proper place.

Chapter XIV

THEN we must carefully consider that although they might have had many of the major sciences, or if they did actually have some of them, yet it was not a time when these sciences were used except in two cases, astronomy for the calendar and music for the divine Office. For the histories make it clear how Eusebius of Caesarea, the blessed Cyril, Victorinus, and Dionysius, the Roman abbot whose instruction the Church now follows, as well as others by apostolic order worked through the laws of astronomy in this matter. But the other major sciences were neglected, and particularly those known to contain the judgments and the sublime works of knowledge. The reason for this was fourfold. For philosophy before the advent of Christ gave laws to the world with the exception of the Hebrews, both concerning divine worship and morals and the laws of justice, of peace among citizens, and of war against the opponents of the commonwealth. Since, moreover, laws of this kind were given as far as human reason could do so, as Aristotle maintains at the end of the Ethics, when he passes to the book of

* An unidentified work.

laws saying, "Let us state now how much of philosophy is possible in human affairs," the leading men of the world under the guidance of philosophers were unwilling to receive the law of Christ, which was above human reason, and for this reason philosophy hindered the entrance of the faith in this particular, that the world under the influence of philosophy feared to agree to a higher law.

But not only in this way did philosophy hinder the faith of Christ, but by right of its own laws concerning the defense of the commonwealth from every opposing view it seemed through decisions regarding matters of the future and the revelation of secrets of the present, and through works wonderful beyond the power of nature and art acting in common,—it seemed, I say, to be contending with preachers of the faith, whose peculiar province it was not through nature and art, but through the virtue of God, to utter prophecy of future events, to bring forth hidden things into the light, and to arouse the miraculous. For that the power of prophecy is able to accomplish great results, which the rank and file not only of the laity, but also of the clergy viewed as miracles, the following will show. Moreover, the rulers of commonwealths everywhere, on the advice of philosophers zealous for their own laws, inflicted on God's saints severe penalties of persecution and death. The art of magic besides was increasing in strength throughout the whole world, and seizing upon men under every form of superstition and religious fraud; and although it was hated by philosophers and fought by them all, as will be shown in the proper place, yet the early sacred writers finding the world occupied with it as well as with philosophy reckoned both as the same art, since both in many ways were hindering the fruit of faith. For as the magicians of Pharaoh withstood Moses, and made the people of Egypt disobedient to God's command, so did the early Church suffer violence through the art of magic. Since magic produced the same effect as philosophy in opposing the work of faith, all censure of the former was directed against the latter, the more important of the two.

Moreover, it pleased God that no human testimony should be given the Church in the beginning, but that the truth of the faith should shine upon the world with such power that it might be proved that on God's authority alone was the truth made known through witnesses chosen by his command. For these reasons, therefore, philosophy was not only neglected by the Church in the beginning and by God's sacred writers, but also hated, not, however, on account of anything contained in it contrary to the truth. For although it is imperfect in comparison with the Christian profession, nevertheless its power is not hostile to the Christian sect; nay, it is altogether in accord with it and most useful and absolutely necessary to it, as all believe and as I shall show clearly. Not therefore because of any evil in philosophy did the Church of God neglect and reject it in the beginning, but on account of its abusers, who were unwilling to couple it to its true end, that is, Christian truth. For this reason, moreover, the primitive Church was not eager in regard to the translation of the chief sciences of philosophy and accordingly the sacred Latin doctors did not have access to the great principles of philosophy, and conforming themselves to the early Church neglected many things most excellent in philosophy, just as they were neglected in the beginning for the reasons given above, not because of anything false or unworthy that can be found in the power of philosophy. These facts will be established more surely in their proper place by means of God's sacred writers themselves. For it is shown that the holy patriarchs and prophets at the beginning of the world received all sciences from God, to whom he gave that great length of life, that they might be able to make trial of what had been revealed to them, to the end that, when the faith of Christ was introduced and the fraudulence of the art of magic was purged away, the power of philosophy might be advantageously applied to divine things.

Chapter XV

Neither the sacred doctors made use of the splendid sciences of philosophy, nor did their successors, Gratianus, the Masters of sentences and of histories, Hugo of Saint Victor, and Ricardus of the same. For they had not been translated in their times, nor were they in use among the Latins, and therefore these men neglected them, and did not know how to judge of sciences worthy of the sacred mysteries, but from prejudice rejected that of which they had no experience, and in many places speak against these sciences, taking occasion to do so all the more from the fact that sacred doctors neglected the same; but they paid no attention to the reasons of the sacred writers, no doubt because these works had not been translated in their time, and because the Church neglected to order them translated for the five reasons touched on previously. Modern teachers of the people, although many important portions of philosophy have been translated, make no use of them, delighting in small and trivial works and neglecting the two better works on logic, one of which has been translated with a commentary of Alpharabius on it, and an exposition of the second made by Averroës has been translated without the text of Aristotle. Far more do they neglect all the rest possessing still greater value, as the nine sciences of mathematics, and the six great natural sciences, comprehending in them many others, and the four very excellent divisions of the morals; and they seek a wretched solace for their ignorance in Gratianus and the other recognized masters who had no knowledge of the parts of philosophy, even as these men have not. They take refuge also in the bald statements of certain sacred writers, since they do not understand the reasons before stated. For the sacred writers after Christ did not take advantage of the great value of philosophy, not because it is contrary to sacred conceptions or unworthy of them, since philosophy can render useful and glorious aid in understanding absolutely theological truths both in regard to the Church of God, the commonwealth of the faithful, the conversion of unbelievers, and the reproof of those foreknown, as will be shown when necessary. And it is so much the more remarkable that a host of modern students neglect important sciences, although they were introduced after Gratianus and others and are used by men of excellent learning; and some are still living who have read them through in the course of their sacred studies.

Chapter XVI

Although I am attacking earnestly those universal causes of all evils, and I could wish that all should be reduced to firm authority and the way of thinking of the learned and experienced, who are few, yet let not your Serenity suppose that I intend to excite the clemency of your Holiness in such a manner that Papal Majesty should attack violently weak authorities and the multitude itself; nor that I in my unworthiness should stir up under the shadow of your glory any trouble over the matter of studies; but that I in my need may gather crumbs necessary to me as they fall from the Lord's table filled with the dishes of wisdom. The greatness of your Power will be able to make provision for itself and its successors in regard to the advantageous fullness of all knowledge not only to be possessed absolutely, but arranged in the four ways mentioned previously. Then when the wisdom of your Fatherhood shall have obtained fuller certainty regarding these matters, the judgment of your authority will be able easily to persuade the studious and wise that, what the rank and file of students cannot receive, those eager for knowledge may rejoice in obtaining; moreover, hope promises as much as suffices the multitude. For Jerome says on Isaiah, "The multitude after learning the truth easily changes its mind." And this is true except when it is in the hands of crack-brained leaders. For although the throng is rather prone to evil, and since it too frequently finds a leader weak, yet unless the leader hinders, it is quite easily directed to imperfect good, because of its own nature it is unstable, and once set in motion cannot preserve a due measure, and for this reason as far as it is concerned easily turns to the opposite view at the direction of its leader; since it is turned by every wind of doctrine, like a reed, and considers that the will of its leader has the force of law. For we see in every gathering of men that the members are moved at the will of the head. For if the leader neglects goodness, those under him fall asleep; if he stirs them to evil, they madly rush into it; if to an imperfect goodness, they likewise hasten without discretion. If, however, he should advise the paths of perfection, the multitude then sniffs at a distance, but is unable to taste nor does it care to do so, nor should this be required of it, as was shown above. But if it should not belong to your time to consummate all universally, your Magnificence will be able to place the foundations, dig out the sources, and make firm the roots, so that the successors of your Holiness may be able more easily to complete what shall have been happily begun.

PART TWO OF THIS PLEA

Chapter I

ACCORDINGLY after the four general causes of all human error have been banished to the lower regions and have been completely removed from this plea, I wish to show in this second part that there is one wisdom that is perfect and that this is contained in the Scriptures. From the roots of this wisdom all truth has sprung. I say, therefore, that one science is the mistress of the others, namely, theology, to which the remaining sciences are vitally necessary, and without which it cannot reach its end. The excellence of these sciences theology claims for her own law, whose nod and authority the rest of the sciences obey. Or better, there is only one perfect wisdom, which is contained wholly in the Scriptures, and is to be unfolded by canon law and philosophy. I make this statement since the exposition of divine truth is made through those sciences. For it is itself unfolded as it were in the palm with these sciences, and yet it gathers within its own grasp all wisdom; since all wisdom has been given by one God, to one world, for one purpose. Therefore this wisdom from its own triple arrangement will obtain unity. But the way of salvation is single, although there are many steps; but wisdom is the way to salvation. For every consideration of a man that does not belong to his salvation is full of blindness, and leads down to the darkness of hell; for which reason many sages famous in this world have been condemned, because they did not have the true wisdom, but an apparent and a false one, whence reckoning themselves wise they became fools according to the Scripture. But Augustine, speaking concerning the Scriptures, says in the second book on Christian Doctrine, if elsewhere there is truth, it is found here; if there is a hurtful thing, it is here condemned. And he wishes the Christian to perceive that, wherever he may have found the truth, it belongs to his Lord, as was said in the beginning. And the truth of Jesus Christ is the wisdom of the Scriptures. Therefore there is no truth elsewhere except that which is contained in that science. Ambrose on the Epistle to the Colossians says, "All knowledge of the science above and of the creation beneath is in Him who is the head and author, so that he who knows Him should seek nothing beyond, because He is the perfect virtue and wisdom." Whatsoever is sought elsewhere is found here in perfection. Since therefore the Scriptures give us this wisdom which is Christ, it is clear that all virtue is here included. But if wisdom elsewhere is so called and is opposed to this, it will be in error, it will have only the name of wisdom; even though it be not called opposing, it is yet different. But difference, although elsewhere it does not produce opposition, does so here, as is shown on Gospel authority, "He who is not with me is against me." So also is it true of this wisdom that what is not connected with it is proved to be against it, and for this reason to be shunned by the Christian.

Chapter II

This moreover becomes clearer as we estimate and consider the division of the sciences. For if we should try to separate the sciences one from another, we cannot say that theology is not both the science of canon law and philosophy; for under one division of philosophy, namely, moral science, which Aristotle named civil, is contained the civil law, as will be shown below. But canon law is named from the canonical Scriptures, not from others, just as the name shows. The books of the Old Testament are called the canonical Scriptures, as is the practice in the first part of the Decreta in the ninth section, and elsewhere; or the Scriptures are called canonical from this same word canon, for *canon* in Greek is called *regula* in Latin; and the law, canonical as well as divine, is acknowledged to give us the way of living according to rule. But canon law is wholly founded on the authority of Scripture and its expositors, as is clearly shown throughout the whole body of the Decretum and of the Decretals. For either in support of the ordinances of the canons there are brought forward the authorities of the expositors of Scripture, as Augustine, Jerome, Gregory, Ambrose, Isidore, Cyprian, Hilary, and others; or the holy and exalted pontiffs in support of their own decrees cite the authorities and examples of the New and Old Testaments; and hence this law is nothing else but an explanation of the will of God in the Scriptures.

Canon law is likewise called ecclesiastical law, by which the Church of God is ruled in spiritual matters both in its head and in its members. But the Scripture has no other interest outside this direction of the Church. Moreover, the natural law is contained in the Scripture, as is taught in the beginning of the Decretum, but whatever has been accepted in customs or included in writings, if it should be opposed to natural law, must be considered vain and without reason, as is set forth in the first part, eighth section. Therefore canon laws cannot be different from divine law, nay, they must be derived from its sources. The common law, moreover, is either divine or human. It is divine, since it was proclaimed to the world by the mind and spirit of God in his Scriptures; it is human, since it was discovered by the mind of man. But we agree that the Church is ruled by the divine mind and spirit, and therefore by the divine law which is included in the Scriptures, and it is certain that the Church is ruled by canon law. Therefore this divine law must be drawn out of the treasure-house of the Scriptures. This, moreover, is clear if we consider the divisions of canon law. For it orders the gradations of ecclesiastical offices, determines the sacraments of God, discusses the forum of conscience, and settles ecclesiastical cases. But the roots of all these matters and the erect stalk itself are found in the Scriptures; the branches belong to the expositors of the same, so that in the body of the canons are to be had the leaves, flowers, and fruit bringing salvation. For the pleasing embellishment of the language of the canon is compared to leaves according to the Scriptures, but the utility of the flowers and fruits is comprised in the four divisions already mentioned under their appropriate metaphors. And for this reason the canons are merely the golden heads of grain, and the branches, the ripeness of the grapes, are to be offered through the virtue in their own Scriptures. Since therefore canon law is thus subject to Scripture, it is contained in one body, just as the body of a single tree is composed of roots and stalk, branches, flowers, and fruits.

Philosophy

Chapter III

We must show, moreover, both in general and in particular, that the power of philosophy is not foreign to the wisdom of God, but is included in it. After this has been made clear by authorities, examples, and arguments of a general nature, a fuller exposition will then be made covering the four or five divisions of philosophy in regard to the power of the separate sciences and arts. For if Christians ought to snatch from the philosophers as from unlawful possessors the useful facts contained in their books, even as in the beginning I quoted from Augustine, it is evident that philosophy is wholly worthy and belongs to sacred truth. In the same book he says that the gold and silver of the philosophers did not originate with them but are dug out of certain mines as it were of divine providence, which is present everywhere, and he shows that this had been prefigured saying, "Just as the Egyptians possessed vessels and ornaments of gold and silver and raiment, which that people on its departure from Egypt claimed for itself on the ground that it would make a better use of these things; so the systems of the Gentiles contain liberal instruction better adapted to the service of truth and moral precepts of the most useful kind, and some facts are discovered in these philosophers concerning the worship of God himself, and this gold as it were and silver of theirs the Christian should take from them and apply to good use in preaching the Gospel."

And he explains this statement in all matters subject to human management which relate to either morals, or history, or arts, or nature, or logic and grammar. For as regards morals, he says, "The raiment also of those men, that is certain human institutions, but yet adapted to human society, which we cannot do without in this life, must be converted to Christian use." Of matters pertaining to history, he says, "The history of the Gentiles aids us greatly in understanding the Scriptures. For both by means of the Olympiads and the names of the consuls we frequently seek an answer to our questions, and ignorance of the consulship in which our Lord suffered has caused some to err in thinking that our Lord suffered in the forty-sixth year of his age, because it was said by the Jews that their temple had been so many years in building, which was a figure of our

Lord's body." This fact is made clear in almost innumerable places in the New and Old Testament. Concerning other human considerations relating to the arts as well as nature he says, "Of all the other arts, by which something is made or remains after the action, as a house, bench, vessel or any object of this kind, or medicine and agriculture and the art of navigation, or of those activities whose whole effect is the action itself, as dancing, running, wrestling"—a knowledge of these, I say, must be employed in forming a judgment, lest we should be wholly ignorant of that which the Scriptures wish to inculcate when they employ certain forms of expression derived from these arts, and we may take the term *naturalia* in a wider sense so as to include medicine and agriculture. For those sciences are based on nature, and are two of the eight sciences dealing especially with nature, as will be explained below. Yet in general he says in regard to all the natural sciences, "That man would indeed do the Scriptures a kind service who should collect the characteristics of times and places, of stones and the rest of inanimate things, of plants and animals." Moreover, in behalf of logic he says, in the first place, that a training in disputation is valuable in regard to all kinds of questions that must be examined and solved in sacred writings. Elsewhere in the same book he says there is a difference in the case of logic from the other sciences. For certain necessary and important things can be inferred from these sciences on behalf of theology, but I do not see, as he says, whether this can be the case with logic, since it, like nerves, is inferred throughout the whole text of the Scriptures. Also in the third book concerning the order of discipline he says that no one should approach the sacred science without a knowledge of the power of logic. His second, third, and fourth books almost throughout exhort us in regard to the application of grammar to sacred things. Jerome, moreover, in his commentary on the Epistle to Titus, speaking of the usefulness of grammar in regard to theology in comparison with many other sciences, says, "The teaching of the grammarians can benefit life provided it be applied to better uses," a subject concerning which many important statements must be made in what follows. But concerning mathematics, Cassiodorus says in his book on this science, "These four sciences, geometry, arithmetic, astronomy, music, we ponder over with an attentive mind, they sharpen our perception, they wipe away the mire of ignorance, and produce that speculative contemplation by the gift of the Lord. Rightly do the holy fathers persuade us to read these four, since in a great measure through them our appetite is drawn away from carnal things, and we are caused to desire those things which we can view in spirit only with the help of the understanding." But these matters will be shown fully in their proper place. If such is the case in regard to these sciences, much more is metaphysics in accord with divine utterances. For with the philosophers metaphysics occupies the place of one part of theology, being named by them together with moral philosophy the divine science and the theology of physics, as is clear from the first and eleventh books of the Metaphysics of Aristotle and from the ninth and tenth of the Metaphysics of Avicenna. For it considers many things concerning God and the angels and divine topics of this kind, and thus it is clear that Scripture lays claim to the power of the whole science. But not only Augustine gives us the above instruction, but he states that many sacred writers have done so when he asks, "Do we not see with how great a load of gold Cyprian came out of Egypt, that very sweet teacher and most blessed martyr; likewise Lactanius, Victorinus, Optatus, and Hilary, to be silent about the living; likewise innumerable Greeks, as in former days Moses himself a most faithful servant of God had done, concerning whom it was written that he was learned in all the wisdom of the Egyptians?"

Chapter IV

Not only the blessed Augustine, but other sacred writers as well make this statement, and show that this truth was expressed under a figure and bear witness that the sacred writers so expressed it. For in the present instance I think that Jerome in his letter to the great orator should be cited, "For what reason do you ask why I should sometimes place in my works examples from secular literature and pollute the purity of the Church with the filth of the nations? My reply is a brief one. You would never ask this question unless Tullius had full possession of you; you would not ask it if you read the Scriptures or the commentators on them, omitting Vulcatius. For who is ignorant that in Moses and in the books of the Prophets certain things are taken from the books of the Gentiles, and that Solomon both proposed some questions to the philosophers of Tyre and made certain replies? Hence he warns us at the beginning of the Proverbs that we should understand the discourses of wisdom and the subtleties of words, parables and dark speech, words of the wise and enigmas which properly belong to dialecticians and philosophers. But even Paul the Apostle used a verse of the poet Epimenides, writing to Titus, The Cretans are always liars, evil beasts, idle bellies. In another epistle also he places a senarius of Menander, Evil communications corrupt good manners. Moreover, disputing before the Athenians in the Senate of Mars, he calls to witness Aratus, For we are also his offspring. This quotation is in Greek the close of a heroic verse. And lest this should be too little, the leader of the Christian army and the invincible orator pleading his cause for Christ skillfully turns even a chance inscription into an argument of the faith. For he had learned from the true source that David wrenched a sword from the hands of the foe and cut off the head of very haughty Goliath with his own weapon. For he had read in Deuteronomy a precept from the mouth of the Lord that the head of the woman captive must be shaved, her eyebrows, all the hair and nails of her body must be cut off, and thus must she be taken in marriage. Why then is it strange if I too wish to change secular wisdom on account of the grace of its speech and the beauty of its words from a maid servant and a captive into an Israelite? and if I cut or shave off whatever there is dead in her of idolatrous pleasure, error, lusts, and by union with her very pure body beget servants born in the household for the Lord of Sabaoth, my labor is profitable in the household of Christ. Julianus Augustus vomited forth on his Parthian expedition seven books against Christ; if I shall attempt to write against him, am I to suppose that you will forbid me to drive back the mad dog with the doctrines of the philosophers and of the Stoics, that is with the club of Hercules?" And in his proof of this kind he cites the prophets and all the famous doctors from the beginning of the Church who by the teachings of the philosophers have secured acceptance of the faith of Christ on the part of princes and unbelievers and have strengthened it in many ways. Moreover, Bede in his commentary on the Book of Kings says that Christians may apply as their own to divine matters whatever is useful in the liberal sciences. Otherwise Moses and Daniel would not have suffered themselves to be instructed in the wisdom and literature of the Egyptians and Chaldeans. Likewise in his book on building the temple he says that Solomon with his servants signifies Christ, and Hiram with his servants the philosophers and wise men of the Gentiles, so that the temple of God, that is the Church, was built not only on the wisdom of the Apostles, but on that of the philosophers also. The Scripture says, "The servants of Hiram were more skillful in hewing wood than the servants of Solomon." Because, as Bede says on this point, the Gentiles, when converted from error and transformed in accordance with the truth of the Gospel, had a better knowledge of the errors of the Gentiles, and the better their knowledge, the more skillfully did they learn how to overcome and drive them out. Paul had a better knowledge of the Gospel he had learned through revelation. But Dionysius was better able to refute the false dogmas of Athens, the arguments of which, along with similar errors, he was familiar with from a boy, and it is for this reason that Solomon says, "For you know that there is not in my people a man that knows how to hew wood like the Sidonians." These statements and many like them the Venerable Bede makes, and many other writers as well, but let these now suffice.

Chapter V

The reasons can be given why the sacred writers so affirm the point in question and declare that it was set forth in a figure and seized upon in effect by the saints. In the first place the truth wherever found is thought to belong to Christ, as evidenced by the opinion and authorities of Augustine quoted above; in the second place, although in some measure the truth may be said to belong to the philosophers, yet for possessing it the divine light first flowed into their minds, and illumined them from above. For it lighteth every man that cometh into this world, as the Scripture says, with which sentiment the philosophers themselves agree. For they maintain that there is an active intellect and a possible intellect. The human soul is called possible by them, because it has of itself capacity for sciences and virtues and receives these from another source. The active intellect is the one which flows into our minds, illuminating them in regard to knowledge and virtue, because although the possible intellect may be called active from the act of understanding, yet in assuming an active intelligence, as they do, it is so called as influencing and illuminating the possible intellect for the recognition of truth. And thus the active intellect, according to the greater philosophers, is not a part of the soul, but is an intellectual substance different and separated essentially from the possible intellect. And since it is necessary for the convincing proof of my position to show that philosophy exists through the influence of divine illumination, I desire to prove this point conclusively, especially since a grave error has invaded the rank and file of philosophers in this particular, and also a large number of theologians, for what a man is in philosophy, that he is proved to be in theology. Now Alpharabius says in his book on the Intellect and the Apprehensible that the active intellect, of which Aristotle has spoken in his third book on the Soul, is not matter, but is a separate substance. Avicenna also teaches the same principle in his fifth book on the Soul and in the ninth of his Metaphysics. Moreover, the Philosopher himself says that the active intellect is separate from the possible and unmixed. He likewise maintains that the active intellect is incorruptible in regard to its being and substance, since he says that it differs from the possible in respect to incorruptibility. But the possible is incorruptible in regard to its substance and corruptible in regard to its being on account of its separation. Therefore the active must be incorruptible in regard to its being and substance, for which reason it cannot be part of the soul, since in that case it would be destroyed so far as its existence in the body is concerned when the separation takes place. And he says that it bears the same relation to the possible as the artificer to his material or the light of the sun to colors. The artificer, moreover, is not a part of the material in which he works, and is separate from it in essence. Likewise the light of the sun driving away darkness from colors and other things is separate from them in essence and comes to them from without. He also says that the active intellect knows all things and is always actual, which is not true of the rational soul nor of the angels, but of God alone. Again if it were part of the soul, the soul would then know the same thing through the active intellect and be ignorant of it through the possible intellect, because the active intellect is in actuality what the possible is in its potentiality, as Aristotle bears witness. Moreover, a thing should be named rather from its more worthy part, therefore the soul must be spoken of rather as knowing through the active intellect than as ignorant through the possible before discovery and teaching. If it should be said that the active intellect, although it be a part of the soul, yet is not the actuality and form of the body as is the possible intellect, and therefore a man has the functions of the possible and not of the active intellect, this is contrary to the definition of the soul in which it is stated that the soul is the actuality of a natural body. For if one part of this only is the actuality, then the form of the body in that case is wrongly granted to differ from the soul as a whole, and the soul is wrongly defined by the expression actuality of the body absolutely and without qualification, and for this reason in that case the part which is not the actuality of the body ought to be excepted, just as Aristotle himself in the beginning of the second book, when he assumes that certain parts of the soul are not only the actuality of the whole body, but of parts of it, for example, the sensitive and vegetative parts of the soul, excepts the intellect, which he says is not the actuality and perfection of a part of the body, because it is not attached to an organ like the other parts of the soul. And to bring out his meaning more clearly he says that the intellect is in the body like the sailor in the ship, since it is not attached to any particular part any more than the sailor is to the ship, although it is the actuality and perfection of the whole. Yet the sailor is not the perfection of the ship, but only its director. Moreover, in that case the soul would be composed of a separate substance and one that is united, but this is impossible. For intellect, or angel, and soul differ according to species as respects capacity or incapacity of uniting with the body, and therefore the soul cannot be composed of something that is the actuality of the body, and of something that is not such. For one species cannot have in its composition something pertaining to another opposite species. Since, therefore, this opinion is in agreement with the truth, and the text of the Philosopher evidently indicates this, and his chief expositors declare it in this form, and these words active and possible have been accepted by the Philosopher and not by the sacred writers, it is far better in accordance with the opinion of the Philosopher to speak of the active intellect as a substance separate from the soul in essence. For the expert in philosophy has no doubt that this is his opinion, and on this point all the learned experts of the past are in agreement. For when the University of Paris was convoked, I twice saw and heard the venerable president, Master William, bishop of Paris, of blessed memory, in the presence of all express the opinion that the active intellect cannot be a part of the soul; and Master Robert, bishop of Lincoln, and brother Adam of Marsh, and elders of this type supported this same view. How possible objections may be refuted will be explained in the principal work when questions on natural topics are discussed. But, however, lest some caviler may arise on the side, alleging that in which the rank and file are deceived, I reply in the first place that although these words are attributed to Aristotle, "Since moreover in all nature there is something that acts and something that is acted on so is it also in the soul," the translation has often been false and more frequently obscure. For although it is stated in the third book on the Heavens and the World that the circle and the circular figure fill out space, the statement is false, as those know who are expert in the natural and geometrical sciences, even as Averroës proves in the same place. And the statement made in the third book of the Meteorologics that the lunar rainbow occurs only twice in fifty years, is likewise false. For experience shows us that whenever the moon is full and there is rain, and the moon is not covered by clouds, the rainbow appears. So also many other things have been falsely translated, the reason of which will be explained in the third part of this work, when the question of faults in translation will be considered; but far more passages have been translated obscurely and unintelligibly, in which any one may contradict another. In this passage both faults occur or at least the second one, as I shall prove by Aristotle himself. For he says in the second book of the Physics the material cause does not coincide with other causes in the same individual thing, therefore in no nature are there at the same time the active cause and the material cause, therefore not in the mind. If, therefore, the badly translated text should be held to the letter, then it is altogether false and contrary to Aristotle elsewhere; and so great an author does not contradict himself. And however it happens, his statement in the second book of the Physics is true and accepted by all; therefore his statement in the third book on the Soul is falsely translated, or is in need of exposition. For his thought is that in the soul and in the functioning of the soul two things are needed, namely, the active principle and matter, just as in all nature. That is, in every operation of nature two things are required, namely, an efficient cause and matter, and this is true; but the active principle is always different from matter and outside of it as regards its substance, although it acts upon it. But we can otherwise relieve this passage. For Aristotle in the fourth book of the Physics says that there are eight modes of being in any thing, one of which is that of the moving force in what is moved, because the moving or acting principle is in the matter moved as far as its force is concerned, although not as regards its substance. Thus the active principle is in every nature on which it works, and so in the soul. Thus in no way does it follow that the active intellect is part of the soul as is commonly imagined. This view is altogether trustworthy and confirmed by the sacred writers; for all theologians know that Augustine says in his Soliloquies and elsewhere that the rational mind is subject to God alone in its illuminations and in all important influences. And although angels may cleanse, illumine and arouse our minds in many ways, and though they may be to our minds like stars to the eye of the body, yet Augustine ascribes to God the principal influence, just as to the sun is ascribed the flow of light falling through the window, and the angel is compared to one opening the window, according to Augustine in his gloss on the Psalm, "Give me understanding." And what is more, he maintains in several places that we do not learn any truth except in the uncreated truth and in eternal laws, and this at least has to be understood to denote the effect and influence of the truth upon us; although Augustine not only maintains this, but hints at something else in his words, for which reason some have believed that he is here thinking of greater matters, as is generally known. All this is evidence of the fact that the active principle illuminating and influencing the possible intellect is a separate substance, that is God himself. Since, therefore, God has illumined the minds of those men in perceiving the truths of philosophy, it is evident that their labor is not opposed to the divine wisdom.

Chapter VI

THE third reason why the wisdom of philosophy is reduced to the divine wisdom is that God not only has enlightened their minds for the acquisition of a knowledge of wisdom, but they have secured it from him, and he has revealed, presented, and given it to them. For all wisdom is from the Lord God, just as the authority of Scripture states, because, as the Apostle says, "What is known of God is manifest in them, for God has revealed it to them." Augustine says on John that he gave it to them, and the greatest philosopher Aristotle in the book of the Secrets asserts that the whole of philosophy was evidently given and revealed by God. Also one of the greatest of philosophers, Marcus Tullius, in the first book of the Tusculan Disputations asks, "What is philosophy, as Plato says, but the gift, as I say, the invention of God?" Whence he also says that not even a poet has poured forth a theme filled with weighty meaning without some divine instigation of his mind. Moreover, Augustine in the eighth book of the City of God teaches and approves of what Socrates, the father of the great philosophers, maintained, that a man cannot know the causes of things except in the divine light and by its gift, and any one can prove for himself that nothing in the first instance is discovered by man which is within the range of philosophy. I offer an example in a very small matter; for although the Universals of Porphyry have been pretty adequately explained by him, and elsewhere have been stated sufficiently in logic, metaphysics, and natural and speculative philosophy, yet there is no man so well grounded as not to need in many ways to have teachers and to listen to them for a long time and to study before he knows the whole truth of the Universals. And no one with difficulty learns enough about these before death, no matter how many teachers he may have, as is clear owing to the disagreement of all in this matter:

since some maintain that these exist only in the mind, others that they exist only outside of it, while still others maintain that in respect to being they are in things, but in their universal aspect they are in the mind. Avicenna shows in his commentary on Porphyry that he lacked the sixth Universal and that he made several false statements. If, therefore, ignorance of these exists in any man you please, although he may study throughout his whole life the books of the philosophers, and though he may have honored teachers, much more will he be ignorant of these weightier matters and never discover them for himself without books and teachers. Wherefore of necessity the truth of these matters was revealed to man in the beginning. If also any one, however well versed in the Universals, had handed over the book of Porphyry to oblivion and all things necessary for a knowledge of Universals, and could not have books or teachers, it would be impossible that he should ever unfold the truth of the Universals. I am speaking of the Universals in respect to their true being, as a metaphysician must consider them, and not merely in respect to the puerile teaching of Porphyry and the method of logic. Wherefore any one can consider for himself that revelation is necessary in this direction; and since these are puerile and very small matters, my position will be much stronger as applied to the whole knowledge of philosophy. But what is from God and what he himself has revealed, presented, and given, should be in absolute conformity with his own wisdom.

Chapter VII

But the whole aim of philosophy is that the Creator may be known through the knowledge of the creature, to whom service may be rendered in a worship that gives him honor, in the beauty of morals, and in the integrity of useful laws, because of the reverence due his majesty and the benefits of creation and preservation and of future happiness, so that men may live in peace and justice in this life. For the end of speculative philosophy is the knowledge of the Creator through the creatures, and moral philosophy establishes the dignity of morals, just laws, and the worship of God, and persuades us of our future felicity in a profitable and glorious manner so far as lies in the power of philosophy. These things are known as facts by those who peruse all the principal parts of philosophy, as what follows will show. Since, therefore, these things are altogether necessary for Christians and are in complete accord with the wisdom of God, it is clear that philosophy is necessary to the divine law and to the faithful who glory in it.

Chapter VIII

Moreover, all sacred writers and wise men of old in their expositions take a literal sense from the natures of things and from their properties, in order that they may bring out spiritual meanings through convenient adaptations and similitudes. This Augustine declares in the second book on Christian Doctrine, taking an example from the word of the Lord speaking to his Apostles, "Be wise as serpents and harmless as doves." For the Lord meant by this, that like the serpent which exposes its whole body in defense of its head, the Apostles and apostolic men should give themselves and their all for Christ their head and for his Faith. For this reason every creature in itself or in its own likeness or in the universal or in the particular, from the highest part of heaven even to the end of it is placed in the Scripture, even as God has made the creatures and the Scripture, so has he willed to place in the Scripture the things themselves that were made for the understanding of its literal as well as its spiritual sense. But the whole purpose of philosophy is to evolve the natures and properties of things, wherefore the power of all philosophy is contained in the sacred writings; and this is especially clear, since the Scriptures far more certainly, better, and more truly comprehend the creatures than philosophical labor would know how to define them. In place of an infinite number of examples let that of the rainbow make this matter clear. The philosopher Aristotle disturbs us by his own obscurities, nor can we get any clear understanding of the subject through him; nor is this surprising, since Avicenna, his particular imitator, the leading philosopher after him, according to the statement of the commentator on Aristotle's chapter on the rainbow in the third book of the Meteorologics, confesses that he himself did not clearly understand the nature of the rainbow. The reason of this is because the philosophers were ignorant of the final cause of the rainbow; and in their ignorance of the end they are ignorant of those things which pertain to the end, because the end imposes a necessity upon those things that are ordained for the end, as Aristotle maintains in the second book of the Physics. In fact, the final cause of the rainbow is the dissipation of the aqueous vapor, as is clear from the book of Genesis, whence always on the appearance of the rainbow there is a resolution of the clouds into an infinite number of drops, and the aqueous vapors disappear both in the air and in the sea and land; since one part of the rainbow falls into the spheres of water and earth. The disappearance of the aqueous vapor cannot take place through the rainbow except by reason of the solar rays that cause it, for through various reflections and refractions an infinite number of rays are assembled, and the assemblage of the rays is the cause of the resolution and disappearance of the waters, and therefore, the rainbow is produced by multiple reflections. For the rays cannot assemble except through refraction and reflection, as will be shown later in its proper place. From the Scriptural statement therefore of Genesis, "I will place my bow in the clouds of Heaven, that there may no more be a deluge over the earth," we learn the final cause of the rainbow itself, from which the efficient cause and the way in which the rainbow is produced can be investigated. The manner of its production was not clearly understood by the philosophers as their books show us. And such is the case in regard to every creature. For it is impossible for a man to know the ultimate truth of the creature as it is employed in the Scripture unless he shall have been especially illumined by God. For creatures are employed there because of the need of bringing out the truths of grace and glory, concerning which the philosophers were ignorant, and therefore did not attain the ultimate power of knowledge in regard to creatures, as the sacred Scripture contains it in its own vitals. Hence the whole excellence of philosophy lies in the literal sense when philosophy has been adorned with the sacred mysteries of grace and glory, crowned as it were by certain very noble pictures and colors.

CHAPTER IX

THIS general proposition can be proved finally by the fact that the full measure of philosophy was given to the same men to whom also the law of God was given, namely, the holy patriarchs and prophets from the beginning of the world. And not only is this necessary for establishing the point under consideration, but also for certifying the whole range of knowledge. For one could not unaided acquire the principles of the sciences and arts, but needed a revelation. If we prove a revelation, we should have no doubt in regard to the secrets of wisdom discovered by the investigators, although we have had no experience in them. But there is no single point within the realm of knowledge so difficult of proof as this, because it is the main foundation of all human comprehension, and objections and doubts frequently present themselves, and authorities and books must be explained more fully than for any other proposition discoverable in the whole range of knowledge. I say, therefore, that the power of philosophy was given by God to the same persons as the sacred Scripture, namely, the saints at the beginning, so that there thus appears to be one wisdom in its entirety necessary for mankind. For the patriarchs and prophets alone were true philosophers, knowing all things, to wit, not only the law of God, but all the parts of philosophy. For sacred Scripture proves this to us with sufficient clearness, stating that Joseph instructed the princes of Pharaoh and taught the old men of Egypt prudence, and Moses was skilled in all the wisdom of the Egyptians. Bezaleel and Aholiab also proved this point, who had a full grasp and knowledge of the things of nature; for with one breath the Holy Spirit illumined them and taught them the whole of nature's power in metals and other minerals. But Solomon, wiser than all preceding or following him, according to the testimony of Scripture, possessed the full power of philosophy. Josephus in the first book of his Antiquities says that since the sons of Adam through Seth were religious men and made by God himself, God granted them a life of six hundred years on account of the glorious parts of philosophy in which they studied, in order that they might know by experience through the length of their life what God revealed to them. He adds that Noah and his sons taught the

Chaldeans the parts of philosophy, and that Abraham entered Egypt and taught the Egyptians. And he adds that Solomon left no part of nature unexamined, but philosophized about all things and clearly set forth his teaching in regard to their properties, and he mentions how Solomon in treating of things singly composed four thousand and five books; and he adds many things which we know nature as she generally functions does not in any way do. Mighty Aristotle, too, under compulsion of truth itself says in the book of Secrets, God revealed all wisdom to his prophets and to just men and to certain others whom he chose beforehand, and illumined them with the spirit of divine wisdom, and endowed them with dowries of science. Subsequent philosophers, Indian, Latin, Persian, and Greek, received from them the beginning and origin of philosophy. For from those men they received and wrote down the principles and secrets of the arts and sciences, because in their writings nothing false is found, nothing rejected by wise men, but only that which is approved. For philosophers and especially the Greeks, because they were more studious, received all from those men. Averroës states in his commentary on the opening portion of his Heavens and World that in the days of the ancients before Aristotle and the other philosophers philosophy was complete—a completion for which Aristotle hoped in his own times. In the introduction to the Construction of the Astrolabe of Ptolemy and in Albumazar in his larger introduction and elsewhere and in others it is stated in many ways that Noah and Shem, his son, enlarged philosophy, and Shem, his first-born, was particularly eminent in this subject. Then after them came the men who were commonly called philosophers. All the philosophers and famous poets both greater and lesser were after Noah and his sons, and after Abraham. For Aristotle and all the others agree in this that the first to take up philosophy were the Chaldeans and Egyptians, whence he adheres to the opinions of the Chaldean fathers in the eleventh book of the Metaphysics. In Egypt scholastic study began first, as he states in the first book of the Metaphysics. Because although Noah and his sons taught the Chaldeans before Abraham taught the Egyptians, yet study in the regular scholastic method was not begun so quickly, but its method and practice developed little by little.

To the end that all doubt in this matter may be removed, let us consider the descent and succession of unbelieving philosophers and of poets and of all those zealous in regard to the pursuit of wisdom, that we may perceive that after Abraham and his predecessors individuals were found who secured honorably some title to wisdom. For however closely we wish to reckon, Zoroaster discovered magic arts according to Augustine in the twenty-first book of the City of God, and according to all authorities this was generally known; but he was Cham, son of Noah, as Clemens in his book and the Master of Histories and the Historical Mirror state. Then Io, afterwards called Isis, gave letters to the Egyptians, as Augustine states in the eighteenth book of the City of God. Before her time the study of wisdom was not treated of in letters and in writings, although the teaching of the spoken word of Abraham instructed them. This Isis, as Augustine states, is said to have been the daughter of Inachus, first king of the Argives, who reigned in the first year of Jacob and Esau, grandsons of Abraham, as Augustine and the histories state; although some have maintained that Isis came from Ethiopia into Egypt as a queen, and gave them letters, and bestowed many benefits, as Augustine states. But, however, she was not before the times of Inachus, nor is she found in the line of the kings of Egypt in the chronicles. At the same time appeared the virgin Minerva, named Pallas, the discoverer as Augustine says of many things, who is called in the poets the goddess of wisdom. She is also called Athena and Tritonia, as Augustine states. Isidore says in the eighth book of the Etymologies and Pliny in the fifth book that Pallas is called Tritonia from a place called Trito in Africa, and that she lived in the time of the flood which happened in Achaia in the reign of King Ogyges and which therefore bears his name. According to Augustine, Eusebius, and Jerome, also Solinus in his book on the Marvels of the World, Ogyges was a contemporary of Phoroneus the son of Inachus. Inachus, moreover, reigned fifty years and Phoroneus, his son, sixty-one years, in whose time the promise was again made to Jacob, just as to his father, as Augustine states. Therefore Ogyges belonged to the time of Jacob, whence Solinus says that the first deluge in Achaia was in the time of Ogyges and Jacob the patriarch. This deluge was six hundred years before that of Deucalion, as

Solinus likewise states. For as Jerome and Eusebius tell us, the deluge of Deucalion was in the time of Cecrops, king of Athens, in whose reign Moses led out the children of Israel from Egypt. Under Phoroneus moral philosophy began among the unbelievers. For Augustine says that under this king Greece became more famous because of the introduction of laws and judgments. But previously there were customs and laws of living, as is shown by the inhibition of blood, and permission to eat flesh after the flood, and from the sale and purchase by Abraham of the cave, and from circumcision and rites of this kind. The holiness of Abraham and of his fathers infers that just and sacred laws of living were taught by them. And since they perfected less useful sciences, the wisdom of such great men must not have neglected the most useful science of morals. Next Prometheus was foremost among men, distinguished by his title to greater wisdom, who the poets say formed men out of clay, because he had the reputation of being a most excellent teacher of wisdom, as Augustine says in the eighteenth book of the City of God. His brother, as Augustine likewise states, was Atlas, the great astrologer; whence fable finds occasion, as Augustine states, to represent him as bearing up the heavens, although a mountain is called by his name, from whose height rather the idea of his bearing up the heavens seems to have been general. This mountain in the extreme maritime parts of Africa near the Gades of Hercules rises as it were to heaven. But still earlier were the sons of Noah and Abraham, who were skillful astronomers, as Josephus states; also Isidore in the third book and Clemens in the first book make this same statement in regard to Abraham. For these men according to Augustine flourished when Moses was born, and Isidore agrees in the fifth book, saying that Atlas was in servitude to the children of Israel before the birth of Moses. Atlas in fact, Augustine says, was the maternal grandfather of Hermes, the greater Mercury, famous for his skill in the great arts. Because he taught these to men, he was venerated as a god after his death; and, as Augustine states in the eighteenth book, he lived at the time when Moses led the children of Israel out of Egypt. His grandson was Hermes Mercury, who was called Trismegistus to distinguish him from the other Mercury. He was a famous philosopher of Egypt and lands to the south, especially in morals and in those things known to pertain to worship and divine matters, as Augustine teaches us in the eighth book of the City of God; and this man wrote to Asclepius, as is clear in the book on Divination, which is pretty well known. Aesculapius, the grandfather of this Asclepius, was the founder of medicine among the unbelieving philosophers. But Isidore says in the third book of the Etymologies that Apollo was the father of Aesculapius, who is said to have been the first among unbelieving philosophers to teach the art of medicine. For medicine is ascribed to the father also as far as the first rudiments are concerned, but more generally to the son, because he enlarged this art and taught it in a more definite manner. For Apollo's method was by incantations and remedies of this kind, but Aesculapius relied on the truth of experience, as Isidore states, and he is believed to be the great Apollo, who is imagined by the poets to be among the gods and to give responses in the temple of Apollo on the island of Delos, whence he is called the Delphic Apollo. And yet the inestimable glory of medicine existed before those men, as Aristotle indicates in his book on the Directing of Life—a glory which he ascribes to Adam and Noah more than to subsequent philosophers. And since medicine is more necessary to man than many other sciences, without doubt the sons of Adam and Noah discovered it, to whom the fullness of wisdom was given, and who were permitted to live so long for the sake of completing its study.

Chapter X

After this in the time of Othniel, judge of Israel, Cadmus the Theban reigned, who first gave letters to the Greeks, as is shown in the Cluniac chronicles. Bede also in his smaller chronological work and other writers agree that under Aoth the judge, Amphion the musician flourished. This Aoth was first after Othniel under Deborah and Barak. There was another Apollo, a philosopher, according to the Cluniac chronicle, the founder of medicine, a contemporary of the second Hercules, whose deeds are famous, as Augustine says in the eighteenth book of the City of God. This Hercules in the time of Abimelech the judge laid waste Troy, set up his pillars in India, erected his columns at Gades, and refusing to bear the pain of his disease cremated himself in the time of Jephthah the judge, as is confirmed by Augustine and the chronicle mentioned. I have spoken of this second Hercules, because there was another Hercules near the time of the greater Mercury who was a little after him, as Augustine states. After him there was a third, who established the Olympian contest, which after an intermission his son Picus restored in the four hundred and eighth year after the destruction of Troy, as Solinus writes. Whence many have been deceived thinking that there was but one Hercules, who did all the things that were written concerning several, as Augustine says. There was a similar mistake in regard to this philosopher Apollo. For all, as Augustine says, reckon him to have been the one who was worshiped as a god in the island of Delos, under the impression that he was one and the same person, whereas the contrary is shown to be the fact by much evidence. For that Apollo who gave responses in temples is found at least to have given a response when the city of Athens was founded, that Athena, who is Minerva, might be worshiped as a goddess, and for this reason this philosopher cannot be he who was worshiped as the Delphic god. But according to Augustine he was the son of Latona, whose sister was Diana. Isidore also makes the same statement in the eighth book. Similarly he does not seem to be the one of whom Jerome writes in the Epistle to Paul which is placed before the Bibles of the Latins. For that one discovers Hiarchus seated on a golden throne and teaching, and this Hiarchus is said to be Abrachis, the astronomer, who lived after the death of Alexander the Great, as Ptolemy shows in the Almagest. And therefore according to this there were three Apollos as well as three Hercules. Then under Gideon Orpheus and Linus won fame, as Bede states. And these men, namely, Amphion, Orpheus, Linus, and such men of their day, were called theological poets, as Augustine says, because they composed hymns to the gods. According to Solinus, Nicostrate, mother of the Roman king Evander, was called Carmentis from her gift of prophecy, who dwelt on the Capitoline hill in Rome, and was first to give the Latins letters. And this, as Bede states, was in the time of Jair, judge of Israel, but according to the Cluniac chronicle it was in the time of the judge who filled that office seventeen years after Jair. But this question is not within the scope of the present discussion.

On account of the Sibyls and especially the Erythraean Sibyl, who far surpassed all the unbelieving philosophers previously mentioned, the date of the fall of Troy must be settled definitely. For Augustine states in the eighth book of the City of God that many authorities have written that this Sibyl lived in the time of the Trojan war, although others held that she lived in the time of Romulus and Ahaz, or of Hezekiah, king of Judah, as Augustine says. Now the fall of Troy was four hundred years before Romulus. For Solinus proves that Rome was founded in the seventh Olympiad, four hundred and thirty-three years after the Trojan war, as he clearly shows by Hercules and Picus his son mentioned above and by others. And Augustine in the eighth book mentioned above maintains that Troy was captured when Labdon spoken of above judged the Hebrews, whence in the Cluniac chronicles it is stated that Troy was captured in the third year of Labdon. Then in the time of Samuel according to the Cluniac chronicle, but more explicitly in the Deeds of the Greater Britons, lived Homer the famous poet. Then Hesiod the philosopher succeeded Homer before the founding of Rome, as Tullius says in his book of Tusculan Questions. And later came Archilochus in the reign of Romulus, as is stated in the same book, and therefore in the time of Ahaz or Hezekiah, kings of Judah. For in the first year of Numitor, grandfather of Romulus, who was last king of the Albans in Italy, as Augustine states, Rome was founded. And therefore Numitor and his grandson Romulus ruled at the same time, and then the kingdom and name of the Albans ceased, and they were called Roman kings. The king at that time in Judah was Ahaz; or, as some think, Hezekiah his successor, a most excellent and pious king, ruled in the time of Romulus.

Chapter XI

Augustine makes these statements; but in the time of this same Romulus, according to Augustine, Thales is reported to have lived who was the first of the seven wise men. For after the theological poets wisdom increased and men devoted to it were called Sophists, that is wise men. But according to Bede in his book of Chronologies and according to Isidore in the fifth book of the Etymologies and according to others also, Thales lived in the time of Josiah, who as a natural philosopher examined nature and was an astrologer at the time when the Hebrew people, as Augustine states, were led into captivity. Another of the seven wise men appeared, Pittacus by name and a Mytilenean by nationality and race; and five others were in the time of the captivity, whose names are Solon the Athenian, Chilon the Lacedaemonian, Periander the Corinthian, Cleobulus the Lidian, Bias the Prienean. Of these Solon gave the Athenians laws, for the transfer of which the Roman people sent ten men, and they are called the laws of the twelve tables, as Isidore writes in his fifth book and Gratianus accepts the statement from him. But the four others left nothing in writing, as Augustine says. All these men, however, were called wise according to Augustine in the eighth book of the City of God, who were distinguished from other men in the nature of their life by certain precepts pertaining to right living, and according to Augustine in the eighth book these were the Ionic wise men, that is Greek. But another class of men devoted to wisdom arose after these in the Greek language, which, however, is called Italic, but they came from that part of Italy which in ancient times was called Great Greece. They were called the Italic school because they studied in Italy although they were Greeks and used the Greek language. Those men did not wish to be called wise, but lovers of wisdom, of whom the first was Pythagoras from the island of Samos. When he was asked who he was, he replied that he was a philosopher, that is a lover of wisdom, as Augustine says in the eighth book of the City of God; but in the eighteenth book of the City of God Augustine says that he appeared at the time of the Jewish restoration. According to Tullius in the first book of the Tusculan Questions in the time of the Roman king Tarquin the Proud, the seventh from Romulus and last of the Roman kings, after whom the consuls arose, Pythagoras came into Italy and swayed Great Greece with his reputation, discipline, and authority. For many ages afterwards the name of Pythagoras had such weight that no others were considered learned. This Tarquin, as Bede writes, began to reign in the time of Cyrus, king of the Persians, who released the Jews from captivity, and reigned in the time of Cambyses, the son of Cyrus, and of his two brothers the Magians, and of Darius, in whose second year the temple was built, and at that time Pythagoras was considered famous, as Bede says, and Zorobabel, Aggaeus, Zachariah, and Malachi were renowned as prophets. Pythagoras was taught by Pherecydes the Syrian, as Tullius says in the book mentioned above, who was the first to maintain that the souls of men are immortal. His date is established only by the date of Pythagoras his pupil, although Isidore also says in the first book that Pherecydes wrote histories in the time of Esdras, which might have been toward the end of Pherecydes' own life and in the youth of Esdras. For from the time when Pythagoras was said to have flourished there passed the thirty-six years of the reign of Darius, the twenty-six of Xerxes, the seven months of Artabanus, and the six years of Artaxerxes Longimanus, before Esdras went up from Babylon to Jerusalem. For in the seventh year of his reign on the first day of the first month, according to the Scripture and the chronicle, Esdras set out.

Chapter XII

THESE two classes of philosophers, the Ionic and the Italic, branched out through many schools and various successors up to the teaching of Aristotle, who corrected and changed the positions of all his predecessors, and attempted to complete philosophy. Archytas of Tarentum and Timaeus named especially among others succeeded Pythagoras. But the leading philosophers, Socrates, Plato, and Aristotle, did not descend from this line, but were Ionians and true Greeks, of whom the first was Thales of Miletus. How the rest succeeded him, Augustine shows in the eighth book of the City of God. For after Thales first came Anaximander his pupil, whose successor was Anaximenes, and these two were in the time of the Jewish captivity, as Augustine states in the eighteenth book of the City of God, and others likewise agree in this. Now Anaxagoras and Diogenes, auditors of Anaximenes, succeeded him under Darius Hydaspes, in whose second year the building of the temple began. Anaxagoras, as Augustine states, had as his successor

Archelaus his pupil, whose auditor Democritus, according to Isidore in the eighth book, and Socrates according to Augustine in his eighth book, is said to have been the pupil of Archelaus. Socrates was born, according to Bede, under Artabanus, who reigned over the Persians for seven years, whom Artaxerxes Longimanus succeeded, in whose seventh year Esdras descended from Babylon. Esdras and Socrates were therefore contemporaries, but Esdras was the older, as is clear from what has now been said. For this reason Augustine says in his eighteenth book that Socrates was after Esdras, that is, later in respect to birth. For when Esdras was influential at the court of the Persian king and among the Jews, Socrates appeared. This is the Socrates who is called the father of the great philosophers, since he was the teacher of Plato and Aristotle, from whom all the schools of philosophers have descended. Now Plato, according to Bede in his larger work on Chronologies, was born under Sogdianus, who reigned for seven months, whom Darius surnamed Nothus, although Bede writes in his larger work on the Chronologies that Plato was born under the same Darius. But in that work he reckons the time of Sogdianus, since it was short, under the reign of Darius. For he joins him in the succession with Artaxerxes Longimanus. Now at Plato's birth Hippocrates was famed as a physician, as Bede states, and at this time we find Empedocles and Parmenides. But Plato, first learning what Socrates and Greece had to teach, as a teacher of Athens sought Egypt and Archytas of Tarentum and Timaeus, and traversed most laboriously that same Italian coast which was called Great Greece, as Jerome states to Paulinus. And against Rufus Jerome writes that Plato, after founding the Academy and teaching innumerable pupils, when he perceived that much was lacking in his system of instruction, came to Great Greece, and there instructed by Archytas of Tarentum and Timaeus of Locri in the doctrine of Pythagoras united his own elegance and grace with teachings of this kind. Plato is placed before all other philosophers according to the sacred writers, because his books reached their hands, and because he wrote beautiful sentiments about God and much about morals and a future life, which accord closely with the sacred wisdom of God, as I shall explain in the part on moral philosophy. For this reason many Catholic writers have thought that he heard

Jeremiah the prophet in Egypt. For he visited Egypt in his search for wisdom and was taught by the barbarian priests, as Tullius writes concerning him in the fifth book of the Academica. But Augustine says that he did not live in the time of Jeremiah. For Jeremiah, as he says in the eighteenth book of the City of God, first prophesied in the time of the fourth king from Romulus, who was called Ancus Martius, and in the time of the fifth king, namely Tarquinius Priscus. Plato, however, did not live at that time, but was born almost one hundred years after the time of Jeremiah, as Augustine states in the eighth book, and he did not, as some thought, find the Seventy Translators in order to receive their instruction. For as Augustine states in the eighth book and Tully in his book on Old Age, Plato died in the eighty-first year of his life, and this was at the end of the reign of Artaxerxes, who was called Ochus, as Bede states; and from the year of his death, as Augustine states, there were almost thirty years to the time of the Seventy Translators. Hence Plato was not instructed in divine matters by them. But Augustine thinks that owing to his desire for knowledge he learned Hebrew and read through the books of the Old Testament, as he shows in his account of the creation of the world which he has represented in conformity with Scripture, and by the name of God which God himself gave in Exodus, "I am that I am," when Moses asked him what his name was. This is the name used by Plato, who states that it is God's name.

Chapter XIII

Before the death of Socrates Aristotle was born, since he was his auditor for three years, as we read in the life of Aristotle. According to Bede he was born under Artaxerxes surnamed Memnon, the successor of Darius Nothus. In his seventeenth year he was an auditor of Socrates and listened to him for three years. After the death of Socrates he became an auditor of Plato, according to Bede, and remained so for twenty years, as we read in his life. After Plato's death he lived twenty-three years. As is clear from the statements made, the extent of his life was sixty-three years. This statement is likewise made in Censorinus' book on the natal day. Çensorinus' book states that

Philosophy

Aristotle maintained a struggle against a mortal disease for three years by the greatness of his soul rather than by the virtue of medicine. Aristotle became the teacher of Alexander the Great and on the authority of his pupil sent two thousand men throughout the world to inquire into the secrets of nature, as Pliny tells us in the eighth book of his Natural History, and Aristotle composed a thousand books, as we read in his life. He purged away the errors of preceding philosophers, and enlarged philosophy, aspiring to that full measure of this subject possessed by the ancient patriarchs, although he was not able to perfect each of its parts. For his successors have corrected him in some particulars, and have added many things to his works, and additions will continue to be made until the end of the world, because there is no perfection in human discoveries, as has been shown in what precedes. Nature made this man strong, as Averroës says in the third book on the Soul, that she might discover the ultimate perfection of man. Aristotle, on the testimony of all great philosophers, is the greatest of them all, and that alone must be ascribed to philosophy which he himself has affirmed; whence at the present time he is called by the title Philosopher in the realm of philosophy, just as Paul is understood by the title of Apostle in the doctrine of the sacred wisdom. But the larger portion of the philosophy of Aristotle received little attention either on account of the concealment of the copies of his work and their rarity, or on account of their difficulty, or unpopularity, or on account of the wars in the East, till after the time of Mahomet, when Avicenna and Averroës and others recalled to the light of full exposition the philosophy of Aristotle. Although only some of his works on logic and certain others have been translated from Greek by Boetius, yet from the time of Michael Scotus, whose translations with authentic expositions of certain parts of Aristotle's works on nature and metaphysics appeared in the year of our Lord 1230, the philosophy of Aristotle has grown in importance among the Latins. But in comparison with the vastness of his wisdom contained in a thousand books, only a very small portion up to the present time has been translated into Latin, and still less is in common use among students. Avicenna in particular, the imitator and expositor of Aristotle, and the man who completed philosophy as far as it was possible for him to do so, composed a threefold volume of philosophy, as he states in the prologue of his book Sufficiency: one part popular in character like the philosophical dicta of the Peripatetics who are of the school of Aristotle; the second part in conformity with the pure truth of philosophy, which does not fear the thrusts of the spears of contradicters, as he himself asserts; and the third part conterminous with his own life, in which he gave an exposition of the earlier parts and collected together the more hidden facts of nature and art. But of these volumes two have not been translated; the Latins have the first in certain parts; which is called the book of Assipha, that is the book of Sufficiency. After him came Averroës, a man of sound wisdom, correcting many statements of his predecessors and adding much to them, although he must be corrected in some particulars and completed in many others. For of making many books there is no end, as Solomon writes in Ecclesiastes.

Chapter XIV

From these considerations the main conclusion is clear, and it is manifest that all unbelieving philosophers and poets and Sibyls and whosoever were devoted to wisdom, appeared after the true, believing, and perfect philosophers, who were the sons of Seth and Noah with their sons, to whom God granted a life of six hundred years to complete the study of wisdom, as Josephus says in the first book of Antiquities, saying that in less time they could not complete philosophy, especially on account of astronomy, in which there is the greater difficulty because mortal men are far distant from the heavenly bodies. But God revealed all things to them, and gave them length of life to complete philosophy through experience. But because of the evil of men, who have abused the paths of wisdom, as in the first instance did Nemroth and Zoroastes, Atlas, Prometheus, Mercurius Trismegistus, Aesculapius, Apollo, Minerva, and the like, who were worshiped as gods because of their wisdom, God darkened the foolish heart of the multitude: and gradually the knowledge of philosophy disappeared until Solomon again recalled and perfected it in its entirety, as Josephus shows in the eighth book of the Antiquities. And again on account of men's sins the study of philosophy vanished by degrees, until Thales of Miletus took it up again, and his successors broadened it, until Aristotle completed it, as far as was possible for that time. But those men learned all things from the Hebrews, as Aristotle says in the book of Secrets, as was shown above. Since, therefore, the first unbelieving philosophers, as Nemroth, Prometheus, Atlas, Apollo, and others were after Seth and Noah and Shem and Abraham and after their sons who completed philosophy; and since following Solomon, who a second time perfected it, were the rest of the unbelieving philosophers, Thales, Pythagoras, Socrates, Plato, and Aristotle, it is clear that philosophy in its perfection was first given to the holy patriarchs and prophets, to whom likewise was revealed the divine law by one and the same God. This would not have happened unless philosophy were wholly conformable with God's saints and holy law, and useful and necessary for the understanding, execution, and defense of that law. Moreover, a belief in philosophy must be secured, philosophy must be demonstrated, diffused, and enlarged, for it is necessary in all these ways, as will appear on going through the separate parts of philosophy. Therefore philosophy is merely the unfolding of the divine wisdom by learning and art. Hence there is one perfect wisdom which is contained in the Scriptures, and was given to the saints by God; to be unfolded, however, by philosophy as well as by canon law.

Chapter XV

Hence it follows of necessity that we Christians ought to employ philosophy in divine things, and in matters pertaining to philosophy to assume many things belonging to theology, so that it is apparent that there is one wisdom shining in both. The necessity of this I wish to establish not only on account of the unity of wisdom, but because of the fact that we must revert below to the lofty expressions relating to faith and theology, which we find in the books of the philosophers and in the parts of philosophy: so that it is not strange that in philosophy I should touch upon the most sacred truths, since God has given to the philosophers many truths of his wisdom. The power of philosophy must be applied to sacred truth as far as we are able, for the excellence of philosophy does not otherwise shine forth, since philosophy considered by itself is of no utility. The unbelieving philosophers have been condemned, and "they knew God, and did not glorify Him as God, and therefore became fools and perished in their own thoughts," and therefore philosophy can have no worth except in so far as the wisdom of God required it. For all that is left is in error and worthless; and for this reason Alpharabius says in his book on Sciences that an untaught child holds the same position with respect to a very wise man in philosophy as such a man does toward the revelation of God's wisdom. Wherefore philosophy by itself is nothing, but it then receives vigor and dignity when it is worthy to assume the sacred wisdom. Moreover, the study of wisdom can always continue in this life to increase, because nothing is perfect in human discoveries. Therefore we of a later age should supply what the ancients lacked, because we have entered into their labors, by which, unless we are dolts, we can be aroused to better things, since it is most wretched to be always using old discoveries and never be on the track of new ones, as Boetius says, and as we proved clearly above in the proper place. Christians likewise ought to handle all matters with a view to their own profession, which is the wisdom of God, and to complete the paths of the unbelieving philosophers, not only because we are of a later age and ought to add to their works, but that we may compel the wisdom of the philosophers to serve zealously our own. For this the unbelieving philosophers do, compelled by truth itself as far as it was granted them: for they refer all philosophy to the divine wisdom, as is clear from the books of Avicenna on Metaphysics and Morals, and from Alpharabius, Seneca, and Tullius, and Aristotle in the Metaphysics and Morals. For they refer all things to God, as an army to its chief, and draw conclusions regarding angels and many other things; since the principal articles of the faith are found in them; for as will be set forth in the morals, they teach that there is a God and that he is one in essence, of infinite power and goodness, triune in persons, Father, Son, and Holy Spirit, who created all things out of nothing; and they touch on many things concerning Jesus Christ and the Blessed Virgin. Likewise also they teach us of Antichrist and the angels and of their protection of men, and of the resurrection of the dead and of future judgment and of the life of future happiness promised by God to those obedient to him, and of the future misery which he purposes to inflict on those who do not keep his commandments. They write also innumerable statements in regard to the dignity of morals, the glory of laws, and concerning a legislator who must receive the law from God by revelation, who is to be a mediator of God and men and a vicar of God on earth, the Lord of the earthly world. When it shall be proved that he has received the law from God, he must be believed in all things to the exclusion of all doubt and hesitation; who must direct the whole race in the worship of God and in the laws of justice and peace, and in the practice of virtues because of the reverence of God and because of future felicity. [We must avail ourselves of their teachings] because they wrote that the worship of idols should be destroyed, and because they prophesied of the time of Christ. From whatever source the philosophers got these statements and similar ones, we find them in their books, as a clear proof will show in what follows, and any one can discover the fact who cares to read through the books of the philosophers. For we cannot doubt that these things were written by them, from whatever source they received them. Nor should we be surprised that philosophers write such statements; for all the philosophers were subsequent to the patriarchs and prophets, as we brought out above in its proper place, and therefore they read the books of the prophets and patriarchs which are in the sacred text.

Chapter XVI

They likewise composed other books, touching on the mysteries of Christ, as in the book of Enoch and in the book on the testaments of the patriarchs and in the third, fourth, and fifth books of Esdras, and in many other books concerning some of which mention is made in the sacred text, as the books of Nathan, Samuel, and Abdon the prophets. For in books of this kind articles of the faith are expressly touched upon, and far more explicitly than in the canon of Scripture. For besides other books the book on the testaments of the patriarchs shows all things that were fulfilled with respect to Christ. For each patriarch at his death made known to his sons and his tribe and predicted those things which must be believed regarding Christ, as is clear from that book. Although these books are not in the canon of Scripture, yet the sacred and wise Greek and Latin writers have used them from the beginning of the Church. For the blessed Judas received authority from the book of Enoch, and Augustine bases much on that book in order to show that the saints possessed philosophy before the philosophers did, and he says that that book lacks authority more because of its excessive antiquity than for any other reason. Moreover, in regard to the other books it is clear that they were used by the sacred writers and wise men of antiquity because they are known to contain clear truths concerning Christ. The philosophers therefore, curious and diligent in the study of wisdom, traversed many different regions in their search for wisdom, and for this reason read through the sacred books, and learned many things from the Hebrews. For Avicenna in the Foundations of Moral Philosophy quotes the words of Isaiah in regard to life eternal, saying that life eternal is that which eye hath not seen nor ear heard, and he states that charity takes away sin, as the prophet of truth, Tobias, says. Augustine maintains in the eighth book of the City of God that Plato had read the book of Genesis because Plato's description of the creation of the world is similar to that given in Genesis; also that he read the book of Exodus because of the name of God found in it, namely, "I am that I am." For Plato employed this name, and he could not have found it elsewhere, as Augustine agrees. In addition to the sacred prophetic books the patriarchs and prophets composed books of philosophy, nay, they twice perfected the whole of philosophy; and that which the philosophers received exclusively from them, as Aristotle says, has been clearly shown in what precedes. And since there is one complete wisdom, which suffices the human race, for this reason the sacred writers have mingled divine matters with others, as far as philosophy was able to receive them. On this account by means of those philosophic books of the sacred writers the philosophers learned much concerning divine truths.

Philosophy

Chapter XVII

Moreover, since the philosophers themselves were devoted to truths and to every good quality of life, despising riches, luxury, and honors, aspiring to future felicity as far as human frailty could do so, nay, rendered victorious over human nature, as Jerome writes concerning Diogenes in his book against Jovinianus, it is not strange if God, who illumined them in these lesser matters, should give them some light in regard to the greater truths; and that he should do this, if not primarily for them, yet for our sake, that through their persuasion the world might be prepared for the faith. For this purpose he caused many Sibyls to appear, namely, ten prophetesses, as all sacred writers agree, according to Augustine in the eighteenth book of the City of God, and Isidore in the seventh book of Etymologies, and likewise other writers. Moreover, histories and philosophers and poets are in agreement universally in regard to these Sibyls. It is certain that they gave utterance to divine truths respecting Christ and future judgment and matters of this kind. Therefore it is much more probable that the wisest and best philosophers received truths of this nature from God. That the Sibyls uttered divine truths is shown by the sacred writers and otherwise; and it is sufficient to quote what Augustine says in the eighteenth book of the City of God. Now those women gave utterance to statements of this kind; "they will smite God in the face with unclean hands, and will spit out from impure lips poisonous spittle, and he will without guile offer his sacred back to the scourge and will receive the blows in silence, and will be crowned with thorns. For food they gave him gall and for drink vinegar. Foolish race, thou didst not perceive thy God playing with the minds of mortals but thou didst crown him with thorns, thou didst mingle the horrid gall. The veil of the temple shall be rent and in the middle of the day there shall be the darkness of night for three hours. And he shall die, for three days remaining asleep." And again the Sibyl said in verse:

"As a sign of judgment the earth shall be wet with sweat, from Heaven a king shall come destined through the ages, present in flesh to judge the world, whence the unbeliever and the faithful shall see God on high with his saints at the end of the world. Thus shall the souls with the body be present, that he himself may judge them. Fire shall burn the earth, the sea, and the sky. Flames shall consume the guilty for ever. Each shall then speak disclosing his secret deeds, and God shall bring to light the secrets of the heart. The brightness of the sun is taken away and the chorus of the stars passes away, the sky shall be rolled up, the splendor of the moon shall depart. He shall cast down the hills and exalt the valleys. So likewise the fountains and rivers are parched with fire. The earth yawning shall disclose Tartarean Chaos. Fire and a stream of sulphur shall fall from the heavens."

If therefore weak women spoke in this fashion, much more should we believe that the wisest philosophers had a taste of truths of this kind. Augustine maintains in the eighteenth book of the City of God that others perceived the truth of God besides those who descended one after another from the stock of Abraham up to Christ. For Job knew of the resurrection and the truths of God. And in the chronicles of Eusebius it is written that in the reign of Helena and Constantine a body was dug up, with which was found this written statement, "I believe in Christ; under Helena and Constantine the sun will again behold me." And now in the time of Pope Alexander the Fourth a Saracen in Bozea, despising the world, and devoting himself under his own law to God and virtue and the contemplation of another life, received an angelic visitation advising him to be converted to the faith of Christ, and was baptized by a priest of the Januensian merchants. The fact was known to Pope Alexander and to many, and at the present time very many recall it.

Chapter XVIII

This same conclusion, moreover, can be proved by two principles of metaphysics. For this science deals with what is common to all things and all sciences, and for this reason shows the number of the sciences and the need of another science beyond philosophy, whose principles it touches upon in the universal, although it cannot assign this science in the particular. For philosophy knows its own imperfection, and is aware that it lacks full knowledge of matters regarding which it is most important that it should know, as Aristotle states frequently in the Metaphysics, and Avicenna likewise as we mentioned above, and shall mention again in its proper place. And for this reason philosophy advances to the discovery of a higher science, and proves that it must exist, although philosophy cannot unfold it in its special function; and this science is in its entirety the divine one, which philosophers call the perfect theology, and for this reason philosophy raises itself to the science of divine things. Moreover, philosophers were anxious above all to inquire concerning the verification of a school in which was contained the salvation of man, and they give very clear methods of proving this, as will be shown from the morals. And they find for a certainty that there must be some school trustworthy and sufficient for the world, whose principles they assign, which cannot be found except in the religion of Christ, as is proved in the appropriate places, and it is shown that it accords with the goodness of God and human necessity that this trustworthy school should be known by men. But this cannot be proved to unbelievers by the law of Christ nor by the sacred writers, because following the rule of disputation they can deny all things in the law of Christ, just as Christians deny what is contained in other laws. And since they deny Christ, it is not strange if they deny the authorities of the Christians. Now persuasion in regard to the faith is necessary, but this can be accomplished in only two ways, either by miracles which are above believers and unbelievers, concerning which no one can presuppose, or by the road common to believers and unbelievers, but this is only by philosophy. Therefore philosophy must give the methods of proof of the Christian faith. But the articles of this faith are the principles belonging to theology; therefore philosophy must enter into the proofs of the principles of theology, although less deeply than into the principles of other sciences; and in this way let the subject rest on this reasonable basis until we come to the examination of the schools. For there we shall show that moral philosophy more efficaciously serves theology in this particular, and therefore although in point of truth such matters belong to theology, they are nevertheless philosophical, but because of theology.

Chapter XIX

Moreover, all speculative philosophy has moral philosophy for its end and aim. And since the end imposes a necessity on those things pertaining to the end, as Aristotle says in the second book of the Physics, therefore speculative science always aspires to its own end, and elevates itself to it, and seeks useful paths to this end, and for this reason speculative philosophy is able to prepare the principles of moral philosophy. Thus therefore are the two parts of wisdom related among the unbelieving philosophers; but with Christian students of philosophy moral science apart from other sciences and perfected is theology, which adds to the greater philosophy of the unbelievers the faith of Christ and truths which are in their nature divine. And this end has its own speculation preceding, just as the moral philosophy of the unbelievers has its own. There is therefore the same relation between the ends in view as between the speculations; but the end, namely, the Christian law, adds to the law of the philosophers the formulated articles of the faith, by which it completes the law of moral philosophy, so that there may be one complete law. For the law of Christ takes and assumes the laws and morals of philosophy, as we are assured by the sacred writers and in the practice of theology and of the Church. Therefore the speculations of Christians preceding their own law must add to the speculation of the other law those things which are able to teach and prove the law of Christ, in order that one complete speculation may arise, whose beginning must be the speculative philosophy of the unbelieving philosophers; and the complement of this must be added to theology in accordance with the peculiar characteristics of the Christian law. And for this reason the complete philosophy among Christians must have a much more profound knowledge of divine things than it has among the unbelieving philosophers; and for this reason Christians ought to consider philosophy as if it had just been discovered, so that they might make it suitable for its own end; and therefore many things must be added in the philosophy of the Christians, which the unbelieving philosophers could not know. And there are reasons of this kind rising in us from the faith and from the authorities of the law and of the sacred writers, who are acquainted with philosophy, and they can form the common points of complete philosophy and theology. And these are recognized by the fact that they must be common to believers and unbelievers, so that they may be so well known, when they are brought forward and proved, that they cannot be denied by the wise and those instructed in the philosophy of the unbelievers. For the unbelieving philosophers are ignorant of many things at present concerning divine matters, and if these were suitably set before them and proved by the principles of the complete philosophy, that is, by the vivacity of reason, which has its origin in the philosophy of the unbelievers, although completed by the faith of Christ, they would receive it without contradiction and would rejoice in regard to the truth set before them, because they are eager for wisdom and are more studious than Christians. I do not say, however, that any one of the special articles of the Christian faith should be received on trial, but there are many common rational truths, which every wise man would easily accept from another, although he might be ignorant of them himself, as every man studious and desirous of knowledge learns many things from another and receives them by rational arguments, although he was formerly ignorant of them.

Those philosophizing should not be surprised, therefore, if they must needs raise philosophy to the level of divine things and theological truths and the authorities of sacred writers, and employ these freely whenever the occasion arises, and prove them when necessary, and by means of these prove other matters; since without doubt philosophy and theology have much in common. The sacred writers not only speak as theologians, but as philosophers, and frequently introduce philosophical subjects. Therefore Christians desiring to complete philosophy ought in their works not only to collect the statements of the philosophers in regard to divine truths, but should advance far beyond to a point where the power of philosophy as a whole may be complete. And for this reason he who completes philosophy by truths of this kind must not on this account be called a theologian, nor must he transcend the bounds of philosophy; since he can handle freely what is common to philosophy and theology and what must be accepted in common by believers and unbelievers. There are many such matters besides the statements of unbelieving philosophers, which belonging as it were within the limits of philosophy the man philosophizing in the right way should collect, wherever he finds them, and he should assemble them as though they were his own, whether they occur in the books of the sacred writers, or in the books of the philosophers, or in sacred Scripture, or in the histories, or elsewhere. For there is no author who does not besides his main theme introduce incidentally some matters which belong rather elsewhere; and for this reason there is a linking together of sciences, because each thing in a manner is dependent on another. But every one who handles a subject in the way he should must assign what belongs to it, both what is necessary and what befits its worth; and therefore wherever he finds these things he knows how to recognize his own, and therefore he must seize them as his own and arrange them in their proper places. For this reason the philosophizing Christian can unite many authorities and various reasons and very many opinions from other writings besides the books of the unbelieving philosophers, provided they belong to philosophy, or are common to it and theology, and must be received in common by unbelievers and believers. If this be not done, there will be no perfecting, but much loss. And not only must this be done to complete philosophy, but because of Christian conscience which must reduce all truth to divine truth that the former may be subject to and serve the latter; also for this reason that the philosophy of the unbelievers is essentially harmful and has no value considered by itself. For philosophy in itself leads to the blindness of hell, and therefore it must be by itself darkness and mist.

PART THREE OF THIS PLEA

On the usefulness of grammar

CHAPTER I

AFTER making the statement, therefore, that there is one perfect wisdom, which is contained in the Scriptures and must be expounded by canon law and philosophy, by which the world must be directed, nor is there needed any other science for the advantage of the human race, for it contains in itself the whole power of law and philosophy; I now wish to take up those parts of philosophy which are especially valuable for the exposition of this splendid wisdom. These are five, without which neither divine nor human things can be known, while a sure knowledge of them makes it easy for us to know all things. First, there is grammar, developed in the foreign tongues from which the wisdom of the Latins has sprung. For it is impossible for the Latins to reach what is necessary in matters divine and human except through the knowledge of other languages, nor will wisdom be perfected for them absolutely, nor relatively to the Church of God and to the remaining three matters noted above. This I now wish to state, and first with respect to absolute knowledge. For the whole sacred text has been drawn from the Greek and Hebrew, and philosophy has been derived from these sources and from Arabic: but it is impossible that the peculiar quality of one language should be preserved in another. For even dialects of the same tongue vary among different sections, as is clear from the Gallic language, which is divided into many dialects among the Gauls, Picards, Normans, Burgundians, and others. A fitting and intelligible expression in the dialect of the Picards is out of place among the Burgundians, nay, among their nearer Gallic neighbors; how much more then will this be true as between different languages? Therefore an excellent piece of work in one language cannot be transferred into another as regards the peculiar quality that it possessed in the former. Hence Jerome in his epistle on the best kind of interpretation speaks thus, "If I translate literally, the result is absurd in sound. But if any one does not think that the grace of a language is changed by translation let him translate Homer literally into Latin. I shall make a further statement; let him translate this same author in his own tongue in the words of prose, he will see a ridiculous order and a very eloquent poet speaking with difficulty." For let any one with an excellent knowledge of some science like logic or any other subject at all strive to turn this into his mother tongue, he will see that he is lacking not only in thoughts, but words, so that no one will be able to understand the science so translated as regards its potency. Therefore no Latin will be able to understand as he should the wisdom of the sacred Scripture and of philosophy, unless he understands the languages from which they were translated.

Secondly, we must consider the fact that translators did not have the words in Latin for translating scientific works, because they were not first composed in the Latin tongue. For this reason they employed very many words from other languages. Just as these words are not understood by those ignorant of those languages, so are they neither pronounced correctly nor are they written as they should be. And, what is bad, owing to their ignorance of Latin they have used Spanish, and other native tongues, to an almost endless extent in place of Latin. For let one example suffice for many from the book on Plants of Aristotle where he says, "*Belenum* which is very harmful in Persia when transplanted to Jerusalem becomes edible." This word is not the scientific one but colloquial Spanish. For *jusquiamus* [henbane] or the seed of the cassilago, is its name in Latin. After being laughed at by my Spanish students, familiar as they were with the words of their own language, when I did not understand what I was reading, I at length learned from them the meaning of this word and of many more besides.

Thirdly, although the translator ought to be perfectly acquainted with the subject which he wishes to translate and the two languages from which and into which he is translating, Boetius alone, the first translator, had full mastery of the languages; and Master Robert, called Grosse-Teste, lately bishop of Lincoln, alone knew the sciences. Certain other ordinary translators, like Girardus Cremonensis, Michael Scotus, Aluredus Anglicus, Hermannus Alemannus, whom we saw in Paris, have failed greatly as well in the languages as in the sciences; even as this same Hermannus has confessed concerning himself and others, as their translation shows. For so great is the perverseness, crudity, and terrible difficulty in the translated works of Aristotle that no one can understand them, but each one contradicts another, and false statements are found again and again, as is clear from a comparison of the different translators and of the texts of the different languages. Likewise in the sacred text false statements are found and many bad translations. For Jerome proves that the translation of the Seventy Interpreters and of Theodotion and of Aquila had many errors; and since these errors were published throughout the whole Church, and all men stood for the translation of the Seventy as for their own life, Jerome was considered a falsifier and a corrupter of the Scriptures, until little by little the truth of the Hebrew became clear when turned into Latin by the sainted Jerome. Lest, however, he should hinder the Latins by too much alteration, for this reason, as he himself writes, he sometimes adapted himself to the Seventy Interpreters, sometimes to Theodotian, sometimes to Aquila; and therefore has left much as it was translated by others, and for this reason many false statements have remained. For, as Augustine proves in the second book on Christian Doctrine, the translation given in the book of Wisdom is a bad one, "Spurious vines will not produce deep roots." For it should be, Spurious plantings, or adulterous plantings, as Augustine proves by the Greek. And yet Jerome let this pass like many other places for the peace of the Church and of the doctors. And it is clearly known that Jerome, subject to human frailty, sometimes erred in his own translation, as he himself frequently confesses. For since he had made a bad translation of the nineteenth chapter of Isaiah, he takes it up again in the original in the fifth book saying, "In this place also which we translated *curving* and *bridling* we can say *curving* and *frisking*. But while hastily translating what was written, deceived by the ambiguity we translated the Hebrew word *acmon* as *bridling*." He also again considered another place which he had translated badly in the same chapter saying, "I think it better to censure my error rather than while blushing to confess my ignorance to persist in the error. In the passage which I translated, 'and the land of Judah shall be a festivity to

Egypt,' the reading in Hebrew is *agga* which can be translated both as *festivity*, whence *aggeus* is translated as festive, and as *fear*, which Aquila has more significantly translated by *girosin*, since any one fearful and afraid turns about his eyes and fears a coming foe. Therefore if we wish to take it in a good sense that the recollection of Judah is a joy to Egypt, festivity is the right word: but if, as I judge, the idea is one of fear instead of joy we translate it terror or fear."

Chapter II

The fourth reason for this condition is the fact that the Latins up to the present time lack very many philosophical and theological works. For I have seen two books of the Maccabees in Greek, namely the third and the fourth, and Scripture makes mention of the books of Samuel and Nathan and Gad the seer, and of others which we do not have. And since the whole confirmation of sacred history is given by Josephus in his books on Antiquities, and all the sacred writers take the fundamentals of their expositions from those books, it is necessary for the Latins to have that work in an uncorrupted form. But it has been proved that the Latin codices are wholly corrupt in all places on which the import of history rests, so that the text is self-contradictory everywhere. This is not the fault of so great an author, but arises from a bad translation and from the corruption by the Latins, nor can it be remedied except by a new translation or by adequate correction in all fundamental points. Likewise the books of the great doctors like the blessed Dionysius, Basil, John Chrysostom, John of Damascus, and of many others are lacking; some of which, however, Master Robert, the aforesaid bishop, has turned into Latin, and others before him translated certain other works. His work is very pleasing to theologians. If the books of these authors had been translated, not only would the learning of the Latins be augmented in a glorious way, but the Church would have stronger supports against the heresies and schisms of the Greeks, since they would be convinced by their own sacred writers whom they cannot contradict.

Likewise almost all the secrets of philosophy up to the present time lie hidden in foreign languages. For as in many instances only what is common and worthless has been translated; and much even of this character is lacking. For lines almost without number, chapters, parts of books, and whole books are omitted in the works on metaphysics, nature, logic, and on other topics, besides great secrets of the sciences and of the arts and the hidden things of nature that have not yet been translated. Such is the second philosophy of Avicenna, which he calls oriental, which is devoted to pure philosophy, and does not fear the thrusts of the lances of contradicters, and the third, conterminous with his life, in which he collected secret experiments, as he noted in the introduction to his first philosophy. Likewise, although Aristotle completed the eight principal parts of natural philosophy containing in them many sciences, we do not have all of the first part, and almost nothing of the others. And in the same way although he himself completed the nine sciences composing mathematics, we have no part of his text. What we have on metaphysics can be reckoned of no value on account of many grave defects. Although there are five great sciences composing morals, we have only the first and a little of the second. Also there is missing from his logic a book better than the others, and the book next in excellence to it has been badly translated, and cannot be known, nor is it in general use, because it has come only lately into the hands of the Latins in a defective and rough translation. Nor is it strange if I call those books of logic better, for of necessity there must be four true arguments; for two stir the speculative intellect or reason, namely, the dialectic, through the feeble and initial habit of the mind, which is opinion, so that we are disposed toward knowledge, which is the complete and final habit, in which the mind is at rest in its contemplation of truth. And this habit of the mind is acquired by the demonstrative argument. But since the will or active intellect is nobler than the theoretical and virtue united with happiness excels infinitely pure knowledge, and is without comparison more necessary for us, we must of necessity have arguments to arouse the active intellect, especially since we are weaker in this particular than in theory. For we all gladly taste of the tree of the knowledge of good and evil, but approach with difficulty the tree of life, that we may embrace the noble virtues for the sake of future felicity. Therefore the active intellect must of necessity have its own aids, and be aroused by its own proper arguments just as the speculative intellect by its arguments. And for this reason it was necessary that the teaching in regard to these arguments should be handed down, which moral philosophy and theology employ freely. For as the speculative sciences rejoice in the speculative arguments of opinion and pure knowledge, so the active sciences, as theology and active moral philosophy, consider arguments by which we are aroused to action, that is, to good works, and are turned to a love of eternal felicity. And here there are two ways of turning us. One is that which influences the mind to belief, agreement, pity, compassion, and acts of this kind; and to their contraries when there is need; and this argument is called rhetorical, and is related to the active intellect in the same way as the dialectic argument to the speculative intellect. For it causes the weak habit, namely, the persuasion of credulity and faith, to be followed by the complete habit, which is the love of what is believed and affection strengthened by opinion. And this habit, which turns us to the love of good works, is secured by the poetic argument, because true poets, like Horace and others, Greek and Latin, attack vices and magnify virtues, so as to attract men to a love of good and a hatred of sin. For as that famous poet says, "Poets wish either to benefit or please. He has won full approbation who has mingled the useful with the sweet."

For he confers no small benefit on his fellow citizens who delights them in the subject of morals. For it is necessary not only to teach, but to delight and urge forward. Hence the poet as well as the orator ought to do these three things, that by teaching he may render his hearers docile, by delighting them he may make them attentive, and by urging or turning them he may force them to work. These arguments are most potent in matters pertaining to salvation, but in purely speculative matters are impotent; just as demonstration is most effective in pure theory but wholly impotent in practical matters and in those pertaining to salvation. Accordingly Aristotle says in the first book of the Moral Philosophy that it is equally a fault for a mathematician to use a rhetorical argument and for a rhetorician to try a demonstration, since, as he says in the second book, this science does not exist for the sake of mere contemplation but for the purpose of making us good. Aristotle therefore composed books on these arguments, and Averroës and Alpharabi expounded them in their commentaries, and Avicenna instructs us in his works on logic about these arguments. Alpharabius in his book on the Sciences states that two parts of logic ought to consist of these two arguments, because logic alone ought to teach the nature of arguments and how they are composed for the use of all other sciences; and then logic serves the speculative sciences by two arguments, the dialectic and the demonstrative. Moreover, it furnishes practical arguments to morals. And because theology and canon law determine morals and laws and rights, therefore these two arguments are necessary to them, and every theologian and jurist and moral philosopher makes use of these arguments of necessity, by custom and practice, whether his plea be made to prelate, or prince, or judge, or people, or private individual, although the Latins do not yet possess the knowledge of these arguments as regards the force of the art of logic. How these arguments are composed we need not state at present; but in the work which your Beatitude has demanded they must be explained. There is nothing, however, in regard to the speculative sciences more useful in proving the faith to unbelievers, that they may be converted to a belief in the Christian faith and to a love for it; and likewise that we may be skillful preachers to all who are in need of preaching, and so also in regard to the other forms of persuasion useful for salvation. We are, moreover, greatly aided by Augustine in the third and fourth books on Christian Doctrine and by the works of Tullius and Seneca, and by their letters and certain other facts that can be collected in the Latin language concerning these arguments, although the text itself of Aristotle is lacking to us.

Chapter III

There is a fifth reason in addition, because the sciences were formed and expounded with the same idea, and therefore since the sciences were given to the Latins from other languages, all the sacred writers and Latin philosophers who expound the sciences have used copiously other languages, and multiply for us words without end in Greek, Hebrew, Chaldean, and Arabic, besides those contained in the texts. We are the sons and successors of the sacred writers and of the wise philosophers, like Boetius, Pliny, Seneca, Tullius, Varro, and other men of wisdom even up to these latest times. For we saw certain men of the past who labored much in languages, like Master Robert, mentioned above, the translator and bishop, and Thomas, the venerable president of Saint David, lately deceased, and brother Adam of Marsh, and Master Hermann, the translator, and certain other men of science. But since we do not imitate them, we therefore lack a grasp of the sciences to a degree past belief, because we cannot understand their authentic expositions and consequently we are unable to gain an understanding of the sciences. I offer two examples to serve for many. Jerome says in the prologue to his commentary on Daniel that Daniel and Esdras are written in Hebrew letters but in the Chaldean speech, also one pericope of Jeremiah. Now all theologians state that this pericope consists of the Lamentations of Jeremiah, because pericope is the same as small part. But all Hebrews know that the Lamentations are written in Hebrew letters and in the speech of the Hebrews. Any one who knows anything about Hebrew can perceive this; hence the boy here is not ignorant of this fact. Then we can mention this pericope from the tenth chapter of Jeremiah, where it is said, "Thus therefore shall ye say unto them, Let the Gods that have not made the heavens and the earth perish from the earth and from those things which are under the heavens." For this passage alone in Jeremiah is written in the Chaldean language, as all Hebrew scholars know. It is certain that the Chaldean and the Hebrew have the same tongue but a different dialect, like the Gaul and the Picard; for dialect is the particular form of a language determined by a nation. Hence the Hebrew says *Heloim* for God or Gods, the Chaldean says *Heloa,* and for heaven or heavens the Hebrew says *Samaim,* the Chaldean *Samaa,* and for *not* the Hebrew says *lo,* but the Chaldean *la;* and thus there are accidental differences in the same language. Such is the case in the pericope mentioned above. Yet before this pericope is written here in Hebrew and Chaldean, the Hebrew alphabet must be given in order that the subject under discussion may be more easily understood. The letters of the

Hebrew alphabet are written first, then in the line above are given their names, and last our letters corresponding to the Hebrew ones, in order that we may know the value of the letters and the sounds indicated, some being vowels and some consonants.

z	v	e	d	g	b	a	m
zain	vaf	he	dalet	gimel	bet	aleph	mem uverte
ז	ו	ה	ד	ג	ב	א	מ

l	ch	ch	i	t	h	s
lamet	chaf	chaf	iot	teis	heis	sazake dreite
ל	ך	כ	י	ט	ח	ץ

s	a	s	n	n
sazake torte	ain	samech	nun dreite	nun torte
צ	ע	ס	ן	נ

m	t	s	r	k	p	p
mem close	taf	sin	ris	kof	pe	pe
ם	ת	ש	ר	ק	ף	פ

There are six vowels, *aleph, ain, he, heth, iot, vav;* the rest are consonants; *he* and *heth* are aspirated, *he* at the beginning, *heth* not only at the beginning, but at the end, and *heth* is produced in the throat, *he* in the mouth. *Aleph* likewise in the mouth and *ain* in the throat. But we must bear in mind that *iot* has only one sound, namely, j, like our j, and becomes a consonant and vowel like j with us. V, as Jerome says in his Hebrew Questions, has a double sound, namely our v and o. The remaining four have the sound of our five vowels, namely, a, e, i, o, u, as is shown by Jerome in the book of interpretations. This difference of sounds they indicate by points and dashes. For if under *aleph* a line be drawn without a point thus, אַ or with a point under the line thus, אָ, the sound is a. But if two points are made lying under the *aleph* crosswise, אֵ, or two vertically, אְ, or three like a triangle, אֶ, or five points in this way, אֱ, the sound is e. But if three points lying under *aleph* descend obliquely thus, אֻ, the sound is u. If one point be placed under the letter, אִ, the sound is i. But if a point be placed above it, the sound is o, thus, אֹ. The same is true of *ain* and *he* and *heth,* which have these five sounds by means of difference in pointing. When *vaf* has the sound of u, there may be the sign of the three points, as stated, thus, וֻ, or one point may be placed in its bosom thus, וּ. The letters for the vowels are not often written, but the vowel signs mentioned above take their place. Therefore we must note that they attach these signs to the consonants to indicate what vowel sound is to be joined in the syllable with the consonant. If I wish to indicate *ba, be, bi, bo, bu,* I must write,

ba	be	bi	bo	bu
בַּ	בֶ	בִ	בֹ	בֻ

There are likewise other signs indicating strengthened or weakened sounds of the consonants. Hence when a dash is placed above the letter it is weakened; when a point is placed in its bosom the letter is strengthened. Hence when a dash is placed above *daleth* thus, דֿ, a weakened sound is indicated like our z, as when I say, *adamas.* But when a point is placed in its bosom thus, דּ, the sound is strengthened, as when I say, *dabo.* And thus we here find in this Hebrew which follows

dii	eis	dicetis	sic	
elaa	lehom	temerun	chidena	Hebrew letters
אְלַהיאַ	לְהוֹם	תֵּאמְרוּן	כִּדְֿנַה	Chaldean tongue

non	terram et	coelum	qui
la	areka ve	semaa	di
לַא	וְאַרְקַא	שְׁמַיַּא	דִי

terra de	pereant	fecerunt
area me	iebedu	ebadu
מֵאַרְעַא	יֵאבְֿדוּ	עֲבַֿדוּ

coelo	sub de et
semaa	thehot mi u
שְׁמַיָּא:	וּסְתְּחוֹת

dii	eis	dicetis	sic	
elohim	lahem	tomeru	co	Hebrew letters
אֱלָאחִים	לָחֶם	תֹּאמְרוּ	כֹּה	Hebrew tongue

fecerunt	non	terram et	coelum	qui
asu	lo	ares ve	samaim	eser
עֲשׂוּ	לֹא	וְאֶרֶץ	שְׁמַיִם	אֲשֶׁר

sub de et	terra de	pereant
thahat mi u	eres me	iobedu
וּמִתַּחַת	מֵאֶרֶץ	יֹאבְדוּ

isto	coelo
ele	samaim
אֵלֶּח:	שַׁמַיִם

There is therefore a manifest and wretched error on the part of all in this matter owing to ignorance of these languages. Another example is taken from Greek. Since many Greek examples will be given in what follows, I wish for this reason to give here the Greek alphabet with the diphthongs which they use in writing: for what I have to say will be made clearer by this means.

a	b	g	d	e
alpha	vita	gamma	delta	e. penti, *i.e.* quintum
α	β	γ	δ	ε

z	i	th	i	k	l	m	n
zita	ita	thita	iota	kappa	labda	mi	ni
ζ	η	θ	ι	κ	λ	μ	ν

x	o	p	r	s	t
xi	o. micron	pi	ro	sima	taf
ξ	ο	π	ρ	σ	τ

y. Greek among Latins	ph	ch	ps
y. psilo	phi	chi	psi
υ	φ	χ	ψ

o
o.mega, *i.e.* magnum
ω

There are seven vowels, counting the different letters, since there is a threefold i and a twofold o. But there are only four vowels possessing a principal sound, namely, a, e, i, o. The diphthong with the Greeks is a union of two vowels with the sound of one, or a vowel with a consonant; and the final letters in the diphthongs of the Greeks are *iota* and *ipsilo. Ipsilo* therefore can follow *alpha* thus, αυ, and then the sound is like a with υ consonantal, a sound somewhat similar to the sound of a itself with f, and therefore we commonly explain the sound as that of *af;* or it can follow e thus, ευ, and then the sound is like e vocal with υ consonant, and like *ef,* as was stated in regard to *alpha* and *ipsilo;* or it can follow *ita* thus, ηυ, and sounds like *if,* as was stated concerning the others; or *ipsilo* can follow *o micron* thus, ου, and then it has the vowel sound of u; and in this case only do the Greeks have the sound of this vocal u. If *iota* follows *alpha* thus, αι, then the sound is e; if it follows e thus, ει, then the sound is i through the *iota;* if it follows o thus, οι, it has the y sound through the *ipsilo.* It can also follow *ypsilo* itself, thus, υι, and then it has the i sound with the *ipsilo.* These eight diphthongs are called proper diphthongs. But three are called improper, and are made by writing this letter *iota* beneath *alpha, ita,* and ω *mega,* thus, ᾳ, ῃ, ῳ. Sometimes, however, *iota* is placed on the line after the letter as in other diphthongs, thus, αι, but in improper diphthongs the sound never alters, but remains that of the principal letter, that is, of the letter beneath which the *iota* is written. For when it is written beneath *alpha,* the sound is a; when beneath *ita,* the sound is *ita;* when beneath ω *mega,* only the ω *mega* sounds.

The Greeks use these three diphthongs always in the dative case of the first and second declensions.

Now the example in the present instance is concerning Jacob, who meeting his brother Esau coming from Mesopotamia said, "For thus I beheld thy face, as the face of God." Augustine asks in the book of Questions, and it is in the gloss, how a holy man could compare an evil one to God and reckon him as God. He answers the question with the statement that the word God is used in Scripture in many ways, sometimes for the true God and sometimes otherwise; but the Seventy Translators, in order to point out that he was not speaking of the true God, added the article to the name of God. For the article has the property of showing the truth of a thing; but this force does not appear in Latin, since it lacks the article. But it does appear in the Gallic tongue; and hence when they say in Paris, *li reis vient,* the article *li* designates the particular and actual king of such a place, since they are speaking of the king of France. This would not suffice to denote the arrival of the English king to the city of Paris. For no one would say of the English king's coming to Paris, *li reis vient,* but would add something, *li reis de Engleterre vient.* Therefore the article alone suffices to designate the truth and property of the thing under discussion. For this reason Augustine says that the Greek runs thus, προσωπον θεου, which in Greek has the sound *prosopon theu* without the article; and not thus, προσωπον του θεου. *Prosopon* here signifies countenance or face, *theu* is the genitive case of this noun *theos,* that is, God, and *tu* is the article in the genitive. There is great need, therefore, that the Latins should know languages on account of the sayings of the sacred writers and of other men of science.

Chapter IV

The sixth reason is the correction of errors and false statements without end in the text of theology as well as philosophy, not only in the letter, but in the sense. That, moreover, correction is necessary I prove by the great amount of corruption. Since the error in the text of God is more serious and dangerous than in the text of philosophy, I shall therefore apply the power of languages to the corruption of the sacred text, that the necessity of these may appear, owing to the endless corruption of the edition in common use, that of Paris. God knows that nothing can be brought before the Apostolic Seat in need of such vigorous correction as this endless corruption. For the letter in the common text is everywhere false or in doubt for the man who has taken this corruption into account, and if the letter be false or in doubt, then the sense, spiritual as well as literal, will contain falsehood and inexpressible doubt. This I now wish to show without possible contradiction. For Augustine says against Faustus, "If there is disagreement in the Latin codices, we must have recourse to the ancient ones and to several of them." For the ancient ones, as he thinks, are to be preferred to the late ones, and the larger number to the fewer. But all the ancient Bibles, which lie everywhere in the monasteries, and which up to the present time have not been glossed nor touched, have the true translation, which the holy Roman Church received in the beginning, and ordered spread abroad throughout all the churches.

But there are many contradictions in the edition of Paris. Therefore this edition needs a thorough correction by means of the ancient texts. But Augustine says in the same work that if a doubt still remains in the ancient Bibles, we must then have recourse to the Hebrew and Greek languages. He also makes this statement in the second book on Christian Doctrine, and shows it in examples. Jerome also imparts this teaching to Frecella and Sunnia and so states in the commentary on Zachariah, and all the sacred writers agree in this, and any one else with reason. But Greek and Hebrew along with the ancient Bibles stand in opposition to the Paris edition, and hence it is in need of correction in many places. But Jerome says to Damasus in this event that where there is diversity, the truth is not known. But those who are striving with all the truth they know to correct the text, namely, the two orders of Dominicans and Franciscans, have now formed from the corrupted text various scriptures and more than one Bible may contain. They contend with each other and contradict without end; and not only the orders, but the brothers of both orders, oppose one another even more than the orders as a whole. For every house contradicts another, and in the same house correctors succeeding one another in

CLEMENT IV

From *Effigies Pontificum Romanorum Dominici Basae*

turn destroy their mutual positions with infinite scandal and confusion. Hence although twenty years before Dominicans effected a correction in the Scriptures, others have come and arranged a new correction, which contains more than the half of one Bible, some of whose flowers when collected are scarcely contained in so great a space as that required for the New Testament. And because they see that they have erred in the old correction, they have now made a statute that no one shall adhere to it, and yet the second correction on account of its dreadful length has along with many truths incomparably more false statements than the first correction, as your Glory will be able clearly to perceive when the proof in particular shall be presented to your authority.

Chapter V

What I have shown in general can be shown in examples. For corruption of the text takes place without limit by the addition, substraction, alteration, union, division of statement, word, syllable, letter, diphthong, mark of breathing, so that not only the letter but the literal and spiritual sense is changed. The faults are found not only in one statement, but in many, nay, they affect many folios. I shall now give one or two examples of each of these faults. For many prologues are placed in the text superfluously, since they are not prologues of the text giving an explanation of the translation of the books to which they are prefixed; but they are either letters sent to friends, like the letter of Jerome to Paul, which in the caption of the Bible is reckoned a prologue and is commonly so called, and yet it is contained in the book of Jerome's epistles; or they are prologues to commentaries on originals not to the text, like the one prefixed to the book of Ecclesiastes. For without doubt it is the prologue of the original itself and this is clear from its purport. And the same is true of many others which are not in ancient Bibles. Of a superfluous statement there is an example in Deuteronomy, "Cursed be he that sleepeth with the wife of his neighbor, and the people shall say Amen"; since neither the ancient codices nor the Hebrew or Greek have this verse. Of a superfluous word there is a horrible and dreadful example in the eighth chapter of Genesis, when the statement is made that the raven did not return to the ark. For the Hebrew has the affirmative and all the Jews hold this view: and the ancient Bibles have the affirmative, and Jerome in the original. The negative was accepted a short time after from another translation, doubtless that of the Seventy, whose falseness Jerome shows in numberless places, and which since the time of Isidore and before has been discredited. For he himself states in his book on the Offices that as a general rule all Latin Churches use the translation of Jerome, because it is more truthful in the expression of thought and clearer in phraseology, with the exception that owing to the frequent practice of singing in the Church the translation of the Psalter following the Septuagint has remained. But before the Roman Church ordered this translation to be used everywhere, Augustine and others and Jerome himself in his time used, as did the Church, the ancient translation. Therefore when Augustine quotes this text in the sixteenth book of the City of God and expounds it, he had to use the translation in common use and received among the Latins, nor could he do otherwise. Since, indeed, a glossarist a hundred years later placed glosses on the text, he accepted the authority of Augustine from the City of God, and placed the negative beside the text, but he did not alter the text nor did he insert the negative. Modern scribes, paying no attention to the difference in translations, nor considering what translation they were using, have inserted the negative on their own authority and in the first instance some one noted among the rest did this. In this way a terrible error was spread abroad, since a contradiction is given for its opposite. But we see in philosophy that there is sometimes a double or triple translation of the same book, and one translation has what is different from another or sometimes what is contrary. But there is no one who has ventured to mix one translation with another.

The error found in ecclesiastical books of reading a negative was introduced by those interested in these books from a corruption of the original. As regards the change of a syllable, and consequently of the whole word, there is a strange example in the case of Joseph, who is said in the common version to have been sold for thirty pieces of silver on account of the example of the Lord. But according to the ancient codices, and the Hebrew,

Greek, Arabic, and Jerome in the original, and Josephus in the first book on Antiquities, the reading should be twenty, not thirty. Likewise in the Psalter by a change of syllable the whole word is changed with great error, when it reads, "My soul has thirsted for God the living fountain." For since the Church uses the translation of the Seventy in the Psalter alone, Jerome has corrected this translation twice, and has placed *fortem* where we put *fontem* through an error due to the similarity of the words, and because in the preceding verse mention is made of fountain, and because thirst corresponds to the idea of water. But as I have said, Jerome has corrected it to *fortem;* and this is the reading in the Hebrew and the Greek, and in Jerome's own translation which he made from the Hebrew. This is also the reading in all ancient Bibles, and in the ancient Psalters of the monasteries. For I have examined this matter carefully, and there is assuredly here merely a very despicable error due to the similarity mentioned.

As regards the change of a letter this a notable example in the first chapter of Judges where we find the words *in monte Hares*. *Hares* is interpreted by *testateo* [of tile, or brick], with the penult letter e not i; but it is generally taken as *testatio* with an i, so that it would be a nominative case and formed like *testificatio* from *teste;* but it must be the ablative derived from *testa,* for in all ancient books it is *testateo* with an e, both in Greek and Hebrew, when *Hares* occurs. Jerome has translated it by tile, or by something derived from tile, or by something like it, as brick or dryness. For *Hares* in Hebrew signifies tile or some one of the aforesaid meanings in Latin. Hence Jerome in the sixth book on Isaiah expounding this word in the sixteenth chapter, "To those who rejoice over walls of baked brick," says, "*Hares* signifies tile or baked brick," and in the eighth book, expounding this word in the twenty-fourth chapter of Isaiah, "The moon grows red," he says that *Hares* signifies tile or dryness. The fact that in the thirty-first and thirty-second chapters of Jeremiah the names *Ananeel* and *Anameel* are erroneously confounded, so that the letters m and n are placed without distinction in the penult, is a great error in the change of one letter. For Jerome says in the original that *Ananeel* written with an n is a tower, with an m is the son of Sellum, cousin on the father's side of Jeremiah; and it is so found in the Hebrew.

As regards the diphthong, there is that example in the sixteenth chapter of Proverbs, *lapides sacculi* [stones of the bag] according to the Hebrew and the Greek and ancient writers, although it is commonly given as *seculi* without a diphthong for *saeculi* in some not very ancient writers. The error in this instance secured a foothold because this noun *seculum* ought to be written with the diphthong, and it is thus correctly written in all ancient books in every instance. Since there is only a slight difference between c and e, some of the old writers of our own times were deceived and changed the first c into an e, thus writing *saeculi;* and since modern writers do not write it with a diphthong, they have therefore retained this noun *seculum* written in their own way, and have neglected the noun *sacculum,* which has the correct letter.

As to the mark of breathing there is an example in the First Epistle to the Thessalonians, when we read, *"Ad tempus ore,"* so that *ore* would be the ablative case of this noun *os oris,* and not the genitive of this noun *hora horae.* It is written therefore in the ablative case, and is glossed, not by a sacred writer, but by the Master of Sentences, who glossed the Epistles; but just as he has failed in many instances elsewhere owing to his ignorance of Greek, so has he failed here, since without question in Greek from which it was taken the genitive case of this noun *hora* is found to be *horas* and is aspirated in both Greek and Latin. But *os oris* is not aspirated, for this noun *hora* is Greek, although it is declined in the Latin manner like *domina.* But Greek declines it thus, *hora, horas, hora, horan, hora.* Hence the nominative, dative, and vocative are alike, the accusative in *an,* genitive in *as;* ablative the Greeks do not have. And this word in the Greek is *horas,* as I have read with care, and any one can prove it who knows Greek, and the aspirate is found in ancient writers.

I have desired to offer these examples, that some proof may be given by way of hint of the necessity of knowing foreign languages, owing to the corruption of the Latin text both in theology and in philosophy. But the method followed in a clear proof, and particularly in one dealing with all the corruptions of the Bible, is postponed to another time, owing to the greatness of the matter, which can, when you wish so to order, be presented to your Holiness, but not by me adequately but rather by another; the importance of which I shall explain to you in what follows.

Chapter VI

The seventh reason why it is necessary that the Latins should know languages is particularly false interpretation, although the text be absolutely correct. For in both theology and philosophy interpretations are necessary, especially so in the sacred text and in the text of medicine and in that of the secret sciences, which are too obscure owing to the ignorance of interpretations. For physicians are confused because of the bad interpretations which they call synonyma. For it is not possible for them to use the established remedies owing to the error in these synonyma, and therefore there is no end of peril in their hands. It is the same with the sacred text; for the chief difficulty in knowing it is due to the variety and obscurity of an infinite number of interpretations, as is clear in a familiar example which will serve for others without number. For the common interpretation of the name Israel for the patriarch is the "man beholding God," and this continued in use up to the time of Jerome, and even up to the time that his translation and exposition were ordered to be used in all the churches. But he himself says in the original, although they are men of great influence, who have interpreted Israel as the "man beholding God," and although their shadow oppresses us, yet we agree with God rather, or the angel who gave this name, than with the authority of a man of secular eloquence. He accordingly proves his assertion in admirable fashion. For those who interpreted it thus believed that this word has the same signification united as when divided, like *respublica* with us. But this is not in general true, nay, there is an instance of it in many cases in every tongue. Now in Hebrew *is* is man, *ra* beholding, *el* God, and therefore they believed that this name of the patriarch must be divided into those three words. But Jerome proves this interpretation false by many arguments; for four arguments can be drawn from his statements based on the word, and four or five based on the fact. For in those three words there are other letters and more in number than in the name of the patriarch and they are found in a different order and syllabication. Therefore from this triple argument taken from the letters Jerome concludes that there cannot be the same signification in this case and in that; since the reason for the same signification rests on the identity of the words. But it is clear that the word and the letters differ too much, since in the name of the patriarch there are these five letters in order, *iod, sin, res, aleph, lamet,* as the Hebrew thus arranged shows יִשְׂרָאֵל, Iserael. But in this triple word these eight letters have the following order, *aleph, iod, sin, res, aleph, he, aleph, lamet,* as the Hebrew shows.

el	ra	is
אֵל	רָאה	אִיש

A fourth argument can be drawn from the pronunciation. For, as the points show, the proper name does not retain in the Hebrew the exact sound of those three words, but it has a greater sound, because Iserael is pronounced in four syllables; but the three words are limited in pronunciation to only three syllables, so that we say *is, ra, el;* since one point under a letter has the sound of *i,* two points that of *e,* and a line with a point beneath it has the sound of *a.* But according to Jerome stronger arguments are drawn from the sense of the word. For it is shown by the text, Hebrew, Greek, and Latin, and by Josephus that Israel ought not to be called "the man beholding God," but "chief or prince with God," since in Hebrew the literal sense is as follows, "And God said, thy name shall be called no more Jacob, but Israel, since if thou wast a chief or prince with God, thou shall be able to be one with men also." Therefore Jerome says that the sense is, "Thy name shall not be supplanter, that is Jacob, but thy name shall be prince with God, that is Israel. For since I am a prince, so shalt thou be called a prince who wast able to wrestle with me. If moreover thou wast able to strive with me, how much more with men, that is with Esau, whom thou shouldst not fear." And the Hebrew itself shows this here written in this way.

Jacob	non	dixit et
iaecove	lo	iomer va
יַעֲקֹב	לֹא	וַיֹּאמֶר
nomen tuum	amodo	dicetur
simecha	oze	ieamer
שִׁמְךָ	עוֹד	יֵאָמֵר

quoniam	Israel	si	quoniam
ki	icerael	im	ki
כִּי	יִשְׂרָאֵל	אִם	כִּי

Deo	simul	principatus
elohim	im	saritha
אֱלֹהִים	עִם	שָׂרִיתָ
poteris et	hominibus	simul et
tuchal va	enasime	im ve
וַתּוּכָל	אֲנָשִׁים	וְעִם

The Greek text has the verse as follows: "Since thou didst prevail with God, with men also shalt thou be strong." And the Latin has: "Since thou wast strong against God, how much more wilt thou prevail against men." Josephus in the first book of the Antiquities says that he was called Israel because he withstood an angel. Therefore all these expressions, namely, to be a prince with God, and to prevail, and to be strong, and to stand against or with God, are reduced to the same meaning, as is evident, but interpreted by different words, no one of which in virtue of its meaning can have the signification of beholding God. Therefore the true interpretation is "a prince with God." And in addition Jerome confirms this by an argument from derivation. For Sarith, which is derived from the name Israel, means prince, as he states. Whence also Sara the wife of Abraham is called princess, just as Jerome says on the seventeenth chapter of Genesis. Wherefore if people in general or some ancient writers, like Eusebius of Caesarea in his book of Hebrew Names, which Jerome translated into Latin, and others, abusing a well-known interpretation, say that Israel is interpreted as the "man beholding God," we may say with Jerome: "The interpretation of Israel as *the man beholding* God, given in the book of Names and generally accepted seems to be more forced than true." If, therefore, any one should argue that his authorities for the statement that the true meaning of this word Israel is "the man beholding God" are Eusebius in the book of Names, translated by Jerome into Latin, and Ambrose and other perchance sacred writers, we must answer that they spoke following the common exposition, before the truth was disclosed, which later the blessed Jerome revealed to the Latins by a true and correct interpretation; even as it is contained in his books, and appears also in the gloss. If, therefore, it be said that it is the custom of modern theologians to take this interpretation, the proper answer is indicated in the statements given above of Augustine, Cyprian, Isidore, and others. For according to them custom should yield to truth when revealed, in order that giving up the error of the throng we may follow the truth; and that which has come from ignorance ought not to be alleged in proof, as is being done in the matter under discussion. And above all we should not oppose a sacred author and teacher when he brings forward in support of his position convincing reasons and authorities. Furthermore, for the assurance of all, any one can consult Hebrew scholars, and he will find the judgment of the blessed Jerome ratified and unshaken. There is the greatest need of remedies against false statement in these interpretations on account of the form of Hebrew speech. For in the common interpretations, which are placed at the end of the Bible, there are infinite occasions for errors because a word is reckoned according to the Latin standard which has many forms in Hebrew. And there is the greater error because to such a word various interpretations are given, as though they belonged to the same Hebrew word, whereas each belongs to a different one, because a Hebrew word written by us without due consideration in a single way has different letters in Hebrew and different ways of writing, according to which it receives different meanings. Jerome gives an example of this in his epistle on Mansions.* For *or,* if it is written with *aleph,* means light; if with *ain,* skin; with *heth,* hole; with *he,* mountain. He says, therefore, that in the twentieth chapter of Numbers some have interpreted it in these four ways; but he eliminates three of these because in the Hebrew this word is written with the letter *he,* and therefore in this place has the meaning of mountain only. But in preaching and reading theologians have recourse to all four expositions in this word; and so also in other words because of their various interpretations.

The last scientific reason for the need of other languages is the fact that Latin grammar was formed from Greek and Hebrew. For we received our letters from the Greeks, as Priscianus shows, and the whole method of treating the parts of speech Priscianus received from the Greeks, as he bears witness, and he mixes Greek freely in all his books. The words themselves of the Latin tongue, both theological and philosophical, were brought in for the most part from other languages; of which words the Latins suspect some to be from another tongue, while others they do not consider as descending from such a source. Many, in fact, are reckoned as wholly Latin when in reality they are Greek or Hebrew, Arabic or Chaldean, in which words error is frequent on the part of the Latins in their pronunciation, writing, and meaning. For it is no small impropriety to make mistakes in words; because as a consequence a man errs in his statements, then in his arguments, and at length in what he reckons as conclusions. For Aristotle says, "Those who are ignorant of the meaning of words often reason falsely." Boetius places the first and principal foundation of learning in an exact and complete knowledge of terms, as he shows in his work Disciplina Scholarium; and we experience this in each of the sciences. For the principal difficulty in a science and its usefulness are found in knowing how to understand the words employed in the science, and to express them in a wise manner and without error. When a man has advanced thus far, he can accomplish the rest by himself without further teaching if he be diligent in study. For the texts of the sciences are plain to him after he has learned how properly and correctly to understand and interpret; and without difficulty he can understand any scientist, and can confer ably with any one, and be instructed,

* *I.e.,* the forty-two stopping-places of the Children of Israel.

if necessary, by any one. Aristotle says in the first book of the Heavens and the World, "A small error in fundamentals is a great one in what is derived from them." For he who makes an error in the foundation of necessity piles up his whole building on the error.

We judge, therefore, that our language is composed of Latin words and contains few words from other tongues, whereas words in common use are from foreign tongues, as *domus, scyphus, clericus, laicus, diabolus, Sathanas, ego, pater, mater, ambo, leo, bos, ager, malum,* and a host of others which with difficulty could be contained in a large volume; especially so if the words used in the different sciences were examined, particularly theology and medicine. Nothing would be more useful than such a volume, if it furnished all the words correctly written and properly pronounced, together with a trustworthy derivation and an accurate interpretation. But as it is we make countless mistakes in these four particulars to the great detriment of all learning, as we can perceive from a few examples. For we do not consider the order of languages, nor the fact that an earlier language does not receive the interpretation of a later one, nor that different languages in that in which they are different do not mutually expound themselves, as Jerome says on the seventeenth chapter of Genesis, "No one in naming a thing in one language takes the etymology of the word from another." Servius also makes this statement in the commentaries on Vergil. Above all, an earlier language cannot have its origin in a later one, as is clear to every reasonable man. Hence Greek does not spring from Latin, nor Hebrew from Greek. Therefore Hebrew must not take its etymology from Greek, nor Greek from Latin. Hence Jerome says in the place mentioned, replying to certain objectors, "*Sarra* ought not to have an explanation from the Greek but from the Hebrew, since it is a Hebrew word." Servius says that *Lenaeus* is derived from *lenos,* wine vat, not from *lenio,* because a Greek noun cannot have a Latin etymology. But we commonly and without distinction disregard this rule. For we say that *amen,* although it is Hebrew, is derived from *a,* the Greek prefix, meaning without, and *mene* a Greek word meaning defection. And although *Parasceve* is Greek, we say that it is derived from *paro, paras,* and *coena, coenae,* which are Latin words.

The statement is also made that *dogma* is derived from *doceo,* and that *jubileum* comes from *jubilo,* and similarly in regard to a host of other words, all of which derivations are false. Not only the rank and file of the Latins but the authorities err in these matters; as, for example, Hugutio and his followers, who think that *jubileum* is derived from *jubilo,* whereas *jubileum* must be Hebrew and *jubilus* is Latin. But the word should not be written *jubileum* with the letter *i* in the second syllable as in *jubilo;* but it should have the letter *e* and be written *jubeleus,* as Isidore and Papias maintain, and all ancient books so have it. For it is derived from *jobel,* which is Hebrew.

Chapter VII

Since, moreover, we know that many words in use among Latins must be interpreted through other tongues, we believe, owing to our familiarity with this fact, that far more words than the fact warrants derive their etymology from a foreign source. For only those words having a Greek and a Hebrew origin and derivation ought to be interpreted by those languages. For words of pure Latin origin can have no exposition except by means of Latin words. For the pure Latin is altogether different from every other tongue, and therefore does not have an interpretation from a foreign source. But the Latins pay no attention to this fact, nay, without distinction they interpret pure Latin words by other languages, like the words derived from the Greeks. Hence in many ways they interpret in Greek this word *caelum,* which is pure Latin, stating that *caelum* is equivalent to *casa helios,* that is, house of the sun, for the sun is called *helios.* But their statement is inconsistent and false. For since *helios* is in the nominative case and not in the genitive *casa helios* is inconsistent; for they should say *casa heliu,* because *heliu* is in the genitive case. Secondly, the statement is false; for as Varro, most learned of the Latins, to quote the words of sacred writers and philosophers, teaches us, and Pliny confirms his statement in the prologue to his Natural Philosophy, that *caelum* is derived from *caelo, caelas,* that is, *sculpo-pis,* because it is carved and adorned with stars, which is manifest from the rule for writing words. For *caelo-las* for *sculpo-pis* is written with the diphthong *ae* in all ancient books; and likewise this noun *caelum* in all the ancient codices is written with this same diphthong, and is therefore derived from *caelo,* which is the same as *sculpo.* Hence it follows that it is not derived from *celo, celas,* which is a synonym for *occulto-tas,* as those claim who give this noun a Latin etymology, stating that it is so called because it is hidden and far removed from us. These interpreters like those preceding have been deceived by a wretched error. Likewise this word *ave,* which is pure Latin, they expound in Greek, stating that it is formed from *a,* meaning without, and *ue,* as though it were *sine ue.* But this cannot be since this word is not taken from the Greek word of cognate signification. For *chere* in Greek signifies *ave* in Latin, but these two words do not agree, and therefore much less will it draw its exposition from other sources. This then is one way in which mistakes are made in an almost countless number of Latin words.

Chapter VIII

Another form of error is due to the fact that we do not pay attention to the many different ways of writing in Greek words, and that words very similar in sound are distinguished in meaning. The Greeks have three forms of *i,* two of *o* and two of *t, p,* and *c;* they have eleven diphthongs and many other ways of showing the variation of their words in meaning. For *cenos,* meaning empty, from which comes *cenodoxia,* that is, vanity, of which we read in the seventh chapter of Deuteronomy, is written with a short *e;* and *cenos,* that is, new, from which comes *encenia,* that is, innovations like a new feast and dedications of churches, of which we read in the tenth chapter of John, is written with the diphthong *ae,* thus, *caenos.* But *cenos,* that is, common, from which come *cenobium* and *epicenon,* is written with the diphthong *oi,* although the Latin has an *e,* but it should have *i,* so as to read *cinos.* Hence from this word is formed *cinomia,* which is, according to Jerome in his correction of the Psalter, the common fly or fly of every kind. Hence Papias says that it is written with the diphthong in the first syllable thus, *coinomia,* and this is proved from the Greek

Psalter. Also the word *cynos,* dog, when it is written with the Greek *y,* whence *cynomia,* that is, dog fly, of which we read in the eighth chapter of Exodus. The word *xenos* with *x* means foreign, from which comes *xenia,* meaning presents or gifts, of which we read in the first book of Maccabees and in the twentieth chapter of Ecclesiasticus. Also *schenos* with *sch* is a rope, from which is formed *schenobates,* one who walks on a rope or over a rope; and *scena* means shade or tent, from which is formed *scenopegia,* that is, setting up a tent or the art of tent-making, at which Paul the Apostle labored. Since, therefore, the derivatives and compounds of those words differ so in their meanings, although they are similar in expression and sound, like the words from which they are derived, it is clear that he cannot escape error in the literal sense who disregards the written form of words of this kind. Hence many famous living expositors have been at times deceived, such as Rabanus, who on the eighteenth chapter of the Acts says that the *schenofactoria ars* [art of tent-making] teaches how to make ropes, because he thought that *schenos* meaning rope is the word from which the noun is derived. But Bede teaches to the contrary, holding that it is derived from *scena* [tent]. And this is shown clearly in the writing of the word, namely, in the eighteenth chapter of the Acts, in the Greek text the word is written with the first syllable lacking the aspirate, and with the vowel called *ita,* that is long *i,* and is so written *scena* for tent; but *schenos,* meaning rope, is written with the diphthong *oe,* and with the aspirate. And thus there is a contention among the doctors about *cynomia* and the other words mentioned. Hence as regards *xenia* the majority believes that it does not exist and corrects it in the sacred text into *exenia.* But in the ancient Bibles it is not so written, nor in Greek, nor can the word be so formed according to Greek grammar, because it would require the placing there of the Greek preposition *ex,* which is impossible, since the word begins with a consonant, as is clear according to the grammar of the Greeks. In this way error arises in almost countless words.

Chapter IX

The third form of error is due to the fact that we do not observe as we should that while the Latins have much in common with the Greeks they yet differ from them in some particulars. For since Priscianus states the fact and all the Latins are aware that the name of a tree is in the feminine gender and terminates in *us,* and the name of a fruit is neuter and terminates in *um,* as *pomus, pomum; pirus, pirum,* etc., they think that this principle holds good in regard to all words in use among the Latins, like the words *malum* and *amigdalum* and others. For the Latin rule is to be understood as applying only to Latin words, not to Greek words nor to others. That this is a fact is clear in the first place because the Latin gives its rules concerning Latin words, and it is not within its province to form rules for other languages; in the second place, Priscianus says that every Greek word passing into Latin retains its own gender that it had in Greek; and therefore since the word *malum* for tree is Greek and of neuter gender, it will remain so in Latin; and therefore whether it has the meaning of fruit or of tree it will be of the same gender and of the same termination. And this is proved by Vergil, who says in the Georgics *mala insita;* trees are ingrafted, not fruits; and in regard to this Servius the commentator, who was greater than Priscianus, for he frequently quotes the authority of Servius, says that this rule, "All names of trees are feminine," is to be understood of Latin, not of Greek nouns. The word *malum,* moreover, is Greek, as he states; and it is certain that it is Greek, although in accordance with Latin custom pronounced somewhat differently, since no word in Greek ends in the letter m, but in n; and Latin is accustomed to terminate its own words in m, as *scamnum, lignum, pomum,* and the like. Likewise the Latin often changes a vowel in a Greek word, as where the Greek says *grammaticos* the Latin has *grammaticus,* and so in many instances. This is the case here, for the Greek says *melon* for tree and fruit; the Latin changes the e into a, and n into m, and says *malum.* But this change does not alter the word as regards the nature and root; because it was taken from Greek, although altered in pronunciation, and to this fact all authorities bear witness. But throughout Latin texts in ancient books on theology and philosophy we always find the word *malum* for tree. For in the first chapter of Joel *malum* for tree is generally found in all Bibles; and up to the present time the correctors have let it pass in the new Bibles, and in all the ancient Bibles it is likewise found in the fourth chapter of Canticles, where we read, "As the appletree among the trees of the wood"; and so Bede expounds it in the original. And in the twelfth chapter of Ecclesiastes is the word *amigdalum* in all ancient Bibles; and *malogranatum* in the singular and *malogranata* in the plural are found in the books of the law, which would not happen if *malum* were not of the neuter gender. Therefore these words are changed in accordance with the form of Latin words inconsistently, and it is particularly strange that in one passage the correctors allow the ancient letter to remain, and in another passage rub it out, a procedure altogether opposed to reason.

Chapter X

Likewise in the pronunciation of the Greek letters there is a great deal of error because the Latins wish to keep their own manner of pronunciation in the Greek words. And in this there is a very grave error; since all the Latin poets and all the ancient Latins pronounced according to the primitive custom. But we of modern times have violated this rule in many ways contrary to the practice of all the ancient Latins and their authors. For example, when Priscianus says that possessive adjectives ending in *nus* are long and are accented on the penult, as *Latinus, bovinus, equinus,* and the like, the rule is to be understood of Latin words, not of Greek, for certain reasons previously touched upon. And therefore since *adamantinum, byssinum, crystallinum, hyacinthinum, bombycinum, onychinum, amethystinum, smaragdinum,* and the like are Greek, they should be short in the penult, as the Greeks make them. Moreover, these are not possessives; for there are only two terminations for possessives in Greek, namely in *cos,* as *grammaticos,* and in *os,* as *uranios,* that is, belonging to heaven. But all the Latin poets make the penult short, and it is not a poetic license, because they all do it in every instance. For what happens rarely and for a reason is to be ascribed to poetic license, but not that which is of common and constant occurrence. Whence Juvenal's *Amethystina convenit illi.* And he likewise says *grandia tolluntur crystallina,* with short penult. Persius also shortens the penult, saying *hyacinthina laena* at the end of the verse, and so do all the poets without exception. Therefore it is not a case of poetic license, but it happens in accordance with natural law. And since in the seventeenth chapter of the second book of Kings the words occur *quasi siccaret ptisanas,* a famous exposition of the words of the Bible, to which all give adherence, tries to prove that the middle syllable of *ptisanas* is long; and the author of this exposition defends himself by a verse of Horace, which reads: *Tu cessas agedum sume hoc ptisanarium oryzae.* But it is an error, for, as can be proved by all authors, only one syllable is elided at the end of a word in meter; and therefore this should be scanned, *hoc ptisanari oryzae,* making the syllable *sa* short and the syllable *na* long. This is clear for another reason, for in all derivatives *a* before *rium* is long, as *contrarium, armarium,* and numberless others of this kind. This rule is observed in the scansion above, but not in the one generally given, *ptisanar oryzae,* requiring the elision of two syllables, because in the latter this syllable *na* is short, as is evident. Therefore of necessity the middle syllable of this word *ptisana* is short and must be grave in accent. Moreover, there is an error in the writing of the word. For in the new Bibles it is given as *tipsanas,* which has no meaning; and therefore the *p* should be placed before the *t,* as in the name of Ptolemy and in many other words. This kind of error occurs over and over again in other words; and we have made such violent changes in the correct rules of accents that there is no help to be found in our teachers; since custom forces all to pronounce badly, as is made clear by a single example in place of the thousands that might be cited. *Butyrum* has a short penult in Latin authors. Hence Statius says in his Achilleid,

Lac tenerum cum melle bibit, butyrumque comedit.

And Macer in his books on Herbs has,

Cum butyro modicoque oleo decocta tumorem.

The Greek also has it short, and the component parts of the word require it. For it is formed from *tyros* and *bos,* and *tyros* is short in the first syllable. The word means milk-food which comes from the cow. But far greater mistakes are made by many, and there is ignorance of the truth in regard to accents on the part of all. But a broader consideration of this matter is required than the limits of the present production permit.

CHAPTER XI

SINCE I have now shown how a knowledge of languages is necessary to the Latins owing to the pure zeal for knowledge, I now wish to state why this should be secured because of the wisdom established for the Church of God, and the commonwealth of the faithful, and the conversion of unbelievers, and the repression of those who cannot be converted. For in four ways is this knowledge necessary to the Church, first on account of the divine Office, because Greek, Hebrew, and Chaldean words are used in the Office, just as they are in Scripture; and we hear many words, which the Scripture does not use, like *agios, atheos, athanatos, iskiros, imas, eleyson, kyrie,* and the like. When, therefore, we are ignorant of the writing and the correct pronunciation and the sense, we shall miss much of truth and devotion in our singing; for we speak like the magpie and the parrot and other brute creatures which imitate human voices, but neither pronounce correctly nor understand what is said. For since we say *alleluia* many times in a year, it is very proper and right that all those who sing throughout the whole Church should know that they are two words, namely *allelu* and *ia.* For *allelu* signifies the same as *laudate* and *ia* denotes *Dominus,* since it is one of the ten names of God, as Jerome writes to Marcella; and in particular it signifies the invisible one, and God is the most invisible of all. Hence it does not designate any one at all who is invisible, but it designates God. In every Mass we say *Osanna,* a word composed of two elements one correct and the other incorrect. For, as Jerome says to Pope Damasus, *osi* is the same as *salvifica,* and *anna* is the interjection of one praying, the first syllable in this case being written with *aleph.* Whence the meaning is the same as *salva deprecor* [save, I pray]. In other cases the first syllable is written with the letter *he* and then denotes a conjunction which the Latin language does not have. And when we salute the glorious

Virgin saying, *Ave Maria gratia plena, Dominus tecum,* it is very essential for a true and devout understanding that each educated person should know the meaning of the word, and especially so since many thinking that they know make many errors in this matter. There is a Syrian word, *Maron,* signifying Master, from which *Maria* comes; and it is the same as *dominatrix* [mistress], as Jerome says in his interpretations. This meaning especially befits the most blessed Virgin, who is mistress over all the uncleanness of sin that she may drive it from us, and over all diabolic guile and wickedness, because she is the terror of sin and of the demons, like the ordered line of battle of a camp; and not she only, but all who really put their trust in her. This interpretation is most correct and without sophistry. Jerome, however, says in his interpretations that many have thought that *Maria* should be interpreted "she that illumines" or "myrrh of the sea." He does not accept this meaning but says that it should be "star of the sea" or "bitter sea" according to the Hebrew interpretation. And she is rightly called Star of the Sea, that she may direct us to a port of safety; also Bitter Sea, because she lived in absolute poverty and temporal bitterness in this world, and at length a sword passed through her soul in the death of her Son, so that she is an example to us of all patience and a comforter in every adversity in this world. It is necessary for us therefore in all our psalm-singing and prayers to know how rightly to pronounce and understand, and that as regards the proper meaning of words we should know how to frame our petitions devoutly, in order that we may obtain by the goodness of God and the saints and by the merits of the Church what we ask for rightly and devoutly. But we cannot do this without the knowledge of the words in another tongue; and therefore it is very expedient and necessary that we should have this knowledge. The second reason is because a knowledge of languages is necessary to the Church of God on account of the sacraments and consecrations. For intention is necessary to a sacrament, as theologians know. Understanding and knowledge of the thing to be done precede intention. And therefore in every way it would be expedient for the Church that her priests and prelates should know how correctly to pronounce and understand all the words of the

Masses and sacraments and consecrations; just as in the beginning the holy and high pontiff, and all the sainted fathers and founders of ecclesiastical orders decided and knew how the mysteries of God consist in words and their meanings. Whence it is not only proper, but expedient and necessary that those who minister the sacraments, beginning with the first exorcisms and purifications and so passing on through baptism and all the sacraments, should know the correct pronunciation and the required sense, to the end that no sacrament may be impaired. But lately throughout the universal Church countless numbers pronounce the words instituted by the Church and do not know what they are saying, nor do they keep the correct pronunciation of the words. This cannot happen without injury to the sacrament. Would that the effect of the sacrament may have full efficacy! And since the Church established these formularies with definite knowledge and all the ancient fathers knew the correct pronunciation and the sense of the words as befitted the sacraments, we have no excuse; but our ignorance is a base and vile one to be excused by no subterfuge. When in the consecrations of Churches letters of another language according to the order of the alphabet are made with the point of the pastoral staff, it is certain that very few make the letters required, as they were originally determined on by the sainted fathers and by the Church, owing to ignorance of the characters of the other language and in particular an error is made in this manner that three figures are made, which in no way ought to be written in the Greek alphabet. For without doubt the figures which are called episemon ϛ, koppa, and the character ϡ are not from the alphabet of the Greeks, nor did the Greeks ever insert them in the order of their letters; but they are figures and marks of numbers. Modern Latins, however, do not bear in mind that the Greeks number with the letters of the alphabet, and that they have inserted to complete their numeration the three figures mentioned above, namely, ϛ ϟ ϡ. But they do this when they count, not when they use the figures as letters and in writing. Hence in writing they never use these three characters, nor do they place them in the order of the alphabet. But the Church determined that letters alone of the alphabet should be written, in the consecration of the Church, and decided to use letters, not the signs of numbers. Therefore it is most improper that erroneous writing of this kind should be practiced throughout the universal Church.

It is a wretched thing that these words $\overline{\text{I H C}}$. $\overline{\text{X P C}}$. are written in Greek letters, and people think that they are Latin, or are in ignorance of the manner in which they are Greek. For without doubt in this word $\overline{\text{I H C}}$. the first letter is *iota,* equivalent to our *j;* the second is *ita,* equivalent to *e* long; the third is *sima,* equivalent to our *s.* And in this word $\overline{\text{X P C}}$. the first is *chi,* equivalent to *ch* aspirated; the second is *ro,* equivalent to our *r;* the third is *sima.*

There is a third reason for the necessity to the Church of God of a knowledge of languages. For many Greeks, Chaldeans, Armenians, Syrians, Arabs, and nations of other tongues are subject to the Church of the Latins, with whom the Church has to arrange many matters, and to give them various directions. But these matters cannot be handled in the right and advantageous way that they should be, unless the Latins have a knowledge of their languages. Of this there is proof in the fact that all the nations mentioned waver in faith and morals, and neglect the orders of the Church pertaining to salvation, because a genuine plea is not addressed to them in their mother tongue. Hence everywhere among such nations there are evil Christians, and the Church is not ruled as it should be.

The fourth cause is due to the progress of the whole Church from the beginning to the end of days. For the Lord says, one jot or one tittle shall in no wise pass from the law, till all be fulfilled. And therefore there is an admirable exposition in the book on the meanings of the Scriptures stating how the individual letters of the Hebrew alphabet had significance respecting the ancient people, and how they show the number of centuries through which the state of that race passed as regards the different periods and ages, in accordance with the special powers and potencies of the letters; and then the progress of the Church of the Latins is shown by the virtues of the Latin letters. A similar examination is made of the Greek Church by means of the letters of the Greek alphabet. In a remarkable examination of this kind the periods of time are distinguished as regards the varying conditions of the Church to the very end, and we learn through how many hundreds of years each change happening to the Church in its history will last. If we should unite to this remarkable examination prophecies and reliable testimonies, we would be able by the grace of God to perceive in advance to our profit those things which the Church shall receive in her prosperity as well as adversity. Therefore nothing would be more useful than an examination in this way of the value of the letters along with other similar examinations. For to secure certainty in such important matters many ways are needed, of which the one at least is not ignoble which employs the letters of the different languages. I cannot sufficiently admire the manner in which this examination was devised, although it may seem to the uninitiated to have a weak basis in the letters of the alphabet, which are the first rudiments of children. But according to the teaching of the Apostle, lesser things are more necessary and are to be accorded greater honor. And as God has chosen the weak to confound the strong, so has supreme power given greater weight to matters which we reckon insignificant than the human mind can grasp. Such is the case in these letters of the three alphabets; whence not without cause the epitaph of the Lord was written in Hebrew, Greek, and Latin, so that we might be taught that the Church redeemed by the cross of Christ must consider the virtues of the threefold alphabet; especially since the Church began among the Hebrews, made progress among the Greeks, and was perfected among the Latins.

Chapter XII

In the second place, a knowledge of languages is very necessary for directing the commonwealth of the Latins for three reasons. One is the sharing in utilities necessary in commerce and in business, without which the Latins cannot exist, because medicines and all precious things are received from other nations, and hence arises great loss to the Latins, and fraud without limit is practiced on them, because they are ignorant of foreign tongues, however much they may talk through interpreters; for rarely do interpreters suffice for full understanding, and more rarely are they found faithful. A second reason is the securing of justice. For countless injuries are done the Latins by the people of other nations, the sufferers being the clergy as well as the laity, members of religious orders, and friars of the Dominicans and Franciscans who travel owing to the varied interests of the Latins. But owing to their ignorance of languages they cannot plead their cases before judges nor do they secure justice. The third reason is the securing of peace among the princes of other nations and among the Latins that wars may cease. For when formal messages along with letters and documents are drawn up in the respective languages of both sides, very often matters which have been set on foot with great labor and expense come to naught owing to ignorance of a foreign tongue. And not only is it harmful, but very embarrassing when among all the learned men of the Latins prelates and princes do not find a single one who knows how to interpret a letter of Arabic or Greek nor to reply to a message, as is sometimes the case. For example, I learned that Soldanus of Babylonia wrote to my lord, the present king of France, and there was not found in the whole learned body in Paris nor in the whole kingdom of France a man who knew how satisfactorily to explain a letter nor to make the necessary reply to the message. And the lord king marveled greatly at such dense ignorance, and he was very much displeased with the clergy because he found them so ignorant.

Chapter XIII

In the third place, the knowledge of languages is necessary to the Latins for the conversion of unbelievers. For in the hands of the Latins rests the power to convert. And for this reason Jews without number perish among us because no one knows how to preach to them nor to interpret the Scriptures in their tongue, nor to confer with them nor to dispute as to the literal sense, because they have both the true letter and their own ancient expositions according to ————* and of other men of wisdom as much as the literal exposition requires, and in general as much as it requires for the spiritual sense. For the text everywhere sounds the spiritual note of the Messiah whom we call Christ, even as the Hebrews themselves are not ignorant, because they expect that he will come, but are deceived in re-

* Blank space in manuscript.

gard to the time of his advent. O unspeakable loss of souls when with ease countless Jews might be converted! What makes the situation as bad as possible is the fact that the foundation of our faith began with them, and we should bear in mind that they are of the seed of the patriarchs and prophets, and, what is more, from their stock the Lord sprang and the glorious Virgin and the Apostles and innumerable sacred authors have descended from them from the beginning of the Church. Then the Greeks and the Rutheni and many other schismatics likewise grow hardened in error because the truth is not preached to them in their tongue; and the Saracens likewise and the Pagans and the Tartars, and the other unbelievers throughout the whole world. Nor does war avail against them, since the Church is sometimes brought to confusion in the wars of Christians, as often happens beyond sea and especially in the last army, namely, that of the king of France, as all the world knows; and if Christians do conquer other lands, there is no one to defend the lands occupied. Nor are unbelievers converted in this way, but they are slain and sent to hell. The survivors of the wars and their sons are angered more and more against the Christian faith because of those wars, and are infinitely removed from the faith of Christ, and are inflamed to do Christians all possible evils. Hence the Saracens for this reason in many parts of the world cannot be converted; and especially is this the case beyond sea and in Prussia and in the lands bordering on Germany, because the Templars and Hospitallers and Teutonic Knights hinder greatly the conversion of unbelievers, owing to the wars that they are always stirring up and because they wish to have complete sway. For there is no doubt but that all nations of unbelievers beyond Germany would have been converted long since but for the violence of the Teutonic Knights, because the race of pagans was frequently ready to receive the faith in peace after preaching. But the Teutonic Knights are unwilling to keep peace, because they wish to subdue those peoples and reduce them to slavery, and with subtile arguments many years ago deceived the Roman Church. The former fact is known, otherwise I should not state the latter. Moreover, the faith did not enter into this world by force of arms but through the simplicity of preaching, as is clear. And we have frequently heard and we are certain that many, although they were imperfectly acquainted with languages and had weak interpreters, yet made great progress by preaching and converted countless numbers to the Christian faith. Oh, how we should consider this matter and fear lest God may hold the Latins responsible because they are neglecting the languages so that in this way they neglect the preaching of the faith. For Christians are few, and the whole broad world is occupied by unbelievers; and there is no one to show them the truth.

Chapter XIV

The fourth reason, the repression of those who cannot be converted requires rather the way of wisdom than the labor of war. For the unbelievers always return to their own provinces, as we see beyond the sea and this side of it in Prussia and in the lands of the pagans bordering on Germany and everywhere else; because Christians signed with the cross although sometimes victorious, yet after making a foreign expedition return to their own lands, and the natives remain and multiply. The faith should first be preached by men wise in all knowledge, but who are well versed in languages or have excellent and faithful interpreters. When we learn that some race will obstinately resist we should not only get ready a military expedition, but men of learning should assemble who should subjugate not temporarily nor a part of the unbelievers, but all of them who are in proximity to Christians, so that at least the Holy Land with Jerusalem may always remain in the possession of Christians without fear of its loss in the future. And although many secrets of the sciences and of the mighty works in the arts are needed in this matter, of which I shall make mention later in many places, not only on account of those now living, but because of Antichrist and his followers, yet the power of languages and of the different letters must not be despised. For such great virtue can consist in words that no mortal can trace it out. And at this I wish to hint in many ways, because the matter is difficult and subject to great contradiction. For we see that the words of the sacraments have infinite virtue. And we know that at the command and at the words of the saints from the beginning of the world the laws of nature were changed and elia [*sic*] and other brutes were obedient, and in this way numberless miracles were performed. But the hand of the Lord is not shortened; and we should believe that if on the authority of the Church and with right intention and from desire many true and wise Christians should utter holy words for the propagation of the faith and the destruction of falsehood, that many blessings might result by the grace of God. Oh, how many tyrants and evil men have been confounded at the words of power and convicted rather than through wars! And not only by the words of the saints or of the faithful, but by the words of philosophers have they been so stunned that they were forced to obey the truth. The histories give us definite information in regard to them, and we have seen many of the people who by certain forms of speech have freed many from very great perils. For by two verses containing the names of three Kings of Cologne it happened that ——————* I know a man who when a boy found a man in the fields who had fallen in epilepsy, and wrote those verses and placed them around his neck, and immediately he was cured. He had no return of the disease, until long afterwards his wife, wishing to confuse his mind because of her love for a certain cleric, caused him to be stripped, in order that he might lay aside at least during the time of his bath the amulet† from his neck to protect it from the water. On this being done his infirmity at once seized him in the bath. His wife frightened by the miracle again bound on the amulet and he was cured. Who will venture to put an evil interpretation on this and ascribe it to demons, even as some inexperienced and foolish people have ascribed many things to demons, which frequently happened by the grace of God or by the operation of nature and the power of the excellent arts? For how has any one succeeded in proving to me that the incident related was the work of a demon, since the boy had neither the knowledge nor the wish to deceive. And the woman who wished to deceive not only her husband, but herself through fornication in removing the writing, after beholding the miracle was stirred by piety and bound on again the amulet. I prefer to view the matter reverently with a view to the praise of God's

* Blank space in manuscript.

† Literally, "writing of attestation."

blessings than to condemn with great presumption that which is true. Likewise in Poland and in many districts charms with which to exorcise are made of iron, which is carried in the hands or is walked over. And likewise of the water in which an accused person is placed, a practice of the Church and her priests. The innocent come forth without danger, but the guilty are committed; just as in the old law a woman accused of adultery drinks the consecrated water and was freed if she was innocent, but if guilty her sin was brought to light. But it is certain that the rational soul, which is superior to creatures lower than angels, has in accordance with the rights of its own dignity great power in respect to creatures of lesser worth; just as we see that heavenly bodies, because they are nobler, have power over what is inferior. And any inferior thing that is nobler in virtue can change a thing that is less noble, so that the ignobler things change to their own natures things more noble, as wine intoxicates a man and fire consumes him. But this takes place in so far as wine or fire is nobler; for everything active is nobler than what is passive, as Aristotle states, and it is a fact that, owing to a defect in any creature, one inferior has a certain prerogative which the nobler one lacks. Since therefore the rational soul is without comparison more worthy than the whole animal soul, there is no doubt that it has great power in its works when it is free from spot of sin or when commanded by the grace of God it acts with strong desire and firm intention. But its especial action is the word, and therefore the saints always performed their miracles by pronouncing words. But the rational soul itself knows how to select the time of the chosen constellations for all its works, as the skillful physician selects the proper time for his medicines and blood lettings and other things, as Hippocrates, Galen, Aristotle, Ptolemy, and other authors state. And such is a fact not only in these things, but in all things in which earthly bodies are altered by the virtues of the heavens; and therefore men of wisdom, not only the pure philosophers, but the saints like Moses and others, performed their works under chosen constellations, as I shall show in the proper place, by the locations of which they changed and excited men to many things, without the loss of freedom of the will. Hence they both waged wars advantageously and carried on many great works. And therefore just as they performed their other wonderful works in their appointed times, so also they formed words, which received great virtue from the heavenly constellation itself, and accordingly they accomplished many things through these words. But these matters will depend on what follows, and therefore I pass them over until we come to the opportune place. If, however, other works can receive virtue from the spotless soul, strong desire, firm intention, and celestial virtue, many sages believe that far more has the word power, which is the principal and the first work of the rational soul, and especially so in the three languages consecrated by the divine mysteries, Hebrew, Greek, and Latin. But let us pause here. For these matters cannot be understood as they should without what follows.

From what therefore has been said in regard to languages, it is evident that the Latins suffer a great loss of knowledge owing to their ignorance of them. Hence in this particular they cannot boast of their knowledge, nay, they are far from glorious and they languish with the loss of knowledge in many fields. Since they have paid no attention to this matter, modern Latins of necessity have been forced to bear the loss along with the censure, from both of which all the sainted doctors, philosophers, and sages remained free.

PART FOUR OF THIS PLEA

In which is shown the power of mathematics in the sciences and in the affairs and occupations of this world.

FIRST DISTINCTION

In three chapters

CHAPTER I

AFTER making it clear that many famous roots of knowledge depend on the mastery of the languages through which there is an entrance into knowledge on the part of the Latins, I now wish to consider the foundations of this same knowledge as regards the great sciences, in which there is a special power in respect to the other sciences and the affairs of this world. There are four great sciences, without which the other sciences cannot be known nor a knowledge of things secured. If these are known any one can make glorious progress in the power of knowledge without difficulty and labor, not only in human sciences, but in that which is divine. The virtue of each of these sciences will be touched upon not only on account of knowledge itself, but in respect to the other matters aforesaid. Of these sciences the gate and key is mathematics, which the saints discovered at the beginning of the world, as I shall show, and which has always been used by all the saints and sages more than all other sciences. Neglect of this branch now for thirty or forty years has destroyed the whole system of study of the Latins. Since he who is ignorant of this cannot know the other sciences nor the affairs of this world, as I shall prove. And what is worse men ignorant of this do not perceive their own ignorance, and therefore do not seek a remedy. And on the contrary the knowledge of this science prepares the mind and elevates it to a certain knowledge of all things, so that if one learns the roots of knowledge placed about it and rightly applies them to the knowledge of the other sciences and matters, he will then be able to know all that follows without error and doubt, easily and effectually. For without these neither what precedes nor what follows can be known; whence they perfect what precedes and regulate it, even as the end perfects those things pertaining to it, and they arrange and open the way to what follows. This I now intend to intimate through authority and reason; and in the first place I intend to do so in the human sciences and in the matters of this world, and then in divine knowledge, and lastly according as they are related to the Church and the other three purposes.

Chapter II

In which it is proved by authority that every science requires mathematics.

As regards authority I so proceed. Boetius says in the second prologue to his Arithmetic, "If an inquirer lacks the four parts of mathematics, he has very little ability to discover truth." And again, "Without this theory no one can have a correct insight into truth." And he says also, "I warn the man who spurns these paths of knowledge that he cannot philosophize correctly." And again, "It is clear that whosoever passes these by, has lost the knowledge of all learning." He confirms this by the opinion of all men of weight saying, "Among all the men of influence in the past, who have flourished under the leadership of Pythagoras with a finer mental grasp, it is an evident fact that no one reaches the summit of perfection in philosophical studies, unless he examines the noble quality of such wisdom with the help of the so-called quadrivium." And in particular Ptolemy and Boetius himself are illustrations of this fact. For since there are three essential parts of philosophy, as Aristotle says in the sixth book of the Metaphysics, mathematical, natural, and divine, the mathematical is of no small importance in grasping the knowledge of the other two parts, as Ptolemy teaches in the first chapter of the Almagest, which statement he also explains further in that place. And since the divine part is twofold, as is clear from the first book of the Metaphysics, namely, the first philosophy, which shows that God exists, whose exalted properties it investigates, and civil science, which determines divine worship, and explains many matters concerning God as far as man can receive them. Ptolemy likewise asserts and declares that mathematics is potent in regard to both of these branches. Hence Boetius asserts at the end of his Arithmetic that the mathematical means are discovered in civil polity. For he says that an arithmetic mean is comparable to a state that is ruled by a few, for this reason, that in its lesser terms is the greater proportion; but he states that there is a harmonic mean in an aristocratic state for the reason that in the greater terms the greater proportionality is found. The geometrical mean is comparable to a democratic state equalized in some manner; for whether in their lesser or greater terms they are composed of an equal proportion of all. For there is among all a certain parity of mean preserving a law of equality in their relations. Aristotle and his expositors teach in the morals in many places that a state cannot be ruled without these means. Concerning these means an exposition will be given with an application to divine truths. Since all the essential parts of philosophy, which are more than forty sciences distinct in their turn, may be reduced to these three, it suffices now that the value of mathematics has been established by the authorities mentioned.

Now the accidental parts of philosophy are grammar and logic. Alpharabius makes it clear in his book on the sciences that grammar and logic cannot be known without mathematics. For although grammar furnishes children with the facts relating to speech and its properties in prose, meter, and rhythm, nevertheless it does so in a puerile way by means of statement and not through causes or reasons. For it is the function of another science to give the reasons for these things, namely, of that science, which must consider fully the nature of tones, and this alone is music, of which there are numerous varieties and parts. For one deals with prose, a second with meter, a third with rhythm, and a fourth with music in singing. And besides these it has more parts. The part dealing with prose teaches the reasons for all elevations of the voice in prose, as regards differences of accents and as regards colons, commas, periods, and the like. The metrical part teaches all the reasons and causes for feet and meters. The part on rhythm teaches about every modulation and sweet relation in rhythms, because all those are certain kinds of singing, although not so treated as in ordinary singing. For it is called "accent" since it is, as it were, song [*accantus*] from *accino, accinis*. Hence these subjects pertain to music as Cassiodorus teaches in music, and Censorinus in his book on Accent, and so too in those on other topics. Authorities on music bear witness to this fact as well as do their books on that science. And Alpharabius agrees with them in his book on the Division of the Sciences. Therefore grammar depends causatively on music.

In the same way logic. For the purpose of logic is the composition of arguments that stir the active intellect to faith and to a love of virtue and future felicity, as we have already shown, which arguments are handed down in the books of Aristotle on these arguments, as has been stated. But these arguments must have a maximum amount of beauty, so that the mind of man may be drawn to the truths of salvation suddenly and without previous consideration, as we are taught in those books. And Alpharabius especially teaches this in regard to the poetic argument, the statements of which should be sublime and beautiful, and therefore accompanied with notable adornment in prose, meter, and rhythm, as befits place, time, personages, and subject for which the plea is made. And thus Aristotle taught in his book on the Poetic Argument, which Hermannus did not venture to translate into Latin on account of the difficulty of the meters, which he did not understand, as he himself states in the prologue to the commentary of Averroës on that book. And therefore the end of logic depends upon music. But the end of everything is its noblest part in every matter and imposes necessity on what is related to it, as Aristotle states in the second book of the Physics; nor have those things any utility of their own which are naturally formed for the end, except when they are related to their end, as is clear in individual cases. And therefore the whole utility of logic is drawn from the relation of all logical arguments to arguments of this kind, and therefore since they depend on the arguments of music, necessarily logic must depend on the power of music. All these facts are in accordance with the opinion of Alpharabius in his book on the Sciences, and they are likewise clearly stated by Aristotle and Averroës in their books, although these are not used by the Latins. But not only does a knowledge of logic depend on mathematics because of its end, but because of its middle and heart, which is the book of Posterior Analytics, for that book teaches the art of demonstration.

But neither can the fundamental principles of demonstration, nor conclusions, nor the subject as a whole be learned or made clear except in the realm of mathematics, because there alone is there true and forceful demonstration, as all know and as we shall explain later. Therefore of necessity logic depends on mathematics.

What has been said is applicable likewise because of its beginning and not only because of its middle and end. For the book of Categories is the first book of logic according to Aristotle. But it is clear that the category of quantity cannot be known without mathematics. For the knowledge of quantity belongs to mathematics alone. Connected with quantity are the categories of when and where. For when has to do with time, and where arises from place. The category of habit cannot be known without the category of place, as Averroës teaches in the fifth book of the Metaphysics. But the greater part of the category of quality contains the attributes and properties of quantities, because all things that are in the fourth class of quality are called qualities in quantities. And all the attributes of these which are absolutely essential to them are qualities, with which a large part of geometry and arithmetic is concerned, such as straight and curved and other essential qualities of the line, and triangularity and other figures belonging to surface or to a solid body; and the prime and non-factorable in numbers, as Aristotle teaches in the fifth book of the Metaphysics, as well as other essential attributes of numbers. Moreover, whatever is worthy of consideration in the category of relation is the property of quantity, such as proportions and proportionalities, and geometrical, arithmetical, and harmonic means, and the kinds of greater and lesser inequality. Moreover, spiritual substances are known by philosophy only through the medium of the corporeal, and especially the heavenly bodies, as Aristotle teaches in the eleventh book of the Metaphysics. Nor are inferior things known except through superior ones, because the heavenly bodies are the causes of things that are lower. But the heavenly bodies are known only through quantity, as is clear from astronomy. Therefore all the categories depend on a knowledge of quantity of which mathematics treats, and therefore the whole excellence of logic depends on mathematics.

Mathematics

CHAPTER III

In which it is proved by reason that every science requires mathematics.

WHAT has been shown as regards mathematics as a whole through authority, can now be shown likewise by reason. And I make this statement in the first place, because other sciences use mathematical examples, but examples are given to make clear the subjects treated by the sciences; wherefore ignorance of the examples involves an ignorance of the subjects for the understanding of which the examples are adduced. For since change in natural objects is not found without some augmentation and diminution nor do these latter take place without change; Aristotle was not able to make clear without complications the difference between augmentation and change by any natural example, because augmentation and diminution go together always with change in some way; wherefore he gave the mathematical example of the rectangle which augmented by a gnomon increases in magnitude and is not altered in shape. This example cannot be understood before the twenty-second proposition of the sixth book of the Elements. For in that proposition of the sixth book it is proved that a smaller rectangle is similar in every particular to a larger one and therefore a smaller one is not altered in shape, although it becomes larger by the addition of the gnomon.

Secondly, because comprehension of mathematical truths is innate, as it were, in us. For a small boy, as Tullius states in the first book of the Tusculan Disputations, when questioned by Socrates on geometrical truths, replied as though he had learned geometry. And this experiment has been tried in many cases, and does not hold in other sciences, as will appear more clearly from what follows. Wherefore since this knowledge is almost innate, and as it were precedes discovery and learning, or at least is less in need of them than other sciences, it will be first among sciences and will precede others disposing us toward them; since what is innate or almost so disposes toward what is acquired.

Thirdly, because this science of all the parts of philosophy was the earliest discovered. For this was first discovered at the beginning of the human race. Since it was discovered before the flood and then later by the sons of Adam, and by Noah and his sons, as is clear from the prologue to the Construction of the Astrolabe according to Ptolemy, and from Albumazar in the larger introduction to astronomy, and from the first book of the Antiquities, and this is true as regards all its parts, geometry, arithmetic, music, astronomy. But this would not have been the case except for the fact that this science is earlier than the others and naturally precedes them. Hence it is clear that it should be studied first, that through it we may advance to all the later sciences.

Fourthly, because the natural road for us is from what is easy to that which is more difficult. But this science is the easiest. This is clearly proved by the fact that mathematics is not beyond the intellectual grasp of any one. For the people at large and those wholly illiterate know how to draw figures and compute and sing, all of which are mathematical operations. But we must begin first with what is common to the laity and to the educated; and it is not only hurtful to the clergy, but disgraceful and abominable that they are ignorant of what the laity knows well and profitably. Fifthly, we see that the clergy, even the most ignorant, are able to grasp mathematical truths, although they are unable to attain to the other sciences. Besides a man by listening once or twice can learn more about this science with certainty and reality without error, than he can by listening ten times about the other parts of philosophy, as is clear to one making the experiment. Sixthly, since the natural road for us is to begin with things which befit the state and nature of childhood, because children begin with facts that are better known by us and that must be acquired first. But of this nature is mathematics, since children are first taught to sing, and in the same way they can learn the method of making figures and of counting, and it would be far easier and more necessary for them to know about numbers before singing, because in the relations of numbers in music the whole theory of numbers is set forth by example, just as the authors on music teach, both in ecclesiastical music and in philosophy. But the theory of numbers depends on figures, since numbers relating to lines, surfaces, solids, squares, cubes, pentagons, hexagons, and other figures, are known from lines, figures, and angles.

For it has been found that children learn mathematical truths better and more quickly, as is clear in singing, and we also know by experience that children learn and acquire mathematical truths better than the other parts of philosophy. For Aristotle says in the sixth book of the Ethics that youths are able to grasp mathematical truths quickly, not so matters pertaining to nature, metaphysics, and morals. Wherefore the mind must be trained first through the former rather than through these latter sciences. Seventhly, where the same things are not known to us and to nature, there the natural road for us is from the things better known to us to those better known to nature, or known more simply; and more easily do we grasp what is better known to ourselves, and with great difficulty we arrive at a knowledge of those things which are better known to nature. And the things known to nature are erroneously and imperfectly known by us, because our intellect bears the same relation to what is so clear to nature, as the eye of the bat to the light of the sun, as Aristotle maintains in the second book of the Metaphysics; such, for example, are especially God and the angels, and future life and heavenly things, and creatures nobler than others, because the nobler they are the less known are they to us. And these are called things known to nature and known simply. Therefore, on the contrary, where the same things are known both to us and to nature, we make much progress in regard to what is known to nature and in regard to all that is there included, and we are able to attain a perfect knowledge of them. But in mathematics only, as Averroës says in the first book of the Physics and in the seventh of the Metaphysics and in his commentary on the third book of the Heavens and the World, are the same things known to us and to nature or simply. Therefore as in mathematics we touch upon what is known fully to us, so also do we touch upon what is known to nature and known simply. Therefore we are able to reach directly an intimate knowledge of that science. Since, therefore, we have not this ability in other sciences, clearly mathematics is better known. Therefore the acquisition of this subject is the beginning of our knowledge.

Likewise, eighthly, because every doubt gives place to certainty and every error is cleared away by unshaken truth. But in mathematics we are able to arrive at the full truth without error, and at a certainty of all points involved without doubt; since in this subject demonstration by means of a proper and necessary cause can be given. Demonstration causes the truth to be known. And likewise in this subject it is possible to have for all things an example that may be perceived by the senses, and a test perceptible to the senses in drawing figures and in counting, so that all may be clear to the sense. For this reason there can be no doubt in this science. But in other sciences, the assistance of mathematics being excluded, there are so many doubts, so many opinions, so many errors on the part of man, that these sciences cannot be unfolded, as is clear since demonstration by means of a proper and necessary cause does not exist in them from their own nature because in natural phenomena, owing to the genesis and destruction of their proper causes as well as of the effects, there is no such thing as necessity. In metaphysics there can be no demonstration except through effect, since spiritual facts are discovered through corporeal effects and the creator through the creature, as is clear in that science. In morals there cannot be demonstrations from proper causes, as Aristotle teaches. And likewise neither in matters pertaining to logic nor in grammar, as is clear, can there be very convincing demonstrations because of the weak nature of the material concerning which those sciences treat. And therefore in mathematics alone are there demonstrations of the most convincing kind through a necessary cause. And therefore here alone can a man arrive at the truth from the nature of this science. Likewise in the other sciences there are doubts and opinions and contradictions on our part, so that we scarcely agree on the most trifling question or in a single sophism; for in these sciences there are from their nature no processes of drawing figures and of reckonings, by which all things must be proved true. And therefore in mathematics alone is there certainty without doubt.

Wherefore it is evident that if in other sciences we should arrive at certainty without doubt and truth without error, it behooves us to place the foundations of knowledge in mathematics, in so far as disposed through it we are able to reach certainty in other sciences and truth by the exclusion of error. This reasoning can be made clearer by comparison, and the principle is stated in the ninth book of Euclid. The same holds true here as in the relation of the knowledge of the conclusion to the knowledge of the premises, so that if there is error and doubt in these, the truth cannot be arrived at through these premises in regard to the conclusion, nor can there be certainty, because doubt is not verified by doubt, nor is truth proved by falsehood, although it is possible for us to reason from false premises, our reasoning in that case drawing an inference and not furnishing a proof; the same is true with respect to sciences as a whole; those in which there are strong and numerous doubts and opinions and errors, I say at least on our part, should have doubts of this kind and false statements cleared away by some science definitely known to us, and in which we have neither doubts nor errors. For since the conclusions and principles belonging to them are parts of the sciences as a whole, just as part is related to part, as conclusion to premises, so is science related to science, so that a science which is full of doubts and besprinkled with opinions and obscurities, cannot be rendered certain, nor made clear, nor verified except by some other science known and verified, certain and plain to us, as in the case of a conclusion reached through premises. But mathematics alone, as was shown above, remains fixed and verified for us with the utmost certainty and verification. Therefore by means of this science all other sciences must be known and verified.

Since we have now shown by the peculiar property of that science that mathematics is prior to other sciences, and is useful and necessary to them, we now proceed to show this by considerations taken from its subject matter. And in the first place we so conclude, because the natural road for us is from sense perception to the intellect, since if sense perception is lacking, the knowledge related to that sense perception is lacking also, according to the statement in the first book of the Posterior Analytics, since as sense perception proceeds so does the human intellect. But quantity is especially a matter of sense perception, because it pertains to the common sense and is perceived by the other senses, and nothing can be perceived without quantity, wherefore the intellect is especially able to make progress as respects quantity. In the second place, because the very act of intelligence in itself is not completed without continuous quantity, since Aristotle states in his book on Memory and Recollection that our whole intellect is associated with continuity and time. Hence we grasp quantities and bodies by a direct perception of the intellect, because their forms are present in the intellect. But the forms of incorporeal things are not so perceived by our intellect; or if such forms are produced in it, according to Avicenna's statement in the third book of the Metaphysics, we, however, do not perceive this fact owing to the more vigorous occupation of our intellect in respect to bodies and quantities. And therefore by means of argumentation and attention to corporeal things and quantities we investigate the idea of incorporeal things, as Aristotle does in the eleventh book of the Metaphysics. Wherefore the intellect will make progress especially as regards quantity itself for this reason, that quantities and bodies as far as they are such belong peculiarly to the human intellect as respects the common condition of understanding. Each and every thing exists as an antecedent for some result, and this is true in higher degree of that which has just been stated.*

Moreover, for full confirmation the last reason can be drawn from the experience of men of science; for all scientists in ancient times labored in mathematics, in order that they might know all things, just as we have seen in the case of men of our own times, and have heard in the case of others who by means of mathematics, of which they had an excellent knowledge, have learned all science. For very illustrious men have been found, like Bishop Robert of Lincoln and Friar Adam de Marisco, and many others, who by the power of mathematics have learned to explain the causes of all things, and expound adequately things human and divine. Moreover, the sure proof of this matter is found in the writings of those men, as, for example, on impressions such as the rainbow, comets, generation of heat, investigation of localities on the earth and other matters, of which both theology and philosophy make use. Wherefore it is clear that mathematics is absolutely necessary and useful to other sciences.

These reasons are general ones, but in particular this point can be shown by a survey of all the parts of philosophy disclosing how all things are known by the application of mathematics. This amounts to showing that other sciences are not to

* The Latin of this sentence is obscure.

be known by means of dialectical and sophistical argument as commonly introduced, but by means of mathematical demonstrations entering into the truths and activities of other sciences and regulating them, without which they cannot be understood, nor made clear, nor taught, nor learned. If any one in particular should proceed by applying the power of mathematics to the separate sciences, he would see that nothing of supreme moment can be known in them without mathematics. But this simply amounts to establishing definite methods of dealing with all sciences, and by means of mathematics verifying all things necessary to the other sciences. But this matter does not come within the limits of the present survey.

SECOND DISTINCTION

In which it is shown that the matters of this world require mathematics. The distinction has three chapters.

Chapter I

In the first chapter it is shown in general that celestial and terrestrial things require mathematics.

WHAT has just been shown in regard to the sciences can be made clear in regard to things. For the things of this world cannot be made known without a knowledge of mathematics. For this is an assured fact in regard to celestial things, since two important sciences of mathematics treat of them, namely theoretical astrology and practical astrology. The first considers the quantities of all that is included in celestial things, and all things which are reduced to quantity discontinuous as well as continuous. For it gives us definite information as to the number of the heavens and of the stars, whose size can be comprehended by means of instruments, and the shapes of all and their magnitudes and distances from the earth and thicknesses and number, and greatness and smallness, the rising and setting of the signs of the stars, and the motion of the heavens and the stars, and the numbers and varieties of the eclipses. It likewise treats of the size and shape of the habitable earth and of all great divisions which are called climes, and it shows the difference in horizons and days and nights in the different climes. These matters, therefore, are determined by this branch of the subject as well as many things connected with them. Practical astrology enables us to know every hour the positions of the planets and stars, and their aspects and actions, and all the changes that take place in the heavenly bodies; and it treats of those things that happen in the air, such as comets and rainbows and the other changing phenomena there, in order that we may know their positions, altitudes, magnitudes, forms, and many things that must be considered in them. All this information is secured by means of instruments suitable for these purposes, and by tables and by canons, that is, rules invented for the verification of these matters, to the end that a way may be prepared for the judgments that can be formed in accordance with the power of philosophy, not only in the things of nature, but in those which take their tendency from nature and freely follow celestial direction; and not only for judgments in regard to present, past, and future, but for wonderful works, so that all things prosperous in this world may be advanced, and things adverse may be repressed, in a useful and glorious way. Nor are these matters doubtful. For the patriarchs and prophets from the beginning of the world had full information respecting these matters as well as all others. Aristotle restored the knowledge of the ancients and brought it to light. All those informed in great subjects agree in this, and experience teaches it. But concerning these matters an exposition will be given in the proper place.

It is plain, therefore, that celestial things are known by means of mathematics, and that a way is prepared by it to things that are lower. That, moreover, these terrestrial things cannot be learned without mathematics, is clear from the fact that we know things only through causes, if knowledge is to be properly acquired, as Aristotle says. But celestial things are the causes of terrestrial. Therefore these terrestrial things will not be known without a knowledge of the celestial, and the latter cannot be known without mathematics. Therefore a knowledge of these terrestrial things must depend on the same science. In the second place, we can see from their properties that no one of these lower or higher things can be known without the power of mathematics. For everything in nature is brought into being by an efficient cause and the material on which it works, for these two are concurrent at first. For the active cause by its own force moves and alters the matter, so that it becomes a thing. But the efficacy of the efficient cause and of the material cannot be known without the great power of mathematics even as the effects produced cannot be known without it. There are then these three, the efficient cause, the matter, and the effect. In celestial things there is a reciprocal influence of forces, as of light and other agents, and a change takes place in them without, however, any tendency toward their destruction. And so it can be shown that nothing within the range of things can be known without the power of geometry. We learn from this line of reasoning that in like manner the other parts of mathematics are necessary; and they are so for the same reason that holds in the case of geometry; and without doubt they are far more necessary, because they are nobler. If, therefore, the proposition be demonstrated in the case of geometry, it is not necessary in this plea that mention be made of the other parts.

In the first place, I shall demonstrate a proposition in geometry in respect to the efficient cause. For every efficient cause acts by its own force which it produces on the matter subject to it, as the light of the sun produces its own force in the air, and this force is light diffused through the whole world from the solar light. This force is called likeness, image, species, and by many other names, and it is produced by substance as well as accident and by spiritual substance as well as corporeal. Substance is more productive of it than accident, and spiritual substance than corporeal. This species causes every action in this world; for it acts on sense, on intellect, and all the matter in the world for the production of things, because one and the same thing is done by a natural agent on whatsoever it acts, because it has no freedom of choice; and therefore it performs the same act on whatever it meets. But if it acts on the sense and the intellect, it becomes a species, as all know. Accordingly, on the other hand, if it acts on matter it also becomes a species. In those beings that have reason and intellect, although they do many things with deliberation and freedom of will, yet this action, namely, a production of species, is natural in them, just as it is in other things. Hence the substance of the soul multiplies its own force in the body and outside of the body, and any body outside of itself produces its own force, and the angels move the world by means of forces of this kind. But God produces forces out of nothing, which he multiplies in things; created agents do not do so, but in another way about which we need not concern ourselves at the present time. Forces of this kind belonging to agents produce every action in this world. But there are two things now to be noted respecting these forces; one is the multiplication itself of the species and of force from the place of its production; and the other is the varied action in this world due to the production and destruction of things. The second cannot be known without the first. Therefore it is necessary that the multiplication itself be first described.

Mathematics

Chapter II

In which the canons of the multiplication of the forces of agents as respects lines and angles are explained.

Every multiplication is either with respect to lines, or angles, or figures. While the species travels in a medium of one rarity, as in what is wholly sky, and wholly fire, and wholly air, or wholly water, it is propagated in straight paths, because Aristotle says in the fifth book of the Metaphysics that nature works in the shorter way possible, and the straight line is the shortest of all. This fact is also made evident by the twentieth proposition of the first book of Euclid, which states that in every triangle two sides are longer than the third.

But when the second body is of another rarity and density, so that it is not wholly dense, but changes in some way the passage of the species, like water, which in one way is rare, and in another dense, and crystal similarly, and glass and the like, through the media of which we are able to see, then the species either impinges upon the second body perpendicularly, and still travels along the straight line as in the first medium; or if it does not fall perpendicularly, then of necessity it changes its straight path, and makes an angle on entering the second body, and its declination from the straight path is called refraction of the ray and the species. This is because the perpendicular is the stronger and shorter, and therefore nature works in a better way on it, as geometrical demonstrations show, of which mention will be made later more particularly in the proper place. But this refraction is twofold, since if the second body is denser, as is the case in descending from the sky to these lower objects, all the forces of the stars that do not fall perpendicularly on the globe of elements, are broken between the straight path and the perpendicular drawn from the place of refraction. If the second body is rarer, as is the case in ascending from water into the air, the straight path falls between the refracted ray and the perpendicular drawn from the place of refraction. And this is a wonderful diversity in the action of nature, but it is not strange, when countless wonders are performed by nature in accordance with the laws of these refractions; and by means of art aiding nature those things can be done which the world cannot receive, as I shall explain in perspective science. But it will be driven by these means in the times of Antichrist to those things which it will itself wish for in great part.

That, moreover, these things are true authorities teach, and all experts know, and instruments can be made so that we may sensibly see propagations of this kind; but until we have instruments we can prove this by natural effect without contradiction, as this figure shows. Let us take then a hemisphere of crystal or a glass vessel, the lower part of which is round and full of water. When, therefore, rays come from the center of the sun to the body of crystal, or of glass, which is denser than the air, those that are not perpendicular to such a body (and these are the ones that do not pass through its center, as is clear from geometric principles) are refracted between the straight path and the perpendicular drawn from the point of refraction, as is the ray *ac,* which after passing through the whole body of the vessel, comes obliquely to the air which is of less density. Of necessity, therefore, it so travels that the straight path is between the path of the ray and the perpendicular drawn to the point of refraction, and therefore the ray will not travel to *c,* but bends to *f,* on the principal perpendicular, which comes from the sun, that is, ray *Of.* And in the same way at any other point, owing to the double refraction, *hf* will pass through the same point *f* on the ray *Of.* But an infinite number of rays come forth from every point on the sun to the body; therefore an infinite number will meet in this same point by means of the double refraction. But a convergence of rays is the cause of heat. Therefore a burning heat will be produced at this point. And this is a fact, as is clear to the sense; for if a combustible be placed at this point, as wool, silk, or a piece of rag, it will be consumed. Since, therefore, there is combustion at this point and this can-

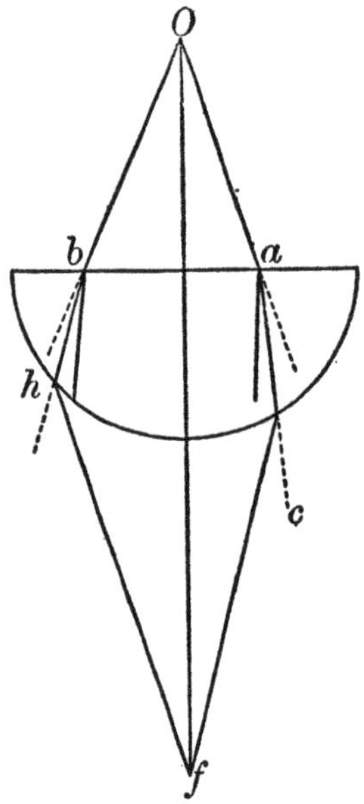

FIG. 1.

not happen except through a convergence of rays, and the rays cannot assemble except through a double refraction, because a single refraction would not suffice and a third refraction is not required, therefore we must assume this kind of refraction, a wonderful thing in the eyes of men of science. For why is it that nature so acts? Surely nothing is pleasing to nature, or to her will, except what is remade by change; but the causes are hidden. We need not now investigate the causes, since we know this marvel by means of a very certain test, and in what follows other tests will be added.

When, however, the second body is so dense as in no way to permit the passage of the species,—I am speaking of a passage appreciated by human vision,—we say that the species is reflected. Yet according to Aristotle and Boetius the vision of the lynx penetrates walls. Therefore the species does actually pass through as a matter of fact; but human vision forms no judgment concerning this, but does concerning the reflection which of necessity takes place. For on account of the difficulty of passage through the dense medium, since in the air from which it came it finds an easy road, it multiplies itself more abundantly in the direction from which it came. And this in the first place can happen in general in two ways; for either the ray falls perpendicularly on the dense body and then returns upon itself wholly by the same path by which it came, and a ray is generated in the same place, as for example the ray *ab* falls perpendicularly, and this is in planes at right angles, as is shown in the eleventh book of geometry, just as in spherical shaped bodies when the ray passes through the center. And the reason for this is that the angles of incidence and reflection are always equal, as a manifold demonstration shows, and authorities maintain, and instruments made for this purpose make clear to the eye. But there are only two right angles at the dense body in the case of *ab*. Therefore by these same angles the reflex ray will return upon itself and therefore in the same place. But the line *ac* which falls at oblique angles and not perpen-

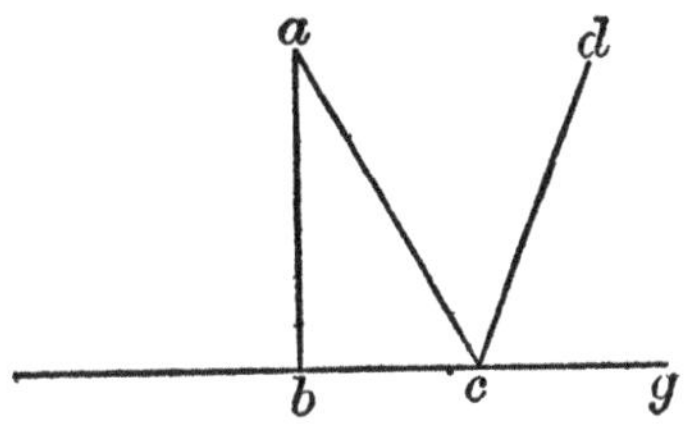

FIG. 2.

dicularly, does not return upon itself, but passes to *d,* because of the equality of the angles of incidence and reflection. Whenever the ray falls at oblique angles, the acute angle is called the angle of incidence; and from the obtuse angle an angle equal to the angle of incidence is cut off by the reflected line. The angle so formed is contained between the reflected line and the dense body, as is the angle *dcg,* and it is called the angle of reflection, which of necessity must equal the acute angle on the other side, and we prove this to the sight in mirrors. For we cannot see things, unless the eye is in the line of reflection, as if the eye be at *d* it will see *a;* and if not, it will not see by means of that reflected ray. These are known facts, and tests will be given adequately concerning this matter in what follows.

Moreover, an infinite number of rays can be assembled by reflection, just as by multiplication, so that strong combustions take place. But from a plain surface rays cannot converge to one point, because one goes to one point, another to another point. Nor can they do so from a convex mirror; but they can converge from a concave spherical mirror, or from one column-shaped, or pyramidal, or ring-shaped, or oval, and so too from others. If, therefore, a concave spherical mirror be exposed to the sun, an infinite number of rays will converge to one point by means of reflection. And therefore of necessity fire is kindled when a concave mirror is exposed to the sun, as is stated in the last proposition of the book on Mirrors, and the demonstration is there given. But an instrument might very well have been made for this purpose, and the phenomenon would then be visible to the eye, as we stated before in the case of refraction. Hence if a mirror should be made of good steel, or of silver, the combustion would occur more easily; but a combustion does not take place by all rays falling on a mirror, but by those only which fall on the circumference of a circle around the axis of the mirror, because all that fall in one circumference fall at equal angles, and therefore are reflected to a point on the axis, because the angles of the reflected rays are equal, and those that fall in another circle are reflected to another point, and those in a third circle to a third point, and the same statement may be made of an infinite number of circles imagined around the axis of the mirror; for of necessity rays falling on different circumferences travel to different points, because they do not fall at equal angles. Those falling in a smaller circle are reflected higher, and those in the greatest circle are reflected to the lowest point, namely to the pole of the sphere, or to the end of the axis. But neither nature nor art is content with combustion of this kind, nay, they wish so to fashion bodies that all the rays falling on the whole surface of the mirror may converge to a single point; and what is more at every distance we desire. This is the ultimate which the power of geometry can do. For this mirror would burn fiercely everything on which it could be focused. We are to believe that Antichrist will use these mirrors to burn up cities and camps and armies. Since if a moderate convergence of rays by refraction or by a concave mirror burns perceptibly, how much more so without limit, when rays without number converge by means of this mirror. Scientists reckon that this is a necessary consequence. And an author in a book on burning mirrors shows how this instrument is made, but without sufficient reason hid in that book much respecting the artifice, and states that he has placed the lacking information in another book, which has not been translated by the Latins. But there are Latins who because of the ill favor of that author in concealing the perfection of his knowledge, have studied this wonderful secret of nature, because that author stimulates greatly men skilled in science to perfect what is lacking, and he shows that the mirror must be nearly ring-shaped or oval, as if the cones of an egg were cut off, the figure would be ring-shaped; if, however, one cone remains, it is oval. Now in such a figure properly constructed all the rays falling on the whole surface must fall at equal angles, and therefore they must be reflected at like angles, and for this reason to one point. Moreover, the most skillful of the Latins is busily engaged on the construction of this mirror, and the glory of your Magnificence will be able to order him to complete it when he is known to you. This triple multiplication is said to be a principal one, because it comes from the agent itself.

But the fourth kind is more necessary to the world, although it is called accidental multiplication. For light is called accidental with respect to the principal light coming from an object, since it does not come from an agent, but from principal multiplications, as in a house the principal multiplication falls through the window from the sun, but in a corner of the house the accidental light comes from the ray of the window. Moreover, the bodies of mortals would not be able to be exposed always to principal species without destruction, and for this reason God has tempered all things by means of accidental species of this kind.

The fifth kind is different from the others, for it does not follow the common laws of nature, but claims for itself a special privilege. This multiplication does not take place except in an animated medium, as in the nerves of the senses; for the species follows the tortuous course of the nerve, and pays no attention to the straight path. This happens through the force of the vital principle regulating the path of the species, as the actions of an animated being require. Concerning this propagation something will be said in the truths of perspective. The first four in respect to which nature delights to work are common to the inanimate things of the world; the fifth is known to pertain to sensation.

Chapter III

In which multiplication is given as respects figures.

We must next consider how multiplication takes place with respect to figures. Multiplication of necessity takes place in the form of a sphere. For the agent multiplies itself equally in every direction, and with respect to all diameters, and all differences of position, namely, above, below, before, behind, to the right, and to the left. Therefore everywhere the lines go forth in every direction from the agent as from a center: but lines traveling everywhere from one place cannot terminate except on the concave surface of a sphere. And this is clear, because the eye sees only by means of the coming species, but if an infinite number of eyes were placed everywhere, all would see the same thing; therefore the species goes forth by means of an infinite number of lines; but an infinite number of lines terminate only on a spherical surface. If it should be said that light entering through a large triangular opening or through one of another polygonal figure does not fall in spherical form, but does so when it enters through a small opening, we must state that the sides of the small opening are not far apart and therefore the light in a short distance is able to regain its figure, but when it passes through a large figure, it cannot do so easily, but it will do so at some sufficient distance, if obstacles are removed. This principle is made evident by the fourteenth and fifteenth propositions of the first book of the elements of Euclid, as the figure shows. For let the rays be drawn as far from the in-

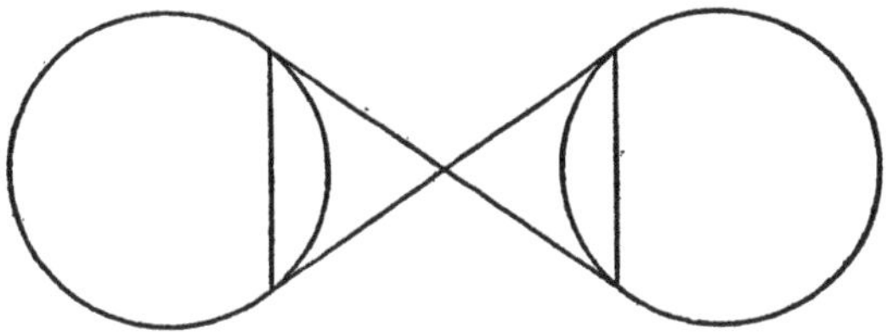

FIG. 3.

tersection as from the intersection to the sun, by the propositions mentioned the bases of the triangles must be equal. But those bases are the diameters of the lights. Therefore the diameter of the species is necessarily equal to the diameter of the sun at some one distance, and consequently the multiplication will be an equal spherical one, and can be varied according to difference of distance, but it will always be of spherical form. Nor is the light of fire a case in point, which ascends in the form of a pyramid; because this is not a multiplication from the proper nature of light, but is owing to the motion of the body of the fire itself, of which light is an accident, and the accident moves in accordance with the motion of its subject, like the light of the sun in the sun. Now fire must ascend in pyramidal form, since the interior parts are removed from the surrounding cold, and therefore they are less impeded and extricate themselves more quickly than the parts on the outside, and for this reason rise higher, and the remaining parts the nearer they are to these, the more quickly do they extricate themselves, and they attach themselves to the ones in the interior, failing somewhat to reach the height of the inmost parts, and so in regular gradation the remoter ones reach a lesser height, because they are more impeded by the contrary force surrounding them; and therefore a pyramid must necessarily be formed. But in the sphere all regular figures can be inscribed, as is clear from the fifteenth book of the elements of Euclid, among which figures is the pyramid.

And although according to the principle of inscribing geometrical figures irregular ones cannot be inscribed, nor round figures, nevertheless all figures can be produced and marked on the sphere. And therefore not only in spherical multiplication shall we find pyramids with sides, which can be inscribed in a sphere, but also round* pyramids, which can be marked and drawn in spherical multiplication. And it is this figure that nature especially selects in every multiplication and action, and not any pyramid at all, but that one whose base is the surface of the agent and whose vertex falls on some point of the surface acted on, because in this way can the species of the agent come to each point of the surface acted on by means of an infinite number of separate pyramids, as is clear in the figure.

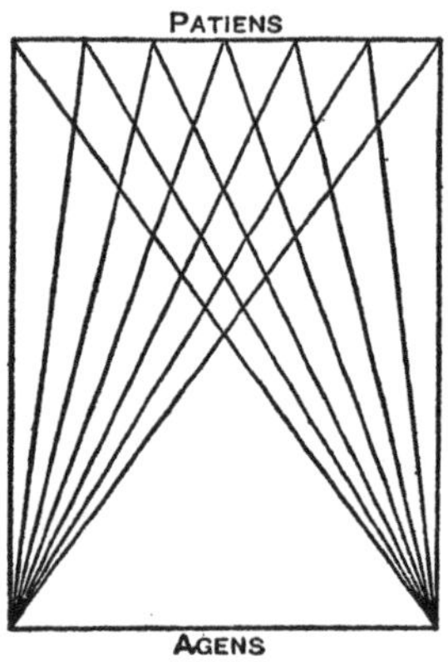

FIG. 4.

For from each point of the surface acted on there are an infinite number of rays and therefore they can be combined infinitely to form an infinite number of round pyramids with one common base, namely, the surface of the whole agent; and to every part of the surface acted on there comes an apex of a pyramid, so that force comes from the whole agent to each point of the surface acted on, and not from some limited part, to the end that the force may come complete and as a whole, not partial and imperfect, so that the action may be complete, because nature acts in accordance with what is better.

* Bacon's expression for a cone.

THIRD DISTINCTION

In which the difference of natural action is made clear by means of geometry. This distinction has three chapters.

CHAPTER I

AFTER considering these facts in regard to multiplication, we must take up some matters in regard to ulterior action. For light by means of its multiplication makes a luminous image, and this action is called univocal, because the effect is univocal, and of one kind and conformable to the agent. But there is another equivocal multiplication, as light generates heat, heat generates putrefaction, putrefaction death, and wine inebriates, and so of every agent, because it produces many effects besides its own species and force univocal to itself. And so the sun and stars cause all things here below, and the angels move the heavens and the stars, and the soul its own body; the force, however, of the agent does all those things, and this is complete action of the agent and of its force and the one desired finally by nature. Concerning this action therefore some canons or rules must be considered. These rules relate chiefly to this action, and yet they have place in a univocal action and contain truth in regard to it.

Nature, therefore, as has been said, works more effectively in a straight line than in a curve, because the former is shorter, and causes the surface acted on to be less distant from the agent, and therefore it receives more of the force of the agent, as he who is near a fire is warmer than one farther away. But the equal is better than the unequal, as Boetius states in the practical part of his Geometry. But in the straight line there is equality. Likewise every united force is of stronger action, as is stated in the book on Causes. But uniformity and unity are greater in the straight line, as Aristotle states in the fifth book of the Metaphysics. For in the curve is an angle, which causes dispersion and irregularity in form and resists unity. Therefore nature acts with more force in the straight line than in the broken or reflected one. But the straight line, which falls at equal angles and perpendicularly either on planes or on spherical bodies, is the one on which nature chooses to work both on account of its equality and greater uniformity, and on account of its shortness. For by the nineteenth proposition of the first book of the elements of Euclid, in every triangle the greater side is opposite the greater angle. But from the seventeenth proposition of the same book, the greater angle in a triangle is the right angle, namely *acb*. Therefore the greatest side, *ab*, is opposite this angle. But this line does not fall perpendicularly. Therefore the perpendicular is shorter; wherefore the force coming along it will act more strongly. Moreover, the effectiveness of a perpendicular force is shown not only by demonstration, but also by experience, whence a stone falling from a height perpendicularly strikes with more force; and if a man falls from a height perpendicularly he is hurt more: for should any one divert from the perpendicular path a man falling from a height, he would not be injured, provided he was near the ground. If this help is not given he will die from the perpendicular fall, and will be shattered completely.

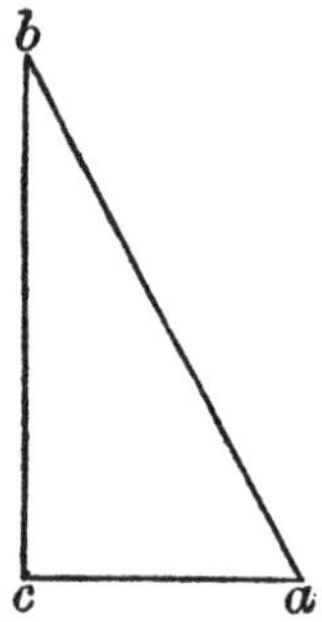

FIG. 5.

And we must consider that if from different points of the agent rays come to the same thing, one perpendicular to the agent and another not perpendicular to it, the perpendicular will always be the shorter. Let *ac* represent the agent, then *ab* is shorter than *cb*, and therefore stronger for this reason. And although *cb* falls on *ed*, a surface acted on at right angles, and *ab* does not, so that *cb* will be strengthened for this reason, nevertheless it will not reach the strength of *ab*, because *ab* is shorter, and because the point *b* of *ed* is likewise a point of another line, as *fg*, to which *ab* is perpendicular, and so the explanation of a body acted on is applicable in its main point to *b*. And the action in particular is most complete on a spherical body, since the force does not fall there at some angle, but

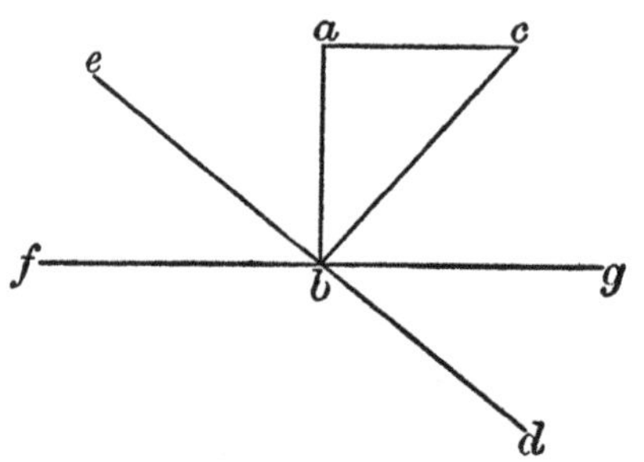

FIG. 6.

passes through the surface and the spherical body, and there is no difference or difformity and completeness of action is there found. For unless circles be marked on the sphere, there is no angle there; but when they are marked the angles of the perpendicular line will be equal, just as in planes, and the action will be complete. But, however, they are not right angles, but obtuse, as is clear from geometry.

In the second place, we must know that nature works on broken lines more strongly than on reflected ones, because refraction is in the direction of the straight path, but reflection advances in the opposite direction, and contrary to the natural path which the nature of the advancing force seeks in continuity and direction unless it is impeded, and therefore reflection weakens greatly the species and the force, and much more so than refraction. But this is to be understood of refraction and reflection as regards the peculiar property of forward motion in them. If, however, we consider that reflection takes place in the same medium, and refraction in different ones, of necessity the double medium impedes more than the single, and this is the case at least when reflection takes place in a rare medium and the refraction in a second denser medium, as in a glass vessel. For if rays should be assembled from a burning mirror and behind a burning glass, the combustion is of necessity greater, as will be explained below in the proper places. Refraction which takes place in a second denser body weakens less than in a second rarer body. For the perpendicular advance is the strongest, and therefore that which approaches nearer to the perpendicular is the stronger. But refraction in a second denser body is deflected toward the perpendicular, which is erected at the same point at which the refraction takes place, as is clear above in the figure, both in planes and in spheres, and therefore that kind weakens less. In the case of reflection at right angles, although the ray is accidentally doubled, and thus the action becomes stronger, yet it is in accordance with the nature of that reflection that by repeating itself it weakens the species more; for it is wholly opposed to the natural effort of the species itself, since the species returns by the same line by which it came. But when the reflection takes place at oblique angles, it is not wholly in the opposite direction, but to the side, and therefore this reflection does not weaken so much as the other. I am speaking in reference to the nature of the reflection, but owing to the doubling of the force in the same place and owing to the equality of the angles and the conditions of the perpendicular, the action is stronger. And yet we must here consider that by the falling of rays at oblique angles more rays can be assembled by intersection than by rays falling at right angles, not only from the property of mirrors, as has been said, but because of rays meeting in infinite numbers in accordance with the law of incidence and reflection at oblique angles, just as happens in the air, because owing to the falling of rays of this kind and their reflection the rays intersect one another at every point in infinite number, and heat is produced. For few rays are incident perpendicularly on any body, because from a point of the agent only one perpendicular ray falls to a point of the surface acted on, and therefore few are reflected. But an infinite number of non-perpendicular rays proceed from each point of the agent, and an infinite number corresponding to these are reflected. Then by the perpendicular falling only two are united at the same point in the air, namely, the incident ray and its equal reflected ray. But by the falling at oblique angles an infinite number intersect one another at every point in the air. And likewise incident rays penetrate reflected ones that do not belong to them, and reflected rays penetrate reflected ones in countless numbers. For at every point of the earth rays in infinite number are incident, and from the same point infinite numbers are reflected, and therefore a stronger action is produced in this way accidentally from the incident and the reflected rays at oblique angles than at right angles. Art, moreover, is able to aid nature in bringing about the action; for it can form mirrors in such a way that a great convergence of forces is produced by means of concave mirrors, and especially by those of oval shape, just as has been stated. But the principal force, namely, the straight refracted and reflected one, is stronger than the accidental, because the accidental one does not come from the agent, but from the species of the agent, and it is the species of a species, and for this reason is weaker.

Mathematics

Chapter II

In which the strength of action according to figures is considered.

AND since the pyramid, as has been said, is required for the action of nature, we must consider that the apex of the shorter pyramid acts more strongly, both because it is less distant from the agent, and because the rays conterminal around the apex of the shorter pyramid are in closer proximity, and proximity of rays and convergence act more strongly; and this is shown in the figure. For by the thirteenth proposition of the first book of the elements of Euclid, all the angles around a single point in a surface are equal to only four right angles, therefore the four angles at the apex of the shorter pyramid equal the remaining four at the apex of the longer pyramid. But by the twenty-first proposition of the same book, the angle in the apex of the shorter pyramid is greater than the angle in the apex of the longer pyramid, namely, *a* is greater than *c;* and by the fifteenth proposition of the same book, angles placed opposite are equal, namely, *a* and *b,* likewise *c* and *d;* therefore *c* and *d* taken together are less than *a* and *b* taken together; therefore since the four taken together equal the other four so taken by the thirteenth proposition, of necessity *h* and *l* are greater than *f* and *e.* Wherefore the rays containing *e* are in greater proximity than the rays containing *h.* And in the same way the rays containing *f* are nearer than the rays containing *l,* and so of an infinite number of rays conterminal in the shorter pyramid, and of necessity all of them are in closer proximity than the rays conterminal in the apex of the longer pyramid. But proximity of forces is the cause of stronger action. But, however, since the forward motion of perpendicular rays is the strongest, and every deflection toward perpendiculars is stronger than deflection away from them, the rays of the

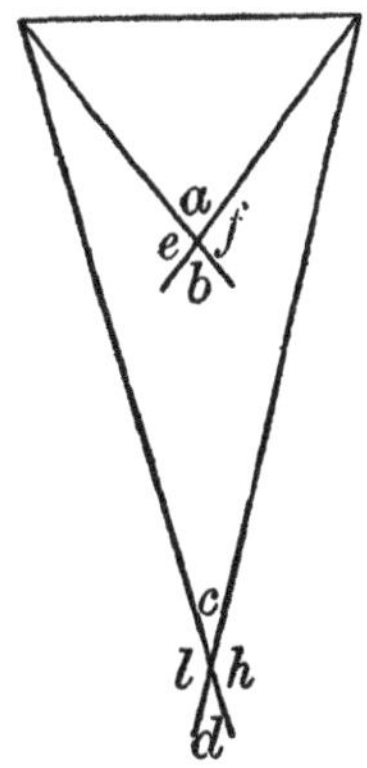

FIG. 7.

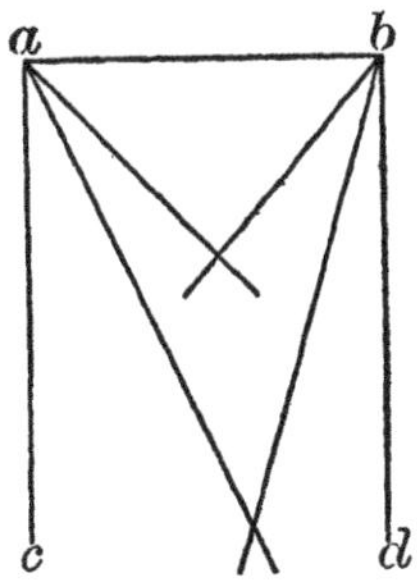

FIG. 8.

longer pyramid, since they deflect more toward the perpendiculars *ac, bd* will be stronger. Likewise as many rays come to the apex of the longer pyramid as to the apex of the shorter, since they are infinite in number on both sides. But the apex of the longer pyramid has the more acute angle by proposition XXI. Therefore its rays are the more united. Therefore they will burn more strongly. And we must state that these reasons are demonstrations for both conditions, but they are stronger for the first condition, and therefore they outweigh the other. Whence, as far as the last reasons can, so far do they draw conclusions, but there are other more potent ones and more effective in action.

Chapter III

Explaining how much of the surface acted on is changed and how much of the agent produces the change.

To these statements we must add that in equal spherical bodies the half of each receives the force of the other, because the extreme rays touch those bodies, and therefore they pass through the ends of the diameters, and no ray reaches any part of the other half. But the lesser body receives the force of the greater in its own larger portion, because the extreme rays are not equidistant always, but converge and are able to embrace more than half of the smaller. For the diameter of the greater sphere is greater than the diameter of the smaller; and therefore rays coming from the ends of the diameter of the greater

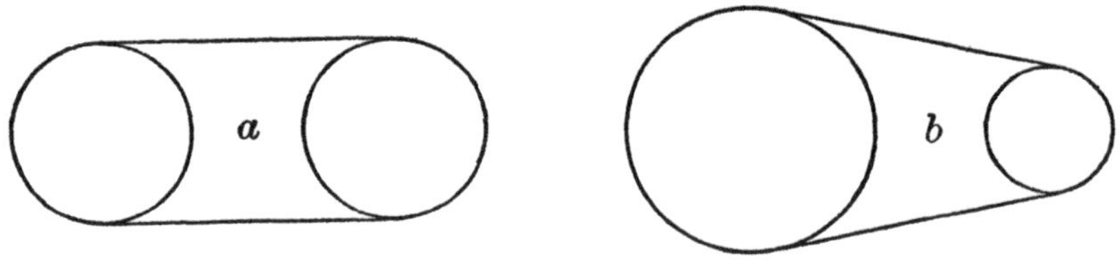

Fig. 9.

body are able to pass beyond the diameter of the smaller body so as to embrace the larger portion of it; and on the contrary the smaller portion of the larger body receives the force from the smaller body, because the diameter of the smaller body does not equal the diameter of the larger, but equals some chord of the smaller portion of the larger body. And no point of a spherical body can exert a force from it except to the space which is separated from it by a line touching that body in that point from which the force acts; since from the point *a* no other line can fall between the tangent line and the spherical body, as is proved in the fifteenth proposition of the third book of the elements of Euclid. And therefore the space which is in the angle of tangency, as well as that back of the whole tangent line on the side of the body, will not receive the force from the point *a,* but all the space beyond the tangent line, in which are the points *bcde* will receive the force. And all rays coming forth from the surface of the spherical body, whose direction falls to the center of that body, are perpendicular to it and such lines come forth everywhere in infinite numbers, as is shown in the figure. And yet from the same point on the surface of a body, at which the ray is perpendicular, there are an infinite number of rays on the same body, as is shown at the point *a,* and this is true of all points; but only one ray, namely, *ab,* is perpendicular to that body, because, it alone, if produced continuously and in a straight line, passes through the center of the body, and therefore it is the strongest and possesses more force by far.

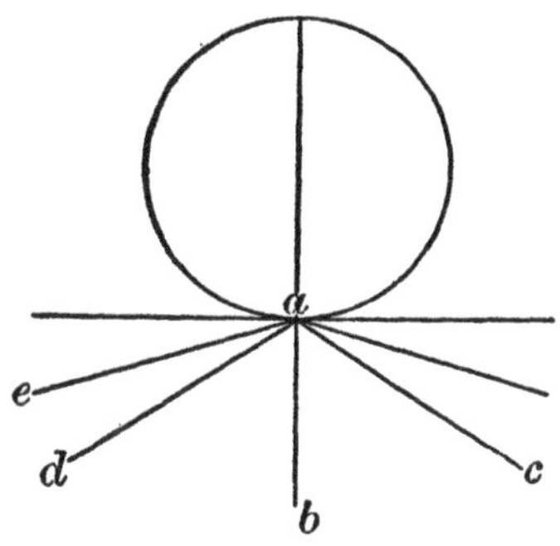

FIG. 10.

And those perpendicular rays, because they meet at the center of the body, are not equidistant; yet the eye can be so far distant from the body that it does not perceive their meeting, and judges such rays to be equidistant, as we experience in the case of rays from the sun and stars, and therefore we judge the shadows of different things equidistant when they are opposite to the sun, but we do not do so in accordance with the fact. For we think that many things are equidistant, because we do not perceive their meeting, as the walls of any house seem equidistant to our sense, but they are not so, because everything with weight tends naturally to the center, and therefore the house would fall if the walls were absolutely equidistant. And the meridian circles of different states seem to be equidistant, and the meridian lines, because we do not perceive their meeting, and yet they do meet at the pole of the earth.

We should also know that rays falling to the çenter of the spherical body from which they come are the ones by which we judge when we consider the stars through the openings of instruments. Hence the astronomer and the man working with perspective, who examine matters of this kind, make use of those rays, because they are perceived by the senses and are strong. For although from some portion of a spherical body opposite to the thing acted on there comes a pyramid with infinite rays, which are from the individual points of that portion, and all rays converge at the apex of the pyramid with the perpendicular ray, yet one alone is perpendicular in a pyramid, and that perpendicular is superior in strength, and is the axis of the pyramid and the whole pyramid is named from it by experimenters, and it is called the ray of the acting body, as is shown in the figure. For let *a* be the center, and *dc* the por-

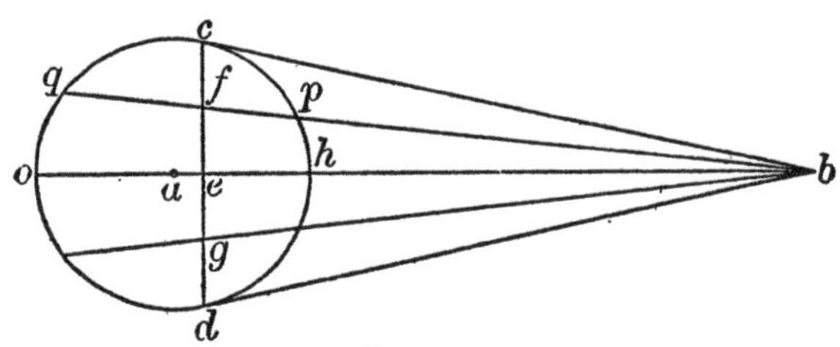

FIG. 11.

tion of the sun opposite to the earth, and *b* the apex of the pyramid falling upon the earth; it is evident that the ray *eb* falls to the center of the sun, and no others that are of the body of the pyramid, although they are infinite in number. For *gb* diverges from the center of the sun, as is clear, and so do all the others, *fb, cb*. And therefore it is perpendicular to the sun's body, and is the axis of the pyramid, and therefore is stronger and has more force, since force comes by this ray from the whole depth of the sun, which is not the case in the others. For the diameter *ho* is longer than *pq,* and longer than all lines falling in the circle to the side of the diameter, and therefore receives more of the sun's substance, and therefore has more force. And the line *hb* is shorter than *cb* and all other lines descending from the portion of the sun to the earth, wherefore it has more force according to the statements made. And that which has now been said is clear from the eighth proposition of the third book of the elements of Euclid.

FOURTH DISTINCTION

Chapter I

In which the rules cited are applied to the light of the stars.

BY these principles and the like given by means of geometry, a man can verify every action of nature, because every truth in regard to the action of an agent in a medium, or on matter in general, or on celestial bodies, and on the whole machine of the world, is derived mediately or immediately from the principles just stated, and from certain similar ones, because I have not been able to place in this plea all that a larger work requires. And I wish to illustrate my statement by some examples in different things of the world, and I shall begin with things above. Aristotle says in the first book of the Meteorologics that all the stars receive their light from the sun; and this is shown by an eclipse of the moon; for when the earth is interposed between the sun and the moon, the moon is eclipsed, and when the earth is not so placed, the moon is illuminated; and the same thing would happen in the case of other heavenly bodies, if they were in the same situation as the moon. But they are not, for the apex of the pyramid of shadow reaches only to the sphere of Mercury, and therefore the moon alone can fall within the earth's shadow. And yet the lower stars eclipse the higher ones, as Aristotle maintains in the second book of the Heavens and the World, when the lower ones fall between the sun and the higher ones, and this happens properly, but is not so visible as in the case of the moon. But since, as is clear from what has already been said, rare and transparent bodies permit the passage of the species, like the air, and since the species of the eye and of the stars pass through the sphere of fire, and through the medium of all the orbits of the seven planets, these media must of necessity be rare and transparent, and it is clear that they do not limit vision. Therefore they are not dense. Therefore they are invisible, because that only is visible, as Avicenna teaches in the third book on the Soul, which can terminate vision, and this is a fact. But if they are not visible, they are non-luminous, because what is luminous is visible. I am speaking of a luminous body with a fixed light belonging to it and not one passing through it, a light which is able to multiply rays from itself, like a star and fire; I am not speaking of a luminous body that receives light passing through it, as the air, which Aristotle calls luminous in the second book on the Soul: this is an equivocal use of the word. Therefore they make a mistake who judge that the sphere of fire is naturally luminous, just as here below, and especially since it is rarer than air, and therefore less visible, and on this account less fitted for light, because density is a cause of illumination, as Averroës states in the second book of the Heavens and the World, and in the book on the Substance of the Sphere. And likewise a worse error is commonly made in maintaining that the spheres in which the stars move are luminous, especially when a false statement is made and imputed to Averroës. For it is said that a star does not differ from the sphere in which it moves except by a greater and lesser aggregation of light. But Averroës does not say this, but teaches and proves the contrary: for this whole statement is correct except the genitive case at the end, namely, *lucis* [of light], in place of which he says *perspicui coelestis* [of the transparent celestial substance]. And because almost all the words of the common statement and of Averroës himself are the same, and people consider transparent and luminous as the same because of Aristotle's use of the word in the second book on the Soul, where luminous is used in an equivocal sense, they impute to Averroës the teaching that the sphere in which a planet moves is luminous with a fixed light of its own, like a star, although less. But he states the contrary, maintaining that because of the strength of action that a star has in this world, it must of necessity have much of the substance of the heavens collected in its own body, and therefore the transparent celestial substance, which is dispersed in parts of the orbit, is condensed in the body of the star, so that it has a strong force in the alteration of the world. And therefore he teaches in this place and in his book on the Substance of the Sphere that the star alone is luminous, and no part of its orbit. And this error is common among those philosophizing and among all theologians when they speak and write about the stars.

Since in the book on the properties of the elements it is stated that the sun is like a candle and the stars are like mirrors, the rank and file of students judge that the light coming to us from the moon and stars is the light of the sun reflected from their surfaces. But this is impossible because of the equality of the angles of incidence and reflection, as is shown in the figure. For as we saw before, if this were the case, the angle *a* of incidence must equal the angle *b* of reflection. Therefore the ray *bc* comes only to a definite portion of the earth and not everywhere, and this is true of all the light coming to the surface of the moon. For all the light is like a single ray, and falls at unequal angles, and is reflected to a definite part. Therefore if the light came to the earth, the moon would illuminate only some definite part of the horizon; but we see that it illuminates the whole hemisphere like the sun. Therefore that light which comes from the moon and stars is not reflected. Averroës in the second book of the Heavens and the World uses this demonstration, and confirms the fact by his own authority, that the light descending to us from the stars is not reflected from their surface, but belongs to them and is innate, yet derived from the potency of the matter in the body of the star through the force of the sun coming to the star, and this force alters and transforms the star and produces light in it; and when it has light naturally generated in it, just as the sun has created light, it is then able to multiply light from itself in all directions like the sun. And then only must we concede that the light of the sun is reflected from the surface of the moon, but the light does not come to the earth, but deflects to some part of the sky among the heavenly bodies in accordance with the equality of the angles of incidence and reflection.

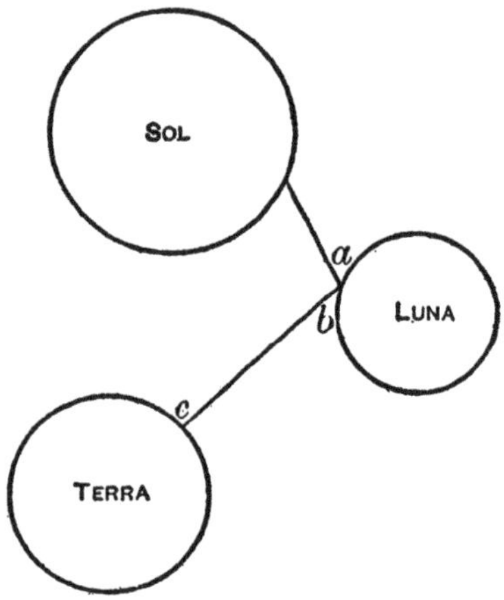

FIG. 12.

From the statements given above it is clear how much of the earth and of the stars is illuminated by the sun. For the greater portions of these are always illuminated, because they are less than the sun. For the sun is about 170 times larger than the whole earth, as Ptolemy shows in the fifth book of the Almagest, and so all things pertaining to illuminations and protractions of rays in the heavenly bodies can be verified, as the difference in the moon's form according to periods, and the reason for her red and pallid appearance in eclipses, and the complete reason of the eclipse; for this is on account of the principal light, which comes from the principal rays below the shadow, and in a measure it is thus imperfectly illuminated by accidental rays. We can not only verify these matters, but the other forces produced by planets and stars on other stars in accordance with all the varying conjunctions and aspects, which astronomers employ in their studies and in which not only the rays of light, but the forces of the stars belonging to their substance in their turn are multiplied, respecting which we say that the moon in Aries is warm and dry, in Gemini warm and moist, and in Cancer cold and moist; and when she is in conjunction with Saturn she becomes cold and dry, and when with Jupiter warm and moist, and so of all such matters. For all these things are proved by the multiplications of species and of forces determined in accordance with the principles already touched upon. And not only these matters pertaining to the qualities and the natural substances of the stars are so verified, but also that which pertains to the figures, magnitudes, altitudes, and number of the heavens and of the stars, and to like things.

Chapter II

In which the rules quoted above are applied to the whole world.

These things are verified by the means stated not only as regards the celestial bodies, but also as regards the elements and the whole world. For although philosophers before Aristotle maintained all things were of one substance, this statement can be shown to be untrue by the laws of refraction. For if any one by means of the instruments used in astronomical investigation, such as armillary spheres or others, learns the position of a certain star about the equinoctial at its rising, and then learns the position of the same when it comes to the line of the meridian, he will find that it is sensibly more distant in its position on the meridian from the north pole of the heavens than it was at rising. Therefore our vision sees the star in different ways at those different times; for if it saw the star in the same way, it would always find it in the same position. But when the star is on the meridian line, it approaches the zenith of the head of the observer, which is the point in the heavens directly overhead, wherefore the rays fall perpendicularly to the vision, and to the center of the world, and therefore they are not refracted, and for this reason the vision sees the star by means of straight lines in its true position. Therefore when the vision errs at the rising of the star, it will not see the star by means of perpendicular lines, because the star is far distant from the zenith above the head, and therefore the rays fall at oblique angles, for which reason they are refracted, and therefore the vision then sees by means of the broken lines, and errs in regard to the position of the star. So, moreover, Ptolemy in the fifth book on Optics teaches us to note this same fact, and Alhazen also in his seventh book, and I have noted it in instruments, and it is an assured fact. Since, therefore, the refraction of rays happens in this world, it is evident that there are many bodies in the world. And the first refraction is found according to the reason mentioned at the surface of fire immediately under the celestial sphere, namely, under the lunar sphere; wherefore the sphere of fire is different from the celestial sphere, although the followers of Plato and Augustine and many ancient authors mention in Platonic fashion that fire and the heavens are of one nature. But this is not possible, because of the demonstration mentioned and because of other natural demonstrations, which Aristotle gives in his book on the Heavens and the World, which no one in modern times contradicts; for this truth is well known in natural phenomena. This demonstration, however, is not known to the rank and file of writers on nature, for Aristotle does not touch upon it nor his expositors. Since, moreover, there is no refraction in the sphere of air, as those same authors teach, and there is the certainty of experience, therefore scientists are much disturbed over the question whether the sphere of air and of fire are two or one. For it seems according to the authors before mentioned and because of the removal of refraction that there is one surface of air and fire, and one body. But this is impossible, because Aristotle says in the third book of the Heavens and the World, that air is heavy in its own sphere, and follows naturally the surface of water not of fire; for if fire mounted in its own sphere, air would not follow, as he states, because when water descends, air follows its surface, as we see with the eye. Wherefore air and fire are not one body; and all doubt is removed by the law of refraction. For three things are required for refraction, namely, that the second body have a surface distinct from the first, and that it be of a different rarity, namely, more rare or less rare, and that the rays fall at oblique angles. But if any one of these conditions is lacking, refraction is impossible. Because of the first condition refraction does not take place in the same body, although it may have one part more rare and another part less rare, as the air is rarer above than below. And because of the second condition there is no refraction in the celestial spheres, because they are of the same rarity, as far as we can perceive. And for this same reason there is no refraction in the sphere of air, because it gradually becomes rarer up to the point where it is equal in its highest part to the rarity of fire in its lower part, and therefore refraction does not take place there. Since, moreover, this refraction is between the straight path and the perpendicular drawn from the point of refraction, as these authors teach, and experience itself shows, therefore it follows that the second body is denser than the first and therefore the body under the celestial sphere is denser than the celestial sphere. Wherefore we must maintain a complete diversity of bodies with respect to the heavens and an element. And when we grasp these things, by the rays and the luminous pyramids of the stars coming to our instruments we shall verify all that concerns the heavenly bodies, namely, the number of the heavens, and the magnitude, altitude, and thickness of the stars, and all things that are in the heavens.

Chapter III

In which by means of the multiplications before mentioned the conditions of the localities of the world about its poles are investigated.

After these matters we shall take up the spheres of the elements and investigate all the conditions of these with respect to the separate parts of our abode, and we shall find that the localities beneath the poles of the world are naturally uninhabitable because of the cold. Nor do the reasons noted above for the longer pyramids stand in our way, for, as was there stated, the demonstrations for the shorter pyramids outweigh the others in this particular, although every man ignorant of the rudiments of this kind of multiplication of light would judge in favor of the longer pyramids. But the fact is that the apexes of the pyramids coming to the places beneath the poles are too far distant from the sun, and are therefore weak, and are capable only of raising vapors from the water and the earth into the air, and are not able to consume them; and for this reason the air is of necessity dense and foggy and congealed by vapors everywhere and at all times, so that plants and animals cannot live there, as Ptolemy teaches in his book on the Arrangement of the Sphere. But if along with the longer pyramids we should add the authority of Aristotle in his second book on Plants saying that "among those where the days are prolonged," etc., and that is, as the commentator explains in regard to this word, where the day is half a year, and the night is half a year, there are neither animals nor plants, because the heat burns up their matter; it will appear that those places beneath the poles are uninhabitable because of the heat, not on account of the cold. For without doubt the day is half a year there, and then it would seem that the heat would be too abundant, while the sun enters the first degree of Aries, until he comes to the first degree of Libra, and this is through half a year. And there is the added fact that the dawn begins for a month and a half before the rising of the sun above the earth, and the twilight lasts for a month and a half beyond the setting of the sun, so that the light of the sun appears above the earth before the rising and setting through the whole time mentioned, just as for us in summer the light of the sun appears after its setting and in the dawn; wherefore because of the length of the twilight and of the dawn the heat will be greatly increased, since they do not have absolute night except through three months of the year, as is certain on the authority of all authors and men of science. Likewise the sun never leaves those parts, except through its greatest declination, which is about twenty-four degrees, because the equinoctial is the horizon of those parts.

But the sun recedes from us through a double declination, namely, forty-eight degrees. Therefore from a proximity of this kind the heat will be increased. Wherefore we conclude in accordance with truth that the localities there are uninhabitable because of the heat. But when Pliny in his Natural History and Martianus in his description of the regions of the world found by certain experience that the regions beneath the poles are the most temperate in this world, as they themselves state, and they allege the experience of men who were there, we cannot deny that the regions there are the most temperate; and who will bring into accord so great a contradiction? Surely no one, unless he be well acquainted with the principles of the multiplication and action of species. Therefore I present an argument for this view that naturally in accordance with its location in regard to the heavens and the sun the place is of necessity uninhabitable because of the cold; and very many authorities agree in this view.

But because of the accidental formation of the locality together with some natural causes, it is possible that some part of this region is consumed with heat, and another part is temperate. For because of the reasons derived from the longer pyramids, and because of the length of the day, and of the twilight and dawn, and because of the fact that the sun recedes from those regions only by its greatest declination, if we add an accidental cause to these, perchance we shall find what we are seeking. For without doubt, according to the teaching of Pliny, Martianus, and others, very lofty mountains are near the fertile districts of the North, such as the Rhipean, Hyperborean, and other mountains, whose height is immense. Owing to their great height they are able to exclude the cold of the North, as is the case in the mountains of Italy near the districts which are between the sun and the mountains. And there is likewise the added fact that mountains are found of stone, and others, hardened into crystal and salt, as we see in many places of the world. These mountains have surfaces more polished and equal, for which reason a greater and better reflection can be produced from them than from rugged mountains. For from a polished, equal, and smooth surface a sensible reflection is made, as is evident in mirrors; and this is because the parts agree in a single action, and the species is not dissipated, but returns intact as it came; but owing to the inequality of the surface of a rough body no part is in accord with another, but the more elevated part first reflects, and then the more depressed, and so the whole species is dissipated and does not come intact, for which reason we do not see by means of rough bodies but by smooth. The excellence, therefore, of reflection that can be found in certain localities about the poles because of the polished surfaces of mountains, is able to coöperate in generating heat along with the height of the mountains. And we must consider, further, that mountains have various shapes, for one can have a shape like a burning mirror, and another like a spherical mirror, or a column-shaped, or pyramidal one, and where the shape of a burning glass is found together with other causes of heat, there must of necessity be strong combustion, so that nothing can live there, and in this sense is Aristotle to be understood with his commentator. Where, moreover, the causes of heat and cold are tempered in respect to the height of mountains, the smoothness of their surfaces, and their shape, the locality of necessity is temperate, and so are Pliny and Martianus and other experimenters to be understood.

Chapter IV

In which the complexion is investigated of the localities in the central portion of the world.*

The parts of the world in which we exist throughout the whole habitable portion up to the limit of the third climate are of bearable heat. Jerusalem is in the third climate, but beyond under the tropic of Cancer begins the torrid zone and the region unsuited for habitation. And because the path of the sun is between the two tropics, it is commonly thought that that whole region is hot and that no part of it is temperate; and this view is held because the sun twice in the course of a year passes over the heads of the inhabitants, and reaches the tropics only once, as at the summer solstice it comes to the tropic of Cancer, and at the winter solstice it arrives at the tropic of Capricorn.

* *I.e.,* the combination in a certain proportion of the qualities, *hot, cold, moist, dry.*

But in its passage from one to the other it passes twice over the equinoctial circle, namely, at the beginning of Aries and of Libra, and this happens at the beginning of spring and of autumn, for which reason many think that the region beneath the equinoctial is especially hot, and reasoning in accordance with the multiplications mentioned above support this view. For there twice at least in the year there are shorter pyramids and rays falling at right angles and perpendicularly, and consequently not refracted; and then they return upon themselves so that the rays are doubled, and therefore it seems on the first view that the whole strength of the whole natural action is there concurrent far more than elsewhere. But it is clear to us that beneath the tropics the Ethiopians are subject to the heat. Therefore it appears that the region beneath the equinoctial circle is the hottest of all; just as is commonly supposed. But without doubt Ptolemy maintains in the book before mentioned that that region is temperate in comparison with the tropics. And Avicenna teaches in the first book on Animals, and in the first book on the Art of Medicine that that region is the most temperate. And for this reason theologians maintain in these days, that paradise is situated there, and therefore the unlearned throng errs in respect to this place. We shall hold the view of Ptolemy at least, whatever may be the truth about the opinion of Avicenna and of the theologians, although the laws of the multiplication of species are here conclusive as far as they can be. But stronger causes are in opposition, which Avicenna excellently states, namely, that the declination of the sun is great there and the parallels are distinct and far separated. For while the total declination of the sun is about twenty-four degrees of colure, about twelve belong to the equinoctial signs, namely, Aries and Virgo, Libra and Pisces, and about eight belong to Taurus, Leo, Scorpio, and Aquarius, and about four belong to Gemini, Cancer, Sagittarius, and Capricornus. Therefore when the sun is in the signs where the declination is about four degrees, it advances, owing to the uniting of the parallels above this same region, for forty days, and burns as it does at the tropics. But less so in the other signs, where the declination is eight degrees, and least where it is twelve degrees, and this is about the equinoctial.

Another reason is due to the equality of day and night, because the air is tempered only during the night, so that there cannot be superfluous heat during the day, and for this reason there is in that region an equality in the air, and excellent arrangements of complexions, as Avicenna shows. There can be no doubt that the region is temperate, but whether it be the most temperate, I have not yet learned. And for this reason it is not settled whether paradise should be there. Since if the eccentricity of the sun is such as mathematicians maintain, it cannot be absolutely temperate beneath the equinoctial. For one part of the eccentric, which is called *oppositum augis* [perigee], descends to the earth by five parts of the semidiameter of the eccentric more than the remaining part, which is called *aux* [apogee]. And therefore when the sun arrives at the *oppositum augis,* it burns the earth in all ways, so that nothing can live there, on account of its nearness and the fall of the rays at right angles, and the uniting of the parallels, on which it delays above this same region, so that it burns it; and this happens when the sun is in Sagittarius and Capricornus and Scorpio. For the *oppositum augis* is in Sagittarius; and in regions near this sign the sun burns in the same manner and distempers. And therefore the sun will distemper Libra and Aries. For it has already passed through the middle portion of the eccentric when it arrives at the beginning of Libra, and has approached sufficiently the *oppositum augis,* so that it can cause an excessive amount of heat, as is clear in the figure. But it is agreed that paradise possesses an absolutely even temperature. But the proof of these matters does not belong to the present argument. And therefore I return to the proposition stating that the complexions of the localities of the world cannot be discovered unless a man knows the laws of the multiplications above stated, since he will neither avoid what is false, nor will he be able to prove what is true.

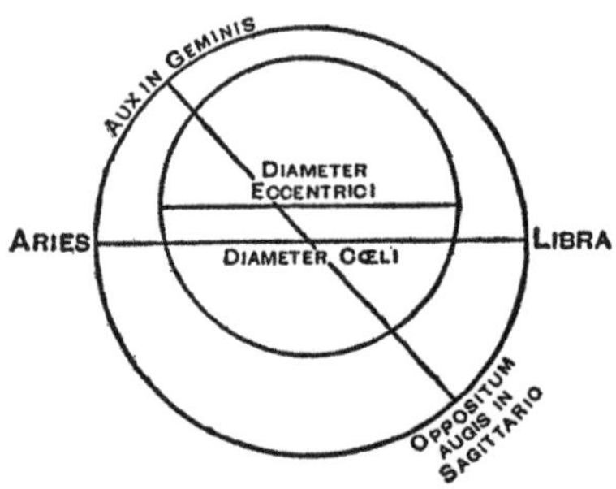

FIG. 13.

Mathematics

CHAPTER V

BUT place is the beginning of our existence, just as a father, as Porphyry says. And we see that all things vary according to different localities of the world not only in nature, but also men in their customs; since the Ethiopians have one set of customs, the Spaniards another, the Romans still another, and the Greeks yet another. For even the Picards, who are neighbors to the true Gauls, have such a difference in customs and language, that we cannot but wonder at such diversity in neighboring localities.

But since the objects of this world located in different places, however near they may be, receive the apexes of different pyramids coming from the whole heavens opposite to them, an infinite diversity happens on this account. For to the separate points of the earth come apexes of separate pyramids, and each point is the center of a new horizon. And therefore we see that two herbs grow out of the ground at the same time without anything in common, and for this reason twins in the womb of their mother receive by lot a difference in nature, so that afterwards they have different manners, and follow different arts and occupations through their whole life. And for this reason the forces of the heavens and stars produce everywhere different things in their properties and natures, and diversity in things generated according to propagation. And not only the multiplication of celestial force is operative, but that of father and mother, since forces are determined in the seeds, as physicians teach. Especially from the vital principle of the mother is there a continual multiplication of force and species over the foetus up to the completion of generation and birth. And when the child at birth is exposed to a new air, another world as it were, he then receives apexes of celestial pyramids as respects his separate members, and thus receives new impressions, which he never gives up, because what the new jug receives it tastes of when old. And then is formed the radical complexion, which remains to the end of life, although the current complexion may be changed for a whole day. And this radical complexion is followed by inclinations to morals and to sciences and languages, and to all the trades and occupations, and to all that diversity which we see in all things. And if the disposition of the heavens is evil at the conception and nativity of the child, the apexes of the pyramids harm the complexion, and consequently the man is inclined to evil morals and perverse arts, according to the diversity of the celestial constellation: and if the constellation is good, the complexion is good, and there follows an inclination toward good morals and useful sciences; and if mediocrity happens in the celestial constellation, the man is mediocre in all things as far as his natural disposition is concerned, although he will be able to change himself through free will, God's grace, temptation of the devil, and good or bad counsel, especially in youth. And now what I have said in general can be proved, if there were time, by examples in individual cases. But concerning combustion made by crystal and glass and concave mirrors and other means visible to the eye, we are sensibly shown that refractions and reflections can produce natural effects. And it is well known that rainbows, comets, and many other impressions of fire in the air, and circles around the sun and moon are made by multiplications of this kind of rays, since they cannot have another cause. And the same is true of all other matters, although it is not evident in individual cases, because not of all agents are the species visible.

Chapter VI

In which is given the reason of the ebb and flow of the sea by means of rays.

And now I shall give an example hidden from all, and cite it in a case where it is less evident that a multiplication according to lines and determined angles is required—a matter as difficult as possible, which, however, is rendered sufficiently plain by means of the multiplication, I refer to the ebb and flow of the sea. Alpetragius in his book on Celestial Motions reasons that all bodies of the world except the earth are moved by the motion of the first heavens, and this is true; but in accordance with the fact that the farther bodies are removed the slower are they moved and with the greater hindrance. Hence water is moved more slowly and irregularly in its sphere than other bodies of the world. He accordingly adds that this motion produces the ebb and flow; but this explanation does not satisfy us in this case, because the ebb and flow are determined and fixed, and they move as the moon varies in the parts of the heavens. But the motion of water is confused by the motion of the heavens, without order, and irregular, because the force of the first heavens is too far removed from its origin, when it is in the water, and therefore the proper force of the water prevails, namely, its own weight, because it strives to remain quiet in its own place, because this motion cannot be so regular and distinct at definite times as respects the ebb and flow as we see in the sea. And therefore Albumazar in the larger Introduction to Astronomy determines all the differences of ebb and flow, and states that they happen every day and night according as the moon is in the different parts of its circuit and with respect to the sun. But he does not tell us the reason, except that the moon is the cause, and that when the moon is in one place there is the flow, and when she is in another there is the ebb. Wherefore we must consider that when the moon rises over the sea of any region, its rays fall at oblique angles, as any one acquainted with the incidence of angles knows. And because they fall at such angles, the rays are necessarily of weak force, as has been shown before. And therefore they are only able to raise vapors from the depth of the sea, like swelling bottles, and overflowing waters of the sea, so that they are driven from their channels. These vapors the rays cannot draw out to the air nor consume them because of their weakness; and therefore of necessity the water flows from its resting places, as long as this kind of ebullition lasts. But when the moon approaches the middle of the sky, the rays fall more and more at right angles and become strong over the body of the sea and draw forth the vapors to the air and consume them, whence the flow grows weaker little by little according to the moon's approach to the meridian; and when she reaches it, the vapors are kept in check and consumed, so that at once while the moon is descending to another quarter of the heavens, the reflux begins, since the effect ceases when the cause ceases. And I shall give a sensible example for this. For in food placed above a fire the fire in the beginning resolves the vapors, and causes them to go forth by the mouth of the vessel; when, however, the fire becomes strong and its action continues, it consumes the vapors, and the liquor settles in the bottom of the vessel. For in general a weak heat resolves vapors and does not consume them, for the strong heat of the sun consumes them, whence at night and in the morning and in the evening there is a greater resolution of vapors than at mid-day, and at that time they are consumed, and likewise natural heat after food and drink in the first place resolves the vapors vigorously and causes sleep. But when it has acquired strength for some time over the food it consumes the vapors, and the man awakes. For resolved vapors obstruct the paths of the spirits and of the heat coming to the senses from the heart and head, and therefore of necessity the senses become quiet and this is sleep. And such is the case with respect to every active force, whose property is to resolve and consume.

Albumazar states and others agree that the moon causes a similar effect in opposite quarters. For while the moon is in the quarter between the east and south the flood is in that quarter and in the opposite quarter, which is between the west and the angle of the earth. And while it is resolving the vapors when it is in the quarter between the angle of mid-heavens and the angle of the west, there is then a reflux likewise in the sea from the angle of the earth as far as the angle of the east. But these authors do not explain this matter, merely stating that the moon has a similar effect in opposite quarters. But how will the moon be active where she is not? And it is agreed that her rays do not pass through the middle of the earth. But in this matter the reflected multiplication aids us. For without doubt the starry heaven or ninth is dense throughout, for our vision continues to the second of these heavens, but vision is terminated only by something dense. At the second of the heavens the multiplied rays of the moon in one quarter are reflected to the opposite quarter, and so the force of the moon is direct in one quarter and its reflection at the same time is in the opposite quarter. But the force of the sun is excessively powerful, and therefore at whatever angles it falls, it is able to consume quickly the vapors that it resolves. But the force of other stars is weak because of their distance from the earth, although they are much larger than the moon, and because of this weakness they do not produce an effect of this kind, although their rays fall at different angles like those of the moon. I do not, however, understand that there is here an ebullition with intensity of heat as in a liquid over a fire, but there is a resemblance as far as regards the resolution of vapors made by the force of the moon, either through rays of light, or rays of her own natural substance, or through both. For when during the night by the rays of the stars, and in the morning and in the evening by the sun there is a resolution of vapors from the earth and from the water, they do not on this account boil with heat, and yet they are raised up in very great volume and become clouds without number. And therefore it is not necessary that the waters of the sea should grow hot in the flow.

Chapter VII

In which the multiplication of forces is considered in reference to the health and the weakness of the human body.

Wonderful, therefore, is the power of this multiplication, since all things hidden and revealed happen in accordance with its laws. And a knowledge of these matters is not only necessary in the sciences, but affords us many advantages in body and soul, if we diligently investigate them, for it is very useful to know these laws in the conservation of health. For since we cannot avoid all impacts of species and forces of things evil and injurious to health, nor are we able always to adapt our bodies to meet more fully the forces coming from things conducive to health, yet we should always be anxious to receive not the principal forces of harmful things, namely, those refracted and reflected in straight lines, but the accidental ones if we can, and if we are unable to avoid all the principal forces, at least let us avoid those in straight lines, let us evade the fall at right angles, and, if we cannot, at least let us beware of the shorter rays of the pyramid. And these considerations have place when a man is exposed to harmful celestial impressions, like the sun in summer and the moon at night, which exhaust our bodies. Hence many have died from not protecting themselves from the rays of the moon. And especially is this true when a man is exposed to the rays of Saturn and of Mars, since those two induce much hurt and corruption in things, as experience teaches. Likewise the species from corrupt and unclean places, and when the species are multiplied of the leprous and of the infirm and especially of those with contagious diseases, and of those with bad complexions, and particularly of a menstruous woman; since if she looks in a new mirror, a bloody cloud appears in the mirror from the force of the menstruation staining it, as Aristotle states in his second book on Sleep and Waking, and likewise from serpents and other venomous things. And especially must this fact be considered when men and animals become angry, and have the desire to do harm and a mind of malignity. For witchcraft is reduced to this source, whence what force it possesses it receives from this cause, since without doubt impression then becomes stronger, because nature obeys the thoughts and the desire of the soul, and she is aroused to stronger action, as Avicenna teaches in the eighth book on Animals and in the fourth on the Soul by examples and various experiences, and this is an assured fact. Whence Solinus tells us in his book on the Wonders of the World that in a certain northern region there are women with two pupils in the eyes, who on becoming angry kill people with sight alone, of whom Ovid says the double pupil does the harm. And we must be especially careful lest the noble parts, as the eyes and the face, be exposed to species of this kind: for I saw a physician made blind while he was endeavoring to cure a patient with a disease of the eyes, because of the multiplication of the species coming from the eyes of the patient. For it is necessary in the case of those harmful species to use great precautions, and especially so when the evil is severe or intolerable. So Alexander, instructed by Aristotle, as the histories state, by means of large polished bodies bent back upon a city the poisonous species of a basilisk placed on the wall to slay the army, so that it was destroyed by its own venom. And on the contrary the adapting of the body to receive the species of things conducive to health, as far as one can do so efficaciously, is very beneficial to both the healthy and the infirm.

Chapter VIII

Concerning the infinity of matter.

Since I have continued for too long a time my argument to show how in the things of the world as regards their efficient and generating causes nothing can be known without the power of geometry, I now wish briefly to consider them with respect to their matter, by showing that it is necessary to verify the matter of the world by demonstrations set forth in geometrical lines, if we wish effectively to purge away numberless errors. For the majority of those philosophizing, not only in philosophy proper, but in theology, state and assert that matter is of one kind only in all things and that there is only diversity as respects their forms.

And they argue to prove this that if we should exclude by the intellect from the matter of the heavens and of a stone their forms, we cannot state by what they differ, because actuality, that is, form, distinguishes, as Aristotle says in the seventh book of the Metaphysics. And in the first book of the Physics he says that all things are one in matter but differ in form: and in the second book of the Metaphysics he says that there is no distinction in matter. And if matter were multifarious like form, he says that it must be common and predicable of more than one thing like form. And they allege that every universal has the nature of form. For the parts of definition, as Aristotle says in the seventh book of Metaphysics, are forms: but the parts of definition are through the genus and difference and compose species: therefore species is form only like genus and difference. And it is evident that the remaining two universals of Porphyry, that is, property and accident, are pure forms, because they belong to the categories of accidents. Therefore every universal is form; and for this reason matter will not have the nature of the universal but must be singular and the same in all things. By such fundamentals and authorities badly translated they strive to persuade us. But this is a very grave error. For there is no greater one in speculative truths, because this being granted it is impossible to keep the generation of things, and the whole course of nature will be unknown. But what is more, if this error be discussed, it will be found to be very close to heresy, or wholly heretical, than which nothing is more profane, because the necessary consequence is that matter is God and creator. This I wish to show for the present only by means of geometrical reasonings, as I promised, although reasonings from nature and metaphysics are abundant and efficacious, concerning which elsewhere a long disquisition can be made. And it is necessary because of popular prejudice, which is violent everywhere.

I say, therefore, that if matter can be the same in two substances, by the same reasoning it can be the same in three and in an infinite number. For Aristotle makes a similar argument in the chapter on vacuum in the fourth book of the Physics, saying that if any one and the same thing can be in two substances, then it can be in three, and so in an infinite number; but that which can exist in an infinite number of substances has infinite power. Therefore matter is of infinite power. Wherefore also of infinite essence, as will be proved, and therefore it must be God. But they cannot contradict the aforementioned consequence, namely, that it follows that the same matter can exist in an infinite number of substances if they were existent, and therefore, they grant to it infinite power but deny to it infinitude of essence, because nothing has infinite essence, since nothing has infinite essence except God. And when they are hard pressed as regards infinitude of power, they begin to vacillate in their words, saying that the power of matter is infinite just like the power of continuous quantity. But this argument amounts to nothing, because this power of continuous quantity is not with respect to an infinite actuality, nor with respect to an infinite number of objects to be considered at the same time and in actuality, and therefore this power is not infinite in actuality but in potentiality only. But infinite power is so far granted to matter according to the position stated, that if there were in actuality an infinite number of things, the same matter could actually exist in those infinite things. Therefore it has infinite power, actually and intensively. But the power of continuous quantity is not similar to divine power: because divine power does not have confinement and limitation as far as regards actual existence in several objects and in an infinite number. To such existence the power of continuous quantity does not extend, but to matter there is conceded that same existence in several substances and in an infinite number by the authority and demonstration of Aristotle. Wherefore the power of matter is not like the power of continuous quantity, but like the divine power.

But they vacillate in still another particular, saying that the power of matter is passive, and the divine power is active. For this is excluded in many ways. For passive power is spoken of with respect to a change suffered by an agent. But the power of existing in several substances does not come under this head, and so they differ. Likewise the power of existing in several substances at the same time, which is not confined or limited, is claimed for the highest excellence, and therefore is granted to God and to the soul, which as a whole is in many parts of the body. But we do not concede this quality to God and to the soul because of any passive essence, and therefore we do not concede it to matter. Likewise whatsoever this power be, whether active or passive, the substance of matter must be infinite if this power is infinite, and if this be true, then matter is God or equal to God, because no thing has infinite substance and power except God.

Let therefore the line A be infinite power and let the line B be its infinite substance, and let the line C be some finite power,

A *Potentia infinita*
B *Substantia ejus finita*
C *Potentia finita pars ipsius A*
D *Substantia potentiae C*
E *Substantia aequalis ipsi B substantiae*
F *Potentia proportionalis ipsi E*
G *Potentia major quam F*

which is part of infinite power (for from every given quantity it is possible by the intellect to remove a part, as Aristotle maintains in the first book of the Heavens and the World), and let line D be the substance of power C; let substance D therefore be multiplied until it becomes equal to B, for this is possible, since D and B are finite, and let the line E be this augmented substance. Likewise let the power C be multiplied to the extent that the power may increase proportional to E and let it be the line F, for this is possible, since C and D are finite, and therefore addition can be made to both to the extent that augmented they shall be proportional, just like C and D. But power F is finite, because it is the result of a finite augmentation and has a substance E equal to substance B, to which A corresponds, which is an infinite power; therefore the finite power F and the infinite power A must have equal substances. Wherefore the finite and the infinite must be equal, because equal powers belong to equal substances, and the part must equal the whole, since F is part of A, just as C is, as this assumption was made in the beginning concerning C, and F increases in accordance with the finite augmentation of C, wherefore F must be part of A, and thus the part must equal the whole.

From this we must conclude further that the part must have greater essence and more essence than the whole. For some finite power greater than F may be taken, and may be cut from the infinite power A, and let it be represented by the line G. Therefore the power G, since it is greater than F, must have more substance or be based on a greater and nobler essence than the basis of power F. But F had an essence equal to A, as was proved. Therefore G must have a greater essence than A; and thus to the part there corresponds more essence and a nobler one than to the whole, which is impossible. And it must also follow that the finite will have more essence corresponding to it than the infinite; and all these conclusions are impossible.

Likewise nothing infinite can have finite power. Therefore, on the contrary, nothing finite can have infinite power. The first statement is proved by a demonstration like the preceding one, for let the line *a* represent that infinite substance, and line *b* that finite power. Let some finite substance or part of *a* itself

a *Substantia infinita*
b *Potentia finita*
c *Substantia finita pars ipsius a*
d *Potentia ipsius c*
e *Potentia aequalis ipsi b potentiae*
f *Substantia potentiae e*

be taken and let it be the line *c*. Therefore it must have a less power; and let line *d* represent its power. Let therefore *d* be multiplied until it equals *b*, which is possible because both are finite; and therefore so much can be added to *d*, that augmented it may equal the power *b;* and let line *e* be that augmented power. In like manner then let *c* be multiplied and addition be made to it until it has the power *e*, and let that augmented substance be the line *f;* for this is quite possible, since substance *c* and *d*, its power, are finite, and therefore if to *d* so much can be added that power *e* becomes greater than *d*, to *c* so much substance can be added that the augmented substance will increase, to which the power *e* will be proportional, and that augmented substance must have that power. But power *e* is equal to power *b,* as has been proved. Therefore *f,* which is finite, because it is produced in accordance with the augmentation of a finite substance by a finite factor, must have a power equal to *a,* which is infinite. This conclusion is clearly altogether impossible, and it must follow from it that the part equals the whole, namely, *f* equals *a,* and also that the part is greater than the whole, as was clear in the former reasoning; and that the finite has greater power than the infinite, as is clear in the terms, and all these conclusions are impossible.

Likewise if the power of matter is infinite let this power be designated by the infinite line A, and if this be divided at the point *a,* then the divided parts are equal, since both parts extend to infinity from the point *a.* Let then the point *b* be marked, likewise *c* and *d.* Then the parts divided are equal, but *baf* extending to infinity is greater than *af* by *ab.* Therefore line *bc* extended to infinity is greater than *af.* Therefore it is greater

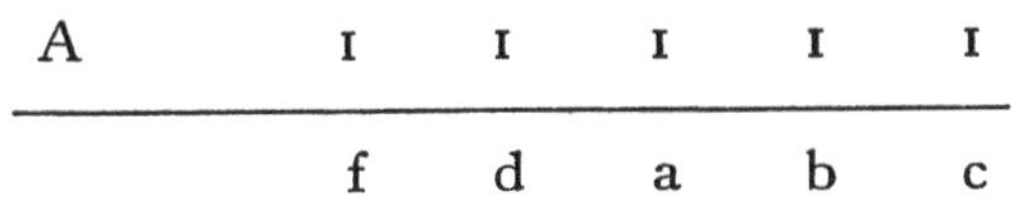

than the equal of *af,* that is *abc.* Therefore the part is greater than its whole. Wherefore it is impossible for the power of matter to be infinite. We must not apply this demonstration to divine power, because it is not divisible like the power of bodily matter.

Chapter IX

Whether bodies touch each other in a point.

Just as I have employed the fundamental principles by way of proof to examples of things in nature, so likewise can some examples be noted that apply to the matter in the things of the world, in which the wonderful power of geometry shines forth. For if the matter of the world were of one kind, of necessity there would be but one form, and all things would be one body, just as Parmenides and Melissus have maintained, opposed to whom is Aristotle in the first book of the Physics and in the book on Generation. For surely from a unity of matter there follows that most false power, which I previously eliminated by the laws of refractions.

Since, moreover, there are false geometrical illustrations in support of this position, which may disturb every man, I wish therefore to introduce certain ones. For if bodies are different in this world, like the heavens and an element, and we imagine a line passing through their middle, they will not cut that line in different points, because there would then be a linear length in the middle, and so a superficial width and a corporeal depth, and so there would be a third body distinct from the heavens and the element between them, which is false. Therefore they must cut that line in the same point. But those are continuous quantities whose terminus is the same. Therefore they are continuous. Many very illustrious men have given various answers to this question, and all gave their adherence to this point of view that they would cut the line in different points, either naturally or mathematically, or in both ways, making a distinction between natural and mathematical continuity in accordance with Averroës in the fifth book of the Physics. But this view is untenable; for the lines must necessarily intersect each other in the same point, so that an iron rod may be imagined; but that point does not continue the cutting bodies, and therefore they are not continuous, but continues the parts of the rod or line passing through, from the continuity of which it cannot be concluded that the bodies are continuous through which the rod passes; but if we should admit a cutting of the rod in different points, then of necessity a part of the rod, which part is a body, would be between; and this is impossible. Likewise it may be argued with more subtility that above as below any bodies assumed in this world, if they are different and contiguous, not continuous, a third body may be placed, to which at equal distance apart two lines may be drawn from the midpoints of the given bodies. If, therefore, those lines terminate at different points in the third body, then a part of that body is between them, and they are always equally distant and approach, and in the direction between the two bodies they will be distant by the body interposed, which is wholly false. Therefore they will continue to one point in the third body, and by the same reasoning to their own extremities at the junction of the assumed bodies, wherefore the assumed bodies will be one body, because one point cannot be in different bodies. And many have been confused by this, nor is there time to review their individual opinions, especially since the truth is evident. If it be said that those lines will terminate at the same point of the body nor would they be parallel, because they have no distance between them, since the bodies are placed together without space between, and since all separation is removed, they must necessarily terminate at one point in the third body, and since two points do not have any length, since they are not quantities, therefore they do not occupy any more space than one point. Hence two consecutive points indicate as much indivisibility as one point taken by itself, because a quantity does not increase by adding point to point, and therefore they can remain different points at the extremities of two lines united in two bodies without space between, although they terminate at one point in a third body. But if in those united bodies two lines be drawn from these points within the bodies, and a line let fall upon their extremities at right angles, by the fourteenth proposition of the first book of the Elements of Euclid the lines extended in the bodies are one continuous line. Therefore the bodies also in like manner, for such is the purport of that proposition.

And because of this proposition Averroës was deceived in the fifth book of the Physics, and all those following him, maintaining that there would be one line mathematically, and consequently that there would be one body mathematically, although they would be different physically. For, as he states there, physical contiguity passes into mathematical continuity; but this is impossible, for a mathematical quantity and a physical one are the same as regards being and as regards reality, but they differ only in the point of view, because the geometer considers a physical line, not as it exists in physical matter, and therefore the line is called a mathematical one. And the natural philosopher considers this same line, as it exists in physical matter, as in iron, or stone, or other physical object. And because the same thing is, as respects being and reality of existence, physical and mathematical, therefore if there should be here one line or one body mathematically, then in the same way would there be one physically. I say, therefore, that there are here lines in reality different and bodies likewise, nor is this statement at variance with the meaning of Euclid. For he does not mean that it is one continuous line, but that as far as making right angles the two lines are as effective as one, since those two lines lie in the same direction of length, just as if they were two parts of the same continuous line: and he has this meaning in many other places where he uses one line for two, since the one does the same thing for his proposition as the two. And this is the case when he speaks of a common difference of surfaces or of a common cutting, for he calls this one line when in reality there are two: but the one line has the function of these two, and therefore we speak more easily of one than of two.

If, however, the objection be raised that if one maintains a separability in bodies, two tablets of plane surface and of circular shape may be taken, and one raised from the other with a simultaneous action throughout the whole surface, and then the air would fill the outer portions more rapidly than the inner. For this reason there would be a vacuum in the inner portion for a time. And in regard to this matter many foolish statements were generally made. For some said that the air moved to the center instantaneously in accordance with a law of universal nature that there should be no vacuum; others said that all bodies touching each other in air or water have moist surfaces, as Aristotle states in the second book on the Soul, and they maintained that this moisture of air or water is rarefied suddenly throughout the whole, in order that there might be no vacuum. But their statement, that there may be no vacuum, is pure negation. No pure negation is a reason of an affirmation; and therefore answers of this kind are false. For this reason we must say that one plane cannot be raised from the other with a simultaneous action throughout the whole surface, but of necessity one tilts when it is raised from the other, and so the air penetrates gradually. Any one can experiment with this plane in a glass cup immersed in a vessel full of water; for it cannot possibly be raised keeping the same formation of its parts. The reason is that the water takes the place of it gradually. And this is an affirmative reason to which consequently vacuum is not admitted.

If, therefore, there are of necessity many corporeal materials and many bodies in this world, since every body is infinitely divisible, it does not follow for this reason that the world is composed of an infinite number of material particles called atoms, as Democritus and Leucippus maintained, by whose position Aristotle and all students of nature have been more hindered than by any other error. Yet this error is wholly eliminated by the power of geometry; for no stronger argument can be used against this error than that the diagonal of the square in that case and its side would be commensurable, that is, would have a common measure, namely, some aliquot part as a common measure, the contrary to which Aristotle always teaches. And the truth is clear by the demonstration from the last part of the seventh proposition of the tenth book of the Elements, where it is shown that if some measure, as a foot or a palm, measures the side, it will not measure the diagonal, nor *vice versa;* so that if the diagonal is ten feet, the side cannot be expressed exactly in feet. And not only does it follow from this position that they would be commensurable, but also equal. For if the side has ten atoms, or twelve, or more, then let the same number of lines be drawn from those atoms to the same number in the opposite side, the sides of the square being equal; wherefore just so many lines will occupy the whole surface of the square; and therefore since the diagonal passes through those lines, and no more can be drawn in the square, the diagonal must receive a single atom from each line, and therefore there will not be more atoms in the diagonal than in the side, and thus they have an aliquot part as a common measure, and the side has just as many parts as the diagonal, both of which conclusions are impossible.

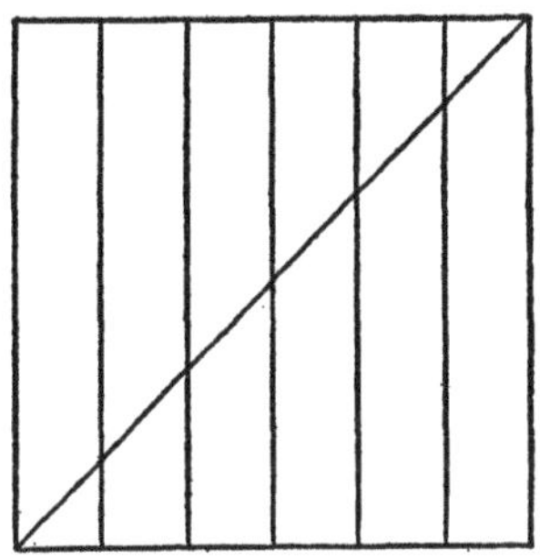

FIG. 14.

CHAPTER X

Concerning the shape of the universe.

SINCE of necessity the bodies of the universe are many and divisible and are quantities, they must have a form required for the existence of the universe. Now form is a property of matter, and is found in things by reason of matter, just like quantity also. For figure in one aspect is quantity inclosed by lines; in another it is called the limitation of quantity. It is necessary, in fact, that the universe on the outside should have a spherical shape. For should any other form be given to it, there would be a vacuum or the possibility for it. But nature does not endure a vacuum nor a possibility in respect to it. For if it were of some angular form, then of necessity there would be a vacuum in its motion; for where at one moment there was an angle, there would be nothing until another angle came to the same place. Other figures especially suitable would be either of oval shape or those like it or of lenticular shape and those formed like it, according to Aristotle in his book on the Heavens and the World. But he states that the heavens do not have a shape of this kind, but he does not give a reason. A lenticular shape is that of the vegetable called lentil. For it has gibbous lateral surfaces, and is not a true sphere, owing to a shorter diameter passing through those sides. If the universe had the shape of an oval, cone, cylinder, or other figure of this kind, and revolved about the shorter diameter, there would actually be a vacuum; but if it revolved about the longer one, there would be no vacuum actually, but there would be the possibility of a vacuum, since there would be just as great a possibility, as far as the figure is concerned, that it would revolve about the shorter diameter as about the longer. If, however, it were of lenticular shape or of the form of a cheese or of similar shape, and moved about the longer diameter, there would actually be a vacuum, and if about the shorter, there would not actually be a vacuum, but the possibility of it would remain. For it is just as possible as a matter of fact that it should revolve about one diameter as the other. Nor can we say that if it were oval in shape it would revolve always about the longer diameter, and if it were lenticular it would revolve about the shorter, in order that there might be no vacuum. We must refute that statement just as we did above, namely, that it is impossible that a pure negation should be the cause of an affirmation; but the statement, that there may be no vacuum is a pure negation. Therefore the universe must be of spherical form, in which solid alone are all diameters equal, so that it can revolve freely on every diameter, and thus no inconvenience results. Likewise within it must be spherical and concave; for it cannot be of a plane figure. Since if three lines be drawn from the center of the earth, and one perpendicular to the surface of the heavens, as *ab,* this last line will be shorter than the others by the eighteenth proposition of the first book of the Elements and the thirty-second proposition of the same book. Therefore the heavens will not be equally distant from the earth: but it must be of the same nature in every part. Therefore every part is naturally equally distant from the earth. Nor can it be convex looked at from within, as

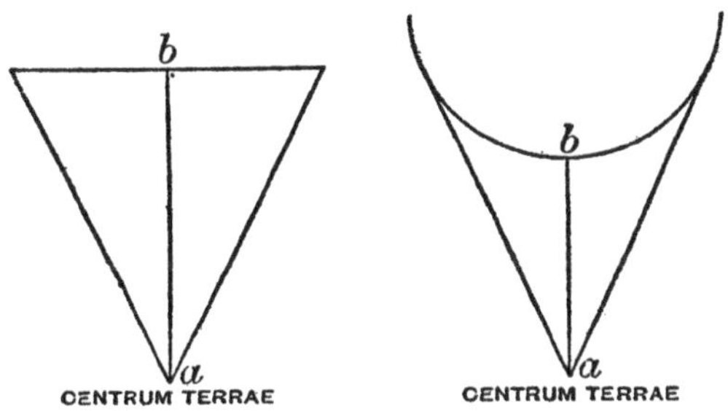

FIG. 15.

is shown by the eighth proposition of the third book, which states that from a point taken without if many lines are drawn to a circle, the line falling on the diameter would be shorter than the others. Therefore it remains that if from the center of the earth three lines be drawn to the convexity of the heavens, one will be shorter, namely, that one which is perpendicular to the sphere, as is clear in the figure, and therefore the heavens would not be equally distant from the earth, which, however, it must be, as has been stated. And again if the universe were of convex figure looked at from within it would not naturally contain all things; but the universe does naturally contain all things. But if it cannot be of a plane figure nor of a convex, it must be concave, since there is no other shape. But it can be concave in many ways, either spherical, or cylindrical, or pyramidal, or lenticular, or of many other shapes. It is not possible, however, that it should have any other concavity than a spherical one, because in this figure alone are all the lines equal which are drawn from a point to the surface. For it is not possible in other figures to take a point from which all lines drawn to the surface may be equal: for the diameters are unequal. But the parts of the heavens must be equally distant from the earth on account of the equality of nature. Therefore of necessity it must be of spherical form.

Moreover, among all isoperimetric figures the sphere has the greatest content, as the eighth proposition of the book on Isoperimeters states. Plane figures are called isoperimetric, as the triangle, quadrangle, and circle, when the sides of the triangle extended continuously and in a straight line have the same length as the four sides of the quadrangle when so extended, and the same length as the circumference of the circle if it were extended, and the same thing is true of all other plane figures. Whence the word isoperimeter is formed from **ἴσον**, equal, **περί**, around, and **μετρόν**, measure, meaning, as it were, of equal circummensuration. And among all these isoperimetric surfaces the circle is the maximum, as the seventh proposition on isoperimeters states. Solids, moreover, are called isoperimetric; as the sphere, cube, cylinder, and any other, when the surface of the sphere extended continuously and in the same plane contains as much in length and width as the six surfaces of the cube and as much as the surface of the cylinder, and so of the others. But among all those figures the sphere has the maximum content, as is demonstrated in the book mentioned above. Since, therefore, the heavens must contain all things, it was necessary that it should be of spherical form. Likewise the nobleness of the universe and the excellence of this figure correspond. For this figure is the first of the solid figures, because it is contained in a single surface, while all the others have more. Therefore it is adapted to the first body, the heavens. Moreover, this figure is the simplest, since it is without angles, vertex, sides, and free from every difference. Therefore it belongs to the simplest body, namely, the heavens. It is likewise most fitted for motion. Therefore it belongs to the *primum mobile*. It is likewise removed from chances and hindrances, because it has no angle against which something might strike. Therefore it is especially fitted for a body that cannot have hindrance or chance of accident. It is, moreover, the most perfect, because nothing can be added to it; but to all others something can be added. Therefore it belongs to the most perfect body.

The fact, moreover, that the bodies contained in the heavens should have a spherical shape, is proved from water, which lies in the middle, so that it is consequently evident as regards the others. If lines be drawn everywhere to the surface of the water from the center of the earth, it is plain that water always runs to a lower place because of its weight, as we see. Therefore if one of those lines should be shorter than another, the water would run to the extremity of that line until equality was established. Therefore all lines drawn in every direction from the center of the world to the surface of the water are necessarily equal. But lines so drawn to a plane cannot be equal, by the eighteenth and thirty-second propositions of the first book of the Elements, as was stated above, nor can they be equal when drawn to a convex surface by the eighth proposition of the third book. Therefore of necessity the surface of the water containing the earth must be concave, and not of any other kind of concavity than that of the sphere, since in that figure alone are all the diameters equal. And this demonstration holds with regard to the water, not only as viewed from within but as well as from without. For on the outside it flows always to a lower place as well as on the inside. And therefore the water must be convex on the outside, for neither to a plane nor to a concave figure on the outside could all lines drawn from the center be equal, according to the form of the previous demonstration, and this is shown by experiment.

For let *gd* be a ship, and *a* a port, and *c* the deck of the ship where the mast is fastened, and *b* the end of the mast, and let the line *ca* be drawn perpendicular from the port to the end of the mast. It is then plain by the nineteenth and eighteenth propositions of the first book of the Elements that line *ab* is longer than *ca*. Therefore if the sea were plane-shaped, the eye at *c* would see the port better than at *b*, since *b* is more distant from *a* than *c*. But we know by experience that he who is at the top of the mast can see the port more quickly than a man on the deck of the ship. Therefore it remains that

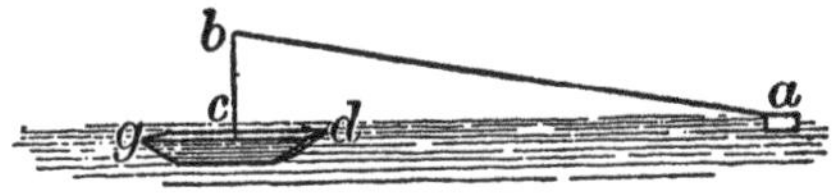

FIG. 16.

something hinders the vision of the man on the deck of the ship. But there can be nothing except the swelling of the sphere of water. Therefore the water is of spherical form. But if this is true, then the earth is of convex spherical figure, for otherwise it would not be equally distant from the heavens, nor would it be equally near the center of the universe; but this must be the case. Likewise there would be a vacuum where they did not touch each other: since if it were a plane or of a concave shape it would not touch the concavity of the water, as is evident, and therefore there would be a vacuum between them.

In a similar way we must proceed in regard to the sphericity of the air concave within and to its sphericity on the outside, as we did in the first demonstration in regard to the water; because the air is heavy in its sphere like the water, as we maintained previously, and therefore it flows to a lower place. And, moreover, there would be a vacuum between the air and the water, if the air were of another shape. For if it were a plane, it would not touch the water except in a point, by the third proposition of the first book of Theodosius; because a sphere does not touch a plane except in a point, as is there stated. If, however, it were of a convex shape, it likewise would not touch except in a point by the twelfth proposition of the third book of Euclid. For if two circles are marked on the convex sphere of water and on the convex air, the circles will not touch each other except in a single point, as is proved in that twelfth proposition. Therefore the same is true of the bodies on which the circles are marked; for if the bodies touched each other in many places, so also would the circles, as is clear to the sense. And so of necessity in either way there would be a vacuum. And if it were of a concavity different from that of the sphere, not all lines could be drawn equal from a point to the surface. Therefore of necessity the surface of the air must be spherical within and without.

Then by these last demonstrations the same fact is evident with respect to the fire as well as the air. But the proof relating to the descent to a lower place does not hold here, because fire is not heavy in its sphere. Nevertheless, it is necessary that all lines drawn from the center of the world to the surface of the fire should be equal, because fire is opposite to the earth and of the utmost lightness, and of the same nature in the whole and in the separate parts. Wherefore it is equally distant from the locality of things with weight, and from the center of the world, and this is true both as it is as viewed from without and from within. Therefore it must be concave within, and convex without, and spherical in form, as we conclude from the arguments given.

Chapter XI

That a vessel placed in a low position contains more water than one in a higher position.

But now as regards the form of the water a great wonder of nature can be noted; since if a cup containing water be put in a low place, it will be able to contain more water than in a higher place, as in a cellar and on a balcony. For on account of the natural inclination of the water to the center of the world, wherever the water may be, either in a higher or lower place, its portions always run to the lower place; and therefore they are always distant from the center by equal lines; and therefore the higher part of the water must be a portion of a sphere described around the center of the world, although in the bottom of the cup the water retains the shape of the vessel, because there only does it touch the vessel and not above. This top part will be shaped according to the law of the gravity of water, and this holds with respect to the center of the world, and therefore the upper part will be a portion of a sphere to be conceived of as around the center of the world. But it is evident that in a lower position it will be a portion of a smaller sphere and in a higher position it will be a portion of a larger sphere, because it will then be more distant from the center; for the higher sphere will contain the lower one, as is evident around the same center. Moreover, the diameter of the cup will be a chord of both portions, if the cup be filled in both cases to its full capacity. Therefore that diameter will cut from the larger sphere a smaller portion, and from the smaller sphere a larger portion. By the twenty-eighth proposition in Jordan's Triangles, in unequal circles the same chord cuts from the larger circle the smaller portion, and from the smaller circle the larger portion, and the same is true of spheres.

For let those circles be marked on those spheres, and the portions of the circles on the portions of the spheres, and it is evident that the result is the same. Therefore the portion of water above the diameter of the cup will be greater when the vessel is in the lower position than when it is in the higher; and therefore the gibbosity is greater and the swelling higher; wherefore there must be more water there, if the cup is completely filled, than in the higher position. Wherefore to the same water more can be poured into the cup when it is lower than when it is higher. For in the lower position the water which is above the diameter of the cup will contract from the sides of the vessel, and will narrow itself into the portion of the smaller sphere, because it is nearer the center of the world, and therefore the chord of the same water will be less than when it was above. And a part of the diameter of the cup will be its chord, not the whole diameter, but on both sides a part of the diameter will be cut off, so that the part left becomes the chord of the portion of water. Wherefore at the sides of the water from the point where the diameter is cut up to the vessel there will be two small spaces where it will be possible to pour in more water than when the cup is in the higher position. This whole phenomenon is due to the inclination of the water following the law of its gravity to form itself into a smaller and larger sphere with respect to the center of the earth; notwithstanding the fact that this seems utterly impossible to the rank and file of students, nay, to great men ignorant of the power of geometry.

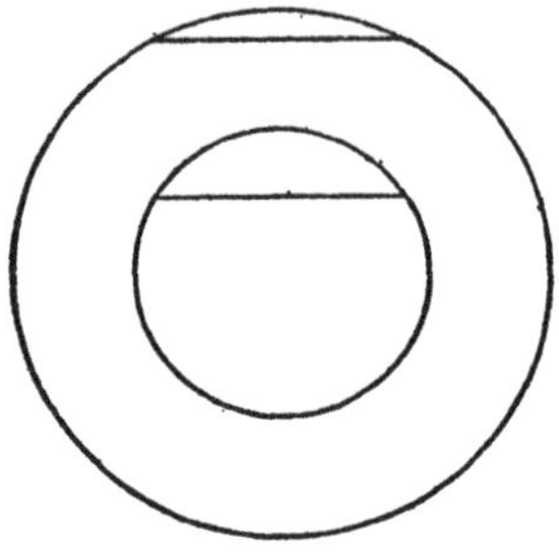

FIG. 17.

CHAPTER XII

Whether the five figures of the regular solids correspond to the world as the Platonic school maintained.

IN consequence of the spherical forms of the bodies of the universe a firm basis is disclosed in the proof of the things of nature, and violent errors are purged away. For the Platonists, in whose time geometry flourished, as Averroës states in the third book on the Heavens and the World, thought that the principal bodies of the world, namely, the heavens and the four elements, had the shapes of the five solid figures called regular, which are equiangular and equilateral, and can be inscribed in a sphere and circumscribed about the same, and there are no others. And for these reasons they are the noblest of figures, with the exception of the sphere, which the author of the present work can easily describe. Three are formed with triangular surfaces, the fourth with square surfaces, and the fifth with pentagonal surfaces; and there cannot be more than these, a fact to be marveled at. The first has four triangular surfaces, and is called a tetrahedron, from τετράς, four, or pyramid of four triangular bases. The second has eight triangular surfaces, and is therefore called octahedron, for *octo* is pure Greek, not Latin. The third has twenty triangular surfaces, and is called icosahedron, from εἴκοσι, twenty. And no more regular figures can be formed with triangular bases. For no angle of a solid can equal four right angles in a plane, as the twenty-first proposition of the eleventh book of the Elements teaches. But six angles of equiangular triangles are equal to four right angles: for three equal two right angles, as is evident from the thirty-second proposition of the first book, which is well known. And therefore six angles of the triangles cannot form an angle of a solid; and therefore a solid figure cannot be formed with triangular surfaces of which the six angles meet to form an angle of the solid. But a solid can easily be formed, because the sum of five, four, or three angles of the triangles is less than four right angles. And if three angles of the triangles form an angle of a solid, it must then have four triangular surfaces. And if four angles of the triangles form an angle of a solid, there must be eight triangles in the solid. But if five angles of the triangles form an angle of a solid, there must be in the solid twenty surfaces triangular on all sides, as is evident to the sense in solid figures. Moreover, from two angles in a plane an angle of a solid cannot be formed, because every angle of this kind is formed from at least three surfaces, as Euclid states in the beginning of the eleventh book; and every geometrician knows this fact. Therefore only three regular solids can be formed from triangles. From squares only one can be formed; for the angle of a square is a right angle, and therefore only three such angles placed together can form the angle of a solid, because if a fourth be added no longer is the angle of a solid possible, because every angle of a solid is less than four right angles. But if three angles of squares are concurrent to form an angle of a solid, in the solid so formed there will be six square surfaces, as in the taxillum; and this figure is called cube and hexahedron, from ἕξ in Greek, which is *sex* in Latin. But if the angles of regular pentagons are taken, three form an angle of a solid, and no more, because if four be taken, the sum would be more than four right angles, since the angle of a pentagon is one and one-fifth right angles, as is evident from the thirty-second proposition of the first book of the Elements, and so there would be four and four-fifths right angles. But every angle of a solid is less than four right angles. Therefore there can be only one solid with pentagonal surfaces, and it must have twelve pentagonal surfaces, as is evident in the formation of that body, and it is called dodecahedron, that is, a solid figure of twelve bases of equilateral pentagons.

All these regular solids are formed with regular surfaces which are equiangular and equilateral. It is not possible to form a regular solid with hexagonal surfaces, because the angle of a hexagon equals one and one-third right angles, as is evident from the thirty-second proposition of the first book of the Elements. Therefore three such angles make three and three-thirds right angles, but three-thirds equal one whole unit. For there can be only three thirds in anything. Therefore the three angles of regular hexagons equal four right angles, but no angle of a solid equals four right angles. Therefore no regular solid can be formed with hexagonal surfaces, and far less with heptagonal and octagonal ones, or with surfaces having a still greater number of angles, because such figures have larger angles than those of the hexagon.

Since the dodecahedron permits all the others to be inscribed in it, as is evident from the fifteenth book of the Elements, the Platonists gave this shape to the parts of the heavens, because the heavens are known to contain all things. Whence they said that the parts of the heavens are concurrent in a point in accordance with this figure, so that the body of the heavens may be established. And because fire ascends in a pyramidal form, they gave the parts of fire such a form. Since, moreover, the octahedron resembles very greatly the pyramid, and air is very much like fire, they give it that form. But since the parts of water revolve in numerous windings and fluxes, they gave it the form of the icosahedron, which in its many-formed sides and angles is whirled around in the sphere. To the parts of the earth they assigned the cube, because that figure is stable and fixed among all, as the earth among the bodies of the universe has obtained stability and fixedness.

But Aristotle opposes them in the third book of the Heavens and the World, and proves that with this formation there will be a vacuum in the sphere of water, in the sphere of air, and in the celestial sphere. For the cube and the pyramid alone are able to fill space, when they are assembled about a point. For it is possible to fill space in two ways; in one, as we commonly say, because every body fills space, unless there be an instance in the last heavens, and the filling of space is not taken thus in the proposition. In another manner space is said to be filled not only solidly but superficially, and more properly so because the fewness of surfaces filling their spaces is the cause of the fewness of solids filling their spaces, as Averroës states in his comment on the third book of the Heavens and the World. Now to fill out superficial space is to complete four right angles, because there cannot be more about a point in a surface, as is evident from the intersection of two lines at right angles in this way. And so squares most properly can fill superficial space, namely, four of them, because the angle of a square is a right angle; and six triangles, because angles of this kind equal four right angles, and three hexagons, because their three angles equal four right angles, as explained before. But pentagons cannot fill space, because three angles of pentagons are equal to three and three-fifths right angles, which in all is less than four right angles, and four angles of pentagons equal four and four-fifths right angles, which is more than four right angles and therefore this regular figure cannot fill space. Likewise neither the heptagon, nor any other, as is evident. And therefore only three surfaces fill space. And for this reason there will be few solids that can fill space solidly, and filling space solidly in this way is not done by a single body, but takes place when several bodies are grouped about a point on all sides, so that they fill a space solidly about that point. And this space has eight solid angles, and twelve superficial right angles separate in point of fact, although there are twenty-four theoretically, since each solid angle is formed of three superficial ones, and therefore we have a total of thrice eight, or twenty-four. But some are frequently repeated, since these angles are united; for if they were divided, there would of necessity be twenty-four separate angles, as each solid angle distinct from another would have three right angles belonging to it. All these facts are evident from three lines intersecting one another at right angles, as in three rods or as in other examples. Since the angle of a cube is formed from three right angles, eight such can most properly fill the space about a point. And therefore in the sphere of the earth, according to the formation now stated, there will be no vacuum, because eight cubical parts of the earth placed together about the center of the world fill necessarily the whole space about that center. Now the angle of the pyramid is formed from three angles of triangles, wherefore it equals two right angles, and therefore six such angles equal four cubic angles, for on both sides they equal twelve right angles, and the other six equal the other four cubic angles. Wherefore Averroës concludes in the third book of the Heavens and the World that twelve angles of pyramids placed together about a point will fill the whole space solidly, just as do the eight angles of the cube, and therefore in the sphere of fire there is no vacuum. But other figures placed together about a point cannot, according to Averroës and Aristotle, fill space. For whatever number are placed together their sum will be greater or less than the eight angles of the cube, and therefore they will not fill space. Therefore in the sphere of water, and of earth, and of the heavens, there must be a vacuum, according to the form of the Platonists. As the cube in filling space solidly corresponds to the square in filling a surface, because the cube is formed from square surfaces, since both figures most properly fill space, so the pyramid corresponds to the regular triangle, because the pyramid is formed from triangles, and both figures fill space. But to the third figure, namely, the superficial hexagon, there is no corresponding solid figure filling space, because from superficial hexagons a regular hexagonal figure cannot be formed, as demonstrated before.

And yet the bee forms hexagonal houses that a vacuum may not exist between them; and nature in the womb of the earth forms crystals all hexagonal united into one. So also stones called irises, and said by authorities to be found in Ireland and India only, are formed together in a hexagonal shape. They are called the stone of iris, because they represent the colors of the iris and of the celestial bow, when they are exposed to the solar rays. The same is true of all things generated in this world which are united by their surfaces to retain hexagonal figures, that vacuum may be excluded, and this is a remarkable fact. But, however, there is no real filling of space as Aristotle uses the term in this place: for such filling of space is with respect to every position of solids and surfaces, as four taxilli superficially with respect to every position fill space, and eight fill space solidly, in whatever way angles or sides may be changed, for there is absolute equality in those angles and sides. And the same is true of pyramidal solids and of superficial triangles, squares, and hexagons. This property does not belong to other figures with respect to every position but with respect to some one position, and therefore it is not considered. For if to some houses erected by the bees others be added in another position, the space is not filled up, but there is a vacant space left, and therefore they do not belong to those figures that fill up space, so that the filling may be spoken of absolutely and without qualification. Because of the formation of things in nature the subject of these figures filling space is important and profound. But in general in regard to this formation in the principal bodies of the world I have told enough for the present.

Chapter XIII

Whether there can be more universes and whether the matter of the universe may be extended to infinity.

I pass further to two examples briefly to be noted in bodies in the universe—examples based on geometrical reasoning and connected, moreover, with the matter of these bodies. For Aristotle says in the first book of the Heavens and the World that the universe contains all of its own matter in one individual of one species, which is true of any principal body in the universe, since the universe is one in number, nor can there be more universes in this species, just as there cannot be more suns or more moons, although many have maintained the contrary. For if there were another universe, it would be of spherical figure, like this one, and there cannot be distance between them, because in that case there would be a vacant space without a body between them, which is false. Therefore they must touch; but they cannot touch each other except in one point by the twelfth proposition of the third book of the Elements, as has already been shown by circles. Hence elsewhere than in that point there will be vacant space between them. There is another proof that the matter of the universe is not extended to infinity, as many have maintained. For geometrical reasoning excludes this. Let two lines *a* and *b* meet at an angle at the center of the universe, and from the point of meeting let them be extended to infinity, and let a third line be drawn parallel to one of these and let it be terminated at the other and let it be *ca*. Then *a* and *b* are equal, and *o* (which is part of *a*) and *c* are equal, because from the same point, namely, *o*, they extend to infinity. But line *c* equals line *b*. Therefore line *o* must equal the whole of *a*, that is, the part must be equal to the whole, which is impossible.

Chapter XIV

Concerning unity of time.

Many other geometrical proofs, moreover, might be adduced in addition, and other truths almost infinite in number might be cited in matters of the world, in which geometrical power is manifest. But these suffice for our argument, and I shall merely explode two fallacies which in accordance with popular notions follow upon the numerical unity of matter. For they maintain that time follows upon matter, and age upon form, and therefore as matter is one in number and not more, so time is one in number, and just as form varies in things, so the age is multiplied everlastingly. Hence there are said to be many ages, and unity of time, so that in accordance with the number of angels is the number of ages. But since it has been proved that matter cannot be one, it is false that time has its unity from the unity of matter. Moreover, time can follow only its own subject. But motion is its subject, not matter; and the subject of motion is not matter, but body composed of matter and form.

Finally, geometrical considerations explain to us the cause of unity in time and furnish a proof respecting it. For a body, since it has on all sides dimensions, does not for this reason permit another body with it: for a body has everywhere that with which it excludes another body in length, breadth, and thickness. Therefore a surface in length and breadth will exclude another surface, but not in thickness, because in this respect it is indivisible and lacks dimension. A line in length excludes another line, but not in breadth and thickness, since it does not have these dimensions. Therefore a point, since it lacks all dimensions, has no means of excluding another point from its own indivisible position; but a first point having been conceived of in its own position, a second coming will have exactly the same position, because there is no distance between, and this is also true of a third point and of an infinite number. Motion, in fact, has only linear dimension from prior to posterior in length of space, and this is from past to future. Therefore only in this course, namely, from prior to posterior, or from past to future, will one motion exclude another; namely, prior will exclude posterior, and the past will exclude the future. But a comparison of motion in the present is different from that in its course from past to future. Therefore in respect to the present no motion has dimension nor divisibility, and therefore will not have the means of excluding another from the present. Therefore it can permit an infinite number to be present with it; and hence one present moment suffices for all present motions, and because of this fact this is considered the real reason of the unity of time, and not because of matter. Moreover, from these considerations can be established the real unity of the age as well as of time. For the age has assuredly only linear dimensions, if we should maintain that the age is divisible and has parts, as many think in opposition to the whole force of philosophy and in opposition to Augustine and Dionysius, although Anselm does not agree. If this be true, then the same reasoning applies to the age as to time; the age must be single and not multiple. Or the age will be indivisible, and in that case it will be perpetual, just like the indivisible position for points and atoms, the same position being common to one point and to many, as we have shown.

Therefore there will be one age belonging to all ages. This is a necessary conclusion and doubted by no one skilled in philosophy. Nor is it opposed to the sacred writers and principal doctors, but is in agreement with their view.

Chapter XV

Whether the motion of heavy and light bodies excludes all strain, and how motion generates heat. Likewise concerning a twofold method of knowing the truth.

SINCE motion is the subject of time, and time is the measure of motion, we can now see the great force of geometry in the motions of the bodies of this world. Physicists consider that the motion of heavy bodies downward is altogether natural, and the motion of light bodies upward is likewise wholly natural, so that as a result they suffer no strain. But a geometrical figure shows us the contrary. For let *dbc* be a piece of wood or a stone in the air, *a* the center of the earth, and *gh* the diameter of the earth. Since then *dbc* are always equally distant in their own body, they will fall toward the center along parallel lines. Therefore *d* will fall along the line *de,* and *b* along the line *ba,* and *c* along the line *co,* wherefore *d* will fall outside the center of the earth on the diameter of the earth *hg* toward the heavens, namely, at the point *e,* and *c* at *o;* wherefore in this fall *d* will deviate from the center *a* toward the heavens by the height *ae,* and *e* by the height *ao*. But every deviation of a heavy body from the center toward the heavens is a strain. Therefore *d* and *e* are moved under a strain, and the same is true of all the parts of *dbc* except *b,* which alone falls to the center. Therefore there will be much strain here.

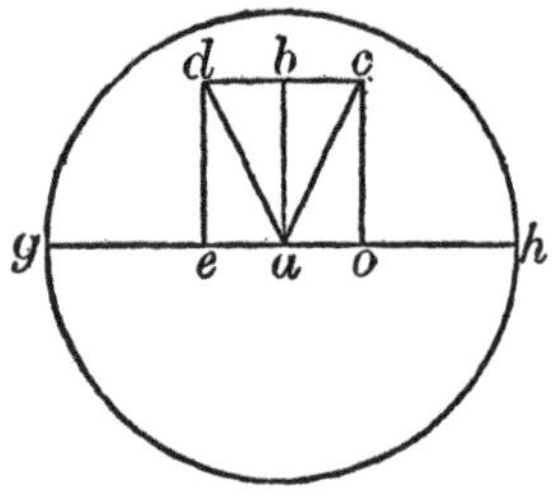

FIG. 18.

But the straight and natural path of *d* is along the line of *da,* whence if *d* be separated from its whole, it would fall to *a* by a straight path, because every heavy body tends toward the center. Moreover, every deviation of a heavy body from the straight path is a strain, but the more *d* is moved along the line *de,* the more it departs from the straight path, as is clear to the perception, because the lines *da* and *de* are more widely separated below than above. Therefore the further *d* falls downward, the more is it moved under strain, and likewise *c* and each of the other parts of the whole weight *dbc,* except *b* which alone is always falling along the straight path. It is clear, therefore, that there is a great and complex strain in the natural motion of the heavy body itself. From this there follows a certain truth in nature, namely, that natural motion generates heat; for since the strain has been proved, and we are agreed that a heavy body's tendency is naturally downward, it is evident that there are two forces in a heavy body downward drawing it in opposite directions. Therefore one of these draws the parts of the heavy body in one direction, the other in another, and in consequence of these pulls the parts of the heavy body are of necessity rarefied. But rarefaction is an immediate disposition to heat, whence by experiment we know that a heavy body falling downward grows warm. Therefore we can state here, as in former cases, that the reasons of things in nature should be assigned by the power of mathematics. A man can see that in the things of nature there are two methods of proof, one by a demonstration that proceeds through the causes, and the other by a demonstration in regard to the effect, as in the demonstrations given it is proved by the cause that strain takes place in a heavy body in its own natural movement, later this same fact is demonstrated by the effect, namely, by the generation of heat. For heat would not be generated except through rarefaction, nor would that rarefaction take place except from forces drawing the heavy body in opposite directions, and these can only be the one natural, the other producing a strain, on account of which a heavy body in its own natural motion is under strain. Thus this conclusion, a heavy body receives a strain in its own natural motion, is proved by cause and effect. But cause alone produces knowledge far greater even than does effect, because Aristotle says in the first book of the Posteriores that we believe we know when we learn the causes. Therefore, when the proof, as he shows in the same book, is a syllogism causing us to know, the proof through cause is of necessity far stronger than that through effect; and Aristotle maintains this in the book of the Posteriores. Wherefore when a proof is given through cause by means of mathematics, and a proof is given through effect by means of nature, the mathematician is more efficient in learning the things of nature than the natural philosopher. And this is particularly clear, that motion purely under strain, as the motion of a heavy body upward, will generate heat, and far more so than natural motion; because in a body moved under strain there are two altogether opposite motive forces, both in the whole body, and in opposite directions, as the natural force of a heavy body tends downward, and the force of strain tends altogether upward. Therefore there is a great difference in the parts of a thing moved under strain, and a greater one than in natural motion.

Chapter XVI

Concerning the movement of the scale.

In addition to what has been said already it is expedient to disclose further the potency of geometry in motions; and I purpose to do this because of a general understanding of the science of weights; a noble one and too difficult for men not acquainted with the causes relating to the motions of heavy and light bodies. Therefore Jordan says in his book on Weights that if a balance is in a horizontal position with equal weights in the scales, it will remain in this position; and if it is disturbed, it will return to the horizontal position. This we see sensibly in both scales, the beam of which is equal in length on both sides, and equal weights are in the pans, and the scale is held horizontally by the support, so that the support is at right angles above the beam of the balance at the center of revolution; for this point is called the center of revolution, from which the support extends at right angles. It is called the center of revolution, because when by the force of a hand depressing one of the equal weights, or owing to the inequality of the weights one of them is lowered, the other will be raised, and this motion of descending and rising will describe a circle of which that point from which the support extends is the center, and therefore it is called the center of revolution. To make this clearer, let us draw a figure. For let the rod or beam of the balance be *ab* and let *cd* be the support, then the center of revolution from which the support extends will be *d,* and on the circumference of this circle the weights will move, for the descending one will describe the lower circle, and the rising one the upper circle. These assumptions being made, the proof is as follows. When one of the arms of the balance with equal weights is lowered by the hand of one depressing it, it becomes heavier, according to Aristotle in the fourth book of the Heavens and the World, because the lower a body sinks, the more weight it acquires, as he himself says. Therefore the descending weight becomes heavier, however little its descent from the position of equality may be, and therefore the greater its descent, the greater will be its weight. Therefore it will become unequal to the other weight and heavier than it. Therefore, although they were equal in the horizontal position, yet when they lose this position they become unequal in weight; wherefore the one which is depressed will always continue to descend, and the other will always ascend, and therefore they will never return to the horizontal position. Just so when two unequal weights are placed on the arms, they immediately lose the horizontal position, and will never return to the same position, but the heavier one always descends. So likewise in the case under consideration, which is contrary to Jordan and to our sense. Jordan likewise says that between any weights there is assumed in the same way a proportion of velocity in descent and of weight, but that the greater the descent of a weight, the heavier it becomes. Therefore it descends with so much greater velocity. Therefore it will never return naturally to the horizontal position. Jordan also says that one weight is less heavy as regards position, because it follows the descent of the other with an opposite motion, that is, because it ascends when the other descends, and *vice versa;* wherefore they will never come to rest in a position of equality. That, moreover, the weight which is depressed is less heavy in respect to position, is clear for this reason, because it takes a less direct descent on

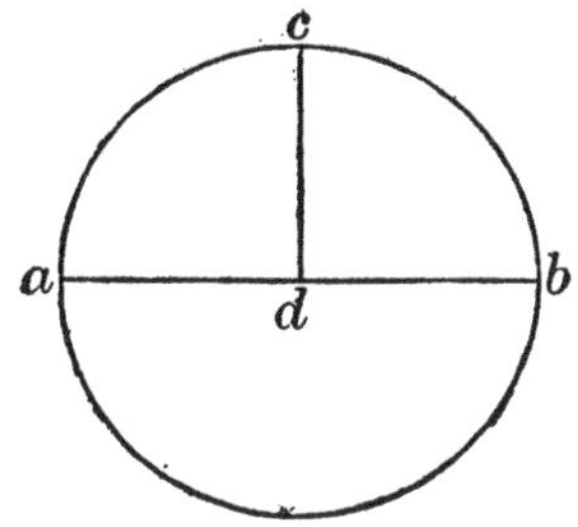

FIG. 19.

the diameter passing through the center of revolution toward the center of the earth; wherefore, according to Jordan, it will be less heavy according to position. This conclusion truth itself requires, as explained by the figure. A drawing of this kind will remove objections, nor can the intellect have assistance except through the figure. Let then a circle be described above the center of revolution, which is *o,* on the circumference of which the weights revolve, and let the horizontal diameter *ab* be drawn, and let another diameter intersecting this one be drawn directed toward the center of the earth, and let it be *dc,* and let equal arcs be marked on both semicircles on both sides of the horizontal diameter, and from both of its extremities, and from the ends of the arcs in both semicircles let lines be drawn parallel to themselves and to the horizontal diameter, namely *fh, gp, tq, sr,* all of which cut the diameter falling to the center of the earth. Of necessity then, according to Jordan and his commentator, those parallel lines will cut from the diameter directed toward the center of the earth unequal parts, so that the parallel which is nearest the horizontal diameter will cut a greater part of the other diameter than will a more distant parallel; as, *tq* will separate a larger part of the diameter *dc* than *sr,* so that the part of the diameter *dc* which is between *ab* and *tq* is greater than the part of the same diameter, which is between *tq* and *sr,* and in the same way the part of the diameter *dc,* which is between *ab* and *gp,* will be greater than that which is between *gp* and *fh.* Accordingly if a parallel is taken in the one semicircle, and another parallel in the other semicircle, which are equally distant from the diameter parallel to them, they will of necessity cut equal parts from the diameter falling to the center of the earth, as *tq* and *gp* will cut equal parts from *dc,* and likewise *sr* and *fh,* just as the twenty-sixth proposition of Jordan on triangles states. If, therefore, the parts of the diameter directed toward the center of the earth divided by the parallels are unequal, so that those parts of the diameter

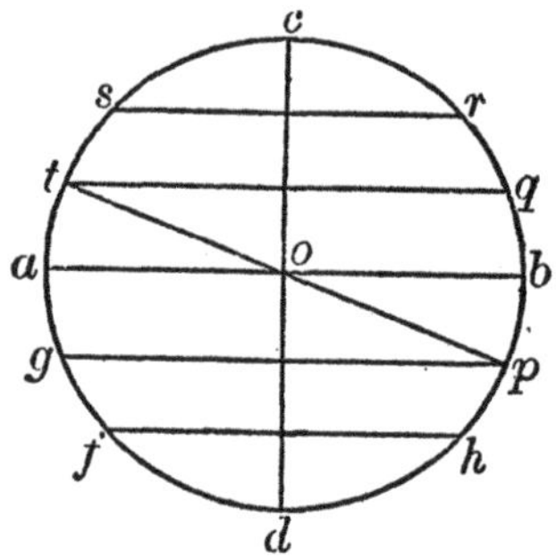

FIG. 20.

which are divided by the parallels nearer the horizontal diameter are greater; then first let us note that the beam of the balance lies on the horizontal diameter, and let the support be upright on the diameter falling through the center, so that the balance and its arms are in a horizontal position; after that let the balance be moved, and one part of the balance be raised to the first parallel on the upper semicircle, the other part depressed to the end of the first parallel on the lower semicircle, so that the beam is in the position of the line *tp* and part of the balance is higher at *t* than the other part at *p*. If, therefore, *p* should descend to the end of the other parallel *h*, it will cross on the diameter drawn to the center the part which is between the parallels *gp* and *fh*, which is less than that part of the diameter which is between *tq* and *ab*, as is clear from what has been said. Therefore if *t* should descend to *a*, it would take a more direct descent on the diameter drawn to the center of the earth than *p* while it is descending to *h;* wherefore *t* is heavier as regards position than *p*. And *t* again descends toward the center of the earth. But *p* on account of the bending of the circle is curved back from the center, and tends less toward the center, as is clear to the sense. It remains, then, that for this reason it will be less heavy. Since this is the case, the first objection is removed; for although the part of the beam *p* descending is nearer the center of the earth, as is clear if a straight line be drawn from it to the center *o*, because that line is shorter than the line which is drawn from *t*, as is clear to the sense, and thus is heavier in so far as it has a lower position, nevertheless since *p* does not move along that straight line *po* toward the center, but along the circular course of the circumference of the circle, and that circular course causes it to take a less direct course on the diameter drawn to the center than *t* takes, and bends and curves *p* from the center, so that it does not tend directly to the center like *t*, when it descends, of necessity *t* is heavier than *p* while they are in such a position. There-

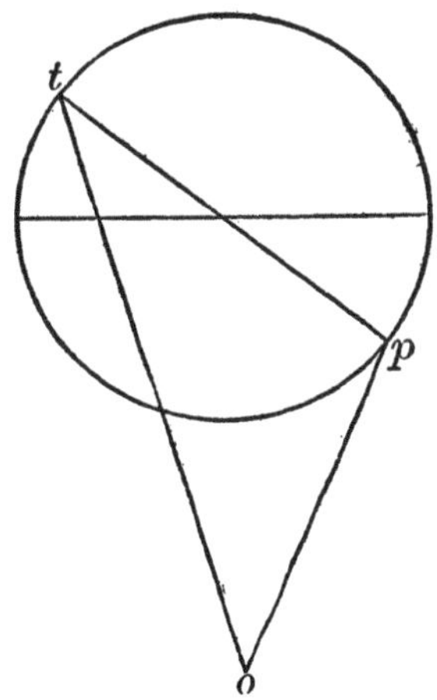

FIG. 21.

fore if they are left to themselves, *t* will descend, and *p* will follow its descent with an opposite motion, namely, an ascent to the position of equality. Therefore *p* will not always descend, but the two reasons for weight here given prevail over that of which the objection has made mention. Again that weight has no force, for it is slight and insensible, and therefore is not operative here, just as if a feather should be added to one of the equal weights, when they are in the horizontal position, it would not cause the weight to descend, and yet as a fact that arm to which the feather was added is the heavier, since the feather has some weight. Therefore the case is similar here; for the descending arm, when it is at its lowest point, is only a short distance from the position of equality, therefore the additional weight it acquires is only trifling and insensible, and therefore the acquired weight is not to be considered.

The other fact is then explained, which Jordan draws as a conclusion from the greater weight, that it will always descend with greater velocity, so that it will never rise. For it is now clear that this additional weight has no sensible effect, and the two causes of weight touched on above are found in the higher arm. When, however, he states in the third argument that it is less heavy because it follows the descent of the other with an opposite motion, namely, an ascent, I gladly grant it. For the lower weight is in this condition because it is ascending when the upper weight is descending, as is the case when they are left to themselves and are equal, as has been set forth. But not on this account does it ascend more and more, nor does the other descend without limit, but they merely reach a position of equality, and there they naturally come to rest, wherefore the objection draws a wider conclusion than it should when it wishes to conclude from this fact, because one ascends and the other descends, that they pass the position of equality.

And if it be said that at this point there is an oscillating motion and therefore the upper arm descends beyond the position of equality, and that the reason why it passes over a small distance and a large one is because this crossing over is of one nature, and similarly in regard to the other arm that it always ascends after it passes the position of equality; we must reply that this descent of the upper arm beyond the position of equality is not due to the nature of the weight itself, but to the impacts of the particles of the air. For when the air is set in motion, it retains the motion readily, and therefore for a long time its particles vibrate hither and thither and do not permit the weight to come to rest at once in the position of equality.

THE APPLICATION OF MATHEMATICS TO SACRED SUBJECTS.

After the necessity of mathematics has been proved in secular things, and in human sciences, the same can now be shown in divine things. And we must give this the more consideration, because human things have no value unless they be applied to divine things. Since then it has been shown that philosophy cannot be known unless mathematics be known, and since all are aware that theology cannot be known unless philosophy be known, the theologian must of necessity know mathematics. But God has placed created things in his Scripture, who alone knows the power of the creatures which he has made, nor can he have false perception, nor does it befit his own truth. Therefore since all things from the highest heavens even to the ends thereof are placed in Scripture by God and the angels, either in themselves or in their similitudes or in their opposites, and since the same knowledge applies to opposites, as Aristotle says, and truth is either in the universal or in the particular, the theologian of necessity must know the things of this world if he ought to know the sacred text.

Moreover, we see that the literal sense is based on a knowledge of the natures and properties of creatures, so that through suitable adaptations and similitudes the spiritual meanings may be elicited. For thus the sacred writers and all the sages of old expound, and this is the true and real exposition which the Holy Spirit has taught. Therefore it behooves the theologian to have an excellent knowledge of the creatures. But it has been shown that without mathematics they cannot be known. Therefore mathematics is absolutely necessary to sacred science.

This, in the third place, can be shown by its properties. Although what I maintain will be proved in many ways, yet I shall attempt to convince by means of the activities of the sacred writers together with a removal of the bad repute of mathematics which many ignorantly allege because they do not understand the testimony of the sacred writers. For the Patriarchs and the Prophets before and after the flood discovered it and taught other men, first, the Chaldeans; then, the Egyptians; and from the Egyptians it descended to the Greeks; and it is not so clearly stated that they so labored in other sciences. But since through those men the divine law was given to us, and they were most holy men, and busied themselves only in the science which is most useful to the divine law, mathematics is for this reason in complete accord with the divine law. The lesser proposition has with it its own proof. The greater one is proved by triple authority. In the first place it is proved by the historians, especially by the chief of these, Josephus. For in the first book of the Antiquities in three places he mentions these sciences, and expressly tells all that our major proposition states. For he says that the sons of Adam discovered geometry, astronomy, arithmetic, and music; and Noah and his sons taught the Chaldeans; then Abraham taught the Egyptians.

In the second place, this is verified by the blessed Jerome and Cassiodorus and other sacred doctors, as even the rank and file of theologians knows; and the sacred writers ratify what Josephus asserts. In the third place, it is proved by the statement of philosophers. For Albumazar in the fifth book of the larger Introduction to Astronomy in the eleventh principle in the eleventh chapter states that Shem, the son of Noah, taught others this science, and in the prologue of the Astrolabe of Ptolemy it is stated that a zeal for this science is affirmed to have been inspired in the world by the son of Shem, who had been put in mind of it by the divine narration or perchance had been stirred by the divine command. The same is indicated by the activity of our sacred writers after the coming of the Lord, as Augustine, Cassiodorus, Isidore, Jerome, Orosius, Bede, Origen, Eusebius of Caesarea. For concerning these sciences have they written, and in them have they exercised themselves and others solely, or at least more than, in other sciences. Since therefore they were teachers of the sacred Scripture and were holy men, it is clear that sciences of this nature are especially valuable to professors of the sacred science. What, moreover, they have written concerning these sciences is shown by Cassiodorus and Isidore, who composed their treatises on all four of these subjects. Augustine also composed different books on numbers and music; Jerome treated of these subjects in different books, and Orosius in writing to Augustine, and Eusebius in writing of the places in the world, definite knowledge of which is derived from astronomical sources, as is certain and as what follows will show. Bede wrote concerning the course of the sun and moon and concerning the whole question of the diversity of time, and both Origen and Eusebius wrote on subjects which are known to be based on astronomy. Not only did they write on these subjects, but they also taught these things to suitable men, in order that they might be sharp-witted against heretics, as is disclosed by Cassiodorus in mathematics and by the sixth book of the Ecclesiastical History; the divisions also of geometry and arithmetic are subjects of instruction, as is stated in the same book, which are more removed from sacred things than are astronomy and music. But not merely did they busy themselves in writing and in teaching others, but they also expounded theological truths through the potency of these sciences; as is shown without contradiction by all the original works of the sacred writers. This is evident in the matter of numbers and of places in the world and in celestial things and in others that are known to pertain to the sciences mentioned, as we shall show below.

But not only have they so expounded sacred things as a matter of fact, but they state that these sciences throughout the whole world have value for divine things, and they extol them beyond other sciences in this particular. Cassiodorus indeed speaks as follows in his preface concerning the arts and discipline of secular studies: "We can call mathematics in the Latin speech doctrinal, and although by this term we are able to designate all subjects of instruction as doctrinal, yet this science has properly claimed for itself the general word owing to its own excellence." And in his treatise on mathematics he thus speaks, "Those are subjects of instruction which never fail, deceived by opinions, and for this reason they are called by such a name. While we turn these over in frequent meditation, they sharpen our perception, they wipe away the filth of ignorance, and by the gift of the Lord they lead us to that speculative contemplation. Rightly do the holy fathers counsel us to read these sciences, since in great measure by them is our appetite drawn away from carnal things, and they cause us to desire those things, which, the Lord alone granting, we can view with affection. A twofold division of mathematics is assigned to numbers in common speech not only by the sacred writers, but also by the philosophers themselves, namely, arithmetic concerning numbers absolutely, and music concerning proportions and other relations which are found in numbers as referring to sound and gesticulation; because sound and gesticulation form a subject apart from music; but they are discussed however by means of the relationships of numbers in the science of music." Augustine, moreover, in the second book on the Order of Discipline says, "We should not aspire to any of the sacred writings without a knowledge of the potency of numbers." He also says in the same book on Christian Doctrine, "Ignorance of numbers causes many things placed in the Scriptures with a transferred and mystic meaning not to be understood"; and he furnishes many examples, adding that "thus under many other forms of numbers secrets are placed in the sacred books, which owing to ignorance of numbers are closed to readers." And therefore owing to the very great utility of numbers Isidore in his treatise on Arithmetic says, "Take from the world computation, and blind ignorance enfolds all things; and men cannot differ from other animals, which are ignorant of the method of calculating."

Moreover, although the praise of music is now manifest in common with arithmetic, because both take into consideration the relations that exist in numbers, yet numbers, as they exist in sounds, Augustine greatly praises, speaking to Memorius. "What importance numbers have in all motions of things is more easily considered in the case of what is uttered by the voice, and this consideration by certain graded roads presses on to the supernal paths of truth, in which wisdom joyfully discloses herself." And in the book of Retractions he says, "The sixth has become especially known, since in it a question worthy of being understood is treated, how from mutable numbers we arrive at the immutable, as if the invisible things of God are perceived through those things which have been made." Likewise in his book on Christian Doctrine he says, "Not a few things in the sacred books an ignorance of the principles of music closes and covers up"; and he gives examples in the case of the ten-stringed psaltery and the cithara and the like, and he adds that "we find music occupying an honorable position in many places in the sacred Scriptures." Cassiodorus, moreover, judges that this science is of importance in the commands of God, in morals, in the sacred Scripture, and in all created things. Hence in his treatise on music he speaks as follows: "Musical training applies to all the acts of our life; first of all, if we are to do the bidding of the Creator and obey with pure minds the laws established by him, music is then the science of proper regulation. If we manage our life correctly we are always shown to have had our share of such training. When, however, we do injustice, we do not possess music. In religion itself there is also a strong admixture of music; whence the ten-stringed instrument of the decalogue, the tinkling of the cithara, the timbrels, the melody of the organ, the sound of the cymbals; without doubt the Psalter itself was named after an instrument of music, because in it is contained the exceedingly sweet and pleasing melody of the heavenly virtues; and to state the whole matter briefly, whatever in things celestial or terrestrial is done fittingly in accordance with the direction of its author is the result of this system. Most pleasing then and useful is the knowledge that lifts our mind to things above and charms our ears with its melody."

Concerning the utility and the knowledge of astronomical matters he says, "If we inquire after astronomy with a pure and disciplined mind it floods our senses, as the ancients say, with great brightness. For such is the result when we turn the mind to the heavens and investigate with inquiring method that whole celestial frame, and with mental acuteness by contemplation gather in part what mysteries of such magnitude have veiled." And he adds, "By these means, as it seems to me, to know the climes, to understand the lengths of the seasons, to note the course of the moon for the determination of Easter, in order that the artless may not be confused in regard to it,—all this does not appear silly. There is also another advantage from such studies which is not to be despised, if we learn from them the suitable time for sailing, the time for ploughing, the dog-star of summer, and the mistrusted storms of autumn." Augustine in the second book on Christian Doctrine, speaking of the utility of this science, maintains that its utility is threefold, namely, "a demonstration of the present, a knowledge of the past, and reasonable conjectures of the future. The demonstration of the present consists of the assignment of the properties of celestial things. Besides a demonstration of the present it possesses something like a narrative of the past, because from the present knowledge and motion of the stars we may revert regularly to their tracks in the past. It possesses also conjectures according to rule of the future, not full of mistrust and ominous, but definite and certain," as he says. Since the sacred writers feel thus concerning the three latter parts of mathematics, they must of necessity have laudatory sentiments in regard to the first part of the science, which to be sure is geometry. For on the knowledge of this the other parts depend, since it is the first of all and the root of the others. Cassiodorus, writing concerning the praise of this subject, speaks thus, "For if it is permitted to speak, sacred divinity since it has given to its own creation diverse species and rules, since by a power to be venerated it assigns the courses of the stars, and causes things that are moved to travel in determined lines, and places in a definite abode things that are fixed, whatsoever is well arranged and complete, sacred divinity, I say, can be attached to the qualities of this subject."

If we wish to consider the qualities of the study of theology, we shall find mathematics wholly necessary for seven important reasons. One is the knowledge of celestial things. For nothing is so fitting to theology and its professors. For theology is celestial by divine will; and therefore no human speculative science will befit it to the same extent as the science of the heavens. And by all Scripture we are called away from the things of this world and urged on to heavenly things. And our conversation according to the Apostle is in heaven if we are really Christians, and if we have the aspiration and the belief that we are to remain corporeally and perpetually in heaven. Wherefore nothing else should be so fully known by us as the heavens, nor should anything else be so desired. And if we rejoice to expound the Scripture, it is proper that the things which are placed in Scripture and cannot be known otherwise should be expounded by the properties of inferior things. Wherefore in like manner since there are in Scripture many difficult things concerning celestial matters, it is necessary for the theologian to know celestial things.

Moreover, since the vastness of things celestial stirs us to reverence the Creator, and since there is no comparison in magnitude of things below with things celestial, the knowledge of things below will have no comparison with that of the things above as regards their end, which is the praise and reverence for the Creator. For Avicenna says in the ninth book of the Metaphysics that "those things that are under the circle of the moon are almost nothing in comparison with those things that are above it." And Ptolemy shows in the Almagest, and all astronomers know, that the whole earth together with all things below holds the same relation to the heavens as the center to the circumference. But the center has no quantity. Wherefore they draw a like conclusion in regard to the earth in comparison with the heavens, although the earth in itself may possess great quantity. The smallest of the visible stars, as Alfraganus states in the beginning of his book, is larger than the earth; but the smallest of the stars in comparison with the heavens has no quantity of any significance. Since from the eighth book of the Almagest and from Alfraganus it is evident that there are six magnitudes of fixed stars; any one of those belonging to the first magnitude is equal to the earth about one hundred and seven times; and any one of those belonging to the sixth magnitude is equal to the earth eighteen times; and the sun is about one hundred and seventy times larger than the earth, as Ptolemy proves in the fifth book of the Almagest. According to him it requires thirty-six thousand years for a star to complete the circuit of the heavens, owing to the vastness of the heavens, notwithstanding the fact that the star moves with incredible velocity. But the earth can be traversed at foot pace in three years. Wherefore the magnitude of things below bears no comparison to things celestial. The same is true of their utility; because the whole utility of things below is drawn from the things above. For the twofold position of the sun under the oblique circle together with the aspects of the planets is the cause of all that happens here below.

If, therefore, we consider things celestial as regards questions of theology, it is evident that theologians inquire in their collections of maxims and in their treatises on these collections whether the celestial orbs are continuous or discontinuous, concerning the number of the heavens, especially as concerns the ninth and tenth, concerning the shape of the heavens, concerning their circles, epicycles, and eccentrics, concerning their motions in these, concerning differences of position in the heavens, as right and left, before and behind, up and down, concerning the properties of the heavens, such as their light, transparency, and the like, concerning the influence of the heavens on things here below, and concerning the species of the heavens and of their nature elemental chiefly on account of fire. For Augustine and others sometimes following the opinion of Plato state that the heavens are of a fiery nature. They also inquire about the localities of the world because of paradise in their effort to determine whether it is under the equinoctial circle or not; and they try likewise to determine the location of hell. They inquire also whether celestial things have power over those things that are generated and corruptible and over the rational soul; concerning fate and like questions, decisions in regard to which are known to belong to astronomy. Other matters are arising daily without number in the inquiries of theology. But not only the treatises on the Sentences, but also the sacred text itself, together with the expositions of the sacred writers, require this aid. For the first chapter of Genesis has many difficulties because of celestial things, as is clear from the text itself and from the expositions of the sacred writers, especially Basil, Ambrose, and Bede in their books that are called Exemeron. In Joshua there is the difficulty owing to the length of the day when the sun stood still, especially in comparison with the length of the day, on which the sun went backward ten degrees at the word of Isaiah. For there seems to be a contradiction in these places. When Solomon says in Ecclesiastes that the sun every day, according to the exposition of Jerome on the original, turns to the north, scarcely any of the astrologers can understand this; since they know that from the winter solstice to the summer solstice the sun advances toward the north about one degree every day, but that during the other half of the year it travels in the opposite direction. There is difficulty also concerning the height of the firmament mentioned in Ecclesiasticus, and how the sun in the south consumes the earth, as is stated in the same book, men lack the astronomical knowledge to understand. The Hyades and Pleiades and Arcturus and Orion and the chambers of the south, of which the blessed Job speaks, possess great difficulty, especially when the blessed

Jerome says in commenting on Isaiah that Orion has twenty-two stars, of which the first nine are in the third magnitude, and nine others in the fourth magnitude, and the four remaining ones in the fifth magnitude, and makes no further statement. But these things cannot be known except from the eighth book of the Almagest, where six gradations of magnitude of stars are given, and the stars that belong in each magnitude are determined. There are also other matters almost countless in number in Scripture and in the expositions of the sacred writers which touch upon the science of the heavens and the decisions of astronomy: wherefore the theologian of necessity must be well informed about things celestial, both on account of the anticipation of questions in collections of maxims and in summaries, and on account of the sacred text itself.

The second head of astronomy, as regards theology and especially as regards the sacred text, is the study of the places in the world. For the whole text is full of these places, and for this reason nothing of importance can be known unless these places are known. For the whole series of Scripture deals with regions, states, deserts, mountains, seas, and other places of the world, certitude with respect to which cannot be gained except through the sciences mentioned, because it is the function of these sciences to distinguish the habitable parts from the non-habitable, and to divide the habitable into three great divisions, Europe, Africa, and Asia, and these three into seven known climates, with many other irregular climates besides. These climates no one can separate with certainty except by virtue of these sciences into provinces and districts and other localities, so that well-known and famous cities may be found, such as Jerusalem, Babylon, Meroë, Alexandria, Antioch, Ephesus, Athens, Tarsus, Rome, and the rest, determined in comparison with others according to the required distance between them, both east and west, north and south. These having been found, it will be possible to find famous regions named from them, also seas and deserts and mountains and all things that are contained in sacred writings. For here these sciences possess great usefulness in sacred Scripture. And perhaps nothing connected with philosophy will be found more useful; since for the man ignorant of the localities of the world the rind of history frequently has no taste because of the infinite number of places, and especially owing to the manifold falsity of the new Bibles; and as a consequence he will be impeded in rising to the spiritual meanings and only imperfectly will be able to explain them. But he who has gained a good idea of places, and has learned their location, distance, height, length, width, and depth, and has tested their diversity in heat and dryness, cold and humidity, color, savor, odor and beauty, ugliness, pleasantness, fertility, sterility, and other conditions, will be pleased very greatly by the literal history, and will be able easily and admirably to gain an understanding of the spiritual meanings. For there is no doubt that corporeal roads signify spiritual roads, and corporeal places signify the ends of spiritual roads and the similitude to spiritual places, since place has the quality of terminating local motion and furnishes the reason for stopping, and therefore the knowledge of those places both causes the literal meaning, as we have said, to be understood, and prepares the way for understanding the spiritual meanings: a fact which is established fully by the words and deeds and writings of the sacred authors.

In the first place, because Jerome says in the prologue to the second book of Chronicles, "He will have a clearer insight into the sacred Scriptures, who has viewed Judea with his own eyes, and has learned the memorials of the ancient cities and the names either the same or changed of the places." In the second place, because the sacred writers have labored in viewing those places and in traveling over them. Wherefore the blessed Jerome says in the prologue just mentioned, "We made it our care to undertake this labor along with the most learned of the Jews, that we might traverse the province of which all the Churches of Christ speak." Moreover, he would not have done this except for the purpose of understanding the sacred writings. In the third place, because he has written many books about places in the world, determining their distance, location, and other facts in regard to them with great certainty. Orosius also gives Augustine a very useful, truthful, and clear account of these places. Isidore, moreover, in many places locates districts and states in a more useful way, if we may say so, than the former writers. Cassiodorus, moreover, does not omit to distinguish them in respect to their climates. Eusebius of Caesarea also, as Jerome states in his book on places, after the geography

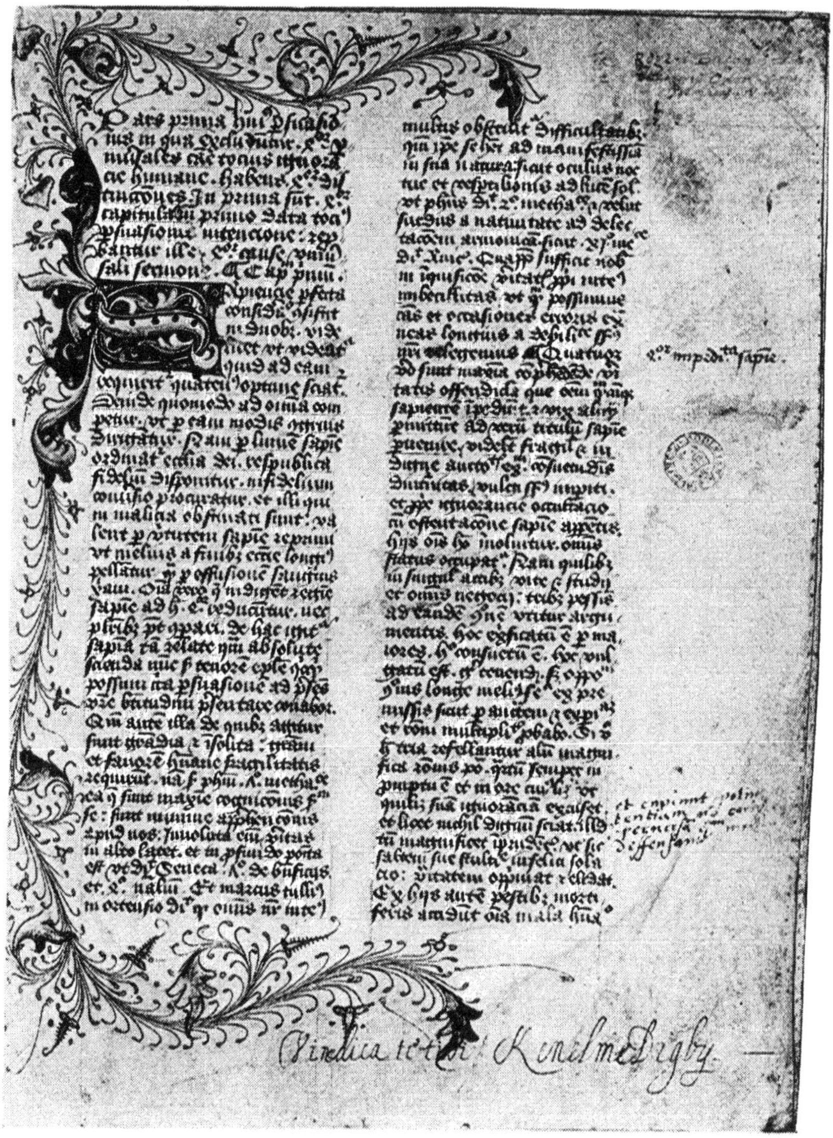

A PAGE OF THE OPUS MAJUS

From the Digby Manuscript in the Bodleian Library

of the land of Judea and the assignments of the tribes, and a picture of Jerusalem itself and of the temple in it with a very brief exposition, labored at the end to collect for us from sacred Scripture the names of nearly all the cities, mountains, rivers, hamlets, and different places, which remain the same or have been changed or corrupted in part. Origen the Adamantine in his commentary on Joshua in the original, in agreement also with the gloss on the eighteenth chapter of Joshua, speaking of the multitude of places mentioned in Scripture and amid other praises of those places, addresses us thus, "Do not read these names with disdain nor think Scripture a worthless texture of many names; but be assured that in these are contained greater mysteries than human speech can utter, or mortal ear hear!" Since then the sacred expounders and doctors have labored so much on those places, and confess that such great mysteries are contained in them, there is no doubt that a knowledge of them is necessary in every way for the sacred Scriptures. But it is the function of astrology and astronomy to give explanations and full certitude in regard to the places in the world. Wherefore these sciences are very necessary in this particular. These statements can be fully verified in examples. For he who hears the accounts of events that took place about Jordan, Jericho with its plain, Mount Olivet, the Valley of Jehoshaphat, and Jerusalem, and has no conception of those places and of their nature, will be ignorant of the literal sense, and will, not undeservedly, have little appreciation of the course of history, and in consequence is blind to the spiritual meanings. But he who knows the lengths, breadths, depths, heights, variety of qualities, as heat, cold, dryness, humidity, and of those following these four, softness, hardness, thickness, thinness, harshness, smoothness, aridity, fluidity, slipperiness, and innumerable other qualities, which are defined in the fourth book of the Meteorologics; moreover, colors, savors, odors, beauties, ugliness, pleasantness, sterility, fertility, matter infected, corruptible, and their opposites, and others which have to be considered in their place,—he who knows all these can understand clearly the literal sense and delight himself in it, and can pass to the spiritual meanings in a calm and glorious manner.

For by considering a few facts relating to the places named, we can express important meanings in a moral, allegorical, and analogical sense. The Jordan flows from the north to the south, and lies to the east of Jerusalem, which is in the west not far from the great sea, and between those two first from Jordan is the city of Jericho with its plain, then Mount Olivet, third the Valley of Jehoshaphat, and then Jerusalem. The world, sacred writers say, is symbolized by the Jordan according to the explanation of its interpretation and because of its properties, since it empties into the Dead Sea, which is a figure of Hades, and the Jordan is so considered for many other reasons. Jericho signifies the flesh, as the sacred writers maintain. Mount Olivet signifies the excellence of the spiritual life, owing to the excellence of the mountain and the sweetness of devotion owing to the explanation of Olivet. The Valley of Jehoshaphat signifies humility owing to the explanation of valley, and true humility before the eyes of majesty, because the interpretation of this name Jehoshaphat is, in sight of the Lord. Jerusalem signifies the vision of peace, and morally it is the sacred soul that possesses peace of heart; allegorically it signifies the Church militant; analogically the Church triumphant.

He therefore who from the beginning of his life or the hour of his nativity in the morning of deliberative reason or with the possession of this faculty desires to arrive at least in the evening of his life and in old age at peace of heart, and this I mean in a moral sense; and he who wishes to be a faithful and perfect member of the Church, under whose shadow he may lie in peace from the insults of the malignant foe, and this I mean in an allegorical sense; and he who wishes that his conversation may be in the heavenly Jerusalem in this life, and that he may pass at death to that holy city, where he shall sit in the beauty of peace in the tabernacles of fidelity or in opulent rest—such a man should first either leave behind Jordan, that is, the world, by subjecting it to him, as the sacred writers have done who remained in the world, or withdraw from Jordan by wholly renouncing the world, as members of religious orders do. For at this point is the first step in the spiritual life and an easier one than the others. When this has been done, he must attack the flesh, because it is not so easy to conquer this as the world. For it is a familiar plague and does not leave its subject. But he must not destroy the flesh and crush it with force, but gradually and discreetly must he conquer its pride. For this reason the flesh is considered Jericho with its plain; and therefore he must proceed as a penitent along the level road, in order that the obedience of the flesh may be a reasonable one, lest if he foolishly overwhelm the flesh he may not be able to reach the higher things of the spirit. For this statement is directed against many who, having been turned to penitence, destroy their bodies in the first or second year, and afterwards become useless, so that they are able to aid neither themselves nor others. After a man has brought the world under subjection, and has conquered the flesh as he should, then and not before is he fit to ascend to the excellency of the spiritual life and to the sweetness of devotion. For then is he able to ascend to Mount Olivet and to reach the summit of perfection and to plunge into the sweetness of prayer and contemplation. When he has been sufficiently exercised in the ascent and passage of that height, he must then cross the Valley of Jehoshaphat, that is, he must conclude his life in perfect humility, that he may be poor and humble in spirit in the sight of God, not in his own eyes or in those of men. For many appear humble to themselves and to others and are before God and the angels most proud. When he has completed his whole life in perfect humility, then is he in Jerusalem, according to its triple sense. For he will have peace of heart, because such peace follows the perfection of the spiritual life. "For there is no peace for the wicked," says the Lord. But to the holy belongs the peace of God, which passeth all understanding, and in the peace of the Church militant such a man rests secure, a peace which unbelievers and sinners living in a state of damnation lack, whom the devil confounds and drives from sin to sin, and from the punishment of sin to fresh punishment. And as has been said, the sanctified man will participate even in this life in certain hope and in revelations by means of that blessed vision of the peace of Jerusalem above, which he shall obtain by the grace of God at death.

But not only these places between Jordan and Jerusalem make history clear and explain spiritual meanings, but other places also without number that are found in Scripture between these two limits. If any one should wish also to consider further the other qualities enumerated, much more and almost incomparably will he be able to bring out the divine meanings, as is evident to him who examines the matter. But it suffices merely to hint how we can succeed in drawing forth many things from few, great things from small, obscure things from those that are more manifest. But the places of the world cannot be known except through astronomy; since it is necessary for us first to know the longitudes and latitudes of such places. Latitude is measured from the equator and longitude from the east, that we may know under what stars places belong and how far they are from the path of the sun. For in accordance with these we see sensibly that the affairs of this world vary, not only in the things of nature but also in things pertaining to morals. It is necessary also to know through astronomy what planets hold sway over the different regions; for in accordance with this principle are regions greatly altered. Many things of this kind must be considered by means of astronomy, that we may know the natures of the places in Scripture; and this information is necessary not only on account of the places, but also because of the things located in these places. A knowledge of all things is necessary for the spiritual as well as the literal sense, as is clear from what has been said.

The third head relates to chronology. For the whole course of history is traced through times and generations and ages from the beginning of the world to Christ the Lord, and all things have been set in order on his account, that no other legislator might be expected, but that he alone may be the Savior of the world by his own law; that the error of the Jews regarding their expectation of the Messiah may be removed; likewise the error of the Saracens regarding Mahomet, who followed Christ; also the error of those who shall still adhere to the proposer of the law of wickedness, who is to come, as Albumazar teaches in the book of Conjunctions, and who in reality will be Antichrist; in order that all sects of pagans, idolaters, Tartars, heretics, and other unbelievers, who throughout the world have been scattered into sects almost without number, may be destroyed through a certification of the time of the Savior, that neither before nor after him may any other be considered, through whom the salvation of the human race may be accomplished. But no one can certify in regard to times except the astronomer, nor can any other science than astronomy certify in regard to them. For all beg for its leavings in this respect, as is clear. And if we consider the matter, we shall find in many ways how necessary astronomy is in this particular. For some of these calendars are lunar, some solar and lunar, some have a fixed beginning, as with the Jewish astronomers. For they begin the year with the moon in October, because they have used from antiquity tables and canons for the setting of the sun at the city of Jerusalem. But they still are glad to use these on account of the land which was given them by God. Some calendars have an undefined beginning, as the times of the festivals among the Jews, and the calendars of the Arabs, and in this they differ much. Certain calendars are solar, and of these certain have always had a fourth besides the full days, like the calendar of the Greeks and Latins; certain have never had this, like the calendars of the Persians; certain have it at one time and not at another, as the calendars of the Egyptians. And the beginning of the year varies among them, as the canons of astronomy teach as well as the Almagest, and there are many other variations. Since, therefore, there are contained in Scripture years lunar and solar as well as those of the Greeks and Latins and the like, and we wish to reduce all the calendars to solar years and to the years of the Latins, which are the years of Christ, it is necessary for us in sacred history to know the differences of these calendars, and also to know what is peculiar to each and how they may be equalized, and how we can draw the greater from the less and the converse, and extract any information desired from any one of them. But this cannot be done, as Scripture requires, except by means of canons, tables, and other astronomical means. But years are computed from the beginning of the world, according to the original Hebrew and according to the translation of the Seventy; but they disagree internally, and all authors of history and chronicles and sacred writers in this matter contradict one another in turn, not only in the matter of the whole time from the beginning to Christ, but also in regard to the particular ages. But this diversity cannot be cleared away except through a sure source. No science can in this matter discover nor has it the ability to ponder on the means of settling such an important question except astronomy, whose function it is to consider the definite and determined revolutions of eclipses, conjunctions of planets, and other celestial revolutions, the order of nature remaining fixed. Wherefore it is necessary that astronomy should here bestow diligence. For Ptolemy in the Almagest certifies us in regard to the time from Nabugodonosor to Alexander and from him to Octavianus Augustus, and from him to Prince Adiran. The Church holds that in the forty-first year of Augustus the Lord was born. Wherefore by this means there is certified much time from Nabugodonosor to Christ. If we consider the views of Albumazar in his book of Conjunctions, we shall see that he fixes the beginning of the world and the first man, namely, Adam, and from him numbers the years to the flood, and fixes the day and hour in which the flood began, and by the revolutions of the planets and by their conjunctions determines the following ages, and the dates of Nabugodonosor, Alexander, the Lord Christ, Mahomet, and of many others. Therefore by the skill of Albumazar and Ptolemy and other astronomers it behooves us to account for the times back to the beginning of the world, so that we know how many years there are from the beginning to Christ, not only as a total, but as regards the separate ages and periods.

In the second place, we can argue for the utility of astronomy because of the determination of the beginning of time, whether it is to be reckoned from the October lunation, the autumnal equinox, or the vernal equinox. For many have maintained according to the general opinion that the world was created about the time of the vernal equinox; but others have held that it was near the autumnal equinox; because according to the Hebrew text, the year, as far as it concerns the natural sequence of time, begins about the autumnal equinox. This fact can be clearly proved by the text of Exodus, where it is said that the Feast of Tabernacles is celebrated at the end of the year, that is in the month after the close of the year. For the twenty-third chapter reads thus, "The feast also in the end of the year when thou hast gathered in all the crops from the field." Therefore the feast begins after the first of the new year. The thirty-fourth chapter of the same book says, "Thou shalt make a feast when at the return of the time of the year all things are stored away." And in the gloss of Jerome on the first chapter of Ezechiel it is stated that October is the first month of the year and January is the fourth. And in the first chapter of Nehemiah, "and it came to pass in the month Caslen, that is December, in the twentieth year"; and in the second chapter, "It came to pass in the month of Nisan in the twentieth year." For if Nisan, that is, April, were the beginning of the year, it would not be stated in the twentieth year, but in the twenty-first year, as is apparent. In like manner we may reason from the commands in regard to sowing and reaping the fields, owing to the rest of the seventh year. For if Nisan, that is April, were the beginning of the year, then the sowing that is made in September or October of the sixth year will not be able to be harvested in the seventh year, since it is holiday according to law, and so the crop will perish. Likewise, since the whole seventh year is holiday, if the year began from April to September, they would not have anything to eat, because the crops of the sixth year will suffice only for the seventh year and for the eighth and for the seed of the eighth, as Josephus states, and not for the ninth year. Likewise by Jerome in his letter concerning Feasts that same matter is made clear. For he speaks thus, "At the end of the solar year among the Jews in the seventh month, when the fruits are gathered into barns and storehouses, then were they directed to keep the feast by law, on the first day of trumpets, on the tenth day of atonement must the sabbath be kept. And from the fifteenth day until the octave of the tabernacles be finished, they are enjoined to keep holiday."

But it is now made clear by Bede in his book on Chronologies, and especially by Josephus and all the Hebrews from the beginning on, so far as concerns the commencement of the festivals, that Moses began the year with April on account of the Passover, which is the first festival, and on account of the mystery of the new chronology, namely, the Christian, the years of which are computed from the incarnation, which was about the equinox, and at which time glorious Easter is now observed. But in selling and buying and in the rest of his government, so far as concerns the beginning of the year, Moses kept the decrees of the prior age, as Josephus states in the first book of the Antiquities. Now the prior age extended from Moses back to the beginning of the world. Therefore since the beginning of the first year of the world and the beginning of the world were the same, those men conclude that the world began about the time of the autumnal equinox, so that after the spoliation of the fruits of the old year the cultivation of the new might begin. Now this dispute is a very serious one, and for this reason wise men have recourse to the science whose function is to give definite information in regard to chronology, I mean the science of astronomy. Those also who hold the first opinion wish to protect themselves by astronomy, saying that the world was created under a better arrangement owing to the generating of man and of things, and for this reason the planets must have been in their more favorable position with respect to their control of the world. Wherefore they maintain that the sun was created in the middle of the world, as it is on the equator, so that it occupied an equal position with respect to the whole world. And they placed the sun in Aries, not in Libra, because the astronomers say that there the sun has its exaltation, which is its major dignity or second after the major. For the sun has five dignities and forces, namely, exaltation, house, triplication, terminus, face. And again they maintain that since planets must have eccentric orbits, it was necessary for the world, that the aux* of the sun should have been in Aries, because the position of aux is far nobler than any other part of the eccentric orbit. For when the sun and moon and other planets are in the position of their aux, they possess a stronger and better control in this world, as astronomers decide and as experience teaches. From all these considerations it follows, according to these men, that the world began at the vernal equinox. As to the objection made to them, that all growing things of the earth were created in the maturity of their fruits, which does not occur in the course of nature, they say that this was accomplished by the power of the Creator and not by the action of nature, although they add that in many southern regions is the heat of spring more favorable for maturing fruits than that of summer. And not only in spring, but again in autumn do they bear fruits, because of the temperateness of both equinoxes. Others, however, who take more things into consideration, namely, the reasons given above concerning the beginning of the world, reasons convincing and incontestable according to the view of those who hold the opinion quoted above, try to prove false the opin-

* Position farthest from center of motion.

ions of their opponents by means of astronomy. For they assert that the ancient astronomers fix the commencement of the year about the beginning of October, as is clear in the explanation of the tables that are called Almanac. And they say that Ptolemy discovered with certainty the aux of the sun in Gemini in his time. But if at the beginning it had been in Aries, as at the exaltation of the sun, namely, in the fifteenth degree or in the nineteenth, in which the exaltation of the sun is placed, the aux would move according to the motion of the planets, namely, according to the signs opposite to the motion of the first heaven. But this cannot be, because in that case the habitable regions above which the aux is would become uninhabitable in the succession of time, when the point opposite to the aux arrived above them: for which reason they maintain that the aux moves according to the motion of the starry heaven, not according to any imagined motion whatever, but by the motion of the Indians and of Thebit, namely, by the descent and ascent of its poles, or by the motion of the heads of Aries and Libra of the starry heaven in a small circle around the fixed heads of Aries and Libra, which are in the ninth heaven. By this motion the heads of Cancer and Capricorn are moved, advancing and receding on the surface of the ecliptic of the immovable Zodiac now to the east, now to the west, as is apparent according to the idea of Thebit, who to the works of Ptolemy adds in this particular the opinions of the Indians. For thus motion is assigned to the eighth sphere. In this view Arzachel agrees in his tables and canons, also Albumazar in his book of Conjunctions, and all astronomers now hold this opinion. But they thus maintain that the aux of the sun moves forward and backward and so does not depart from the sign of Gemini, for which reason it does not move in a circuit about the earth, so that the point opposite to the aux should at some time touch the habitable lands.

But those who maintain in the way mentioned that the sun was created in Aries, say that paradise is under the equinoctial, owing to its maximum temperateness, because both are the noblest places of the world, as they maintain. But if the aux were there at the beginning of the world, then was the point opposite to the aux in the position of Libra, wherefore at the autumnal equinox of the same year the sun was above the same places. Therefore at that time was the temperateness removed, and the excellence of the climate of paradise and of the whole equinoctial region was destroyed yearly, which is contrary to the nature of those places in view of their position. For they were thus quickly rendered uninhabitable because of the point opposite to the aux, because at that time the sun is near the earth and burns it. But they themselves maintain that the more temperate part of the world is beneath the equinoctial, and particularly in the case of paradise, which cannot be made to agree with what has been stated. And therefore those who thus object maintain that the sun was in Libra, so that punishment might follow immediately on the sin of Adam, in order that, having been driven from paradise, he might not be able to attain to it because of the consuming heat of the region beyond the equinoctial. And for the production of this combustion, immediately after the ejection of Adam from paradise, continuously this side of the equinoctial, it was necessary that the sun should be in Libra, because there it would have been placed toward the point opposed to the aux beyond the middle of the eccentric circle, so that combustion will begin at once, which would not have happened in Aries. Many astronomical considerations are here present. For this reason the theologian in this particular should be well acquainted with the principles of astronomy.

In the third place, the theologian needs the viewpoints of astronomy now touched upon because of the longevity of Adam and his sons even after the days of Noah, of whom Scripture speaks. For it is strange how the age of man has declined, when in the beginning it was of such great length and spanned so much time. Therefore many impute it to the excellence of the celestial constellation at the beginning of the world, and about this period. They investigate this excellence by means of the dignities of the planets, maintaining, as I have said, that the sun was at the equinox and in Aries and at aux, and that the other planets were in good positions and in favorable aspects, so that human nature might be fortified by the excellence of the constellation at the beginning, to the end that it might be able to live so long. But gradually, owing to the failure of the aforesaid excellence because of the changes of the position of the planets and the stars, the life of man was shortened, so that it arrived at a limit beyond which it does not succeed in passing; because of all things consistent with nature there is a reasonable explanation and an end, as Aristotle maintains, which limit Scripture fixes around eighty years in vigor, but beyond this is labor and pain. But we have already touched upon the way in which other astronomers object to these arguments. The dispute in regard to this matter is an important one; and perchance it will be discovered that the decline in the age of man is not due to a receding from the favorable disposition of the world at the beginning, but for other prescribed reasons, as we shall set forth below in its proper place.

And in the fourth place, it is necessary that the beginning of the year should be settled because of the deluge. For, as Josephus says, the deluge was in November. For these are his words, "Moreover this event happened in the six hundredth year of Noah's life, in the second month, which is called Dios by the Macedonians, but Maresvan by the Hebrews." But Dios, as Bede says in his book on Chronologies, is November, not May, as the rank and file of theologians consider it, and therefore Maresvan is November, just as Bede likewise states. And it is the second month, because October is the first in the natural order of time, as has already been proved. Hence, Bede says, For Tisseri, which is October, precedes Maresvan, because of the gathering of the crops and because of very celebrated feasts, and the Hebrews call this Tisseri the new year. For thus in Egypt they arranged the year so that Tisseri, that is, October, should begin it. And for this reason the Master of Histories and certain commentators did not understand Josephus when they believed that Dios and Maresvan were May. For this reason all those following them have been deceived because of the Greek and Hebrew names of the months of the Greeks and Hebrews, which they did not understand, as is clear to one examining the view of those who prove that the beginning of the year and of the world was about the time of the autumnal equinox of the year. For this reason the same perplexity occurs here as before. Hence the deluge must have happened in November, in accordance with the proof given above that time naturally began with October. Josephus shows this clearly to every man acquainted with the Greek and Hebrew names of the months of the year.

Not only concerning the beginning of the world and the year is there doubt among the theologians, but also concerning the beginning of the natural day, whether in fact night preceded the artificial day or the converse. This matter must be introduced here as the fifth point in regard to the substance of time. Many say that the day preceded the night, and expound Scripture, as they are able. But according to Jerome's comment on John and on Matthew, night preceded day. For, as Alfraganus says in his Astronomy, "All nations that use lunar months begin the day at sunset." But the Hebrews and the Scripture use lunar months and years, as can be proved in many ways. Therefore the Hebrews and Scripture use the natural day, the night of which precedes the day. Therefore the astronomical tables of the Hebrews, which they used in settling questions of chronology, were arranged with reference to sunset at the city of Jerusalem, just as the tables of the Latin astronomers have been arranged with reference to noon at the city of Toledo or of another city. Wherefore it is settled in the law that the day begins at evening. For in the twenty-third chapter of Leviticus it is said, "From evening to evening shall ye celebrate your sabbath."

The fourth head of mathematics in its relation to theology deals with the accidents and occurrences belonging to chronology, of which kind are the first and other phases of the moon, intercalations, and the like. For the text and the expositions of the doctors require a vast knowledge of these matters, and especially so respecting questions in regard to the Hebrews, pertaining to astronomy as well as to usual topics. But this division of the subject differs from the one treated above, because the former deals with the substance of chronology, but the latter with its properties and accidents. Therefore the basic truths regarding occurrences of this kind must be considered, before our discourse takes into consideration the Scriptures, because otherwise our argument would be unintelligible. I say, therefore, that the beginning of lunation according to astronomers is not reckoned from the moment when the new moon is visible among the Hebrews, as some theologians have said, since this time is not equal, as Alfraganus shows. But one lunation is made equal to another. For sometimes the horns of the old moon in its wane are seen in the morning and on the same day in the evening the horns of the new moon are visible, and sometimes these phenomena are separated by a space of three days, as observation shows and Alfraganus states. Therefore the Hebrews in ancient times determined by means of astronomy the beginning of lunation, and since this did not occur at the moment when the new moon became visible, nor could it be known by sight, they lit torches on a high mountain at Jerusalem, that it might be known that at that moment the lunation began, so that men might be ready to perform the solemn acts and keep the festivals which they had to observe. The lunation is not reckoned with respect to the true conjunction of the sun and moon, since this time is not equal, as the astronomers know. But the moon must be reckoned with respect to the mean conjunction of the sun and moon, as Alfraganus says. For this period of time is always equal. It is not, however, said to be the first of the moon at the conjunction, but after the conjunction, when the moon is so far separated from the sun that it is as regards itself visible, although it cannot be seen. For then occurs the first illumination of the moon, although the moon is not seen at that hour. This diversity happens owing to the latitude of the moon differing from the Zodiac, and in accordance with the fact that it is in the signs of an oblique or straight descent, and according to the difference in the regions in the north and in the south, as Alfraganus teaches in the twenty-fifth chapter of his book. Now the period of the moon runs from the first to the twenty-ninth and a little more. From these periods is reckoned the time of mean or equal lunation, which the astronomers of the Hebrews and of the Arabs call a lunar month, although it may be called a lunar month in many other ways. Although the most skillful astronomers in their tables and canons maintain that the time of equal lunation is twenty-nine days and thirty-one minutes of a day and fifty seconds, as Arzachelem shows in his Toledo tables; yet the Jewish astronomers have given the subject more delicate and accurate consideration. For the time mentioned contains twenty-nine days and twelve hours and forty-four minutes of one hour, as the algoristic work will explain. But the Hebrews divide one hour into 1080 parts, and every minute of the hour contains eighteen parts of an hour, as is clear from the reduction of fractions of one kind to fractions of another. Therefore the time of equal lunation among the Hebrews, corresponding exactly to the lunation of the Arabs, cannot be more than twenty-nine days and twelve hours and 792 parts of an hour. But the Arabs in their tables and canons make too small a computation, and reckon each lunation less than it should be by three seconds and fifteen thirds and forty-four fourths, as is evident on exact examination. For this reason the Hebrew astronomers, wishing to complete the lunation, have added one part, because they could not add less according to this scale which they use. Therefore they reckon even up to the present day in one lunation twenty-nine days and twelve hours and 793 parts of an hour. Their reckoning is far surer than that of the astronomers using the tables and canons current among other nations, although their computation is slightly in excess of the exact length of a lunation. For they exceed it by four thirds and sixteen fourths of an hour. But this is far less than the error of the Arabs mentioned above. Wherefore the computation of the Hebrews is considerably better. We need not take into consideration the excess mentioned above in the Hebrew computation; since in a very long period of time the error amounts to very little and is negligible. The months, moreover, of the Hebrews and those occurring in Scripture are not uniform. For one is of thirty days and another of twenty-nine, because people in general cannot compute except in entire days.

Having considered these matters, therefore, because they enable us to understand the Scriptures and the sacred writers, we are met with the further consideration that we should know that the Jews use a lunar cycle, a consideration of which is necessary in expounding Scripture. For just as we use a cycle of nineteen years, so do they use a lunar cycle, the first year of which begins in our fourth year. Therefore those men speak falsely who have not maintained that the Jews use a cycle; for they have intercalations in their cycle, just as we have in ours; nay, we have taken ours from them; and by means of intercalations they make the lunar years equal to the solar years by collecting eleven days thrice, in order that they may make in the third year an intercalated month, that is, one in excess. And we have taken this method of intercalations from them. Therefore famous men make a wrong statement who say that the Jews did not use a cycle. The Jews collect thirteen lunar cycles and make a table and canon in accordance with it. These thirteen cycles contain 247 years, because in so long a period of time all the observances of their legal festivals recur at the same moment of time. Therefore legal observance in many ways proceeds in accordance with this, and many other matters besides. For new moons and firsts of the month, on which falls the observance of sacrifices and of solemn feasts, of which it is said in the first book, twentieth chapter, of Kings, "Tomorrow will be the first of the month, and thy seat will be required," make it necessary that we should know that the lunar month in general begins at sunset. But the lunation itself has no fixed beginning. Wherefore if the new moon should happen at sunset or before it at some hour of the preceding natural day, the new moon will be reckoned as occurring on the evening following, likewise the first of the month and the new month, because it is already new moon. If, however, it occurs after sunset, as, for example, at the second hour of the day or later, neither new moon nor first of the month, so far as its commencement is concerned, will be reckoned on that natural day. We must consider, however, that the first month lasts from sunset of the first day until sunset of the thirtieth day, and yet the lunation does not last except from the beginning of the night to the morning of the thirtieth day, counting in whole days, although some fractions remain over. The second month, therefore, does not begin before sunset of the thirtieth day, but its lunation begins in the morning of the thirtieth day, and therefore the two first days are assigned to the second month, on which occurred the feasts and sacrifices, namely, on the artificial day of the thirtieth day of the first month and on the thirty-first natural day, because those two days are from the lunation of the second month, although the second of these alone is part of the second month. Wherefore in the first book, twentieth chapter, of Kings it is said that the seat of David on the second day after the first of the month appeared vacant. Whence it happens that even months have always two days of feasts, but odd months have only one. From these facts the statement of Ecclesiasticus becomes clear, "From the moon is the sign of the feast day; the month is called after her name." From these facts it could be determined whether it was new moon at the beginning of the world or full moon, as many have stated. For the Jews and Scripture use lunar months. Therefore the beginning of the first month and the beginning of the world were identical. But the lunar month begins with the new moon. Wherefore at the beginning of the world there was new moon.

What has now been said is necessary in considering Noah's exit from the ark. For the glosses involve us in grave doubt, wherefore the Master of the Histories was deceived, when he maintained that Noah's exit occurred on the twenty-eighth moon on the same day as regards the month on which he entered. For if this were true, then he was in the ark not only for a year, but for a year and a day, because a solar year consists of 365, which are completed from the first day of January to the last day of December, which is the end of the year. This is Bede's statement in his work on Chronology, "Noah with his family entered the ark on the seventeenth day of the second month and on the twenty-seventh day of the same month after the deluge is stated to have gone forth from the Ark. It is clear, therefore, that a whole solar year, that is, 365 days, was described, because doubtless the moon, which in the present year, let us say, appears on the Nones of May as the seventeenth, will occur the year following on the day before the Nones as the twenty-seventh." Such is Bede's statement. Because whatever moon it is, if you add eleven you will have the total on the same day, a year later. For example, if today is the first, the same day a year later will be the twelfth. Now this is true as a rule. However, sometimes its revolution is only the eleventh, sometimes the thirteenth. Concerning the eleventh, for example, if the first takes place today, the day before the Nones of April, on the same day a year hence it will be the eleventh. Concerning the thirteenth, let us suppose that the first occurs today, the fifth before the Nones of May, on the same day a year hence it will be the thirteenth. Let any one count it up and he will find it so for himself. This whole perplexity of the Master arises from a gloss of Strabus. In order, therefore, that we may reconcile the gloss of Strabus with Bede, we say that Strabus' statement in regard to the same day ought to be understood as the same day of the week; for example, if Noah entered on Sunday, he came out on Sunday. It follows that if eleven be added to the day present in his thought, that day being counted with the eleven, he is correct in his statement, and this is proved by what he appends, therefore after a year the eleven being added, it was the twenty-seventh day, or twenty-seventh moon. For if the twenty-seventh day itself were excluded, it would be proved to have been not the twenty-seventh, as he himself maintains in his gloss, but the twenty-eighth. And this is made clear as follows. For since he entered the ark on the seventeenth moon of the second month, it is evident that the year following by the sixteenth moon of the second month the lunar year of 354 days was completed: moreover, from the seventeenth to the twenty-seventh are eleven days, and if these are added to the aforesaid 354, it will become a solar year of 365 days, just as Bede computed above. Thus, therefore, he came out on the same day of the week on which he had entered, but not on the same day of the month. And thus both spoke the truth. But the Master, however, did not give the correct explanation of the statement of Strabus. That, in fact, the seventeenth moon should be included within the eleven is made clear by another gloss, which by the addition of ten exclusive of the seventeenth counts up twenty-seven. Therefore if Strabus reckons up with the addition of eleven only twenty-seven, and the other man by the addition of ten likewise reckons up twenty-seven, it is evident that the former understood it inclusively, the latter exclusively. When, however, the Master wishes to excuse himself on this point and gives the reason why he said that the exit was on the twenty-eighth, whereas Bede says it was on the twenty-seventh, the former's statement being, "for it was possible that he came out of the ark in the evening of the twenty-seventh moon, with the twenty-eighth close at hand, but intermediate times are frequently expressed by any designation you please of their limits";—an explanation that amounts to nothing, because although a true or mean conjunction of the sun and moon may take place every hour of the day as well as of the night, yet the Jews and Arabs compute according to lunar months, and the moon belongs to the night, just as the sun belongs to the day, therefore days and months all begin in the evening. Bede makes these statements concerning chronology, "At whatever hour the moon is illuminated, before evening comes, it will not be called the first." If, however, it is illuminated after sunset, it will be reckoned not the first on the preceding evening, but the thirtieth.

The Master's statement in his comment on Exodus regarding the arrival of the sons of Israel in the wilderness of Sinai before receiving the law of God is also false. His statement is, "In expounding Scripture we take thirty days as a month." For if every month is reckoned as thirty days, since there are twelve months in a year, there will be 360 days in a lunar year. But it is evident that this number is six days in excess. For it is agreed that in the lunar year there are only 354, since the solar year exceeds the lunar by eleven days, having 365 days. Therefore in two months there are only fifty-nine days. This is what Bede says in his book on Chronology, "I am disturbed by some anxiety in regard to the manner in which our ancestors compute the day on which the law was given, which is the third of the third month, the fiftieth from the slaying of the lamb. They maintain that there were seventeen days remaining of the first month, the thirteen preceding ones having passed before the passover, and an additional thirty in the second month together with three in the third, the total of which is fifty days. And let us rest assured that two lunar months contain not sixty but fifty-nine days; but that owing to this fact in the whole of the time mentioned there are found to be not more than forty-nine days." Such is Bede's statement. Therefore according to Bede the first day of June was not the forty-seventh but the forty-sixth, and thus if the law was given on the fiftieth day from the Passover, it was not given on the third day of the month of June, but on the fourth. But the Jews say that on the sixth day was the law given, and on the fourth day of their arrival in the wilderness of Sinai. For the Lord said, "Go to the people, and sanctify them today and tomorrow, and let them be prepared for the third day." And thus that third day was the sixth of their arrival; and according to this on the fiftieth day was the law given, not beginning from the Passover, nor from the first day of the feast of unleavened bread, but from the second day of this feast. This computation agrees with that made in the twenty-third chapter of Leviticus, "Ye shall count from the second day of the Sabbath, that is of the feast of unleavened bread, in which ye brought the sheaves of the first fruit seven full weeks to the second day of the completion of the seventh week, that is, fifty days." Here, therefore, great perplexity is found among readers due to the error of the Master in the length of the month and in his own computation, as is clear to one giving this matter more careful consideration and from the fact that the reckoning of the Hebrews is unknown in these times. A table on lunar cycles harmonizes greatly the observances of the law. Since in accordance with that table it is necessary that, although the beginning of the year is, as a matter of fact, the lunation of October, yet due to the requirements of the law respecting feasts, they have a triple year; one the common year with 354 days, and one a lessened year with 353 days, and one a year in excess with 355 days; and thus the beginning of the year varies accidentally. This is necessary, because the year cannot begin on a Sunday. Since if it did begin on a Sunday, then on the fifteenth day of the seventh month it would be Sunday, and on the vigil of that feast branches of trees are collected, which is unlawful on the Sabbath. In like manner the beginning of the year cannot be either on Wednesday or on Friday. For if it began on Wednesday, then the tenth day of the month would fall on Friday, on which it is not lawful to do anything, since they are not permitted to prepare food, as it is equal to the Sabbath. Wherefore in that event they must needs prepare food on Thursday for Friday and for the Sabbath, which would be burdensome and tiresome, particularly in a warm region and in the hot season such as prevails in the land of the Hebrews. In the second place, if any one should die on Thursday, he would not be buried until Sunday, which would not be endurable in that country. If, however, the year began on Friday, then the tenth day would be Sunday, and the same inconvenience just mentioned would ensue, because there is no difference whether the tenth day precedes the Sabbath or follows it. Wherefore one must know thoroughly the table and the other matters connected with it if he wishes to have an understanding of the law.

But matters of greater importance are here in doubt, unless we are aided to the fullest extent by the power of astronomy. For questions arise contrary to the views of all Latin theologians. But owing to the importance of the matters I shall proceed in opposition to both sides, and let him rejoice in the solution who is able to give it. I shall not, however, make any statement in opposition to the general opinions of the Latins except that which I know not how to answer. May I find some one to solve the difficulty, if the conclusion is false. But if, however, it should prove to be the true one, then a solution would not be necessary. But no one is of sufficiently great authority in the Church, except the Pope, to venture to give an opinion in opposition to the received opinions of the Latins in this matter, although they were false. The majority of Latins then hold that the Lord was born in the second year of a nineteen-year cycle and in the tenth year of a solar cycle, and in these matters there is no doubt, and that he suffered on the eighth day before the Calends of April [March 25] and that the moon was the fifteenth on the day of his Passion, matters concerning which there is a great dispute, and the Latins contradict the Greeks, who have maintained that the Lord suffered when the moon was at the fourteenth day. All the masters state this, and Augustine, Jerome, and Bede give authorities for this view. Opposed to it there is a strong argument. For if he suffered on the eighth day before the Calends of April and the moon was the fifteenth, this cannot be a fact, as Bede writes in his book on chronology, unless it was the thirteenth year of a nineteen-year cycle; and this is true. Because in accordance with this fact the golden number must have been thirteen, so that the moon may be named the first in the Calendar, where the thirteen are written, in order that from that place the age of the moon may be computed, so that it may be found to be the fifteenth on the eighth day before the Calends of April; as any one can find in the calendar. But since the Lord was born in the second year of a cycle, then at the end of that first cycle, he had lived eighteen years, according to the cycle, to which if we add from the second cycle the thirteen which passed up to his Passion, there will be thirty-one years according to cycles. But these years according to cycle are only twenty-nine years of his life, and of the thirtieth year the space of time from the nativity to the Passion; since he was born near the end of the second year of the cycle, that is, only seven days before the beginning of the third year, because the year begins at the Circumcision of the Lord. Wherefore at the end of the first cycle Christ had lived only seventeen years and seven days of his life, to which if there be added thirteen of the second cycle, there will be thirty; so that his Passion was in his thirtieth year. Therefore the Lord lived only twenty-nine years, and so much of the thirtieth as remained up to his Passion; which is contrary to the Gospel of Luke, who states, John the Baptist baptized Jesus "beginning to be about thirty years of age." And again according to the word of the Evangelists it is certain that he preached for several years.

Thus Bede tries to prove that in three years and a half past the thirty, that is, in his thirty-fourth year according to him, he suffered, namely, in the eighteenth year of Tiberius Caesar. And another stronger objection follows; for if he suffered on the eighth day before the Calends of April and the moon was the fifteenth, then, as stated above, the golden number was thirteen; but the Latins in general follow the blessed Dionysius, the Roman abbot, who was the first to reckon years from Christ, while formerly they were reckoned from the time of the sacrilegious Diocletian, as Bede writes, and this is a fact; so that at the completion of the five hundred and thirty-first year from the Incarnation or Nativity he began to form his great cycle, which contains 532 years, resulting from the multiplication of the nineteen-year cycle and the solar one. He begins his cycle from the five hundred and thirty-second year after the Incarnation, and not from the five hundred and thirty-third, because the Lord was born in the second year of a cycle, and therefore of necessity the new cycle began from the five hundred and thirty-second year, that is, at the completion of the five hundred and thirty-first year from the Nativity. Bede shows that it was so much. For he says, "Dionysius writes paschal cycles, beginning with the five hundred and thirty-second year of the Incarnation of our Lord." If, therefore, we find the letter in the table, and revolve from this the cycle of Dionysius twice and as much beyond this as there is up to the present year from the Nativity of our Lord, we shall find that the eighth day before the Calends in the Passion is on Sunday, as any one can find out. But it is clear that he did not suffer on a Sunday, but on a Friday. Wherefore many men diligent in chronologies, but particularly Bede, Marianus Scotus, and Geraldus, famous among them all, leave us a very great doubt as regards the computation following the cycles of Dionysius. For Bede says to the computer, "Return thanks to God, if in this way you shall find the year of our Lord's Passion. Wherefore if you do not discover it, ascribe your failure to the negligence of the chronographers or to your own stupidity." Men oppose him in two particulars; first, because he maintained that the Lord was born in the second year of a cycle; and, secondly, because he maintained that exactly the five hundred and thirty-first year had passed since the Incarnation or Nativity of the Lord.

Marianus, therefore, in his Chronicles concedes that it can happen and is a fact that the fifteenth moon may fall on the eighth day before the Calends of April, and the year be the thirteenth in the years of the cycle. But in that event it will be either the twelfth year of the life of Christ, or in the two hundred and fifty-ninth year; but he suffered neither in the twelfth year of his life nor in the two hundred and fifty-ninth year. Wherefore it does not seem that Dionysius can be proved correct. Therefore Marianus, carefully keeping in mind the failure of Dionysius, yet wishing to safeguard the view of Augustine and Jerome, says, because he suffered on the eighth day before the Calends of April, and likewise because the moon was the fifteenth, but not in the second year of a nineteen-year cycle, but in the eighteenth, two years of Christ must be taken from that cycle, so that the eighteenth year of the cycle is Christ's first year and the nineteenth is his second year, and then another whole cycle, which added together make the twenty-first year. Then thirteen from the third cycle should be added and the result will be thirty-four according to cycles. But the years of Christ's life are fewer than this, namely, thirty-two years and three months, and agree well with this number. Therefore Marianus added twenty-two years to the time of Dionysius. Therefore in our computation from Christ, if we wish to follow Marianus, we shall always add twenty-two years more than the Church reckons according to Dionysius. But according to Bede and according to the certain computation of the years of the emperors, it is evident that Marianus is in excess. For according to him from the birth of Christ to the cycle of Dionysius are 553 years; because to 531 he adds 22. But it was shown above that from the Incarnation of Christ to Dionysius there were not so many years, but only 531. For however liberally the years of the emperors are reckoned, we shall find that he is in excess 22 years, or 21, or at least 20, and therefore is not to be imitated. Marianus is shown also by the blessed Augustine to be in excess. For Augustine says in the eighth book on the City of God, "Thus are reckoned 365 years from the passion of Christ to the consulship of Honorius and Euticianus." Therefore that year, namely, the consulship of Honorius and Euticianus, is the three hundred and ninety-seventh from the Nativity. This number is found by adding the years from the Nativity to the Passion, namely, thirty-two years according to Marianus and three months according to the computation above; but according to the computation of Marianus, as he himself has maintained in his Chronicles, it was the four hundred and nineteenth year. It is manifest then that he is in excess 22 years. For if you take 22 years from 419, there remain 397.

Moreover, the famous Gerlandus, whom all computers and astronomers follow in handling matters relating to computation, seeing that Dionysius cannot be proved correct nor his followers, being unwilling also to adhere to Augustine and Jerome as regards the day of the month on which the Passion occurred, but adhering to the blessed Theophilus, a contemporary of the Apostle, in his statement that the Lord suffered on the tenth day before the Calends, took from the computation of Dionysius seven years, and maintained that the Lord suffered in the ninth year of a cycle. For while Dionysius maintained that the Lord was born in the second year and took one year of the cycle from the Nativity, Gerlandus takes away that one and two to eight inclusive. Therefore the Lord was, according to him, eleven years of age at the completion of that cycle, to which if there are added one whole cycle and five years, because five is the golden number so that it may be the fifteenth moon at that point, there will be thirty-five years according to cycles, and these agree well with the years of our Lord's life. But just as Marianus was in excess, so is Gerlandus deficient. For by the computation submitted above according to Bede, and according to the years of the emperors, from the Nativity of the Lord there are only 531 or about that number, so that at the least Gerlandus is deficient five years, however liberally we compute the years of the emperors.

The Master in the Histories adds to the doubts of the others a new question of doubt. For when he says at the end of his chapter on the Lord's Supper, "If we revolve the table of reckoning, we shall find the moon to be the twenty-second on the Calends of April and Friday at the time of the Passion; therefore the eighth day before the Calends of April was the fifteenth moon and a Friday," he is in manifold error. For if his opinion were the correct one, the golden number would be thirteen, and in that case, just as we have shown above, he would have suffered before the completion of thirty years, which is false. Likewise if we revolve the table, as the Master says, we shall find G to be the dominical letter at the time of the Passion and G to be on the eighth day before the Calends. Wherefore he suffered on Sunday, which is not true and contrary to the Gospel. But many, as Bede writes, and in particular Victorius, as is evident in his letter to Pope Hilary in regard to the paschal observance, say that Christ suffered on the seventh day before the Calends of April and rose on the fifth. But then, as Bede says, it must have been the second year of a nineteen-year cycle, because the golden number will be binary; on the supposition that the Lord suffered on the day of the fifteenth moon, as has always been assumed in the former calculations. But because he maintained that he suffered in the fifteenth year of Tiberius, as is manifest from his works, he is rightly refuted by Bede and by others.

These matters then have been discussed in accordance with the point of view of the writers on chronology, all the conclusions of whom are based on the assumption that it was the fifteenth moon on the day of the Passion. But astronomers anxious in this matter are not able to find the fifteenth moon nor the eighth day before the Calends at the time of the Passion, nor the tenth nor the seventh, so that at the same time it may be found to be a Friday, as it must be according to the Gospels, nor also from the thirtieth year of the Lord to the end of his life can they find it on any day of the month. Besides, there is the gravest doubt in this matter. For I have discussed these matters diligently both independently and in accordance with the consensus of those expert in astronomy. But with respect to the fourteenth moon it is readily found. Wherefore many things according to Scripture are to me and to many incontestable in this matter to the contrary, by which it is shown that he suffered on the day of the fourteenth moon, just as the Greeks maintain and the Jews agree. For it is said in Matthew "Not on the feast day." But the feast day is the day of unleavened bread, which was at hand, and the day of unleavened bread is the fifteenth. Therefore he was slain before it. Likewise John in the eighteenth chapter, "They did not enter into the praetorium, that they might not be defiled but might eat the Passover." Therefore they ate the Passover on the next evening; if then Passover is here understood as the paschal lamb, on that evening began the fifteenth day, and the moon was the fifteenth, and then was so reckoned. Therefore before that evening was the fourteenth moon. Wherefore the Lord was slain on the fourteenth day of the moon, like the paschal lamb in the law. The statement, moreover, that Passover is not to be understood in this passage in the sense mentioned but otherwise, can have no authority from Scripture, and therefore is rejected with the same ease according to Jerome as that with which it is proved. But when they say that passover is to be taken here as unleavened bread, it is impossible. For unclean persons, although forbidden to eat the Passover, that is, the paschal lamb, were not, however, forbidden to eat unleavened bread, if after eating the lamb they became unclean. Nay, if any one should eat leaven, the law declares, Exodus XII, that he should perish from the assembly of Israel. Moreover, neither was leaven found in their houses during those days, wherefore they would not in the case supposed eat bread for seven days, which is wholly absurd. Wherefore this interpretation has no weight. Likewise John XIX, "It was moreover the preparation of the Passover." Therefore on the same day at evening they prepared the Passover. But when they prepared the Passover, the fifteenth moon was beginning. Therefore since he suffered before that day, it was the fourteenth. Likewise in the same chapter John says, "There laid they Jesus therefore because of the Jews' preparation day; for the sepulchre was nigh at hand." For on this account did they hasten to bury him lest they should bury on the fifteenth day, for they would not have buried him on the day of unleavened bread, because they buried no one on noted feast days like the Passover, Pentecost, Feast of Tabernacles, and the like. So too Luke XXIII, "And women returning," namely, on the day of the Crucifixion, "prepared spices and rested the Sabbath day according to the commandment." Therefore that day was not the day of unleavened bread, but was the fourteenth; for it was unlawful for them to prepare spices on the day of unleavened bread. For in Exodus XII concerning the first and the last day of unleavened bread, it is said, "No work shall ye do in them save that which every man must eat." For as they would have rested on the Sabbath day because of the commandments, so would they have rested on Friday if it had been the day of unleavened bread. For the injunction refers to both, although the Sabbath is the more holy. Augustine in his book on Questions of the New and the Old Testament says that he suffered on the day of the fourteenth moon. These and many other statements can be here adduced together with the exclusion of false arguments.

But let these statements now suffice to arouse us to two efforts, namely, that we should know on what day of the moon the Lord suffered, whether the fourteenth or the fifteenth, and if one of these be settled upon, then that the day of his Passion should be discovered by means of astronomical tables, as has been indicated. But all these matters have very great difficulty, yet more so for this reason, because theologians are ignorant of astronomy and computation and the like, rather than owing to the difficulty of the thing in itself. For if they were expert in these subjects, they would easily discover with certainty the age of the moon and the day of the Passion, and they would change many opinions which they hold. For the most skillful in the consideration of those matters hold that the Lord suffered on the day of the fourteenth moon. This being verified, it is easy to find the Calends by special tables compiled for this purpose. To the intent that our mind may be aroused to do this, I shall here place a table in which, according to the tables of new moons, is found the opposition of the sun and moon through all the years of our Lord up to 38, so that we may see in what year it happens in March near the Passion of our Lord and in what year in April. Not, however, because of a certification of this matter am I giving this table, but as an example, in order that the method of argument in this particular may be made plain; for a certification is very difficult, for this reason, because the motions of the heavens have not been completely ascertained, nor do any tables whatever suffice

	Anni Domini.	Literae sextarum feriarum.	Dies proximus ante oppositionem.	Dies oppositionis [et ideo passionis signum, et similiter in alia tabula].	Dies proximus post oppositionem.	Dies sequens proximum.	Dies.	Horae.	Minuta.	Secunda.	Tertia.	Quarta.
	1	g	b	c	d	e	27	14	31	2	0	0
	2	f	e	f	g	a	16	23	19	41	8	48
	3	e	b	c	d	e	6	8	8	20	17	36
	4	c	f	g	a	b	24	5	41	2	42	8
	5	b	b	c	d	e	13	14	29	41	50	56
	6	a	e	f	g	a	2	23	18	20	59	44
	7	g	c	d	e	f	21	20	51	3	24	16
Principium cycli alia via, sc. rationis.	8	e	f	g	a	b	10	5	39	42	33	4
	9	d	c	d	e	f	28	3	12	24	57	36
	10	c	g	a	b	c	18	12	1	4	6	4
	11	b	c	d	e	f	7	20	49	43	15	12
	12	g	g	a	b	c	25	18	22	25	39	44
	13	f	c	d	e	f	14	27	11	4	48	32
	14	e	g	a	b	c	4	11	59	43	57	20
	15	d	e	f	g	a	23	9	32	26	21	52
	16	b	g	a	b	c	11	18	21	5	30	40
	17	a	d	e	f	g	1	3	9	44	39	28
	18	g	b	c	d	e	20	0	42	27	4	0
	19	f	e	f	g	a	9	9	31	6	12	48
Principium cycli secundum computistas.	20	d	b	c	d	e	27	7	3	48	37	20
	21	c	e	f	g	a	16	15	52	27	46	8
	22	b	b	c	d	e	6	0	41	6	54	56
	23	a	f	g	a	b	24	22	13	49	19	28
	24	f	a	b	c	d	12	7	2	28	28	16
	25	e	e	f	g	a	2	15	51	7	37	4
	26	d	b	c	d	e	20	13	23	50	1	36
	27	c	f	g	a	b	20	22	12	29	10	24
	28	a	c	d	e	f	10	19	45	11	34	56
	29	g	g	a	b	c	28	4	33	50	43	44
	30	f	c	d	e	f	18	13	22	29	52	32
	31	e	a	b	c	d	7	10	55	12	17	4
	32	c	c	d	e	f	26	19	43	51	25	52
Hic.	33	b	g	a	b	c	14	4	32	30	34	40
	34	a	e	f	g	a	4	20	5	12	59	12
	35	g	a	b	c	d	23	10	53	52	8	0
	36	e		d	e	f	0	19	42	31	16	48
	37	d	a	b	c	d	19	17	15	13	41	20
	38	c	e	f	g	a	9	2	3	52	54	8

Dies. Horae. Minuta. Secunda. Tertia. Quarta.: Post quot dies de Martio et horas et minuta fuit media oppositio, quantum ad meridiem Novariae.

Anni Domini.	Dies proximi ante oppositionem [mediam].	Dies oppositionis mediae.	Dies proximi post diem oppositionis mediae.	Secunda dies post diem oppositionis.							
1	d	e	f	g	26	2	55	5	15	44	April.
2	g	a	b	c	15	12	3	44	24	32	April.
3	c	d	e	f	4	20	52	23	33	20	April.
4	g	a	b	c	22	18	25	5	57	52	April.
5	d	e	f	g	12	3	13	45	5	40	April.
6	g	a	b	c	1	12	2	24	15	28	April.
7	e	f	g	a	20	9	35	6	40	0	April.
8	g	a	b	c	8	18	23	45	48	48	
9	d	e	f	g	26	15	56	28	13	20	
10	b	c	d	e	17	0	45	7	21	48	
11	e	f	g	a	6	9	33	46	30	56	
12	b	c	d	e	24	7	6	28	55	28	
13	e	f	g	a	13	15	55	8	4	16	
14	b	c	d	e	3	0	43	47	13	4	
15	f	g	a	b	21	22	16	29	37	36	
16	b	c	d	e	10	7	5	8	46	24	April.
17	e	f	g	a	30	15	53	47	55	12	Mart.
18	c	d	e	f	18	13	26	30	19	44	April.
19	f	g	a	b	7	22	15	9	28	32	
20	c	d	e	f	25	19	47	51	53	4	
21	g	a	b	c	15	4	36	31	1	52	
22	c	d	e	f	4	13	25	10	10	40	
23	a	b	c	d	23	10	57	52	35	12	
24	b	c	d		10	19	46	31	44	0	
25	g	a	b	c	1	4	35	10	52	48	
26	d	e	f	g	19	2	7	53	17	20	
27	a	b	c	d	9	10	56	32	26	8	
28	e	f	g	a	27	8	29	14	50	40	
29	a	b	c	d	16	17	17	53	59	28	
30	e	f	g	a	6	2	6	33	8	16	
31	b	c	d	e	34	23	39	15	32	48	
32	e	f	g	a	13	8	27	54	41	36	
33	a	b	c	d	2	17	16	33	50	24	
34	f	g	a	b	21	14	49	16	14	56	
35	b	c	d	e	10	23	37	55	23	44	April.
36	e	f	g	a	30	8	26	34	32	32	Mart.
37	c	d	e	f	18	5	59	16	57	4	April.
38	f	g	a	b	7	14	47	56	5	52	

Post quot dies de Aprili et quot horas et minuta et tertia fuit media oppositio, quantum ad meridiem Novariae; tamen in duobus annis, 17. sc. et 36. ponitur post quot de Martio; et secundum hanc computationem incipit dies in medio diei praecedentis, et terminatur in medio sui.

in this case. For many skilled in astronomy have labored in this matter, that they might find these oppositions of the sun and moon, and have not been able to find the year of the Passion from the thirtieth to the thirty-fifth, when there would be in March an opposition on a Friday, nor a day before the opposition nor a day next after it, so that it would agree with the Passion. Nor have I been able to discover it up to the present time. Where, however, an opposition together with a Friday can be found according to the present table, will be disclosed by an explanation of it.

Explanation of the Table.

The first line, then, occurring in the table below contains the years of our Lord to the thirty-eighth; because it is certain that he suffered within those years. The second line has all the letters of the sixth week day, which happened in those thirty-eight years. The third line contains the days next before the opposition of the sun and moon. The fourth, the day of the opposition. The fifth, the day immediately after the opposition. The sixth, the day following, and here is found the mean opposition of the sun and moon. The seventh line, with those joined to it, contains the time passed in March before the day of the Passion. The rest of the table deals with April, so that the opposition may be found in April around the day of the Passion, with the exception that it occurs in the seventeenth and thirty-sixth years in March. We must bear in mind, however, that these tables were made for the meridian of the city of Novaria, although constructed at Paris; but there was a reason, for Novaria is more out of the way, and noon there precedes the noon of Paris by twenty-five minutes of one hour. If, therefore, from the given time we subtract twenty-five minutes, there will remain the time of the opposition after noon at the city of Paris. According then to this table the Lord suffered three days before the Nones of April on Friday at the mean opposition of the sun and moon in the fifteenth year of a nineteen-year cycle, and in the fourteenth of a solar cycle, in the thirty-third from the Incarnation according to the cycle of Dionysius, and that is, in the thirty-second year, according to the true age of the Lord. This in the second table occurs beyond the letter *b* in the line of the thirty-third year after two days of April and seventeen hours and sixteen minutes and thirty-three seconds, fifty thirds, twenty-four fourths, and so the most expert in these calculations have reckoned, who have labored hard to prove this. Whence according to them the opinion of the Church that he suffered on the eighth day before the Calends of April is not verified by facts, but is the popular view, as is the case in many other matters which lack stronger verification. If, therefore, mean opposition and the fourteenth moon from its illumination occur together at the Passion, the matter is clear according to this table and according to the reckoning of learned men. If, however, the opposition precedes at the Passion the fourteenth moon by a whole day, it would be necessary to have recourse to a table of the illumination of the new moon constructed similarly to this one, and then doubt would be the more removed. But authorization of this view, as well as other matters of which I write, requires the assent of your Holiness, that those expert in such matters may firmly establish the truth. I have shown in this chapter how we can attest this matter, but I do not claim in the present treatise to have solved so great a difficulty myself.

It has been said that there are seven heads under which we must employ mathematics. One is in regard to celestial matters, another concerning places in the world, the third concerning chronology as regards its substance, the fourth concerning the qualities and accidents of chronology, of which mention has been made. Now I wish to present the fifth head, which concerns geometrical forms as regard lines, angles, and figures both of solids and surfaces. For it is impossible for the spiritual sense to be known without a knowledge of the literal sense. But the literal sense cannot be known, unless a man knows the significations of the terms and the properties of the things signified. For in them there is the profundity of the literal sense, and from them is drawn the depth of spiritual meanings by means of fitting adaptations and similitudes, just as the sacred writers teach, and as is evident from the nature of Scripture, and thus have all the sages of antiquity handled the Scripture. Since, therefore, artificial works, like the ark of Noah, and the temple of Solomon and of Ezechiel and of Esdra and other things of this kind almost without number are placed in Scripture, it is not possible for the literal sense to be known, unless a man have these works depicted to his sense, but more so when they are pictured in their physical forms; and thus have the sacred writers and sages of old employed pictures and various figures, that the literal truth might be evident to the eye, and as a consequence the spiritual truth also. For in Aaron's vestments were described the world and the great deeds of the fathers. I have seen Aaron thus drawn with his vestments. But no one would be able to plan and arrange a representation of bodies of this kind, unless he were well acquainted with the books of the Elements of Euclid and Theodosius and Milleius and of other geometricians. For owing to ignorance of these authors on the part of theologians they are deceived not only in matters of the greatest importance, but also in those of very little consequence. For they state along with the Master in the Histories that the small balls of the candlestick were circular bodies not possessing full sphericity. The use of the diminutive in speaking of them is not for this reason, but because they were small balls, yet satisfying completely the definition of the sphere in Euclid and Theodosius. For the Jews in their Gallic speech call those small balls *pumeus* from the roundness and sphericity of the *pumel,* and so too concerning other things without number. Oh, how the ineffable beauty of the divine wisdom would shine and infinite benefit would overflow, if these matters relating to geometry, which are contained in Scripture, should be placed before our eyes in their physical forms! For thus the evil of the world would be destroyed by a deluge of grace, and we should be lifted on high with Noah and his sons and all animate creatures collected in their places and orders. And with the army of the Lord in the desert we should keep watch around the Tabernacle of God, and the table of the shewbread, and the altar, and the cherubim overshadowing the mercy seat, and we should see as though present all the other symbols of that ancient people. Then after the unstable tabernacle swaying to and fro had been removed, we should enter the firm temple built by the wisdom of Solomon. And with Ezekiel in the spirit of exultation we should sensibly behold what he perceived only spiritually, so that at length after the restoration of the New Jerusalem we should enter a larger house decorated with a fuller glory. Surely the mere vision perceptible to our senses would be beautiful, but more beautiful since we should see in our presence the form of our truth, but most beautiful since aroused by the visible instruments we should rejoice in contemplating the spiritual and literal meaning of Scripture, because of our knowledge that all things are now complete in the Church of God, which the bodies themselves sensible to our eyes would exhibit. Therefore I count nothing more fitting for a man diligent in the study of God's wisdom than the exhibition of geometrical forms of this kind before his eyes. Oh, that the Lord may command that these things be done! There are three or four men who would be equal to the task, but they are the most expert of the Latins; and rightly must they be expert, since unspeakable difficulty lurks here, owing to the obscurity of the sacred text and the contradictions of the sacred writers and differences of the other expounders.

But in another manner the usefulness of geometry is disclosed as regards sacred knowledge, both in the text and in the questions: and not only in these, but in beautiful comparisons respecting grace and glory and future punishment, and precaution against vices. Concerning each I shall give an example. And for the sake of all things in general let us recall to mind that nothing can be known concerning the things of this world without the power of geometry, as has already been proved. Also a knowledge of things is necessary in Scripture on account of the literal and spiritual sense, as has been set forth above. For without doubt the whole truth of things in the world lies in the literal sense, as has been said, and especially of things relating to geometry, because we can understand nothing fully unless its form is presented before our eyes, and therefore in the Scripture of God the whole knowledge of things to be defined by geometrical forms is contained and far better than mere philosophy could express it. Nor is it strange, since God himself, the author of all wisdom, has ordained his own Scripture. Wherefore in place of an infinite number of examples I desire at present to cite one. For Aristotle more than all other philosophical writers has involved us in obscurities in dealing with the rainbow, so that we secure no valuable information through him, nay, many false statements are contained in the translation of the Latins, as we maintain from the differences of the expounders. For the statement made in the editions of the Latins that the rainbow occurs only twice in fifty years in the rays of the moon, is manifestly false, as any one can discover in full moonlight when it rains provided the brightness of the moon is not lessened by the density of the clouds. Avicenna, moreover, the greatest authority in philosophy since Aristotle, as all insist, humbly confessed that he himself was ignorant of the nature of the rainbow. Thus it is certain regarding all philosophers that no one of them has been able to gain a knowledge of the rainbow. Nor is it strange, since they have not examined Scripture with the necessary diligence. For all philosophers have been ignorant of the final cause of the rainbow. But the end places a necessity on those things concerned with the end, as Aristotle says in the second book of the Physics, and it is a fact in all matters. Now the end for which the rainbow exists the text of God alone explains clearly, namely, when it is said, "I shall set my bow in the clouds of heaven," etc. From which the fact is established that God's bow was provided against a deluge and an overflow of the waters. Therefore of necessity, whenever this bow appears in the heavens, there must be an active consumption of the aqueous moistures; and this is a fact. For the clouds are freely resolved, and dew falls are produced without number, as the philosophers say, and we see in great part. But the consumption of the aqueous moisture does not happen except by means of something possessing the power to consume. But we find nothing in the production of the rainbow except the rays of the sun and the clouds. The collecting of the clouds is the material cause; therefore the projection of the rays is the efficient cause. But the incident rays cannot effect great and wonderful results, because they do not mutually converge; a convergence, moreover, of forces is required that a vigorous action may result. But convergence can occur only through reflection and refraction. Wherefore the rainbow must be produced by infinite reflections or refractions in numberless drops of water falling without an interval, so that the truth is thus discovered regarding both its colors and form by means of multiplications of this kind in respect to figures, angles, and lines, and not by means of a diversity in the matter forming the cloud, as is stated in the text of the Latins and as all believe, even as I shall explain by certain experiments when I mention the subject of experimental sciences. Just as philosophers, then, owing to their ignorance of the sacred Scriptures have been unable to know the truth regarding the rainbow, so in the same way it is impossible for the unbelieving philosopher to reach full certainty in regard to any creature owing to ignorance of the Scripture. For in truth each creature is entered there according to its ultimate dignity, namely, according to its true definition and description, because God made the creatures he has placed in the Scriptures, and he himself alone knows them as they are. Since, therefore, the power of geometry is required for the knowledge of every corporeal creature, there is no doubt but that in an inexpressible manner it is effective for sacred knowledge because of its aid in understanding things.

But returning to the proposition considered in a spiritual sense I offer an example from Scripture, which says, "The sun in a threefold manner burns up the mountains," etc. For infinite rays fall upon every point of the mountain because infinite rays come forth from every point of the sun, and light is the cause of heat, especially when there is convergence, as we sensibly know, and an infinite number of rays is reflected from the surface of the mountain, because reflection takes place from a dense body, and the rays converge at every point in the air, and draw apart at every point of the air, rarefying the air near mountains, and thus, secondly, do the mountains grow hot. Through the medium of the clouds the rays are doubly refracted, first at the surface of the clouds toward the sun, then in the air between the clouds and the mountains. By this double refraction all rays of the sun coming from one point must converge to a point on the mountain or in the air near the mountains, particularly in very lofty mountains and in the Alps.* Thus heat is produced on mountains, although not on the loftiest mountains; since mountains rising to the clouds or nearly so, like the mountains of Italy and Spain and the Caucasian and Caspian range and innumerable other mountains, are very cold and have almost perpetual snows; because they approach too near to the mid-space of the air, which is the coldest place in this world. Thus from the separate points of the sun rays proceeding and doubly refracted increase and augment the heating of the mountains. All these things are made clear by means of mathematical forms, as is the case in the round beryl, or crystal, or water container, or any transparent body thicker

* *I.e.,* lower mountain pastures. See Bridges' note on this passage.

than air. For when these bodies meet the solar rays, owing to the double refraction a point in the air sensible to sight and touch can be selected, at which dry tow or some other very combustible substance may be set on fire, as has been stated before. This triple heating can be called a threefold combustion, of which the Scripture speaks; or otherwise we can say that it causes a single combustion of the mountains from the three sources of which Scripture speaks when it states, "The sun in three ways consumes the mountains." For in another way combustion is considered with regard to the different kinds of angles, because the multiplication of light is either at right angles, or oblique, or at no angles. If at oblique angles, there is vigorous action; if at right angles, the action is stronger by the eighteenth proposition of the first book of Euclid's Elements and by other methods; if at no angles, there is the strongest action because of the equality in every way, as has been stated above. But from every point of the sun to every point of the mountain exposed to the sun the rays come, and they fall in infinite numbers at right angles, and in infinite numbers at oblique angles, as is clear. Although, moreover, the surface of the mountains is not everywhere polished and smooth, nor spherical, yet in many parts of it is it of such a character, and infinitely so throughout different small portions and therefore by taking into account sphericity of this kind an incidence of rays takes place at no angles, and for this reason there is a triple incidence. Wherefore there takes place a triple falling of rays as regards the angle and the absence of angle. In all ways is there heat produced by the incidence of light and its convergence. Wherefore the second manner of combustion of which the Scripture speaks can be explained in this way now mentioned. In the third place, the falling of light varies as regards its figures, for it falls in a spherical form and in a pyramidal one, of which the apex is on the sun and the base is on the mountain, and in a pyramid of which the apex is at a point on the mountain and the base on the sun. In all these three ways there is formed a multiplication of light from the sun on the mountain, as is certain from what has been said above, although the spherical multiplication of light can be strong, yet the pyramidal multiplication, of which the apex is in the sun, is stronger; and the pyramidal one, of which the apex is at a point of the mountain, is strongest, because, as has been said, by that pyramid from the whole surface of the sun opposite to the mountain light comes to every point of the mountain, since the base of that pyramid is the surface of the sun, and an infinite number of such pyramids come from the same surface of the sun, as is evident to one drawing a figure. Nature does not require more figures in the multiplication of light, or in the multiplication of any force or species, in order that by any agent soever there may be formed a species to be multiplied, as he knows for a certainty who is acquainted with the paths of nature through the potency of geometry. By these three multiplications pictured there arises the third combustion of which the Scripture speaks. But these matters cannot be known, unless a man has an excellent knowledge of the potency of geometry.

Owing to the questions arising I offer one example instead of a thousand, where the power of geometry is required, although the rank and file do not perceive it. For concerning light and its multiplication theologians ask many questions, namely, whether it is a substance or an accident, whether it is a body in a medium, whether suddenly or successively it travels in a medium, etc. But these matters in no way can be known without what has been said concerning multiplication according to lines, angles, and figures. For the multiplication of light is just like the multiplication of every other species of any agent whatsoever. For light in a medium is the species of the light that is in the luminous body. Therefore the rules that were stated concerning the multiplication of species taken generally are understood in the case of light and in the case of any determined species whatever. Since, therefore, the multiplication of species taken absolutely requires the greatest geometrical power, as has been abundantly shown previously, it is manifest that the same power of geometry is necessary to understand the multiplication of light, although theologians do not employ it.

Then regarding the expression of spiritual matters by means of geometrical truths, I offer an example in grace and in glory, both in the case of those to be saved and those to be damned, in order that we may see how lines straight, broken, and reflex, may be adapted to spiritual matters of this kind. Since the infusion of grace is very clearly illustrated through the multiplication of light, it is in every way expedient that through the corporeal multiplication of light there should be manifested to us the properties of grace in the good, and the rejection of it in the wicked. For in the perfectly good the infusion of grace is compared to light incident directly and perpendicularly, since they do not reflect from them grace nor do they refract it from the straight course which extends along the road of perfection in life. But the infusion of grace in imperfect, though good men is compared to refracted light; for owing to their imperfections grace does not continue in them an altogether straight course. But sinners, who are in mortal sin, reflect and repel from them the grace of God, and therefore grace with them is compared to light driven away or reflected. But as of bodies from which light is reflected, some are rough, from which the reflected light is dissipated and does not appear; and others are polished and of an equal and smooth surface, from which there is produced a sensible reflection, as in the case of mirrors; so sinners living in mortal sin are of two kinds. For there are some who so repel grace that nothing good appears in them, but respecting them and those nearest it is manifest that the whole effect of grace is dissipated in them, and these men are the openly evil who do not hide their sins. There are others, however, who, although they are not good, yet hide their sins, and appear good in the sight of men, as do the hypocrites. This comparison can be made also in another way. For sinners in this life are compared to reflecting bodies, as has been said, and men in grace in this life are compared to those bodies in which there is a refraction of light because of the imperfections of human frailty, because however perfect a man may be in this life, yet has he many obliquities, and imperfect in him are the love and knowledge of God. But those living in the kingdom are compared to those things that receive the light within in a straight course without deviation, as far as is possible in a creature, and as far as the order of divine justice exacts, which returns to every man according to his deserts. Since, however, the damned wholly lack the grace and glory of God, and the glorified are in a measure imperfect on the score of imperfection in merit in many things and in accordance with the law of creation (since every creature has imperfection with regard to the glory of its Maker), there can be a third comparison according to the status of the damned and of the glorified and of God, so that to the damned the comparison of reflected light is wholly applicable, and in regard to the glorified the simile of refracted light is appropriate, and the utmost rectitude and perfection of the divine glory are symbolized by the direct course of light, in which way a better proof is given of the trinity and unity of the persons of the Godhead. Such a comparison we make according to our ability and to the extent that we are permitted to set forth by means of creatures the attributes of the Creator.

I shall add what should be carefully noted, that it is impossible for the blessed trinity and unity of essence to be represented more aptly by us in an example of a creature perceptible by the senses than by means of matters pertaining to geometry. For in the triangle alone among all things made is there found a unity of essence with the distinction of three embracing the same essence. Since each of the three angles occupies the same identical space and the whole of it, as is evident to the sense, and yet they are in fact distinct angles, which is wonderful in a creature, nor is it found elsewhere except in the supreme Trinity. And when on a given line it is required to construct an equilateral triangle, as the first proposition of Euclid states, what can be assumed more properly in order that we may perceive that if the person of God the Father be granted, a trinity of equal persons presents itself?

These matters and many like them can be made available as salutary wisdom from the truths of geometry, and especially so from the tenth book, where the rationality and commensurability of quantities are set forth. For these subjects can be adapted in a useful and excellent way to sacred matters and to the figurative representations of the virtues with the exclusion of the vices. For whatever is contained in that book regarding rational quantities can be applied to a life which consists in virtue; and that which is explained regarding irrational quantities can be likened to the irrational and bestial morals of sinners. Whatever also is said concerning the commensurability of quantities can be fittingly applied to the acts of love; and what is set forth concerning incommensurability is known to relate to hate and the separation of souls. But these matters require a longer discussion. Moreover, in precaution against sins we are greatly aided by the consideration of this kind of geometrical multiplications. For since the delectable things of this world are to us like mousetraps, by which we are made captives to sin, as is written in the book of Wisdom; and like bait on the devil's hook, with which we would be strangled except in so far as God deigns to guard us, and our senses carried away in regard to the delectable things announce to our intellect an occasion of sin, the first and principal remedy existing in man is not to receive according to their principal multiplication into his five senses the species of delectable things, as of women and foods and riches, things in which human desire is at boiling heat and proud ambition glories; for the accidental multiplication suffices a man for damnation when it is received eagerly and copiously. But still more to be avoided is the straight sensible multiplication than the refracted or reflected because of the stronger action previously explained. But most of all must we avoid letting the direct delectable species fall at equal angles into our sense, because then is it strongest, especially if the vertex of the short pyramid meets it. So Eve received the species of the voice of the serpent and of the visible apple and its sweet odor; and Adam was tempted so that he damned himself and the whole human race because of the multiplication of species perceptible to the senses. So David the holy prophet deceived by the species of Bathsheba fell from adultery into murder. So the elders whom Daniel judged were deceived by the species of a woman. And it is certain that every man is deceived by the species of the things of this world, because, according to John, "all that is in the world is either the lust of the flesh or the lust of the eyes, or the pride of life," from the love of which no one can withhold himself, who delights in the sense of the species which come from things. For he is taken captive like a beast. Therefore men devoting themselves to holiness turn away their senses as far as possible from all species of delectable things, and especially from those pertaining to the sense of touch and taste, which are gluttony and excess, so that not only do they not touch, but do not see nor hear mention made of these things, lest the species multiplied into the senses should compel the spirit to serve carnal allurements; and particularly are they on their guard against proximity to them, so that they may avoid the shorter pyramids, and multiplications, principal, direct, and at equal angles, as I have said before. These are most of all to be avoided when it is necessary either in confession or for some other reason to talk with women. For all men, however holy, strong species in this particular would disturb, although such species could not overcome those whom God deigns to guard.

We have now spoken of five heads of mathematics as regards Scripture; two others remain for us, namely, numbers and matters relating to music. Since, however, sacred writers abundantly show the utility of these subjects, and express more fully comparisons and praises of these with respect to sacred matters than they do with regard to the heads already considered, I have for this reason touched upon the others more fully, and in regard to these two I shall merely hint at certain methods, the explanation of which is found in the books of the sacred writers given in full. Numbers, in fact, are valuable for the sense of Scripture in four ways. For number was not placed in Scripture with any other purpose than that it should be taken in a literal sense according to all its arithmetical properties, to the end that through fitting similitudes the spiritual meaning may be elicited, just as the sacred writers make clear in almost endless ways. In the second place, for a knowledge and a certifying of chronicles it is necessary for the theologian to have ample ability in computing, so that he may know all the algoristic methods, not only in whole numbers, but in fractions, in order that he may know how to compute, add, subtract, halve, multiply, divide, and extract roots, both fractions and whole numbers: and again that he may not only know vulgar fractions as regards halves, thirds, fourths, fifths, and so on to infinity; but that he may know fractions used in astronomy, namely, minutes, seconds, thirds, fourths, fifths, and so on to infinity, because in these historical matters we must have recourse to the motion of the sun and moon, in which such fractions are in the main considered, concerning which sufficient explanation has been given above. One must know not only the fractions of the Latins and the Arabs, but also of the Hebrews, who divide the hour into 1080 parts, of which mention has been made before. But the man who has to use these must know how to reduce different kinds of fractions in turn; because if between whole numbers there occur seven fifths, and ten sevenths, and twenty eighths, and the like in endless ways, a man cannot extricate himself, unless he reduces these different kinds of fractions to one kind of fraction, so that they may thus be reduced to whole numbers. But in using these fractions and whole numbers there is great subtlety, and the fair lights of wisdom shine out, especially since through these all knowledge sacred and human is directed and harmonized, in accordance with which two authorities, Cassiodorus and Isidore, have expressed themselves clearly, proving us guilty in a cruel fashion, because all ignorant of the potency of numbers, they assert, do not differ from brutes, which know neither sacred things nor human. In the third place, a consideration of numbers is valuable in Scripture, namely, in making division and distribution of the shares of booty and of portions to priests, to Levites, to princes, and to individuals, according to which Moses, most skillful in algoristic computations, distributed to each one what was his, distributing an almost countless number of things into parts most definitely determined by the nicest algoristic operations both in fractions and in whole numbers. No one can elicit the literal sense nor in consequence the spiritual, as the dignity of Scripture requires, unless he knows the reason for these operations and learns the methods in counting by which Moses and others in Scripture proceed.

In the fourth place, one must have an excellent knowledge of the methods of computation because of the corruption of numbers in the Scripture, since they are corrupted in almost endless ways; for almost all the numbers have been corrupted. This corruption cannot be reduced to truth except by means of the ability to compute in every way, both in fractions and in whole numbers. For nearly all numbers, or at least the major part of them, in the sacred text and in the books of the sacred writers have been perverted through the fault of scribes and amenders, of which I shall furnish a few examples to serve for many. For in the eleventh chapter of Genesis it is stated, "Arphaxad lived after he begat Salah three hundred and three years." But in the Hebrew the years are given as 403. From the Greek moreover the truth cannot be ascertained because between Arphaxad and Salah it places Cainan, whom we do not have, nor does the Hebrew. And in the same chapter, Reu lived thirty-two years according to the Hebrew and ancient codices of the Latins. But in the vulgar edition the years are given as thirty-five, so that the true account of the chronicle cannot be a continuous one. In the eighteenth chapter in the part near the end, "Peradventure there shall lack five of the fifty righteous; wilt thou destroy for the lack of five?" Thus read the Hebrew and the ancient books. Modern correctors have inserted forty-five for the sake of clearer sense. But it is not lawful without approval of the apostolic seat to change the text which it has received and ordered distributed through all the churches. And in the thirty-seventh chapter, "Joseph, when he was seventeen years old," according to the Hebrew and the Greek. But modern amenders have sixteen, and that in the same chapter he was not sold for thirty pieces of silver, but for twenty, has been proved above on ample testimony. And in the fourteenth chapter of Exodus, "and he took six hundred chariots," according to the Hebrew and the Greek and Josephus and the ancient codices. But part of the gloss mutilated in nearly all books causes the vulgar edition to read three hundred chariots. For in general this much only is given in the gloss, "in order to fight against the faith of the Trinity." Whence owing to the name of the Trinity they rubbed out the six hundred and inserted three hundred. But in many ancient glosses the statement is found as follows, "to fight against the faith of the Trinity he armed himself with the number six." Therefore according to the gloss six hundred can be read in this passage. In the thirty-second chapter of Exodus the Hebrew and the Greek read, "about three thousand." But the vulgar edition has twenty-three thousand. The Master in the Histories says that this is from an ancient translation. But the ancient translation is that of the Seventy, and is from the Greek, but the Greek does not read thus, nor does the Hebrew. This error has its origin in the tenth chapter of the first epistle to the Corinthians. But the statement there is not to be understood as referring to this passage, as is clear in the text and in the gloss. Since when he says, "Neither be ye idolaters," etc., he does not give the number of slain, because the words do not refer to this passage. But immediately when he says, "Neither let us commit fornication," etc., he continues, "and fell three and twenty thousand," which statement is understood according to the gloss in regard to those who committed fornication with the Midianites, as we read in the twenty-fifth chapter of Numbers. I do not wish to heap up more examples at present, since it does not accord with my main purpose to prove corruption of the text. But I have introduced these examples, so that the necessity of computation in the Scripture may be seen, to the end that we may know how to correct the errors in numbers.

And still a fifth reason can be noted for the usefulness of numbers. For unity multiplied into itself cubically, that is, thrice, as once one taken once, does not multiply essence, but remains the same although it is produced equally in three directions. And so by a familiar example theologians designate the blessed Trinity. Moreover, although a manifold perfection of number is found according to which ten is said to be perfect, and seven, and six, yet most of all does three claim for itself perfection; since an especial perfection that is assigned to other numbers is that all the aliquot parts taken together should give its total, as is evident in the number six. But in the number three alone does it happen that the aliquot part and the non-aliquot part taken together give the number itself, namely, unity and two, a fact which cannot be discovered in any other number. And since in every number there are two things, namely, discreetness of parts and oneness from their union, because of which the number is said to be one: nor does anything else make a number except these two things; these two are found exactly in the number three. For by means of the two is there discreetness, and by means of unity is unity itself obtained. Therefore complete perfection of number is found in the number three and not in any other. For this reason this number befits the Creator more than any other. And as every large number proceeds from unity, so from God himself. Arithmetic at its end teaches how all the proportions are investigated in the ratios of numbers. For neither the geometrician, nor the musician, nor the astrologer treats of these, but applies what arithmetic, which is the science of numbers, teaches. Theologians, moreover, strive to exalt these proportions to the properties of the divine persons, and in particular Richard of Saint Victor in his book on the Trinity, which offers great difficulty on account of the ignorance of these proportions on the part of the majority. Many for this same reason abuse the comparisons of these proportions to sacred things. Now a proportion is of three kinds, namely, arithmetic, geometric, musical. The arithmetic proportion consists in the identity of the differences of the first to the second, and of the second to the third, as four, three, two. For unity alone is the difference from different directions, and a similitude to this is found in the divine persons. For the difference of the Father respecting the Son is that the Father gives to the Son and does not receive from him; and the Son conversely. Similarly between the Son and the Holy Spirit there is a like difference. For the Son gives to the Holy Spirit and does not receive from him; the Holy Spirit conversely receives from him and does not give to him. And since there is a geometrical proportion, when there is the same ratio of the first to the second and of the second to the third, an application of this mean is found by means of the perfect equality and unity of essence in the three persons, and in all their essential qualities, which are power, wisdom, goodness. For we believe and firmly hold that as is the Father, such is the Son, and such the Holy Spirit. The harmonic or musical proportion consists in the identity of the ratio of the first to the third, and of the difference of the first and the second to the difference of the second and the third, as six, four, three. For there is a double ratio between the first and the last; likewise between two, which is the difference of the first and the second, and unity, which is the difference of the second and the third. And thus is there found a similitude in the divine persons. For the same relationship exists of Father to Holy Spirit, since there is equality, as there is between the difference of Father and Son, and the difference of Son and Holy Spirit. For between Father and Son is the difference that the Father gives and does not receive, the Son receives and does not give; and this same difference exists between the Son and the Holy Spirit, as is evident. Thus then briefly are these proportions assigned, although in other ways also are they assigned by many whom I cannot understand; because I do not think that they can be justified by the principles of arithmetic. But the present occasion is not a fitting one to explain the opinions of individuals.

The sixth reason for the utility of arithmetic can be found in Scripture regarding things of this world which it employs. For it touches upon the height and magnitude and thickness and number of the heavens and of the stars and requires certification in regard to them. The sacred writers in their expositions say many things regarding them. For Jerome says on Isaiah that Orion has twenty-two stars, of which the first nine are in the third magnitude, and nine of the remainder are in the fourth, and the last four are in the fifth magnitude, and does not explain himself. The theologian therefore ought not to be ignorant of these things. But especially in these matters does arithmetic give us correct notions, and I therefore wish to disclose these great truths by means of numbers, since the comparison of numbers appears to do so in an excellent and useful manner. Now for this consideration it is necessary to take some known base. This base is the length of the arc of the earth, which corresponds to one degree in the heavens, according to the teaching of Alfraganus in chapter VIII.

Averroës also agrees to this at the end of the second book of the Heavens and the World. These writers, in fact, give the method of verifying this in the number of miles and parts of a mile. Ptolemy, moreover, in the fifth section of the Almagest proceeds by means of an extended demonstration in regard to the altitudes of the sun and moon and in regard to their magnitude. But he does not give a definite quantity in number of miles; nor does he determine from other celestial bodies the quantity. It is necessary, therefore, to assume that a uniform and geometrical cubit contains a foot and a half, and that a mile contains 4000 cubits, and so Alfraganus takes them in his computation. Moreover, every circle of a sphere can be divided into 360 parts, which are called degrees. We are to understand, then, the largest circle in the celestial sphere, which passes through the center and divides the sphere into two equal parts. I say then that to one degree of this circle will correspond many miles on the earth, experimental knowledge of which Alfraganus hints at in the fact that if we take the elevation of the pole star above the horizon and walk to the north or to the south to a point where the pole star appears a degree more elevated or depressed, we discover the number of miles on the earth corresponding to one degree in the heavens. For if on a clear starry night one views through the apertures of a quadrant or an astrolabe or other instrument the star used by sailors, namely, the pole star itself, and notes the degrees that the end of the rod on the back of the astrolabe or the filament on the quadrant reaches and then proceeds on the earth to the north until on another starry night he sees the same pole star to be elevated one degree higher above the horizon, that arc of the earth which he has traversed will correspond to one degree, and will be similar to it, as similar arcs are taken on different spheres according to Theodosius, but they will not on this account be equal.

Since, moreover, he teaches in the twenty-first chapter that the diameters of the celestial orbits and the distances of the auges and of the points opposite are measured by the semidiameters of the earth, which contain 3250 miles, it behooves him to take the true length of the semidiameter. For otherwise a great error would occur in the distances of the auges from the fact that the length of the semidiameter of the earth repeated many times would make a great error in those distances unless the exact length was taken. Since, therefore, we must grant that Alfraganus perceived this error, it is manifest that he himself takes the true length of the semidiameter, namely, 3250, and the true diameter, namely, 6500. Therefore he must of necessity assume the true and complete basis of calculation, which is found in the length of the arc on the earth with respect to a degree of the heavens, although he does not express this perfectly. Wherefore he himself assumes that it is fifty-six miles, and two thirds of a mile, and 27 ninetieths, and one six-hundred-and-thirtieth, or fifty-six miles and 2984 cubits and five sevenths of one cubit. Therefore if one considers the matter carefully, Alfraganus in his eighth chapter in this basis of computation, which is the length of the arc on the earth with respect to a degree in the heavens, as regards the diameter and semidiameter used by him, is found to omit 50 six-hundred-and-thirtieths of one mile, or, what is the same thing, 317 cubits, and a third of a cubit, 28 six-hundred-thirds of one cubit, since he merely states there that this arc of the earth is fifty-six miles and two thirds of one mile. But although he assumes the full basis of computation, he yet omits some fractions, owing to the tediousness of numbers. For it is his custom in his books frequently to omit fractions, and other authors do likewise. If, therefore, we wish to adapt the basis of computation to this length of diameter, namely, 6500, used by him, we shall say that an arc of the earth corresponding to one degree in the heavens contains fifty-six miles, and two thirds of one mile, 27 ninetieths of a mile, and one six-hundred-and-thirtieth. And if we wish to compute by cubits the arc of the earth corresponding to one degree in the heavens will be fifty-six miles, and 2000 cubits, 984 cubits, and eight sixty-thirds of one cubit, which eight sixty-thirds are more than one-eighth of a cubit by one five-hundred-and-fourth of one cubit, as is easily shown. For eight sixty-thirds equal 64 five-hundred-fourths, and 63 five-hundred-fourths are an eighth of the whole: therefore, 64 five-hundred-fourths exceed an eighth by one five-hundred-fourth; and so eight sixty-thirds of one cubit exceed an eighth of it by one five-hundred-fourth. If we triple this diameter, 6500, and add a seventh of this, we shall have the circumference of the whole earth, and it will be exactly 20,428 miles, and 2285 cubits, and five sevenths of one cubit; or in other fractions, this number will be 20,428 miles, and four sevenths of one mile. According to this the whole surface of the earth will be 132,600,000 miles. According to this basis correctly computed a quarter of the earth will have 33,150,000 miles in its surface; an eighth will have 4,143,750 miles. For we are in need of these two quantities, as well as of the others mentioned before.

After looking into these matters we must consider the altitude of the heavenly bodies, and likewise their size and thickness. For Alfraganus says in the twenty-first chapter that Ptolemy and other scientists took the semidiameter as the length with which they measured distances from the center of the earth, and took the magnitude of the earth as the quantity with which they measured the magnitudes of the stars. This is quite clear from the demonstrations of Ptolemy in the fifth section of the Almagest. Therefore Alfraganus is of the opinion, from a comparison of the semidiameter of the earth with that of the starry sphere, that the distance of the starry sphere from the center of the earth is 20,110 times the semidiameter of the earth, 65,-357,500 miles, which if doubled will give the diameter of the whole starry sphere as 130,715,000 miles. When this is multiplied into three and one seventh we shall have the circumference of a great circle in the starry sphere, namely, 410,818,-571 miles, and three sevenths of a mile, that is, 1714 cubits, and two sevenths of a cubit. If we divide this number into 360 parts, one part, which will be the length of one degree of the starry sphere, will have 1,141,162 miles and 251 three-hundred-sixtieths of a mile, that is, 2788 cubits, and eight ninths of one cubit. And if we reckon the diameter into rotundity, the surface of the whole starry sphere will be 53,700,149,508,265,000, namely, fifty-three thousand thousand thousand thousand thousands of miles, seven hundred thousand thousand thousand thousands of miles, one hundred forty-nine thousand thousand thousands of miles, five hundred eight thousand thousands of miles, two hundred sixty-five thousands of miles. Moreover, the semidiameter of the starry heaven is the longer distance of the sphere of Saturn, because they join without intermediate space. But its nearer distance to the earth is 46,816,250 miles, which is the longer distance of the sphere of Jupiter, whose nearer distance is 28,847,000 miles, which is the longer distance of the sphere of Mars, whose nearer distance is 3,965,000 miles, which is the longer distance of the sphere of the sun, whose nearer distance is 3,640,000, which is the longer distance of the sphere of Venus, whose nearer distance is 542,750, which is the longer distance of Mercury, whose nearer distance is 208,541 and two thirds of a mile, and this is 2666 cubits and two thirds of a cubit, and this is the longer distance of the moon, and this, as Alfraganus says, is sixty-four times and a sixth of one time equal to the half of the diameter of the earth, and the nearer distance of the moon is 109,037 and a half of a mile, that is, 2000 cubits, and this distance is thirty-three times and a half of a tenth, that is, one twentieth of one time equal to half of the diameter of the earth. The diameters of the separate spheres are obtained by doubling the semidiameter; the circumference of each is found by tripling the diameter with the addition of a seventh part, and the whole surface of each sphere is found by multiplying its diameter into its circumference, as was explained in the case of the earth and of the starry sphere. Any one can find these dimensions by computation, and for this reason I omit them to avoid prolixity. Since after substracting the nearer from the longer distance there remains the thickness of the sphere, it is therefore evident that the thickness of the sphere of the moon is 99,504 miles, and the thickness of the sphere of Mercury 334,209, and Venus 3,097,250, and of the sun 325,000, and of Mars 24,882,000, and of Jupiter 17,-

969,250, and of Saturn 18,541,250. These dimensions are derived from the principles of Alfraganus and of Ptolemy in the Almagest.

We must note that in all these altitudes distance is taken from the center of the earth. Hence, although scientists sometimes speak of these as being distances from the earth, they understand them, however, as reckoned from the center of the earth, because half of the earth makes no sensible difference. Since, then, the nearer distance of the sphere of the moon is 109,037 miles and half a mile, if half of the diameter of the earth is substracted, that is 3250, from the number of whole miles stated above, omitting the half mile, the remainder of the distance of the sphere of the moon from the earth will be 105,787. Now let twenty miles be taken as a day's march. Then if the moon's distance is divided by twenty the number of day's marches is found to be 5289 with seven miles remainder. If these day's marches are divided by the 365 days of one year, the result will be fourteen years, with a remainder of 179 day's marches, which if divided by thirty, the number of days in a full month, the result is seven full months with a remainder of one incomplete month, that is, a month of twenty-nine days. It is evident then that by taking twenty miles as a day's march a man could pass through the space to the sphere of the moon in fourteen years and five full months and one incomplete month, and there would still remain seven miles and the half-mile which was omitted at the beginning of the calculation.

Concerning the ninth and the tenth heaven nothing can be known by means of instruments belonging to the senses in the matter of their altitude, thickness, and size; also concerning the thickness of the eighth heaven, as we are able to do in regard to the other heavens, since all these matters are hidden from our sense, and therefore in regard to them certification of their magnitudes, altitudes, and thicknesses is lacking. By means of the nearer distance of the moon is ascertained the longer distance of the sphere of fire, but the longer distance of the air is not known. For philosophers have neglected this because it is not a matter of notable utility. Yet they have verified the distance to the position of the clouds, since it is demonstrated in the book on Twilights that their altitude is fifty-one miles and two thirds. Pliny, however, in the second book of

Natural History, says that the clouds have an elevation of four hundred and nine hundred stadia according to the different statements of philosophers. Therefore at the least the air extends so far and more; although Albumazar and certain others estimate that vapors are not higher than two miles and a tenth and a third of a mile. Ptolemy, moreover, in his book on the Arrangement of the Sphere says the greatest elevation of vapors is ten stadia, and Martianus in his Astronomy agrees. For they assume that the highest mountain is ten stadia high, as Olympus, on whose summit is found neither wind nor vapor, which fact men prove by experience. But since Aristotle in his book on Meteorologics maintains that vapors reach to the boundary of air and fire, therefore we must accept the opinion at least of the book on Twilights, and other views must be explained by the difference in quantity between the stadium and the mile. For sometimes we find that the stadium is an eighth part of a mile, and at other times is far more, and likewise there is great diversity regarding the length of the mile; for it is known that the mountains of Italy and other ranges, like the Caucasus, on whose summit the rays of the sun are apparent up to a third part of the night, have an altitude of more than eight times 125 feet, and therefore are more than eight stadia high, according to which the stadium is said to be 125 feet. Elsewhere I found in an exposition of the book Ormesta Mundi that a stadium contains two Gallic miles and two parts of one Gallic mile, and thus one stadium contains five miles and a third of a mile. Hence through differences of this kind the difficulties mentioned above are solved. The maximum height of mountains is with greater certainty considered to be eight miles, as is shown in the book on Twilights, and therefore in Ptolemy's book there may be falsity in translation or in writing, since the greatest height is placed at eight stadia.

After we have the altitudes of the heavens, it is evident that we have the heights of the stars according to the views of Ptolemy and of all mathematicians. For the nearer distance of the moon is the distance of its sphere, and the longer distance likewise is an altitude, and so of the others. But concerning the magnitudes and thicknesses of the stars our knowledge is not gained in this way. He, therefore, who could find their diameters could easily find their circumferences and their total surface by the method I touched upon above in regard to the spheres. Certain ancient writers, as Martianus in his Astronomy, have striven to find the diameter of the sun and of the moon by the flow of water through an aperture in a vessel while the star is rising and while the whole sphere of the heavens revolves; and thus by means of those two quantities of water they found by how great an arc the diameter of the mass of the star is subtended.

Now these were water instruments like certain clocks, which are now made. For experimenters take a caldron with one orifice in the bottom, and fill it with water, and regulate it in such a way that at the exit of the first drop of water through the orifice the sun begins to rise, and at sunset they reckon the amount of water that has escaped, which they divide into twelve parts according to the number of the twelve hours, and therefore the flow of each part through the orifice mentioned marks a revolution of fifteen degrees, which makes one hour of the day. Therefore those wishing in this way to learn the size of the sun or moon allow the first drop of the diurnal water to escape when the first part of the sun rises, and when the last part has risen close the orifice. Therefore the ratio of the diameter of the sun, or of the arc of the heavens corresponding to this chord, to the total arc of the day is the same as the ratio of the water now escaped to the whole diurnal water. Others, however, by means of the astrolabe or the quadrant have reckoned how much the rule is moved while the extremities of the sun and moon are under observation, and then reckon that the ratio of the star to the circumference of the heavens is the same as that of the arc measured on the instrument to its whole circumference. But these methods do not possess sufficient certainty. For this reason Averroës says at the end of the second book of the Heavens and the World that the size of the diameters of stars is investigated by means of an instrument, which is mentioned, with two rules. For the former instrument determines the angle which the diameter of the star passes through, and the latter instrument consists of two rules extended in the manner of a pair of compasses for the space of four cubits, and at their ends are perforated feathers, so that the two extremities of the sun or moon may be seen. For by means of the angle between the two rules, when it is known, the diameter of the star is known, since it is the base of that angle. This instrument was discovered by Abrachis, as Ptolemy maintains in the fifth section of the Almagest in the fourteenth chapter. Ptolemy, however, rejected the first methods on account of their insufficiency, and found this method of Abrachis had a defect. For according to this instrument of Abrachis he estimated the diameter of the moon at every distance to be equal according to aspect to the diameter of the sun because of the equality of the angles. But Ptolemy reckoned that the diameter of the moon is not equal according to aspect to the diameter of the sun except when the moon is at its greatest distance. This happens when the moon is at the aux of the epicycle, and the epicycle at the aux of the eccentric circle, and this again is when it is full. For at that time the diameters of the sun and moon are subtended according to aspect by equal angles. But when the moon is at any other distance this is not the case, and therefore the size of the angle is not determined when the moon is in any position whatever. The reason for this is found in the fact that the table is raised from the rule and the eye is above the surface of the rule, and for this reason the eye is not directed straight to the hole in the table but somewhat obliquely, and in that case what is between the two holes should be perceived, and what is between the rules is actually perceived; for this reason is there doubt regarding the size of the angle. On this account Ptolemy wished to determine the magnitudes of angles by lunar eclipses, in which he makes clear his point. Therefore by means of the instrument and by eclipses the magnitude of the angle is determined which the diameter of the moon subtends when it is at its greatest distance and is full. Therefore by means of the magnitude of the angle is determined the diameter of the moon, and by tripling its diameter with the addition of a seventh part the size of the moon's circumference is determined, and by multiplying the diameter by the circumference the total surface of the moon is determined and its whole size.

Since half of the diameter of the earth is the basis of this calculation, as used by Ptolemy in the fifteenth chapter of the section mentioned before, therefore we perceive the size of the diameter of the moon by its comparison with the diameter of the earth, and the size of the moon by comparison with the size of the earth. Therefore Ptolemy, in the sixteenth chapter, concludes by means of his demonstration and drawing that according to a scale in which the diameter of the moon is one part, the diameter of the earth will be three parts and two fifths approximately. The diameter of the earth therefore will be three times the diameter of the moon and two fifths of it in length. Therefore if from the diameter of the earth, which is 6500 miles, we take the diameter of the moon, by computation the diameter of the moon will be exactly 1911 miles, and three quarters of one mile, and one seventeenth of one quarter; and if we triple the diameter with the addition of a seventh part we shall have its circumference, which will be 6006 miles, and six seventeenths of a quarter part of one mile: and by multiplying the diameter into the circumference we shall have the whole size of the moon, which will be 45,927,882 miles, and this will be about one of the thirty-nine parts of the whole earth, as Ptolemy shows. For according to the measurement by which the size of the moon is one part, will the size of the earth be about thirty-nine and a fourth times as large; since when the size of the diameter of the moon was taken as one part, the size of the diameter of the earth was three parts and two fifths. If, therefore, those quantities are multiplied into themselves cubically and as in solids, it is evident that one multiplied into itself cubically is only one, but three and two fifths multiplied into itself cubically make approximately thirty-nine and a quarter. Wherefore the size of the earth will be approximately thirty-nine and a quarter times as large as that of the moon. This is evident, because the quantity before mentioned respecting the size of the earth was found above as approximately thirty-nine and a fourth times as great as that of the moon, as is clear to one giving the matter consideration.

Moreover, the size of the diameter of the sun and of the sun itself depend, according to Ptolemy, on two things, namely, on the size of the diameter of the moon and on that of the semidiameter of the earth. For he finds the distance of the sun from the earth by means of the distance of the moon already known, and by the angles which the diameters of the moon and of the sun subtend in aspect when the moon is at its maximum distance at the time of full moon. For then the moon seems to cover completely the sun and no more, and for this reason the sun is then said to subtend in aspect the same angle as the moon subtends. Nor does he take into consideration a different distance of the sun, because there is no difference in aspect, owing to the remaining distance, whether the sun is at its nearer or longer distance. He finds by these calculations that according to the scale by which the diameter of the moon is one part, and the diameter of the earth is approximately three and two fifths parts the diameter of the sun will be eighteen parts and four fifths of one part. The diameter of the sun will then be eighteen times the diameter of the moon, and four fifths of it; and the diameter of the sun will be approximately five and a half times the diameter of the earth. If, therefore, by means of the known diameter of the earth, which is 6500, we find the diameter of the sun, it will be 35,941 miles, and 12 seventeenths of one fourth. If this number is tripled with the addition of a seventh part, the size of the circumference of the sun will be 119,803 miles and one third and 12 seventeenths of one quarter. If we multiply the diameter into the circumference the total surface of the sun is 292,783,785,375 miles. If the number representing the diameter of the sun taken with respect to the diameter of the moon be multiplied into itself cubically, this cube will be approximately six thousand six hundred and forty-four and one half times as large; and therefore according to the scale by which the size of the moon is one part, the sun's magnitude with respect to the moon will be approximately six thousand six hundred and forty-four and one half times as great. But if the diameter of the sun compared with the diameter of the earth be multiplied into itself cubically, the cube will be about one hundred and seventy times as great. Therefore the sun is about one hundred and seventy times as large as the earth. This statement of his accords pretty nearly with the explanation of Thebit in his book entitled, Concerning those matters in need of expounding before the Almagest is read. But there is not an exact accord. For he says that the sun is equal to the earth one hundred and sixty-six times. Alfraganus in the twenty-second chapter of his book adds to the quantity mentioned by Thebit a fourth and an eighth of the earth, which fractions I previously determined to make certain of this passage. The sun is therefore one hundred and sixty times with the additions of a fourth and an eighth of the earth larger than the earth. This is the exact size of the sun according to the views of authors.

Ptolemy, however, determined only the size of the sun and of the moon. But Alfraganus says in the twenty-second chapter that the method is similar in the case of other stars to that which Ptolemy employs in these two. For in accordance with his discovery that the diameter of the moon, when it is at its maximum distance and is full, is equal in aspect to the diameter of the sun, so likewise when other planets are at the mid-point of their distances, their diameters have a fixed ratio to the diameter of the sun. For the diameter of Mercury in aspect* is one fifteenth part of the diameter of the sun, and of Venus it is one tenth part of the diameter of the sun, of Mars it is the one twentieth, of Jupiter the one twelfth, of Saturn the one eighteenth part of the diameter of the sun; since the diameter of Mercury is one twenty-eighth part of the diameter of the earth, and the diameter of Venus is a third part of the diameter of the earth and the third of a part, and the diameter of Mars is one and one sixth times that of the earth, and the diameter of Jupiter is four and one half times and one sixteenth of one time as large as the diameter of the earth, and the diameter of Saturn is four and one half times as large as the diameter of the earth, which is 29,250 miles. From these facts it follows that the size of Mercury is one part of the twenty-two parts of the earth, and Venus is one thirty-ninth part of the earth according to Alfraganus, but according to Thebit in the book mentioned it is one part of the thirty-two parts of the earth, and Mars is one and one half times and the half of an eighth of one time as large as the earth, and Jupiter is ninety-four and Saturn ninety-one times as large as the earth. If any one wishes to find by means of the diameter of the earth the size of the diameter of any one of the planets in miles, and by means of the size of the diameter desires to find the circumference and by means of both to find the total surface in miles it is easy for him to do so in the ways mentioned above.

Moreover, concerning the 1022 fixed stars Alfraganus gives the same method. For so many of these are known to scientists, which extend from the north to the remotest point which was visible to them in the third climate. For by instruments they are able to determine their sizes. But of these 1022 there are

* "In aspect" is used for the apparent diameter of the planets in terms of the sun's apparent diameter. See Bridges' note on p. 235, Vol. I.

six groups and differences of magnitudes, according to the teaching of Ptolemy in the eighth chapter of the Almagest, and of Thebit in the book mentioned, and of other scientists. There are, then, six magnitudes, and the first is greater than the others, and so consequently is its order of position. In the first magnitude, then, are fifteen fixed stars, as the Dog Star and the falling Vulture, and the Heart of the Lion and the like, the diameter of each of which is in aspect one twentieth part of the diameter of the sun, and the diameter of each of them is equal to four and one half times the diameter of the earth. Therefore each of these is one hundred and seven times the size of the earth, according to Alfraganus. From the measurement of these stars the size of the others is determined. There are, then, in the second magnitude forty-five stars, for example, six stars of the Greater Bear and two stars of the Lesser Bear, and each one of these is equal to ninety times the earth. In the third magnitude there are 208, for example, some stars of the Lesser Bear, according to Ptolemy, and many others, and each of these is equal to seventy-two times the earth. In the fourth magnitude there are 474, for example, four stars of the Lesser Bear and many others, and each one of these is equal to fifty-four times the earth. In the fifth magnitude are 217, each of which is equal to thirty-six times the earth. In the sixth magnitude there are sixty-two, each of which is eighteen times larger than the whole earth. Then there are other stars in infinite number, the size of which cannot be ascertained by instruments, and yet they are known by sight, and therefore have sensible size with respect to the heavens, like the part with respect to the whole. But the earth does not possess any sensible size with respect to the heavens, as Ptolemy proves in the beginning of the Almagest, and this is a certain fact, since the earth has the same relation to the universe as the center to the circumference of the circle. Wherefore each of those stars known to sight is larger than the whole earth, and Alfraganus states this in the fourth chapter. From all these facts, then, that have been mentioned in regard to the magnitudes of the heavenly bodies it is evident that greater than all, with the exception of all spheres other than that of the earth, is the sun: then in the second place are stars of the first magnitude; in the third place, Jupiter; fourth, Saturn; fifth, all the remaining fixed stars according to their grades and orders; sixth, Mars; seventh, fixed stars known to sight; eighth, the earth; ninth, Venus; tenth, the moon; eleventh, Mercury.

According to the judgment of the sacred writers matters pertaining to music are necessary to theology in many ways. For although it is not necessary for an understanding of Scripture that the theologian should have a practical knowledge of singing and of instruments and of other musical things, yet he should know the theory of them, in order to grasp the natures and properties of these things and of the writings on this subject in accordance with their teachings of music theoretical and practical. For Scripture is full of musical terms like rejoice, shout for joy, sing, play upon the cithara, cymbals, and the like of different kinds. Moreover, Scripture contains in it many kinds of songs both in the New and in the Old Testaments. Likewise many kinds of meters are contained in the sacred Hebrew text, of which the sacred writers in their expositions take note in many ways. But it belongs to music to give the reasons and theories of these, although the grammarian teaches the practical rules regarding them.

Moreover, the whole pronunciation of Scripture consists in accents, longs, shorts, colons, commas, and periods; and all these belong causally to music, because of all these matters the musician states the reason, but the grammarian merely the fact. For so the authors of philosophy decide, also Augustine so decides regarding them in dealing with music. For one part of music deals with what is audible, the other with what is visible, as authorities decide. That which deals with the audible has two divisions, of which one concerns the human voice, the other instruments. That which concerns the human voice is fourfold. For one part concerns melody, as in singing; the second concerns meters, and considers the nature and properties of all songs, meters, and feet; the third concerns rhythm, and considers every variety of relations in rhythms; the fourth concerns prose and considers accents and other aforesaid things in prose discourse. For accent is a kind of singing; whence it is called accent from *accino, accinis* [I sing, thou singest], because every syllable has its own proper sound either raised, lowered, or composite, and all syllables of one word are adapted or sung to one syllable on which rests the principal sound. Thus length and shortness and all other things required in correct pronunciation are reduced to music, and concerning them the authorities on music decide, as is evident in Censorinus on accents and in Martianus and in many others. These matters are explained by Isidore and Cassiodorus in their statements on music. Augustine reduces meters and feet and matters of this kind to the subject of music. The perfect theologian therefore should have the mastery of all these things, because in the literal and spiritual sense their natures and properties are required, as the sacred writers decide in many ways. Now music on instruments determines the structure of the instruments and their use. Therefore since Scripture is full of instruments of this kind, the perfect theologian ought to know, at least in general, the structures of instruments, and to know how use has to be made of them on account of their numberless mystical meanings besides their literal ones. Music, moreover, consisting in what is visible, is necessary; and that it is such is evident from the book on the Origin of the Sciences. For whatever can be conformed to sound in similar movements and in corresponding formations, so that our delight may be made complete not only by hearing, but by seeing, belongs to music. Therefore dances and all bendings of bodies are reduced to gesture, which is a branch of music, since these are conformed to sound in similar movements and corresponding formations, as the author of the aforesaid book maintains. Therefore Aristotle says in the seventh book of the Metaphysics that the art of dancing is not complete without another art, that is, without another kind of music to which the art of dancing is conformed. Dancings, therefore, and other flexions of the body, which Miriam, sister of Moses, Deborah, and other women singers used as well as David and other male singers, of which Scripture treats in many ways, ought to be known to theologians, in order that they may know how to express all their properties, so that they may give utterance to all the spiritual senses of an angelic devotion.

Although it is now sufficiently apparent that these sciences must be necessary in respect to theology and philosophy, yet their far greater utility must be considered by means of the other important ways which pertain to wisdom not only absolutely, but as it ought to direct the Church of God, and care for the other three matters touched upon several times, as will be explained below.

Now that we have shown how mathematics is necessary to wisdom, both sacred and human, we are still under the necessity of proving what has preceded, in order that certain sophistries to the contrary may be exploded and certain statements of the sacred writers may be expounded, so that all doubt may be removed regarding the utility of mathematics. Mathematics is most subject to attack owing to the judgments of astronomy. Since, then, the contention of many ignorant of the potency of philosophy and its very great usefulness to theology, and condemning both relatively and absolutely the views of mathematicians, impedes the pursuit of knowledge and does it very serious harm in this respect; I therefore wish at the present to verify their purpose, and to remove the infamy attaching to true mathematics. Theologians then have found many statements scattered abroad by the sacred writers against mathematicians, and some of these men, owing to their ignorance of the true mathematics and of the false, do not know how to distinguish the true from the false, and therefore condemn the true with the false as though by authority of the sacred writers. For the word for the true mathematics is said by many authors to be written with an aspirated *t,* and to be derived from this word *mathesis* with short middle syllable, meaning knowledge, and it is certainly derived from the Greek; because the verb *matheo* is the same as *disco* [learn], and *mathetes* is the same as *discipulus* [learner], and *mathesis* is the same as *disciplina* [instruction]. Whence *mathematica* [mathematics] is instructional and theoretical knowledge, as Cassiodorus said above. But the word for false mathematics is said by the same authorities to be written without an aspirate and to be derived from *mathesi* with long middle syllable, meaning divination, or with greater certainty, from *mantos* or from *mantia,* which are the same as *divinatio* [divination], as Jerome states in the original, Isaiah, chapter IX. But whatever it may be as regards its writing and derivation, false mathematics is the art of magic. For there are reckoned to be five kinds of magic art, namely, prescience, mathematics, charms, illusions, fortune telling.

Mathematics, then, is the second division of magic. This subject usurps a consideration of the heavenly bodies which is marred by characters, incantations, conjurations, superstitious sacrifices, and various frauds. It maintains that all things by virtue of the constellations happen of necessity, that nothing may happen in one of two ways, that nothing happens by chance nor by fortune, nothing by design, yet from the excellence of its essence and in aid of the constellations it has made a more efficacious arrangement of the forms of the characters appropriate to each of the constellations and of the other things aforesaid. These matters are expressly stated in books on magic, whence this science claims that all these things happen of necessity by means of the heavens, and presumes by this necessity to judge infallibly concerning all future events. But this form of mathematics has been condemned not only by the sacred writers, but by the philosophers, as Isidore states in his tractate on Astrology, asserting that one branch of astronomy is full of superstition, namely, that which is magic, and is called prophetic mathematics. Whence Aristotle and Plato, on the testimony of Isidore, condemned it: and Pliny, attacking it frequently in different places in his Natural History because of the errors which this fantasy has written in matters relating to nature and to medicine, at length in utter abhorrence of it discloses in his thirtieth book its origin, and shows clearly how it has defiled the whole world. Tullius also in his book on Divination, more particularly treating of its evil, shows that it destroyed divine worship, did violence to the republic, and infected medicine and natural philosophy, and subverted all the good arts. Ptolemy also and Aristotle, Avicenna, Messehalac, Hali, and Albumazar, who in comparison with others have spoken with greater authority in regard to these matters, do not maintain that there is an absolute necessity in things below due to the influence of the heavens, because free will is not subject to the things of nature, nor do they think that the decision must be infallible, nay, they do not even place any necessity on free will, since they do not ascribe it to the things of nature, as will be evident. Therefore, philosophers universally condemn the madness of those false mathematicians.

Not only are they condemned on the principal count, namely, because of the error they hold regarding the heavenly bodies, but also because those mathematicians summon to their aid demons of celestial natures by means of conjurations and sacrifices, which is wholly wicked, and, moreover, defile their studies in regard to the heavenly bodies by circles and figures and very silly characters and very foolish incantations, and irrational speeches in which they trust. Moreover, they have recourse to fraud in their acts, perpetrated by means of collusion, darkness, fraudulent instruments, sleight of hand, in which they know there is deception, and they do many things to be wondered at by the foolish by these means, in which matters the influence of the heavens is not operative, and therefore contradicting themselves, they know in their hearts that their statements to others attributing influence to the heavens have no truth. Likewise although elsewhere in certain conjurations and sacrifices and incantations and various figures do they trust as though coöperating with the constellation according to their judgment, yet frequently they arrange these things in a fraudulent manner according to the kinds of fraud mentioned above, and in the presence of those believing in them they ascribe very many things to the constellation in respect to which it has no power. This fact does not escape them. Therefore on account of these follies of false mathematics philosophers have condemned it, and the sacred writers and catholic men perceiving these facts together with the philosophers have rejected it.

But the special reason of the sacred writers for their position was the fact that such mathematicians hindered in the beginning the entrance of the faith into this world, because not only imbued with this fantasy did they err in the faith and fashion their morals in accordance with the heavenly bodies, under the impression that they were made of necessity by the heavenly bodies and by the other agencies mentioned either irascible or mild-tempered, chaste or wanton, and so also concerning other matters; but they also ascribed the miracles, proving the faith of Christ, to that kind of mathematics, saying that the Christians are mathematicians and magi seducing the people. For just as by means of this demoniacal cunning they were able to do many things in the presence of an ignorant people by which they chained them to their errors and ruled over them, so did they bring the charge against the apostles and martyrs and other preachers of the faith that they did not do real miracles by the power of God, but by means of the magic art which they themselves employed.

But true mathematicians, whom in this field we call astronomers or astrologers, because they are so called indifferently by Ptolemy and Avicenna and many others, are not confuted in regard to sacrifices, conjurations, incantations, and characters, as even the rank and file of students are aware, but are censured only in regard to infallibility of judgment and the necessary happening of events. Now we cannot detect their view unless we extract their own statements from their books, so that thus from their own words we may condemn them if in error, or snatch them from the teeth of the ignorant throng and set them free if they are guiltless. Since, then, the most serious charge against them is the error in regard to judgment as though they wished in an infallible way to contend with divine certitude, as they are charged by people in general, let the nobler philosophers be cited, namely, Aristotle, Avicenna, Ptolemy, Hali his commentator, Messehalac, and Albumazar, on whom more than on others rest this burden. In general, then, in our examination of what is to be considered as the view of the philosophers in this matter we note that Ptolemy states in the beginning of the Centilogium, "The astronomer should not state a matter specifically but in general terms, as one who views something from a distance," and he adds, "The judgments that I give you are between the necessary and the impossible."

Hali says on this passage, "He made this statement, because this science does not exist except through demonstration and opinion since the subject to which the whole work of the stars pertains is convertible to the one and to the other." Wherefore in the fifty-ninth chapter of the Centilogium according to one translation he says, "When you have been asked about one that is absent, you should not decide the question of his death, until you remove that of his intoxication; nor should you settle the question of a wound, until you remove the possibility of blood letting; nor the question of a man's acquired wealth, until you remove the possibility that the money was merely entrusted to him; for in all these cases the judgment is the same." Hali also says, "He who gives judgments, judges by the forms which are near the truth, because the ability is not found in judgments to separate or distinguish between an unconscious person and one dead, nor between a wounded person and one who has submitted to blood letting, nor yet between the man to whom money has been entrusted and him who has amassed it": and he adds an example, "When his father was hiding along with others from the emperor, an astrologer visiting him daily directed him to take a very large bronze vessel full of water and place above it a stool, and advised him to sit thereon the greater part of the day. He gave him this advice, in order that he might cause the astrologers of the emperor to err in regard to the state of those in hiding. For they said that he was in the middle of the sea owing to the similarity of the water in the vessel above which he had been seated. For they were unable to distinguish between the one and the other because of the likeness of the container and that contained. For he was contained on all sides by the water, and the concavity of the vessel indicated the concavity of the bed of the sea."

Likewise Ptolemy in his book on the Arrangement of the Sphere says, "The judgments of the astronomer concerning matters here below are not drawn from an artificial science, by which their boundaries are necessary ones, but from those things that are most general." Again he says, "Astronomers have made that statement not in an artificial way, not according to a fixed innate quality, but by means of a conjunction falling upon the thing more nearly related." And in a third passage he says that "it is clear that what they affirm is affirmed in the most general manner not in an artificial way and does not involve necessity." For this reason, therefore, error will often happen in these matters, as he states. Ptolemy also in the first chapter of the Quadripartitum says, "But the hidden profundity of this art so excellent and its practice so majestic and incomprehensible, that it cannot be wholly grasped by human genius, seem to be the result of its ineffable subtlety and of a certain divine quality as it were belonging to it." For it is far removed from human sense, and transcending corporeal nature it rises as it were far beyond man. And in his third chapter he says that "from the excessive profundity of this art error happens at times." Wherefore according to Ptolemy, to whom at the present time especially is ascribed the method of pronouncing judgments, it is clear that the astrologer cannot give full certitude in regard to his judgments, especially in particular judgments. This is clear not only from these, but from his other discourses, in which, although he hints at the possibility of judging about many things with reasonable certitude, yet he unqualifiedly states that such great difficulty is inherent in the art, that it is easily apparent that he himself places the definite limit, that the astronomer ought not to boast of sufficient certitude in particular judgments. Whence in the first chapter of the Quadripartitum he says, "For although this thing is to such a degree difficult, that by no means can it be brought absolutely to a conclusion, yet it does not seem fitting that it should be given up, but rather have we deemed it proper to set forth those things that are not beyond our ability." And after a few statements he continues, "The prosecution of such an art is so difficult and laborious that without a daily practice wholly free from secular affairs and exempt from unprofitable cares no one can by any means aspire to it." And because for those free from other cares and filled with zeal there is an excellent prospect of becoming proficient in this science, he says in the second chapter, "By no means does it seem proper to me, if error happens elsewhere to banish the art wholly from us." Moreover, he cites an example, saying, "For if sailors, exercising too little caution in regard to the rocks in the sea, suffer shipwreck due to some error, they go aboard again when their ships are refitted with no loss of their former feeling of safety, alacrity of mind, hope, and confidence. In this manner also let us eagerly acquire, as far as the possibility of our human talent permits, this science of the stars, firm and certain and unchangeably constant in its truth, so that we may try to win for it the kindly interest both of students and of others, to the end that keeping in mind the dignity of the art itself, they may secure under its guidance a knowledge of the future; but let those matters prohibited by the art itself owing to their difficulty be abandoned." And in the third chapter he argues, "Although error happens here, we must regard the incongruity as passed over by negligence or blotted out by forgetfulness." After a few statements he adds, "But since it is most evident that this preconception of the future is not held with error, and that not all kinds of this art are involved in errors, it remains to disclose with greater care what is to be derived from this efficacious art, also in what ways innovations can be rejected. But if it has not the power to give particular judgments, it should not however escape showing all that has been placed within the range of possibility. If this shall be found to be very small, we do not judge it of less profit that it is reduced to narrower limits." And he gives us an example, "For although physic is not the cause of every man's health, yet since it is of some men, we think that it is desirable because of its utility."

From these statements and from similar ones it is clear that it is not Ptolemy's thought that the astrologer should give in particular a fixed judgment and one sufficing in individual cases; but that his judgment should be a general one and a mean between what is necessary and what is impossible, and the astrologer is not able in all cases to give a final judgment. For this reason Avicenna, who completed the works of Ptolemy, as he states in the prologue of his book entitled Sufficiency, shows in the tenth book of the Metaphysics that the astrologer is not able to give certain assurance in all cases nor should he, owing to the instability of generated and corruptible matter, which does not in all cases obey the celestial force, as Messehalac states, giving as an example the magnet, because it exercises its power over iron only when it is adjusted at the necessary distance and under the other conditions requisite for attraction.

Moreover, they know that the divine rule can change all things according to its will, and for this reason they add always at the end of their decisions the qualification, "Thus shall it happen if God wills it." But they themselves know and testify that the rational soul is able to change greatly and impede the effects of the stars, as in the case of infirmities and pestilences of cold and heat, and famine, and in many other matters, as Ptolemy states and teaches in the Centilogium. Hali expounds him; and Isaac in his book on Fevers sets forth clearly this same principle; Aristotle, moreover, in his book of Secrets confirms and declares it. For when men see beforehand these evils, they are able to prepare remedies. Whence Isaac says, "Evil does not happen to a man unless he is restrained by ignorance of the celestial science," and he gives an example in the case of pestilences and infirmities and matters subject to the will, in which this science can give remedies, if it is thoroughly known. From these statements, then, it is clear that philosophers do not maintain that there is an inevitable happening of events in all cases due to celestial influences, nor is their judgment infallible in particular instances, but in accordance with the possibility of this science; particularly so since they also add that another science, which is called experimental, gives a still more certain judgment than the ordinary astronomy. So teaches Ptolemy in his book on the Arrangement of the Sphere, and this matter is made clear in what follows.

Therefore, after a careful consideration of these statements and the like, it is clear that true mathematicians and astronomers or astrologers, who are philosophers, do not assert a necessity and an infallible judgment in matters contingent on the future. Therefore any persons attributing these erroneous views to these men are clearly proved guilty of ignorance of philosophy, and reprobate the truth of which they are ignorant. Whence their sin is twofold, namely, because they handle these matters of which they are ignorant, and nevertheless blaspheme against the truth. But those who so view things which are true that they reject what is false condemn mathematicians dealing in magic, who are not philosophers but contradict both philosophy and the faith, as I explained at the beginning. Therefore the sacred writers have spoken against them and not against true mathematicians. This is made clear by the opposition of the sacred writers. For they reprobate nothing except the principle that through the stars a necessity is placed on contingent things and particularly on morals and human acts, and that there is an infallible judgment in all cases. But mathematicians practicing magic alone maintain these doctrines, and true mathematicians make no such statements. Wherefore whatever Basil in the Hexaemeron and Ambrose in the Hexaemeron, and Augustine and Gregory in the homilies on the Epiphany and Augustine in his comments on the sixty-first and ninety-first Psalms and on John, and whatever he and other learned men say in many other places against mathematicians, is wholly directed against those who imagine that necessity is placed upon those things in which there is choice, and particularly in matters which proceed from free will, who alone are lying or fraudulent mathematicians, full of superstition, not philosophers, as is manifest. Therefore they err greatly who strive to defend their madness against true mathematicians on the authority of the sacred writers, and making a wrong use of the words of the sacred writers presume to condemn the truth.

Whence they strive to establish their views although false, which they form from an ignorance of mathematics true and false. Since, moreover, they can bring no reason to bear, because their statements against the true mathematicians are opposed to all reason and to the force of philosophy, they seize and purloin the authorities of the sacred writers which have nothing to do with the matter in hand, as is clear to the investigator.

We must bear in mind, moreover, that if we consider the words of the sacred writers we shall clearly discover that not only do they not reprobate true mathematicians, but approve in the matter of their knowledge of the future; since Augustine says in the second book on Christian Doctrine that mathematics has regular conjectures regarding the future, not rousing distrust and ominous, but direct and certain concerning the future as well as the present and the past. Cassiodorus agrees with him, as was said previously. He shows to Januarius in many ways how the future can be learned by means of the heavenly bodies. Basil in the Hexaemeron declares this likewise in many ways; from this source he declares there is a most certain judgment in some cases. Ambrose in the Hexaemeron agrees, and Isidore in his work on Astronomy. The sacred writers confirm their statement by the Gospel, "from the red sky in the evening or in the morning," and again "there shall be signs in the sun and in the moon," etc. Mathematics is not therefore infamous from the knowledge of the future which it gives in a trustworthy manner; although in the beginning the sacred writers made less use of it and commended it less because of the error of false mathematics, to which in name and in a certain consideration of the heavenly bodies respecting the future true mathematics is allied and in accord, as Augustine decides in his book on Christian Doctrine.

But after the falsity of mathematics dealing with magic was cleared away in the Church, the study of true mathematics was taken up by the Catholic doctors, and was continued until the advent of certain theologians, who were ignorant of the potency of philosophy and the fallacies of magic. For this reason they have condemned it in their lectures, preaching, and collations public and private, in the first place to the very great destruction of philosophy, and in the second to the injury of the dignity of theology and to the detriment of the Church and of the whole commonwealth of the faithful and of the conversion of unbelievers, as we have previously shown in great part, and as we shall show later respecting the remainder; because where there is an ignorance of mathematics the whole potency of philosophy will be unknown. If the truth of philosophy is impaired, damage is inflicted on theology, whose function it is to use the power of philosophy, not only absolutely, but in ordering the Church, directing the commonwealth of believers, and aiding the conversion of predestined unbelievers and the reprobation of those foreknown. But not only do they err in this that they ignorantly condemn knowledge of the future secured by means of mathematics, but because for the sake of a part, which they abhor as a result of their own ignorance, they condemn the whole. For although in the parts of philosophy there are very many things that are useless and not so well discussed and some erroneous, yet nevertheless by no one is the whole of philosophy condemned, nor should it be nor can it be condemned. Wherefore since in all the other divisions of mathematics apart from that pertaining to judgment all advantages are offered in respect to theology, the Church, the conversion or reprobation of unbelievers, and in respect to philosophy as a whole, and nothing reprobated is found even in that part to which judgment is ascribed, and since many excellent things, besides the art of judging, respecting the properties of celestial things and the favorable influences of the stars and respecting other matters of this world are discussed, it is manifest that those are wholly in error who have ventured on account of one part which they falsely condemn to reprobate all the remainder, which possesses very great advantages and in no respect is subject to calumny. Again although the part relating to judgment concerning human affairs should be deserving of blame, yet the other part dealing with natural sciences and the heavenly bodies brings no calumny on the faith.

But in human affairs true mathematicians do not presume to certain knowledge, but they consider how the body is altered by the heavens, and when the body is changed the mind is aroused now to private actions and now to public ones, yet in all matters is the freedom of the will preserved. For although the rational soul is not compelled to its acts, yet it can be strongly influenced and aroused, so that it gratuitously desires those things to which the celestial force inclines it, just as we see men, owing to association, advice, fear and love and the like, change greatly their intention, and gratuitously wish for those things which they previously did not wish for, although they are under no compulsion, like him who in the hope of safety casts into the sea his most precious wares. But we see that the species or forces of things below altering our senses, also the species of things visible and audible which feebly change the body, so strongly excite men to wish for those things for which they did not formerly care, that sometimes they do not take into account death, disgrace, or fear, provided they may satisfy their desires, like those who see and hear that their enemies are coming to meet them, and they rush along heedless of risk in order to revenge themselves on them. In the same way voluptuaries, when the chance is secured of indulging their desires respecting things seen and heard, like brute beasts are influenced against the judgment of reason, gratuitously choosing those things to which they are excited.

But to a far greater degree are the forces of the heavenly bodies and the strong species of the stars able to influence the body and its organs, and when these have been greatly altered the man will be strongly excited to actions for which he did not previously care, the freedom of his will remaining unaltered; since the forces of the heavens are stronger than those of the visible and audible things below and of many other objects of sense perception, and are able to alter substance, not merely accidents, and to corrupt and destroy all things here below; since in accordance with the sun's position on the oblique circle generation and corruption in things occur, as Aristotle says. Not only is this position of the sun considered absolutely, but in connection with the forces of the other planets and stars; and for this reason there is a great change in our bodies, due to celestial forces, and consequently the mind is strongly excited to its actions, although it is not under compulsion, and in accordance with this principle the judgment of the astronomer is given, and does not imply infallibility or necessity. The astronomer relies greatly on this, because he sees that men in their actions follow for the most part the temperament that they have, as, for example, the choleric man is easily moved to anger, and is unable to restrain all the first promptings, and so in regard to other characteristics in respect to which men differ in their temperaments. Therefore in the case of the astronomer, where he sees that men follow their complexions, which spring from celestial influence, like all generation, it is not surprising if he extends this influence to the sphere of human actions.

But he sees clearly that the manners of those dwelling under different parallels of the heavens differ according to their position, as, for example, those dwelling toward the pole, as the Scythians, have different manners from those dwelling toward the south, as the Ethiopians; and those dwelling in the fourth climate have different manners from those just mentioned; nay, according to diversity of each climate and even of the parts of a climate do the manners of the inhabitants differ, as, for example, in the seventh climate we clearly see that the Picards, French, Normans, Flemings, and English differ in their manners, notwithstanding the fact that they are neighboring one to another. But this diversity does not have its origin in the men themselves due to a difference in their rational soul, but is due to the complexions of their bodies innate from the nature of the heavens, under the different parallels and stars of which they are situated, and in accordance with the diversity of their location with respect to the planets. Moreover, this diversity is not only in accordance with the latitude of the regions measured from the equinoctial circle toward the pole, but in accordance with longitude also, although the reasons are more occult. By experience we note that under the same parallel regions differ according to their greater distance from the west or east, and not only is this true of the regions themselves, but also of their parts. The chief cause of this difference cannot be found in the earth or in men, but is sought for in the heavens according to all scientists; whence, as has already been stated, on every point of the earth there is incident the vertex of a pyramid full of force from the whole heavens. These virtues are different in nature, and the pyramids likewise, because they have different bases owing to the differences in their horizons, since every point of the earth is the center of its own horizon. Therefore of necessity a great difference in all things arises from this cause, no matter how close they are, like twins in the same womb; and such is the case in all things, just as we see that from two points very close to each other on the earth plants spring differing according to species. On this principle the astronomer rests the foundations of his judgment, and rightly so, because the complete diversity in things is thus discovered to be due to the heavens. Wherefore the skillful astronomer is able not only in the things of nature, but also in human affairs to consider many things regarding the present and the future and the past, and therefore at least as regards kingdoms and states is he able to judge by means of the heavenly bodies and of the secondary members of the same, which are renewed by special forces of the heavens, such as comets and the like, because it is an easier judgment in regard to a community than in regard to an individual. For a judgment respecting a community is a general one, and the astronomer is well equipped to give general judgments. Because the whole province surrounding any famous city is affected by its manners and customs due to the fact that the city serves as a refuge and a central point for transacting the affairs of life, and to the additional fact that cities have power over the neighboring districts (and it is true also of a more powerful kingdom in respect to the surrounding kingdoms), both on account of communication and violence; the prudent astrologer is able to consider many things in this particular advantageously in regard to customs and laws and parties and wars and peace and the like, which pertain to the commonwealth of states, provinces, and kingdoms, although he encounters greater difficulty in judging of the acts of individuals.

If he wishes to consider carefully and without error the hours of conception and nativity of individuals, in order that the domination of the celestial force may be known at those hours, and if he carefully considers when the heavenly bodies will reach these arrangements in accordance with the separate parts of the age of each, he is able to give a satisfactory judgment in regard to all things in nature, such as infirmities, health, and the like, at what time they must happen and how they are limited, as authorities not only in astronomy, but in medicine, like Hippocrates, Galen, Hali, Isaac, and all other authorities, decide. But few physicians of this time know astronomy, and therefore many of them neither understand their own authors nor can they understand them, and for this reason they neglect the better part of medicine. But concerning them a long and useful discussion is possible when the opportunity comes. Since, moreover, in accordance with the complexions, weaknesses, and health vary the wishes of men and their longings and views, although they are not compelled but strongly influenced, as is manifest, the prudent astronomer is able to give a wise judgment regarding the moral actions of an individual, in all cases, however, with a preservation of the freedom of the will, and he will be able in many instances to form a sure judgment in accordance with the possibility of the subject which he handles. For since this is contingent and not necessary, he cannot say that of necessity these or those things will happen, yet he can say in many cases they will happen, and that truth exists concerning the future although not of necessity. For it is one thing to be true, and another to be rendered true of necessity. This is the judgment resting between what is necessary and what is impossible. In cases where he cannot form a judgment of this kind, he will easily form a general judgment, or one lying between the universal and the particular; yet by means of a general judgment and in accordance with what is possible regarding a public personage, as a prince and a prince's counselor in a state or district, he is able more frequently to form a particular judgment concerning the deeds of the commonwealth; because, as has been said, it is easier to judge concerning a community than concerning an individual, and in accordance with the judgment of the prince are states and kingdoms ruled; for that which pleases the prince has the force of law.

Therefore the astronomer is influential respecting the affairs of famous cities, provinces, and kingdoms, provided he be well informed about the domination of the heavens at the conception and nativity of the prince, also as regards the fact that complexion is altered by this means, and that morals are shaped in accordance with the quality of the complexion. Then let him note when the constellation comes which stimulates and strengthens the complexion, so that the mind is stimulated and aroused to similar actions; and likewise let him note when the complexion is altered into the opposite by arrangement of the heavens, so that the mind of the prince is aroused to opposite courses of action. For in accordance with variations of this kind in the constellations and complexions and wishes of princes and prelates innovations in habits and change in laws and customs arise in a people on higher authority. Moreover, there easily arise at times discords and dissensions, which are followed by wars, or at other times, due to the excellence and utility of the laws, concord springs up between citizens and others, and peace is made. Therefore the skillful astronomer is able easily to give judgment regarding the common affairs of this kind in states and districts, because he has the means of proceeding not only by ways proper to them, but by means of the natures of personages in authority. So likewise are they able to judge by the qualities of those who assist princes and prelates, and on whose counsels they rely, because wicked princes are influenced for the good of the state by good counselors, and good princes are perverted by evil counselors. If, therefore, from the constellation of the nativity and conception the complexion of any prince, or of another on whom he relies, is found disposed to perversity of morals and to discords and wars, and if the astronomer at the same time with this sees that men are influenced in this way in the formation of habits and are more strongly aroused when a like constellation happens, he is able to give a rational judgment regarding the woes of the state and kingdom over which they rule, when the arrangement of the heavens and of those things which are renewed by celestial forces, like comets and the like, happen conformably. The best judgments can be made in accordance with the opposite natures of princes and of those whom they trust in all matters, when likewise there is found a celestial arrangement in conformity. Since such personages in one kingdom are few, and are public characters and known to all, whose manners are revealed to the whole kingdom, the astronomer is well able to make certain concerning them, and to give a useful judgment regarding public matters by means of the qualities of such personages. These, then, are the statements I have desired to make in order to remove the infamy resting on mathematics in judgments of this kind, from which it is clear to every man of wisdom that true mathematics is not to be blamed in this matter, but to be embraced warmly and loved owing to the glorious utilities which can come out of the judgments of true mathematics, which in no way is contradictory to the truth.

Opus Majus

Since the potency of mathematics has been shown in respect to the sciences of philosophy and to the things of this world and to theology, and so in respect to all knowledge, just as knowledge itself is considered absolutely in reference to itself, I now wish to show that same thing with reference to the influence of this knowledge on the Church of God, the commonwealth of the faithful and the conversion of unbelievers and the repression of those who cannot be converted. Since in many ways which cannot now be enumerated it is necessary to the Church, I wish now to set forth three cases that are of infinite wonder and ineffable utility. The first consists in the certification of the faith which the Church holds. For we can have great consolation in our faith, since the philosophers who have been led by the exercise of their reason alone agree with us, and strengthen the body or profession of the Christian faith and agree with us in the stability of this body; not because we seek reason before faith but after it, so that rendered certain by a double confirmation we may praise God for our salvation which we possess without doubt. By this means offered by mathematics not only are we made certain respecting our profession, but we are fortified in advance against the sect of Antichrist, about which at the same time with the Church of Christ mathematics is concerned. A very excellent examination of this kind is made by considering all the principal sects from the beginning of the world, which are six in number, nor can there be more, namely, Jews, Chaldeans, Egyptians, Agarenians or Saracens, who descended from Agar and Ishmael, the Church of Christ, and the sect of Antichrist. Nor is it strange if the philosophers have spoken regarding these, since they were after the patriarchs and prophets and were instructed by their sons and books, as we have previously shown.

I shall then at the present time state to the best of my ability the opinions of the mathematicians in which authorities agree. They say, then, that Jupiter and Venus are benevolent and fortunate planets, Saturn and Mars malevolent and unfortunate ones. Mercury, they say, is in a middle position, because he is good with the good, and evil with the evil, since he is of a changeable nature. Of the benevolent and fortunate planets they say that Jupiter is the better and that greater good fortune is owed to him, and less to Venus. Therefore since there are two lives, the present and the future, and the future is more important than the present one, just as that which is eternal is more important than that which is temporal, they say that Venus has significance regarding the fortunes of this life, as far as pertains to games, pleasures, joy, and the like, and Jupiter has respect to the blessings of the other life, which are greater. He has signification in regard to wisdom and intellect, interpretation of dreams, divine worship, faith and doctrine of the law, veneration and fear of God, fitness of morals, and many such things, as astronomers state.

Moreover, they divide the whole heavens into twelve parts, called houses, which are separated by the meridian circle and the horizon, and by four other circles intersecting one another at their points of meeting, so that the first house begins at the horizon and is formed beneath it, then there succeed it the second and the third up to the angle of the earth, which is the point in the heavens beneath the earth opposite to the point of the meridian above the earth. From that point there are three other houses up to the line of the west. Then above the horizon begins the seventh house and the eighth follows and the ninth up to the point of the meridian. Then come the tenth, eleventh, and twelfth to the east. The first house they assign to Saturn, the second to Jupiter, and so on according to the order of the planets, so that the eighth again is assigned to Saturn, and the ninth to Jupiter. All have agreed in considering the ninth house as that belonging to religion and faith. Ptolemy, Albumazar, Altavicus, Messehalac, and all others have assigned to these houses their properties, because the first is the house of life, the second the house of substance, and so on with the others, according to their properties and natural dispositions. Whence the ninth house, as they say, is that of peregrinations and journeys of faith and deity and religion, and the house of the worship of God, of wisdom, of books, letters, and of the accounts of ambassadors and reports and dreams. Therefore rightly, as they say, is the house assigned to Jupiter, who is significant with regard to the blessings of the other life, because for those blessings there are needed faith and religion and the worship of God and the study of wisdom, and a multitude of books and of letters, as is evident from the sacred law; and a large number of ambassadors, such as prophets and apostles and preachers, making suitable reports regarding the noble state of that life and having frequent revelations in dreams and ecstasies and visions concerning this life.

They say, then, that the planets are in conjunction and embrace one another in turn, and this happens when they are in the same sign and especially when they are in the same degree and in the sixteenth minute of that degree or below it. Therefore the philosophers maintain that Jupiter, from his conjunction with other planets, has signification regarding the division of religions and faith. Since there are six planets with which he can be united and in conjunction, they therefore assert that there must be six principal sects in the world. Whence the skillful authorities aforesaid and others say if Jupiter is in conjunction with Saturn, he signifies the sacred books and of the sects that of the Jews, because it is more ancient than the others and prior to them, just as Saturn is the father of the planets and more remote and prior in the egress of the planets and in their order in existence. All faiths acknowledge it, and it acknowledges no other, just as all the planets are in conjunction with Saturn and he with no one of them because of the slowness of his motion. Because when a planet is before him to the east, Saturn never overtakes it, but the other planet is so far superior, that at some time or other it overtakes Saturn and is in conjunction with him. All sects, in fact, trace themselves back to the sect of the Jews, because this was the first and is the root of the others, from which all have taken evidence of some kind and have found a basis for their sect: whence philosophy has received much evidence and many ways of establishing its cult, as we have previously shown. If, however, Jupiter is in conjunction with Mars, they say that he has significance with regard to the Chaldean law, which teaches the worship of fire, of which nature is Mars in natural potency and effect. If he is in conjunction with the sun, his reference is to the Egyptian law, which maintains the worship of the host of heaven, of which the sun is chief. If he is in conjunction with Venus, his reference is said to be to the law of the Saracens, which is wholly voluptuous and lascivious. Although Mahomet reduced this law to writing, yet through long ages was it regarded by its votaries as the rule of life. Whence the book De vitae sua mutatione, attributed to Ovid the poet, speaking of the lascivious sect, which the book stated was the law of the people in the poet's time, says,

> "In which anything pleasing is considered lawful,
> Although a written law regarding it is not yet found."

This law more than six hundred years later Mahomet reduced to writing in a book called the Alcoran. For Ovid was before Christ and lived in his times, and the sect of Mahomet began more than six hundred years after the Incarnation of Christ, as is evident from the difference between the Christian years and those of the Arabs, namely, 631 years and 195 days. But the years of the Arabs are computed from Mahomet, as Alfraganus states and others likewise.

If he is in conjunction with Mercury his reference is then to the law of Mercury. For Mercury, as they say, has reference to deity and to the oracular utterances of prophets, and to belief and to prayer, and particularly so when Jupiter is in conjunction with him; since he then signifies the number of the psalm and the number of the sacred books. They also say that the law of Mercury is harder to believe than the others, and contains many difficulties beyond the human intellect. This is in keeping with the difficult motions of Mercury, whose circuit is in an epicycle and in an eccentric circle and in a concentric one. In these motions we have to consider his motions in longitude and his curving and backward curving in his motions in latitude, owing to the declination of the eccentric circle from the Zodiac to the north and south, and of the epicycle from the eccentric circle in the direction of north and south, and they are more wonderful and difficult than all the other motions of the planets, as is clear from the statements of Ptolemy, and clearer still from those of Albategnus, Thebit, Archaselis, and Alfraganus. For this reason he has reference, as they say, to the law that contains difficult articles and hidden truths, of which kind is the Christian law. But because Mercury signifies writing and writers, and depth of knowledge contained in profound books, and eloquence or sweetness of speech and tongue, oratory and its rapid flow, and the explanation of sentences, he indicates that this law will be defended by such authentic scriptures and by so many profound sciences and by such potency of eloquence, that it will always remain firm in its own strength, until the final law of the moon shall disturb it for a time. They say also that this is the law of the prophet who shall be born of a virgin, in accordance with the teaching of all the ancient Indians, Chaldeans, and Babylonians that in the first face* of the sign Virgo [the Virgin] there ascends a very pure virgin who shall rear a son in the land of the Hebrews, whose name is Jesus Christ, as Albumazar states in his larger Introduction to Astronomy. In morals his authority will be quoted among other authorities on matters pertaining to moral philosophy, because for the matters here mentioned mathematics gives preparation in the service of that philosophy, as will be set forth more expressly in that subject. The virgin birth of the prophet is in full accord with the law of Mercury, because Mercury was created in the sign Virgo, and the dignities or potencies or testimonies or virtues or five fortitudes which are due to the planets in relation to the signs, Mercury has in the sign Virgo, to wit, house, exaltation, triplicity, terminus, face.

Both in the present and in the former instance house is used in a double sense; since these houses are called essential and natural, those mentioned previously are called accidental and relating to position, since these houses are the twelve signs of which there is a natural division, because the sections of the Zodiac and of the heavens remain in their positions respecting the celestial circle, that is, the firmament, which division of the signs is made by six circles intersecting one another at the poles of the Zodiac, and dividing the whole heavens and the world into twelve equal parts. These parts can be considered on the Zodiac only, and then they are properly signs, like Aries and the others, or those circles in imagination can be extended to the poles, at which they intersect one another, and then they divide the whole heavens into twelve parts, narrow at their extremities around the poles and broad in the middle like the bottom of a skiff, so that the broad part contains the extremities of the parts of the Zodiac, which we call by the general name signs, as Aries [the Ram] or Taurus [the Bull], etc. It is taken also in another sense for the whole portion of the heavens contained between two circles crossing, let us say, through the boundaries of Aries and meeting at the poles of the world; and that sign is said to be the sign of Aries, because its breadth

* Face means one third of a zodiacal sign.

consists in the extension of Aries, and thus the stars outside the body of Aries are said to be in the sign of Aries, although they are near the poles of the world. But the other houses are called accidental, because their division is accidental, and the sections do not remain in the same place in the heavens, because they do not follow the motion of the heavens, and for this reason their positions change on the circle or in the heavens hourly. These sections are made, as has been said, by the meridian circle and the horizon together with four other circles. The house, moreover, that is called natural is twofold, one that is principal and one that is not principal. Moreover, the principal house of a planet is that in which it was created, as follows; Leo [the Lion] is the house of the sun, Cancer [the Crab] of the moon, Virgo of Mercury, Libra [the Scales] of Venus, Aries [the Ram] of Mars according to some, Scorpio [the Scorpion] according to others, Sagittarius [the Archer] of Jupiter, Capricornus [the Goat] of Saturn. These are, moreover, inferior [*minus principales*] houses, as follows; Aquarius [the Waterman] is assigned to Saturn, Pisces [the Fishes] to Jupiter, Scorpio to Mars according to one opinion, according to another Aries, Taurus [the Bull] to Venus, Gemini [the Twins] to Mercury; so that each of the five planets has two houses but the sun and moon only one each. So have the sages from antiquity decided.

Now these are the exaltations. The sun has its exaltation in Aries, the moon in Taurus, Saturn in Libra, Jupiter in Cancer, Mars in Capricorn, Venus in Pisces, Mercury in Virgo. Moreover, as the sun has its exaltation in Aries, so has it its descent in Libra, and likewise in the case of the others; and similarly the depression of Mercury is in Pisces, and therefore the exaltation of Mercury is in Virgo, just like his house, and this exaltation is in the fifteenth degree of Virgo. A planet is said to be in its triplicity when it is in the sign in which it was created, or in one of the same nature as the sign in which it was created. Whence we must note that there are four triplicities of signs. One is hot and dry, containing three hot and dry signs, of which kind are Aries, Leo, and Sagittarius. Hence when the sun is in any one of these three he is said to be in his triplicity. The second triplicity consists of Taurus, Virgo, and Capricornus, and is cold and dry; and Mercury when he is any one of these is in his triplicity. Since although the owners of this triplicity are in the day first Venus, then the moon, and in the night first the moon, afterwards Venus, and Mars shares them in the night and in the day, yet Mercury has a share in them properly, as the astronomers say, and therefore has his triplicity in Virgo as well as his exaltation and house. The third triplicity consists of Gemini, Libra, and Aquarius, which is hot and moist. The fourth consists of Cancer, Scorpio, and Pisces, which is cold and moist.

More famous are the boundaries of the Egyptians; Jupiter has the first six degrees of Aries, Venus the following six, Mercury eight, Mars five, Saturn five, Venus in addition the first eight of Taurus, Mercury the following six. Thus with a wonderful diversity do these boundaries vary, as is evident in the table of boundaries, so that Mercury has the first seven degrees of Virgo for a boundary, not only according to the Egyptians, but also according to Ptolemy, and this is our present question. Faces of the signs, moreover, are determined by the division of each sign into three equal parts; and each one consists of ten degrees, which are called faces, and otherwise *decani.** The first face begins at the first degree of Aries and ends at the tenth degree of the same sign and is called the face of Mars. The second face reaches to the twentieth degree, and is called the face of the sun, because the sun succeeds him in the order of the orbits. The third extends to the end of Aries and is called the face of Venus, and so on with the others according to their sequences, as is shown in the table of faces, so that Mercury has the last ten degrees of Virgo for his face. Thus it is clear that Mercury has all these potencies in Virgo. They are called potencies through a similitude. Hence a planet in its own house is compared to a king in his royal house and in his domination; and when the planet is in its exaltation it is like the man in his honor and among his servants and ministers; when the planet is in its boundary, it is like the man in the midst of his parents, relatives, and family; and when it is in its face, it is like the man in his official position of command. House, moreover, is said to have five fortitudes, exaltation four, triplicity three, boundary two, face one. Whence house has in

* Chief of ten parts of a zodiacal sign.

itself the fortitudes of five faces, and exaltation has the fortitude of four faces and so on.

From these statements, therefore, it is clear that these potencies of Mercury essential and principal are all in Virgo. We must add that each planet has in addition a certain accidental potency in the sign belonging to it, which is called *gaudium* [joy]. Hence Saturn, when he enters Aquarius, is said to rejoice, as does Jupiter in Sagittarius, Mars in Scorpio, Venus in Taurus, and Mercury in Virgo. Therefore Mercury nowhere else has so much influence as in Virgo. Nor has any other planet so many rights of ownership in her, wherefore Mercury is assigned properly to Virgo. Therefore for this reason they say that the law of Mercury must be the sect of the prophet to be born of a virgin; and therefore this sect of Mercury is assumed by them to be the Christian law. If, therefore, he be in conjunction with the moon, teachers of astronomy say that the law of the moon will follow and be the last one, because the sphere of the moon is the last one, and this will be the law of corruption and a foul one, which shall do violence to the other laws and suspend them, even the law of Mercury for a time. For the moon, as they say, has reference to necromancy and deceit, and therefore the law of the moon will be devoted to necromancy, magic, and deceit. On account of the corruption of lunar motion and of lunar shapes it signifies the corruption of that law, which will be corrupt in itself and a corrupter of the others. It will not, however, last long, as they say, because the moon changes quickly in its shape, light, and motion, owing to the shortness of its orbit. This dominion, as they say, will be established by some one great and powerful, who shall prevail over others, and astronomers of our faith both modern and ancient consider that this is the law of Antichrist, because he finally at the end of the world will appear, and will bring in the law of corruption, and will infatuate the world by means of magic and his lies.

Thus, then, the astronomers distinguish the sects and in particular Albumazar in his book on Conjunctions, and especially in the first and in the second books, so that there are found to be six principal sects, in which are included the inhabitants of this world. By this means it is established that the sect of Christ is one of the principal sects. If we compare it with the others, it is manifest by reason of the noble nature of its lawgiver and of the sect itself, that no other sect is worthy, but all others are the figments of men. This fact is at once apparent respecting the last law, because there is no truth in it. In the law of Venus, which is that of the Saracens, a delight in sin abounds according to its tenor mentioned above. But philosophy excludes sin from law. Likewise the law of the Egyptians is nothing, nor is that of the Chaldeans, because they teach men to worship a creature, and this philosophy denies. For worship is due to God alone, as will be explained under the topic of morals. The sect of the Jews, however, is less removed from the truth. But the lawgiver was not the son of a virgin, as in the Christian dispensation; and it is not confirmed by so many authentic Scriptures, nor has it such noble articles as we showed previously according to the philosophers. Therefore of necessity the law of Christ must obtain the leadership. But in moral philosophy this will be plain from its properties. For not only thus in general do astronomers investigate these laws, but they determine the times of their inception and the end of some of them. This investigation is made by means of the conjunctions of the planets and the revolutions of their motion. Albumazar, then, in his book on Conjunctions and other astronomers determine that there are three conjunctions of Saturn and Jupiter, namely, a great one, a greater, and the greatest. A great one happens at their conjunction every twenty years, in whatever sign it may take place. For Jupiter completes his orbit in twelve years, and Saturn in about thirty, and therefore it happens that after twenty years they are in conjunction in the ninth sign from that one in which they had previously been in conjunction; and after another twenty in the fifth from the first; and after the third twenty again in that first sign. This is a great conjunction, which occurs in this threefold manner twelve times, or sometimes thirteen times. For the first sign, the fifth, and the ninth make a triplication. This conjunction is said to have reference frequently to the elevation of kings and potentates and to dearness in the cost of provisions, and to the rise of prophets. After they have been in conjunction in this triple manner so many times that they change to another it is then called a greater conjunction. This happens every two hundred and forty years or thereabouts, and has reference to a sect and to its change in certain regions. When the conjunction has changed from this triple form to another one, as from the end of Cancer to the beginning of Aries, it is then said to be the greatest one, through the revolution of Saturn thirty-two times, and happens every nine hundred and sixty years, and has reference to changes in empires and kingdoms, to impressions of fire in the air, to flood, earthquake, and dearness in the price of food. A conjunction of the greater or almost of the greatest type occurred in the twenty-fourth year of Augustus Caesar which wise astronomers said had reference to the coming law of Mercury. In the book concerning the change in Ovid's life which is named Concerning a Little Old Woman, because of whom the change had been made, Ovid is said to have spoken about this conjunction, and from its nature to have broken forth in admiration of the sect of Mercury destined to be introduced into the world by a prophet without human father, born of a virgin. He predicted that this sect would come into being six years after this conjunction, so that according to him the prophet would be born in the thirtieth year of Octavianus Augustus. For twelve years passed from the death of Julius to the battle of Actium, in which Octavianus Augustus obtained full power. For previously he worked to acquire the kingdom rather than actually possessed it. Christ then was born in the thirtieth year of Octavianus. If, however, those twelve years are reckoned as part of his reign, then was Christ born in the forty-second year of Augustus, as some count. But the result is the same. For Augustus waged five great civil wars in those twelve years, as the histories state, and in particular Orosius in his book Ormesta Mundi. The last one was at Actium, in which he defeated Antony and Cleopatra, and the empire rested in peace. Ovid, therefore, speaking of a conjunction of the greater or almost of the greatest kind, writes the following verses: "Such a conjunction occurred lately at a momentous time in the twenty-fourth year from the beginning of Augustus Caesar's reign, which signified that six years later a prophet must be born, without a human father, of a virgin, of whom there is thought to be a type when the influence of Mercury is increased. The early character of the future sect will accord with this influence, for nowhere else has Mercury so much power in respect to the signs as in the sign of the Virgin."

The first face of Virgo was ascending in the east when that

conjunction happened. The conjunction was near the head of Aries. For if we revolve the motions of Saturn and Jupiter to that time, we shall find that they were in conjunction by their mean courses six years, five days, and three hours before the birth of Christ; and the mean course of both was in Aries ten degrees, fifty-six minutes, fifty-two seconds. Now the mean motion of the eighth sphere was ten degrees, five minutes, fifty-one seconds, twenty-seven thirds of the signs of the small circle, and had to be subtracted from the positions of all the planets; whence there remained from Aries two degrees, fourteen minutes, forty-two seconds. Since, therefore, the difference between two conjunctions by their mean courses adds eight signs, two degrees, twenty-five minutes, seventeen seconds, it follows that the preceding conjunction had been in Cancer nine degrees, fifty-one minutes, twenty-five seconds, and thus was the triplicity changed from a watery to a fiery sign. If, however, this conjunction had been nearer the head of Aries, it would have been one of the greatest kind, and then were completed 305 years and nine months, and about twenty-three days, of the Greeks, as can be proved by the tables of years.

By means of the revolutions of the motions of the planets they calculate that same matter. For Albumazar in the eighth distinction of the second book on Conjunctions says that the retarding of sect and of empire and changes in them occur particularly according to the space of ten revolutions of Saturn, especially if for Saturn a change should occur at the signs bringing change, which are Cancer, Libra, and Capricornus, provided Jupiter is setting away from him. But if Jupiter is with him, or is looking toward him, he will greatly lessen the evil owing to his own goodness. For when ten revolutions of Saturn were completed in the days of Darius, Alexander the Great appeared, and the destruction of the kingdom of the Persians took place. After approximately ten other revolutions were completed Jesus, son of Mary, appeared, in regard to whom there are rescripts with change of faith. When ten others were completed Manes appeared, coming with a law which is in force among the Pagans and the Nazarenes. After ten others came Mahomet, an event perhaps before the completion of ten revolutions, as in the ninth revolution, and perhaps after them, as in the eleventh. This is in accordance with the space required by the conjunctions preceding, which are stronger than those revolutions. Similarly a sect is changed more quickly or slowly according to the properties of the planets bearing sway over different regions, as follows: Saturn controls India; Jupiter, Babylonia; Mars, Thrace; the sun, the Romans and their empire; Mercury, Egypt; the moon, Asia.

Concerning the destruction of the law of Mahomet they speak clearly and with certainty. For according to what Albumazar says in the eighth chapter of the second book, the law of Mahomet cannot last more than 693 years. But so long is it able to last and will last, unless owing to some coincident cause, the time shall be shortened as we explained before, since a shortening can take place greater or less from different causes. It is now the six hundred and sixty-fifth year of the Arabs from the time of Mahomet, and therefore it will be quickly destroyed by the grace of God, which must be a great consolation to Christians. Wherefore God is to be praised, who has given the light of wisdom to philosophers, by which the law of truth is confirmed and strengthened, and by which we perceive that enemies of the faith must be destroyed. With this view the Apocalypse agrees in chapter XIII. For it says that the number of the beast is 663, a number less than that predicted by thirty years. But Scripture in many places takes something from a complete number, for this is the custom of Scripture, as Bede says. Perhaps God willed that this matter should not be explained fully, but should be somewhat veiled, like other matters which are written in the Apocalypse. Whence before the final time which is determined for that sect, according to its principal cause, as Albumazar determined it, it may happen that the Saracens will be destroyed by the Tartars or Christians. Already the greater part of the Saracens have been destroyed by the Tartars, as well as the capital of their kingdom, which was Baldac, and their caliph, who was like a pope over them. These things happened twelve years ago.

Although they make these statements regarding sects, and sects depend on the freedom of reason, yet they do not place any necessity on freedom of the will, saying that the planets are signs hinting to us those things which God arranges from eternity to take place either through nature, or through human will, or by his own plan in conformity with the good pleasure of his will. Such is the statement in the book on the Courses of the Planets. Moreover, they state that the will is not under compulsion, but the body, however, is altered by the forces of the heavens, and then the soul united to the body is excited strongly and influenced effectively, although it is in no way compelled, so that it desires gratuitously to follow the inclinations of the body toward actions private or public, and to good ones as well as to evil ones, so that thus opinions and sects and changes in customs are introduced by some person famous among the people and powerful, in accordance with what was seen and known beforehand by God; so that the planets are thus not only signs, but do something in the way of excitation.

Since they have maintained that Jesus Christ is God and man, as the astronomer Ethicus clearly states in his Cosmography, and likewise Alchimus, and, moreover, in that book inscribed, *Ovidius de vetula,* the conclusion is drawn that God is incarnate in Christ, attributing to him, as they do, what is denied to man alone, their contention is that a celestial arrangement could have taken place for a sign of the conception of the Virgin and of the nativity of that Man, so far as he was man. In the same way a star gave a sign at his Nativity, in accordance with which the statement is made in the book on the Courses of the Planets, that all the planets and other stars serve in one way God made man, in another way pure man; in one way the Creator, in another creatures. Therefore God has willed so to order his affairs, as to show by means of the planets to rational souls certain things which he saw beforehand and predestined to happen, with the intention doubtless that the human mind, recognizing the wonders of God, might grow, inflamed to a love of its Creator. Therefore they maintain that the heavenly bodies are able to presage and signify this divine work of the conception and nativity, so that the creature may bear witness to its Creator coming in the flesh. Just as the sun was darkened, contrary to nature, at the Passion, wherefore the philosophers seeing these things said that either God conceded something to nature or the whole frame of the world would be dissolved. Without doubt they maintain that it is impossible for God to be subject to a creature, and that that divine work, in so far as it was from an infinite virtue and above nature, in no way was subject to a celestial arrangement; but that it had merely served for a sign. Yet in so far as the most pure Virgin was the true and natural mother of our Lord Jesus Christ, and functioned according to the force of nature in preparing in advance substance and in fostering it after conception and in like actions, they have maintained that the force of the heavens coöperated with the natural force of the glorious Virgin and aided her in so far as she functioned according to nature, because man begets man. For if there was anything of a natural character in that conception through preparation of substance and nourishment in the womb and the like, in so far as she was true and natural mother, they do not consider it improper to assume that the celestial arrangement is more than a mere sign, taking into account simply the things of nature. But whatever statement they make in this matter, must be reduced to the rule of faith, so that it may not be out of accord with catholic truth. Although all things do not suffice to show fully the secrets of that sect, yet they give convincing evidence on the question whether this is a sect, and also in regard to its quality in general, so that in our admiration of the wisdom granted to them we easily excuse their ignorance in falling short of full certification of Christian rite, since they had not been instructed in it. We should praise them because they agree with us and confirm our profession. But in the third part of the moral philosophy there will be a fuller discussion concerning this confirmation.

Since after the law of Mahomet we do not believe that any other law will come except the law of Antichrist, and astronomers likewise agree in this, that there will be some powerful one who will establish a foul and magical law after Mahomet, which law will suspend all others, it would be a very useful thing for the Church of God to take thought concerning the time of this law, whether it will come quickly after the destruction of the law of Mahomet, or much later. The philosopher Ethicus in his Cosmography states expressly that a race which has been shut within the Caspian gates,* will burst forth into the world, and will meet Antichrist and will call him god of gods. Without doubt the Tartars were within those gates and have gone forth. For already the gates have been broken, as we know. For Franciscans, whom the present lord king of France,

* Narrow passes in the Taurus.

Louis, sent, crossed with Tartars through the midst of the gates far beyond into the mountains where they were shut up. It is known not only to all nations of the East that the Tartars have gone forth from those localities, but also to those who are well acquainted with the geography of the world, and know its habitable parts and the differences in regions by means of astronomy and through authors like Pliny and Martianus and others who describe the regions of the world, and also by means of the histories. I do not wish in this matter to be presumptuous, but I know that if the Church should be willing to consider the sacred text and prophecies, also the prophecies of the Sibyl and of Merlin, Aquila, Seston, Joachim, and many others, moreover the histories and the books of philosophers, and should order a study of the paths of astronomy, it would gain some idea of greater certainty regarding the time of Antichrist.

But these matters may now be dismissed. I shall now introduce a subject not only in accordance with its advantage, but also its appropriateness to the Church, a subject without which great peril and confusion cannot be avoided; although for long periods there has been manifold abuse in this matter. Since all this error proceeds from pure ignorance and negligence in its study, it is so much the more contemptible in the sight of God and of holy men, and in the sight of all men as well as of learned astronomers. But even ordinary computers are aware of the manifold error, and write in regard to it as well as do astronomers. The writings of both these classes are circulated throughout the Church of God, in which these errors are noted and advice is given in regard to their removal.

The matter I have in mind is the correction of the calendar used by the Church. Julius Caesar, instructed in astronomy, completed the order of the calendar as far as he could in his time, and, as the histories state, he disputed against Achorius the astronomer and Eudoxus his teacher in Egypt regarding the length of the solar year, on which our calendar has been based. Hence, as Lucan states, he himself said, "My year shall not be conquered by the calendar of Eudoxus."

But Julius did not arrive at the true length of the year, which he assumes to be in our calendar 365 days, and one fourth of a day. This fourth is summed up through four years, so that in leap year it is reckoned as one day more every fourth year than in other ordinary years. But it is clearly shown by all computers ancient and modern, and rendered certain by astronomical proofs, that the length of the solar year is not so great, nay, less. This deficiency is estimated by scientists to be about the one hundred and thirtieth part of one day. Hence at length in 130 years there is one day in excess. If this were taken away the calendar would be correct as far as this fault is concerned. Therefore, since all things that are in the calendar are based on the length of the solar year, they of necessity must be untrustworthy, since they have a wrong basis.

In the second place, there is another greater error, namely, regarding the determination of the equinoxes and solstices. For this error arises not only from the length of the year, but has in itself grave errors. Moreover, the equinoxes and solstices are placed on fixed days, as if they had always occurred on them and must continue to do so for ever. But astronomers are certain that they are not fixed, nay, they ascend in the calendar, as is proved without doubt by tables and instruments. At the beginning of the Church the winter solstice was placed on the eighth day before the Calends of January on the day of our Lord's Nativity, and the vernal equinox on the eighth day before the Calends of April on the feast of the Annunciation of the Glorious Virgin, and the summer solstice on the eighth day before the Calends of July, namely, on the day of the nativity of the blessed John the Baptist, and the autumnal equinox on the eighth day before the Calends of October. Of this opinion was Hippocrates the physician; with whom Christians easily agreed, because the blessed John said, "He must increase, but I must decrease." Hence some expounders of sacred Scriptures have maintained that the Lord was born when the day increases, and this happens at the winter solstice; and that John Baptist was born when the days begin to decrease, that is, at the summer solstice. Isidore held this view of Hippocrates, as is clear from the eighth book of the Etymologies, also Saint Anatolius, who at the beginning of the Church discussed matters of this kind; but later ecclesiastics changed these dates, and decided that the vernal equinox is on the twelfth day before the Calends of April, as Bede states in his book on Chronologies, and this date is still the accepted one. This is clear to all who know anything about computation, and who consider the practice of the Church. For as under the old law the Passover was celebrated after the vernal equinox when there had been the fourteenth moon, so the Church decreed that Easter should be reckoned from that date, and be observed on the Sunday following; because of necessity this feast among Christians must fall on a Sunday, and for this reason we are not able to celebrate Easter on the day of the fourteenth moon, but on the Sunday following. They have maintained that the earliest Easter is on the eleventh day before the Calends of April, that it cannot happen before that date, because the equinox, they said, is fixed on the twelfth day before the Calends when the moon can sometimes be the fourteenth. For as Bede says in his computation, the Church does not consider the fourteenth moon which precedes the equinox, but that which is on the equinox or after it; and the equinox itself, as he says, is on the twelfth day before the Calends of April, and from this date is Easter reckoned, as all computers know: because if it falls on the Sabbath then on the day following Easter can be kept, because this day is Sunday. Therefore because of the fixed date of the equinox on the twelfth day before the Calends of April, the Church says that the earliest Easter is on the eleventh day before the Calends of April.

But although the Church in its practice in the beginning held that the equinox is on the eighth day before the Calends, and later changed the date, and now considers the equinox as fixed on the twelfth day before the Calends of April, yet it is certain that the equinox is not on those dates, but is ascending in the calendar far from them, and likewise the solstices and the other equinox. For this year the winter solstice was on the Ides of December, twelve days before our Lord's Nativity, and the vernal equinox on the third day before the Ides of March, and the summer solstice is on the seventeenth day before the Calends of July, and the autumnal equinox on the sixteenth day before the Calends of October. This fact can not only the astronomer certify, but any layman with the eye can perceive it by the falling of the solar ray now higher, now lower, on the wall or other object, as any one can note. They will be changed from these dates in subsequent times, because after about ninety-four years, that is, in the year of our Lord MCCCLXI, the winter solstice will be on the day before the Ides of December, and the vernal equinox on the fourth day before the Ides of March, and the summer solstice on the eighteenth day before the Calends of July, and the autumnal equinox on the seventeenth day before the Calends of October; that is, each of these will be one day earlier than it is at present. For after about 125 years it comes a day earlier. This happens from the error in the length of the year, because the sun lacks by one day in about so much time the length of a year. And this is because it accords better with the years which we reckon from the Incarnation. For Ptolemy, in the one hundred and fortieth year, found the vernal equinox on the eleventh day before the Calends of April and the winter solstice on the eleventh day before the Calends of January, as is clear from the Almagest. But from this position in the calendar to the Ides of December, when the solstice now happens, are nine days by which this solstice has risen in the calendar. But from the year of Ptolemy's verification there are now 1127 years of our Lord, because the present year is the twelve hundred and sixty-seventh, from which if there are subtracted the 140, which passed from the Incarnation to the verification of Ptolemy, there will remain 1127 years. But in this time just stated 125 years will be found nine times and two years over. Wherefore this time, namely, 125 years, is in sufficient agreement with the number of years of the Christian era, so that always one day in such a period of time should be substracted from the length of the year, and denote the change in the solstice and equinox.

I protest, however, that in so great a difficulty I am not speaking in precise terms, but there is much at hand to attest the truth, until a final determination is reached in regard to the length of the year and the change in the solstices. By what has now been said, however, according to the verification of Ptolemy the winter solstice could not have occurred on the eighth day before the Calends at the time of the Nativity of our Lord, but must have been on the tenth day preceding the Calends; because in the 140 years from the time of the Nativity to the verification of Ptolemy the solstice could not have changed three days, nor two, but one and a small part of another. Therefore the tenth day before the Calends of January could have been, according to Ptolemy's discovery, the eleventh before the Calends. Since by this same verification of Ptolemy the equinox could have taken place in the first year of the Nativity on the eleventh day before the Calends of April; it could not have been on the eighth, as was at first the belief of the primitive Church, and still more was it impossible for it to happen on the twelfth day before the Calends of April, as is now the belief according to the usage of the Church. Because since the equinox is always mounting in the calendar, and in the time of Ptolemy was on the eleventh day before the Calends of April, previous to that time it was after that date and nearer to April. According to these facts, therefore, neither are the equinoxes and solstices fixed, nor are they fixed on those days set by the Church. Nor was Hippocrates far from the truth, since he was more than three hundred years before Christ, and therefore the equinox could have happened in his times on the eighth day before the Calends, or nearly so, namely, on the seventh.

But the third disadvantage is a far greater one. For, as we have shown, the truth is that without error Easter ought to be celebrated on the Sunday after the fourteenth moon which is found either on the equinox, or after the vernal equinox, because of the conformity of the Christian law to the old law, after the observance of the Passover, which in the first place was in the old law, and preceded as a figure of the new Passover. Since, therefore, it is true that the equinox is on the third day before the Ides of March, and it is possible that on that date occurs the fourteenth moon, namely, in the fourteenth year of a nineteen-year cycle, so that the first moon is reckoned on the day before the Calends of March above the letter C, Easter must of necessity be on the Sunday next after that day in accordance with the truth. But from this fourteenth moon Sunday cannot be farther distant than up to the thirteenth day before the Calends of April, as is manifest in the Calendar. But this is before the eleventh day before the Calends of April on which the Church celebrates the earliest Easter. Wherefore in that fourteenth year of the cycle glorious Easter will be observed at the wrong time. The same thing occurs in the third year of the cycle, as is shown by the golden number on the Calends of March. For Easter will happen on the twelfth day before the Calends of April or short of it. Since Sunday in these years can happen on the day before the Ides of March, and on the

Ides, and so on, up to the twelfth day before the Calends of April, therefore, for various reasons should Easter be observed on those days, which cannot be if the earliest Easter is on the eleventh day before the Calends of April. Since, moreover, the true equinox mounts more and more in the calendar, about the fourteen hundred and eighty-first year it will be on the fifth day before the Ides of March. The equinox will in this way approach continuously the beginning of March, and will pass out of March, due to the error in the length of the year. Easter will then have to be observed around the first of March or in February, and in this way its observance will always come at an earlier date in accordance with the earlier date of the equinox. But this is a most serious disadvantage; because thus not only Easter, but Lent and all the movable feasts will recede in a shocking manner from their positions and the whole order of the ecclesiastical office will be confused. Moreover, since as a matter of fact Easter can be observed before the eleventh day before the Calends of April in accordance with the truth respecting the equinoxes, and this by many days, and by as many as we wish, according to the earlier date of the equinox, of necessity the true beginning of Lent likewise ascends in the calendar in like fashion before it begins according to the usage of the Church, and thus in the real Lent meats will be eaten for many days; and to such an extent is it possible for the date of the equinox to be advanced that the whole of the true Lent will be at a time when Christians will be eating meats, which is very absurd. In this way the Easter feast, by which the world is saved, and the feast of Pentecost, by which the grace of God is diffused in the Church, and the other movable feasts suffer violence; and consequently the other immovable feasts are disturbed also, because they yield to Easter and to the other feasts of its class, as is clearly known. Since these are dreadful things in themselves, they are the more foolish and worthy of derision; since owing to ignorance and negligence the devil has brought about this state of affairs in the Church of God. For it would be necessary only to verify the length of the year and the equinox. If the verification of the length of the year is conceded to the most skillful astronomers, yet is it easier to verify the equinox, because in approximately 120 years it changes only by one day according to the more approved opinion in these days. Scientific skill would be available in this matter, and the equinox could be found for all future ages to the end of the world and beyond, an easy task, and tables could be made and a canon regarding it, and published throughout the whole Church of God, and placed with the calendar, and then would there be no obstacle in this particular. Likewise the other equinox could be found, and the solstices with ease, and all error in these matters would be eliminated to the praise and honor of God, and evil report and stumblingblock be removed, which are at present numerous among all computers and astronomers and scientists owing to the errors which in this matter are permitted by the Church to be spread abroad.

But a greater disadvantage comes from the beginning of lunation, as given by the golden number in the calendar. For any one can see with the eye, if he looks at the sky, that the new moon occurs as a matter of fact three or four days before it is marked in the calendar, and every seventy-six years the beginning of lunation recedes from its place in the calendar sixteen minutes* of one day and forty seconds, and this is more than a fourth of one day, and nearly a third of a day, because sixteen minutes and forty seconds are six hours and forty minutes of one hour. In every 304 years it recedes from the place of first lunation in the calendar one day and six minutes of a day and forty seconds. After 4256 years the moon, according to the calendar, will be called new when it is full of light. After 7904 years there will be an error of one whole lunation, with the exception of a small part, namely, thirty-eight minutes and thirty seconds. This error can reach to one hundred lunations, and then again the first error will return, so that successive ones may follow in order, and thus these errors roll around forever. Skillful astronomers have no doubt that all these statements are facts. Moreover, every computer knows that the beginning of lunation is in error three or four days in these times, and every rustic is able to view this error in the sky.

As briefly as I can, I shall give a general exposition of the errors mentioned. Our calendar assumes a cycle of nineteen years to be equal to nineteen solar years taken with a whole fourth, and accordingly the calendar uses a solar year of this kind, as we have explained previously. This cycle contains all

* See Bridges' note, Vol. I, p. 275.

the lunations which happen in nineteen lunar years. These are reckoned as 235, because in any lunar year, which is called common, we have twelve lunations, and to these we add seven lunations in a whole cycle of nineteen years to make up deficiencies, which occur in twelve lunations with respect to the twelve solar months in a solar year, because the lunar year consisting of twelve lunations has only 354 whole days, and the solar year eleven more, for it has 365. Therefore the lunar year ends eleven days sooner than the solar year; and then those eleven days in excess are collected together up to the third year, and one lunation or lunar month is formed. Thus in the third year of a cycle a lunar month is placed, and is called intercalary and the year an intercalary one, that is, an excess year, because it has thirteen lunations; and so on through the whole cycle are collected these intercalary months, so that they become intercalary years, in order that a nineteen-year cycle of lunations may be equal to nineteen solar years. For according to Ptolemy in the Almagest and all other astronomers the lunar month is not properly reckoned from the sight of the new moon, since this time is unequal: because sometimes in the morning of the same day there is the crescent form of the old moon, and in the evening the crescent of the new moon, and sometimes there is a space of two days between them, and at other times three days intervene, as is plain to the sense, and the reasons are given by astronomers; but months should be equal. Likewise, it does not begin with the conjunction of the sun and moon according to their true course, because this time is unequal; and therefore it will be reckoned with respect to the conjunction of the sun and moon according to the mean and equal course of each, because this time is equal and uniform. According to the proof of Ptolemy in the Almagest, this time is twenty-nine days and thirty-one minutes of one day, and fifty seconds, and eight thirds, and nine fourths, and twenty fifths. Arzachel agrees with this who based his tables on the length of a lunar year containing 354 days and a fifth and a sixth of one day, that is, twenty-two minutes. According to this the time of an equal lunation will be twenty-nine days, and thirty-one minutes, and fifty seconds, which multiplied by twelve make 354 days and twenty-two minutes. But Arzachel omitted the thirds and fourths and fifths of Ptolemy, because in a very long period of time they amount to but little.

If we assume, then, the time of equal lunation is twenty-nine days, and thirty-one minutes, and fifty seconds, it happens that the least time reducing whole lunations to the same beginning of time is thirty years of the Arabs, which contain 360 whole lunations; and this period contains 10,631 days exactly. Since as one lunation is twenty-nine days, and thirty-one minutes, and fifty seconds, twelve lunations, which make one year of the Arabs and one lunar year, will contain 354 days and twenty-two minutes of one day. But 354 multiplied by thirty make 10,620, and twenty-two multiplied by thirty make eleven whole days, which added to the preceding make 10,631, reducing first equal lunations to a similar beginning of time. Since, therefore, thirty years of the Arabs form the least time which reduces whole lunations to a similar beginning of time, it is not possible that any other time should do this, unless it be equal to this time or a multiple of it. But twenty-nine years of our calendar are not equal to thirty years of the Arabs, nor a multiple of them, because thirty years of the Arabs contain twenty-nine solar years, and one month, and eight days. Therefore it remains that nineteen solar years cannot equal a true cycle of first lunations, nor consequently the nineteen years of the nineteen-year cycle. Therefore the nineteen-year cycle is not a true cycle of first lunations.

Moreover, this fact can be made clear in greater detail, so that the disadvantages mentioned above may be apparent. For successive periods of nineteen solar years are not of the same length, because in the first cycle there are only four leap years, and in three others there are five, because the fourth year in the first cycle is a leap year, and the eighth, twelfth, and sixteenth, because the fourth year is always a leap year. Therefore in the second cycle the first is a leap year and so on, so that in this cycle there are five leap years, as is clear to any one considering the matter, and likewise in the other two; and then the cycle returns again which has four leap years, and the three follow which have five and thus continually does the order of cycles proceed. The first cycle of nineteen solar years has exactly 6939 days, and in these years are assigned 235 lunations. But if we multiply the period of an equal lunation, that is twenty-nine days, and thirty-one minutes, and fifty seconds, into 235, the result will be 6939 days and forty minutes and fifty seconds, which are more than two thirds of one day. Wherefore after the completion of nineteen years containing only four bissextile days, 235 lunations are not yet completed, but lack forty minutes and fifty seconds of one day. But any period of nineteen years having five bissextile days contains 6940 days. Hence since 235 equal lunations have 6939 days and forty minutes and fifty seconds, nineteen years with five leap years exceed 235 lunations by the space of nineteen minutes and ten seconds, which is about one-third of a day. This is manifest, if from the one day by which the nineteen years first mentioned exceed the 235 lunations there are subtracted the forty minutes and fifty seconds, by which those lunations exceeded the nineteen years with only four bissextile days. Thus is it evident that periods of nineteen solar years are not equal to one another in turn. But four times nineteen years, making seventy-six years, return to the same beginning in time, and are always equal to other periods of seventy-six years. If, then, there are accumulated three times the nineteen minutes and ten seconds, which every period of nineteen years with leap years adds in excess of 235 lunations, there will accrue for us fifty-seven minutes and thirty seconds, by which those thrice nineteen years with five leap years taken together exceed their lunations. But nineteen years with only four bissextile days are less than their lunations by forty minutes and fifty seconds. If, therefore, these forty minutes and fifty seconds are subtracted from the aforesaid fifty-seven minutes and thirty seconds, there will remain sixteen minutes and forty seconds, by which four times nineteen solar years, making seventy-six years, exceed their lunations.

This is the first notable error, which I noted at the beginning, from which follows the second. For if we take four times seventy-six years, making 304 years, they will exceed their lunations by four times sixteen minutes and forty seconds. But these taken four times make one day and six minutes and forty seconds. There remains then the second error that 304 years exceed their lunations by one day and six minutes of one day and forty seconds. Therefore it happens that after 304 years according to the calendar we call the moon the first, although the existing one is of greater age to the amount of one day and six minutes and forty seconds. If these are facts, it is manifest to the inquirer that with the multiplication of time will be multiplied the inconveniences. Therefore after 4256 years according to the computation of the calendar the moon will be denoted as new when it is filled with light. And further, one whole lunation will superabound; and at length a hundred lunations in a very great period will be in excess, as is manifest by the principles just given. Nor is it necessary to make greater computations for the present plea, since proofs are ready for this matter, when the fitting time arrives. Very manifest, then, is the error of first lunations according to the calendar, and not a single one but multiplex, nor small but well-nigh infinite.

Since, moreover, the error of the nineteen-year cycle is in the separate years, it follows that likewise will there be an error in the cycles of epacts. For the eleven days in a solar year in excess of a lunar year are called the epact of the following solar year, as it were, *epiaucta,* that is, additionally increased; and by those eleven days is the age of the moon increased on the first day of the following year, and by the same days is the age of the moon increased at the beginning of each month of the coming year beyond its age at the beginning of each moon of the year prior. Because of the existence of these errors, of necessity the true age of the moon must be ascertained by other means than cycles of this kind.

The remedy for all these ills is found in the fact that we are able to learn the primation of the moon according to astronomical truth if we count time according to the years and months of the Arabs; because the first day of each month of the year of the Arabs is the day of the conjunction of the sun and moon according to the mean course of each. Hence if we desire to call this day the first day of the age of the moon, then when the beginnings of the months are known the beginnings of lunations are also known. But if we wish to begin our first lunations on the first day of the visible moon or on the second or third, we shall begin our lunations with the second or third day of each month of the Arabs, and we shall proceed uniformly in the computation of lunations and we shall encounter no error. Because the knowledge of the years of the Arabs and the beginnings of their months will give us the true knowledge of the beginnings of lunation, all we need do then is to have recourse to the tables and canons of the years and months of the Arabs. I make this statement in accordance with the scientific principles which are the common property of Latin and Arabic astronomers. If, however, we should wish to proceed in another way, we can with the same certainty but on greater authority prove what we wish, namely, by means of the tables of the Greeks and especially of the Hebrews. For from the beginning the Hebrews have been very skillful in the knowledge of astronomy; and all nations have obtained this science as well as the other sciences from them, as I proved in what preceded. Therefore, if any one will consult the tables of the Hebrews for the setting of the sun at Jerusalem, he will find the truth in such matters.

But although the errors I have mentioned are terrible in themselves, yet they bear no comparison to those which follow from the facts now stated. For the whole order of Church solemnities is thrown into confusion by errors of this kind respecting the beginning of lunation according to the Calendar, as well as by the error in determining the equinoxes. And not to refer to other years for evidence of this error, I shall state the case in this present year. For not only the mean conjunction of the sun and moon this year was on the sixth day before the Calends of April, above the letter B, but also the moon's first illumination and appearance. Therefore the moon was the fourteenth on the fifth day before the Ides of April, above the letter A, and the fourteenth moon is the terminal date of Easter, and the Sunday following is Easter. Wherefore at the letter B following, namely, on the morrow, that is, on the fourth day before the Ides of April, should be Easter as a matter of fact. But it will be transferred eight days farther on owing to the beginning of lunation taken according to the golden number. For the golden number of this year is fourteen, which is placed on the third day before the Calends of April above the letter E, and there the beginning of lunation is said to take place according to the calendar. Therefore according to this the fourteenth moon, which is the terminal date of Easter, is the day before the Ides of April, and on the Sunday following is Easter placed, namely, on the fifteenth day before the Calends of May, and thus eight days beyond the true date. Wherefore the feast of

Easter, by which the world is saved, will not be celebrated at its proper time, but there is fasting this year through the whole true week of Easter. For the fast continues eight days longer than it should. There follows then another disadvantage that the fast of Lent began eight days too late; therefore Christians were eating meats in the true Lent for eight days, which is absurd. And again then neither the Rogations nor the Ascension nor Pentecost are kept this year at their proper times. And as it happens in this year 1267, so will it happen the year following. For according to the golden number Easter will fall on the fourth day before the Ides of April in the year of our Lord 1268. But it should be on the third day before the Nones of April, eight days earlier, because the first moon is three days earlier than it is given by the golden number. Therefore on the fourth day before the Calends of April, or at least on the third, will occur the fourteenth moon, which is the terminal date of Easter; wherefore the first Sunday following will be the day as a matter of fact. Just as it happens this year, so also more frequently can it occur through the earlier date of the equinox and of the beginning of lunation, that Easter will be celebrated not only eight days, but many more before it should, also the other feasts, just as we explained before in regard to the equinox. For after a long period of time it will happen that the moon is called new when it is at any distance whatever from the sun, as we explained before in the errors in the beginning of lunations. Therefore a very great inconvenience and an intolerable one will here follow.

Since, then, all astronomers and computers can see these errors, and every man who is willing to investigate the truth can see them, it should be necessary and required and well pleasing to God and desired by all men of science that the remedy be applied. The remedy would be an easy one; for the true equinox would be found by the tables of astronomy and by instruments, and the beginning of lunation would be attested by the same means, so that all kinds of the aforesaid errors would be eliminated, and the fourteenth moon from the equinox would be taken, whether it fell on the day of the equinox or after it, and there would be the terminal date of Easter, so that Easter day would fall on the Sunday following. Therefore the golden number would not have to be followed as something fixed regarding the Calends. Tables, moreover, could be made for the beginning of lunations and for the equinoxes, and in accordance with these could the Calends be arranged in the year and in the months, following a similar device used by the Hebrews.

But in opposition to this it may be said that the Council of Nice decided that the paschal beginnings of lunation should be found according to the nineteen-year cycle. The blessed Pope Leo, when this dispute was under discussion, decided to adhere to the opinion of the Council of Nice. Bede also in his book on Chronologies in chapter XLIII strives as far as he can to prove correct this cycle and the beginnings of lunation occurring through the course of this cycle. For he introduces the authority of the Council of Nice and of Pope Leo and cites a miracle. For when many maintained at a certain time that Easter should be celebrated on the eighth day before the Calends of April, and others on the tenth day before the Calends of May, as the order of the nineteen-year cycle required, the dispute was settled by means of a certain baptistery, in which on the night of Easter yearly the sacred font was filled with water of itself, and after the people had been baptized the water went away as it had come. This happened on the tenth day before the Calends of May as the reckoning of the cycle required, and not on the eighth day before the Calends of April.

But if these things are rightly understood they do not contradict the truth. For when Eusebius, bishop of Caesarea, first arranged this cycle, he was very little prior to the Council of Nice, so that as his arrangement of this cycle was in correct accordance with the course of the moon at that period, there could not have been a sensible change at the time when that sacred Council was held. Therefore the holy fathers decided that the cycle be followed, because it did not at that time contain error, nor for a long time thereafter was it notably in error. Hence since Pope Leo was subsequent to the Council of Nice about one hundred and twenty years, under the Emperor Martian, it is manifest that the beginning of lunation cannot have receded from its place in the calendar from the time of Eusebius more than one day. Since the Pope himself discussed this error, he at least gave posterity the opportunity of considering the truth in this matter. But the error was a new one and not very great, and the Council of Nice was of great authority, for which reasons it was not expedient at that time to go contrary to the decision of this Council.

Moreover, he did not find astronomers in the Church equal to this task, because from the beginning astronomy was hateful to Christians for the reasons stated above, namely, in the second part of this treatise. Nor yet up to the present have any been found to give the remedies in this matter. For there were many who knew how to prove those errors by astronomical means, and to what extent there is error; nor is this surprising, since our vision shows many of them to us; and in general they have stated what the remedies were, namely, that the length of the year should be attested, that the equinoxes and solstices should be accurately determined, and the beginning of lunations likewise; and they discuss the general methods applicable to these matters. But no one has yet given us the true length of the year, with full proof, in which there is no room for doubt. Nor have they given us what we require in respect to the other questions involved. Therefore, it was not strange if these things were not attested in the times of the early Church, when mathematics was reprobated among arts under suspicion and useless to the Church of God, like many other noble sciences, owing to the five reasons noted in the first part of this work. The Council of Nice decided that the cycle should be followed while it contained the truth, and because in those times it was not in error. But it decided that this should be observed, in order that the contention arising from individual judgment might be avoided, until there should be in the Church of God the potency of mathematics, by which all things in dispute can be attested. For it is an incomparably lesser evil that a single inconvenience should be borne for a time owing to the impossibility of a remedy, than that one should promulgate his own opinion equally as false as the one held commonly by all. So speaks Bede. For he says, "When they thus determined the moon according to the reckoning of the cycle, they avoided another greater peril by this means." And as to the fact that Bede adduced a miracle, it must be conceded that in those times there was no error in the cycle, as is manifest, but later an error grew and appeared sensibly.

Also the fact that he adduces the case of the stone selenite, whose splendor increases with the moon in the beginning of the month and then decreases, by which in ancient times they learned the beginning of paschal lunations, is not an argument in favor of the cycle except while it was correct, but rather the reverse; because that stone shows us the true equinox, and the fourteenth paschal moon at or after it to be taken as the terminal date of Easter, which this cycle in these times cannot do, nor will it ever do so, nor has it done so for a long time in the Church. But Bede lived about the time of the introduction of the cycle. For he himself states that in his computation he at that time had reached the seven hundred and first year of our Lord. From the Incarnation up to the times of Constantine, under whom the Council of Nice was held, they were, according to Bede, 332 years. And from this time to the Emperor Martian there were 120 years, under whom was Pope Leo, which are in all 452 years. From this it is evident that Bede was not 300 years later, but even from the Council he was distant only 359 years, and for this reason the lunation could not have receded much from its position. For the recession could not have been more than one whole day and part of a second, because in 304 years there is a change of one day. Therefore for the reasons discussed above the cycle could have been followed properly in the time of Bede, just as well as in the time of Pope Leo. But, however, from the time of Leo a scruple of doubt has arisen, and this doubt was increased in the time of the blessed Pope Hilary, who directed Victorius to compose a new cycle, in which, since there was no certainty, even as there was none in the nineteen-year one, posterity has adhered to ancient custom and to the decree of the Council of Nice.

It is evident from these facts that, without impairing the full authority of the Council of Nice, this cycle can be changed; because at that time there was no error, and the cycle was kept, that a greater danger might for the time be avoided, until the Church should be able to have the aid of astronomy, by which alone the remedy can be applied. For the early Church did not have the advantage of astronomy. Therefore subsequently, up to the present, has the correction of this matter not been made because of long custom, and because the study of astronomy was not taken up by prelates and by the great body of students, nor is it pursued up to the present; although there are some quite able and skillful in this particular. But at the present time the remedy must be applied because of those manifest and palpable errors, and because of the great discredit brought upon the Church. For all those trained in computation and all astronomers know these things, and deride the ignorance of prelates who maintain them. Unbelieving philosophers also, Arabs, Hebrews, and Greeks, who dwell among Christians, as in Spain, Egypt, parts of the East, and in many other regions of the world, abhor the folly which they behold in the arrangement of the chronology followed by the Christians in their festivals. At length Christians possess skill in astronomy, by which attestation can be made. Therefore your Reverence has the power to command it, and you will find men who will apply excellent remedies in this particular, and not only so in the aforesaid defects, but in those of the whole calendar. For there are thirteen principal ones with an almost infinite number of branches. If then this glorious work should be performed in your Holiness' time, one of the greatest, best, and finest things ever attempted in the Church of God would be consummated.

Since we have stated how urgently mathematics is needed for philosophy, theology, and the Church of God, we must now show how necessary it is for directing the commonwealth of the faithful. It is important in two principal ways: in one, because of the knowledge it gives us of the future, present, and past; in the other, in useful works. Since the human race is exposed to countless dangers respecting the future, it is very necessary that it have the means of learning the future. Since God has granted man the greater things, namely, soul and body, and promises eternal life, he must not have denied things that are less. For the sun rises upon the wicked and the seas lie open to the pirates. Much more, therefore, will God give to the good a useful knowledge of things, and especially to the multitude, because in it is found the general good. Moreover, since there are always found in the world some good men and pleasing to God, therefore has God given by this means a manifold knowledge of the future, nor can the world stand without it, as Avicenna shows in the fourth book on the Soul, and in the tenth book of the Metaphysics. But the principles concerning the knowledge of the future have been touched upon above in that distinction in which I argued in excuse of mathematics, and showed that a sufficient judgment is possible in all things, which rests between what is necessary and what is impossible, and between the universal and the particular. For by these means is the human mind illuminated, so that it is able to discourse wisely on all topics and to provide advantageously for itself and for others. I took up later, as the matter required, particular judgments in human affairs, namely, in the distinction on sects. If, moreover, in human affairs and especially in matters of this kind a true and useful judgment may be made, much more can this be done in the things of nature, in a particular as well as proper consideration.

Although in my defense of mathematics, and above in my comparison of celestial forces with those here below, I touched upon the knowledge of the places in the world and of the things generated in them by means of the heavenly bodies, yet now shall I set forth this matter more fully in passing to the art of healing the human body, a knowledge of which is more necessary to man than that of anything else in this world. Not only do I show how we gain knowledge of things in different localities of the world, but also how they are caused in those same places by the different courses of the seasons. But an effect is known only by means of its cause, as all know; but celestial things are the causes of things below, whence these things that are generated must be known through those that are not generated, namely, celestial things. That, moreover, celestial things not only are universal causes, but the proper and particular causes of things below, is proved by Aristotle, who says in the second book on Generation, that the elements are less active than the tools and instruments used in an art are in comparison with the artificer. But the whole action in an art is attributed in the main to the artificer, as, for example, to a builder, not to his tool, as, for example, an axe. Therefore, it is manifest that the principal action of all terrestrial things is attributed to the heavens, since the only agents are the heavens and the elements, which are the instruments of the heavens. Moreover, this is manifest inductively. For in all inanimate things the heavens are the particular cause without contradiction; because inanimate things do not generate anything, nor do they produce individuals of their own species, because a stone does not generate a stone, as a human being does a human being and an ass an ass. It is manifest then that the force of the heavens incorporate in the matter of the elements produces all inanimate things, and animate things likewise through putrefaction; for in the generation of these things there are only the heavens and the elements. Moreover, Averroës says in the seventh book of the Metaphysics that the force of the sun does the same thing in putrefied matter as the paternal force in the seed, wherefore the heavens are the particular cause up to the generation of things by propagation. But that the same principle holds there, I shall prove. For in the Vegetabilia Aristotle says in the first book on Plants that the sun is the father of plants and the earth the mother; and in the Animalia he maintains this same principle, because in regard to man, concerning whom the principle seems less applicable, he says in the second book of the Physics that man and the sun generate man out of matter, and it is evident that a father does not continue nor terminate generation, but only begins it by letting the seed fall; wherefore of necessity that which continues and perfects generation is the sun or the force of the heavens. Moreover, not only are the heavens the cause in things correctly generated, but also in the faults of nature and in monstrosities. For Avicenna in his eighteenth book on Animals says, "If the embryo is not able to receive human characteristics, it will receive those of an animal, as in the case of monstrosities: for example, when the son of a man has had the head of a ram, and a lamb the head of a bull: since the force in it has produced a form in accordance with celestial shapes, which happen to each one. If, moreover, we proceed further, we are able to investigate the causes of terrestrial things more properly by means of celestial things. But the first principle in this is that every point of the earth is the apex of a pyramid filled with the force of the heavens.

But in addition, that my purpose may be seen more surely and plainly, we must consider what the diversity of the regions of the world is, and how the same region varies in different seasons, and how the different things of the same region receive different qualities at the same time. But these things cannot be known unless we form clear notions of the size and form of the habitable earth and of its climates. But in order that we may obtain such ideas, we must assume that the world is of a spherical form, as we proved above. We shall imagine three lines drawn from the terminals of the world intersecting each other at the center of the world at right angles, so that one is from right to left in the heavens, and this is due to the fact that it is drawn from east to west through the center of the world; the second up and down, that is from south to north; and this is from the Antarctic pole to the Arctic pole; and the third line forward and backward, that is, from the point in the middle of the heavens above us to the opposite point in the heavens beneath the earth. By a certain transumption in the word it is called the angle of the earth. Thus Aristotle teaches us to imagine six differences of position in the heavens in the second book of the Heavens and the World.

If, therefore, we imagine a circle passing through the east, the middle of the heavens, the west, and the angle of the earth, this circle divides the heavens into two equal parts, leaving one half with respect to one pole and the other half with respect to the other pole, and is called the equinoctial circle, both because the dwellers under it have a perpetual equinox, and for all inhabitants of the earth it is the equinox when the sun comes to that circle and describes it in a natural day. This happens at the beginning of spring and at the beginning of autumn when the sun enters the heads of Aries and Libra. If, now, we imagine another great circle, which crosses through the poles of the world and through the limits of the east and west, intersecting the former circle at right angles, which is called the colure passing through the equinoxes, then will the heavens be divided in four quarters, two of which will be above the earth in one location and the other two beneath the earth. One fourth to the north will be above us, namely, the fourth contained between half of the equinoctial circle and two quarters of the colure mentioned, terminated at the Arctic pole in one direction and at the points of east and west on the equinoctial circle, as is shown in the figure, and this is the fourth part above. In a similar manner we must imagine that the earth is spherical, and those aforesaid three lines will pass through the center of the earth, intersecting one another at this point at right angles; for they are perpendicular to the earth, and since at their intersection the center of the heavens and of the whole world is located, of necessity the same point is the center of the world and of the earth; because that intersection is only at one point by the fifth proposition of the first book of Theodosius on Spheres. For if a straight line falls perpendicularly from the heavens to a plane surface tangent to the sphere of the earth, the center of the earth will be on that line by the fifth proposition, which says, if a plane is tangent to a sphere, and from the point of contact a straight line is drawn perpendicular to the tangent plane, the center of the sphere must be on this same line; but that same line will be perpendicular to the plane tangent to the sphere of the heavens, therefore on that line will be the center of the heavens, and this line is one of the three mentioned. Similarly each of the others, by the same reasoning, will pass through the center of the heavens and the center of the earth; but any one of these lines intersects another only in one point, therefore at that same point there will be the center of the heavens together with the center of the earth, and for this reason the earth lies at the center of the world. Therefore if we imagine two circles on the earth corresponding to those described in the heavens, one beneath the equinoctial circle passing through the east and west and the point on the earth beneath the point in the middle of the heavens; and the other through the east and west points and the points on the earth corresponding to the poles; then by these circles will the earth be divided into four quarters, of which two will be on the surface of the earth in our location; and the other two in the other direction of the earth. One will be to the north, namely, from the middle of the earth under the equinoctial circle up to the point of the earth beneath the Arctic pole contained below the lines which run from the east and from the west to the pole or to the point of the earth situated under the pole. This is the fourth part, about which we are inquiring, in which the habitation is known, and it is located beneath the quarter of the heavens noted above.

Now the subject of habitation is considered in two ways; in the first, with respect to the heavens, namely, how much can be inhabited on account of the sun and how much not. This question was mentioned previously in a general way, and will be touched upon later. In the second way, the size of the habitable earth in comparison with the water is considered, namely, how far the water hinders. We are now to consider this latter question. Ptolemy in his book on the Arrangement of the Sphere maintains that about a sixth part of the earth is habitable, due to the water, and that all the rest is covered by water. Therefore in the Almagest in the second book he states that habitation is not known except in a quarter of the earth, namely, in that in which we dwell; whose length is from east to west and is one half of the equinoctial circle; and its width is from the equinoctial circle to the pole, and is one fourth of the colure. But Aristotle maintains at the end of the second book of the Heavens and the World that more than a fourth is inhabited. And Averroës confirms this. Aristotle says that the sea is small between the end of Spain on the west and the beginning of India on the east. Seneca in the fifth book on Natural History says that this sea is navigable in a very few days if the wind is favorable. And Pliny teaches in his Natural History that it was navigated from the Arabic Gulf to Cadiz; whence he states a man fled from his king in fear and entered a gulf of the Red Sea which is called Arabic, which is distant about the space of a year's voyage from the Indian Sea according to Jerome in his letter, as will be explained below. Therefore the width of the earth through which the Red Sea extends is very great; from which fact it is clear that the beginning of India in the east is far distant from us and from Spain, since the distance is so great from the beginning of Arabia toward India. From the end of Spain beneath the earth the sea is so small that it cannot cover three quarters of the earth.

This fact is proved by the weight of another consideration. For Ezra states in the fourth book that six parts of the earth are habitable and the seventh is covered by waters. That no one may lessen this authority by saying that that book is apocryphal and of unknown authority, we must state that the sacred writers have used that book and confirm sacred truths by its means. Many times in the divine Office they use authorities from that book. Therefore, whether Ezra or another composed this book, it must be taken as an authority. Wherefore I say that although the habitable portion known to Ptolemy and his followers is reduced to less than a quarter, yet the habitable portion is more. And Aristotle was able to know more, because on the authority of Alexander he sent two thousand men to investigate the things of this world, as Pliny states in the eighth book of the Natural History. Alexander himself journeyed as far as the boundary of the east, and as is clear from the History of

Alexander and from the letters which he wrote to Aristotle, he always sent him word about all the wonderful and unusual things he found in the East. Therefore Aristotle could attest more than Ptolemy could. And Seneca likewise; because the Emperor Nero, his pupil, in similar fashion sent him to explore the doubtful things in this world, as Seneca states in the Questions on Nature. Therefore according to these facts the extent of the habitable portion is great and what is covered by water must be small. For toward the poles of the world the water must abound, because those parts are cold, owing to their distance, but cold multiplies moisture, and therefore from pole to pole the water runs down into the body of the sea and extends for no great width between the end of Spain and the beginning of India, and is called the ocean; so that the beginning of India can be far beyond half of the equinoctial circle beneath the earth quite close to the boundary of Spain. But that the truth in this matter may not be condemned as false, we must know that Spain in this locality is considered not as hither but as farther Spain, of which certain authors speak, as Pliny in the Questions on Nature, and Merlinus in his prophecy, and Orosius in his book Ormesta Mundi, and Isidore in the fourteenth book of the Etymologies. Since he teaches that between the Spain so called at present and the Africa so named at this time water did not flow, but the land was continuous in times past, but later the ocean broke through in the low parts of the earth and joined the Tyrrhenian Sea, which flows by the coast of the province of Arragone and of Italy. Therefore Hither Spain extends from the Pyrenees Mountains to Carthage: but Farther Spain crosses the Strait of Gibraltar up to the provinces of Africa. Hence it extends beyond Gibraltar and reaches the Atlas range. I have cited these statements from the authors mentioned, of necessity, in order that Aristotle and his commentator should not be derided through ignorance of Farther Spain, when they state as a proof of the smallness of the sea between Spain and India that elephants are found only in those two places. For it is a fact that elephants abound near the Atlas range, as Pliny states, likewise Aristotle, so too in India, and therefore in Farther Spain there is an abundance of elephants; but Aristotle says that there cannot be elephants in those places unless they are of a similar character, and if the distance were great they would not be of similar characer, and therefore elephants would not be in those places only. Wherefore he concludes that these places are nearer, and that the sea for this reason must be small between them.

The sea will not therefore cover three quarters of the earth, as is estimated. For let the upper half of the earth be *abcd,* in one quarter of which, namely, *abc,* is the habitable portion known to us. It is now clear that much of that fourth will be beneath our habitable portion, because the beginning of the east and the beginning of the west are near, since a small sea separates them from the other part of the earth. Therefore the habitable portion between east and west will not be half of the equinoctial circle, nor half of the circumference of the earth, nor twelve hours, as is estimated, but much more than half of the circumference of the earth, and more than the revolution of one half of the heavens. But how much this may be has not been measured in our times, nor have we found it attested, as it should be, in the books of the ancients; nor is this strange, since more than half of the quarter in which we are is unknown to us; nor have the states been included by the philosophers, as will appear from what follows. Similarly, if we speak about the two other quarters, and consider natural processes in accordance with natural philosophy, we shall find that those quarters are not covered by water, as the majority of mathematicians estimate. For since the poles and the regions near them are of the same remoteness from the sun and the planets in accordance with the relative position of the poles to the paths of the planets at the middle of the world, between the two tropics, there must, in accordance with these facts, be equal dispositions of land and water in our quarter and in the quarter beyond the equinoctial circle toward the other pole; and likewise in the fourth under our feet up to the equinoctial circle: and in the fourth beyond the equinoctial circle there will be a similar arrangement according to the tenor of the preceding statements. Therefore if our quarter is not covered by waters at least up to the latitude and distance from the equinoctial circle of 66 degrees, as at the end of the islands of Scotland

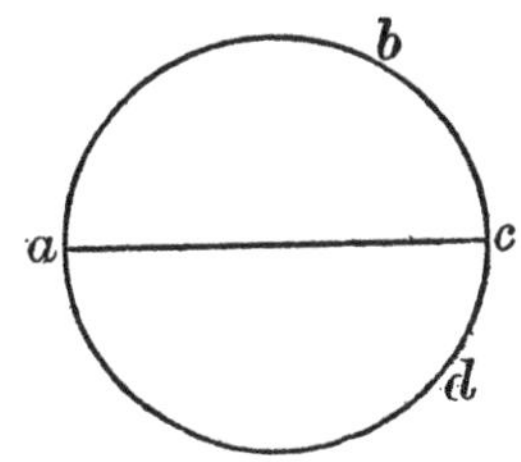

FIG. 22.

and in the kingdom of Norway, it is manifest that a similar natural cause will exist in the other quarter beyond the equinoctial circle in the upper part of the earth as in the one in which we are, because remoteness from the path of the sun induces cold, and cold multiplies moisture, and for this reason about the poles there will be a natural gathering together of waters, and in the regions that are near. Therefore in the other quarter beyond the equinoctial circle there must accordingly be much that is habitable; at least as far as the regions whose latitudes reach 66 degrees, just as is the case here.

But according to mathematicians a larger habitable portion can be assumed there than in our quarter due to the lack of water, since in that part is the point opposite to the aux of the sun, and the sun draws much nearer there to the earth. Whence it must necessarily parch that quarter in some portion of it, and render the remaining portions as far as the pole hotter than the portions of our quarter in which we dwell. We can argue in a similar way in regard to the remaining quarter under that one. And again, Aristotle in the first book of the Heavens and the World and Averroës are cited in support of the view that the remaining half of the earth beyond the equinoctial circle is a locality elevated in the world and nobler and for this reason most fitted for habitation. Wherefore it will happen according to the arrangement of nature that hindrances to habitation are to a greater degree excluded, at least in a large part of that half, namely, in that portion further removed from the point opposite the aux of the sun, if we assume an eccentricity, and everywhere if an eccentricity be not assumed, and this fact is due to nobler stars in that part, as Averroës maintains in the first book of the Heavens and the World. Moreover, Ptolemy states in his book on the Arrangement of the Sphere that nature requires that there should be two races of Ethiopians, namely, under the two tropics. From which some argue that there is habitation beyond the equinoctial circle just as there is this side of it; and according to these views the habitable portion will not have the shape of a quarter of the sphere nor of a semicircle drawn in a plain, nor will the water encompass the world through the poles and the east and west covering three quarters of it, as is believed; but rather will the form of the water be of this kind or similar to it, so that this sea may be called ocean, having most of its water about the poles, extending in length from pole to pole between the beginning of India and the end of Spain, which is known to the mathematicians.

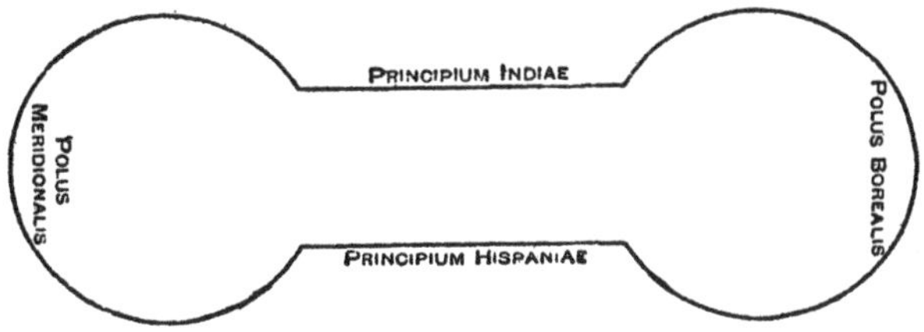

Fig. 23.

Since the habitable portion is known to mathematicians only in the quarter in which we are, and also the whole quarter is unknown which is contained within half of the equinoctial circle and half of the colure crossing through the poles and the extremities of the equinoctial circle beneath this figure, therefore, in discussing the opinion of the mathematicians we must omit what is uncertain and consider that which is known to the great philosophers. Among all since the Incarnation of our Lord has Ptolemy given the most accurate information in this particular, and has in the second book of the Almagest defined that quarter in which we are. His division and that of Alfraganus and of others is better known by means of the seven famous climates. Climate with them is the space of the earth in which the longest day exceeds by half an hour the longest day of the second division of the earth or is exceeded by the same. It is certain, however, that a more natural and truer distinction would be according to the quarter hour, as Ptolemy at first made the divisions; but because those parts are small, the greater philosophers have united two of them together and consider them as one climate. Since these climates and the famous cities in them cannot be clearly understood by means of mere words, our sense must be aided by a figure. In the first place, then, I shall give a drawing of this quarter with its climates, and I shall mark the famous cities in their localities by their distance from the equinoctial circle, which is called the latitude of the city or region; and by the distance from the west or east, which is called the longitude of the region. In the divisions of the climates and in the latitudes and longitudes of cities I shall follow the authority and experience of learned men. But in marking the position of a city by means of its longitude and latitude found from the authorities on this subject, I shall give an additional device by which the position of the state may be known by its distance from south, north, east, and west. This device consists in intersecting a straight line parallel to the equinoctial circle, marked on a plain and coinciding with a straight line drawn from the number of the degrees of the region marked on the fourth of the colure drawn from the equinoctial circle to the pole of the world—consists, I say, in the intersecting of such a line with the arc of a great circle passing through the poles of the world and through the number of the longitude of the city marked on the equinoctial circle. This is a better and easier method and suffices for a study of the places in the world in a drawing of this kind which appeals to the sense.

I shall also show, together with the latitude of each climate, how many miles each one contains, how many degrees in the heavens correspond to each and how many hours the longest day contains. Moreover, the elevation of the pole above the horizon in each climate is its latitude from the equinoctial circle, and the zenith distance from the equinoctial circle is the same as the latitude and the elevation of the pole; and I give the number of miles of the whole space of the seven climates. But although the philosophers note only seven climates, they nevertheless mark the other spaces of the earth both before the climates and after them. For Ptolemy says in his book on the Arrangement of the Sphere that a journey was made with the help of the kings of Egypt as far as the equinoctial circle. By few men, however, and rarely, has this space before the climates been crossed, owing to its distance, but still more to the negligence of princes who should have aided philosophers in this matter. I mark, then, three spaces before the known climates, which contain more of the earth's surface than a single climate, and I give the number of miles in the width of that space between the equinoctial circle and the first climate, and I state how many miles there are from the equinoctial circle to the end of the seventh climate; then I divide the space which is beyond the climates. Ptolemy in the seventh book of the Almagest distinguishes that space by adding the excess of a quarter hour beyond the length of the day in the preceding region up to the point where he reaches the latitude of the region which is 61 degrees. After that he proceeds by the half-hour period up to the latitude of 64 degrees; and from that location he divides the spaces by one hour up to the latitude of 66 degrees, where the night is continuous at the winter solstice, except that half of the sun comes suddenly to the horizon, and the day is continuous at the summer solstice, and this is far beyond Scotland. Then the sun is always visible in the north toward the pole.

But the space beyond is then divided notably by the length of a day of one month, or of two, three, four, five, or six. For dwellers beneath the pole have day for half the year, that is, they have the sun above the horizon for six months and below the horizon for the other six; but the evening twilight lasts for seven weeks and one day, namely, from the sixteenth day of September inclusively, on which the sun enters at the present time the sign of Libra up to the sixth day of November exclusively. For so long a period does the brightness of the sun continue above the earth, just as with us in the summer after sunset; because on that sixth day the declination of the sun beneath the horizon is 18 degrees and 6 minutes; and twilight lasts up to the end of the 18 degrees and no more. From that same sixth day of November inclusively up to the twenty-first day of January inclusively there is dark night for ten weeks and five days. For on that twenty-first day the sun is beneath the horizon 18 degrees and 6 minutes, and therefore morning twilight cannot begin on that day, but the sun must first pass through those six minutes. But from that day, namely, from the twenty-first of January exclusively, the dawn begins, and lasts up to the entrance of the sun into the sign Aries, which happens on the thirteenth day of March at the present time, and lasts for seven weeks and one day. For the remainder of the time while the sun advances from the first degree of Aries to the first degree of Libra, the sun is always above their horizon, that is, for half a year; because their horizon is the equinoctial circle. Therefore the six northern signs are always above the horizon, as is apparent sensibly on the sphere. Therefore while the sun is in those signs they have clear day. Nevertheless the two twilights, in which the brightness of the sun appears above the horizon, taken together continue for three months, two weeks, and two days, and therefore in comparison with the twilights and the day the dwellers in that locality under the pole have little night throughout the year.

All these things I have written, following principally Ptolemy and Alfraganus, and the table of longitudes and latitudes of states. For in the latitudes of the climates and of the spaces before and after the climates I have followed the view of Ptolemy in the Almagest. But Alfraganus has been my chief authority in what I have written in regard to the extent in miles of the climates and of the spaces before and after the climates, and in regard to the states and regions contained in those climates and spaces, with these exceptions, that the computations of which he does not treat in exact terms I present by means of a fuller examination; and in following other authors I sometimes make a change and an addition, as the need arises for greater certainty; as in the case of the city of Syene.

But if it be objected that in the canons of astronomy and in the tables the longitudes and latitudes of cities are found to be different, as is clearly evident concerning Toledo, for the meridian of which the Toledo tables are made, we must say that first in one way and then in another do they understand the east and the west. For the east and the west of which we are speaking here are the remotest points of the habitable earth. But in one way east and west are understood as points beneath the equinoctial circle at the middle of the earth; so that the beginning of India beneath the equinoctial circle is the east of the habitable earth; and the boundary of Farther Spain, if it extended to the equinoctial circle, would be the west; but it does not extend so far, nay, there is a great stretch of country from the southern part of Farther Spain to the equinoctial circle. The boundary therefore of that stretch westward is the west of the habitable earth. But since the earth westward may have great breadth, namely, from the equinoctial circle as far as the Atlas range and Gibraltar, and on this side through the whole circuit of all Spain and Ireland, the term west can be understood by different people in different ways; whence some understand it from Cadiz, some from the Atlas range, some from the boundary of the habitable world on the equinoctial circle. But when it is taken on the equinoctial circle it is defined better, because it is a single method and a better one, since this point is in the middle of the world between the two poles, and therefore is the true west; and the same is true of the east. But the table of latitudes and longitudes does not take longitudes from the west on the equinoctial circle, as we know. For reckoned thus the longitude of Toledo is 29 degrees west, and according to the table it is only 11. The author of that table took the west known to him and fixed, and with respect to the situation of his own region. We must here consider, however, that the division of the earth cannot and must not be taken in any other way than with respect to the farthest points of the earth, where the sea called ocean ends, which lies between India and Farther Spain and the other regions west and east that follow India and Spain. Whence these terms must not be understood as referring to the limits of some horizon in the way in which we speak sometimes of east and west with reference to the rising and setting of the sun. For there are an infinite number of horizons, both inclined and straight. Therefore east and west in the division of the earth are not taken with reference to the horizon: for then would the east of one horizon be the west of another horizon and the middle of it.

We must note that the true west and east are on the equinoctial circle, as has been said; the remotest habitable point of Farther Spain is considered the west and the remotest point of Farther India as the east. If, therefore, we wish to take the distance of a city from the west as now stated we shall draw a line from this west point parallel to the city, and the line intercepted between the city and the parallel mentioned will denote its longitude westward. Similarly let a line be drawn from Arym, a city at the middle of the world, to the Arctic pole, and from this line let a straight line be drawn to the other city. This line will mark the distance from the middle of the world. But sometimes distance is taken from the west with respect to the boundary of the habitable earth in a straight line with that city; and because this varies in infinite ways among different inhabitants, the taking of longitudes varies. But it is better to take it from the west on the equinoctial circle, because this is done in only one way. Moreover, because the longitude from Toledo and the latitudes of other cities are found collected only in this table, I have for this reason followed it in this particular, although greater accuracy is required, since the longitudes and latitudes of cities and regions have not yet been attested among the Latins; nor will they ever be attested except by apostolic or imperial authority or through the aid of some great king who will furnish the means to men of science. In accordance, then, with the statements made I am offering the present description on the whiter part of the parchment, where the cities are marked by red circles: for in another part of the parchment it will be possible to give another description for greater clarity with respect to the places in the world. This second description I am adding on account of the very great importance of the places. Since, then, there is a very great advantage in knowing the places in the world, the other description must be presented. For the things of the world cannot be known except through a knowledge of the places in which they are contained. For place is the beginning of the generation of things, as Porphyry says; because in accordance with the diversity of places is the diversity of things; and not only is this true in the things of nature, but in those of morals and of the sciences, as we see in the case of men that they have different manners according to the diversity of regions and busy themselves in different arts and sciences. Since then philosophy introduces itself to the things of the world, there is much still lacking in it among the Latins, because it does not possess attestation respecting the places in the world. But this attestation consists in a knowledge of the longitude and latitude of every place; for we should then know under what stars each place is, and how far it is from the path of the sun and the planets, and from what planets and signs places receive their control, all of which things cause the different characteristics of places. If these things were known, man would be able to know the characteristics of all things in the world and their natures and qualities which they contract from the force of their location.

Not only philosophy requires this, but also theology, of which the whole sequence deals with places in the world. Whence the literal sense rests on a knowledge of the places in the world, so that by means of suitable adaptations and similitudes taken from things the spiritual meanings may be elicited. For this is the proper way to expound Scripture, as I have shown above in the example. This knowledge of the places in the world is very necessary to the state of the believers, and for the conversion of unbelievers and for opposing unbelievers and Antichrist and others. For because of different advantages to the state and because of the preaching of the faith men are sent to the different places in the world, occupations in which it is very necessary for those setting forth to know the characteristics of foreign places, so that they may know how to choose temperate places through which to pass. For the most vigorous men sometimes through their ignorance of the places in the world have destroyed themselves and the business interests of Christians, because they have passed through places too hot in the hot seasons or too cold in the cold seasons. They have also met with countless dangers, because they did not know when they entered the regions of believers, or of schismatics, Saracens, Tartars, tyrants, of men of peace, barbarians, or of men with reasonable minds. He who is ignorant of the places in the world lacks a knowledge not only of his destination, but of the course to pursue. Therefore whether one sets forth to convert unbelievers or on other matters of the Church, he should know the rites and conditions of all nations, in order that with definite aim he may seek the proper place; lest, if he should wish to visit pagans he fall among idolaters, or if he has those as his goal, he invade schismatics, or pick out as schismatics those obedient to the Roman Church, or those indifferent to both sides, of which kind are the tribes who are called Aas; in order also that he may avoid Nestorians when he wishes to reach Nicolaitans; and so among the many races of divers sects he may not choose one instead of another through error. For very many have been foiled in the most important interests of Christian people because they were ignorant of the distinctions in regions.

Moreover, no small necessity of knowing the places in the world arises from the fact that the Church should have excellent knowledge of the situation and condition of the ten tribes of the Jews, who will come forth in the days to come. For Orosius in his work, Ormesta Mundi, addressed to Augustine, says in the third book that Ochus, who is Artaxerxes, compelled very many of the Jews to emigrate, and directed them to dwell in Hyrcania on the Caspian Sea. It is the opinion that these tribes exist up to the present day with great increase in their numbers and that some time or other will burst forth from that region. The Master in the Histories adds that Alexander the Great found them confined there, and because of their wickedness restrained them more narrowly, who, however, according to his testimony, will come forth near the end of the world and will cause a great massacre of the people. Moreover, in his Cosmography Ethicus the astronomer says that various races must come forth around the days of Antichrist, and they will call him God of Gods who is destined to lay waste the regions of the world. Jerome, moreover, confirms this in the book which he translated on the wise sayings of this philosopher.

Moreover, Alexander the Great fought with these nations, but could not conquer them, as this same Ethicus testifies and Jerome recounts, and therefore he lamented and said, "Nations composed of reasonable and wise people have I destroyed, and I have overthrown a famous and noble people and an honest race. Of what advantage or necessity was it, when we have left here hiding in the human species all the demons of hell and phalanxes of adversaries? Oh, that they may never hear or learn of the earth flowing with honey and of the very great glory of the world, lest perchance they rush forth over the whole surface of the earth and snatch away and devour all like bread. O earth, mother of dragons, nurse of scorpions, pit of serpents, den of demons, it had been better for hell to be in thee than to give birth to such races. Woe to fruit-bearing and honey-flowing earth when so many serpents and beasts shall rush forth into it. Woe to the inhabitants of the world, when those races begin to triumph. If Alexander had not taken measures effective for the time being against these tribes, no race could have endured their oppression." So writes Jerome. Since, then, these nations shut up in certain localities of the world will come forth to desolate regions and meet Antichrist, Christians and especially the Roman Church should study carefully the location of places, that it may be able to learn the ferocity of nations of this kind and through them to learn the time and origin of Antichrist; for these races must obey him. Therefore if they come from one part of the world, he will come from that opposite, when, however, the bars of Alexander have been broken. For some were shattered before Isidore, since he writes about them. Friar William, moreover, whom the lord king of

France sent to the Tartars in the year of our Lord 1253, when he was beyond sea, wrote to the king aforesaid that he crossed with Tartars through the middle of the gates that Alexander constructed. For when he was not able to conquer these races, then, as Ethicus writes and Jerome confirms, Alexander offered victims to God, and prayed a whole night and day for God's mercy and counsel, and by the divine power a great earthquake happened, and opposing mountains came together and approached through the distance of a stadium, so that space was left only for a single chariot, and he himself then erected gates of wonderful size and smeared them with an unknown cement, which can be dissolved neither by fire nor iron nor water nor by anything else except a severe earthquake.

Since, then, there is a boundless advantage in a knowledge of the places in the world for philosophy, theology, and the Church of God, I wish to compose still another dissertation on places of this kind and to assign more clearly the divisions of the regions; and I shall follow Pliny rather copiously, whom all the sacred writers and men of science have followed. But when I shall have discovered anything definite through other authors, both the sacred ones, as Jerome, Orosius, and Isidore, and through others, I shall not neglect to give those facts that are necessary. But the particular divisions of the regions known to us it is unnecessary to note, likewise the individual places in other regions, but rather those more notable and famous in Scripture and philosophy; from which tyrannical nations will come and have come that are reported from the past to have ravaged the world or as destined to do so at some time. I shall state the rites and sects of the races, for example of those who are pagans, idolaters, or Tartars, and so concerning others, that a clearer understanding of places may be in the mind of the reader. This path by which I shall proceed is not that of astronomical attestation, namely, by the true longitudes and latitudes of places with respect to the heavens, because the Latins do not have this knowledge as yet; but the information is derived from authors who describe the regions of the world according to the ability of each to describe the places of his native soil and to be instructed by others concerning foreign parts.

Sometimes, however, many things are found written which

authors have gathered from reports more than from experience. For Pliny was not accurate in saying that the Caspian Sea rises from the ocean, and Ptolemy was plainly in error regarding the location of Greater and Lesser Britain, as is manifest to everybody, and so are those authors in error in regard to many other matters, and other authors likewise. Wherefore I shall have recourse to those who have in great measure traveled over the places of this world. Especially in the northern regions I shall follow the friar mentioned above, whom the lord king of France, Louis, sent to the Tartars in the year of our Lord 1253, who traversed the regions of the East and of the North, and wrote these facts to the aforesaid illustrious king. I have examined this book with care, and I have conferred with its author, and likewise with many others who have explored the places of the East and South. Yet just as I composed the former description rather to serve as a model and an inspiration to your Glory, that it might be completed at the proper time by the scientific men of this world, rather than as an attestation, so also am I writing this division, that your Wisdom may recognize that greater labor is here needed than the present plea ought to contain. For the perfected treatise you require must complete both descriptions.

Now in setting forth the views of writers on nature and of experimenters and of the sacred writers regarding the habitable parts of the world we must not confine ourselves to those facts which mathematicians among the Latins have attested, for such facts are few; but going more broadly afield, supported by authority and manifold experience, we state that not only are the seven climates inhabited, but a fourth of the earth, and much more than a fourth, contains nations of men. For we find in Pliny and in others that there are certain localities in our habitable portion that are called *ascia,* that is without shadow, from *a* meaning without, and *scia,* shadow, and these places vary in many ways. For in some places things do not have a shadow north or south at the summer solstice. For when the sun is over their heads at noon, there is no shadow to the north or south, nor to the east or west, and this happens on an island of the Nile which is called Syene, in the remotest part of Egypt at its boundary and that of Ethiopia, as is manifest from Pliny in the second and in the fifth book; and Lucan says, "Syene

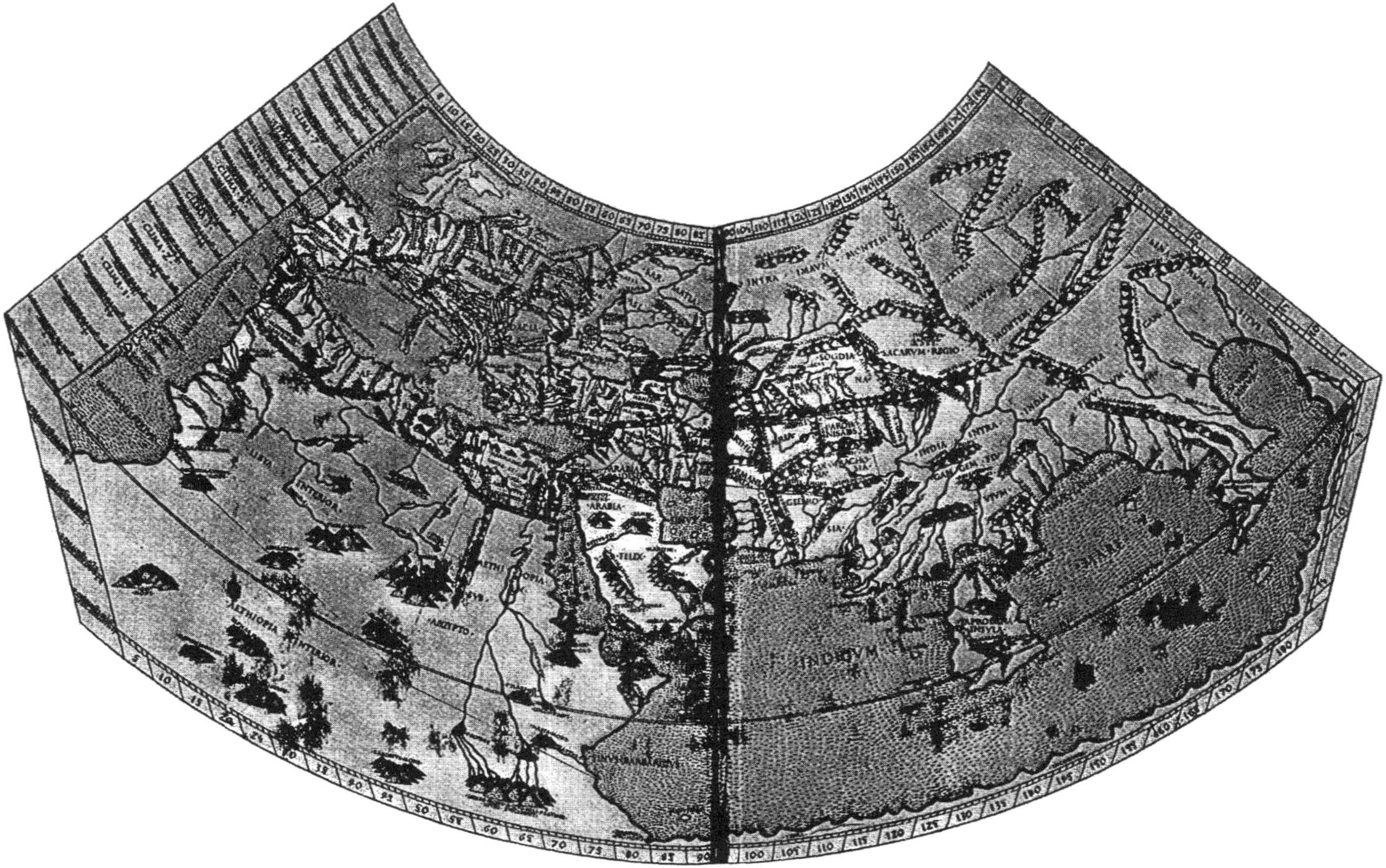

THE KNOWN WORLD IN THE MIDDLE AGES

From the Berlinghieri Paraphrase of Ptolemy's *Geographia*

nowhere casting shadows"; that is, at noon of the summer solstice, and this happens toward the boundary of the second climate, of which mention is made many times in Ezekiel. Other places are called *ascia,* because twice a year they do not have shadow. For certain places in our summer cast a shadow to the south when they have the sun to the north. In our winter they make a shadow to the north since the sun is in the south, according to Pliny in the second book; and alternation of shadow is through six months, as Pliny states in the sixth book, chapter XIX. This cannot be except on the equinoctial circle, since although those who dwell between the tropic of Cancer and the equinoctial circle have a manifold variety in casting shadows, now to the north, now to the south, yet they cannot have this variety through six months, but they will cast the shadow more to the north than to the south, because they have the sun more to the south than to the north. But those who are on the equinoctial circle have the sun equally to the north and to the south, namely, both for six months. This race in India is called the Orestes and the Monedes and the Simari, among whom there is a mountain, Malcus by name, on which shadows thus vary for six months, as Pliny states in the second and in the fifth book. But what is more, we discover through him that there is habitation on the tropic of Capricorn or beyond it. For a region in India is called Pathalis with a very noted port, as he says, where the shadows fall only in the south; therefore dwellers here have the sun always to the north. He states the same thing in the sixth book respecting the island of Taprobane in India, from which men coming to Rome in the reign of Claudius were surprised that their shadows fell to the north and that the sun rose from the south; and therefore among these people shadows fall always to the south and the sun always rises to the north.

Therefore the statement of Ptolemy in his book on the Arrangement of the Sphere is correct, that nature requires that there be two races of Ethiopians on the two tropics. But if the sun has eccentricity, then although as far as the natural arrangement of the heavens is concerned a place on the surface of the earth will be uninhabitable owing to the heat when the sun comes to Sagittarius and Capricorn, because of his nearness to the earth and on account of the uniting of the parallels and on account of the falling of the rays at right angles; and it will be uninhabitable owing to the cold when the sun is in Gemini and in Cancer, because the sun then recedes too far from them, and his rays fall at oblique angles; yet owing to other accidental configurations of the localities, because of the height of mountains keeping off the heat of the sun, and for other reasons, some places there can be habitable, and especially in places beneath the earth when the sun is near the point opposite to the aux. Elsewhere there may be a plain of such a kind between these mountains and the sun, and behind it mountains formed like concave burning mirrors of such smoothness that it is possible for the localities to be habitable when the sun is at the aux without hindrance from the cold, just as I stated above in regard to the localities about the poles of the world. But if we assume the circle of the sun as concentric together with an epicycle, as is possible, just as Ptolemy says in the Almagest, then habitation in that locality can easily be maintained, because in that case the sun does not approach so near to the earth as wholly to consume that part beneath the winter tropic, nor does it depart so far from it as to abandon it in the cold. But if we assume neither an epicycle nor an eccentric circle, according to the view of the writers on nature, there follows no difficulty in the matter of habitation. Therefore whatever opinion we hold in this matter, Pliny can be sustained. And of necessity so, since he learned the certainty of the fact from experience in the case at least of those who came to Rome from Taprobane and of those who were conducted to the island. But although the locality beyond the tropic of Capricorn forms an excellent habitation, because it is the upper part in the world and nobler according to Aristotle and Averroës in the first book of the Heavens and the World, yet we do not find in any author that land described, nor the inhabitants of those localities named, nor the information given that they visited us nor we them. Therefore it is the opinion of some that paradise is located there, since it is the noblest place in this world, according to Aristotle and Averroës in the second book of the Heavens and the World. But not only the philosophers, but also the sacred writers, like Ambrose in the Hexaemeron and Basil, agree in the matter of this diversity of shadows. For in the fourth book Ambrose says, "There are those who through two whole days of the year are without shadow in southern parts, because having the sun above their heads they are illuminated on all sides; whence they are called *ascii,* that is, without shadow; and *amphiscii,* that is, shaded round about: These people dwell on the equinoctial circle and about it on either side, who when the sun is not above their heads cast sometimes the shadow to the north, sometimes to the south, according to the sun's position now to the north of them and now to the south." He adds, moreover, "There are people located to the south in this world we inhabit who are seen to cast their shadow to the south." Because he speaks expressly of shadows to the south, he can be understood to have reference to those who have shadows falling to the south only, namely, dwellers on the tropic of Capricorn and beyond it; because the sun is always north of them except at one time when it is over their heads on the tropic of Capricorn.

How far habitation extends north Pliny shows through actual experience and by various authors. For habitation continues up to that locality where the poles are located; and where the day lasts six months and the night for the same length of time. Martian, moreover, in his description of the world, agrees with this statement; whence they maintain that in those regions dwells a very happy race, which dies only from satiety of life, attaining which it casts itself from a lofty rock into the sea. These people are called Hyperboreans on the European side and Arumphei in Asia. These statements have been made, therefore, according to the latitude of the regions which is on this side or on the other of the equinoctial circle, in order that we may see that the inhabited portion of the earth exceeds a quarter according to the latitude of the regions.

This fact can in like manner be made apparent regarding longitude, which is reckoned from east to west. For as Pliny in the sixth book of his Natural History writes, "India alone is a third part of the habitable earth." Jerome, moreover, says to the monk Rusticus, "Those navigating the Red Sea with many difficulties and dangers arrive at a very great city. The voyage is a lucky one if they reach after six months the port of the aforesaid city, from which the ocean begins to open, through which scarcely at the end of a whole year does one arrive at India." Therefore from the port on the Red Sea toward us a voyage of a year and a half is required to reach India. Jerome states in his book on places that the fleet of Solomon for the space of three years carried wares from India, so that the voyage thither required a year and a half, and the return voyage the same time. But the distance of the Red Sea to the end of Farther Spain about the Atlas range is immense. It is manifest, therefore, that from the limit of the west to the end of India above the earth will be much more than half of the earth: so that of necessity we adopt the opinion of Ezra, Aristotle, and Averroës, which was discussed above in regard to the size of the habitable portion of the world between the east and the west, which exceeds a quarter of the earth according to longitude. When Pliny says that Europe is larger than Asia, he does not include in this division India owing to its size, since it is a third part of the habitable earth according to him, as has been stated.

Since we have considered the facts with reference to the size of the habitable earth as a whole, it is now worth our while to take note of some more famous parts in Scripture and in philosophy, about which it is useful for Christians to know for the conversion of unbelievers, and because of diverse affairs to be transacted with the different races, and because of the advantages to the Church against the rage of Antichrist and of those who are believed to precede his times, in order that they may lay waste the world first until the very great tribulation comes through Antichrist. And here not only is there necessary a depicting and a representation of the places, but also a narrative of those things which must be depicted; for neither suffices. I shall follow authors and investigators with such diligence as I can as far as suffices at present until complete information respecting places is required. I shall begin with the parts to the south and to the east, particularly on account of Scripture, which is more concerned with those localities. I say, then, in accordance with the former statements that the southern frontier of India reaches the tropic of Capricorn near the region of Patale and the neighboring lands which are washed by a great arm of the sea flowing from the ocean which is between India and Farther Spain or Africa, concerning which Aristotle was quoted above. For Pliny expressly states that the sea touches the southern part of India, and this fact is confirmed by Jerome and Alfraganus. That sea stretches along the southern coast of

India and extends for a distance equal to a year's voyage until it joins the Red Sea, as is clear from Jerome, Pliny, and others; and in that sea to the east with respect to India is the island of Taprobane distant across the Sea of Nadosius a voyage of seven days. On this island the Great Bear and the Pleiades are not visible. The inhabitants have gold, silver, and precious stones in abundance, and their resources are greater than those of Rome. But the Romans make a greater use of their wealth, as Pliny states. They choose as king an old man of a mild disposition and childless: but if he has sons later, the royal power does not pass to them. To the king are given thirty governors, whose advice he follows in governing the people. If he is guilty of a crime he is condemned to death. The sentence, however, is carried out in such a way that no one lays hands on him; but food and all other things are denied him, also no one talks with him, whence his death is due to himself. Human life among them is counted short at a hundred years.

The southern coast of India passes from the tropic of Capricorn and cuts the equinoctial circle near the Malcus range and the regions conterminous with it, and passes through Syene, which is now called Arym. For in the book on the Courses of the Planets it is stated that there are two Syenes; one beneath the solstice, of which I spoke above, the other on the equinoctial circle, of which I am now speaking, distant 90 degrees from the west, but more from the east, because the longitude of the inhabited earth is more than half of the heavens or of the earth. This is true eastward. Therefore Arym is not distant from the east only 90 degrees. But mathematicians assume that it is in the middle of the inhabited earth on the equinoctial circle equally distant from the west and east, north and south. Nor is there a contradiction, because mathematicians speak of the habitable earth known to them according to the true conception of longitudes and latitudes of the regions; and this is only so far as is known through experience derived from journeys by land and sea, as given in Pliny and other writers on nature. According to the information given by authors, and especially by Pliny, the Indian Sea, passing along the coast of India from the tropic of Capricorn and cutting the equinoctial circle, crosses the southern coast of India; then embracing an immense space of the earth turns west by south, until it takes in the straits and mouth of the Red Sea, and continues southward beyond the southern boundary of Aethiopia to the sea in the west.

Between the straits of the Red Sea and the Sea of Aethiopia, Aethiopia begins. Where the latitude of the region is about 16 degrees, and the length of the longest day about thirteen hours according to Ptolemy in the Almagest—Pliny also is in practical agreement in his sixth book and in his second book—is Saba, an island formed by the Nile, the royal city of Aethiopia, concerning which we read in Isaiah "the labor of Egypt, the traffic of Aethiopia and of Saba," in regard to which in the thirteenth book Jerome says that it is the Sabaean tribe which is across Aethiopia. This is Meroë, which is at the remotest point in Aethiopia and at the end of the known habitable portion of the earth, as was before stated, and it is mentioned in Ezekiel XXVII, which Josephus in the first book of the Antiquities states was given the cognomen Meroë by King Cambyses from the name of his sister, and Jerome is witness to this fact in his book on Places. This city is distant from the Sea of Aethiopia about 700 miles according to Pliny in the sixth book. This city is on the first climate, from which the climate receives the name Diameroës [through Meroë]. Candar, a woman, was ruler there, whence the name of Candax for many years now has passed to the queen, as Pliny states, and he infers that when the Aethiopians were in control of affairs this city was of great renown. They state that it was accustomed to furnish 250,000 armed men and support 400,000 artificers. Moreover, the Acts of the Apostles make mention of a eunuch of Candace, queen of the Aethiopians, whom Philip baptized; whence Candax is a name of rank, like Caesar, Ptolemy, Pharaoh, Antiochus, and Abimelech. For the Abimelechs among the Philistines, the Antiochi in Syria, the Ptolemies in Egypt after the death of Alexander, and the Pharaohs in the same country in ancient times were like the Caesars and the Augusti in the Roman empire, as Jerome states in the ninth book on Ezekiel. But where the latitude of the region is about the same on the shore of the Red Sea is the city of Ptolemais, founded by Ptolemaeus Philadelphus for the earliest hunting of elephants, where for about forty-five days before the solstice and for as many after it there are no shadows at noon, and in those ninety days or thereabouts the shadow falls to the south, because the sun is to the north, and after these days the shadow falls for the rest of the year to the north. Men dwell here beneath the circuit of the half of Taurus and the half of Leo. Whence the sun crosses twice a year over their heads in those halves of the signs.

Then beyond, in the same latitude but toward the west, between Ptolemais and Meroë 4820 stadia distant from Ptolemais, as Pliny states and Bede confirms in his book on Chronologies, is Berenice, the city of the Aethiopian Troglodytes, where the sun passes similarly twice a year, and the shadows are like those in Ptolemais. For the region of the Troglodytes must bend in part to the west, as will be explained below. It will not, therefore, be in eastern Aethiopia, but more in the central part. Moreover, Scripture mentions these Troglodytes in the second book of the Apocrypha in the twelfth chapter, who came with Selsac, king of Egypt, as an aid. These, as Pliny states in his fifth book, hollow out a cave, which serves them for a house; their food is the flesh of serpents, their voice a harsh noise, and they lack the intercourse of speech. In the sixth book he says, "The race of Troglodytes of marvelous speed from hunting, said to be swifter than horses." Whence Isidore in the ninth book, "The Troglodytes, a tribe of Aethiopians named for this reason because they possess such wonderful speed that they overtake wild beast by swiftness of foot." After these to the east are the Aethiopians from Nubia, and lastly those who are called Indians because of their proximity to India, with whom Pliny begins his description of the race of the Aethiopians. For according to Isidore in the ninth book, there are three principal tribes of Aethiopians, the Hesperi, Garamantes, and Indi. The Hesperi are in the west, the Garamantes in the middle, the Indi in the east. He identifies the Troglodytes with the Garamantes, who were united with them. Meroë, which is the chief city of the tribes, is located, according to Alfraganus, between the Nubians and the Indians and the Garamantes. The Garamantes, named from Garama, the chief city of their kingdom, free from marriage ties, live at random with women. The Hesperi live near Spain. For Spain is called Hesperia: whence those dwelling above Farther Spain are called Hesperi. There are, however, many other Aethiopians joined to these three tribes in different places, much degraded also from that which human nature should be, whose names, localities, and manners it is not within the scope of the present plea to give. All these statements from the books of Pliny and of others are sufficiently clear, and should be noted in the Scriptum Principale.

Aethiopia terminates below at the Red Sea eastward, with Africa to the west and Egypt in the middle between these extremes. In this middle division is the city of Syene, of which Ezekiel speaks expressly in chapters XXIX and XXX, saying, "From the land of Syene up to the ends of Aethiopia shall the foot of man not pass." Syene is the lower limit of Aethiopia and the remotest part of Egypt, as Jerome states on Ezechiel in the ninth book. But Meroë is the upper limit of the known inhabited portion of the earth, according to Pliny. For to the south he places Meroë as the beginning of the known inhabited portion of the earth. Pliny says in the sixth book that from Syene on either side, east and west, there did not remain a town, fort, or house as far as Meroë. For all were destroyed in the constant wars, as sacred Scripture attests. From Syene to Meroë, according to Pliny in his second book, are 5000 stadia. In the sixth book, however, he gives 972 miles. Moreover, the latitude of this Syene has been stated. For it is located on the tropic of Cancer, and from it the second climate is called *clima Diasyenes* [parallel through Syene].

What follows cannot be explained unless there is here given a description of Egypt and of Africa and of the course of the Nile. The capital of Egypt is Syene, as has been said, but Egypt is twofold, namely, upper and lower. That named lower Egypt is enclosed within the Nile in the form of a triangular island, like the Greek letter called delta; and for this reason Egypt was called Delta in ancient times. This has to the east the land of the Philistines, to the north the Mediterranean Sea, to the west Africa, to the south upper Egypt. In the direction of Palestine is a mouth of the Nile called Pelusium, where falls one side of the triangle, that is, one branch of the Nile into the sea.

This is the Pelusium referred to in Ezekiel XXX, where it is said, "I shall pour out my fury on Pelusium the strength of Egypt"; and Jerome says in the ninth book, "It is called the strength of Egypt because it has a very safe harbor, and maritime interests center there." Another mouth is called Canopium, where the second arm of the triangle falls into the sea toward

Africa. Between these mouths of the Nile is the base of the triangle on the sea coast, containing 170 miles, as Pliny says in the fifth book. From the junction of the branches of the Nile at the vertex of the triangle to the mouth Canopium is a distance of 146 miles, and to the mouth Pelusium 256 miles. Upper Egypt is conterminous with Aethiopia, as Pliny says, and is called the Thebais, and begins at Syene, which is a city of the Thebais, as Jerome says in his book on Places. It has to the south, Aethiopia, to the east, part of Arabia, as is shown more clearly below; to the west, the upper part of Africa. This is the region of the Thebais, in which is the city of Thebes. Egyptian Thebes was built by Cadmus, as Isidore states in his fifteenth book. Among the cities of Egypt Thebes is considered famous by reason of the number of its gates. The Arabs bring their wares to this city from all directions. Then Cadmus journeyed to Greece and founded Grecian Thebes in Achaia, which is now called the land of the Prince of Amorea [Morea].

Lower Egypt has in the direction of Africa Alexandria on the sea, a noble city founded by Alexander the Great, which from that time became the capital of Egypt. Alexandria is on the third climate, which is named from it *clima Dialexandrios* [parallel through Alexandria], and is distant from Syene, according to Pliny in the second book of the Natural History, 5000 stadia. From this city eastward beyond the seacoast about a hundred leagues, as the experience of travelers shows, is the city of Memphis, once the stronghold and capital of Egypt, which is now called Damiata. From this city a day's journey distant is Tampnis, where Pharaoh dwelt and Moses performed his miracles, as Jerome states in the ninth book on Isaiah. On the extreme confines of Egypt, as Jerome says in his letter on Mansions is Ramises, a city toward the east, which the children of Israel built. Formerly, Jerome states in his book on Places, the whole province was so named, in which Jacob dwelt with his children; and this is the land of Gessen, as the book of Genesis states, and Jerome says, in the aforesaid book, that in it Jacob dwelt with his sons near Memphis. Not far from Tampnis is Heliopolis, city of the sun, bordering on Arabia, as Pliny says, and a city of great renown, in which, as Jerome states in the book mentioned, was the priest Phutifar, whose daughter Joseph took, as we read in Genesis, chapter XLI.

Tana is a city of Egypt, as Jerome says in his book on Places, in which the Jews dwelt who from fear of the Babylonians fled with Jeremiah, and not only in this city did they dwell, but in Memphis and in the land of Phatures and of Magdalon, as is stated in Jeremiah, chapter XLIV. Phatures, as Jerome states in his book on Places, is the part of Egypt in which exiled Jews dwelt. But Socoth and Ethan and Phiaroth and Magdalon, of which Exodus XIII and XIV speak, are not in Egypt but just outside of it, as is clear from the letter of Jerome on Mansions, and this is eastward toward the Red Sea. For these places the children of Israel set out from Egypt, before they crossed the Red Sea, as Exodus states.

I have now given, as far as is here necessary, a description of Egypt and I must now turn my attention to Africa. Although Pliny and many others write fully on this subject, yet Sallust's statement in the Jugurtha is more certain and clear, which I will give in the main, since Jerome says in his book on Places and Hegesippus in his History of Jerusalem that Sallust is a most accurate author. I shall devote greater care to this province, because although it is near us, yet it is less known than Europe and Asia; and sacred Scripture and the statements of the sacred writers and the histories require much information regarding this region. Africa is named from a descendant of Abraham named Affer, as Jerome states on Genesis, who is said to have led an army against Libya, and to have settled there after conquering his foes, and to have called his descendants Africans and the country Africa, which, however, formerly was called Libya and first of all bore the name of Phuth, from the sons of Chem, as will be explained below.

In the beginning, as Sallust states, Africa was possessed by the Getulians and Libyans. The Getulians, as Isadore once stated, also Hygucio, came by sea from the north from the Getae or Gotae. But Jerome on Genesis is author of the statement that they came from Evila, the son of Chus, the son of Chem, the son of Noah; nor is it likely that foreigners in the first instance inhabited a land destined for a nation, whence Africa was destined for the sons of Chem, like Egypt and Aethiopia. The Libyans were the children of Labaim, the children of Mesraym, the children of Chus, the children of Chem, as Jerome states on Genesis. From this Labaim Libya was called, although according to Jerome on Genesis and on the last chapter of Isaiah, Libya was formerly called Phuth or the region of Phuth, from the son of Chem named Phuth. For there is still a river in Libya called Phuth, and the whole region was so named. The Getulians dwelt more toward Egypt and the Libyans toward the west, both roving more widely owing to the broad extent of the regions. Sometimes the whole of Africa was called Libya, so named from an important part, and the tribes of the country were called Libyans, who are mentioned in the second book of the Apocrypha, chapters XII and XVI, and in the third chapter of Nahum, and frequently elsewhere. But, as Sallust says, after Hercules died in Spain, his army, composed of various races, disbanded. Of this number Medes, Persians, and Armenians, passing over to Africa in ships, seized the localities bordering on our sea; but the Persians settled more in the interior and nearer Egypt and Italy than those dwelling near the sea under the Getulians. For the Getulians were nearer to the sun and were close to Aethiopia. These gradually united with the Getulians by marriage. He infers, moreover, that because in their frequent trials of new fields they had sought other places thereafter, they called themselves Numidians, that is, a people roving and wandering without a town, as Isidore says in the ninth book. The Medes and Armenians beyond the coast of our sea dwelt from the Numidians as far as Gibraltar under the Libyans, who were confined beyond them to the south toward the Aethiopians. But gradually the Libyans corrupted the name of the Medes, calling them Mauri in their barbarous tongue. All these dwelt from the ocean and Gibraltar as far as the province of the Carthaginians. For according to Sallust in later times Phoenicians coming from Tyre and Sidon in their desire for empire invaded parts of Africa, and pressed together Numidians and Getulians and other Africans, and made Carthage or the province a stronghold, in which there are famous Punic, that is, Phoenician, cities, namely, Hippo, the city of the blessed Augustine, Utica of the great Cato, and Carthage, which was like another Rome. The empire of Carthage extended toward Egypt as far as Arae Philenorum, which the Seventy-two Translators have placed in Ezekiel, where the Hebrew has Tharsi, as Jerome says in his book on Places, and not only have they done so there, but in Isaiah XXIII and frequently elsewhere. Ezekiel XXVII is understood as referring to the Carthaginians.

Then follows the Tripolitan region, belonging to those who inhabit Byzacium, which Tyrians and Sidonians seized, whence it is called Africa or Libya Phoenices, because Phoenicians, namely, Tyrians and Sidonians, inhabited it. This country is very fertile, for it yields fruit a hundred-fold, as Pliny states. There famous Leptis is situated between the Lesser Syrtis in the direction of the Carthaginians and the Greater Syrtis in the direction of Egypt; which, as Sallust describes them, are shallow places full of sand, and when stirred by the winds and waves of the sea belch forth dust and sands copiously, whence they are called syrtes from drawing or pulling sands and dust. For *syrma* in Greek is *tractus* [drawing] in Latin; and *syro* is the same as *traho* [I draw]; whence they annoy and disturb those inhabiting the adjoining localities.

Then follows the province of Pentapolis, which is called Cyrenean in the Scripture, where there are five great cities, of which the chief is called Cyrene, of which mention is made in the Gospel of Luke, also in Mark and in Matthew. When the Lord was being led to his Passion "they laid hold upon one Simon a Cyrenian, etc." And in the Acts of the Apostles, "And there arose of the synagogue which was called that of the Libertines and of Cyreneans, etc.," and in the fourth book of Kings, chapter XXVI, it is stated that the king of the Assyrians transferred the people of Damascus to Cyrene; and in Amos V and IX there are references to this city. Since, however, Arae Philenorum is found in many writings of the sacred authors and in the histories, and frequently is read corruptly, so as to be called Arae Philistinorum, for this reason to avoid error it is worth while to consider what Sallust says on this subject. For after the Cyreneans and Carthaginians had waged many wars, and since the boundaries between the kingdoms had not been settled, for the sake of peace they decreed that on the same day and hour envoys should be sent from both cities, and that where they met should be the boundaries of the kingdoms. But the envoys of the Cyreneans, hindered by chance, were not able to advance as far as they wished. They pretended, therefore, that the Carthaginians had departed from their place sooner than they should have done, and said that they should be buried alive at the place which they had reached, if they wished to keep it as the boundary of their kingdom; or that the Carthaginians should allow them to pass through to the point they wished to reach; and that they themselves would there choose death. The envoys of the Carthaginians agreed, and the chief men were two brothers, who were named Philenes or Phileni. These two with their own consent were buried alive for the sake of their country, in memory of whom the Carthaginians erected altars which are called the altars of the Phileni to this day. Under the province of the Cyreneans is reckoned the whole country as far as Egypt, according to many authors; but Pliny defines it as a small province by itself, which is called by him Libya Mareotis. And thus is terminated the stretch of the whole of Africa from Gibraltar to Egypt, with the distinction of its provinces.

Above Egypt and Africa, to the south extends Aethiopia from east to west as far as the Aethiopic Sea, and the principal regions are those, as I have said, of the Indians, Sabaeans, who belong to Meroë, Nubians, Troglodytes, Garamantes, and Hesperi. Part of the country of the Troglodytes bends toward the west beyond the Greater Syrtis and neighboring parts, from which it seems to be distant, according to Pliny in the fifth book, a journey of eighteen days. Therefore although the greater part of the race of the Troglodytes lies toward the Red Sea, yet some portion of it bends toward the west beyond the regions of Africa. Beyond these westward is the region of the Garamantes in the direction of the Lesser Syrtis and of the Carthaginians; the eastern portion, however, of the Garamantes lies in the direction of the district of Cyrene, according to Pliny in his sixth book, so that they reach at length the Hesperides to the west in the region of Mount Atlas.

Since the Nile waters Egypt and Aethiopia, and separates in many ways their provinces, and is frequently mentioned in Scripture, and is a familiar topic in philosophy and in the histories, it is quite fitting that some account of it should be given. It rises in paradise, as Scripture states, but where it breaks forth into our inhabited world is reckoned differently by different authorities. It is, however, likely that it rises in Aethiopia on the shore at the beginning of the Red Sea, as Orosius affirms in his book Ormesta Mundi to the blessed Augustine, and Seneca in the third book of his Natural Questions practically agrees. For he states that the emperor Nero sent two centurions to explore the source of the Nile, and when they came to the first king of the Aethiopians they were instructed and aided by him, in order that the other kings of the Aethiopians might grant them safe conduct. At length they came to shallow marshes filled with vegetation, the extent of which was unknown to the inhabitants; and they despaired of ascertaining it. For they could explore neither in a boat owing to the shallowness of the water, nor could the muddy ground support the weight of a man. The inhabitants therefore believed that the head of the Nile was in this place. Therefore Pliny's statement that the Nile rises at the confines of the west near Mount Atlas not far from the sea is not to be believed. For the double testimony is stronger here than the single one, and the experience of the emperor Nero is much to the point.

An African river flows into the region which is called Egyptian Libya, to an immense lake in which it ends, as Orosius says. There is the additional fact that paradise is in the east. Therefore it is more likely that the Nile breaks through in the east than in the west; nor is the African river one and the same as the Nile, although they do nourish similar fishes and like monsters and crocodiles, as Pliny argues, because we see rivers in different regions nourishing creatures similar in species; and according to Pliny himself and others rivers in India nourish crocodiles like the Nile. But his statement that the Nile of Egypt is augmented by rains and snows through the African river must be accepted, because it is possible that the lake mentioned, into which the river empties, drains by a hidden passage into the bed of the Nile, a phenomenon familiar to us in many different regions.

The course of the Nile from its source through Aethiopia and Egypt is described by Pliny and others, although in regard to its source he does not agree with others. It flows, then, from its source, as Orosius says, for a long time to the west, and crosses through the middle of Aethiopia, forming many islands, of which Meroë is the most famous, which is also called Saba. Then bending to the north between Meroë and Syene, as Pliny explains, and inclosed by mountains it forms cataracts amid the opposing rocks, so that it does not seem to flow but to rush, where it deprives the inhabitants of the sense of hearing by its excessive noise. For this reason they emigrated to quieter localities, as Seneca says in the eighth book of his Natural Questions. I mention this fact, because Macrobius the Pythagorean, wishing to show how we uninjured endure with our ears the endless noise from the motion of the heavens, gives a pointless example regarding the race which calmly endures from habit the roaring of the Nile. But his example is an erroneous one, as Pliny and Seneca show, and cannot be cited, as Aristotle shows in the second book of the Heavens and the World. This place is near Syene, according to Jerome in the ninth book on Ezekiel; and he states that as far as Syene the Nile is navigable from the Italic Sea. The Nile continues on to the north and at length projects its mouth into the sea between Egypt and Italy. It has two mouths, namely, that at Pelusium and that at Canopus. Moreover, as Jerome says in the fourth book, on Isaiah, chapter XIX, the Nile before Augustus Caesar had a single channel, but was then divided into seven, whence one branch flows down to Pelusium, and flows to Memphis, which is Damiata; and another branch extends upward to the south as far as Kayr and Babylonia, from which it is now called the Soldanus of Babylonia, about three days' journey from Damiata. From Damiata a branch of the river extends laterally about southeast to a village called Lancassor, where a Christian army was defeated, when Lord Louis, son of Louis the son of Philip, famous king of France, first took the Cross to parts across the sea. Other branches of the Nile flow down near Tampnis and Alexandria and other places of Egypt.

Now according to Pliny and others a peculiarity of the Nile is the fact that it inundates at certain times and waters the plains of Egypt. According to the overflow of the river is fertility granted or denied to Egypt. For if it passes beyond its natural limits by only twelve cubits, Egypt feels hunger; if for thirteen cubits, Egypt is no longer hungry; fourteen cubits bring joyousness, fifteen security, sixteen delight. If its overflow is still greater yet in moderation, it rouses the inhabitants to an abuse of pleasure. If its overflow is excessive, harm is done, as Seneca states. The Nile begins to increase, authorities state, at any new moon after the solstice sensibly, that is, by degrees and moderately, while the sun is passing through Cancer, and most abundantly while he is in Leo, and the river rises no further while the sun is in Virgo. In the same way in which it rose is it called back within its banks while the sun is in Libra and on the hundredth day from the beginning of its flow. It is difficult to assign the reasons of this inundation and increase, since it is very strange occurring in the heat of summer as it does when waters are consumed more than at other times. But no other river has an inundation of this kind, according to Aristotle in his treatise on the Nile; according to Pliny no river except the Euphrates. We can, however, name a third one, namely, the Ethilia, which is larger than the Euphrates and forms the Caspian Sea, of which I spoke above. This fact is attested by those who were among the Tartars, as Friar William and others. Aristotle, however, and Pliny spoke in accordance with their experiences.

A natural peculiarity of this kind, which is found in very few rivers of the world, is quite remarkable. Moreover, the almost endless disagreement of learned men over the causes of this increase induces perplexity in us, so that it is not clear what view we should adopt, especially since many authors reject theories as probable as those which they affirm. Seneca also, more definite than all other authors, with the possible exception of Aristotle, in every matter on which he fixes his attention, offers on the present question merely rejected theories in his book on the Nile, which is the eighth of the Natural Questions. Seneca is defeated by this difficulty, although gloriously victorious elsewhere. Aristotle, too, although he intersperses opinions, yet can always be doubted, owing to many contradictions. I shall state with tolerable brevity the conflicting views which I consider the more important ones to give for the sake of this plea which is preliminary to the principal treatise.

Latin scientists, neglecting experience in this matter, adhere to the theory of Thales, who was first of the seven wise men. His theory was that yearly settled winds blow opposite to the mouths of the Nile rolling back the waves and sands of the sea, by which the mouths are obstructed, and the waters of the Nile flow back to the interior, and thus the river overflows its banks. But this theory is found to be false by authority and experience. For according to Aristotle and Seneca and in accordance with what the experience of those who have been in Egypt shows, the waters of the Nile begin to flow from the upper part of Egypt from Aethiopia. For the Egyptians exulting for joy go aboard boats and hasten to meet the down-flowing Nile with singing and many kinds of musical instruments. Hence the river does not begin to flow from the mouths previously mentioned, but from the upper portion toward the mouths.

Anaxagoras the philosopher offers a theory more probable than all the others, and states that the snow melts in summer in the mountains of Aethiopia, and that the Nile is thus augmented, just as the Rhone, Po, Danube, and rivers of this kind near the Alps are increased from the melting of the snow. But this theory Aristotle and Seneca deny and reprobate. Aristotle refutes it by the fact that only a little water is produced from a great quantity of snow, but the increase in the Nile is very great, because it inundated very wide districts sometimes to a depth of thirty cubits. Again he argues in another way. Waters flowing from a distance are stronger at the end, just like winds coming from a remote quarter, while those from a point near at hand are more impetuous at the beginning. This reason is applicable to the waters, because in a great distance many different waters flow together into one body and there are many rains and many vapors come from the earth, and therefore all rivers are strengthened toward their end and are larger than at their source. This is true also of the winds, owing to the multiplication of vapors flowing together into one body from different places on account of the great distance. But the inundation of the Nile is an augmented one at the beginning and grows stronger and stronger for a time, until it diminishes at the end, as Aristotle and Pliny state, and experience teaches. Therefore those waters do not come from a remote locality, but even the mountains of Aethiopia, where it might be said with greater probability that snows abound, are distant from the Nile a journey of five months. Therefore the inundation of the Nile is not caused by snows. The major proposition of this argument is quite sound; for it contains much scientific knowledge, whatever may be the case with the minor proposition. Again he states that at full moon all congealed substances melt and are more resolved, but the Nile increases at the end of the month, an increase not due, therefore, to snows. Again, the Nile flows more when the north wind blows than when the south wind blows; but the south wind to a greater extent dissolves snows because it is a warm wind. But Aristotle says that in Aethiopia, owing to the very great heat which consumes all things, snows cannot exist, a statement that must be readily believed. Seneca agrees in these matters and adds that snows dissolve in the spring and melt when touched by moderate heat and cause an increase in the rivers; but that the heat is never moderate in Aethiopia before the winter. The Nile, however, rises after the summer solstice. In this preliminary plea it is not necessary to state the theories of Pythagoras, Diogenes, Democritus, or of many other philosophers.

But let us quote in place of them all the theory of Aristotle, who says that in Aethiopia in our summer there are many rains and none in winter, and that the Nile rises in those regions, whence the swamps and lakes are augmented; and he adds that yearly settled winds blow in summer from the east and drive the clouds to the regions whence the Nile flows, which are dissolved into its lakes. The reason why the river is augmented at the end of the month is given by Aristotle in the second book of the Posteriores, because the end of the lunar month is colder, and cold multiplies moisture and is increased by the north wind, since that wind drives the clouds naturally because of its impetuosity. It has this quality since it comes from near by, for its habitation is in the northern quarter of the earth, and it is impetuous in this quarter, as Aristotle teaches, and for this reason drives the clouds before it, so that they can collect over the swamps of the Nile, which are of almost incomprehensible size, as was explained above. They receive, therefore, much from the region of the clouds and their waters can thus be multiplied through the resolution of the clouds into rains.

But as in the case of the other theories, objections can be urged against this one. For since the country is uninhabitable owing to the heat, that is, very unsuitable for habitation since it is parched, how can there be an abundance of rain and especially in summer, in the same way that there can be no snows, as Aristotle himself argues in opposition to the second theory? Against the first theory he says that the same thing would happen in other rivers, and that annual winds do not always arrive at their accustomed times. Since, then, there are rains in many regions where there are great rivers and annual winds, and yet we do not see an increase of this nature; and the north wind likewise drives the clouds with greater force in regions near by, because the wind is less distant from its origin; hence, rivers in our localities would be the more increased and likewise at the end of the month; but we see no such increase. Wherefore it is very difficult to give the reasons for this strange increase, because it occurs only in the Nile, according to Aristotle, or at least only in the Nile and Euphrates, according to Pliny. It happened, however, in still a third river mentioned above, namely, in the Jordan before the destruction of Sodom and the neighboring cities, on the testimony of the Scripture in Genesis. Therefore owing to the difficulty of this subject it suffices in this plea that we learn the views of the philosophers, in order that, stimulated by these ideas, we may inquire with greater certainty into the truth in the more important treatise.

We must return, then, to the description of the regions. We shall find in Pliny in the sixth book, Alfraganus and Lucan also agreeing, that the ancients called part of Arabia all the land which is inhabited from the Aethiopic Sea and the south passing through Meroë and Syene, so also that the Heliopolis of Egypt, of which mention has been made, is reckoned in Arabia; and therefore all that country which is inhabited from Meroë, Syene, and Heliopolis toward the east between the Red Sea and the Aethiopic Sea is included under Arabia. Hence Alfraganus placed on the first and second climates the island of the Arabs. This island is in the Aethiopic Sea or near the beginning of the Red Sea. Lucan says, "You Arabs have come to a world unknown to you who wonder that the shadows of the forests do not fall to the left." He says this respecting the Arabs who came to Rome in aid of Pompey, who were surprised that shadows to the left, namely, to the north do not pass, that is, do not change to the right or south. For in their country, which lies between the tropic of Cancer and the equinoctial circle, they have in one part of the year shadows to the south when the sun passes beyond them toward the tropic of Cancer, because the sun is then in the north with respect to them: and when it passes beyond them toward the equinoctial circle, then of necessity they have shadows to the north because the sun is to the south of them. All this part of Aethiopia around Meroë, Syene, and Heliopolis toward the east is included under Arabia; and not only this section, but whatever there is around the tongue, that is, the extremity of the Red Sea and beyond its shore eastward from the point of its tongue to its Persian Gulf. It extends from the Red Sea as far as Pelusium in Egypt to the west, and spreads to the north through the whole of the desert in which the children of Israel wandered as far as the land of the Philistines above our sea bounded by Egypt, and extending eastward until the region of the Amalechites is reached, which lies to the east of the land of Philistia, and as far as the land of Edom, or Idumaea, lying to the east of Amalech and reaching as far as the land of Moab. Then it turns more to the north through the land of Sehon, king of Esebon, and of Og, king of Bashan, as far as Mount Galaad and Lebanon, and turns still more to the north as far as Cilicia and Syria Commagene, and as far as the Euphrates.

Whence taken in its widest meaning it is a very great land, and contains in it first the Desert of Sur or Ethan, for Ethan is called a desert on both sides of the Red Sea and at its extremity is united to Egypt and Palestine, since it is stated in Exodus that the children of Israel encamped in Ethan; and that they crossed the Red Sea and came again into Ethan. For the Scripture says that after the passage of the Red Sea they came into the desert of Sur, and there pitched their tents in Mara, and marched for three days before they pitched their tents, and first they pitched their tents in Mara and then in Helim. But Jerome says in his letter on the Mansions that the desert of Sur and Ethan are the same. In this Arabia near Sur toward the east, beyond where the children of Israel crossed, beyond the shore of the Red Sea, is the country of the Elamites according to Pliny and Jerome in his book of Interpretations. Here is the city of Elam, the remotest city belonging to the people of Palestine. For in this part near the desert of Sur an angle of Palestine turns toward the Red Sea, according to Jerome. Pliny says Stagnos* is near there, an island of the Red Sea, which dogs do not enter, and if landed die wandering over all the shores.

After the desert of Sur toward the east follows the desert of Sin, where there were five sojournings of the children of Israel according to Jerome in his letter on the Mansions, of which

* See Bridges' note, Vol. I, page 326.

the first is not given in Exodus, but in the thirty-third chapter of Numbers, where it is said, "and they proceeded from Helim to the Red Sea which is called Jamsuph." Jerome asks how they returned to the Red Sea, and he gives a double solution of the difficulty, namely, that there might have been a gulf of the Red Sea extending into the desert of Sur from the main body of the sea, for *Iam* means sea and *Suph* red. But his second solution is better. *Suph,* he says, can signify red or bulrush, and here it does not have to be taken as red but is to be understood as bulrush. Hence he says that we can imagine that they came to a certain swamp and lake, which was full of sedge and rushes. That, moreover, sacred Scripture calls every collection of waters a sea there is no doubt. Here, therefore, according to the correct interpretation of the Hebrew the word signifies a marsh of rushes. But since the former rendering became too widely known in the ancient translation, where this sojourning is called the Red Sea, Jerome allowed it to stand in our version, as it had been in the ancient one, just as he did in many other passages throughout the whole body of Scripture. The last of those five sojournings, which was the eleventh from the exodus of the children of Israel from Egypt, is Raphidim directly to the north belonging to the country of the Amalechites, which fought with the children of Israel and was defeated by them. Then more toward the east is the desert of Sinai, where Mount Sinai is situated, which is the Mount of God Oreb, as Jerome says in his book of Places. But the rock Oreb is not at Raphidim, from which Moses drew water. For Horeb meaning Mount Sina is written with the letter *heth,* but Oreb without it. Then come the sepulchers of concupiscence and Asseroth, which are two sojournings after Mount Sinai in this desert of Sinai.

Then more to the east is the desert of Pharan, where the country of the Ishmaelites begins toward the Red Sea to the east. But to the north of Pharan is Ebron, the city of David, and the place where mighty Adam and Abraham and Isaac and Jacob were buried. By the way of the desert between Pharan and Ebron Moses sent Joshua and Caleph and other spies. In this desert of Pharan, as Jerome says in his book on the Mansions, the children of Israel made eighteen sojournings, from the twenty-fifth to the thirty-second, by counting the extremes with the means, so that the last is Asyongaber; whence the desert of Pharan is very extensive. Here were they smitten by the Amalechites and the Canaanites, and here the Lord gave judgments, and the sedition of Chora arose, the rod of Aaron budded, and many other things happened, as is manifest from the thirteenth chapter of Numbers to the twentieth.

After leaving the desert of Pharan they passed more to the east to the desert of Sin, which is Cades Barne, where the people murmured at the waters of strife. But this is not that former desert of Sin. For it was much more distant from the Red Sea, with the land of Edom joining it to the northeast. For from this place the children of Israel sent envoys to Edom saying, "Lo, in the city of Cades, which is in your remotest confines encamped we beg permission to pass through," as the eighteenth chapter of Numbers states. But if any one contends that the desert of Sin mentioned above can extend to this place, it is manifestly not so, as shown by Jerome in his letter on the Mansions, by interpretation, and by Scripture, since the first Sin is written with a *samech* and is translated bramble bush or hatred; but this one is written with a *sade* and is translated command. Therefore the children of Israel turned aside by the way which leads to the Red Sea, passing around the land of Edom, and came to Mount Or in the remotest confines of that land, on which mountain Aaron died.

In the territories of the land of Edom they made three more sojournings, according to Jerome, until they reached the boundaries of Moab, whence Moab is east of Edom. For departing from the territories of Edom they pitched their tents in the desert which looks toward Moab facing the east, as the twenty-first chapter of Numbers states. Then, as Deuteronomy, chapter II, states, they passed a city of Moab called Ar and came to the confines of Ammon. In these localities there begins to the east the land of Sehon, king of the Amorites, and the land of the Ammonites, and therefore these places must be carefully considered. For here begins the land of the children of Israel, and concerning these places much is said in Scripture and in the sacred writers. There is, then, on the border of these places a cliff rising on high called Arnon, and this cliff marks the limits of the children of Ammon and of Moab, and of King Sehon of the Amorites. Hence the land of the children of Israel begins here. But beneath this cliff is a valley called Arnon, near which on the southern side is the city of Ar, capital of the kingdom of the Moabites, which was afterwards called Ariopolis, compounded from the Hebrew word and the Greek, that is, city of the adversary, as Jerome says in the fifth book, commenting on the prophet Isaiah. Moreover, from this cliff a torrent flows toward the west, called the torrent of Arnon, beyond the bank of which is situated a town called Aroer near the Arnon. These facts are attested from the book on Places of Jerome, from the text in the book of Numbers, chapter XXI, of Deuteronomy, chapters II and III, of Joshua, chapter XIII, of Judges, chapter XI, and from many other passages of Scripture.

The land, then, of Moab ascends from the Arnon westward to Edom and to the Dead Sea and to the region where the submerged cities were located, and to the Jordan opposite Jericho, as is clear from the aforesaid passages of Scripture and of Jerome. Below the land of Moab near the Arnon and Ariopolis is Madian, the city of Jethro, father-in-law of Moses, as Jerome states in his book of Places. Here necessarily must be the Midianites. For as is clear from Numbers, chapters XXII, XXIII, XXIV, and XXV, Balac, king of the Moabites, called Balaam the seer to curse Israel, who counseled to offer them the daughters of the Midianites, and they sinned with them and many of the Israelites were slain. Later the Midianites were destroyed and wiped out by the children of Israel. On the other side of the torrent Arnon, the land of the children of Ammon began toward the north and east, stretching toward the Euphrates, and toward the west a corner of it toward the Jordan reaches the ford or torrent of Jacob, which Jacob crossed coming from Mesopotamia in Syria. The angel wrestled with him after his passage of this stream, as Genesis, chapter XXXII, records.

That at this torrent of Jacob is the boundary of the children of Ammon is made clear by Deuteronomy, chapter III. Here is the boundary of Ammon and of Sehon, king of the Amorites, and of Og, king of Basan, as is clear from Judges, chapter XII. For from Jacob the land of Sehon began, as is there stated; and where his land ends, the land of Og, king of Basan, begins and extends also nearly to the torrent of Arnon and the confines of Esebona, city of King Sehon of the Amorites. Therefore the land which belonged to Sehon is bounded to the south by the Moabites and to the east by the Ammonites, and to the west has the river Jordan, and to the north the land of Og, king of Basan. But Sehon, becoming more powerful, invaded the countries of Moab and Ammon, and took away their lands. For that he took away the land of the children of Ammon is established from Judges, chapter XI; and that the children of Ammon lost half of their country is shown by Joshua, chapter XIII. Moab also lost much, as is clear from Numbers, chapter XXI.

Since these regions in the direction of the tongue of the Red Sea have been discovered by means of the sojournings of the children of Israel, we must consider besides that in the deserts lying between the Red Sea and the lands named are other vast regions which extend around the aforesaid lands, namely, the land of the children of Ammon and of Moab, and the desert of Pharan, as far as the land of Elamites, which I said above was situated above the coast of the Red Sea from the passage of the children of Israel on toward the east. In this region, then, and likewise in Pharan, dwelt the children of Kentura and Agar, descendants of Abraham, of whom mention is made in Genesis, chapter XXV. And first from the Euphrates begins the Nabathaean country, so named from the first son of Ishmael, who was called Nabaioth, as Jerome states on Genesis, chapter XXV, and Pliny agrees with the statement in his first book, except that he calls one part of the Nabathaeans Nomads, who wander about the Euphrates near the Chaldeans. After these toward the desert of Pharan is the region of Cedar, which is named from the second of Ishmael's sons, who was called Cedar. Although other regions belonging to the children of Ishmael are named as far as Sur, for he dwelt from Evila as far as Sur, as the Scripture states, yet all these regions are called Cedar, as Jerome maintains in his fifth book commenting on that Burden upon Arabia in Isaiah, chapter XXI. He says, "Here he is speaking for Kedar, which is the country of the Ishmaelites, who are called Agareni and Saracens by a wrong name." In the seventh book, commenting on Isaiah, chapter IX, he says in regard to these regions of Cedar and Nabathaena that Cedar is the country of the Saracens, who in Scripture are called Ishmaelites, and Nabaioth is one of the sons of Ishmael, from whose names the desert is called, which lacks crops, but is filled with cattle. Evila is a part of the country of Ishmael distant from the town of Pharan in the desert of Pharan a three days' journey, as is stated in the book on Places. There is also another region called Evila in India near the river Ganges, of which mention is made in the first chapter of Genesis.

Between Cedar and the region of the Elamites mentioned above extends the region of Saba beyond the shore of the Red Sea, according to Pliny, book V. This region produces incense and is filled with spices and has three divisions. One is called Arabia Eudaemon, which is included between the Persian Gulf of the Red Sea and the Gulf of Arabia, according to Orosius in his book Ormesta Mundi, and according to Isidore, book XIV. The second is Madian, called from a son of Abraham and Ketura. The third is Epha, a region called from a son of Madian, as is clear from Genesis XXV. That, moreover, these two regions are part of the kingdom of Saba, Jerome maintains expressly in his comment on Isaiah XVII, saying "Madian and Epha are regions abounding in camels and the whole province is called Saba; whence was the queen of Saba," as he states. That Arabia Eudaemon is part of Saba is manifest, because it joins immediately the Chaldeans, as Orosius states. The Chaldeans and the Sabaeans their neighbors attacked at the same time the cattle of the most holy Job, as is clear from the first chapter of his book. This is also clearly stated by Isidore in his fourteenth book, who speaks as follows, "Arabia is called sacred, because the region produces incense and perfumes." The Greeks have called this land Eudaemon, and our writers Beata [blessed], in whose pastures are produced myrrh and cinnamon; and there is the phoenix born. This land is also called Saba from the son of Chus, son of Chem, son of Noah, and this son of Chus was called Saba. This fact is stated by Jerome in his Hebrew Questions. Therefore when Madian and Epha are enumerated in Isaiah, chapter IX, Saba also is added, when he says, "All they from Sheba shall come." For Arabia Eudaemon in particular is called Saba chiefly, although that whole region and Madian and Epha are called Saba. Therefore the whole region from the Chaldeans beyond the Red Sea as far as Elam is called Sabaea.

We must here note the fact owing to some contrarieties that Arabia taken broadly includes all the regions mentioned on both sides of the Red Sea according to Pliny, Alfraganus, and the ancient philosophers. Taken, however, in the more restricted sense it is understood to be only the region which extends from the tongue of the Red Sea as far as the Euphrates and the Persian Gulf toward the east, and as far as Idumaean Palestine toward the north in one of its parts, of which the other part to the north and more to the east extends to Mount Lebanon, and embraces the whole region of Moab and of the children of Ammon and the kingdoms of Sehon and of Og, king of Basan, and some regions connected with these. Such is the usage of Scripture, as, for example, in Isaiah when he says, "The burden upon Arabia," where Kedar is included under it; and Mount Sinai is thus in Arabia, as the Apostle states in Galatians, chapter IV. In a still more restricted third sense it is taken so as to exclude Pharan, Cedar, Madian, Epha, and Saba Eudaemon. For thus in the time of Jerome and thereafter was Arabia understood, since he himself says in his book of Places that Pharan is beyond Arabia. Also in the fourth book and in the seventeenth, commenting on Isaiah, he says that Madian, Cedar, and Nabathaea are regions beyond Arabia. That incense-producing Saba Eudaemon and all the region of Saba are distinct likewise from the Arabia mentioned in that third sense is clear from Jerome in his Hebrew Questions. For he says that when it is said in the Psalms, "The Kings of the Arabs and of Saba shall bring gifts," the fragrant and incense-producing Saba is meant, concerning which he cites the authority of Vergil. "To the Sabaeans alone," he says, "belongs the frankincense shrub." For as Jerome says and as is clear to any one acquainted with Hebrew, in the Hebrew the statement is, "The Kings of Saba and Saba shall bring gifts." But the first Saba is interpreted Arabia and is written with the letter *sin,* the second is written with the *samec,* and is the famous incense-producing Saba, and from this Saba came the magi, who worshiped Christ, not from the Saba which is in Aethiopia. For according to the Gospel the magi came from the East, and those are the kings of Saba, or the kings of the Arabs and of Saba.

After these divisions there follows a very large region called Syria, which according to Scripture, Pliny, and other ancient authors contains all the provinces from the Tigris River on the east as far as Arabia on the south and as far as our sea on the west or the great sea, dividing Italy and Syria and Egypt, and from Cilicia and the very lofty Taurus range on the north. Its first and principal region is Mesopotamia or Assyria, for they are the same according to Pliny. Jerome, moreover, states in the third book, commenting on Isaiah, that the whole region between the Tigris and the Euphrates is the country of the Assyrians. Similarly Mesopotamia is included between the Tigris and the Euphrates. Whence it is named from *meson,* meaning middle, and *potamus* meaning river, as if contained between two rivers, the Tigris, namely, and the Euphrates. Therefore, in ancient times Mesopotamia and Assyria were identical. This Mesopotamia or Assyria on the east has the Tigris, on the west the Euphrates, on the south the Persian Sea, which is the Persian Gulf of the Red Sea, on the north the Taurus range. Its length is about 800 miles, and width 300, according to Pliny. In Mesopotamia are Nineveh and Babylon, and all the land of the Chaldeans, and the tower of Babel constructed in the land of Senaar. In Mesopotamia, moreover, are the cities which Nimroth built, namely, Arad, that is, Edissam, and Archad, which is now called Nisibis, or commonly Nisibin, and Calampne, which was afterwards named by King Seleucus Seleucia, as Jerome explains in his comment on the tenth chapter of Genesis. Moreover, in Mesopotamia is Aram, as is stated in Genesis, which still retains its name. Aram is distant two days' journey from the Euphrates, and Jerome says that Aram is beyond Edissa. Therefore Edissa is between Aram and the Euphrates, and Nineveh is about ten days' march from Aram, that is, eastward beyond the Tigris River, according to the statement of Scripture that the Tigris flows opposite the Assyrians. For the Ninevites were generally called Assyrians. From Aram to Baldac toward the south is a journey of about twenty-six days. Baldac is a royal city, in which the caliph lord of the sect of the Saracens has established the seat of his dignity. In those parts is the tower of Babel, and the mighty ruins of

Babylon, which was the capital of the kingdom of the Babylonians and Chaldeans, since they had been Mesopotamians and Assyrians from the beginning, as the whole land between the Tigris and Euphrates was called Mesopotamia and Assyria, and since Babylon, capital of the Chaldean races, possessed the greatest renown among cities in the whole world. The remaining part of Mesopotamia and of Assyria was called Babylonia, as Pliny states, so that at length the Babylonians prevailed. For as is clear from the books of Kings and the Apocrypha at first they were called the kings of the Assyrians, as Salmanasar and Sennacherib and others. Then Nabugodonosor, king of Babylon, and his successors ravaged the Assyrians, and became masters through the whole region between the Tigris and the Euphrates. Noah and his sons at first after the flood dwelt in Babylonia, as Albumazar says in the fifth book in his larger Introduction to Astronomy. For since they themselves were clever astronomers and were first to teach the Chaldean astronomy, as he states in the same book, they knew that the fourth climate is the most temperate, in which Babylon is, and for this reason turned aside to it.

Since the Tigris and the Euphrates are two of the four principal rivers of the world and numbered along with the Nile, something must be said about them. Their source is variously given. For as a matter of fact their origin at first is in paradise, as Scripture maintains. Then, according to Pliny, the Tigris bursts forth in greater Armenia and later empties into a lake supporting all weights placed on it and exhaling mists. The lake has only one kind of fish which do not enter the channel of the water flowing through it, nor do the fish from the Tigris swim into the lake. Then when the Taurus range meets it the river plunges into a cavern, and on the other side of the mountain bursts forth into a lake, and afterwards returns to the form of a river and joins the Euphrates, and flows through Nineveh, and after a long course empties into a gulf of the Red Sea which is called the Persian Gulf. The Euphrates, according to Pliny, book V, rising in greater Armenia, separates Cappadocia from it, then it is met by the Taurus range after flowing to the west. Again it turns southward and divides into two branches. One branch empties into the Tigris inclosing on the north Mesopotamia, the other branch waters it on the west and flows through the middle of Babylonia, as Orosius says to Augustine, then flows into swamps, and finally into the Persian Sea. For the Chaldeans are in Babylonia southward toward the Persian Sea, and the Euphrates waters them, like the rest of the Mesopotamians and Assyrians, on the west, separating them from the other regions of Syria and from Arabia. The Euphrates, as Pliny says, swells like the Nile, differing little from it. For it inundates Mesopotamia when the sun reaches the twentieth part of Cancer, and begins to fall when the sun is in Virgo after Leo has been traversed. The inundation is over at the thirty-ninth part of the Virgin. The statement of Boetius in the fifth book on Consolation and that of Sallust that the Tigris and the Euphrates flow from the same source can be understood of their source in paradise; for this is a fact according to Scripture, with which Boetius at least was well acquainted, and Sallust might have believed from a study of the history in the Scriptures. This is also true as regards their origin in Armenia, since both rivers have their source there, according to Pliny; or it can be understood of their source this side of the Taurus range, for when they meet it they sink into the ground, and burst forth on the other side.

From the Euphrates which flows in the east extends Arabia, of which mention has been made, toward the south and the Red Sea. Toward the north are the remaining regions of Syria, namely, Syria-Comagena, Syria-Coele or Coele-Syria, and the Syria of Phoenicia, and the Syria of Palestine, which includes the provinces held by the Jews, namely, Judea, Samaria, and Galilee this side of the Jordan, and the region across the Jordan which the tribes of Reuben and Gad possessed, and the half tribe of Manasseh, where the region of Decapolis lies and Iturea or the region of Trachonitis.

In these provinces are found all the sacred places that first the holy patriarchs and prophets trod, then the lord himself and his Mother and the holy Apostles, and in which the primitive Church grew. With these places the Gospels resound, in which greater mysteries are contained than mortal ear can hear or human mind understand, as Origen maintains, commenting on Joshua, chapter XVIII. Wherefore I must speak with more care regarding them.

First, then, we must locate the famous cities on our sea or near it, which divides Italy and Egypt and Syria, so that further to the east the places that we wish may be easily grasped by the understanding. First, then, there is the famous city of Gaza in Palestine, not on the sea, but near it, about three leagues distant, on the confines of Egypt, Palestine, and Judea, as Jerome says in his book on Places. Then nine leagues distant is Ascalon, the metropolis of Palestine, situated on the sea. Joppa is twelve leagues distant. Then comes Acon, distant twenty-four leagues, or a journey perhaps of two days. For it is two or three leagues to Assur, formerly called Azotus, and then nine or ten to Caesarea in Palestine, in ancient times called Stratonis turris, where Peter baptized Cornelius, as Jerome states in many places. Five leagues further is Castrum perigrini, and three more to Caiphas and then four to Acon. Tyre is nine leagues distant in the heart of the sea. Four or five leagues distant is Sidonian Sarepta, where the widow fed the prophet Elijah; then three or four leagues further is Sidon, and eight or nine more is Berithum, which is called Barut. Then nine leagues further is Biblium, which is now called Gibeleth, in regard to which Ezekiel, chapter XXVII, says, "The ancients of Biblium and the wise men thereof had mariners." Tripolis is nine leagues further; then Tortosa, formerly called Radum, one day's journey distant. To Laodicea is a journey of about three days. For it is about ten leagues to Valania. Others, however, say that from Tortosa to Margat is one day's journey; and from Margat to Laodicea one day's journey, and from Laodicea to Antioch two days' journey. But Antioch is five leagues inland from the sea. From Antioch to Tarsus, made glorious by the Apostle Paul, and the metropolis of Cilicia, is a journey of about three days. It is a journey of one and a half or two days to the borders of Cilicia.

We must now pass to the places inland. Abraham and Isaac frequented the parts of Gerara. For Gerara, from which the region derives its name, as Jerome states in his book on Places, was formerly the limit of the Palestinians between Cades and Sur, and there Beer-sheba is situated, which is called the well of the oath, where Abraham and Isaac made a covenant with Abimelech. From this place the land possessed by the Hebrews begins. They did not possess anything south of this place, as

Jerome states in his letter to Dardanus on the Promised Land; although the land promised them by God began at a torrent in Egypt, as Jerome states in the eighth book, commenting on Isaiah, and in the first book likewise. For that torrent, as Jerome says, is a turbid stream on the borders of Egypt toward Palestine and Judea, which is dry at times. It is not far from the Nile, but is near a stronghold called Rinocorura, which the Seventy Translators have placed in the locality of the torrent, as we find in their translation in Isaiah. From Beer-sheba twenty miles northward is, according to Jerome in his book on Places, Hebron, once the metropolis of the Philistines, but honored as the burial place of four noble patriarchs, namely, Adam the greatest, Abraham, Isaac, and Jacob. It is easy then to turn our attention to the neighboring places, namely, the vale of Mambre and the oak of Abraham and the field of Damascus to the south of Hebron, which is so named from Damascus, a slave of Abraham. Hence this field is not near that great and famous city Damascus, the capital of Syria. For it is distant from that place about five days' journey, but it is near Hebron, where Adam was formed, and where Cain slew his brother, as the Master of the Histories states in his comment on Genesis. Carmel, where Nabal once was, is now the village Carmela by name at the sixth milestone from the town of Hebron to the east, as Jerome says in his book on Places. Near Carmel to the east eight miles from Hebron is the village of Ziph, where David hid, near which is a rugged mountain with the same name, Ziph, on which David abode near Carmela, as Jerome states. Fourteen miles to the east is Bethlehem, the city in which the Lord was born. According to Jerome, six miles from Bethlehem to the north was Jerusalem, the most famous by far of the cities of the East, as Pliny states. This city is distant from Joppa twelve leagues, and from Acon about three days' journey. Jericho is distant eastward from Jerusalem nine leagues. Jericho and the Jordan are distant two leagues. Tekoa, the village of the prophet Amos, is twelve miles distant to the southeast, as Jerome says in the second book on Jeremiah.

According to Jerome in his letter concerning the epitaph of the holy Paula, the region of Pentapolis, containing five accursed cities, Sodom, Gomorrah, and others, was near Tekoa.

For he describes Paula's return from that place to Jerusalem and states that she first passed through Tekoa, which was next to it. Orosius, moreover, in the first book of the Ormesta Mundi states that the region of Pentapolis is situated on the confines of Arabia and Palestine, and that the valley in the middle which the Jordan had watered the sea has now flowed over and covered. This is the Dead Sea and the sea of salt, and is called the sea of salt pits, and the lake of bitumen, and the salt valley, and the valley of salt pits, and the sea of Araba, that is, desert. Whence in the books of the Kings it is written, "From the entrance of Emath as far as the sea of Araba and the sea of Asphalt," that is, bitumen, according to Jerome in his book on Places and in his comment on Genesis. For in the valley of the salt pits were wells of bitumen before the overthrow of the cities, but after the rain of sulphur it was turned into the Dead Sea, which is called the lake of bitumen. Four cities were submerged, and the fifth, which was afterwards called Bale, remained at the prayer of Lot, in order that he might die in it after the destruction of the others. Later it was called Segor, and now in Syriac is called Zoara, as Jerome says in his comment on Genesis and in many places. This city, although it was not consumed with its sister cities by the fire of sulphur, yet after a lapse of time was destroyed by a third earthquake, as Jerome says. It is said to have been restored and called Zoara by the inhabitants, who are called Zoari. This city is at the end of the Dead Sea to the west, from which not far beyond the Dead Sea on the west is the city of Engaddi abounding in palms, whence come balsam and opobalsam. For there is a tree distilling balsam in the vineyards of Engaddi, of which Solomon makes mention in the Canticles. This city is called in Genesis Asasontamar, which in our tongue means city of palms. Tamar means palm, as Jerome states. Although many things are written by many authors on the characteristics of this sea and of the destroyed places, yet I shall here cite chiefly Hegesippus in his fourth book on the Destruction of Jerusalem, because he writes more fully than others, and many of the others have taken their statements from him and narrate them as if they were their own.

Regarding the Dead Sea he says that all living bodies spring back and are ejected at once and cannot be submerged in it;

that the water itself is bitter and sterile, containing no form of life, and finally that it permits the presence of neither fishes nor of birds accustomed to the water and delighting in diving. A lighted lamp, it is said, floats on its waters, and without being overturned is submerged with the extinction of the light, and although submerged with difficulty remains beneath the surface. It is also said that the Emperor Vespasian ordered men ignorant of swimming, with their hands bound, to be cast into the depths, and that they all forthwith floated as though borne up by some spirit of the wind, and when forced down sprang upward with great force. It is certain that lumps of bitumen float to and fro on the waters in a black liquid. These lumps are collected by men approaching in skiffs, who make this their business. Bitumen, it is said, coheres so that it can be cut away in no manner by iron or by any other kind of sharp metal. It yields indeed to the blood of women, who are said to be aided by bitumen in their menses. At the touch of this blood, as those allege who have had experience of its use, the bitumen is reported to be severed. It is said to be serviceable for the seams of ships, and healthful for the human body mixed with medicines. The width in stadia is 150 in the vicinity of the Sodomites, who once inhabited a very fertile region. Four cities, accordingly, were burned, a kind of semblance of which and form are seen in the burnt ruins. The lands burned, the waters burn, in which there still remain the remnants of the celestial fire. You can see there what look like green apples and formed bunches of grapes, so that they produce in the beholders a desire to eat them. If you pluck them, they fall to pieces and are resolved into ashes, and give out smoke as though they were still burning. Whatever, then, Isidore and Solinus state concerning the marvels of the world, and likewise Jerome in his fourteenth book commenting on Jeremiah, and Pliny also and many others, is included in the statement of Hegesippus. For if he summarizes briefly some things, saying that nothing living can be nourished in this sea nor sink in it, the other writers are specific. Pliny and Solinus say that it contains no form of animal life: that bulls and camels float on the water. Jerome says that owing to the bitterness of the water nothing with breath is found in it, whence there are neither fishes nor serpents, but even the fishes carried down by the floods of the Jordan into this sea at once die. Isidore in his book of Etymologies, explaining the words of Hegesippus, says that it is not stirred by winds, since the bitumen resists the whirlwinds. All the water is made stagnant by the bitumen, and supports no other matter except that in the form of bitumen. The lake extends between Jericho and Zoara, which is Segor.

Since the Jordan flows into the Dead Sea and there loses its name and identity, and since the regions of the Hebrews and many others are identified by means of the Jordan, I must now speak concerning it. Although Jerome speaks in a polished and truthful manner regarding the manifest source and course of the Jordan, also Pliny, Isidore, and others, yet Hegesippus is rightly to be placed before them all, who in his third book explains more definitely and fully the origin of this river. For all authors except Hegesippus, who proceeds by experience, thinks that the Jordan has its origin in two sources at the roots of Mount Lebanon near Paneades, which is now called Caesarea Philippi. Of these sources one is called Jor and the other Dan, which flowing in separate streams at length unite and form the Jordan. For some distance it preserves its identity and then passes into a lake called Gennesareth, to which is contiguous the lake of Tiberias, and then emerging, the Jordan flows to the east of Jericho and empties into the Dead Sea mentioned above.

That whole statement is correct with the exception of that part respecting its origin. Its manifest origin and the one popularly accepted is, as has been stated, from the two sources. But Hegesippus proves that it does not begin there, but from the fountain of Phiala, which is on the other side of Jordan in the region of Traconitis distant 120 stadia from the city of Caesarea. From this source, then, the water flows by subterranean passages and bubbles out again at Caesarea. For Philip, tetrarch of the region of Traconitis, put chaff into the Phiala, which the subterranean stream brought to the surface at Caesarea. Whence it is clear that the origin, not of the Jordan, but of the two branches, is at Caesarea. He adds also, respecting its course, that from Paneades or Caesarea it flows no longer in a hidden and concealed passage through hollows in the earth, but in a visible and open stream through the lands and beginning to spread out intersects Lake Semeconitum and its marshes. Thence directing its course for 120 stadia it proceeds to a city called Julias. It crosses later the lake called Genessareth, from which, after traversing much desert land, it is received by Lake Alfacius and is lost to sight in it. Therefore leaving the two lakes as a victor it is held fast by a third one.

Now that the Jordan is known, we must describe the cities and regions. The district of Jericho extends to the city of Scythopolis to the north of Jericho, according to Hegesippus. The former city is called in the Bible, Bethsan, as Jerome states in his book on Places, and is a town in the tribe of Manasseh, from which the children of Manasseh were unable to expel the original inhabitants. Westward and to the north of Jerusalem is the priestly city noted as the birthplace of the prophet Jeremiah, which is called Anathoth, distant three miles from Jerusalem, as Jerome states in his fifth book on Jeremiah. Thence more to the north, twelve leagues from Jerusalem and twelve miles from Caesarea in Palestine in the direction of Caesarea, is the famous city of Samaria, the metropolis of the ten tribes, which is now called Sebaste. Moreover, beyond the region of Caesarea in Palestine from the great sea to the boundaries of the Ptolemaean region is Mount Carmel extending in length about two days' journey, on which the prophet Elijah prayed, thickly covered with olive trees, orchards, and vineyards, as Jerome states in the fifth book, and in the first in commenting on Jeremiah. After the district of Samaria to the northeast comes the plain of Saba, so named at the present time, but anciently called the great field of Estrelon, mentioned in Judith, chapter I. The plain of Megiddo follows, in which excellent King Josiah was slain.* And through its borders to the north flows the torrent Fison to the great sea between Caiphas and Acon. Then to the north of that plain and distant seven leagues eastward from Acon is Nazareth, blessed city of our Lord Savior. Then two leagues further to the east is glorious Mount Tabor, on which the Lord showed his glory to his three disciples and Moses and Elias.

Then to the east is the city of Tiberias, which in ancient times was called Cenereth, as Jerome states in his fourteenth

* A hiatus occurs here in all the MSS.

book on Ezekiel, and near this the sea of Tiberias, also called the Lake of Cenereth, on which the city is located, which, according to Isidore in the thirteenth book, "is more healthful than all other waters of Judea, and is 160 stadia in circuit, which the lake of Gennesareth joins, the largest of the lakes in Judea. It extends in length 160 stadia, in width 40 stadia, the waters of which are unruffled by the winds with their blasts, but the lake ruffles itself, whence also Gennesareth is called by a Greek name indicating that the lake produces its own blasts of wind. Then through more extended spaces it is roughened by frequent gusts. For this reason its waters are purer and sweet and fit for drinking." These are the statements of Isidore, who distinguishes these lakes by their size and natural quality, although the gloss on Matthew VI says that the same lake is called the lake of Gennesareth and the sea of Tiberias and the lake of salt pits. But the lake of salt pits, according to all authorities, as stated above, is called the Dead Sea, and therefore this gloss of the master was based more on rumor regarding the places than on authority of the sacred writers or on experience. The statement, however, that the sea of Tiberias and Lake Gennesareth are one can be referred to their vicinity. For they are contiguous and join, and for this reason are reckoned as one. The Gospel of John shows that they are different lakes, because it states in the sixth chapter that Jesus went away across the sea of Galilee, which is the sea of Tiberias, and then came to the desert of Bethsaida according to Luke; and later the disciples came to Bethsaida according to Mark; then they put to sea that they might go to Capernaum, according to John. This cannot properly be the sea of Tiberias because they had crossed it before. This is therefore the sea of Gennesareth, and hence they are distinct but adjoining bodies of water. Then toward the north with the desert between, in which our Lord fed the five thousand from five barley loaves and two fishes, is Bethsaida, the city of the prince of the Apostles and of Andrew and Philip. Then comes Capernaum. It is clear from the Gospels that this is the order of those places. For before the miracle of the loaves, John says, Jesus went away across the sea of Galilee which is the sea of Tiberias; and then the multitude he fed met him; and after this the disciples went aboard ship, in order that they might cross to Capernaum, as John says, but before they arrived there, and before he fed the multitude, he came to a desert place, which is Bethsaida, with his disciples, as Luke states. Mark also states that they came to Bethsaida. Wherefore first is Tiberias, then beyond the sea of Tiberias toward the north is the desert of Bethsaida, and near it is Bethsaida, and then the lake of Gennesareth and finally Capernaum on its shore. All these facts are shown in the large gloss on the sixth chapter of Mark. Then after Capernaum is Julias, a town of which mention was made above; then Caesarea Philippi, at the foot of Mount Lebanon.

Further from Acon than Nazareth is Cana of Galilee, in which the Lord changed water into wine. Cana is distant from Acon five leagues. The distance from Cana to Nazareth is two leagues. Moreover, from Acon to the northeast nine leagues is Sapheth, the city of Tobias, about five leagues beyond Cana. Then at the distance of a league and a half is the city of Corazaim, distant from Tiberias about two leagues. In these places, Tiberias, Bethsaida, Corazaim, Capernaum, Cana, and Nazareth, the Lord spent most of his time preaching and doing miracles, as the Gospels state. The Lebanon range extends from Paneades or Caesarea from the region of Tyre, and Sidon, and Baruch, and Biblium, and Tripolis for a thousand and fifty stadia, as Pliny says, and opposite Tyre water flows down from Lebanon underground, and runs near Tyre for a league, and there bursts forth into a very wide well and like a tower in depth, which passes by means of a canal into the country and irrigates the neighboring localities. This well is the well of living waters which flows from Lebanon with force, of which mention is made in the Canticles. Moreover, opposite Tripolis is a fountain of gardens near the hills of Lebanon, flowing as far as a foreign mountain, from which a river flows and empties into the sea between Tripolis and Tortosa, yet near Tripolis. From that same fountain of foreign origin runs an aqueduct into Tripolis. The fountain of gardens and Lebanon are distant from Tripolis three leagues. Sidon is distant from the hills of Lebanon about three leagues, from the great mountain about five leagues.

Since now the cities and mountains this side of the Jordan and toward our sea have been discussed, we can now add some places toward the Euphrates. First beyond the Dead Sea is Macheron, formerly a stronghold second to Jerusalem, as Pliny states, at which the tribe of Reuben begins to the south; and in the north beyond Jordan near Mount Lebanon is the city of Pella, which is the farthest limit of the land of the Hebrews beyond Jordan in the north near Caesarea Philippi. According to Jerome in his book on Places the limit of that land to the east is the city of Philadelphia, which is called in the Bible Rabath of the children of Amon. On the boundary of the kingdom of Sehon and of Og, king of Basan, is Jacob's stream, as has been mentioned, where Ramoth-Gilead is located, of which frequent mention is made in the wars of the kings of Israel. Near Pella and Caesarea Philippi is Mount Hermon, towering above Caesarea and over against Lebanon to the east of the land of the children of Israel, as Jerome says in his book on Places, from which snows in summer are brought to Tyre as a luxury. But, as Pliny states, behind Lebanon with a valley lying between there rises a range of mountains equal to Lebanon, which is called Anti-Lebanon, and is situated toward the eastern part of Mount Lebanon, as Jerome states in his book on Places. Mount Galaad, according to Jerome in his book on Places, at the boundary of Phoenicia and Arabia joins the hills of Lebanon to the south, and extends through the desert to that place where, across the Jordan Sehon, king of the Ammonites, once dwelt. This mountain fell to the lot of the tribes of Reuben and Gad and the half-tribe of Manasseh. Jeremiah says, "Galaad, thou art to me the head or beginning of Lebanon." Anti-Lebanon extends, according to Jerome in his book on Places, around the regions of the city of Damascus, which fell to the lot of the tribe of Manasseh, and is the Damascus called in the books of the Kings the capital of Syria, which is distant from Jerusalem about four days' journey, from Acon about three, from Tripolis about two, from Baruch one. Then from Damascus about a seven or eight days' journey to the north is the famous city of Alap, which in ancient times was the abode of Abraham, distant from the Euphrates about two days' journey and from Antioch a journey of a day and a half. Then at the end of the promised land to the north in the east is the city of Amath, as mentioned in Numbers, chapter XXXIV, in the second book of Kings, chapter VIII, in the first book of the Apocrypha, and in many other places. Jerome says in his book on Places that he inquired diligently regarding this city and found that it is called Epiphania. Then comes the city of Comaga near Cilicia, where glowing fossil tar is found, which if cast on an armed soldier burns him up, nor can it be extinguished by water nor by any other liquid, but must be put out by placing earth on it. For a long time was the Roman army disturbed and confused by this bitumen cast upon its soldiers, until they learned the expedient of casting dust on the place touched by the bitumen. Pliny states this fact in his second book.

After the cities, mountains, bodies of water, and other particular places have been given, the provinces and regions near these can now more easily be considered. This whole Syria this side of the Euphrates is of immense length, but is narrower in width, as Isidore states. Pliny says that its length from Cilicia to Arabia is 470 miles; it has, moreover, many provinces all bearing the name of Syria. For in it are Syria-Comagena, Syria-Coele, Phoenician-Syria, Palestinian Syria, Galilee, Samaria, Judea. For these regions belong to Palestinian Syria, according to the authorities. Syria-Comagena, as Isidore states in his fourteenth book, is called from the name of the city Comage, which was once considered the metropolis in that region. This province has the Euphrates on the east, Cilicia and Cappadocia on the north, our sea on the west, Syria-Coele on the south, which is written with the diphthong, and is called Coele-Syria. Its capital and chief city is Antioch in the west, with which on the sea are connected Laodicea, Ateradum, and neighboring states as far as the province of Phoenicia, and in the east is Emath. For Jerome says in his book on Places that investigating diligently he found that this city of Emath was in Coele-Syria, and Pliny states this fact. So also is Alap, which is near Antioch and far distant from Damascus, which is in Phoenician Syria. This province has then on the west the great sea, on the north Syria-Comagena, on the east the Euphrates, on the south Phoenician Syria, which begins at the northern terminus of Mount Lebanon. For Pliny says that this range extends to Coele-Syria and this is around Tripolis; in this division are Tripolis, Tyre, Sidon, and Acon as far as Caesarea in Palestine.

For Pliny says that on the coast of Phoenicia is Ptolemais, which is called Acon, and Caesarea Philippi is in the province of Phoenicia, as Jerome states, and all this side of the Jordan as far as Palestine; and it contains Mount Lebanon and Anti-Lebanon and Damascus with its region, and all beyond Jordan to Pella, Mount Hermon, and Mount Gilead and those lands of the children of Israel beyond Jordan. That Damascus is in Phoenician Syria is manifest. For Jerome says in his book on Places, Damascus is a famous city of Phoenicia; and in his comment on Genesis he includes Damascus in Phoenician Syria, when he states that Hus, son of Aran, possessed Damascus and as far as Coele-Syria. The chief cities, however, of the Phoenicians are Tyre and Sidon. For, as Isidore states in the fourteenth book, Phoenix, brother of Cadmus, coming from Thebes in Egypt to Syria, reigned about Sidon, and called this province Phoenicia from his own name. This people likewise founded Tyre, from whom the whole land is called Phoenicia. This, however, is divided into two main parts, namely, the region of the Tyrians, Syrians, Aconensians, and the whole country between Lebanon and Tripolis; and the other main part is the Syria of Damascus. The city of Damascus as far as respects the kingdom of Syria between the Euphrates and Mount Lebanon was called the capital of Syria. For the name of Syria in the time of the kings of Israel was given to Damascus and its region. This province, therefore, of Phoenician Syria has the land of the Hebrews to the south and the land of the Philistines; but the land of the Philistines begins at the confines of the territory of Acon as far as the turbid stream of Egypt, and in ancient times contained nearly the whole country of the Jews this side of Jordan.

Since, however, the Jews occupied much of the region of the Philistines and shut them up in their maritime cities, namely, Caesarea, Joppa, Ascalon, Gaza, and others; we must therefore here note that this side of the Jordan there are three principal regions of the Jews, namely, Galilee, Samaria, and Judea named in a special sense, as stated in the gloss on Matthew XIX. The whole province of the Jews to distinguish it from other nations is called Judea, but in a special sense the southern district is so named to distinguish it from Samaria, Galilee, Decapolis, and other regions of the same province.

Moreover, this whole country of the Jews across Jordan and on this side Josephus in his book of Antiquities divides into parts and arranges, whom Hegesippus has followed in his third book, and explains those matters which are found somewhat obscure in Josephus. By these writers the whole region across the Jordan is called Pera. Its length is from Macheron beyond the Dead Sea to Pella near Caesarea Philippi and Mount Hermon, its width from Philadelphia to the Jordan. We find two main divisions in it; one is Decapolis, containing ten cities, of which one is Pella, as Pliny says, and the others connected with it to the south of Lebanon and Anti-Lebanon, according to Pliny, toward Philadelphia, which, as he says, two tetrarchies surround, namely, Paneas or Caesarea Philippi on the west and the region of Traconitis to the south beyond Jordan. Therefore this Decapolis is near Lebanon and Caesarea Philippi, and joins the territories of Tyre and Sidon, as Mark states in chapter VIII, "Jesus departing from the coasts of Tyre came through Sidon unto the sea of Galilee through the midst of the coasts of Decapolis." After this division to the south beyond Jordan is Ithuraea or the tetrarchy of Traconitis, in which is Phiala, the source of the Jordan, as has been stated above, not far distant from Caesarea, and also the region of the Geraseni, whose capital is Gerasa. The gloss on Mark, chapter V, says, Gerasa is a city of Arabia across the Jordan, adjoining Mount Gilead in the tribe of Manasseh not far from the lake of Tiberias, into which the swine rushed. Here, therefore, dwelt the Geraseni or Gergasenes, as we learn from this place. Then toward the south is another part of Ithuraea or of the region of Traconitis, of which Jerome speaks in his book on Places. The region of Traconitis or Ithuraea, of which Philip was tetrarch, according to the Gospel of Luke, is beyond Bosra, a city of Arabia, in the desert facing the south, and to the north it looks toward Damascus. This Bosra is in the desert beyond the Jordan, which fell to the tribe of Reuben to the south of Jericho and extends toward Macheron and the confines of the Moabites. But that there may be no question about Bosra, it must be stated that there is another city of this name in Idumaea, of which Isaiah says, "Who is this who comes from Edom, with dyed garments from Bosra?" as Jerome points out in his book on Places. Moreover, that Ithuraea extends almost to Decapolis and Caesarea is clear from the statement of Pliny that Ithuraea surrounds Decapolis.

The regions this side of the Jordan are thus divided. For according to Josephus and Hegesippus first there is entire Galilee, which has to the north the confines of Tyre and Sidon, to the west the territory of Acon with Mount Carmel, to the east Decapolis, to the south Samaria and Scythopolis given above. Galilee, however, has two divisions, the upper one being that of the Gentiles, which adjoins the territories of Tyre and Sidon. This, therefore, is called Galilee of the Gentiles, because Solomon gave to Hiram, king of Tyre, twenty cities in it. Therefore since Gentiles in this Galilee have mixed with the Jews, and because it is near nations of Gentiles to the north and to the east and to the west, it is called Galilee of the Gentiles. This ends near Tiberias and below the great plain of Esdrelon. In these places begins lower Galilee, called Galilee of the Jews, which is in the tribe of Zebulun. But we must be careful not to think that it is across the Jordan, as many have done owing to the statement of Isaiah and the Gospel of Matthew, "The land of Zebulun and the land of Nepthalim by the way of the sea across Jordan, Galilee of the Gentiles." But the gloss on Matthew says, Galilee of the Gentiles which is in the tribe of Nepthalim bordering on the Tyrians; and therefore since the tribe of Nepthalim is this side of Jordan, this Galilee is also. Jerome states this fact in his book on Places; and Josephus, Hegesippus, and all others attest it. But we must consider the form of expression in this matter. For we frequently find this form of expression in the Gospels. In the sixth chapter of Mark it is said, "And they departed into a desert place by ship. And the people saw them departing, and many knew him and ran afoot thither out of all cities, and outwent them and came together unto him." Where the gloss says, Jesus and his disciples did go to the other bank of the lake or of the Jordan, but crossed some sound or arm of the lake, and the inhabitants came to the neighboring localities, which they could reach on foot. From this we learn that across Jordan here means to pass over some portion of it, not the whole river. In the same way when it is said in the sixth chapter of John, "After these things Jesus departed across the sea of Galilee, which is the sea of Tiberias," he did not cross to the other shore, where the country of the Gergasenes is situated, but crossed an arm of the lake on the same side, namely, this side of Jordan, and therefore in this passage as in the one above a part is put instead of the whole. And when in the same chapter the evangelist says, "They came across the sea to Capernaum," they were still in the same region and were always this side of Jordan. Therefore the whole sea is not understood, nor is there a crossing to the other shore, but the part is understood instead of the whole on the same side of Jordan; and so here when it is said across the Jordan, Galilee of the Gentiles, a part is taken instead of the whole. For from the place where Isaiah said these things there was a long distance on the Jordan to Galilee of the Gentiles, which must be crossed in going from the one place to the other, and therefore he says across the Jordan, that is, across a great part of the Jordan which extended from Isaiah's locality to upper Galilee. Then to the south of Galilee of the Jews is Samaria, which is not only the name of a city but of a district, beginning in the great plain and extending to Judaea. The width of Judaea is from the Jordan to Joppa, according to Josephus and Hegesippus, and its length extends to Beer-sheba.

Now finally we must settle the question of the size of the promised land and the amount of it possessed by the Jews. But Jerome determines this matter definitely in his letter on the land of promise, saying that neither David nor Solomon ever possessed the land except from Dan to Beer-sheba, although after their victory they received many enemies and tributaries. The length of the land between Dan and Beer-sheba is scarcely 160 miles, as he himself states and infers. It is a shame to mention its breadth, for from Joppa to our little village of Bethlehem the distance is forty-six miles, and from Bethlehem to the Jordan about one day's journey. Hence the portion possessed by the Jews was small. We must bear in mind, however, that the portion just discussed was only their possession this side of Jordan.

Two tribes and a half-tribe, it is evident, had their possessions beyond the river, as Jerome explains in this letter. But the land promised them was from the Euphrates on the east to our sea on the west, and from Cilicia and the Taurus range on the north to the turbid river of Egypt, and to the land of Edom, and of Moab, and of Ammon on the south. For in his eighth book Jerome says, commenting on Isaiah, that the country from the Euphrates to the river of Egypt was promised to the Jews, nay, as far as the Nile. For that river is near the Nile. The Euphrates is to the east of this land. Moreover, the river of Egypt with our sea, into which it empties, is to the west. In his first book also he makes this same statement. He adds that on the north the land was promised them from Cilicia and Mount Taurus, and in the fourteenth book on Ezekiel he says the northern region begins at our sea stretching as far as Zephirus, a town of Cilicia, and to the very lofty Taurus range, and to Emath, a city of Coele-Syria now called Epiphania, and in the west continues from the torrent of the city of Rincocura flowing into the great sea, to that point on the sea opposite Emath, a city of Syria, of which mention was made above. The southern part begins at the torrent of Egypt, where it falls into the great sea, ascending through the desert of Sin and of Cades, and through the land of Edom, of Moab, and of Ammon, as far as the Euphrates. For if the Euphrates is in the east, and the sea into which the river of Egypt falls is in the west, then the southern part extends between that sea and the Euphrates. This follows of necessity. But this fact does not rest on a single authority, but is verified from many sources and follows from what has been said. For since in chapter XXXIV of Numbers and in the fourteenth book on Ezekiel in many places it is stated that the sea of Cenereth and the Jordan and the like are in the east, this fact is established with respect to the land possessed by the Jews this side of the Jordan. But beyond Jordan they possessed much, as is clear in the case of the two tribes and the half-tribe, and more was promised them, since the land promised reached to the Euphrates.

In the middle of Judaea is Jerusalem, rich in varied resources, whence, thanks to the elements, the Jews thought this the promised land flowing with milk and honey, since God from this source promises them the privilege of the resurrection. The separation of the ten tribes gave the Jews their name, for before that they were called Hebrews or Israel. But from the time when the people of God was divided into two kingdoms, the two tribes which had kings from the stock of Judah have been called Jews. The remaining ten tribes which appointed a king for themselves in Samaria were called Israel.

In considering other regions we must describe the Taurus range, since it separates countless regions. It begins in the east at the Indian sea and crosses to the west through the confines of India and the kingdoms of the Parthians, and through Mesopotamia and Syria. It leaves these provinces toward the south, and to the north the whole region of the Scythians and part of greater Armenia and Cappadocia, and crosses into Cilicia. In different regions it bears different names. For sometimes it is called the Caucasus, where it is higher, because of the abundance of its snows, for in the tongue of those people over whom it towers Caucasus means dazzling white, in another locality it is called the Caspian range, in still another the Taurus, and sometimes the Hyrcanus; and by many other names, more than twenty in number according to Pliny. For it has eight names from the Indian Ocean in the east before it is called by its own name, Taurus, then Caucasus, and after that it takes three foreign names, and is again called Taurus. But where it opens and the Caspian gates are formed, it is called the Caspian and Hyrcanus, and by many other names, concerning which we need not bother at present. These are Pliny's statements, although Orosius differs, and many others are at variance with the latter. Whence that whole range is called by many the Caucasus in the direction of India and then the Taurus. Others reverse these names, but we need not care since the different ideas and the different names in this case refer to the same thing. The more common practice of learned geographers is to call the range the Caucasus in the east, then the Caspian or Hyrcanus, after that the Taurus, and again the Caucasus, since where it reaches its greatest height it is called Caucasus, and then again the Taurus. The whole range is also called the Caucasus, and the whole is called the Taurus, following different views.

We must return then to the eastern regions beyond Mesopotamia, Assyria, and Babylonia, and state according to Pliny and all others that in that region are the kingdoms of the Medes and Persians and of the Parthians. These kingdoms have on the west the Tigris River, on the east the Indus River, on the south the Persian Sea or the Persian Gulf of the Red

Sea, on the north Armenia and the Taurus and Caucasus mountains, and the Caspian gates, or the Hyrcanian, and the land of the Hyrcanians and the Hyrcanian Sea, which is the Caspian. For the Caspian is the same as the Hyrcanian Sea, as Pliny states. We understand that the kingdoms of the Persians now belong to the Parthians, as Pliny states. Yet in fact that part which is on the Persian Sea is near Persia; for from it the Persian Sea is named. For there are eighteen kingdoms of the Parthians, as he states. Eleven are called upper and northern, which begin at the confines of Armenia and the shores of the Caspian or Hyrcanian Sea, and these peoples are properly Parthians, and extend to the Scythians, with whom they live on terms of equality. The Scythians dwell in the regions of the Caspian or Hyrcanian Sea and mountains. The other seven kingdoms of the Parthians are to the south, and adjoin the Persian Sea, and are properly called kingdoms of the Persians, and those people are Elamites, that is, princes of Persia, as Jerome states on Genesis and in his book on Places. For Elam is the chief city of the Persians, in which was Susis or the fortress of Susa, of which mention is made in the eighth chapter of Daniel, where was located the capital of the kingdom of the Persians. Pliny states that near the Tigris and 250 miles from the Persian Sea is Susa, capital of the Persians, founded by Darius, son of Hystaspes, on the northern channel of the Tigris. Near there is the town where the inhabitants unlike all other mortals cause gold to be hated, and dig it in order that no one may use it. The Medes are neighbors of both the Parthians and the Persians. For one part of the Medes, namely, that in the north, is next to the Parthians and Caspians, and begins directly at the Caspian Gates and on the confines of Armenia. Therefore they have the Parthians to the east, and to the north Armenia and the Caspian gates, to the west the Tigris, because the Parthians are beyond them toward the Indus River. The southern part of the Medes turns between the upper and lower kingdoms of the Parthians, so that it has the lower kingdom of Persia not directly to the east, but more to the south tending toward the west, according to the view of Pliny. For Media includes both the kingdoms of the Persians and embraces them, according to Pliny.

Beyond the Indus River toward the east is the whole of

India as far as the Scythian Sea, which is to the north, and the Himanus, Hemodus, and many other mountains which are parts of the Caucasus; and extends to the Mare Eoum, which is the sea in the east, and to the Indian Sea in the south, into which empties the Indus River, as Pliny states, because the Red Sea has now disappeared. Hence India has the Indus River on the west, and the kingdoms of the Persians and Medes; and it has the Scythian Sea, the Caucasus and Taurus mountains on the north and the kingdoms of the Scythians, and the Indian Sea to the south, and the Mare Eoum to the east, the configuration of which at the beginning was touched upon in many places, because there begins the habitable world. Therefore we had to begin with this country that the pen might proceed on through the longitude of the habitable world to the west through the regions of the Aethiopians, and again return from the west according to longitude.

The regions succeeding the former ones have been discussed as far as India, of which some things must still be said. For it has very great rivers, among which especially are the Indus and the Ganges, of which Scripture speaks. Regarding the size of the Indus Pliny says Alexander the Great on no day sailed less than 600 stadia on the Indus, nor was he able to sail over the whole of it before five months with the addition of some days. Yet the Ganges is greater, as he says; and this river, as Scripture states, compasseth the whole land of Havilah, where there is excellent gold. For rising in the mountains of the Caucasus in the north it divides India, flowing to the east, where are situated its great mouths at which it disappears in the Mare Eoum, that is, the sea in the east.

The Brahmins, of whom mention is made in the epistle of Jerome prefixed to the Bible, are in India. Because the sacred writers and philosophers and the histories relate more wonderful things about them than about other races, I shall here insert something about them, and in particular I shall cite what I wish from the writings of the blessed Ambrose for greater certainty. He states then in his letter to Palladius on the life of the Brahmins that they dwell by the river Ganges, where it flows into the Mare Eoum, but the men dwell beyond the river toward the sea, and the women on this side because of their remarkable chastity. For only because of their desire for offspring do the men consort with the women at certain times, namely, July and August, as the aforesaid sacred writer states. But when they have completed forty days with their women, they then return to their own abodes. When the wife of a Brahmin has travailed and brought forth a first and a second child, her husband does not cross again to her, for the sons one at a time take their fathers' place and the fathers refrain from intercourse of this nature for the rest of their lives. If, however, it happens that any one selects a barren wife, he crosses to her and sleeps with her for five years, and if she does not become pregnant at all within that period, he then abstains altogether from her. As it appears from this letter and from the principal work written by the blessed Ambrose on the life of the Brahmins, they have a very temperate climate, so that they do not use garments but cover themselves with the leaves of trees. They do not cultivate their lands, nor do they have trees or bread or wine; but they live on plants and leaves and fruits growing without cultivation, and they slake their thirst with their excellent waters. They are healthy without infirmity and live to a great age.

North of India, as has been said, are the Scythian Sea and those great mountains which are called Caucasus and Taurus and by many other names according to the diversity of places and races. To the west is Persia or Parthia, and Media. Then next them to the west is Mesopotamia and all Syria, as has been said. But on the confines of Media and Parthia is the iron gate of Alexander, which is a city named from the gates, and these gates are called the Caspian gates, not the Caucasian, as Pliny states. For the gates of Caucasus are different ones, as will be stated later, because these gates are on the shore of the Caspian Sea. For there is a certain sea, which is formed from the union of very great rivers flowing from the north and is called the Caspian Sea and the Hyrcanian Sea, according to Pliny. For the Caspii and the Hyrcanii dwell on the shores of that sea; nor does this sea enter from the ocean, as Isidore and Pliny and all other authors write. For in this case they did not have definite experience, either personal or through others, but wrote from hearsay. But in books on the manners of the Tartars and by men worthy of belief who have been in those regions, it is made clear that this sea is formed from the union of rivers, and is quite large. For a journey round it requires four months. Hyrcania is near at hand beyond the southern shore of that sea on the confines of Parthia; and where Parthia joins Media at its gates, extends from the Caspian gates toward the east, as Pliny states; then facing the remainder of Media to its north and to the west of Hyrcania is greater Armenia, divided by the Euphrates from Cappadocia, as Pliny states, wherefore Cappadocia is to the west of greater Armenia.

Then toward Syria and our sea is Cilicia, which is called lesser Armenia. Hence it lies partly south, partly west, of Cappadocia and begins less than a journey of two days from Antioch. Comprised under Cilicia to the north on the sea is Pamphylia, as Pliny states. The Isaurian tribe is passed over or not counted separately because of its small size, and is included with these states. In Cilicia is Tarsus its metropolis in which the blessed Apostle Paul was born. Cilicia extends from the south northward through Tarsus in width about four days' march toward Turkia; for to the north of Cilicia is Lycaonia, where is Iconium a very noted city, from which Lycaonia is designated as Iconia. Hence the ruler is called the sultan of Iconium and Turkia, for Lyconia is now called Turkia. From the confines of Armenia to Iconium is a journey of eight days. The names of provinces in these regions have been much changed owing to wars. For Turkia occupies many lands, which have ancient names in the authors, such as part of Asia Minor, Phrygia, and Lydia. Asia Major contains more than half of the world and all except Europe and Africa, and hence includes Asia Minor. Asia Minor is now called among the Greeks Anatolia, that is Eastern Greece, in which is Galatia, whence the Galatians to whom the Apostle writes, and Ilium, also called Troy, that very famous city. And there are many others, as Ephesus, and the seven churches of the Apocalypse, and Nicaea, whence the Council of Nice, and many other cities. In summer it is a journey of twenty days from Iconium to Nicaea; and from it to the Arm of Saint George, called by the ancients the Hellespont, is a journey of about seven days. This arm extends from the sea between Italy and Antioch. There Asia Minor terminates on the west. On the south it has that sea which is between Italy and Greece and Antioch and Egypt. To the east it has Phrygia. For, as Pliny says with reference to the divisions, Phrygia, situated above Troas, on its northern side joins Galatia, on its southern side Lycaonia, and to the east has Cappadocia. He also states that Lydia joins Phrygia on the east, whence came Croesus, a very rich king of the Lydians. The Arm of St. George is very narrow and has Constantinople on the west in Europe, and extends from the great sea between Asia, Egypt, Syria, and Italy, about one hundred leagues to the north to that sea which is called Pontic, and greater sea. That sea has the shape of a Scythian bow, and separates many regions.

Hence the regions of the north begin here, of which southern philosophers have little knowledge, as the astronomer Ethicus states in his book; but he traveled over all these regions and sailed the northern ocean with its islands. I wish, therefore, to follow him and also the books on the manners of the Tartars and particularly Friar William, whom the lord king of France, Louis, then in Syria, sent to the land of the Tartars in the year of our Lord 1253. Friar William wrote to the king of the location of the regions and seas.

This greater sea extends in length from the west, namely, from Constantinople, eastward for 1400 miles, and at its middle on both sides contracts into bays, and on the southern bay is the stronghold and port of the sultan of Turkia which is called Sinopolis. On the northern side he has another stronghold on a bay. This place is called Soldaia, and is in a province now called Cassaria or Cessaria. It is 300 miles from Sinopolis to Soldaia, the width of the sea between those bays. These strongholds are two famous ports, from which men cross from the regions in the south to those in the north and vice versa. From those strongholds toward the west or Constantinople the sea extends 700 miles in length and in width, similarly to the east 700 miles. This province of Cassaria is surrounded by the sea on three sides. For on the west it has that part of the Pontic Sea where is situated the city of Kersona, in which Saint Clemens became a martyr. Near it is an island, on which is a temple said to have been built by angelic hands, in which the body of the saint was buried. From Kersona to Soldaia there are 400 settlements, nearly every one of which has its own dialect. There are many Goths in that region, all of whom speak the Teutonic language.

The Pontic Sea extends from the southern part of Cessaria,

and to the east of it empties the river Tanais into the sea where it has a width of twelve miles. Here is located the city of Matrica. This river toward the north forms a sea of 700 miles in length and width, with a depth nowhere exceeding six feet. This sea is the very famous lake of Maeotis, of which philosophers, the histories, and poets speak. The river Tanais extends beyond that lake toward the north to the Riphaean mountains, which are in the extreme north. This river rises in these mountains and flows down through a vast stretch of country into the aforesaid lake, which it forms. The river then leaves this lake and flows into the Pontic Sea, as I stated above. This famous river divides Europe from Asia in those parts, and the lake mentioned and many others are contiguous, but are reckoned as one; and are called the lakes of Maeotis, or Maeotic in adjectival form. Those swamps then which bear the name of that shallow sea are in the east of Cassaria, and are part of the river Tanais, which is between the marshes and the Pontic Sea.

This province of Cassaria has on the north a vast desert, extending from the river Tanais in the east as far as the Danube in the west, a journey of two months if made swiftly on horseback, at the rate at which the Tartars ride, who cover a distance equal to that from Orleans to Paris in a single day. Hence this land extends for a distance equivalent to a journey of about four months as other men generally ride. This land belonged wholly to the Cumani, who were called Captac, but the Tartars destroyed it completely and slew the Cumani except the part which fled to the kingdom of Hungary, becoming tributary to it. This tribe is called Volana by the Teutons, but Alania to the west by Pliny, Isidore, and others. This province has the Danube, Poland, and Hungaria to the east.

On the north of this province is Great Russia, which likewise extends from Poland on its one side to the Tanais; but on its great side has Leuconia on the west, which is a land as large as Alemannia. On the western side of this country are many lands in the circuit of a sea, formed by many arms of the ocean which enter through the middle of Dacia; and beyond this country it widens into a great sea which on the west has Dacia and Sweden. Sweden is to the north of Dacia, sloping somewhat to the east beyond Dacia. Beyond these lands to the north is Norway. Then across a great intervening sea are Scotland and England and, separated by a small sea, Ireland. These regions are known, but I touch upon them to give a notion of the others. If then from the limits to the west we go on the north toward the east, first comes Ireland, second Greater Britain, containing England and Scotland, then Norway, Sweden, and Dacia. After these toward the east is the great sea mentioned above, called the Eastern Sea, because the ocean does not extend beyond that sea. But beyond the northern side of that sea is Esthonia; then Livonia toward the east of that sea; then Curonia or Curlandia as we pass to the southern side; after that Prussia, a large country, on the southern side; then Pomerania; after that Lubec, a great and famous port on the confines of Dacia and Saxonia. In the middle of that sea is an island called Gothland. Beyond Livonia to the east is Semi-Gallia. These lands, namely, Esthonia, Livonia, Semi-Gallia, Curonia, are surrounded by Leucovia mentioned above and it is surrounded by Great Russia on both sides of the aforesaid sea. Russia terminates to the south at Prussia and Poland. Poland lies to the south of Prussia, and to the south of it is Bohemia, then Austria. To the west of these lands is Germany, and then France and Spain. But these countries are known, yet I mention them on account of the others. To the east of Austria and Bohemia is Hungary, to which the western part of Albania extends. For it lies beyond the Danube, which flows through the middle of Hungary and beyond it into the Pontic Sea by twelve great mouths. At the end of eastern Hungary on the north is Albania, opposite which to the south of the Danube are the Blachi and the Bulgari and Constantinople, lands that were called in ancient times Thrace. Albania, therefore, to the west extends from the Danube after the limit of Hungary toward the east, as far as the river Tanais, with Cassaria on the south and Blachia, Bulgaria, and Constantinople; on the west Hungary, Poland, and the extremity of Russia; on the north it has the whole length of Russia.

Beyond Russia to the north is the Hyperborean race, so named from great mountains called Hyperborean. This race, owing to the healthfulness of the climate, lives in the woods, a race long-lived to a degree that they disdain death, of excellent habits, quiet and peaceful, harming no one and molested by no other nation. But others flee to this race as to a refuge. How it is possible for this region to be very temperate, I explained previously in treating of the characteristics of the localities of the world. Thus we have before us the notable northern regions in Europe. The rites of these races are of different kinds. For the Pruseni, Curlandi, Livonii, Estonii, Semi-Galli, and Lencovi are pagans. The Alani no longer exist, because the Tartars invaded that country and drove the Cumani to Hungary. The Cumani are pagans, so likewise were the Alani, but they have been destroyed. The Rusceni are Christians but schismatics following the Greek rite, but they do not speak the Greek tongue but the Sclavonian, which is one of the tongues spoken in many regions. For it is the language of Russia, Poland, Bohemia, and of many other nations. The Tartars inhabit the land of the Alani or Cumani from the Danube almost to the remotest parts of the east; and the other nations bordering on them to the north and south have they subjugated for the most part. For some tribes are in the mountains in very well-protected places which they are unable to conquer, although these tribes are their neighbors, because they are unconquerable.

The Tanais River flows down from the very lofty Riphaean mountains in the north, nor is there any habitation beyond them to the north. At the end of Russia and Alania, where congregate merchants and others who have come from Hungary, Cassaria, Poland, and Russia, there is a village, where the river Tanais is crossed by boat. The Tanais at that place is about the width of the Seine at Paris. Beyond that river is upper Albania as far as another great river called the Ethilia, four times larger than the Seine, and one of the larger rivers of the world, swelling in summer like the Nile. To the north this river is distant from the Tanais ten days' march, but toward the south they are far apart. For the Tanais falls into the Pontic Sea, and the Ethilia into the Caspian, and with many other rivers forms this sea. These rivers flow from Persia and from other localities. For the Pontic Sea, according to Pliny, is distant 380 miles from the Caspian.

In this land dwelt the Cumani, but the Tartars destroyed them all, just as they did on the other side of the Tanais as far as the Danube, as has been stated. The Tartars have countless herds of cattle and dwell in tents, having neither houses nor fortified places except very rarely. One chief with his army and his herds roams between the rivers, as, for example, between the Danube and the Tanais, and a second chief between the Tanais and the Ethilia, and so on eastward, because they have always been divided by the pastures and the waters. From January they begin to travel to the northern parts within the rivers until August, and then they return toward the south on account of the cold in winter. Toward the north the Ethilia is distant from the province of Cassaria one month and three days as the Tartars ride.

This land of the Tartars has between the Tanais and the Ethilia certain tribes to the north. First is the Arumphean tribe near the Riphaean mountains, which is similar in all respects to the Hyperboreans. These two tribes are near the pole in the north; but not so far north beyond the Tanais there is first a tribe called Moxel, subject to the Tartars. They are still purely pagan without law; they have no city but inhabit huts in the forests. Their chief and many of them were slain in Poland by the Poles, Alemanni, and Bohemians. For the Tartars led them to war with the Poles. They strongly favor the Poles and the Germans, still hoping to be freed by them from the Tartar yoke. If a trader comes among them he must give to the master of the first house in which he is entertained his expenses for the whole time he expects to remain there. For this is the custom of that region. Next to these eastward is a tribe called Merduim, subject to the Tartars. They are Saracens, following the law of Mahomet. After these comes the Ethilia, a river before mentioned, which flows down from greater Bulgaria, a land of which mention will be made later.

To the south of this land of the Tartars beyond the Pontic Sea are the Hiberi and the Georgians. In Georgia there is a metropolitan city called Thephelis, in which the preaching friars have a house. Beyond toward the east is the land of the Corasimini, but they have been destroyed by the Tartars. In these localities the Amazons used to dwell in ancient times, according to Pliny and the astronomer Ethicus. For the Amazons, as Ethicus states, were women leading a great army collected from women and without men. The Amazons, calling men to them at certain times of the year, conceived; but they slew the male children when born, reserving the females, whose right breasts they cut off by the art of surgery in their youth, that they might not be hindered by their breasts in shooting arrows. From their youth they nourished at their own breasts minotaurs and centaurs, most savage monsters. Hence these creatures used to precede the Amazons as though the Amazons were their mothers, and the Amazons overwhelmed every army more by means of these monsters than by arms. Likewise from their youth they nourished elephants and trained them for battle; and thus for a hundred years they laid waste the southern parts of Asia and Greece, until they were carried off and destroyed by Hercules. These districts of the Georgians and of the Corasimini have the land of the sultan of Turkey and Cappadocia to the south. For on the southern side of the Pontic Sea is the land of the Sultan as far as Sinopolis, of which mention has been made. Next to it on the same side of the sea toward the west is the terra Vastachii, namely, Greece to the east. For it is called Greece to the west, where Constantinople is situated, and the regions joining it on this side of the Arm of Saint George in Europe.

Greater Armenia is beyond Cappadocia to the east, and therefore that Armenia, although south with respect to Georgia, yet lies to the east and extends as far as Media and Mesopotamia. This whole land is considered by many to be the land of Ararath, because in Isaiah it is stated that the children of Sennacherib, after their father had been slain, fled into Armenia. But Jerome in his second book commenting on Isaiah solves this problem with the statement that the region of Ararath is Armenia in the plain through which the Araxes flows, a land of incredible fertility up to the foot of Mount Taurus, which extends to this point. Wherefore Ararath is not the whole of Armenia, but a limited yet large region. For the Araxes River, from which the region is called Ararath, flows from its source for a distance equivalent to a journey of three months and more. Its source is a fountain in the mountains of Armenia, where near by rises the Euphrates on the north and the Tigris on the other side of the mountain toward the south. On the mountains of Armenia, as Scripture testifies, the ark of Noah rested; but it did not rest on the mountains of Armenia without restriction, since the sources of those three great rivers are not in those mountains, but on the very lofty summit of Mount Taurus, where is situated the region of Ararath; as Jerome states, commenting on Isaiah. The ark in which Noah was saved, when the flood receded, rested not on the mountains of Armenia in general, but on the very lofty mountains of the Taurus, which tower above the plains of Ararath.

Near these mountains is a city that was very great before the Tartars destroyed it. For there were 800 churches of the Armenians in it; and in the time of Friar William, when he passed through it, there were only two small ones. Near there the blessed Bartholomew and the blessed Judas and Thaddeus suffered martyrdom. Two prophecies were uttered there. One is that of the blessed martyr, Methodius, who was of that race, and clearly prophesied of the Ishmaelites, a prophecy which had its fulfillment in the Saracens. Their other prophet is called Akaton, who prophesied concerning the Tartars and their destruction. For he says "from the north an archer race shall come, which shall subjugate all the nations of the East, and shall come into the kingdom of the West, namely, to Constantinople, and there shall they be destroyed by the princes of the West. Then shall all races be converted to the faith of Christ, and there shall be everywhere a peace so great, that the living shall say to the dead, Woe to you in that you did not live to the present time. A Christian Emperor shall place his throne in Taurinus in Persia." The Armenians consider that prophecy as gospel. This city just mentioned is now called Naxuam, which formerly was the capital of the kingdom, and is situated toward the northern parts of Armenia. For Friar William from the feast of Saint Clement journeyed up the Araxes, where it ends toward the north. On the Feast of the Nativity he came to that city, and in the octave of the Epiphany departed from it, and traveled along the Araxes to its source, at which he arrived on the second Sunday in Lent. Therefore it is further from that city to the confines of Armenia on the south than to the north.

Then further to the east are the mountains of the Alani and of the Aas, who are Christians, and welcome without distinction all Christians, both Latins and Greeks, whence they are not Greeks, and they fight with the Tartars; and so do the

Alani. After these to the east are the Saracens, called Lelgi who because of the strength of their country fight with the Tartars. After these to the east are the Caspian gates on the Caspian Sea, which Alexander the Great constructed where the mountains meet. For when he wished to conquer the northern race, he was unable owing to its ferocity and great numbers. And, as Ethicus says, he remained for a year and three months, that he might defend himself from them, and groaned because there was such a very wicked race in the regions of the north, and cried out to God praying that he would give relief lest the world be destroyed by them. Although he was not worthy to be heard, yet God of his goodness for the preservation of the human race commanded that there should be a great earthquake, and the mountains distant a stadium approached each other so that the width of only one gate separated them.

Alexander then built at the bottom of the pass bronze columns of wondrous size, and erected gates and covered them with bitumen, which could be dissolved neither by fire, water, nor iron. He sought this bitumen from the islands of the sea. These gates could be destroyed in no manner except by an earthquake; and now they have been destroyed. For Friar William passed through them with Tartars. Here is located the city called the Iron Gate of Alexander; at which toward the east begins Hyrcania beyond the Hyrcanian Sea, which is the Caspian, as was stated above. For Hyrcania lies on the southern shore of that sea and extends to the confines of India, on the southern side of which are Media and Parthia, as was noted above. These gates are not the Caucasian gates but the Caspian, as Pliny says, nor are the Caucasian gates the Caspian. For the Caucasian gates are distant from the Caspian 200 miles toward the Pontic Sea; and from the Pontic Sea they are distant 180 miles in the parts about Hiberia and Georgia. Those places with the intervening mountains are called the bars of Alexander, by which he restrained the northern nations from breaking through and laying waste the lands to the south. For Alexander waged many wars with them, as Ethicus states, and sometimes within three days there fell on both sides ten hundred thousand men. Alexander, however, conquered more by his art and genius than by force of arms. When they were aroused like bears from their caverns, he was not able to restrain them by violence, but God aided him by an earthquake and the closing together of the mountains. But now the gates have been broken, destroyed long ago either by an earthquake or by age.

We must give these places careful attention. For Gog and Magog, of whom Ezekiel prophesied and also the Apocalypse, have been shut up in these places, as Jerome states in the second book on Ezekiel. The Scythian race of Gog stretches across the Caucasus and the Maeotic and Caspian seas as far as India; and all who have been made subject are called Magog from prince Gog, and the Jews likewise, who Orosius and other sacred writers state will come forth. Alexander, as Ethicus states, shut up twenty-two kingdoms of the stock of Gog and Magog, destined to come forth in the days of Antichrist. These nations will first devastate the world and then will meet Antichrist, and will call him God of Gods, as also the blessed Jerome attests. Oh, how necessary it is for the Church of God that prelates and catholic men should consider these regions, not only for the conversion of the races there, and consolation of Christian captives in the same, but because of the persecution of Antichrist, so that we may know whence he is to come and when, by studying this matter and many others. At the Caspian gates the Caspian Sea begins to extend in length to the east, and in breadth to the north, and it is not smaller than the Pontic Sea, as Pliny states, and to make its circuit requires four months. Friar William on his return from the emperor of the Tartars traversed its western side, and on his journey to him walked along the northern side, as he himself related to the present lord king of France, in the year of our Lord 1253. To the north it has a vast desert, in which are the Tartars. Beyond these are many regions of the north before one arrives at the ocean. Therefore that sea cannot be a gulf of the ocean, which, however, almost all authors state it to be. But the experience in these times of Friar William and other believers shows that this Caspian and Hyrcanian Sea does not come from the sea, but is formed by the confluence of many large rivers. This whole land of the Tartars from the Tanais as far as the Ethilia belonged to the Cumani, who were called Canglae, and were all destroyed by the Tartars. That whole land was called Albania in ancient times. In this region are dogs of such size that they kill lions and pull down bulls. Men hitch them to chariots and plows.

Then beyond the Ethilia is the third principality of the Tartars, by whom the indigenous people, the Cumani Canglae, were destroyed, as stated above. That principality extends from the river to the east a journey of four months on the south until one arrives at the principal country of the emperor. But on the north it extends for a journey of two months and ten days. From this it is evident that Cumania was a very great land. For from the Danube as far as this land, in which the emperor resides, the Cumani used to dwell, who were all destroyed by the Tartars except those who had fled to the kingdom of Hungary. This principality has on the north first greater Bulgaria, from which the Bulgarians came who are between Constantinople and Hungary and Sclavonia. For this Bulgaria in Europe is lesser Bulgaria, having the speech of those Bulgarians in greater Bulgaria, which is in Asia. These Bulgarians of greater Bulgaria are evil Saracens, a strange fact, since that land is distant from the iron gate or from the Caspian gates thirty days and more across a stretch of desert; and it is situated in the extreme north. Hence it is very strange that the religion of Mahomet has reached them so far removed from the Saracens. From this Bulgaria comes the Ethilia, of which mention has been made. Next to this land to the east is the country of Pascatyr, which is great Hungaria, from which the Huni came, who later were called Hungri and now are called Hungari, who uniting with them the Bulgarians and other northern nations broke, as Isidore states, the bars of Alexander. Tribute was paid to them as far as Egypt, and they destroyed all lands as far as France. Hence they have become more powerful than are the Tartars up to the present time. A great part of them live in the land now called Hungaria, beyond Bohemia and Austria, which is now among the Latins the kingdom of Hungaria. Near the land of Pascatyr are the Blachi from greater Blachia, from which came the Blachi in the land of Assani between Constantinople and Bulgaria and lesser Hungary. For that people is called now Ilac by the Tartars, which is the same as Blac. But the Tartars do not know how to sound the letter *b*. To the south of that desert of the Tartars is the Caspian Sea, and then the Caucasian mountains to the east.

This principality extends from the Ethilia to Black Cathay, whence it is called Caracathaia. Cara is the same as black; and it is called Black Cathay to distinguish it from the other Cathay, which is far beyond this to the east many regions distant from this Black Cathay, of which mention will be made later. This land with those adjacent to it is the principal land of the emperor of the Tartars, in which he is always roaming with his court, traveling in summer to the cold regions, and in winter to the warm ones. This Black Cathay was the land of Prester John or King John whose fame used to be so great and regarding whom many false things have been said and written.

Here must be explained the origin of the Tartars, not only for greater evidence in distinguishing regions, but because of the race itself, which is now very famous, and has the world at its feet. We must note, then, that at the time of the war at Antioch, Coir Cham ruled in that land. For we read in the history of Antioch that the Turks sent for help against the Franks to King Coir Cham, who held the monarchy in the regions of the north at the time when Antioch was captured, and who was from Caracathaia. Coir is his own name and Cham his title, meaning diviner. For princes there rule the people by divinations and sciences instructing men in regard to the future or such as form the divisions of philosophy, as astronomy and experimental science, or the arts of magic, to which the whole East is devoted and with which it is imbued. All the rulers of the Tartars are called Cham, just as among us rulers are called emperors and kings. At the time of Coir's death, there was a certain Nestorian shepherd in that land, a man of power and master of a tribe called Naiman. These people are Nestorian Christians, who are wicked Christians yet say they are subject to the Roman Church. These Nestorians are not only in the land of Naiman, but are dispersed through all the regions to the east. The shepherd raised himself to the rank of king, and was called Presbyter and King John. This John had a brother, a powerful shepherd, named Unc, possessing pastures a journey of three weeks beyond his brother, and he was master of a certain village called Caracarum, which is now the imperial city, and larger than any other in the land of the ruler, yet it is not so good a one as Saint Dionysius near Paris in France, as Friar William wrote to the lord king. Beyond the pastures of this man about twelve days' journey were the pastures of the Moal, who were poor people, foolish and simple, without law. Near these and similar to them were other poor people, called Tartars. At the death of John his brother Unc raised himself to the throne, and called himself Cham, whence he was called Unc Cham. He sent his herds toward the confines of the Moal.

Among these Moal was a certain smith, Cingis by name, who pillaged and carried off the cattle of Unc Cham himself. When the latter collected an army, Cingis fled to the Tartars, saying to them and to the land of Moal, "Because we are without a leader do our enemies oppress us." He was then made their leader, and collecting an army he attacked and conquered him. He became ruler in the land, calling himself Cingis Cham, and took the daughter of Unc and gave her to his son in marriage. She became the mother of Mangu Cham, who divided the kingdom among those Tartar chiefs who now rule and are at discord with one another. For to this Mangu Cham Friar William was sent. Cingis Cham himself everywhere sent the Tartars ahead in battle, whence came the fame of the Tartars, who have been almost annihilated by frequent wars. Although for this reason we call that race Tartars in whose hands are the chief power and dominion, yet are their rulers and princes always from the race of the Moal. They do not wish to be called Tartars but Moal; because their first emperor, namely, Cingis Cham, was of this nation. From this nation they have had up to the present, previous to those who now rule, only three kings, namely, Cingis Cham, Keu Cham, and Mangu Cham. Keu Cham was the son of Cingis, and Mangu was his son.

This race, then, of the Moal from its beginning has been a very foolish and poverty-stricken one, which, however, by divine permission little by little has subjugated all the neighboring nations, and in a short time has subverted the world in its whole breadth. If this race were a harmonious one, it would lay waste Egypt and Africa at the first assault, so that it would in this way surround the Latins on all sides. For at the present time on the north they rule as far as Poland, because all Russia is subject to them; and all the land on the east as far as the Danube and beyond the Danube, namely, Bulgaria and Blachia, is tributary to them. Thus their empire extends to the land belonging to Constantinople. The sultan of Turkey, the king of Armenia, the prince of Antioch, and all princes in the east as far as India are subject to them with the exception of a few who are either too far distant or possess very strongly fortified places in the mountains which cannot be taken. First then in the land where the emperor dwells is Black Cathay. This was the land of Presbyter John. Next to it is the land of his brother, a journey of three weeks further, then still further the land of the Moal and Tartars, a journey of about twelve days. But this whole land is that in which the emperor dwells roaming through different places. The land, however, in which the Moal first dwelt, is called Oznam Kerule; and in that land the senate house of Cingis Cham still remains; but because Caracarum with its district was their first acquisition, they consider that city as the imperial one, and near it they choose their Cham, that is, their emperor. Then next to the Moal and Tartars to the east are brave men called the Tangut, who in the first place captured Cingis Cham in war; but peace being made, he again subjugated them. Their tribe has very spirited cattle with hairy tails like horses, the cows of which do not allow themselves to be milked unless they are sung to; and if these cattle see a man clad in red garments they spring upon him in their desire to kill him.

Next to these eastward are men called Thebeth, who were accustomed to eat their own parents as a filial duty, so that they might not make for them other sepulchres than their own vitals, of whom philosophers, as Pliny, Solinus, and others, write. Friar William also attests this fact in his book, and Friar Johannes de Plano Carpini likewise in the book he composed on the Tartars, among whom he was in the year of our Lord 1246, having been sent by our lord, the Pope, on an embassy to the emperor of the Tartars. But since they were from this practice an abomination to every other nation, they at length changed this rite. Yet they still make cups out of the bones of their parents' heads, from which they drink to their memory. Next to these to the east is a race of small people, swarthy like the Spaniards, called the Solangi, whose envoys when they come to the senate of any ruler have in their hand an ivory tablet, on which they gaze when they state their mission as though all things were written thereon. Beyond these is a people, whose animals do not belong to any individual in particular nor have they a herdsman, but if any native wishes to have an animal, he stations himself on a hill and cries out as he pleases, and they come at his voice, and he takes what he wishes. But if a foreigner should come, he would put them all to flight by his odor and make them wild. Therefore when foreigners arrive, they shut them up in a house, and give them the necessities of life, until they receive a reply from those on whose behalf they come; nor do they permit them to go at large through the country.

Beyond these is Great Cathay, which is called Seres by the philosophers. It is in the extreme east, to the north with respect to India, divided from it by an arm of the sea and by mountains. Here excellent Seric garments are made in great quantities and are exported to other lands. This people breathes much through the nostrils. They are excellent workmen in every art, and there are good physicians among them in all matters except as regards the urine, about which they form no judgment, but make an excellent diagnosis by means of the pulse and by other symptoms, and are well acquainted with the virtues of herbs and the potency of the whole medical art. Many of them are among the Tartars. The common money of these people of Cathay is a card of mulberry bark on which are stamped certain lines. Nor is this strange, since the Ruscani who are near us have as money the face of squirrels. This Cathay is not twenty days' distant from the land in which the emperor lives. In that land are lofty cliffs on which certain creatures dwell with a human form in every respect; they do not, however, bend their knees, but walk by leaping. They are not over a cubit in height, have the whole body covered with hair, and do not speak. Hunters bring beer and make holes in the rocks like cups, and those animals come and drink the beer, fall asleep, and so are captured. The hunters bind their hands and feet, and opening a vein in their necks draw three or four drops of blood, and then untie them and let them go. This blood is most valuable as a purple.

We must note that from the beginning of Cathay as far as

the eastward limit they are for the most part idolaters, but intermingled with them are Saracens and Tartars, and Nestorians who are imperfect Christians, with their patriarch in the East, who visits the districts and ordains infants in their cradles to holy orders, because he alone ordains, and cannot visit a place more than once in about fifty years. He says he has authority from the Roman Church dating from the remote past and is ready to obey this Church if the way were open. They teach the sons of the noble Tartars the Gospel and the faith and others also when they have the opportunity; but because of their scanty knowledge and their evil morals they are despised by the Tartars. In the Mass they consecrate a piece of bread as wide as the palm, and they divide it first into twelve parts corresponding to the number of the Apostles and then they divide those parts according to the number of the people. The priest places in the hand of each the Body of Christ, and then each one takes the Element from his own palm reverently. But idolaters are most numerous in all these regions, and all agree in having temples as we do and large bells. Therefore the Churches of the Greeks, of Armenia, and of the whole Orient are unwilling to have bells on account of the idolaters. The Rusceni, however, have them, also the Greeks in Cassaria. All their priests shave the whole head and beard, and observe chastity from the time of shaving the head, and they live a hundred or two hundred in a society. On the days when they enter their temples they place in position two benches, and sit in a straight line, choir facing choir on the ground, with books in their hands which sometimes they lay on those benches. They keep their heads uncovered as long as they are in the temple, and they read in silence, nor would they in any manner speak in the temple except the words of their office. Wherever they go they have in their hand a string of one or two hundred nuts, like our Pater Noster, and they continually repeat these words, "On man baccan," that is, God thou knowest. These things are common to all idolaters. But the Ingeres who dwell in the land where the emperor lives differ from the others. For the others do not maintain that there is only one God but that there are many, and they adore the creature. But the Ingeres, owing to the proximity of Christians and Saracens, hold that there is but one God. They are excellent writers, whence the Tartars have adopted their letters, and they are the great writers of the Tartars. They write from above downwards, and increase the lines from left to right and read. The Thebeth write as we do, and have characters similar to ours. The Tangut write from right to left, like the Arabs, but increase the lines ascendingly. The people in Cathay to the east write with the same instrument with which painters paint, forming in one character groups of letters, each group representing a sentence. By this method characters are formed with many letters together, whence reasonable and natural characters have been composed of letters, and have the meaning of sentences.

The whole land from the Danube to the east is called by the ancients Scythia, whence the Scythians and all the regions of the Tartars belong to Scythia as well as Russia and all as far as Germany.

I have described, then, all the regions of Asia and Africa and the northern ones of Europe. I shall now note briefly the eastern, southern, and western ones of Europe. For nearly all of them are known to all men. It has been stated that western Albania terminates at the Danube on the Pontic Sea, and extends to lesser Hungary. But on the other side of the Danube on the same sea the first country met was called Thrace in ancient times, in which Constantinople is situated. Beyond the Danube to the west it joined Moesia, but in those regions are at present Blachia and lesser Bulgaria. Then comes Hungary to the west, then Moravia, which is next to the kingdom of Bohemia. To the south of this is Histria; in the west Bohemia succeeds Moravia, and Austria, Histria. Then comes all Germany to the west, and then France, countries which are known.

After Thrace toward the south is Macedonia, famous because of its very great kings, Antigonus, Philip, Alexander the Great, which was called by another name, Emathia. Next to it to the south is Magnesia; then Thessalonia, to the people of which the Apostle wrote. Then still to the south is Boeotia, where the famous city of Thebes is situated, eighteen miles to the east of which is a well-known city named Niger Pons. Next to those lands to the west first is Sclavonia on the borders of Macedonia, Thessalonia and Bulgaria. Next to Boeotia is Attica, called from the city of Athens, nurse of philosophers, where Socrates, Plato, Aristotle, and other famous men taught. This Attica is part of Arcadia, according to the philosopher Ethicus, which is a very famous province not only on account of Athens, which Ethicus calls the navel of Greece, but because of its military power. For this land was called Sicyonia from King Sicyon, whence it is called the kingdom of the Sicyonians, one of the four principal kingdoms of the world which have been from the beginning, namely, the kingdom of the Scythians, originating under Reu, father of the great-great grandfather of Abraham; the kingdom under Seruch, great-grandfather of Abraham; the kingdom of the Assyrians; and that of the Sicyonians under Nachor, grandfather of Abraham; as Bede writes in his Chronologies. This last is the kingdom of the Greeks. For, as Ethicus says, all Greece has combined under the name of Sicyonia, since the kingdom of the Greeks was called the kingdom of the Sicyonians because the chief military strength was among the Sicyonians, who are Arcadians. I have treated these matters more fully, because all the histories have a great deal to say about the kingdom of the Sicyonians. In the philosopher Ethicus alone can be found clearly stated the explanation of the word and of the signification of the name. According to Pliny there is Achaia, then the province of the Peloponnesus, in which is famous Corinth. Then comes Locris and then Epirus, and so Greece ends.

Next to Greece westward is Dalmatia, in which is Duracium. Next to it is Illyria, from which the Illyrian Sea is named, which extends from the river Arsia to the river Dirinus, called as a whole by another name, Liburnia. But the Illyrians belong properly around the river Dirinus, and this country extends in length from the Arsia to the river Dirinus, and lies beyond the Adriatic Sea, which is the sea of Venice. At its western end is Venice. Illyria has many small islands, and has a length of 530 miles, a width of 315 miles. I have quoted these statements word for word from Pliny, because we of modern times do not know how to understand the word of the Apostle when he says he thoroughly disseminated the Gospel as far as Illyricum. In many of the histories we find Illyricum and the Illyrians and we do not understand the statements made. Whence the Illyrians lie between Dalmatia and Histria. Therefore, where now are located Sclavonia and Forum Julii and the districts of Venice, the Illyrians used to dwell. All these nations are included between the Arm of Saint George and the great sea on the east; and between the Danube, which is called the Hister for much of its length to the north; and between the Adriatic on the south. Distances in miles and days' journeys can be given in some of them. For it is more than 400 miles along the seacoast from Venice to Duracium. From there the distance to the famous city of Patras is forty miles, thence to Corinth sixty miles, thence to Athens forty miles, thence to Thebes forty, thence to Niger Pons eighteen, from which the distance by sea to Constantinople is 500 miles. From Niger Pons to the island of Crete is 300 miles. Then on the other side of the Adriatic Sea between it and the sea which runs from the Adriatic to Spain, lies all Italy, then Provence, and finally Spain. Since these countries are well known, nothing more need be said about them. This then, is the history which I have wished here to compose in accordance with the experience of writers on nature and of travelers respecting the places and races of the habitable world, until your Reverence requires the principal work.

After the description of places four other topics should follow, namely, a certification of the natures of the fixed stars and of the planets, that by these means a surer certification may be made of all the characteristics of the places and things located, to the end that a judgment may be made respecting the present, past, and future, so that at length, in the fifth place, actions may take place promoting all things advantageous to the state and excluding all things harmful. But since I have not been able owing to hindrances to complete the description of places in a form in accordance with the methods of writers on nature, and these four topics follow, it has been necessary to abandon the treatise on them; I wish, however, here in a summary to explain the design of the treatise on these four topics, as though the treatise had been written, just as I have done concerning other topics which I have handled; to the end that your Wisdom may see what is needed to secure advantage for the state, and how you are to seek writings and wise assistance from each man.

It was stated above that there are 1022 fixed stars, the size of which can be learned by astronomical instruments. These stars possess different forces in heat, cold, moisture, dryness, and all other qualities and natural changes. Among these stars are the principal ones of the twelve signs, by which all other things are especially subject to change. The signs are Aries, Taurus, Gemini, Cancer, Leo, Virgo, Libra, Scorpio, Sagittarius, Capricornus, Aquarius, Pisces, which are so named because the stars in the heavens have the quality of the things thus named. These signs Aries, Leo, and Sagittarius are in effect fiery; Taurus, Virgo, and Capricornus have reference to earth; Gemini, Libra, and Aquarius have reference to the air; Cancer, Scorpio, and Pisces are aquatic signs. Aries, Cancer, Libra, and Capricornus are changeable signs, because in them are renewed the four principal complexions of all things, namely, warm and humid in Aries; warm and dry in Cancer; cold and dry in Libra; cold and humid in Capricornus. Taurus and its followers are fixed signs, because the qualities named are fixed in them and are completed. Gemini and its followers are called common signs, because the quality in them turns to a new one which is renewed in the sign following. Moreover, these signs have many other differences and properties which must be determined by other treatises.

The first difference of planets is in their proper forces. For Saturn is cold and dry and the cause of all sloth and death and destruction of things through the egress of dryness and cold. Mars is destructive because of the egress of heat and dryness. These two planets never do good except by accident; just as a poison sometimes is an accidental good, like scammony, which clears away the matter of disease, but yet of itself injures nature. These planets are called unfair, and unfortunate, and malevolent. Jupiter and Venus have heat and humidity; but Jupiter more so and better; and these planets are called fair in fortune and benevolent. Mercury is in a middle state between good and bad and is of a changeable nature. For with the good he is good and with the evil he is evil. The moon is cold and humid. The sun has generative and vital heat, because he is the cause of life and of generation in all things, whence although he is hot and dry, yet his heat does not cause corruption, but is generative, and his dryness is not death-dealing, and for this reason different from that in Mars.

The planets, moreover, possess other virtues from the signs;

for when they are in a warm sign they have the power of producing heat, and so, too, with respect to the other signs. But besides these virtues the planets have special ones, of which mention was made above, which are house, exaltation, triplicity, boundary, and face; and in accordance with these dignities they have wonderful effects. And again there are forces with respect to their aspects, which are conjunction, opposition, etc. Planets are said to be in conjunction when they are in the same sign, in opposition when one is in the seventh sign from another. There is a third aspect when they are four signs apart; a fourth when they are three signs apart; and a sixth when they are two signs apart. Opposition and the fourth aspect are evil of their own nature. The third aspect and the sixth are good, and conjunction likewise. These aspects, moreover, are considered in the five planets, especially with respect to the moon and sun. When, therefore, an evil planet is in a bad aspect, the condition is detestable, because it then doubles the evil; but when a good planet is in a bad aspect, the condition is tolerable; and if an evil planet is in a good aspect, its evil is mitigated. But when an evil planet is in opposition or conjunction with an evil one, then is the evil a great one.

Planets have also a great variety of actions as regards their eccentrics and epicycles. For when they are in the upper portions of their orbits, they cause very vigorous actions, but in the lower positions weak ones: since when they are in the upper portions of their circles, which positions are called auges, they move around the world in a daily motion in great circles, and then they move rapidly, and speed of motion induces vigorous action in things which have been created to move, of which kind are the stars. For they have various effects with respect to the parts of their revolutions and their whole revolutions. For in accordance with their transit through a fourth of the heavens, the half, three quarters, and the whole circle, they have sensible differences in their effects. And not only so, but this is also true with respect to any one of their revolutions within fixed limits; as, for example, Saturn, when he has completed ten of his revolutions, causes a great change in the world, as we showed above in the matter of the sects. The twelve houses also, into which the whole heaven is divided, and which were discussed above, are especially studied with respect to the forces of the planets. For the planets draw various potencies from these houses, and by means of the forces of these houses the planets effect different results in this world: and therefore mathematicians teach us to give very great attention to these houses. These, then, are the basic principles in the actions of the stars, the roots, as it were, which have branches and flowers, and fruits without number. This, then, is the design of the treatise, I proposed to compose in this work on the virtues and actions of the stars.

In the third place, we must consider how the places in the world and the things are altered in their characteristics by the principles just stated. Certain general discussions have been presented on this matter in what precedes, as was necessary, but in this part of the work more things should be attested in particular. It is a fact, according to Aristotle, that the heavens are not only the universal, but the particular cause of all terrestrial things. For Aristotle says in the second book on Generation that the elements act more weakly with respect to the heavens than tools with respect to the workman, as, for example, a hatchet and pickaxe with respect to a carpenter; and therefore although all things are done here below through the qualities of the elements, yet their relation to the heavens will not differ from that of tools to the workman. Therefore if the workman is the active principle universal and particular with respect to the work, much more the heavens with respect to generating a thing. Averroës says in the seventh book of the Metaphysics that the force of the heaven has the same action on putrefied matter, as the force of a father in the semen. Therefore the things generated by putrefaction, although they are animate, are made directly from the heavens, and much more vigorously so other inanimate things. This holds true not only in these things, but in things generated by propagation in plants: because Aristotle says in this book on Vegetation that the sun is the father of plants and the moon their mother; and in the case of men and of animals he says that man begets man with the help of the sun. But Averroës says that the sun does more than man in producing a thing. For the force of the sun continues in the seed from the beginning of generation to the end, while that of a father does not, but is confined to one act only, namely, sowing of the seed; and therefore it would accomplish nothing, unless the force of the sun were continuously multiplied and infused, regulating the whole generation.

The complexions of all things are due to the heavens, and not only are regions diversified by the heavens, but things in the same region and parts of the same thing. This holds true not only in regular, but in monstrous generation and in the faults and errors of nature. For owing to differences of horizons, as the separate points on the earth's surface are centers in new horizons, it is manifest that all things vary, as we noted above with regard to plants of different species, and with regard to the difference in twins in the same womb: because the vertices of different pyramids containing the forces of the stars and of the parts of the heavens above the heads of the inhabitants come to the individual points of the earth, so that complete diversity takes place in things. But another difference is due to the distance from the poles and from the middle of the world. The cause of this is twofold. One is the universal cause, namely, the distance of the sun or its nearness as we stated above in the section on the places of the world. The second is the particular cause, namely, the diversity of the fixed stars above the heads of the inhabitants. For in particular by means of these stars do natural things vary in different regions, and men also, not only in their natural qualities, but in their morals, sciences, arts, languages, and in all other things. The third cause arises from the predominant force of the twelve signs. For the different signs are dominant in different regions, either because at the beginning of the world they were in the direction of those creatures when they received their first forces, and what the jar receives when new it retains the savor of when old; or because they are similar in nature to the stars which revolve over the heads of the inhabitants. The fourth cause is due to the planets. For the planets are assigned to the different regions owing to their domination, as in the case of the signs, and this follows from the twofold cause stated. Now regarding the signs, in what regions they are to be assigned and likewise the planets, it is difficult to attest, for authorities differ. A treatise must unfold these matters.

But there is no full attestation regarding these as well as many other difficulties except from the books of the Hebrews, which they composed as the first astronomers who had a special revelation from God in all matters. Not only is there a diversity in quality in the different regions due to the stars, but also in the things of the same region. For certain things are affected by the particular quality of the sun and certain by that of the moon, and this is true also regarding the other planets. According to the five dignities previously mentioned of their planets do things become vigorous, greatly strengthened, and augmented. When, however, the planets are in positions opposite to their dignities, things deteriorate and suffer loss of their natural vigor. The same is true of the signs, for different things follow the characteristics of the different signs. For in accordance with the fact that certain signs are fiery and certain planets also, that is, hot and dry, some things are of a fiery nature, and things of this kind are called martial from the planet, and are of the nature of Aries, Leo, and Sagittarius, which are hot and dry signs. The same principle holds true with regard to other characteristics of things, of signs, and of planets. To name, however, and mark things individually with respect to planets and signs, is a matter quite difficult of attestation and impossible except through the books of the Hebrews, that we may have recourse to the fountain of certainty. The difficulties just stated regarding the assignment of the signs and planets to the things of different regions and of the same region are among the greatest of philosophy. In this matter there is great diversity in Latin authors.

Not only do things of the same region vary thus, but parts of the same thing and especially so in man, since all things are on man's account. For the head is of the quality of Aries; the neck of the quality of Taurus; the shoulders and arms of the quality of Gemini, and so on. Regarding this division authorities are in practical agreement, and an adequate reason dictates it as well as experience, which is more important. For if the moon is in Gemini, the sign corresponding to the arms and shoulders, it is dangerous to touch such members with iron, as in blood letting, scarifying, or cupping and especially with knife or sword. And not only so, but this holds true of all surgical remedies; for they cause great difficulty and languor, and sometimes death; which they would not do, if the moon were in another sign. For Ptolemy says in the Centilogium, "If the moon is in a sign corresponding to a member, there is danger in touching the member with iron." Hali the physician in this matter states that due to the presence of the moon in such a sign humors flow together to the corresponding member and cause trouble. Experience proves this daily. For in the year preceding, a somewhat famous physician in France had a surgeon treat the tibia of his brother. But a skillful astronomer forbade his doing so, because the moon was in Aquarius, which is the sign corresponding to the tibia. And it happened that his infirmity and languor continually increased until death. Cases of this kind happen often, but are not understood, owing to ignorance of astronomy.

But the things of the same region not only thus vary in their complexions through the diversity of the signs and planets at the same time, but at different times. For a planet is dominant over every hour in particular, and half of a sign rises above the horizon every hour, from which the hours always vary, and the quarters of days likewise, and the days themselves, since as in every tongue the days are named from the planets, so have they diversity from them, as all learned men agree. This is by divine ordinance from the beginning of the world. Whence during the first hour of the Sabbath day Saturn is dominant and continues chiefly so over the entire day, because by his quality is the first hour affected, which is the chief one and the beginning when the force of Saturn is derived for the whole day. Then Jupiter is assigned to the second hour, Mars to the third, the sun to the fourth, Venus to the fifth, Mercury to the sixth, the moon to the seventh, and again Saturn to the eighth, and so on, so that he is dominant in the fifteenth, and in the twenty-second; Jupiter in the twenty-third and Mars in the twenty-fourth, and thus is completed the natural day of the Sabbath, so that the sun begins to be dominant in the first hour of Sunday, which is therefore called the day of the sun among all nations. On this theory proceeds the variation in all the days of the week and in the hours. From these facts the characteristics of the quarters of the day and of the night are obvious; so also we note which are the hours of blood, which of jaundice, which of phlegm, and which of melancholia. But physicians ignorant of astronomy do not judge respecting the quality of the hours except by the increase or diminution of the heat of the sun or by its absence. For they merely consider that six hours of the natural day are hot and moist, and the remaining six are hot and dry; still another six are cold and dry, and the last six cold and moist. But they do not consider of what quality the sign is that rises every two hours, and which planet dominates at the hour according to the rule given above; moreover, which planets rise above the horizon, although these matters and certain others should of necessity be considered. Yet this diversity in days and hours should be well known to every man, and especially to physicians, and to every one who wishes to preserve his health. For according to the diversity of the complexion of each one and according to the diversity of age and occupation, should each choose different hours for blood lettings, for medicines, and for all other actions of life, and for business affairs and occupations and for undertaking tasks, or he should be guided by the knowledge of others, because the forces of the heavens in these hours and days in various ways influence the different complexions of the healthy and the infirm. And further, as the complexions vary, the minds are aroused and influenced to desire gratuitously to follow the motion of the complexion, even in all voluntary actions, the liberty of the will, however, in all cases being preserved. Whence on the Sabbath day and especially during the first hour, a man ought not to be bled, nor should anything else of importance be begun which has reference to the complexion, either first in matters pertaining to nature or second in matters of the will. Such caution must be observed owing to the bad quality of Saturn, who in all matters causes misfortune as far as he is concerned, unless something else prevents. Yet all men, due to error, have themselves bled on Saturday, which they do owing to the enjoyment of rest on Sunday from their occupations and labors.

Not only in hours and days does a great variation occur to be considered as due to the heavenly bodies, but also in the week. This is due particularly to the age of the moon. For while the moon is crescent, all things increase; while it is waning, all things decrease or suffer diminution. This fact is considered not only by astronomers, but by physicians; as, for example, Galen in his book De Dynamidiis, and others in the light of former experience, as in the case of the seas, the brains of human beings, in marrow, and in shellfish and in all other things. For they are augmented and are full when the moon is full and diminish with its waning. For the moonstone called selenite increases and decreases according to the moon, by means of which the sacred writers in ancient times proved the changes in lunation of the paschal feast, as was stated in its proper place in this work. For a certain kind of mallows grows up at the new moon, and increases until full moon, and then decreases with the moon. It is a plant of wonderful virtue; with the new moon two leaves sprout forth on both sides of the stock and in the second quarter two others, and so on until full moon, and then in a similar way they fall as the moon wanes. But as Ptolemy says, and Hali explains in the fifty-sixth proposition of the Centilogium, the humors in our bodies in the first and third weeks of the moon recede from the interior parts to the external ones, just as rivers flow from their channels. But in the second and fourth weeks they flow back from the external parts to the interior, and in this there is a great and wonderful change in things. Hence it follows, as a matter of fact, that in the first week and in the third we should employ external evacuations, like blood letting, and in the second and fourth weeks laxative medicines; because Hippocrates in the first article of the Aphorisms says that we should draw from the direction found in nature through the proper region, that is, through the part of the body to which the natural humors flow. But physicians ignorant of astronomy pay no attention to this fact, frequently to the injury of their patients.

Another great change in things is brought about by the lunar mansions, which number twenty-eight. A mansion is the space of the Zodiac which the moon traverses in a day. Moreover, these mansions are distinguished, because certain are temperate, certain are distempered in dryness, cold, heat, or moisture. Accordingly on any day if one sees in the morning in which mansion the moon is, he will be able to judge in regard to rain and atmospheric changes, as Albumazar shows in his larger Introduction, and as is obvious by experience. We must consider especially the disposition of the moon at new moon; for of such a disposition must be at least the first week or fortnight, and at different times the whole month; similarly at the beginning of the second week, of the third, and of the fourth the same observation holds. For at those quadratures the action of the moon is strongest, according to Galen in his book on Crisis and Critical Days, as the moon crosses the quarters of its circle. Not only weeks, but whole months vary especially owing to the motion of the sun in the different signs, according to its recession from us or approach to us. Thus the quarters of the year change obviously in their complexions, since from these commonly the four principal complexions have their origin, namely, heat and moistness in spring, heat and dryness in summer, cold and dryness in autumn, cold and moistness in winter. For the mutation of the stars is strongest at the quadratures of their circles, as authors show and experience attests.

There is a greater mutation of years through these other planets, which according to their revolutions at the quadratures of their circles and at the revolution of the whole circle change years; and especially the three superior ones, Saturn, Jupiter, and Mars; as in the case of very great matters, which come slowly, because these planets are slow in their motion, and therefore their effects are not daily or monthly, as are those of the moon or sun. Moreover, effects of this kind are floods, earthquakes, pestilence, severe famine, appearance of comets and of other fires in the air; as Albumazar teaches in his book on Conjunctions, and as all scientists agree. Hence when a dreadful comet appeared in the year of our Lord 1264 in the month of July, the fact is established that it was generated by the force of Mars. For although Mars was then in Taurus and the comet began in Cancer, it did not cease speeding to its cause, namely, Mars; just as iron hastens to the magnet. For just as a magnet attracts iron so the planet, by the force of which the fiery vapor is raised and assimilated to the celestial nature, attracts the comet, and therefore since it moved toward Mars and there disappeared, of necessity it was caused by Mars. Then since the nature of Mars is fiery, whose nature is to increase jaundice, and consequently to excite men to anger, discord, and wars, therefore it happens that that comet portended angry passions, human discord, and wars, as learned astronomers teach us. But the experience of the whole Church too truly has attested the fact through the wars of England, Spain, Italy, and other countries, which happened about that time and have happened from that time. Oh, how great an advantage might have been secured to the Church of God, if the characteristics of the heavens in those times had been discerned beforehand by scientists, and understood by prelates and princes, and transferred to a zeal for peace. For so great a slaughter of Christians would not have occurred nor would so many souls have been sent below.

Not only do I make this statement on account of the revolutions of planets generating comets and the like, but more so because of the conjunctions of the superior planets from the time mentioned. Conjunctions, moreover, of the planets increased greatly in these times; but in the chapter above on the sects and on confirming the faith of Christ the remarkable knowledge of astronomers regarding these conjunctions was cited, showing how the usual laws of nature are changed, and the minds of men are aroused to seditions and to changes in customs and laws, to discords and wars, and how dominions and kingdoms are changed, and the vicissitudes of princes renewed. Albumazar and other learned men clearly teach this. If these examples should prove to be on too high a plain, let us turn our pen to common examples and to such as are useful for men's bodies. For Ptolemy says in the nineteenth proposition of the Centilogium, "If any one take a purgative when the moon is in conjunction with Jupiter, its action will be shortened and its effect lessened." Hali states that the reason of this is found in the fact that nature is so strengthened from the beneficial action of Jupiter that it hinders the effect of the medicine. If the moon should be in conjunction with Saturn, the one taking the potion will suffer gripings and be in danger owing to the evil influence of Saturn, nor will he be able to relieve himself, owing to the dryness of Saturn, by which the medicine will be retained in his body, because a dry substance is easily retained.

But how diseases are to be learned in conjunctions of this kind and in the other aspects and dignities of planets, medical authorities show excellently; and especially so Hippocrates in his book on the changes in human bodies which happen from the motion of the moon when subject to the conjunctions and other aspects of the planets. Galen, moreover, states in the third book on Crisis and Critical Days, "I have returned to a matter which I have studied with zeal and careful investigation, and I have found the same to be true and certain which is imperishable, a fact which the Egyptian astrologers discovered, I mean that the moon marks on days what the condition of a man will be in his sickness and in his health; that is, if certain equal stars are connected with the moon, and these are stars called fortunate ones, those days will be beneficial to him. And if certain stars departing from equality are connected with it, those days will be days of injury and sadness." He here uses the term "connected with" as referring to conjunctions and other aspects, and he has here much to say on this subject.

Hali the physician, also, in his book The Royal Regimen shows clearly that physicians need study of this kind, and hence he says, astronomy is needed by medicine, which uses selected remedies at selected times, when the moon is tempered by good and fortunate planets and by favorable signs of the Zodiac. All authors agree in these matters. Because of these conjunctions and reciprocal revolutions of the planets, according to Avicenna in his second book on Animals, all generation is augmented and regulated, also the beginning of life and of death. Whence he says it is necessary that the limits of life and of death should have been placed in a species, as long as life is concordant with the revolution of some star or of many stars. The first limit is the day with its night: then the second is that established by the moon at the quarter of its circle; then there is the month, which is the circle completed at the conjunction of the sun and moon; then the year of the planets; then the year of mean conjunctions and reversions to those forms in which these conjunctions took place, which tend to the augmenting of natural things and the diminution of unnatural things. Not only do these things happen in regular generation, but in that which is monstrous and in the faults of nature, as Avicenna himself teaches in the same book, saying, "If the embryo is able to receive humanity, it is able to receive animality, as in things monstrous, for example when the son of a man has the head of a ram, and a lamb the head of a bull, since the force in it has induced forms and celestial figures, things happening to anybody."

Not only in these ways do changes occur in the conditions of terrestrial things, but they are also due to the mounting of planets to the apsides of their orbits, and then to their descent to the opposite point, since their motion is rapid at their apsides and slow at the points opposite, as we see by experience. For all things produced on the earth grow more about the summer solstice, when the sun is at its apsis, and gain more vigor in one day than at other times in a week. And when the moon is at the apsides of her circles, as at new moon and at full moon, her actions are more vigorous, as is obvious in the tides of the sea and in fishes; for then are fishes better, then are shellfish more healthy, as authors teach and experience shows. This same is true of all the planets.

But the last point I wish to touch upon is regarding the twelve houses, which arise from the division of the heavens into twelve parts, the study of which is especially recommended by the philosophers. For Albumazar in his book on Flowers says that these houses must most certainly be discussed and examined. For by means of these houses we first consider which planet controls the complexion of the whole year. This study is made at the beginning of spring, when the sun enters the first minute of Aries at the vernal equinox. By these attestations we know how the planets affect the complexions of individual days and hours and all things, as they rise above the horizon. For astronomers show this fact clearly and experience manifests it. These then are the basic principles—roots, as it were,—regarding change in the complexions of things due to the forces of the stars. In a treatise on this subject the branches and fruits and flowers must be displayed along with tables and excellent drawings, whence the potency of figures and computations is more obvious; moreover, the utility and beauty of this knowledge are very great.

After these matters a fourth follows, namely, consideration regarding judgment and knowledge of the past, present, and future. For if the cause of the complexions of things is the celestial constellation, an effect of this kind can be known through this cause. The method or time is attested by the sacred writers, and especially by the patriarchs from the beginning of the world, from whom the philosophers learned the possibility of forming judgments. We must consider that all philosophers have agreed on this point, and Aristotle has attested it in his book on this subject, and experience teaches it. Nor is there any contradiction except on the part of boundless ignorance, which is rife in the common throng and in those rulers who reprobate and neglect all things of which they are ignorant. The method of forming a general judgment by means of the planets consists in this, that a man know by the tables and instruments how to equalize the motions of the heavens and find the positions of the planets; and how to judge which of them has the more numerous and stronger influences from its position in accordance with those five famous ones, namely, house, exaltation, triplicity, boundary, and face; and along with this how to view the influences which happen from the aspects of those planets and from their motions at the apsides of their orbits and from the twelve houses.

When he has carefully examined and considered these matters and those connected with them, then if the sun shall be found to have more evidences and influences of this kind, he must decide according to the solar complexion in things, because it is dominant in them; and he must proceed in this way with respect to the other planets. But a particular and special consideration in individual matters has its own fixed rules according to the condition of affairs. One can examine history at past periods, and study the effects of the heavens from the beginning of the world, as in the case of floods, earthquakes, pestilences, famines, comets, prodigies, and other things without number, which have happened both in human affairs and in nature. After he has discovered these facts, he should consult the tables and canons of astronomy, and he will find that there are constellations corresponding in an appropriate way to the effects in each case. He should then study with the help of tables similar constellations in future time, either near or remote according to his wish; and he will then be able to express judgments on the effects, because they will be similar to those in the past, since if we assume a cause the effect is taken for granted.

Instruction of an introductory character regarding these judgments is given in sufficient fullness in the work of Albumazar dealing with this subject, and in the book of Alkabiz and in that of Hali Abenragel. But the roots, as it were, of forming such judgments are given in the hundred words of Ptolemy, who has more weight in speculative philosophy. The branches are developed in the fourfold divisions of Ptolemy, as those same authors state. The flowers and fruits are collected in many books, but particularly in the book of Conjunctions by Albumazar, through which a knowledge of this science has come into the hands of the Latins. I make this statement, because they do not yet possess complete knowledge on these matters, since the Latins have nothing of value except from other tongues; and the translators have been few and bad, as is obvious. For the books of the Hebrews give definite information on these topics, a certification which Aristotle studied and reproduced in Greek. For Averroës says in the fourth book of the Heavens and the World that Aristotle attested this matter in his book on celestial impressions, a book which is superior to all the philosophy of the Latins and which can be translated by your order.

The fifth consideration stated above is the principal one of them all, for all those named lead up to it; since it does not suffice a state to know all things, but useful measures must be known and taken in advance, and harmful things avoided. But in this matter, although geometry, arithmetic, and music give a great deal of assistance, yet astronomy regulates all things, because every splendid work ought to be done at times selected. Therefore the activities of other sciences require proper times, the selection of which is based on the knowledge of the astronomer, and for this reason astronomy is mistress of all other sciences in this particular, although it has special remedies also without number, in which other sciences possess no potency. Because the potency of this science is so general, Aristotle, the wisest of philosophers, teaches Alexander in the book of Secrets that he should neither eat nor drink, nor do anything without the advice of an astronomer, because there are times selected for all things; for all things have a time, as Solomon, a wiser man than Aristotle, states. Aristotle raises an objection to his own statement, saying, "God has seen beforehand all things from eternity, therefore the astronomer cannot hinder or change these things." He replies that those things which God saw beforehand are immutable. But the effects of those things which God has placed in the power of man in accordance with his eternal foresight, man can change as he wills, because in contingent things there is a choice. God has not imposed necessity on human actions, although he has known from eternity how a contingent matter must terminate, and there is freedom of the human will. Therefore man can take thought beforehand for all his advantages, and remove obstacles, if he is skillful in this science. He gives an example. For if, as he says, there must of necessity be excessive cold at a future time, the astronomer foreseeing it can arrange for relief, so that the cold can be borne without injury, which others without warning have been unable to endure. For the astronomer can prepare in advance warm places, warm foods, warm clothing, and many other remedies, so that the cold will harm him in no way, although others who have not foreseen it are dying from the cold. It is therefore in the power of a man skillful in this science to avoid what is harmful and to secure what is useful.

Isaac in his book on Fevers, in the chapter on pestilence, teaches this truth excellently, not only in the matter of diseases, but in all things in general, saying that the rational soul is worthier than the stars, for their action does it very little harm unless it is possessed with ignorance. For provided he himself knows their action, he will be able wholly to be on his guard; and he says, "This fact is attested by what we see done by very excellent mathematicians, who since they know with the greatest certainty the days of death in the case of those who are putting to sea, are able effectively to guard them; similarly in the case of those who are to die in war." He adds, moreover, "Man differs from the animals, for it is the function of reason and discretion to inquire and long for what is good and to reject what is evil unless a man is possessed with ignorance." Therefore Ptolemy says in the fifth book of the Centilogium, "An excellent astrologer will be able much more to prevent what is destined to come according to the stars"; and in the eighth book he says, "A wise soul will aid the work of the stars, as a sower does the forces of nature," whence it will be able to drive away what is harmful and promote what is useful. Isaac, moreover, gives an example regarding a coming pestilence, and proposes a question to himself on the ignorance of physicians, just as Aristotle did. He asks them whether a rotting of this kind must come owing to the heavens, since in that case the physician cannot prevent, and therefore foreknowledge of this event seems useless. But he answers the question, saying that corruption of the air harms only bodies ready for it; and therefore when the physician has seen the sign of a corrupted air at any future time of the year, he will be able to cleanse nature, and carry off the evil disposition of the body, so that he knows how to resist the coming corruption. For if he sees by the motion of the stars that heat is being generated in the air and too much dryness and burning, he should hasten to purge away the jaundice of those in whom there is a choleric habit, and he should give cold and moist foods and drinks which resist the coming heat. Hali expresses a like opinion in expounding the fifth proposition of the Centilogium. From these facts it is evident that the astronomer is able to offer remedies against things harmful and to promote things useful, not only in diseases but in other matters generally, by the correct method of selecting the time suitable for all actions, and by the removal of things harmful. This truth holds good especially in the case of our human bodies, than which nothing is more important as far as the particular good of our citizens is concerned.

But nevertheless, as regards the common good of the state and of kingdoms, more works are wrought through the potency of astronomy than any one can tell, and more than any one can explain in writing; for these are many of the most secret works of science. But owing to the importance of the matters and the ignorance of the majority of students, who care nothing about the works of science, and owing to the frauds and evil practices of many who have abused these matters, they are always kept secret from the commons and from their rulers. Concerning these matters, however, Moses and Aaron, Solomon, Aristotle, Ptolemy, and other scientific men have labored very admirably. Hence since this subject is one of the most important ones and in a measure potent in all matters, it is not proper that it should be kept hidden from your Glory. Since you have commanded me to write on the wisdom of philosophy, I shall cite to your Clemency the opinions of sages, especially since this knowledge is absolutely necessary to the Church of God against the fury of Antichrist. When Moses in his youth was the leader of the Egyptian army against the Aethiopians, and for the sake of peace had married an Aethiopian princess, on account of whom

Aaron and Miriam spoke against him, as recorded in the twelfth chapter of Numbers, being unwilling to dismiss her that he might return to Egypt owing to the depth of his love, he made, since he was a skillful astronomer, two images on rings, one of forgetfulness which he gave to the woman, and the other of memory which he kept for himself, and thus he freely departed from her with his army and without war. Josephus in the first book of the Antiquities states these facts, and the Master in the Histories and many others attest them. This was a wonderful thing that changed the heart of the woman. Solomon, moreover, made regulations respecting many things of this kind, which happened contrary to the ordinary course of nature, as Josephus states in the eighth book. Solomon, who was wiser than all who preceded him and all who followed, could not neglect these matters, and therefore left many things written in an enigmatical form, which later by the magi were turned to evil uses and were interpreted in the wrong way, in which have been mingled many hideous things by fraudulent people. But the wise know how to separate the grains from the chaff and the antidote from the serpent.

Aristotle, the greatest of the philosophers, shows how these works are performed in his book of Secrets, to the end that all wonderful things may be usefully promoted and all harmful things destroyed. For by these means Aristotle caused Alexander the Great with an army of less than 40,000 to conquer the world. When he lay on his death bed, and was already on the threshold of death, he rescued his city and country from enemies. The first of these facts is noted in the deeds of Alexander, and the second in Valerius Maximus, book V. But it is evident that he could not have done this by bodily power but by the great force of the wisdom which he has left in the books of the Secrets to be investigated only by the most wise. When owing to the burden of old age he handed the world over to Alexander after the defeat of Darius, he withdrew to his own land and bade Alexander write him on any matters he wished, and promised to give him fitting advice. When Alexander found tribes with very bad morals and wrote to Aristotle asking what he should do with them, that prince of philosophy replied, "If you can alter their atmosphere, allow them to live: if not, slay them all." Oh, how occult is the reply, yet how full of the power of wisdom! For he understood that in accordance with a change of air, which contains the celestial forces, are the morals of men changed; for which reason the customs of Gauls, Romans, and Spaniards differ from one another, and the same is true concerning all other countries. Aristotle meant then that Alexander should change for the better the quality of the air of those tribes, so that in accordance with that change their morals might be changed and they might be influenced to adopt a high moral standard, without, however, losing freedom of will; just as each nation is influenced to adopt its moral standards by its own atmosphere, which contains the forces of the stars which are over the heads of the inhabitants, and in accordance with the signs or planets dominant over the particular regions. For I show in the treatise which I am sending that any one can be influenced to good and to evil, both public and private, by the forces of the heavens without constraint, just as by way of example we see that men change their desires through masters, friends, associates, occurrence of new conditions, and in countless ways without restraint. This matter I explain sufficiently in the chapter on the power of judging with regard to human affairs by means of the heavens, in which chapter there is an explanation of all matters relating to this subject. The philosopher wished then that he should perform deeds of wisdom by means of the necessary constellations in the manner of Moses, who stirred the mind of the woman by means of the celestial forces received in the material. For as that woman could be changed to purity and to a forgetfulness of her husband by means of images, so could she have been influenced also to adopt other morals, and not she only, but any other woman. Ptolemy in the ninth proposition of the Centilogium teaches that visages in this world are subject to the celestial visages. Hali says that Ptolemy in this chapter wishes to disclose the secrets of images. The purpose of these in general is obvious, since if one happens at chosen times to carve these images after the faces of the heavens, all harmful things can be repelled and useful things promoted. Thebit, the greatest philosopher among all Christians, who in many things has added to the works of Ptolemy and of other astronomers both in theory and practice, in particular has broadened this science, and other very wise men were in the habit of stressing these matters.

But since these works seem to the rank and file of students to be beyond the human intellect, because the throng with its teachers has no leisure for the works of science, scarcely any one has ventured to speak about these works in public. For they are straightway called magicians, whereas they are the very wisest who know these things. Undoubtedly theologians and judges not instructed in such matters and seeing at the same time that evils as well as blessings can be produced in this way, neglect and abhor these things and reckon them as magic. They see also that magicians and those who abuse the teachings of science employ these means, and they therefore judge them unworthy of Christians. But truth must not be condemned as ignorance, nor utility as an evil, although they may become such; for in that case men should do without knives at table, since they are able to slay their table companions with the same piece of iron with which they are cutting their food, and weapons should be destroyed by the Church, and also the secular arm should be destroyed, because many evil things can be done by these means, and are accomplished daily. Good men must have recourse to the laws although many lawyers take advantage of the laws by means of sophistries and frauds. And now we see that students make more use of what has been written concerning sophistries than of the correct use of demonstrations. The study of science is not, however, on this account to be condemned. For things from which good men produce various blessings are always turned to evil by evil men.

Since, moreover, the rational soul has especial need of words formed efficaciously and by design, the astronomer is able to form words for chosen times which will have inexpressible power. For when the purpose, desire, and force of the rational soul, which is nobler than the stars, are in harmony with the force of the heavens, of necessity either a word or something else is produced of wonderful force in altering the things of this world, so that not only the things of nature, but human minds are drawn toward those things which the skillful adept wills, the freedom of the will remaining unimpaired, since the mind can follow the celestial forces fully without compulsion, as we showed and stated in the proper place.

From this source the use of characters and incantations began. For characters are like images, and incantations are words uttered in accordance with the intention of the rational soul, which receive in the mere act of pronouncing them the force of the heavens: whence I make mention in the third part of the wonderful potency of letters. For by this power our bodies are cured, venomous animals are driven off, all brutes are summoned to one's hand, likewise snakes from their caverns and fishes from the depths of the waters. For the matter of the world is changed to many wonderful forms, if those means are correctly employed, and therefore they can be wonderfully effective against malevolent men and against the enemies of the state, just like other actions due to stellar influence, as necessity requires. But accursed magicians have brought the greatest discredit on this branch of science, since not only in their evil practices have they abused characters and incantations written by the wise against harmful things and intended to be very great blessings, but they have added false incantations and worthless and fraudulent characters, by which men are seduced. Moreover, demons have tempted many, and both women and demons have taught many superstitious practices, with which the nation is filled. For old women of their own volition everywhere make characters, incantations, and enchantments, and the magicians use invocations of demons and conjurations of them and perform sacrifices to them. But all these things are accursed and outside of the paths of the philosophers, nay, opposed to their expressed opinions; and by these means the potency of philosophy is defamed. Therefore theologians of the present day and Gratian and the sacred writers have reprobated many useful and noble sciences along with magicians, not noting the difference between magic and true philosophy owing to the five reasons of which mention was made in Part One; a chapter necessary for every man aspiring to the great things of wisdom and desiring to separate the true from the false. For Gratian, even as he has written many laws which are now abrogated, a wiser opinion prevailing, so in speaking of the sciences has said many things which should be altered to another purport, as I shall explain more fully below.

I return, then, to the words and acts of the wise, formed by the force of the stars and the power of the rational soul; and

I shall give an explanation in summary regarding them in accordance with what the wise have taught. For just as a child born and exposed to a strange atmosphere, as it were to a new world, receives the impression of the celestial forces, from which he has a radical complexion which he can never lose, because what the new jar receives it retains the savor of when old; so is this true in regard to everything newly made, since it receives the force of the heavens at the beginning of its existence, and that force which it received at the beginning it never loses until it is deprived of its natural being and is corrupted. Therefore in these images, incantations, and characters, composed by means of the necessary constellation, the forces of the stars are received and retained, so that through them they can act on the things of this world, and when the constellation recedes in which the things of this kind were composed, they recede. And since the rational soul is nobler than the stars, therefore just as the stars and all things impress their forces and species on things external, species and forces of which I have written adequately in Part Four in discussing the potency of geometry, so then is the rational soul, which is the most active substance after God and the angels, able to impress and does continuously impress its species and force on the body, of which it is the moving impulse, and on the things external, and especially so when it acts with strong desire, definite purpose, and great confidence. Concerning these things Avicenna speaks ably in the sixth book of his Naturalia. Therefore things and words of this kind of which I am speaking receive not only a force from the heavens, but also from the rational soul, which is nobler, and for this reason they can have a great influence in altering the things of this world.

If it be said that just as things of this kind receive the force of the heavens, so do all other things which are in the same region, and existing at the time of the composition of words and things of this kind; and thus all things must possess these forces, men, cattle, horses, and trees, for the rays of the celestial constellations touch an infinite number of objects at the same time; we must reply that the objection has no weight, since the things are not in the same horizon; for all points on the earth's surface are the centers of different horizons, to which the vertices of different pyramids of the celestial forces come, so that they are able to produce plants of different species on the same very small spot of the earth's surface, and to cause twins in the same womb to differ in complexion and habits, and in the use they make of the sciences and languages, in their occupations, and in all other things. Moreover, because other things were made earlier before the composition of the image, although they exist with it, yet they at their beginning received their own basic influence in accordance with which they act, and therefore the force of the heavens at this hour of which we are speaking does not have a natural effect on things of this kind, made previously, as it has in these things and words just now newly formed; and again since they were not formed by the rational soul, or at least not at that time; or if at that time, many other things are done by man, yet not with this purpose, nor desire, nor confidence, nor are they intended for actions of this kind, and therefore things of this kind will not have the power to alter, or at least not such noteworthy activities. If it be said that at all events everything will alter things external to itself and will sensibly change them, and especially so at its beginning by the force it receives at its beginning from the celestial constellation, since the basic complexion according to the force of the constellation remains in it; and that through the process of time, and through the other continuous forces of the heavens, the first force is lessened and weakened little by little, until it fails; we must reply that this is a fact; and in accordance with the origin of such things do great changes at times occur, although we do not consider whence such alterations happen, just as in the case of comets and some other things.

Again it is possible for us to bring out the true meaning of fascination, although it is a term under suspicion and can be understood in the way old women, fortune tellers, and magicians take it; yet it is certain that many men are of an evil complexion and of a corrupt and diseased nature, so that they are full of contagion like men diseased with many contagious sicknesses. Especially is there given to a nature of this kind an infected and contagious force by an evil constellation under which the child was conceived or born. For from the healthiest parents and possessed of the best complexion are born countless numbers who are of the very worst complexion; and these people, like all other things producing their own species and force, contaminate things near them, especially things tender in age and complexion, and particularly have they this power through the eyes; since they are porous and not dense, and vapors and corrupt spirits come out of them and infect things. Hence Vergil says, "Some one or other is fascinating with his eyes my tender lambs." Just as a menstruous woman, if she look at a new and polished mirror, stains it with a bloody mist, as Aristotle says in his second book on Sleep and Waking, and experience teaches us, and thus she infects other things, although not so apparently, thus also lepers spread infection. All these things are in the course of nature. But if further some malign soul should think strongly on infecting some one else, and should ardently desire it, and definitely design it, and earnestly consider that he is able to harm, there is no doubt that nature will obey the thoughts of his mind, as Avicenna teaches in the eighth book on Animals and in the fourth on the Soul, so that there will be a more vigorous multiplication of species and a more violent infection; just as a leper, if he purposes and desires to harm some one whom he hates, and is full of confidence in his ability to do so, does that man far more harm than one against whom he plans no evil. For every action of a man is stronger and more impetuous when thought and will are directed to it, and with fixed purpose he forms his resolution, and firmly hopes that he can accomplish his purpose. For the wise say that as scammony sharpens a medicine and gives it vigor, so do intentions, desire, confidence, affect human actions; and still this is not magic.

For Pliny shows in the seventh book of the Natural History that many men are of such a complexion, who infect others and do harm from their presence, both by sight and by word. For since a word is generated from the natural members within and is formed with thought and care and a man delights in it, and it is the most ready instrument of the rational soul, therefore it has greater efficacy than any other thing man does, especially when it is uttered with definite intention, great desire, and strong confidence. A proof of this is found in the fact that nearly all the miracles performed by the saints from the beginning were effected by the force of words, whence there is a very great potency in words, just as I have explained. If the multiplication of species of this kind and the utterance of the word take place under the necessary constellation, vigorous action must follow; and in all these things there is nothing magical or foolish. If this is called fascination, we can change the name if we wish. But with the wise such action is not subject to calumny. I have made mention of fascination; but it is explained here more fully. But since magicians and accursed old women do not view fascination in this way, but assert that sudden changes are brought about indifferently by any men whatever when the thing is seen early or late, or when they are talking about it, therefore the opinion of these people is worthless and reprobated by the wise.

I am writing these facts not only for scientific consideration, but because of the perils which happen and will happen to Christians and to the Church of God through unbelievers, and most of all through Antichrist, because he himself will employ the potency of science and will convert all things into evil. By means of words of this kind and by means of actions induced by the stars, and composed with a desire of doing harm and with a most definite intention and strong confidence, he himself will render unfortunate and will fascinate not only individuals, but also states and countries. By this wonderful method without war he will accomplish what he wishes, and men will obey him just as beasts, and he will cause kingdoms and states to fight against one another on his behalf, so that friends may destroy their friends, and thus he will accomplish his desires regarding the world.

Thus have the Tartars and Saracens been able to accomplish what they have done. For it is agreed that the Tartars give more time to astronomy than others, since although there are learned astronomers in many nations the rulers of their state are directed only by such advisers. Astronomers hold the same position among the Tartars as prelates do among us. Mangu Cham, emperor of the Tartars in the year of our Lord 1253, when the lord king of France, Louis, sent Friar William of the Franciscan Order to the Tartars, said to the Christians assembled before him including the friar mentioned, "We have a law from God interpreted by our diviners, and we do all things as they direct. You Christians have a law from God interpreted by prophets, but you do not follow it." This he said because in the East there are many evil Christians, like the

Nestorians and many others, who do not live according to the law of Christ. Friar William wrote to the lord king that if he had known a little about the stars, he would have been well received by them; but they despised him because he was ignorant of the terms of astronomy. The Tartars accordingly proceed in all things by means of astronomy, both in foreknowledge of the future and in the works of science. A proof of this is the fact that they have already subdued the whole world in its extent from north to east and from east to south, and now lack only two corners of the land of the Christians, namely, Egypt and Africa, although the Tartars are small, weak men, and hardly eat or drink anything to strengthen their nature, not swift of foot, and properly speaking unarmed except for arrows with which to frighten those whom they pursue, and although they never fight at close range in opposing lines of battle. For unless the Lord checked them and allowed frequent discords to be sown among them, they would have already seized upon the whole world. Hence their success must be due to the wonderful works of science by means of which they tread the world under foot. For the friar mentioned states in his book on the Manners of the Tartars which he sent to the lord king, that 14,000 Tartars defeated the sultan of Turkey, who had 200,000 horsemen without foot soldiers. But they could not have done this by force of arms, as is obvious, and hence they must have succeeded by means of science and especially by means of astronomy, by which they profess to be ruled and directed in all things. Similarly the Saracens make a great deal of use of astronomy and the wise among them know how to employ these means. It is greatly to be feared that the Tartars and the Saracens dwelling in their countries may send men to the Christians to bring misfortune on them by means of astronomy and to sow discords among Christian rulers, because the enemies of Christians strive especially to cause wars and discords among them.

For many things of this kind have been done although the foolish multitude does not consider their source. Perhaps you saw or heard for a certainty that the children of the kingdom of France once followed in countless numbers after an evil man, so that they could not be restrained by fathers, mothers, friends, and were placed on board of ships and sold to Saracens. This event happened less than sixty-four years ago. Likewise in our times the shepherd leader stirred up all Germany and France, and drew to him a multitude of men, and had favor in the sight of the whole body of the laity in contempt of the clergy and to the confusion of the Church. He stated to Queen Blanche that he was going to her son beyond the sea, with such words deceiving a very wise woman. The wise should not doubt that they were emissaries of the Tartars and Saracens, and that they had some means by which they fascinated the people. I saw him with my own eyes carrying openly in his hand something as though it were a sacred object, and in the way a man would carry relics, and he went with bare feet, and was always surrounded by a host of armed men, yet so dispersed in the fields that he could be seen by all who met him, making an ostentatious display of that which he carried in his hand. Whatever the case may be regarding the Tartars and the Saracens, it is certain that Antichrist and his followers will employ these means. Moreover, unless the Church meets the situation by similar means which will hinder and destroy activities of this kind, she will be intolerably burdened by these scourges of Christian people. All wise men believe that we are not far removed from the times of Antichrist, as is obvious in the chapter on the sects viewed together in the light of astronomy. If then Christians knew that these means must be employed by papal authority to hinder the ills of Christianity, the result would be sufficiently laudable, and not only should such means be used to drive away evils, but also to promote blessings of all kinds. Since, moreover, individuals, states, and regions in accordance with what precedes can be changed for the better, life will be prolonged to a fitting length, and all things will be provided for in a useful way, and much greater things accomplished than should be entrusted to the present writing, not only in the things of nature, but in morals, sciences, and arts, as was revealed by Moses and Aristotle. These results are especially possible when the species and force of the rational soul of the worker, which is nobler than the heavens, is in accord with the force and species of the heavens, in such a way that there are present intense thought, ardent desire, definite purpose, full confidence, and especially sanctity of life; since nature obeys the thoughts and affections of the soul, and sanctity especially. For in the eighth book on Animals Avicenna gives as an example the case of a vigorous hen which conquered a wretched cock, and in consequence of the glory of her victory a spur grew for her on the leg. Avicenna here remarks that in this instance we learn that nature obeys the thoughts of the sensitive soul. Also in the fourth book on the Soul he gives the example of a man who walking on a plank over water falls, because he is thinking of falling and loses confidence, and nature obeys the thoughts and affections of his mind. Hence he says that heat is not produced from heat nor cold from cold, but from the thought alone of the mind, and thus disease is caused and every change in the body belonging to it. Without doubt many effects can be produced in the body of another; since there is a certain kind of wolf that renders a man hoarse if it sees him first, as is commonly stated by authors. Much more vigorously can the rational soul, because of the nobility of its species, work many changes under the five conditions mentioned above, and especially when aided by sanctity, because sanctity is nobler than thought or desire, and for this reason the matter of the world is more obedient to the sanctified soul than to the other four conditions. We see this in full measure in the case of saints who have performed miracles, to whom the elements of the world were obedient. Avicenna in the fourth book on the Soul teaches that the soul sanctified and cleansed from sins is able to change the universe and the elements, so that by its virtue are produced rains, storms, and all changes in bodies in the world. It is, moreover, a fact that the grace of God does much, but the sanctified soul coöperates with grace, working the thing that is pleasing, so that a man is not saved by grace alone, but the soul must coöperate with such grace; much more vigorously then will it be able to cooperate with the grace given freely, which is the grace of miracles. But on other wonders it is better to ponder than to write, until by apostolic authority a greater attestation is required, and these things are possible. If you and the successors of your Holiness wish, it will be possible for all things to be completed. These are the basic principles in summary of the treatise which I proposed to write on such matters, but owing to obstacles I have been unable to write more.

THE OPUS MAJUS OF ROGER BACON

A Translation by

ROBERT BELLE BURKE

PROFESSOR OF LATIN AND DEAN OF THE COLLEGE
UNIVERSITY OF PENNSYLVANIA

VOLUME II

PHILADELPHIA
UNIVERSITY OF PENNSYLVANIA PRESS
LONDON: HUMPHREY MILFORD
OXFORD UNIVERSITY PRESS
MCMXXVIII

BACON'S STUDY AT OXFORD

From a Gough Print

CONTENTS

VOLUME II

ILLUSTRATIONS

VOLUME II

THE OPUS MAJUS OF ROGER BACON

PART FIVE OF THIS PLEA

CONCERNING OPTICS, IN THREE PARTS

The first part treats of matters common to the other two; the second deals mainly with direct vision; the third with reflected and refracted vision. In Part One there are twelve distinctions.

FIRST DISTINCTION

This distinction deals with the properties of this science, with the parts of the mind and brain and with the organ of vision, in five chapters.

CHAPTER I

Concerning the properties of this science.

HAVING explained the fundamental principles of wisdom, both sacred and human, which are found in the tongues from which the sciences of the Latins have been translated, and likewise in mathematics, I now wish to discuss some principles which belong to optics. If the consideration just mentioned is noble and pleasing, the one in hand is far nobler and more pleasing, since we take especial delight in vision, and light and color have an especial beauty beyond the other things that are brought to our senses, and not only does beauty shine forth, but advantage and a greater necessity appear. For Aristotle says in the first book of the Metaphysics that vision alone reveals the differences of things; since by means of it we search out experimental knowledge of all things that are in the heavens and in the earth. For those things that are in the heavenly bodies are studied by visual instruments, as Ptolemy and the other astronomers teach. So also are those things that are generated in the air, like comets, rainbows, and the like. For their altitude above the horizon, their size, form, number, and all things that are in them, are verified by the methods of viewing them with instruments. Our experience of things here in the earth we owe to vision, because a blind man can have no experience worthy of the name concerning this world. Hearing causes us to believe because we believe our teachers, but we cannot try out what we learn except through vision. If, moreover, we should adduce taste and touch and smell, we assume a knowledge belonging to beasts. For brutes are busied with the things pertaining to taste and touch, and exercise their sense of smell because of taste and touch, but the things are of little value, few in number, and common to us and to brutes concerning which these senses give verification, and therefore they do not rise to the rank of human wisdom. But because of necessity, utility, and difficulty, sciences are formed, since art has to do with the difficult and with the good, as Aristotle says in the second book of the Ethics. For if what is sought is easy, there is no need for the formation of a science. Likewise although a matter be difficult yet not useful, no science is developed concerning it, because the labor would be foolish and vain. Also unless a subject were very useful and possessed many excellent truths, it does not require the formation of a separate science, but it suffices that this subject be treated in some particular book or chapter along with other matters in general science. But concerning vision alone is a separate science formed among philosophers, namely, optics, and not concerning any other sense. Wherefore there must be a special utility in our knowledge through vision which is not found in the other senses. What I have now touched upon in general, I wish to show in particular by disclosing the basic principles of this very beautiful science. It is possible that some other science may be more useful, but no other science has so much sweetness and beauty of utility. Therefore it is the flower of the whole of philosophy and through it, and not without it, can the other sciences be known. We must note, moreover, that Aristotle first treated this science, of which he says in the second book of the Physics that the subject is placed under another head. He also mentions it in his book on Sense and the Sensible, and has proved Democritus in error, because he did not name refractions and reflections of vision with reference to the optic and concave visual nerves. This book has been translated into Latin. After him Alhazen treats the subject more fully in a book which is extant. Alkindi also has arranged some data more fully, likewise authors of books on vision and mirrors.

Optical Science

Chapter II

Concerning the internal faculties of the sensitive soul, which are imagination and the common sense (sensus communis).

Since the optic, that is, the concave nerves causing vision have their origin in the brain, and writers on optics ascribe to a distinct function through the medium of vision the formation of judgments concerning twenty species of visible things, which will be considered later, and since it is not known whether that distinct function is among the functions of the soul, the organs of which are distinct in the brain, and since many other things to be treated of later suppose a definition of the functions of the sensitive soul, therefore we must begin with the parts of the brain and the functions of the soul, in order that we may discover those things that are necessary for vision. Writers on optics give us a means to this end by showing how the visual nerves descend from the membranes of the brain and from the lining of the cranium, but no one explains all things necessary in this matter. I say, then, as all writers on the subjects of nature, all physicians and authorities on optics agree, that the brain is enfolded by a double membrane, one of which is called the pia mater, which enfolds the brain by direct contact; and the other, the dura mater, which adheres to the concave side of the bone of the head called the cranium. For this latter membrane is harder, so that it may resist the bone, and the other is softer and tenderer owing to the softness of the brain, the substance of which is like marrow and ointment, with phlegm as the chief constituent, and with three distinctions, which are called chambers, cells, parts, and divisions. In the first cell there are two faculties: the one in the anterior part is the common sense, as Avicenna states in the first book on the Soul, which is like a fountain with respect to the particular senses, and like the center with respect to the lines extending from that same point to the circumference, according to Aristotle in the second book on the Soul. This common sense judges concerning each particular sensation. For the judgment is not completed in regard to what is seen before the form comes to the common sense, and the same is true in regard to what is heard and to the other senses, as is clear from the end of the work on Sense and the Sensible, and in the second book on the Soul. This common sense forms a judgment concerning difference of impressions on the senses, as, for example, that in milk whiteness is different from sweetness, a distinction which sight cannot make, nor taste, because they do not distinguish things in other categories, as Aristotle maintains in the second book on the Soul. It judges concerning the operations of the particular senses, for vision does not perceive that it sees, nor hearing that it hears, but another faculty does, namely, the common sense, as Aristotle maintains in the second book on Sleep and Waking. But the final action of this faculty is to receive the forms coming from the particular senses and to complete a judgment concerning them. But it does not retain these impressions, owing to the excessive slipperiness of its own organ, according to the statement of Avicenna in the first book on the Soul. Therefore there must be another faculty of the soul in the back part of the first cell, the function of which is to retain the forms coming from the particular senses, owing to its tempered moistness and dryness, which is called imagination and is the coffer and repository of the common sense. Avicenna cites as an example a seal, the image of which water readily receives, but does not retain owing to its superabundant moistness; wax, however, retains the image very well, owing to its tempered moistness with dryness. Wherefore he says that it is one thing to receive and another to retain, as is clear from these examples. Such is the case in the organ of the common sense and of imagination. The whole faculty, however, composed of these two, namely, that which occupies the whole first cell, is called phantasia or the *virtus phantastica.* For according to the second book on the Soul and that on Sleep and Waking and the book on Sense and the Sensible, it is evident that phantasia and the common sense are the same according to subject but differ according to being, as Aristotle says, and that phantasia and imagination are the same according to subject but differ according to being. Wherefore phantasia includes both faculties and does not differ from them except as the whole from the part. Therefore since the common sense receives the form, and imagination retains it, a complete judgment follows regarding the thing, a judgment formed by phantasia.

Optical Science

Chapter III

Concerning the attributes that are apprehended by special senses, by the common sense and by imagination.

We must note that imagination and the common sense and any particular sense form judgments of themselves only concerning twenty-nine properties; as sight concerning light and color; touch concerning heat and cold, moistness and dryness; smell concerning odors; taste concerning savor. These are the nine special properties that belong to their own senses, as I have named them, of which no other particular sense can form a judgment. There are, moreover, twenty other properties, namely, distance, position, figure, magnitude, continuity, discreetness or separation, number, motion, rest, roughness, smoothness, transparency, thickness, shadow, darkness, beauty, ugliness, also similarity and difference in all these things and in all things composed of them. Besides these qualities there are some that are placed under one or more of these qualities, as order under position, and writing and painting under figure and order. Further examples are straightness and crookedness, and concavity and convexity, which are placed under figure; multitude and fewness, which are placed under number; equality, augmentation, and diminution, which are placed under similarity and difference; eagerness, laughter, and sadness, which are apprehended from the figure and form of the face; lamenting, which is apprehended from the form of the face together with the shedding of tears; and moistness and dryness, which are placed under motion and rest, since from the sense of vision moistness is not apprehended except from the liquidity of the moist body and from the motion of one part of it before another, and since dryness is apprehended from the retention of the parts of the dry body and from the absence of liquidity. Here, however, we must consider that Aristotle maintains in the second book on Generation that moist and dry are in one way primary attributes which naturally belong to the elements, and that through these attributes spring the non-elemental moistness and dryness, which are reduced to the primary attributes and are caused by them. Concerning, then, these primary attributes it has been stated that they are proper objects of sense and perceptible by touch alone. We are speaking, however, concerning the others. For primary moistness is that which passes over easily into all forms, possessing no definite bounds of its own, but conforming readily to the limits of something else, as in the case of air particularly and next in order in the case of water. Dry is the opposite, particularly in the case of earth, secondarily in the case of fire. But here moistness is used in the sense of liquid and slippery, and dry in the sense of arid and coagulated. The same is true concerning many other attributes, which are reduced to the forms and principal modes enumerated above belonging to visible things. All these matters are explained in the first book of Ptolemy on Optics and in the second book of Alhazen on Aspects, and in other authors on optics. There are, moreover, common qualities, some of which Aristotle defines in the second book on the Soul and in the beginning of his work on Sense and the Sensible, as, for example, magnitude, figure, motion, rest, and number. These are not the only common qualities but also all those mentioned before, although most writers on the subjects of nature do not consider this fact, because they are not expert in the science of optics. For common properties are not so called because they are perceived by the common sense, but because they are determined by all the special senses or by several of them, and particularly by sight and touch, since Ptolemy states in his second book on Optics that touch and sight participate in all these twenty. These twenty-nine, with those that are reduced to them, are apprehended by the special senses, and by the common sense, and by imagination, and these faculties of the soul cannot judge of themselves concerning other qualities except by accident.

Chapter IV

Concerning the investigation of the estimative, memorative, and cogitative faculties.

But there are other sensibles *per se,* for animals use sense alone, since they do not possess intellect. The sheep, even if it has never seen a wolf, flees from it at once; and every animal experiences fear at the roaring of a lion, although it has never heard a lion before nor seen one. The same is true in regard to many things that are hurtful and contrary to the constitution or complexion of animals. The same principle holds good as regards what is useful and in conformity with their natures. For although a lamb may never have seen another lamb before, it runs to one and willingly remains with it, and the same is true concerning other animals. Brutes, therefore, have some perception in things advantageous and in things harmful. There is then something sensibly in them besides the twenty-nine properties mentioned above and besides those that are reduced to them. For there must be something more active and productive of change in the sentient body than light and color, because it not only causes apprehension but also a state of fear or love or flight or delay. This is the property of the complexion belonging to each object by which it is assimilated to others in a nature special or general. Through this quality things agree, are strengthened and invigorated, or differ and oppose one another and are mutually harmful. Wherefore not only do light and color produce their forms and impressions, but to a far greater degree do the properties of complexions, nay, the very natures of things as regards their substance, agreeing or disagreeing with one another, produce strong impressions, which change greatly the sensitive soul, so that it is moved to states of fear, horror, flight, or the opposites. These forms or impressions coming from things, although they change and alter special senses and the common sense and imagination, just as they do the air through which they pass, yet no one of those faculties of the soul judges concerning these impressions, but of necessity a far nobler and more powerful faculty of the sensitive soul does, which is called estimation or the estimative faculty, as Avicenna states in the first book on the Soul, a faculty which he says perceives the insensible forms connected with sensible matter. Sensible matter is spoken of here as that which is apprehended by the special senses and by the common sense, as are the aforesaid twenty-nine. We call insensible form that which is not taken cognizance of by those senses of themselves, since they are commonly called senses, although other faculties of the sensitive soul may equally well be called senses, should we wish so to name them, because they are parts of the sensitive soul. For every part of the sensitive soul can be called a sense, because it is in truth a sense, and a sensitive faculty. The statement, therefore, that qualities belonging to complexions are not apprehended by a sense must be understood as applying to a special sense and the common sense and imagination; but they can readily be apprehended by estimation, which although not called a sense is, however, a part of the sensitive soul.

But estimation does not retain a form, although it receives it like the common sense, and it therefore requires another faculty in the remotest part of the posterior cell to retain forms coming to the estimative faculty and to be its storehouse and repository, just as imagination is the storehouse of the common sense. This is the memorative faculty, as Avicenna states in the first book on the Soul. Cogitation or the cogitative faculty is in the middle cell and is the mistress of the sensitive faculties. It takes the place of reason in brutes, and is therefore called the logical, that is, the rational faculty, not because it employs reason, but because it is the ultimate perfection of brutes, just as reason is in man, and because the rational soul in man is united directly with it. By this faculty the spider weaves its geometrical web, and the bee makes its hexagonal house, choosing one of the figures that fill out space, and the swallow its nest. The same is true of all the works of brutes that are similar to human art. Man by means of this faculty sees wonderful things in dreams, and all the faculties both posterior and anterior of the sensitive soul serve and obey it, because they all exist on account of it. For the forms or species that are in the imagination multiply themselves into the cogitative faculty, although they exist in the imagination according to their nature primarily because of phantasia, which uses those forms; but the cogitative faculty holds those forms in a nobler way, and the forms of the estimative and memorative faculties exist in the cogitative faculty in accordance with a nature nobler than that existing in those faculties, and therefore the cogitative faculty uses all the other faculties as its instruments. In man there is in addition from without and from creation the rational soul, which is united with the cogitative faculty primarily and immediately, and uses this faculty chiefly as its own special instrument. Species are formed in the rational soul by this faculty. Wherefore when this faculty is impaired the judgment of reason is especially perverted, and when it is in a healthy condition the intellect functions in a sound and rational way.

Chapter V

An exposition of authorities at variance regarding the faculties mentioned.

But the Latin text of Aristotle does not disclose to us this division, for there is no mention expressly made except in regard to the common sense, imagination, and memory. But since the text of Aristotle owing to faulty translation cannot be understood in this part of his work, and the same difficulty is found in other passages, and since Avicenna was the perfect imitator and expositor of Aristotle in nearly every respect, and was the greatest philosopher subsequent to him, as the Commentator states in the chapter on the Rainbow, we must for this reason hold to the opinion of Avicenna, which is clear and perfect, although the translators of the works of Avicenna, as in his book on the Soul, and in that on Animals, and in his books on Medicine, have translated differently and have changed words, so that Avicenna's meaning has not been translated to the same purport throughout, since in his book on Animals we find that the estimative faculty takes the place of reason in brutes, and so too we sometimes find elsewhere contradiction with respect to the aforesaid matters; but the fact is of little moment that different translators use different words, and sometimes in part of a subject show some diversity. But we must hold to his view as expressed in his book on the Soul, because he there discusses the faculties of the soul as his principal topic, but elsewhere his mention of this subject is rather incidental. Moreover, that work has been translated far better than his other books, a fact that is evident, since it has few or no words belonging to other tongues, while his other books have them in countless numbers. If one really considers what has been said above, he must assume that there are three faculties wholly different, according to the three cells. For a diversity of objects shows us a diversity of faculties. For there are two kinds of properties, one external, as the twenty-nine mentioned above, the other internal hidden from external sense, as the quality of a harmful or useful complexion, or rather the essential nature itself whether useful or harmful. There must then for this reason be two kinds of senses, one containing the special senses and the common sense and the imagination, which are influenced by the first class of qualities; and the other containing the estimative faculty and memory, which are referred to the second class of qualities. But the cogitative faculty, owing to the excellence of its functions as compared with other faculties, is distinguished from them. In the generally accepted translation of Aristotle the whole faculty is called memory, which has the power of retaining forms, and therefore not only the storehouse of the common sense, but also that of the estimative faculty, is called memory. Therefore what is here called imagination is comprehended under memory in the translation of Aristotle that is in common use. But without doubt memory must be a double faculty with two quite distinct parts, so that one shall be the storehouse of the common sense and the other that of the estimative faculty; and these will differ in species, subject, organ, and operation. Moreover, although these faculties have been placed in the brain, we must understand that the marrow-like substance of the brain is not itself sensitive, as Avicenna teaches in the tenth book on Animals, respectfully correcting Aristotle on this point.

For the marrow in other parts of the body is not sensitive, and therefore it is not so here. But it is the container and storehouse of the sensitive faculties, containing slender nerves in which sense and sensible forms are located. But in order that all doubt may be removed, we must consider that the sensitive soul has a twofold organ or subject; one is basic and the seat of life, namely, the heart, according to Aristotle and Avicenna in the books on Animals; the other is that which is first changed by the forms of qualities, and in which the operations of the senses are more manifest and distinguished, namely, the brain. For if the head is injured an evident injury is suffered by the sensitive faculties, and an injury to the head is more evident to us than one to the heart, and frequently occurs. For this reason in accordance with the more obvious consideration we place the sensitive faculties in the head; and this is the opinion of physicians who do not think that the origin of the faculties is in the heart. But Avicenna in the first book on the Art of Medicine says that, although the opinion of physicians is the more obvious to the sense, yet the opinion of the Philosopher is the truer one, since all the nerves and veins and all the faculties of the soul have their origin first of all in the heart, as Aristotle proves in the twelfth book on Animals, and Avicenna in the third book on the same subject.

SECOND DISTINCTION

Containing three chapters. The first deals with the nerves leading to the eye.

Chapter I

CLEARLY, therefore, in order that there may be no scruple of doubt in what follows, the structure of the eye must be studied, because without this nothing can be known concerning the method of vision. But some writers say less, some more, and in certain things they are at variance. For authors on perspective more commonly pass over these matters, assuming a knowledge of the natural philosophers and of the writers on medicine, as though every one who reads the science of perspective has already consulted the writers on medicine and on nature. Therefore their discussion is in itself obscure, nor can we understand it unless we have recourse to the fuller treatment of the subject by the writers on medicine and on natural philosophy. For this reason something more must be said here on this topic than is found in the writers on perspective. Although it may be too difficult to attest these matters, and to explain those that are attested, yet I hope that these facts can be made clear by certain authors. But that I may not draw upon sources of individual opinions to too great an extent, I shall confine my description of the structure of the eye mainly to three authorities—Alhazen in his first book on Perspective, Constantinus in his book on the Eye, and Avicenna in his books; for these writers are sufficient and they treat more definitely the matters in which we are interested. I cannot, however, give the exact words of each, because they are sometimes at variance owing to faulty translation, but from them all I shall form a single statement of the truth. All, then, are agreed that there are two parts in the anterior cavity of the brain, which are called ventricles, or concavities, or cells. These ventricles cannot be organs of the common sense and of the imagination, of which mention was made previously. For those faculties are arranged anteriorly and posteriorly, while these ventricles are placed on the right and on the left, as Constantinus states. For the entire brain cell can be divided into an anterior part and a posterior part, as was stated above. There is, however, a division according to right and left. The anterior part of it, namely, the place of the common sense, has a right and left, where are the two ventricles in a measure distinct. From the pia mater, which covers both ventricles, comes a double nerve, one from the right ventricle, and the other from the left, and these are the optic nerves, that is, the concave ones, according to the authors mentioned. The hollow does not begin at the middle of the anterior part of the brain, because there is located the organ of smell. This is a single nerve with two small bits of flesh on the sides like the tips of the teats, as Avicenna teaches in the tenth book on Animals. But according to Avicenna, and the author* on Perspective, and Constantinus, these visual nerves of which we are speaking come out at the bottom of the ventricles of the anterior part. This is the explanation of Constantinus, who teaches us to examine the heads of large animals, when they are killed not in the summer or in the heat, and we shall find a small opening in the skull through which the nerve passes, and then we should examine the membrane of the pia mater cautiously lest it be broken, and we shall find that a nerve comes out at the bottom of the anterior part of the brain. But the two nerves, as we have stated, from the two directions right and left, meet, according to all authorities, and after meeting again are divided. It was, moreover, better that they should meet in the opening than before it or after it. For in either of these cases they would make two openings in the bone of the head, but it is better that the nerve should pass through one opening than through two. The bone is firmer the less it is perforated. Therefore, since nature acts in the better way possible, the junction will be in the opening of the skull, at which place they are again divided. The nerve that comes from the right goes to the left eye and the one from the left to the right eye, so that there is a direct extension of the nerves from their origin to the eyes. For if the nerve coming from the right part of the anterior cell of the brain passed to the right eye, there would be an angle in the common nerve where they meet and the nerve would be bent and would not extend directly to the eye. But this would hinder vision, because vision always selects straight lines as far as

* Alhazen.

possible. Moreover, since the bone of the eye is concave with an opening toward the head, the nerve enters the opening of the eye and spreads out in the hollow of the bone like the utensil with which wine is put in jars. Let, then, *abc* be the skull, and *d* be the right part of the anterior cavity of the brain, and *e* the left, enveloped by the pia mater, from the bottom of which let two nerves come forth on the right and left, meeting in the opening in the skull, and afterwards dividing, so that the nerve coming from the right goes to the left eye, which is *f*, and the nerve from the left to the right eye, which is *g*, and let them enter the openings of the concave bones, so that they spread out in that hollow, as is shown in the figure. But we must understand that just as the two nerves are derived from the pia mater, so are they also derived from the dura mater, and similarly from the lining of the skull, in which externally it is wrapped, and these three nerves are concave and meet in the opening, and one nerve is formed with three nerve coats, and this nerve so formed goes to each eye, and each eye has naturally a position similar with respect to the meeting of these nerves in the opening and at an equal distance from it, so that vision may be completed with greater certainty.

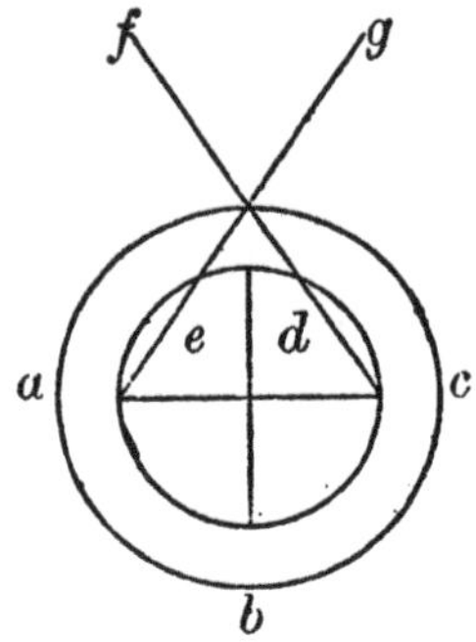

FIG. 24.

CHAPTER II

Concerning the coats of the eye formed from the three nerves mentioned.

THE eye, therefore, has three coats or membranes, and three humors, and a web like that of a spider. Its first coat is formed from the interior coat of the nerve, which comes from the pia mater, according to all authorities, and spreads out from the end of the nerve where it enters the opening in the bone, and is ramified like a concave net in its first part. It is therefore called rete or retina, according to Avicenna in the third book on Medicine and according to Constantinus, and is supplied with veins, arteries, and slender nerves. Its second part is thicker, as Avicenna says, and spreads out spherically up to the fore part of the eye with an opening in the middle of its own fore part, in order that the impressions of light and color and other visible things may be able to pass through the middle of the eye to the nerve coming from the brain. For this opening is placed directly opposite the extremity of the nerve at which the retina expands, and therefore in this whole coat Alhazen states there are two openings, one in front and the other behind, which is the extremity of the concave nerve. This second part is called the uvea, since it is similar to a grape [*uva*], owing to the fact that it has an opening in its anterior part like that left in a grape when its stalk is removed, as Avicenna says in the third book on Medicine. From the coat of the nerve coming from the dura mater, the second coat of the eye is expanded, according to all authorities. This coat has two parts. For the first part is composed of veins, nerves, and arteries, and is called secundina, because it is similar to the after-birth [*secundinae*]; and the second part spreads as far as the anterior of the eye, and there part of it is apparent, namely, the portion of a sphere, which forms a circle above the extremity of the uvea, and is like transparent horn and is therefore called cornea. It is composed, as Avicenna says in the book mentioned, of four thin rindlike coats, and they are like rinds, so that if one of them should be stripped off, the others would not be injured because of it. The result is that this coat is strong because of the injuries and attacks coming from the air, and yet it is very transparent, so that the large number of its coats does not hinder the passage of the impressions of visible things. The third coat of the eye is formed from that third skin of the nerve that comes from the membrane of the skull, and its first part is joined to the bone of the eye, and is, in fact, hard and solid, and for this reason is called sclerotic, but the remaining part extends to the cornea. For this coat is not a complete one but lacks a portion of the sphere, and is filled with fat and white flesh, as we see externally in eyes, and it is called consolidativa or conjunctiva. We must bear in mind that in one way there are said to be only three coats, but in another way six; and both views are true and rational. For if we consider the coats as whole ones, there are only three. If, however, we consider the parts behind as distinct from the parts in the front both in name and fact, there are six. For three parts are posterior and three are anterior. Some also have maintained that they are fewer, and that, too, with elaborate argument; but we should not pay attention to them, because their interpretation is forced and deviates from sound reason. Some also have maintained that there are seven coats; but this is false, because they reckoned as a coat the spider-web structure, although it is not. Those who say that there are three coats call the whole first coat the uvea, and the whole second coat the cornea, and the whole third one the consolidativa. Hence authors on perspective call the whole first coat the uvea, and so I wish chiefly to employ the term in describing the method of vision. Therefore Alhazen says that the uvea has two openings, one anterior, and the other posterior, which is the opening of the nerve, at which the expansion of the concavity of the uvea begins, whence the extremity of the nerve with the whole cavity following as far as the anterior opening is the uvea in fact.

Chapter III

Concerning the humors of the eye and the weblike structure.

This coat contains in itself three humors and one web. For from the anterior part of that coat there begins a web small and fine like the web of a spider, and in this is contained a body that is called glaciale, crystallinum, or grandinosum [like ice, crystal, or hail], and this body is formed directly above the extremity of the nerve. But it has two parts; one is interior at the extremity of the nerve and is similar to melted glass, and on this account is called the vitreous humor. The other part is anterior and is like ice and hail and crystal, is whiter than the vitreous humor, and is called anterior glacialis, not having any other name of its own in the author* on Perspective. But it is called in other authors the humorlike crystal, ice, or hail, because it is similar to them, and the whole body contained beneath the web is so called from this part. Then toward the anterior part of the eye outside of the web a humor like the white

* Alhazen.

of egg fills the anterior cavity of the uvea, and on one side touches the anterior part of the glacialis, and on another side enters the opening of the uvea, and reaches to the cornea, so that the spherical convexity of this humor touches the concave figure of the cornea, and is called humor albugineus. There will then be the cornea, the humor albugineus, the humor glacialis, and the humor vitreus, and the extremity of the nerve, so that consequently impressions of things will pass through the medium of them all to the brain. Avicenna says in the fourth book on Animals that the retina conveys nourishment as a matter of fact to the parts of the eye, and contains the vitreous humor, as Constantinus states, and the author on Perspective agrees, maintaining that the lower part of the uvea contains the vitreous humor, and at the end of it carries blood in its veins and arteries well digested, with which the vitreous humor is formed and nourished, in order that the vitreous may be able to nourish the crystalline humor. For, as Avicenna says in the third book on Medicine, the vitreous humor is the nourishment of the crystalline humor. Constantinus also makes this statement. Hence, since the crystalline humor is quite white and clear, blood is not suitable for it as direct nourishment, but it needs a nourishment intermediate between blood and the crystalline humor, and of this kind is the vitreous humor, which is whiter than blood, and less white than the crystalline humor. Avicenna says that the humor albugineus is the overflow of the crystalline humor, and therefore is opposite in position as regards its nourishment, which is the vitreous humor, and for this reason the crystalline humor is between these. The vitreous humor fills the whole cavity of the nerve as far as the common section, and is thicker than the glacial humor. Each, however, is transparent, so that the impressions of objects may pass through in them. The crystalline humor is called the pupil, and in it is the visual power, just as in a subject that at first is changed, although not radically, since the common nerve is the radical organ, and there vision is completed, as far as the visual power can, as what follows will show.

THIRD DISTINCTION

In three chapters. The first is on the sphericity and central points of the vitreous humor and of the glacial humor, of the cornea, of the humor albugineus, and of the uvea.

Chapter I

WE must consider next the form of the eye and of its parts, and the location of the centers of its coats and humors: for these facts are quite necessary; without them the method of vision is not understood. The whole eye, then, approaches the form of a sphere, and the coats likewise and the humors owing to the admirable qualities of the spherical form, because it is less subject to obstacles than a figure with angles, and is simpler than other figures and is more spacious than other isoperimetric figures, as Alhazen, the author on Perspective, states. But in what precedes these properties and others were touched upon. The anterior glacial humor is the portion of a sphere different from the sphere of which the vitreous humor is a portion; for they are different bodies and of different transparency. Moreover, the anterior glacial humor is the portion of a larger sphere than is that of which the vitreous humor is a portion. Hence the bodies are not complete spheres, but are portions of different spheres, and therefore since these are intersecting spheres, they must have different centers. Since the cavity of the vitreous humor is toward the glacial humor, its center is beyond the center of the eye toward the anterior of the eye. Similarly the center of the anterior glacialis is in the depth of the eye. These centers are, however, on the same straight line that enters through the anterior opening of the uvea and through the opening that is at the extremity of the nerve, where the retina begins to expand. For these bodies are so arranged, according to the authors on Perspective, that from the opening of the bone, where the nerve enters, the nerve extends for some distance and is constantly more dilated, until it comes to the circumference of the whole glacialis, and is consolidated with its circumference. Then above the extremity of the nerve the whole glacialis is placed and is contained in the inferior part of the uvea, which

Alhazen calls the point of the cavity of the uvea, in the last part of which is the opening that is the extremity of the nerve, where the uvea begins. But the middle of the whole glacialis, namely, the vitreous humor, is in the orifice of the opening. For the extremity of the nerve contains the middle of the sphere of the whole glacialis, as Alhazen says, and this middle is the vitreous humor; and the uvea is consolidated with the circumference of the glacial sphere. The humor albugineus contained in the uvea touches the sphere of the anterior glacialis, and it fills the opening up to its contact with the cornea, not so as to touch the cornea in a point, but by the junction of surfaces, just as an interior sphere is contained by an exterior one. But since the convex surface of the cornea is continuous with the surface of the whole eye, and with the whole eye, as Alhazen states, they must have the same center. Since, moreover, the concave surface of the cornea is equidistant from the exterior convex surface, both surfaces of the cornea and the whole eye must have the same center, according to the book of Theodosius on Spheres: for equidistant spheres, one containing and the other contained, have the same center, for example, the spheres of the universe, as the starry heaven, and the sphere of fire, and similarly in the others; for the center of the universe is the center of them all. Since, moreover, the concave surface of the cornea and the convex surface of the humor albugineus, which is in the opening, are like an interior and an exterior sphere, the convex surface of the humor albugineus must have the same center with the aforesaid. But since the concave surface of the cornea does not touch the uvea in a point, nor is united to it as an exterior sphere to an interior one, but is united to it on the circumference of its opening, the cornea must cut the uvea, and therefore they will have different centers. Since the cornea is a larger sphere than the uvea, because it is continuous with the surface of the whole eye, and the uvea is contained below the sphericity of the cornea, the center of the cornea, therefore, must be further in the depth of the eye, as is clear to the sense in the case of spherical bodies so united. This fact is shown by Theodosius on Spheres, and Alhazen so states it.

Chapter II

In which is explained a difficult doubt regarding the aforesaid.

But then there is serious doubt as to what fills the space between these bodies, where the smaller sphere deviates from the larger, and many for this reason think that the humor albugineus spreads itself below the cavity of the cornea, where the uvea is separated from it. But the objection is made that since the humor albugineus is concentric with the cornea, it will be contained in the cavity of the cornea, like a sphere joining it or equidistant; but it is not equidistant, since it touches it; therefore it will join this in its cavity, and will fill the space that is between the uvea and the cornea. But opposed to this statement is the fact that the author on Perspective does not so state it, but always says that it is within the uvea. The objections, therefore, are removed by the fact that the parts of the eye are not complete spheres, but portions of spheres, as is clear in regard to the parts of the glacialis and so, too, in regard to the others that precede these and serve them principally, as are the anterior of the cornea, and the humor albugineus in the opening of the uvea, and the anterior of the uvea. Hence we must take into account here only the sphericity of the portions; and therefore when mention is made of the sphericity of the cornea the statement has reference only to that portion which is necessary for vision, namely, that which is in the anterior part of the eye, but elsewhere is not spherical. Moreover, the uvea, although in the upper part spherical, is not so in the lower part; similarly the humor albugineus does not have a sphericity concentric with the cornea, except in the opening of the uvea, where it is joined to the cornea, for below the uvea it has concentricity with the uvea. Moreover, since such is the case, the humor albugineus cannot flow between the cornea and the uvea. For if the bodies were of complete sphericity, this result would follow, but such is not the case. But where the cornea and uvea are apart the consolidativa spreads itself, and fills up whatever there is to be filled between the consolidativa and the cornea, or the cornea and uvea giving up their complete sphericity dilate, and externally unite or internally or in both ways, and fill all that should be filled.

Moreover, since the anterior glacialis in its convexity cuts the uvea, similarly its center must be different from that of the uvea and farther within the depth of the eye: and since the whole eye, and the cornea, and the humor albugineus have a different center from the uvea and in the depth of the eye, just like the anterior glacialis, and these are required, that vision may take place in the glacialis, it is better, as Alhazen says, that the anterior glacialis should have the same center with them. Therefore the whole eye, the cornea, the humor albugineus, and anterior glacialis are concentric. But concerning the anterior glacialis it will be more definitely shown in what follows that it must have the same center as the cornea and the whole eye, when refraction in the vitreous humor is pointed out. In the meantime let what has been said suffice.

CHAPTER III

Concerning the center and sphericity of the consolidativa.

THE consolidativa is thought to have a different center from all these farther back in the depth of the eye. But the author on Perspective does not say this, but merely states in regard to the uvea and the vitreous humor that neither mutually nor with the others have they the same center, and in arguing that the center of the cornea and the center of the uvea are not the same he says that the sphere of the uvea is not in the middle of the consolidativa, but extends forward to a part of the visible surface of the eye itself, and the visible surface of the eye is a part of a larger sphere than the sphere of the uvea, wherefore the center of this visible surface will be farther back in the depth of the eye than the center of the uvea. But the surface of the cornea and that of the visible eye, as he assumes here and later explains, are the same. Therefore the center of the uvea and that of the cornea are not the same. From this it is argued by some that the center of the concave surface of the consolidativa and that of the cornea are the same, because through the elevation of the uvea from the middle of the consolidativa it is shown that the uvea has a different center from that of the cornea and of the whole eye. But it is to be noted that the exterior surface of the consolidativa is not concentric with the interior surface, as is evident; nor, moreover, is the interior surface completely spherical, because it fills the space where the cornea and the uvea are separated, for which reason it loses its true sphericity, and goes farther into the eye in that part than elsewhere, and therefore the cornea and the interior consolidativa are not concentric. But if both spheres, namely, the cornea and the consolidativa, were complete, the cornea would lie in the cavity of the consolidativa, and the interior of the consolidativa and the exterior of the cornea would be concentric. Moreover, since the exterior and interior surfaces of the consolidativa are not concentric, the exterior of the cornea and the exterior of the consolidativa will not be concentric. Since also the consolidativa does not have perfect sphericity on the exterior, as Alhazen states, for it tends in its anterior toward pointedness and therefore does not strictly have a center from which all lines drawn to the circumference are equal, for this reason neither on the outside nor on the inside is the body really spherical, and therefore a center is not assigned to it by Alhazen. It can, however, have a point similar to a center farther back in the depth of the eye than the center of any other part, as will be shown in the figure. If, however, we wish to remove all contention, we can say that the exterior surface of the consolidativa is not wholly spherical, just like that of the whole eye, because it is somewhat pointed in the anterior part, and thus the whole eye will not have the center of a sphere, nor will the exterior surface of the consolidativa. If the interior surface of the consolidativa were spherical, it would not fill the space that had to be filled between the cornea and the uvea. But either the cornea contracts itself at the surface of the uvea, and is deepened by losing its true sphericity, except in the anterior part that is opposite the opening, or the uvea elevates itself and passes into a protuberant form, externally losing its true sphericity. But although the centers are different in the parts of the eye, yet all are in the same line, which is perpendicular to the whole eye and to all its parts. This line passes through the middle of the opening of the uvea, and through the middle of the opening in the extremity of the nerve, on which the eye is formed. This line is the axis of the eye, by which the eye sees with certainty and by which it passes over the separate points of a visible object, so that it may make certain of each in succession, although it grasps at the same time and at once with full certainty a visible object or a part of a visible object. For since this line is perpendicular to the eye and to all its parts, an impression coming along it is strongest, because approach along the perpendicular is the strongest, as was demonstrated in what was said on the multiplication of species. This arrangement is a necessary one, in order that the eye may grasp most strongly and with the utmost certainty what it should.

I shall draw, therefore, a figure in which all these matters are made clear as far as is possible on a surface, but the full demonstration would require a body fashioned like the eye in all the particulars aforesaid. The eye of a cow, pig, and other animals can be used for illustration, if any one wishes to experiment. I consider this figure better than the one that follows, although the following one is that of the ancients. For it is impossible that the center of the vitreous humor should be below the sphere of the anterior glacialis, because in that case right will appear left and the reverse, as will be shown below: nor yet on the surface of its body, because in that case an impression on the right would go too far to the right, and one on the left too far to the left, and they would never meet in the common nerve, wherefore the center will lie outside toward the anterior part of the eye. Since, moreover, the opening of the uvea is small, and the anterior glacial opening is less, since it is farther in, concentric with the humor albugineus in the opening, the vitreous must cut a very small portion from the glacialis, *cfd* in the figure, so that it can scarcely be drawn by the hand of man, for to this portion corresponds *mo,* the opening of the uvea, which is here larger than the opening of the uvea in the eye of any man of this day. The size of the opening is determined by the boundary lines of the visual pyramid, which is *abl.* For let *al* be the base of the pyramid, which is the visible object, the impression of which penetrates the cornea under the pyramidal form and enters the opening, and which tends naturally to the center of the eye, and would go to it if it were not met first by a denser body by which it is bent, namely, the vitreous humor, *chd.* For this reason, therefore, I have so placed the center of the vitreous humor and have so drawn its sphere; and it is evident that its sphere is less than the sphere of the anterior glacialis. But in the eye there are not whole spheres, but only small portions of them, like the portion of the sphere of the anterior glacialis, *cfd,* and the portion of the vitreous humor, *chd;* I have completed, however, the spheres in the figure to show the centers and the portions. The uvea, moreover, is drawn in a complete sphere, except in its opening, *mo,* and yet it is not in the eye completely spherical, as has been

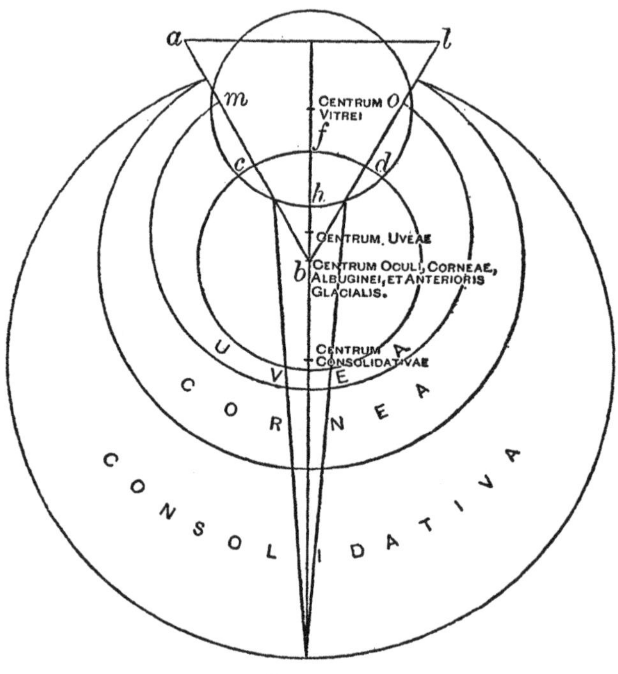

Fig. 25.

stated, but is so in its anterior part, in which is the opening, because sphericity is not required for vision except in that place; elsewhere it is of irregular form, in order that a void may not exist between the uvea and the cornea; and the lines *c* and *d,* which are drawn toward the interior of the eye, are in the sides of the nerve of the uvea, and between these lines is the opening of the uvea, which accordingly is toward the interior of the eye. Above this opening the vitreous humor is formed, as is apparent; for the aperture of these lines terminates at the extremities of the portion of the vitreous humor, and its distance is between *c* and *d,* and this distance between the sides of the nerve is filled with the vitreous humor as far as the common nerve in the surface of the brain. This nerve, however, in which is this path of the vitreous humor, spreads and expands in the circuit of the humor vitreus, glacialis, and albugineus, as far as the anterior opening of the uvea, *mo,* which is opposite to its own interior opening, which is *cd.* Then follows the cornea, and then the consolidativa, as indicated in the figure.

FOURTH DISTINCTION

In four chapters. The first chapter is on the properties of the cornea, the humor albugineus, and the uvea.

Chapter I

THE coats and humors, according to Alhazen, have their admirable qualities, from which follow the benefits of vision, as he himself shows. The first function of the cornea is the closing of the opening in the uvea, preventing the escape of the humor albugineus; it is, moreover, transparent, so that the impressions [species] of light and color may pass through it, as was verified before in the multiplication of species. It has strength, moreover, so that it is not quickly destroyed. Since it is exposed to the air, and could easily be destroyed by smoke and dust and the like, it has for this reason four coats, as explained above. The humor albugineus is transparent, so that impressions may pass through it and beyond. It possesses moisture, in order that it may always keep moist the humor glacialis and the spiderlike web, which is very thin and for this reason could be destroyed by excessive dryness. The uvea is usually black, in order that the humor albugineus and the glacialis may be obscured, so that feeble impressions of light and color may appear in it, since feeble light is very apparent in dark places, and is concealed in places full of light. It is somewhat strong, in order that it may retain the humor albugineus and prevent any of it from exuding. It is also thick so as to be obscure. In the eyes of human beings, however, it is sometimes found to be gray, and frequently so in the eyes of horses. This is due to the fact that the natural heat was not able sufficiently to assimilate the substance of the uvea and of the humors, and for this reason they are somewhat white, because the weak action of the assimilating heat in moisture is the cause of whiteness. Or it sometimes happens from the complete assimilation of the moisture and the victory of the dryness, as is evident in the leaves of the trees in autumn. This grayness either can be due to the uvea, for if the uvea is gray, the eye is gray; if black, the eye is black. Or the grayness can be caused by the humors, since if the moistures are placed near the exterior, and the crystalline is of large size, and the albugineous is moderate-sized, the eye will be gray, unless the opposite happens from the coat. If the moistures are dark, and the crystalline lies toward the interior of the eye and is small, and the albugineous is large, so that it causes darkness, just like very deep water submerging and covering things, the eye will be black. Aristotle held this view, and Avicenna in the nineteenth book on Animals.

Chapter II

Concerning the properties of the anterior glacialis.

THE anterior glacialis has many properties. For the first and chief of these is the fact that in it alone is the visual power, according to Alhazen and the other authorities. For if it is injured, even though the other parts are whole, vision is destroyed, and if it is unharmed and injury happens to the others, provided they retain their transparent quality, vision is not destroyed; for while the transparent quality through the glacialis remains continuous with the transparent quality of the air, vision is not destroyed, provided the anterior glacialis itself is uninjured. Moreover, the anterior glacialis is moist, so that it may more quickly respond to the impression of light and color, for very dry substances do not easily receive impressions; and it is of delicate structure, since such a structure in a body causes delicacy of perception. It is somewhat transparent, in order that it may receive the forms of light and color, and that they may pass through it to the common nerve; and it is somewhat thick, so that it may retain in it impressions for some time, in order that they may become apparent to the visual power and permit the exercise of judgment. For if the transparent quality were too great, the impressions would pass through, and would not remain so that a judgment could be formed. Hence it must be somewhat thick, in order that it may experience a feeling from the impressions [species] that is a kind of pain. For we observe that strong lights and color narrow vision and injure it, and inflict pain. But every action of light is of one nature and of color likewise, except that one is stronger, another weaker. Therefore vision always experiences a feeling that is a kind of pain, although it does not always perceive this, that is, when the impressions are moderate. But a painful feeling would not exist in a body unless it be fairly dense, because if it were of too great rarity the impression would not remain until it could cause the action of pain. Its surface forms part of a larger sphere than that of the vitreous, so that its surface is equidistant from the anterior surface of vision, that they may have the same center, which is the center of the whole eye and of the cornea and of the humor albugineus. These bodies aid it mainly in the act of vision, and more so than the uvea. The surface or portion of the anterior glacialis is less than half of its sphere, for otherwise it would not follow that its center is farther within the depth of the eye, as has been assumed.

Chapter III

On the properties of the vitreous, the weblike structure, the optic nerve, and the consolidativa.

THE vitreous humor is thicker than the anterior glacialis, since the ray that is not perpendicular must be refracted between the perpendicular drawn from the point of refraction and the direct path. Concerning this refraction sufficient mention was made in a preceding part on the multiplication of light. Unless refraction took place thus, vision would never be completed, as will be explained below. The color of the vitreous is whiter than that of blood, and not so white as the anterior glacialis, because the vitreous is its nourishment, as explained above. It is the portion of a smaller sphere, so that the center of this sphere is different from that of the anterior glacialis, since this is necessary owing to the refraction noted above. Its noblest property is the fact that the perception that is in the anterior glacialis is continued in it through the whole optic nerve to the point of ultimate perception, which is in the anterior part of the brain, as Alhazen says. These two humors are wrapped around by a web, since humors unless retained would flow somewhere and would not retain one shape. This web is very thin, so as not to hide the ray, and is spherical, because it contains the portion of a sphere, although there are other reasons for this, as for the whole eye and its parts. The nerve, moreover, as Alhazen says, on which the eye is formed is entirely the optic, so that the impression may be transmitted in it to the brain, and the impression of the visible object and the natural heat due it and the force of first perception may come freely to the eye, and therefore the term optic is the same as concave. The consolidativa is exterior, so as to assemble and preserve all the parts; and is somewhat moist, in order that it may adapt itself better to the positions of the coats in it, because more quickly and easily they assume the form of their position in it owing to its softness than if it were hard; and in addition it is moist, so that dryness may not quickly happen in the coats. It is somewhat firm and strong, so as to preserve the positions and forms of the coats, in order that they may not be altered quickly; and it is white, that by its means the form of the face may be beautiful.

Chapter IV

Concerning the eyelids, the lashes, and the whole eye.

The eyelids are to protect the eye during sleep and to enable the eye to rest when fatigued by a strong impression. And although the impressions are moderate even, the eye needs the means of closing furnished by the eyelids, in order that it may not work continuously. In a similar way smoke and dust and other things harm the vision, for which reason it needs a means of closing, and therefore the lids have a quick motion, in order that they may swiftly close over the eye when harmful things approach. The lashes are to moderate the light when vision takes place, and for this reason when looking one unites them and draws them together, so as to look from a narrow opening when the strong light would harm him. Creative goodness has supplied two eyes, so that if accident or injury should happen to one of them, the second might remain. Another purpose is to add beauty to the form of the face. Both eyes, moreover, are alike in their arrangements, in their coats, and in the forms of their coats, and in the position of each coat with respect to the whole eye; and both have a similar position with respect to the common nerve and the brain. Moreover, although the general reasons for the roundness of the eyes and of their parts were given above in the properties of the sphere, yet it was especially necessary that they should be round for two reasons, namely, owing to the swift motion of this form, enabling the vision to pass at our will from one visible object to another, and from one part of the same visible object to another part, in order that anything may be perceived with full certainty by a swift motion of this kind. But of all figures the sphere is best adapted to motion. It was necessary also that the eyes should be round and the parts likewise, for if the surface of the eye were plane, the impression [species] of an object larger than the eye could not fall perpendicularly on it, because perpendicular lines to a plane tend to different points and at right angles in each case, as shown in the figure. For the lines can fall perpendicularly on the eye *fg* that come from the visible object *cd,* which is equal to the eye, but from the points *a* and *b* the ray cannot come perpendicularly, but at oblique angles. But sensible action, and such as is required in seeing, does not take place except when the ray falls perpendicularly on the vision. Since, therefore, the eye sees large bodies, as almost a fourth part of the heavens in one view, it is evident that the eye cannot be of a plane figure nor of any other figure than that of a sphere, since on a small sphere perpendiculars infinite in number coming from a large body can fall, and they tend to the center of the sphere, and thus a large body can be seen by a small eye, as is shown in the figure.

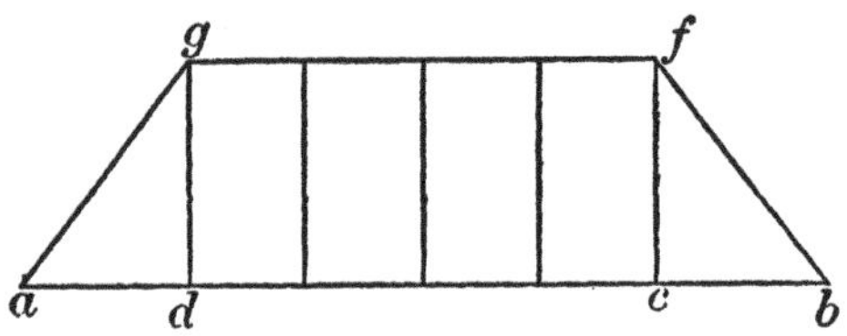

Fig. 26.

FIFTH DISTINCTION

In three chapters. The first chapter shows that impressions [species] of light and color are required for perception.

Chapter I

AS we now understand those things that need explanation on account of the mode of our vision, we must pass on to a consideration of the mode itself and how vision takes place. First we must determine what is required for rectilinear vision. The first requisite here considered is, that vision needs the impression of a visible object, for without this there can be no vision. Accordingly Aristotle says in the second book on the Soul that in every case the sense receives the impressions of sensible things, so that there may be an act of sensation. Moreover, the passive must resemble the active through the active; but vision is a passive force, and for this reason must resemble the active, which is the visible object. But there is no similitude of the active except the impression, as all know. Moreover, an object makes an impression [species] everywhere along all diameters. But when there is an obstacle between the impression of the object and vision, vision does not take place. But when every impediment is removed, so that the impression comes to the eye, the object is seen. Wherefore vision must happen by means of an impression; but especially by means of the impression of light and color. For it is evident that colors have their effect on vision from the fact that when one views a thick growth of vegetation on which the light of the sun falls, and continues gazing at it, if he then removes his vision and turns it to a dark place, he will find in that dark place the form of that light tinged from the green of the vegetation. And if under this condition he looks at visible objects that are white in the shadow, and in a place where the light is weak, he will find the colors of the objects mixed with green. If, moreover, he closes his eye and looks in it, he will find in his own eye the form of the light with green. In the same way he will find a change of this kind if he looks at the color blue or purple, or at any other strong color, as any one can prove by experiment. Therefore color must have an effect on vision. But light more so, since in accordance with the diversity of light falling on the vision and on visible objects is the diversity of the image in individual objects. For a very bright light hides other visible objects, and even injures and oppresses vision and weakens the action of sight. But a very weak light does not change vision as it should, nor does it reveal objects. A moderate light greatly strengthens vision in its action and reveals objects sufficiently. Wherefore the species of light is especially essential for vision. And again we see that in accordance with the diversity of the fall of the same light on the same object the aspect is changed, and the color appears different to the vision, as in the case of the dove's neck when it turns the neck in different positions to the light, and so too in the case of the peacock's tail. In the same manner many things, as, for example, the scales of fish, or decayed oak, and certain worms, and the bird that is called noctiluca [shining by night]; when the light shines on them, their light is hidden and a color is seen; but when they are in darkness, their light is apparent. It is clear, then, that the species of light is especially essential in vision. Without contradiction we find by experiment that without light nothing is seen; for the extrinsic light of the sun, or of the stars, or of a fire must be present in the air, or the light belonging to the eye and multiplied by it, as is the case in the eye of the cat. Wherefore the species of light must always be necessary for vision.

Chapter II

Showing that vision is not completed in the eye, but in the common nerve.

We must understand that vision is not completed in the eyes, according to the teaching of the authors on Perspective. For two different species come to the eyes, and a difference in the species causes a difference in the judgment, since by the different species will one object be judged to be two. A similar result follows through a difference in the one judging. For in two eyes there are different judgments. Therefore one object as viewed will be considered as quite different. Therefore there must be something sentient besides the eyes, in which vision is completed and of which the eyes are the instruments that give it the visible species. This is the common nerve in the surface of the brain, where the two nerves coming from the two parts of the anterior brain meet, and after meeting are divided and extend to the eyes. Therefore the visual faculty is located there, as in a fountain, and since in this case the fontal faculty is a single one, to which the faculties of the eyes are continued through the medium of optic nerves, an object can appear as single, as far as concerns this cause. But it is, moreover, necessary that the two species coming from the eyes should meet at one place in the common nerve, and that one of these should be more intense and fuller than the other. For naturally the two forms of the same species mingle in the same matter and in the same place, and therefore are not distinguished, but become one form after they come to one place, and then since the judging faculty is single and the species single, a single judgment is made regarding one object. A proof of this is the fact that when the species do not come from the two eyes to one place in the common nerve, one object is seen as two. This is evident when the natural position of the eyes is changed, as happens if the finger is placed below one of the eyes or if the eye is twisted somewhat from its place, both species do not then come to one place in the common nerve, and one object is seen as two. This happens in the case of him who does not have the position of his eyes similar with respect to the common nerve, and for this reason the species of his eyes, unless he takes diligent care and rectifies their position, will come to different places in the common nerve, and therefore one thing frequently appears to be two. But naturally eyes that are well formed and healthy have a position similar with respect to the common nerve, and therefore the two species come to the same place in it and become one, so that thus through one species and one perception one judgment is formed regarding a single object. Experience teaches that when the common nerve is injured vision is destroyed in the eyes, and that the faculty is not destroyed in the common nerve owing to an injury to the eyes. When the common nerve is cured, vision is restored in healthy eyes; and when the eyes become well, vision is again restored, because the faculty in the common nerve was preserved, and is not destroyed in the common nerve.

Chapter III

On the ultimate perception.

But since Alhazen says that this ultimate perception is in the anterior part of the brain, some one might think that it is the common sense and the imagination or phantasy which are in the anterior brain, as was stated above: especially as it was there said that a judgment concerning any sensible object is not completed before the impression [species] reaches the common sense. But it is necessary to state that the ultimate perception can be the origin of all the senses, and in this signification we are not speaking here of the ultimate perception: for this is the common sense in the anterior part of the brain. The ultimate perception is taken otherwise in the case of the special sensations of seeing, hearing, smelling, or the others, when some particular sense is mentioned; and thus the ultimate perception is the common nerve in vision with respect to the two eyes, which are the instruments that are first affected by a visible object. The same is true of the small pieces of flesh similar to the nipples of the breasts; for they are the instruments that are first affected by an odor, and the nerve to which they are continued in the anterior part of the brain is the radical and original instrument of the sense of smell. As to his statement that the ultimate perception is in the anterior part of the brain, we must note that this anterior part of the brain is not taken as the first cell of the brain, but as the place near it, namely, in the opening in the skull where the nerves meet; for since it is before and near the brain it is called the anterior part of the brain. The same is true regarding the olfactory process; for it begins at the middle of the anterior brain and extends between the two nipples [olfactory bulbs] near the brain, and closer than the visual nerve, because it is very necessary for an animal that the brain be strengthened by odor. This is especially true in the case of man, because he has a larger brain in comparison with his body than any other animal, as Aristotle says in the book on Animals.

Thus it is manifest that the eyes not only judge concerning a visible object, but that the judgment is begun in them and is completed by the ultimate perception, which is the visual faculty with its source in the common nerve. In a similar way it is evident that the eyes perceive and not only the common nerve. But since the eyes are adjusted to the radical faculty, and from it flow the forces to the eyes, and the sensitive force is continued through the whole nerve from the common nerve to the eyes, as Alhazen says, therefore the visual act is a single and undivided one, which is performed by the eyes and the common nerve. And although he says that the eye is the instrument of the ultimate perception and is the medium between it and the visible object, yet the eye of necessity has judgment and the power of sight, although its judgment is incomplete. For the angle by which the size of an object is known does not exceed the glacialis, and the arrangement of the object seen is made to accord with the object itself on the surface of the glacialis. It is by this arrangement that the object is distinguished and known. This is the statement that is forthwith made.

SIXTH DISTINCTION

On the removal of difficulties in the theory of vision, in four chapters.

Chapter I

In which is removed chiefly the difficulty that seems to arise from the smallness of the pupil.

WE must consider, then, as was verified in what precedes, that the natural action is completed by a pyramid whose vertex is in the patient [object receiving the action] and whose base is the surface of the agent. For thus does the force come from the whole agent opposite to the patient, as has been stated before, so that the action may be vigorous and complete; and therefore in vision there is the requirement that the impression [species] should come from the whole surface of the agent. But although in the natural change produced in the objects acted upon it is required that separate pyramids come to the individual parts of the body acted upon, because each point of this body must be affected, yet in the change produced the chief requirement is that one pyramid come from the agent, and the vertex fall on the eye. This pyramid falls perpendicularly on the eye, so that all of its lines are perpendicular to the eye. For the chief requirement is that vision should perceive distinctly the object itself both with certainty and sufficiency, and this can be done by means of one pyramid in which there are as many lines as there are parts or points in the body seen. Along these lines the separate impressions come from the individual parts to the anterior glacialis, in which is the visual force. And those lines will terminate at the individual parts of the glacialis, so that the impressions of the parts of the thing seen are arranged on the surface of the perceiving member just as the parts are arranged in the thing itself, so that the judgment in regard to the separate parts may be distinct and not confused. Those lines, moreover, are perpendicular to the eye, in order that stronger impressions may come, so that it may be possible to see with perfect vision and judge concerning the thing as it really is. For the eye either does not judge at all, or forms an imperfect judgment by means of those lines alone that are not perpendicular, owing to the weakness of the impression coming by their means, although those lines converging with the perpendiculars will act more fully in securing a perception of a visible object, as will be shown below. The requirement of perpendicularity in lines and species for effective action has been expained at sufficient length in what precedes on the multiplication of species.

But we must now verify the fact that on the surface of the glacialis, although it be small, the distinction of any visible object whatsoever can be made by means of the arrangement of the species coming from such objects, since the species of a thing, whatever be its size, can be arranged in order in a very small space, because there are as many parts in a very small body as there are in a very large one, since every body and every quantity is infinitely divisible, as all philosophy proclaims. Aristotle proves in the sixth book of the Physics that there is no division of a quantity into indivisibles, nor is a quantity composed of indivisibles, and therefore there are as many parts in a grain of millet as in the diameter of the world, as is shown in the figure. If a triangle or pyramid has a large base, *abc,* and a very short line, *ed,* subtends its vertex, it is evident that from any point of the line *ab* a line can be drawn to *c,* likewise from its other points and from all its parts, because an infinite number of lines can terminate at a point. This fact is well established. If, therefore, all those lines extend to *c,* then they pass through the points of the line *de;* since, therefore, they do not meet before *c,* they will pass through different points in the line *de,* because if all or some of these lines passed through the same point, the lines would meet before *c,* but it has been assumed that they do not meet. For if there were a meeting of all the lines or of some of them in some point of the line *de,* then without doubt after the meeting they would separate one from another to infinity, and would never meet in *c,* as is evident to the sense in this shorter pyramid, *fgh.* Since, therefore, the species of an

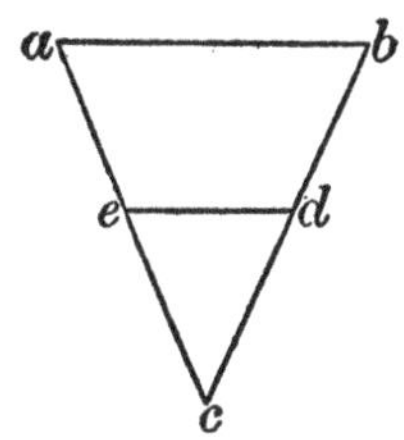

FIG. 27.

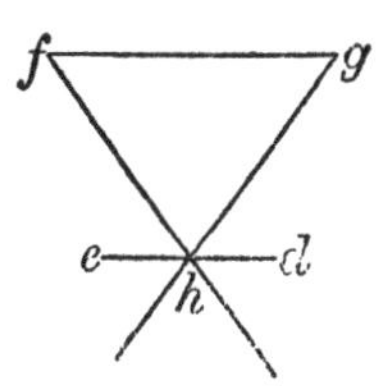

FIG. 28.

object seen, of whatever size it may be, can be arranged on the surface of the glacialis owing to the divisibility of quantity which continues to infinity, and which assumes as many parts in a small body as in a large one, no confusion happens when a large species comes to the small surface of the glacialis.

CHAPTER II

In which there is removed a second difficulty due to the meeting of oblique rays with the perpendicular ones.

BUT we must remove another difficulty that can be imagined from another source. For from any part of an object seen come forth species in infinite number, as explained in the laws of multiplications. Therefore to any part of the glacialis there comes the species from the whole object, and separate pyramids, the vertices of which are in any point of the eye and of the cornea and of the opening of the uvea, and the base of all these pyramids is the object seen. Therefore any point of the cornea and of the opening of the uvea will have species of all the parts confused in it; wherefore the judgment will become confused. We must not say that any point of the eye is infinitely divisible, in order that we may fall into the sophistry given above, since we here accept a point of the pupil or a part as the least sensible portion in the division of the parts of the perceiving member in accordance with the division of the parts of the thing seen. And therefore, excluding this sophistry, we can say that although in reality the vertex of one pyramid from the whole thing comes to every point of the eye and the cornea, and that although the species of all parts are there mingled, yet to one point of the eye or cornea and opening of the uvea a species does not come perpendicularly except from one point of the thing seen, although to the same point an infinite number of species come falling obliquely at unequal angles. Therefore since the body of the eye is denser than the air, according to the laws of refraction determined above, all lines falling obliquely are refracted at the surface of the cornea. And since the species is weakened by falling at oblique angles and likewise by refraction, and since the species falling perpendicularly is the strong one, therefore the perpendicular species conceals all the oblique ones, just as a larger and stronger light hides many weak lights, as, for example, the light of the sun hides the countless lights of the stars, whence to the point *b* there comes a perpendicular from *c* itself, and to the same *b* comes *ab* not perpendicular, since it does not fall to the center of the eye, and therefore the species of *a* itself is concealed, although from the point *b* the species of *a* itself may be able to come to the glacialis by the refracted line *bd;* and therefore the judgment is in accordance with the perpendiculars. And since perpendicular species are distinct and arranged on the surface of the vision, therefore a distinction takes place. The pyramid, therefore, coming perpendicularly produces both a strong vision, so that by reason of its strength forms falling obliquely are hidden, and nevertheless a distinct vision, so that it excludes the confusion that seems to happen from an infinite number of rays falling obliquely which meet each point of the pupil. This pyramid is called the visual and the radiant one, by which vision is chiefly caused.

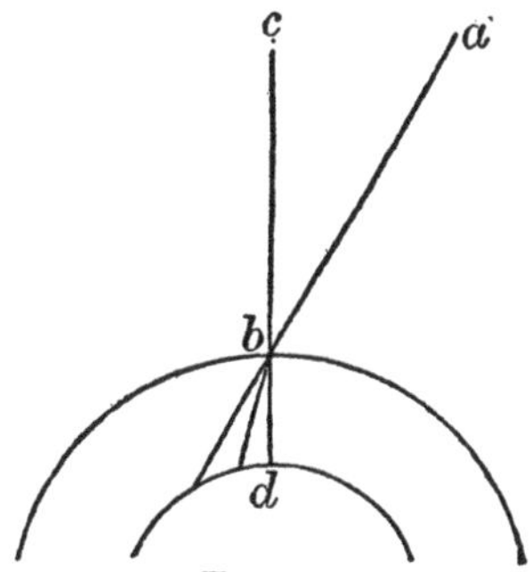

FIG. 29.

I say this because rays coming from a point of an object from which a perpendicular ray comes to a point of the eye, although they do not fall to that point directly, but to other points, yet can extend from the other points to which they fall, by refraction in the coats of the eye, to the same place in the glacialis and the common nerve to which the perpendicular species comes from the same point in the object, from which those oblique rays come; so that there is a fuller vision of any part of a visible object, since it is seen by means of its own direct rays and by means of the refracted ones. But concerning this matter mention will be made in what is to be said in regard to refraction. Moreover, for another reason I have said that vision takes place chiefly by means of the radiant pyramid; for since this pyramid alone is perpendicular to the eye, and falls to the opening of the uvea, and is directly opposite to the center of the eye, it produces for this reason a good vision and the principal one; and yet species can come outside of this pyramid to the eye which will fall not perpendicularly on the cornea, and will all be refracted, so that in this way vision is produced by them; but the vision will be weak, because things seen not perpendicularly are not clearly apparent to the eye. Therefore we can here consider two pyramids: namely, the principal one, which falls to the opening of the uvea, or the greater one, composed of this pyramid and of the rays coming from either side of the opening on the cornea. This whole pyramid so formed is not called the visual pyramid, or the radiant pyramid, although the eye sees by means of it, but it sees principally and clearly all that falls under the visible pyramid, and other things obliquely and weakly, namely, those things that fall outside this pyramid. Hence a thing can be so large that some part of it will fall in the visual pyramid and be well seen, and other parts on the sides will fall without the pyramid on the eye, and will be seen imperfectly. Or it may happen that a thing of moderate size falls in the visual pyramid and other different things will be seen on the sides. Or it may happen that several small objects will fall in the visible pyramid and others fall in like fashion at the side. But that which falls in the visible pyramid and nothing else will always be seen principally and clearly. But with greatest clearness will that be seen at which the axis of the visual pyramid terminates. For that axis is perpendicular to all the coats and humors, and passes through all the centers; and therefore the species which comes along it is strongest and fullest, and causes certainty of vision. But a discussion of this matter will be given below.

Chapter III

In which a third difficulty due to the mixing of species in the air is removed.

But up to the present time there is no small doubt in regard to the removal of the third difficulty, which requires full discussion in the treatise on the generation, multiplication, corruption, and action of species. Without this treatise perspective cannot be understood. But, however, this doubt ought here to be set forth owing to the requirement of vision somewhat more briefly, that it may not increase too greatly our confusion in the mode of our vision. For as a matter of fact the species of colors mingle at every point of the medium, since from extreme colors an intermediate is formed, and from two species of the same specific nature one is formed. For opposites, that is, extremes, according to the statement of Aristotle in the tenth book of the Metaphysics, produce an intermediate, as, for example, white and black; and two white colors blend into one when they are in the same subject; for in the same place and subject they cannot be distinguished, but become one. But what is true of colors is also true of the species of colors, for the species is of the same nature as that which produces it, and therefore any species of color is of some genus of color, since the species of white cannot be derived from substance or from any predicable [of substance] other than quality; nor of this, from any genus other than color; nor of this, from any species, however specialized, other than white, not from black, say, or from green. Hence it remains true of white [as formerly of color] that any particular species of it, being like white, will be an individual case within the special class [of colors] to be called white. From which follows that just as in the same subject white is commixed with black, so species of white with species of black. And if this is true, then a mixed species comes perpendicularly from every point of the air to the eye, and the whole radiant pyramid will be mixed from its point of mingling in the air, and this is a necessary conclusion.

Many philosophers hold that in this view there is difficulty, and say that species have a spiritual existence in the medium and in the sense, and Aristotle and Averroës maintain this in the second book on the Soul. And because they have a spiritual existence and not a material one they therefore do not observe the laws of material forms, and for this reason do not mix, because due to a material existence material forms mix. Therefore they maintain that different species of light in the medium and countless sources of light are found at the same point of the air and are distinguished, so too the species of color and all similar species of things, and for this reason vision is able to see things distinctly. This error is a very serious one, for it contains many false and absurd things, and arises from the fact that they believe that they must maintain the distinction of vision, which they think cannot exist unless the species are wholly distinct in the air. First, then, I shall preserve the distinction of vision, so that it may be seen that it is not necessary to err in this way. Then I shall remove the error more easily, and I shall explain the authors who seem to be at variance.

I say, therefore, that species have a material and natural existence in the medium and in the sense; and that opposite species mingle in a real mixture, as, for example, the species of white and black and of intermediate colors, and that the species of two white colors and of two lights become one, and the same is true of the other species of the same categorical species. A mixed species will come to the eye from the place where the species mix, and the whole pyramid will be mixed. But the species of a visible object has a principal and primary multiplication, but the others have an accidental. Moreover, the principal or primary multiplication is straight, refracted, and reflected, and comes from the agent, as we have shown above. But the accidental or secondary does not come from the agent, but from the principal species; just as is the case with light which comes to the corners of a house from a ray of the sun through a window; and this is so weak that it has no comparison with the principal one, nor does it direct the eye to the object from which the multiplication comes. Hence a man in the corner of the house with the secondary species of the solar light in his eye does not see the sun but the ray falling through the window. If, however, he applies his eye to the principal ray, he will see the sun distinctly. I say, therefore, that just as the perpendicular ray hides all the oblique rays that terminate with it, so the principal ray hides all the accidental rays. Hence at the point *d* there is a real mixture of the white, black, and red, and from this point comes a mixed species to the eye along the line *de*. But in the line *de* there is no principal multiplication except from the visible point *b* itself, nor is there one from *a* or *c*, but merely an accidental and secondary one; because a multiplication similar to *a* and *c* does not come except from their species, not from the points themselves. But the principal multiplication hides all accidental ones, just as the perpendicular ray hides all oblique ones conterminous with it. Thus the whole pyramid is mixed everywhere, but no mixture as regards the principal multiplication comes to the eye.

This conclusion is confirmed by the fact that when different colors have the same principal multiplication, a mixed color appears to the eye; as when glass or crystal or some other transparent colored body is placed before the vision, and another dense body is behind the transparent one in a direct line with it and the vision, the species of both bodies enter the sight in the same place as regards the principal multiplication, and therefore a mixed color appears. Therefore, on the contrary, a single color will appear, when one color multiplies itself along the principal line, and another accidentally, although in the same place. If, therefore, philosophers should consider this distinction of vision, they would never maintain that species do not mix in a medium, because owing to the distinction of vision, which they do not know how to preserve, they fall into that error. And if it should be said that just as those species mingle in a mixture of principal multiplications at any point in the air, so do they also in the eye; therefore there will then be complete confusion, because a mixture will take place there real in character and principal as regards multiplications; the reply to all this must be that a real and principal mixture can take place at any point of the eye; but only one species at one point will be perpendicular, which falls to the center of the eye, and all others will be oblique, as are *ah, cg,* and therefore will be hidden, and a judgment will not be formed concerning these but concerning the perpendicular one, which is *be,* and by means of it concerning the thing itself, which is *b.*

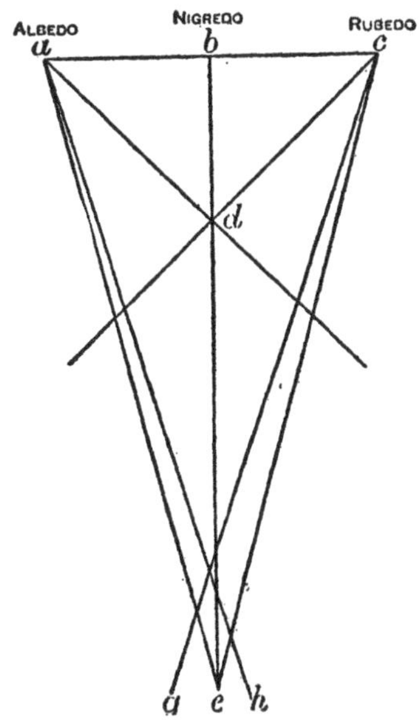

FIG. 30.

Opus Majus

Chapter IV

In which a real mixture of species is proved at any point of the medium, with the removal of cavilings to the contrary.

But when they say that species has a spiritual existence in a medium, this use of the word spiritual is not in accordance with its proper and primary signification, from spirit, as we say that God and angel and soul are spiritual things; because it is plain that the species of corporeal things are not thus spiritual. Therefore of necessity they will have a corporeal existence, because body and soul are opposed without an intermediate. And if they have a corporeal existence, they also have a material one, and therefore they must obey the laws of material and corporeal things, and therefore they must mix when they are contrary, and become one when they are of the same categorical species. And this is again apparent, since species is the similitude of a corporeal thing and not of a spiritual; therefore it will have a corporeal existence. Likewise it is in a corporeal and material medium, and everything that is received in another is modified by the condition of the recipient, as the book on Causes states, also Boetius in the fifth book on Consolation. Therefore it must have a corporeal existence in a corporeal medium. Moreover, it produces a corporeal result, as, for example, the species of heat warms bodies, and dries them out, and causes them to putrefy, and the same is true of other species. Therefore since this produces heat, properly speaking, and through the medium of heat produces other results, species must be a corporeal thing, because properly speaking a spiritual thing does not cause a corporeal action. And in particular there is the additional reason that the species is of the same essence as the complete effect of the agent, and it becomes that when the agent affects strongly the patient [thing acted on]. Since at first, when wood becomes warm, while it still remains wood it has the species of fire, and later the action becomes stronger, and the species is changed into complete fire, when the fire has destroyed the specific nature of the wood, and flame is produced and charcoal. The species does not, therefore, differ from the charcoal and the flame, except as the incomplete differs from the complete, the embryo from the child, and the child from the man. But it is agreed that the complete is material; therefore also the incomplete, wheresoever we observe these things, because the incomplete becomes complete. It is manifest, therefore, that the species of corporeal and material things will have always a material and corporeal existence, whence it is madness to think otherwise.

When, therefore, Aristotle and Averroës say that the species has a spiritual existence in the medium and in the sense, it is evident that spiritual is not taken from spirit nor is the word used in its proper sense. Therefore it is used equivocally and improperly, and this is the truth. For it is taken in the sense of imperceptible; for since everything really spiritual, as, for example, God, angel, and soul, is imperceptible and does not fall under the sense, we therefore convert the terms and call that which is imperceptible spiritual. But this is homonymous, and outside the true and proper meaning of a spiritual thing. Hence the species of things do not fall *per se* under the strong and distinguishing sense; for since nothing is visible *per se* unless it be dense, because this alone can terminate vision, light or the species of color in transparent air is not visible *per se,* but becomes so *per accidens,* because no doubt there is something denser than the air, at which vision terminates, and thus perceives that there is a transparent quality in the medium, at which vision is not terminated, and consequently the clearness of light is apparent in it. And similarly when a ray falls through a window, it is seen *per accidens* owing to the fixed shape of the window, by which the light is shaped, and owing to the opaque places everywhere, so that thus an opposite placed beside its contrary becomes more easily apparent. Similarly when a solar ray passes through glass or through a highly colored cloth, the species of the color is apparent on an opaque body. But this takes place doubly *per accidens,* both because of the excessive clearness of light with respect to color and with respect to an opaque body, which intercepts the light. And in the bodies of the stars the species of the solar light is seen, not because of this light, but owing to the density of the body of the star, a density that terminates vision; and a dense body is the cause of illumination, as was noted above. In these cases, therefore, the species is seen *per accidens,* and likewise at times because of excessive weakness of vision and because of negligence in looking, as in certain cases will be explained below. And since it is only *per accidens* (as from defect of vision or negligence in looking) that species can sometimes, as it were, by chance be perceived in a measure, they are not for this reason said to be visible and perceptible in the simple and absolute use of the word. The same principle applies to the species of objects perceived by taste, smell, and the other senses; perception of them is gained neither *per se* nor *per accidens,* and therefore the species are imperceptible. And because they are imperceptible they are called spiritual; but this spirituality does not contradict corporeality nor materiality in material and corporeal things.

That, moreover, species do unite into one, and that one species is really formed from several, is shown by Alhazen the author on Perspective and by Ptolemy, who makes these statements, and by what has been said concerning the first part of the act of perception. For in it the two species coming from the eyes must become one, so that the object seen may appear as one and not as two. Alhazen says in the first book that lights mix in the medium; and Ptolemy in his third book clearly teaches a mixing of the species. But as to the fact that Alhazen wishes to prove by experiment that lights do not mix in the air, when three candles are placed opposite an opening; for then the lights appear distinct beyond the opening, and therefore also in the opening, as he seems to say; we must state that in one way there is understood to be a real mixing, and in another way there is said to be a distinction. For in fact they do mix in the opening: but since light travels along a straight path, while it is being multiplied in the same medium, therefore the light of each candle, just as it passed along different straight lines before the opening, so must it continue to do beyond the opening as regards the principal multiplication, and therefore the primary and principal paths are divided beyond the opening just as they were before it. But an accidental multiplication of the two candles occurs with the principal multiplication of the third candle, and thus there is a mixing beyond the opening. But since an accidental multiplication is not taken into account with a principal one, nor does vision form a judgment in regard to the former, because it is hidden by the principal one, therefore no confusion is apparent to us, nor is there a mixing in the places where the lights of the candles fall. There is then a mixing in the case of the lights, namely, of the accidental light with the principal; and the author [Alhazen] says that there is no appearance of mixing, because it is hidden, and he says that there is no mixing of the principal multiplications in the places where they fall; and I grant this; and yet there is a mixing there, which I mentioned in regard to the opening. I say that light considered without restriction as it is in the opening must mingle in a natural mixing and become one undivided light, and he does not deny this. But if we consider the lights as regards their relation to the straight principal paths divided after the opening, just as they were before it, they are thus said to be divided and not to mix. Hence speaking without restriction they do mix, but with respect to the different principal paths, into which it is stated that they must be now distributed in the opening, they are divided and distinct. But distinct existence in this way is not understood in the ordinary sense of the term, nor is it opposed to a real mixing without restriction, because this has to do with effect and not with mode of existence; for they are said to be distinct in the opening solely because they make after the opening distinct paths, just as the sun is said to be hot because it produces heat, not because it is formed by heat. Whenever, therefore, any of the sacred writers, or of the philosophers, or of the men of science in the past say that the species of light and color or other species at the same time are distinct in a medium, this statement must not be understood without qualification, but it is made because the species make distinct principal paths beyond the place of mixing, just as they did before it.

SEVENTH DISTINCTION

In four chapters. First there is removed the error of vision which would result if the vitreous and the lens [glacialis] were of the same nature.

Chapter I

HAVING removed the difficulty in seeing, we must now show how other inconsistencies may be avoided. For if the rays of the visual pyramid meet at the center of the anterior glacialis, they must be mutually divided, and what was right would become left and the reverse, and what is above would be below, and thus the whole arrangement of the visible object will be changed, as is easily apparent in the figure: and thus the species of the right part of the object will not come to its place, but to the opposite side, and the same is true with regard to the left, and with regard to other differences of positions. In order, therefore, that this error may be avoided and the species of the right part may pass to its own side, and the left to its side, and so too of other positions, there must be something else between the anterior of the glacialis and its center to prevent a meeting of this kind. Therefore nature has contrived to place the vitreous humor before the center of the glacialis, which has a different transparency and a different center, so that refraction takes place in it, in order that the rays of the pyramid may be diverted from meeting in the center of the anterior glacialis. Since, therefore, all rays of the radiant pyramid except the axis, which passes through all the centers, are falling at oblique angles on the vitreous humor, which is of a different transparency, all those rays must be refracted on its surface, as was shown above in refractions. Since, moreover, the vitreous humor is denser than the anterior glacialis, it follows, therefore, that refraction takes place between the straight path and the perpendicular drawn at the

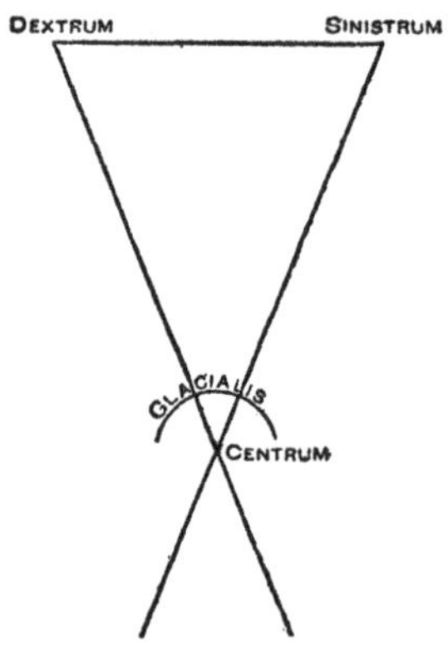

Fig. 31.

point of refraction, as has been shown in the multiplication of species. Wherefore of necessity the ray *mq,* when it reaches the point *q* on the surface of the vitreous humor, which is *gdf,* does not pass by the straight path to *a,* center of the anterior glacialis, which is *ghf,* but will be refracted at the point *q,* between the straight path, which is *qa,* and the perpendicular drawn from the point of refraction, which is *q,* into the vitreous humor. This perpendicular is *bl;* for *bl* goes to the center of the vitreous humor, which is *b.* Thus the right species will always go according to its own direction until it comes to a point of the common nerve, which is *c,* and will not go to the left. In the same way the ray *pu* will not pass to *a,* the center of the anterior glacialis, but will be refracted between the straight path, which is *ua,* and the perpendicular *bs* drawn from the point of refraction, which is *u,* and thus the ray *pu* will pass to a point of the common nerve at *c,* and will always be to the left. The same is true of the species coming from all other parts, because they will always travel along the paths that are due to them, and through the location they should have, so that no error may happen. Since the nerve is filled by a similar vitreous humor as far as the common nerve, there is, therefore, no other refraction, but the species travels uniformly without refraction, nor does it change its straight path in any manner except in accordance with the tortuosity of the nerve. And in this fact we must wonder at the power of the soul's force, in that it causes the species to follow the tortuosity of the nerve, so that it flows along a tortuous line, not along a straight one, as it does in the inanimate bodies of the world. For while it is in an inanimate medium it always travels along straight paths, as stated above: but owing to the necessity and nobility of the processes of the soul, species in an animated medium keeps to the path of the medium and

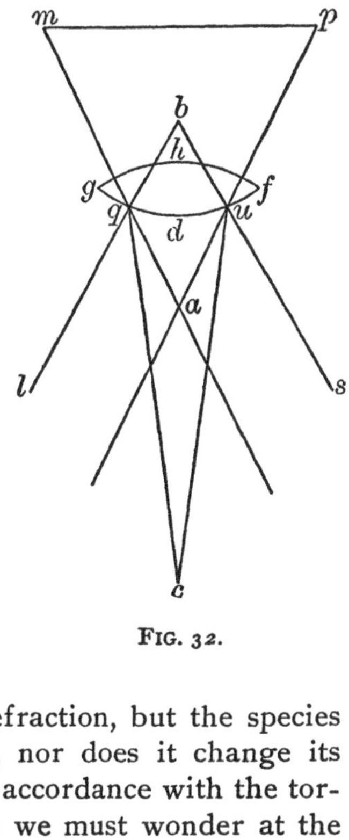

FIG. 32.

disregards the common laws of natural multiplications, rejoicing in the special privilege of the soul. Thus, therefore, we must consider that the species of the thing seen is necessary to vision, and we must also consider how it falls upon the vision and all its parts.

Chapter II

In which it is shown that the species or force of the eye radiates to the visible object because of the act of seeing.

We must now consider whether the species of vision is necessary for the act of seeing. It is clear, moreover, that a species is produced by vision just as by other things, because accidental qualities and substance inferior to vision are able to produce their own forces; much more therefore has vision this power. This fact is evident also for this reason, because the eye is seen by itself, as by means of a mirror, and can be seen by another. But nothing is seen except by means of a species coming from the object seen. But whether this species, or force of vision, or visual rays come from the eye to the thing seen has always been a doubtful matter with scientists. But Aristotle settles this question on his own authority in the ninth book of the De Animalibus, saying that vision is nothing else than the visual force touching the object seen. Ptolemy also in his book on Optics, that is, on vision or in his theory of perspective, previous to the teaching of Alhazen, which Alhazen expounded as received from Ptolemy, maintains throughout his whole work that visual rays come from the eye to the object seen. Tideus also in his book on Aspects affirms this, and states that sight never determines the distance between itself and the object seen, nor the size of the object seen, nor the position and situation of it, unless the visual rays pass to the object seen and rest upon it, and grasp its surface and contain its extremities. This likewise Jacobus Alkindi asserts in his Science of Optics, also Euclid and all other authorities. If we wish, moreover, to confirm this by the sacred writers, we shall state that they agree in this view, Augustine in particular; for he maintains in the sixth book on Music that the species of vision comes and is propagated in the air to the object. Hence just as an inanimate object produces its own inanimate species, so does an animate thing produce a species that has in a measure the force of the soul [*anima*]. For just as an inanimate thing has a relationship to its species, which is similar to it, so is an animate thing related to a species similar to it. A medium, however, which is inanimate will not because of this fact be animate, but will be made like an animate one through its likeness now received.

Chapter III

In which objections are removed.

If, moreover, Alhazen, and Avicenna in the third book on the Soul, and Averroës in his work on Sense and the Sensible are cited as opposed to this view, I reply that they are not opposed to the generation of the species of vision, nor to the part it plays in producing sight; but they are opposed to those who have maintained that some material substance as a visible or similar species is extended from the sight to the object, in order that vision may perceive the object itself, and that it may seize upon the species of the object seen and carry it back to the sight. For this was the opinion of some of the ancients in this matter, who did not yet possess definite knowledge of vision. We must state, then, that the philosophers mentioned above, Alhazen, Avicenna, and Averroës, are opposed only to this view, as is evident from their text. But nevertheless the majority is imbued with the contrary notion, owing to a passage of Aristotle in the Topics, because what one hears from his youth he accepts as a matter of habit, so that he is unwilling to receive anything else. For Aristotle in the book of the Topics, since he is giving the art of proof for every problem, states examples, which are the arguments of philosophers on matters concerning which there was doubt and which are discussed in common among them, as is evident from that book. Therefore that famous instance, that vision is the result of internal reception and not of external transmission, he cites in accordance with well-known opinions. For the Stoics held this view, as is shown by Boetius in the fifth book on Consolation in that verse, "Quondam porticus attulit obscuros nimium senes, etc." And in the book of the Priores he says that we do not always cite examples because they are true, but that he may assent who learns. Aristotle, therefore, does not assert that vision is not the result of external transmission, but makes the statement in accordance with the general opinion, and as an example, not as a truth. As to the fact, also, that in the second book on the Soul he strives to show that in general sensation belongs to the class of passive faculties and does not teach that sensation is active, we must state that this was necessary owing to the position of his master Plato and of many Platonists. For it was the common belief among them that vision was only active, and that it emitted a visible species for viewing all visible objects, whence according to them vision sends suddenly to the stars a visible species which views them and returns to the sight their species. Therefore Aristotle, who wished to verify all things as far as the possibilities of his age permitted, rejects both opinions regarding vision; namely, that of the Stoics, who maintained that it is passive only, and that of the Platonists, who held, and erroneously so, that it was only or principally active. But one of these views, namely, that of the Stoics, he refutes in his book on Animals, and the other, that of the Platonists, in his book on the Soul, just as it suited his purpose. But those versed in the philosophy of Aristotle and particularly in perspective think that vision is active and passive. For it receives the species of the thing seen, and exerts its own force in the medium as far as the visible object. Since the multiplication of species is instantaneous for every distance, as most people reckon, or rather it does require time, but an insensible amount, this time escapes perception owing to its brevity.

Chapter IV

In which the true theory is given.

The reason for this position is that everything in nature completes its action through its own force and species alone, as, for example, the sun and the other celestial bodies through their forces sent to the things of the world cause the generation and corruption of things; and in a similar manner inferior things, as, for example, fire by its own force dries and consumes and does many things. Therefore vision must perform the act of seeing by its own force. But the act of seeing is the perception of a visible object at a distance, and therefore vision perceives what is visible by its own force multiplied to the object. Moreover, the species of the things of the world are not fitted by nature to effect the complete act of vision at once because of its nobleness. Hence these must be aided and excited by the species of the eye, which travels in the locality of the visual pyramid, and changes the medium and ennobles it, and renders it analogous to vision, and so prepares the passage of the species itself of the visible object, and, moreover, ennobles it, so that it is quite similar and analogous to the nobility of the animate body, which is the eye. But since this theory is doubtful to many, therefore besides the verifications now given I shall adduce diverse true and certain experiences, as they shall appear in different places below in regard to other conclusions, which necessarily accompany this view of the subject. Concerning the multiplication of this species, moreover, we are to understand that it lies in the same place as the species of the thing seen between the sight and the thing seen, and takes place along the pyramid whose vertex is in the eye and base in the thing seen. And as the species of an object in the same medium travels in a straight path and is refracted in different ways when it meets a medium of another transparency, and is reflected when it meets the obstacle of a dense body; so is it also true of the species of vision that it travels altogether along the path of the species itself of the visible object. And although the species of the eye lies in the form of a pyramid, whose vertex is in the eye and whose base rests on all parts of the object seen, yet from the surface of the glacialis there are pyramids in an infinite number, all of which have one base, and the vertices of these pyramids fall upon the separate points of the thing seen, so that thus all parts of the visible object are seen with such intensity as is possible. Nevertheless one pyramid is the principal one, namely, that one whose axis is the line passing through the center of all parts of the eye, which is the axis of the whole eye; for that pyramid attests all things, as was stated above and will be explained more fully.

Although the species of visible things, as those of light and of color, mix in a medium, that is, several lights unite to form one, and several colors mix, as has been stated, and this species of an object and the species of vision lie in the same undivided place; yet is there no confusion of these species, or mixing, nor is one thing formed out of them, since they are not of the same species or of the same genus; because the pupil does not have color, nor do color and light have the force of the soul. Moreover, the species of the eye is the species of an animate substance, in which the force of the soul holds sway, and therefore it bears no comparison to the species of an inanimate thing, so that one thing should result from them, just as there is no such result from whiteness and sweetness in milk; and much less so in the case in point, because there is a much greater distinction between the animate and the inanimate, than between the two inanimate.

EIGHTH DISTINCTION

In which we are shown that besides species there are nine conditions required for vision. The distinction contains three chapters. In the first it is shown that light and a proper distance are required for vision.

Chapter I

AFTER these things we must consider that besides species nine things are required for vision, as the authors on Perspective show. One of these is light, because nothing is seen without light. For light in the first place is the visible thing, then color and the rest of the twenty things which I enumerated before, and all other things are seen with the participation of these, but no one of them is seen unless it is bathed in light. The reason for this is thought to be found in several ways, namely, either because color does not have a real existence in darkness, according to Avicenna in the third book on the Soul, or if it has, it is not able to produce its species in darkness, according to Alhazen, or if it can, it will not affect vision, nor will it change it, so that an act of vision takes place, according to the same Alhazen. The first supposition Ptolemy proves false in the second book on Perspective as follows: For if such were the case, any two things having the same position with respect to light and vision would seem of like color, the reverse of which we see in different things quite generally, and in the same thing at different times, as in the chameleon, which changes its color in accordance with the difference of those things that approach it, and in the case of him who turns red from shame and becomes pale from fear; although the thing always has the same position with respect to the light. The second supposition is shown to be false by comparison with every other active thing that produces species in darkness and in light. The third supposition is the true one, and with reason, because the first and principal visible thing is light, and therefore nothing can be seen without its help; just as in the other senses we smell nothing without the aid of the sense of smell, nor do we touch without the aid of the first four qualities, which are hot, cold, moist, dry.

The second thing that is required for vision is distance. For in general a sensible object placed on the organ of sense is not perceived, as Aristotle states in the second book on the Soul. The reason for this is that every sense acts by external transmission, that is, by emitting from itself its own force into the medium, so that the sensible species is returned more fitted to the sense, and receives a nobler essence from the species of the sense, so that it may be more conformed to the sense. Moreover, we find this in all the senses. For we have shown this above in regard to vision; and the like is true in the other two senses that have an extrinsic medium, namely, the sense of smell and hearing, since Aristotle says in the nineteenth book on Animals that forces are produced by the sense of smell and by hearing, just as water from canals. Similarly also regarding the senses that do not have an extrinsic medium, but an intrinsic one, namely, touch and taste. For concerning touch Aristotle says in the second book on Animals that its medium is flesh and its instrument is nerve. But in the twelfth book on Animals he maintains that the flesh perceives in touch, just as the eye in vision. Avicenna, moreover, in the first, second, and third books on Animals maintains that the skin and the flesh perceive. Therefore the sensitive force which is in the nerve scatters its force in the medium of touch, which is the flesh and the skin. But taste is a kind of touch, as Aristotle says in the second book on the Soul, and has an intrinsic medium, as touch has. Therefore the force of taste, which is in the nerve, emits its species into the flesh and skin of the tongue, nay, what is more, into the palate and other parts of the mouth, so that in a measure those parts seem to perceive the savor.

To what distance we can see on a plain of the earth's surface and on mountains the author on Twilights shows, saying that we see on a terrestrial plain about three miles; and on a very high mountain, the greatest height of which is eight miles, we shall see even on a plain of the earth's surface only about 250 miles; and the gibbosity of the earth restricting vision causes this.

CHAPTER II

Concerning the third condition which requires that the visible object confront the eye.

THE third condition is that the visible object confront the eye. For this is required in vision effected along straight lines, the kind of vision here under consideration, although by reflection and refraction a thing can be seen without confronting the eye. But this is a remarkable fact, since we hear and smell in all directions, and feel in front of us the heat of a fire placed behind us, if it is large and strong; and vision, which is a nobler sense, does not act in this way. Moreover, the cause of this thing is quite hidden and still strange and unperceived by scientists. For Aristotle in his book of Problems should have informed us in regard to this matter, for he touches upon it there among his other secret problems. But either a bad translation or error in the Greek edition, or some other reason hinders us in this particular.

It is certain, however, that we hear in every direction sounds without refraction and without reflection, as, for example, one speaking hears his own voice; but it is impossible that this happens by an accidental multiplication, because an accidental species does not cause us to perceive an object, as has been explained above. Nor does it take place by refraction, since there is only one medium; nor is there present a dense body, from which reflection takes place; wherefore the action must be along straight lines produced to the ear, therefore there must be a real sound and not merely the species of sound opposite the ear, and there must be a multiplication into it; but the first sound cannot strike the ear. Therefore a real sound must be produced opposite the ear; and the way in which this is true I shall describe. For sound is produced because parts of the object struck go out of their natural position, where there follows a trembling of the parts in every direction along with some rarefaction, because the motion of rarefaction is from the center to the circumference, and just as there is generated the first sound with the first tremor, so is there a second sound with the second tremor in a second portion of the air, and a third sound with the third tremor in a third portion of the air, and so on. Moreover, because these tremors are violent and especially so in the air, which is easily moved, and when it is put in violent motion retains the impression of the motion, therefore the second sound and successive ones for a good distance are not merely the species of sound, but a real sound possessing at any rate more than the species. Moreover, since this is a fact, for this reason a sound similar to the voice multiplied from the mouth to the surface of the ear can be generated by tremors of this kind in the air produced in every direction to the ear, not by an accidental generation but by a principal one, since the reason for this is found in the tremor mentioned, which causes the air to vibrate to the ear, and produces a real sound in it, and in every direction. Moreover, a sound made in the ear or near it, and in a straight line with it, is not the species of a species, but is sound itself produced by the tremor. Moreover, a proof that sound in the parts of the medium of the air from the first place of its production is not merely the species of the sound, but has more of the nature of the object, is the fact that sound is more violent than other sense perception, because sound suddenly confuses the hearing and destroys it, when it is very loud.

Odor not only produces a species, but from the odorous body a vapor goes forth, which is a subtle body diffusing itself everywhere in the air, and when it comes opposite to the nostrils it multiplies its species to the organ of smell, and therefore that vapor has real odor, just like the first odorous body: and for this reason not only the species, but the real odor is found here in the opening of the nostrils, not, however, the first odor, but the second, which is in the vapor.

Now the four qualities, as stated above in the laws of multiplications, are able to complete their species like the four elements, because of the necessity of generation. For we see that fire not only generates its species, but also a real and perfect fire, in flame and charcoal; and thus the heat of fire, having more than does mere species, is able to generate real heat, and this can take place everywhere in the air. Therefore it can reach to the opposite part of an animal which is not exposed to the fire, namely, when it comes outside the shadow of the fire, which the exposed object makes as far as the point *a,* and

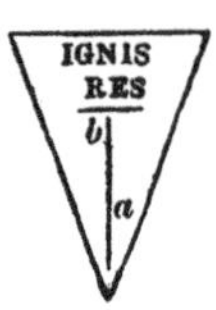

FIG. 33.

then is able to produce a species on the surface of the object not exposed to the fire by the line *ab*. But cases of this kind do not happen in those things ordinarily seen, and therefore an object must confront the sight.

CHAPTER III

Concerning the sensible magnitude of the visible object.

THE fourth condition required for vision is that the object be of a magnitude perceptible by the sense. For the object can be so small that it will not be visible. The reason for this is that the species coming from the parts of the visible object must be arranged distinctly on the surface of the lens, and in addition sensibly as regards the sentient faculty. But when the object is too small, the species coming from the individual parts of the object to the parts of the sentient organ, although distinguished according to the position of its magnitude, as its surface is infinitely divisible, are not, however, distinguished as regards sensation, but are confused because of too great proximity on the small part of the sentient organ which the visual pyramid occupies.

Joined, moreover, to this consideration is that of the maximum magnitude that we can see by means of the visual pyramid. This question is one of great doubt, namely, How much in extreme magnitude can be seen by the eye? Men versed in the problems of perspective estimate that the eye on the surface of the earth, as we now see, cannot see a fourth part of the heavens by means of the radiant pyramid. But if the eye were at the center of the earth, it would see a fourth part of the heavens under that pyramid; since they maintain that the pyramid contains in the eye a right angle, because by that angle is subtended the side of a square that can be described in the sphere of the uvea, namely, because the portion of the uvea in which the opening is can take the side of a square, and to the side of a square a right angle corresponds. Since this is a fact according to the opinion of those men, they think that in the portion seen of the heavens the side of a square can be contained, and for this reason the whole fourth part would be seen if the eye were at the center of the earth. But if this be a fact, then without doubt a fourth part cannot be seen by an eye that is on the surface of the earth, by the twentieth proposition of the Elements of Euclid. For according to this proposition, if from the extremities of the base of a triangle two lines are drawn beneath the triangle they will contain a larger angle, as is evident in the figure. Therefore the vertex of a pyramid coming on a fourth part of the heavens to an eye on the surface of the earth would contain an obtuse angle. Since, therefore, the angle of the pyramid according to these men is a right angle, the pyramid itself will not have a quarter of the heavens as a base, since the eye is on the surface of the earth, but will have less than a fourth.

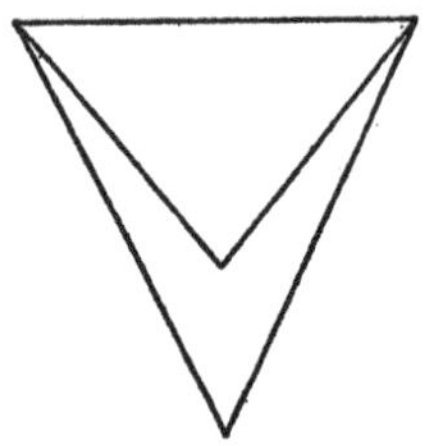

FIG. 34.

But the supposition that the radiant pyramid contains a right angle has neither authority nor experience nor proof up to the present time, and therefore is rejected with the same facility with which it is proved. Moreover, the reason for this assigned in the side of a square described in the sphere of the uvea cannot stand; because the sphere of the uvea and the anterior glacialis are not concentric, and therefore a right angle in the sphere of the glacialis will not correspond to the side of the square in the sphere of the uvea, nor will the portions of the glacialis and the uvea be similar, and therefore the portion of the sky will not be similar to the portion of the uvea in which the side of the square is drawn; which, however, would be necessary. Moreover, I am here calling portions similar which have portions proportional with respect to their spheres when they are cut by the same diameters of the earth, as is evident in the figure. For let the sphere of the heavens be *aci,* the sphere of the glacialis concentric with it *do,* and let the sphere of the uvea, whose center is different toward the anterior of the eye, namely *t,* be *efg.* It is evident, therefore, that *ac,* a fourth of the heavens, and *do,* a fourth of the glacialis, are similar, and will have the same right angle, namely, *ohd;* but *eg,* the fourth of the uvea, does not have that angle, since it faces the angle *ghe,* which is part of the right angle *ohd,* but the portion is larger than that fourth is, namely, *ef.* Therefore the side of the square described in a quarter of the uvea cannot face the right angle of the pyramid and of the glacialis, and therefore the angle of the pyramid cannot be closed by that side. But if it should be so closed, it will be acute, not right, as is evident to the sense in the angle *ghe,* since it is part

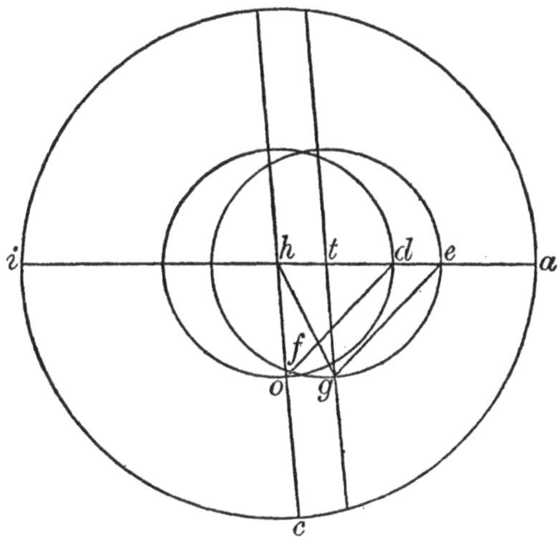

FIG. 35.

of the right angle *fhd* and *ohd,* which are identical. But it would be better to say that the portion of the anterior glacialis is a fourth of the sphere; for thus the angle of the pyramid would be a right angle, because that whole portion, however small it may be, is occupied by the angle of the pyramid, because it as a whole is the organ of vision, and not merely some part of it. Then, moreover, the side of the square described in the glacialis is subtended by the right angle of the pyramid; and similarly in the portion of the albugineous humor, which is in the opening of the uvea; and in the same way in the portion of the cornea, for they will then be fourths of their spheres, because they are concentric with the glacialis. Moreover, since the heavens would be concentric with them, if the center of the eye were at the center of the earth, the fourth part of the heavens would then correspond to them, together with the side of the square described in it, and thus a fourth part of the heavens would be seen. But there has been no verification of the statement that the portion of the anterior glacialis is a fourth of its sphere, and therefore we have no proof that an eye at the center of the earth would see a fourth of the heavens, nor by this method or proof can we be certain that an eye on the surface of the earth will see less than a fourth, nor how much less.

But experiment proves that the eye cannot see a fourth part of the heavens on the surface of the earth. For if one views a star that is above his head, and stands in a level space, he will not be able to see as far as the earth, however much he tries; but a fourth part of the heavens extends from the zenith to the earth. Therefore he will not see a fourth part. But he will see, however, a little less, since if on fixing his eye he bends his head a little he will see the earth, and he will not see the star. There is no other reason for this than the arrangement of the eye, because the pupil is so situated, and the opening of the uvea so arranged, that he cannot see more. This is manifest, because one person owing to the arrangement of his eye will see more of the quarter and another will see less. For he who has the opening of the uvea small, and the pupil deep set will see less of this fourth, and where the arrangement is the opposite, one will see more, as is shown in the figure; so that if the pupil is at the point *a* and the chord of the opening is *bc,* one will see less than if the pupil is at *d,* because the lines *db* and *dc* diverge more than *ab* and *ac.* Similarly, if the opening is greater with nearness of pupil more will be seen, as is evident by the lines *df* and *dg,* which separate more widely; and this is the definite reason in this case. Therefore those suppositions given above have no basis in fact.

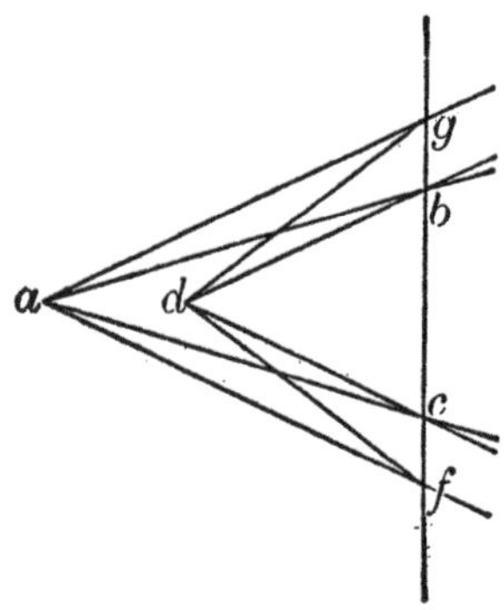

FIG. 36.

NINTH DISTINCTION

In four chapters. The first deals with the density and rarity of the object.

Chapter I

THE fifth condition is that the visible object, which we commonly call the object of vision, must exceed the density of the air and of the heavens, as Alhazen shows. For this reason we see clearly water, because it is denser than air, and vapors, and clouds, and glasses and transparent objects of this kind, which have little density as compared with those things that have perfect density. But, however, we should know that Ptolemy says in the second book of the Perspective that we see the air or the celestial transparent body far off and within a considerable distance although not within a distance near at hand; for much of a transparent body is accumulated in a great distance and has the same effect on vision as that which is perfectly dense within a small distance. Much, therefore, of a transparent object accumulated in a great distance becomes shaded, just as we see in the case of deep water, through the medium of which we cannot see the ground as we can through water not too deep. For the parts of deep water cast forward a shadow on those that succeed them, and a darkness is produced that absorbs the quality of the rarity, so that in this way the whole body of water appears like some dense body, and the same is true of the air or celestial transparent medium at a distance, for which reason it is rendered visible, but not so at close range.

But there is also another reason, namely, that vision is terminated at a distant transparent object. For as Alhazen says in the seventh book, the rarity of bodies in the world is limited, and for this reason all of them have some density and rarity, although not very perceptible especially when near; for the species of vision is emitted from the sight to the visible object, and is weakened in distance, so that although on account of its strength it penetrates the air near at hand, but does not penetrate, however, the transparent celestial medium at a distance; and therefore vision is terminated at it. But Avicenna says in the third book on the Soul that that is really visible which terminates vision, and therefore this transparent celestial medium is really visible at a distance. I offer an example in the case of air filled with vapor and cloud in winter, which is seen at a distance but not near at hand, and when the air filled with vapor is uniform in density and rarity. But from a great distance it is seen because it can terminate the species of vision and resist it; and thus it becomes visible, a visibility it cannot have near at hand owing to the strength of the species of vision. Nor is there deception of vision, because that air filled with vapor becomes an object of vision. The explanation of this according to Avicenna lies in the fact that it is able to terminate vision. Likewise I say here that air, or the sphere of fire, or the heavens, near and remote, is of similar rarity as far as perception is concerned; but it has, however, some density of its own nature, and this density is able in a great distance to terminate the species of vision, which it cannot do in a short distance, and therefore it will be quite visible at a distance, but not near at hand. Alhazen's statement, therefore, that a visible object must exceed the density of the air, and the fact that he calls air wholly transparent as far as the stars, are to be understood in the case of the usual visible objects which can be seen within the required distance and strength of vision. Why a color appears approaching black, namely, blue, is explained in the same way as in the case of deep water, where in a like manner that color appears owing to shadows projected by particles. Darkness is caused by these shadows, which is similar to blackness. This is what takes place in the air or medium between us and the last heaven. If some celestial body were a dense body, we could then say that that body would be visible chiefly because it terminates vision, and that the whole transparent medium between the object and the eye would not be perceived, except for the fact that vision perceives that it is not terminated before its species reaches that dense body. Moreover, it is thought by learned astronomers, who are ignorant of theology, that the starry heaven is everywhere dense. But there is another view, namely, that that which is seen beyond is the heaven of water. For it has a color like the water of the sea, and we see that heaven through the medium of all the eight heavens that are on this side of it. For there is no doubt in the minds of theologians and those philosophizing in accordance with theology that the ninth heaven is of water; and then beyond it is the tenth. But concerning these the discussion belongs elsewhere.

But an objection is raised in regard to the ray or luminous air falling through a window that it is quite visible, and yet that portion of the air is rare, and rarer than that outside the path of the light, because light rarefies air as it generates heat; for heat is not generated except from rarefaction. Here, moreover, some have attempted to show that density is not required in luminous bodies, whence they judge that the stars are not dense, nor are the spheres. But owing to the excessive brilliancy of the light they say that vision cannot penetrate them, but is beaten back or fails from the excess of splendor. But the example of the ray falling through the window is clearly explained, as stated above. For it is not of itself a visible object nor does it terminate vision, but the dense body surrounding it terminates vision, and not the ray; for if there were not a dense body beyond it, it would not be seen, just as the luminous air outside is not seen, unless it be terminated by the density of some portion of the heavens. The air in the window is rendered more perceptible, because its shape is determined by the parts of the window and by the ground on which it falls. It is not, therefore, a visible object except in a secondary and not a primary sense, nor is this of which we are speaking here a visible object, which, namely, has the power of itself by means of the necessary distance to terminate vision. Moreover, just as the wrong understanding of this example is now removed, so must we now have the right view regarding that which they attempt to prove by the example just quoted in regard to luminous bodies. For after the antecedent is shown to be false, we need not believe the consequent. We have stated above in the sections on mathematics that the spheres are not luminous, but that a star is so solely on account of its density, as Averroës says, and as was stated in that place. Moreover, although too great brilliancy confuses vision, as is evident when we gaze on the sun in its full strength, yet this does not exclude density in the sun, because density is the cause of illumination, as Averroës states, and the moon and the fixed stars when viewed do not confuse vision from excessive splendor. Therefore vision would penetrate them, unless their bodies possessed density.

Now from this chapter it is clear to him who gives it due consideration that vision must take place by means of its species emitted to the visible object. For if a single continuous transparent medium in extreme distance terminates vision, and does not terminate it owing to a perfect density, which independently is perceptible in every distance, but terminates it because of the weakness of the species of vision, which fails at a too great distance, then vision must be produced by sending forth its species; that is, by emitting its species owing to the act of seeing.

Chapter II

Concerning the rarity of the medium.

The sixth condition required for vision is rarity of the medium. For if a dense body is placed between the sight and the visible object, the species cannot pass through from one side to the other, and thus of necessity vision is cut off. But some, possessing an excellent understanding of many things in the science of Perspective, raise an objection in regard to a flame placed between the sight and a visible object impeding vision; and the flame, as they say, is a very rare body, because Alhazen says in the third book that the rarity of flame deviates from the mean. Therefore a rare body hinders vision more than a dense one. This, however, is a great error, for of necessity flame is denser than air, because Aristotle says in the second book on Generation that flame is a glowing terrestrial vapor; and therefore the deviation of rarity from the mean can be understood, either through a change to an extreme deficiency, or to an extreme excess, because the mean is midway between the two, just as liberality is between prodigality and avarice. And therefore just as avarice deviates from the mean of liberality through a lack of this quality, so does flame deviate from the mean of rarity by a lack of it, not by an excess. This, moreover, is Alhazen's meaning, for he so explains himself in what follows. If then an objection be raised regarding the lynx, which sees through the middle of a wall, as Boetius in the third book on

Consolation imputes to Aristotle, we reply that although this be true of the sight of the lynx, it is not true, however, of human vision with which the science of Perspective deals. For we are here discussing the latter vision.

But if we assume that there is a vacuum between heaven and earth, there would be neither a dense body nor a rare one. And yet Democritus thought that an eye on the earth could see an ant in the heavens, as Aristotle states in the second book on the Soul. Moreover, vacuum does not possess some nature by which it should impede species, or resist it, because no form of nature exists in a vacuum, as Aristotle says in the fourth book of the Physics. Therefore species will not pass through from heaven to the eye, and thus we should see the stars without a rare medium and without a dense one. But we must here state that we should not see anything if there were a vacuum. But this would not be due to some nature hindering species, and resisting it, but because of the lack of a nature suitable for the multiplication of species; for species is a natural thing, and therefore needs a natural medium; but in a vacuum nature does not exist. For vacuum rightly conceived of is merely a mathematical quantity extended in the three dimensions, existing *per se* without heat and cold, soft and hard, rare and dense, and without any natural quality, merely occupying space, as the philosophers maintained before Aristotle, not only within the heavens, but beyond.

The seventh condition required for vision is a sensible time. For Aristotle says in his book on Memory and Reminiscence that our whole intellect is concerned with continuity and time; much more so, therefore, is sensation. For the action of the intellect is spiritual, and the act of sensation is corporeal. But if a thing is brought suddenly before the eyes it is not seen distinctly and perfectly, and therefore a sensible time is required for vision in which the judgment of vision may be made. Whence Ptolemy in the second book of the Optics says that things which pass through the visual pyramid are thought to move swiftly, like the embers of fire and things passing through holes and narrow places, to which the vision penetrates. For since they pass through the pyramid in a short time, they are thought to move swiftly. But such is not the case.

Chapter III

Showing that the species of vision and of the visible are produced in time.

But here a very great doubt arises in regard to the species of sight and of the visible object, whether they are produced suddenly and instantaneously, or in time; and if in time, whether they require a sensible and perceptible time or not. Alkindus, moreover, tries to show in his book De Aspectibus that the ray passes through in a wholly indivisible instant, and cites in support of his view quite a curious and probable reason, when he says, "If the species, as for example the light of the sun when it rises, is produced in a particle of time in the first section of the air, then if the time is doubled in the second section of the air, and tripled in the third, and so when the light reaches the west, there would result a time made up of many parts which would be large in comparison with the first particle of time; and although the first particle would be imperceptible, yet the whole time owing to its magnitude incomparable, as it were, with respect to the first particle, will be perceptible." Aristotle, moreover, says in the second book on the Soul that although in a small distance the multiplication of light might be able to escape our sense, yet it would not be able to do so in so great a distance as that which separates east and west. Therefore, if the species required time this would be perceptible to the sense. Therefore the species will not be produced in time but instantaneously. Aristotle says in the book De Sensu et Sensato that the explanation in regard to light differs from that in regard to other things of perception, and in regard to these others he teaches that their multiplications take place in time. Therefore the multiplication of light is instantaneous.

All authorities make this statement except Alhazen, who attempts to prove this view false in the second book, arguing as follows: "Let us take the last instant at which the light is at the terminus *a quo* and the first at which it is at the terminus *ad quem.*" Since therefore the instants are different, as he strives to prove by experiment, there will be a time between them; and he says that every change takes place in time, but the medium and the eye are changed by species. But these reasons of Alhazen do not have any weight, because the first is explained elsewhere. For it is not necessary to give always a last instant of an existing thing at the terminus *a quo,* just as universally happens in the generation of permanent things, but it is necessary to give the first instant of the terminus *ad quem,* as Aristotle teaches in the eighth book of the Physics. Whence when Socrates becomes white from non-white, it cannot be said now at last is he not white, taking *now* as an instant, but now first is he white; for he is not-white during the whole time measuring the change, and he becomes white at the end of this time, namely, at the instant which is its terminus, as Aristotle teaches, and as is certain, although it is too difficult to understand unless well explained; but this is required elsewhere. His second reason has no weight; for all holding the opposite view say that the multiplication of light is not a successive and temporal change.

A sound argument, however, for the view of Alhazen can be drawn from the statements which he makes in the seventh book. For he there teaches that from the same terminus the perpendicular ray reaches more quickly the terminus of the space than the ray that is not perpendicular. But quicker and slower belong only to time, as Aristotle says in the fourth and in the sixth book of the Physics. And this is demonstrated without possible contradiction. For no finite force acts instantaneously, as Aristotle says in the sixth book of the Physics; and he proves this, because in that case a greater force would act in less than instantaneous time, which is impossible. But the force of the eye, and of its species, and of everything created is finite. Therefore no force can act instantaneously. Moreover, in the eighth book of Physics at the end he maintains that a finite force and an infinite one cannot act in the same and equal period of time, since in that case they could have equal results, and thus in turn the forces themselves would be equal. But it is the property of an infinite force to act instantaneously. Therefore a finite force cannot produce any result in an instant, wherefore it must require time. Moreover, an instant has the same relation to time as a point to a line. Therefore, interchanging terms, an instant has the same relation to a point as time has to a line; but the passage through a point is in an instant. Therefore the passage through the whole line is in time. Therefore species passing through linear space, however small, will pass through in time. Moreover, before and after in space are the cause of before and after in making translation over space and in duration of time, as Aristotle says in the fourth book of the Physics. Therefore since space through which species is carried has a before and after, passage of the ray must have a before and after both in itself and in duration; but before and after in duration exist only in time, since this condition cannot exist in an instant. If, moreover, it be said that this is true of those things that have a corporeal existence in a medium, not concerning those things that have a spiritual existence, as it is assumed here, evidently the objection has no weight, owing to what has already been said. Again if it be said that this is true in regard to those things that are measured by parts of space, but that species is not so measured, as they assume, this objection also has no weight, because that second statement is not made except on account of a spiritual existence. Since, therefore, the species of a corporeal thing has a really corporeal existence in a medium, and is a real corporeal thing, as was previously shown, it must of necessity be dimensional, and therefore fitted to the dimensions of the medium.

But if at the same instant it were throughout the whole medium, it would then be at the terminus *a quo,* and at the middle point of the space, and at the terminus *ad quem,* namely, at one and the same time. But this is in many ways impossible. For, in the first place, it follows from this that a created thing would be at one and the same time in several places, and by this reasoning if in several places, then also in an infinite number, as has already been shown in the chapter on matter. It would therefore have an infinite power and would be God, or equal to God. In the second place, the argument is drawn from this that while the thing is at the terminus *a quo* it is wholly quiescent, nor does it suffer change in any way; and when it is at the terminus *ad quem* the change has already been effected, and the change takes place between these termini. Therefore at the same instant the species would be quiescent before its passage, and the change would be effected, and would take place actually through the whole space. Therefore at the same time it would be changed, and would not be changed, which are contradictory, as Aristotle argues reducing the matter to an impossibility in another case in the sixth book of the Physics.

But there is still another reason for this, namely, since the multiplication of light does not depend on some other motion, we may assume, therefore, that the heavens are at rest and that there is no motion, for in a stationary heavens the multiplication of light can take place excellently, and will be accomplished at the end of the world, if the heavens shall be stationary, as is assumed. If, therefore, the multiplication of light is instantaneous, and not in time, there will be an instant without time; because time does not exist without motion. But it is impossible that there should be an instant without time, just as there cannot be a point without a line. It remains, then, that light is multiplied in time, and likewise all species of a visible thing and of vision. But nevertheless the multiplication does not occupy a sensible time and one perceptible by vision, but an imperceptible one, since any one has experience that he himself does not perceive the time in which light travels from east to west.

Chapter IV

Concerning the removal of objections to the truth.

Moreover, to this statement made by Jacobus Alkindi we must reply that, just as the first period of the time is imperceptible, so also is the double of it, and the triple, and the thousandth multiple: whence the whole time is imperceptible, although it has many parts which taken together make an imperceptible whole, for this motion of the species is of such great velocity that it can traverse in an imperceptible time a very great distance. Aristotle's statement is true according to his understanding of it, for he is arguing in that place against Empedocles, who maintained that light is a body and the flow of a body, just as water flows from a spring; and it is not possible that a body should change its position wholly from east to west, so that it would not be perceived owing to the great distance. But the species is not a body, nor is it changed as regards itself as a whole from one place to another, but that which is produced in the first part of the air is not separated from that part, since form cannot be separated from the matter in which it is, unless it be soul, but the species forms a likeness to itself in the second position of the air, and so on. Therefore it is not a motion as regards place, but is a propagation multiplied through the different parts of the medium; nor is it a body which is there generated, but a corporeal form, without, however, dimensions *per se,* but it is produced subject to the dimensions of the air; and it is not produced by a flow from a luminous body, but by a renewing from the potency of the matter of the air, as we stated above when the question was discussed in regard to the generation of species. Moreover, if we inquire still more carefully why we do not perceive this generation of light take place successively in the particles of the air, the answer can be given that light in the air is not an object, but a species with a weak and in a manner imperceptible existence as regards itself, and its subject between the east and the west is imperceptible, namely, the air itself, and for this reason the sense is not able to perceive a successive generation of this kind.

Moreover, as regards Aristotle's statement that there is a difference between the transmission of light and that of the other sensory impressions, we must say that many are deceived in this particular; for his statement is true, but this difference is not to be understood as consisting in the fact that light is transmitted instantaneously and the other impressions require time; nay, we must understand that, although light has a succession in its transit, it does not, however, have so great a one as sound and odor, of which he is speaking in this place. For sound has the motion of the displacement of the parts of the body struck from its natural position, and the motion of the following tremor, and the motion of rarefaction in every direction, as was stated before, and as is evident from the second book on the Soul; and these are the three local motions of the particles of the air, as well as of the body struck, no one of which motions takes place because of the multiplication of light. For although the air must be rarefied for light to produce heat, yet rarefaction is not necessary because of the multiplication of light itself, since in celestial spaces light is multiplied, where rarefaction and the generation of heat are not possible. Since, therefore, there is no succession on the part of light except the succession itself in multiplication, but in the multiplication of sound a threefold temporal succession takes place, no one of which is present in the multiplication of light, for this reason there is a great difference between light and sound. However, the multiplication of both as regards itself is successive and requires time.

Likewise, in the case of odor the transmission is quite different from that of light, and yet the species of both will require time for transmission, for in odor there is a minute evaporation of vapor, which is, in fact, a body diffused in the air to the sense besides the species, which is similarly produced. There is also a strong attraction of the nostrils for this kind of vapor and of species, so that sensation may be produced, as Avicenna shows in the third book on the Soul and as we know by experience. This results in the removal of the covering that is over the organ of smell, according to Aristotle in the second book on the Soul. Therefore in the sense of smell there is a twofold local motion, one from the resolution of the vapor, and the other from its attraction, besides a succession in the multiplication of the species. But in vision nothing is found except a succession of the multiplication. The fact that there is a difference in the transmission of light, sound, and odor can be set forth in another way, for light travels far more quickly in the air than the other two. We note in the case of one at a distance striking with a hammer or a staff that we see the stroke delivered before we hear the sound produced. For we perceive with our vision a second stroke, before the sound of the first stroke reaches the hearing. The same is true of a flash of lightning, which we see before we hear the sound of the thunder, although the sound is produced in the cloud before the flash, because the flash is produced in the cloud from the bursting of the cloud by the kindled vapor. Therefore his statement that there is a difference in the transmission of light and of the other sensory impressions can be understood, because this difference is not one of instantaneousness and time, but of less time and more time. For all authors, whether sacred or others, who state that light is multiplied instantaneously, are to be understood in regard to a divisible instant, which is imperceptible time, and not in regard to a real instant, which is the indivisible terminus of time, just as the point is of the line. If, therefore, the vision of the species and the visible multiplication take place in imperceptible time, how was it said above that vision will take place in a perceptible time? It is evident that besides the multiplication there is a judgment of vision in regard to the object seen, and this judgment must be known to the sense; wherefore the judgment must be made in a perceptible time.

The eighth condition required for vision is a healthy state of the eye together with its natural arrangement; for an eye that is torn out, or blind, or much injured, or clouded from some flowing humor, or from a resolution of vapors confusing the pupil cannot judge regarding objects, as is evident, and therefore need not be considered here. For if anything need be said further on this point, it will be explained in what follows. Concerning position, which is the final condition in vision, an explanation cannot be given here, because it coincides with other matters of which mention must be made later.

TENTH DISTINCTION

In three chapters. The first shows with greater precision than above what things are perceived per se *and what* per accidens.

CHAPTER I

HAVING stated these eight conditions without which vision cannot function, we must consider of what things vision takes cognizance, and in what ways they are perceived and certified to, and how and why vision errs in taking cognizance of visible objects, although vision has been produced by direct rays. We must note, then, that when these nine conditions are present in just degree, that is, they neither exceed nor fall below what is required, a reliable vision is assured. When, therefore, they differ from the due measure either by excess or deficiency, error is present in vision. Moreover, the twenty-two qualities enumerated above, as light, color, remoteness, etc., are determined by vision; and besides this, vision perceives man and horse and the other things of this world. For by light and color these twenty things are recognized. Then through the medium of light and color and those twenty qualities other things are known, as is possible for the sense, for the particular senses are not able to certify regarding all things; but those twenty-two the particular sense and the common sense and imagination are able to comprehend without error, provided the eight conditions previously mentioned are present in proper measure. Those qualities, moreover, are called the sensibles *per se,* certainty in regard to which can be secured from the senses; but the sensibles *per accidens* are those that cannot be apprehended in this way.

Although we previously touched upon these sensibles *per se* and *per accidens,* yet in order that no mistake may be made they must be explained more fully. I say, then, that the sensibles *per accidens* are twofold: certain can be determined by other faculties of the sensitive soul, as, for example, by the estimative and memorative faculties, as previously stated; certain, however, are called sensibles *per accidens* with respect to particular senses and the common sense and imagination, since such senses do not perceive things of this kind independently, but because they are found in the same things together with their own sensibles *per se*. As, for example, since enmity with respect to a lamb is present at the same time with the form and color of a wolf, the lamb in looking at a wolf saw something hostile and colored, but the eye *per se* forms no judgment concerning the hostile object, but judges it such, because it is found in connection with the colored object. Moreover, since a particular sense and the common sense are commonly called senses, for this reason the sensibles in regard to which they certify are called sensibles *per se,* and those concerning which they do not certify are called sensibles *per accidens,* although some of them can be recognized by other inner faculties of the soul. For the estimative, cogitative, and memorative faculties are not called senses in the ordinary meaning of the term, although they are parts of the sensitive soul, and therefore the sensibles determined by them are called sensibles *per accidens,* because the sensibles are referred to the particular senses and the common sense.

But there are other sensibles besides those that are recognized by the faculties of the sensitive soul. As, for example, when I see a strange man I cannot perceive by the sense whose son he is, nor at what hour or in what place he was born, or what name he bears, Peter or Robert; and there are countless things of this kind that are accidental to each person, in regard to which no faculty of the sensitive soul can inform us, nor can one know the truth except through information given. And yet on looking at that man vision falls upon all his qualities. For if he be the son of Robert, and a Frenchman, and born in Paris during the first hour of the night, in looking at him the eye sees Peter, a Parisian, born in the first hour of the night, son of Robert, because these things are coincident with his color, and form, and other visible characteristics. In like manner the natural substances of things, both in things animate and inanimate, are not perceptible by some faculty of the sensitive soul except *per accidens.* Those things are excepted that are harmful or useful, which the estimative faculty grasps, and sense, however, falls upon this *per accidens.* Hence when I see a man I see substance and an animated object, and therefore vision falls in a measure on his natural substance, and on his soul also, which is a spiritual thing; but this is surely *per accidens.* Moreover, the sensi-

LOUIS IX

From the Seurre Statue at Versailles

bles belonging to one sense are sensibles *per accidens* of other senses, whence hot and cold, moist and dry, odor, sound, savor, are sensibles *per accidens* with respect to vision, and thus the sensibles belonging to each sense are sensibles *per accidens* with respect to other senses. Sensibles, then, *per se,* as has been said, are twofold: certain are proper, namely, nine, and certain are common, namely, twenty, because they can be perceived in common by several senses, especially by sight and touch. For as Ptolemy says in the second book all things that vision perceives touch discerns, except light and color; and all things of which touch certifies can be certified to by vision, except four proper sensibles, namely, hot, cold, moist, and dry.

Chapter II

Concerning those things that produce species in vision.

Moreover, that we may know how sensibles of this kind are recognized *per se,* we must know in the first place whether all things propagate their species to the sense. The determination of this is difficult, but Ptolemy in the second book of the Optics settles this question with the statement that only light and color propagate their species to the eye. This view Alhazen maintains in the fourth book; whence other things are not active upon the sense or upon the medium. The reason why they are not active is that they are all magnitudes or properties of magnitudes, as is evident, and magnitude is not active, because it is a quality of matter which is not active but passive, as Aristotle states in the first book on Generation, and Avicenna agrees with him in the second book on the Soul. For the medium or instrument that takes up the sound is soundless, and the instrument that takes up color is colorless, as he states. Therefore his opinion is that the medium and the senses do not have to have the natures of the sensibles whose species they must take up, in order that they may judge concerning the sensibles perceived by means of these species. Whence the humor glacialis does not possess some nature of light or color to the degree possessed by objects outside it. For although the eye has light, this is in respect to the perception of color, not in respect to light, because the object acted upon does not possess in actuality but in potentiality the ability to become like the agent. Nor does the anterior glacialis possess some degree of color, by which it may be made really like colored objects outside itself, of which it has to form judgments, although the eye has in its humors and coats some weak substance of color, by means of which imaginary colors sometimes appear, as will be explained when we consider the iris. But the eye does possess form and magnitude and corporeality, and other common sensibles that are appropriate to it, and therefore it is not fitted naturally to receive the species of these things, nor are they themselves active. And although Aristotle says in the second book on the Soul that the final perfection of every sense consists in its organ being the medium of sensibles, yet that degree of mediumship does not exist in things pertaining to the senses; since if it did exist and its species were produced on the sense, the sense would not judge concerning that mediumship, and therefore vision would not be able to receive the species of things so as to judge by means of these species concerning things whose natures are similar in vision. When, therefore, the objection is made that these are sensibles *per se,* that they therefore act on the sense, just like proper ones; we must reply that they are not called sensibles *per se* because of their action on the sense, but because the sense is able to certify concerning them. And if the objection should be made that vision will not certify concerning them, since they do not propagate their species to vision, we must reply that such is the case; for its own species is not required in all things, but the species of vision is sufficient along with the species of light and color and with certain other considerations. When these have all been combined certainty can be secured, as will be explained later. If it be said that authors on perspective state frequently that shape and quality and the like of the thing seen are arranged on the surface of the sentient member, and that they cannot reach there except through their species, we must reply that the statement is not their own, or has not been correctly translated; for they do not mean to say more than that from the whole magnitude and form the species of light and color come which are arranged on the surface of the sentient member, and this suffices.

When it was stated above that we see the air or the heavens at a distance, and that it is certain that in the water of a river or in some other body of water that has great width the heavens are seen by reflection, just like an object in a mirror; and an object in a mirror is seen by this means, because its species is reflected from the mirror to the vision; some one might then say that the celestial transparency of air at a distance produces a species by means of which it may be seen in water. But we must state that no species of the thing seen exists there, but the species of vision, which in the air without water is multiplied along a straight line to the celestial transparency at a distance, and apprehends that transparency with the aid of the species of light illuminating that transparency; and when it is seen by means of water reflexively, the species of vision is reflected from the water to the air seen at a distance, and the species of the thing is not reflected. Hence not everything that is seen by reflection is seen by means of its own species. But this happens as in the case of several familiar visible objects, but this must not be understood generally, because I can see water by means of a mirror, but the species of water is not propagated there, but the species of vision is, and this is the case here.

Chapter III

Distinguishing three universal modes of perception through vision.

In the next place, we should know that besides the particular modes of perceiving the sensibles *per se,* there are three universal modes, according to the authors of Perspective. But in particular Alhazen explains these modes and he does so adequately, except that sometimes an inept translation of words occurs. There is, then, first an impression by the sense alone without any faculty of the soul, and it is thus that light and color in general are perceived. For vision is able to decide that there is color or light without error when it beholds an object, provided the eight aforesaid qualities are in their proper proportion; and therefore these two are perceived by the intuition alone of the sense. But the species and modes of color and of light cannot be so easily recognized; for if a strange color appears which we have not seen before, we will not know what species of color it is. Likewise if we see some bright object in the air, which has a light different from other objects with which we are familiar, as, for example, a star with a train or some other object, and we have not seen it before, we cannot decide by the vision what that light is. Likewise if we have seen some color before, and later have forgotten it, then when it again appears to the sight we can judge it a color, but we will not perceive what color it is. Likewise when in infancy we saw the full moon we did not perceive whether it was the light of the sun or of the moon until we became accustomed to it and the fact became fixed in our minds that such a light belongs to the moon, not to the sun. This is also true of the stars, because many men at some time or other see Jupiter, or Venus, or Mercury, and because of the beauty of these stars they look at them with pleasure, and they are then told by astronomers that the light of one particular star is of this kind, and the light of still another is of that kind, and they are made sensible of the difference. But after time has passed, when on another occasion they see one of those stars, they do not distinguish whether the light is that of Mercury or of Venus or of some other star, because they have forgotten the impression of the particular light of each of those stars seen formerly, but they recognize the lights of these stars by means of a second perception which is by similitude, according to Alhazen. This mode of cognition concerns not only color and light, but all things in which we distinguish the universal from the particular, and particulars from one another. For example, when I see a man whom I have seen before, if I have the impression that I have seen him before, I recognize then not only a man in general, but that particular individual whom I distinguish from others by means of this kind of cognition. If, however, I have forgotten, I see a man, but I do not know who he is.

We must bear in mind, however, that when the statement was made that vision perceives color or light as a universal by the sense alone, and not as a particular, the particular is excluded which is lower in the line of predication, as, for example, the species of color. Again, particulars designated by the species of something are excluded, as the light of the sun, and the light of the moon, because perchance light does not have species but modes, because all lights of stars have their origin in the rising sun. But the indefinite particular is not excluded, for it is as general as its own universal and is convertible with it, as some color, some light, some man, some ox. Therefore the recognition of universals from one another and from particulars, and particulars from one another by a comparison of a thing seen with the same thing seen before, by recollecting that it was seen before and known to the beholder, produces here a second mode of comprehension through vision.

But there is still a third perception, which cannot take place by the sense alone, and does not depend on a comparison with previous vision, but without limitation considers the thing present. For its perception several things are required, and the process is like a kind of reasoning. As, for example, when one holds in his hand a transparent stone, and does not perceive its transparency, but if he exposes it to the air, and if there is some opaque body behind it at the required distance and if the light is sufficient, he will see the light and the opaque body beyond the stone; and then, since he is not able through the medium of the stone to see the object that is behind it, unless the stone be transparent, he reasons that it is clear and transparent. But in ordinary things we employ this perception instantaneously, and we do not perceive that we are reasoning, although we are doing so. For a man reasons naturally without difficulty and labor; as, for example, when a boy is offered two apples, one of which is finer in appearance than the other, he chooses the finer one, but for no other reason than because it looks finer to him, and for this reason should be chosen rather than the other. Whence he follows this reasoning; that which is finer in appearance in so far is better, and what is better is rather to be chosen, therefore the finer in appearance should be chosen; and yet he does not perceive that he is reasoning, because of the swiftness of the reasoning process innate in man, as Alhazen teaches.

And now in metaphysical and logical terms I have established the view of Alhazen, and I have proved that our knowledge of the science of reasoning, which is logic, is derived from nature, but we are ignorant of the proper terms at the beginning, and these through their zeal for discovery the first authors on logic have found, but we learn them by study. Because of these terms the formal subject of logic exists, not because of the potency of the science itself, because this is innate in every one, as Alhazen here maintains and I have proved elsewhere. Thus are comprehended the twenty common sensibles, with their species, for they could not be determined except through this perception, and it is evident that the sense alone is not able to determine in these two modes; and this is called the sense alone when vision is in the pupil and in the common nerve reaching to the common sense. For unless imagination and the memory of the previous sight of an object be present, there will be no comprehension in the second mode. But imagination and memory are beyond the common sense. The third mode is still further removed from the sense pure and simple, because in this mode more things are taken into consideration than in the second mode, and there is more recourse to reason because of the process of ratiocination. But these modes do not have names properly translated. The first Alhazen calls perception by the sense pure and simple. The second he chiefly calls perception by science. The third he calls perception by syllogism, because of the mode of reasoning. But these names are not proper, because the faculties of the sensitive soul have these cognitions, for which neither science nor syllogism is needed as they are generally accepted. But concerning this we shall learn more definitely when we inquire what are the faculties of the soul which form these judgments through the mediumship of vision. This cannot be accomplished before it becomes more apparent, in the case of examples from different kinds of visibles, how they are determined.

Second Part of Perspective

FIRST DISTINCTION

Here begins the second part of this treatise, which is on the particular modes and causes of vision along a straight line in the main. It has three distinctions. The first is on vision as determined by the structure of the eye and is in three chapters. The first chapter is on those things that men see at a distance and on what they see near at hand.

Chapter I

SINCE, moreover, the knowledge of opposites is the same, as Aristotle says in the Topics, and in the first book on the Soul asserts that that which is straight is the test of itself and of that which is not straight, because by the privation of health sickness is perceived, therefore I shall designate particular modes of determining visibles at the same time with the defects of vision and with its errors, so that at once the defect itself and the error may be apparent by the determination of the contrary truth. In the first place, then, we must give an explanation of vision in relation to those things that relate to the structure of the eye. Those whose eyes are deep-set for this reason must see further than those whose eyes are prominent. The first reason for this is the greater strength that a deep-seated eye has owing to its greater proximity to the common nerve in which is located the visual force, as in a fountain. The second reason is because a deep-set eye is more protected from ills and is more remote from injuries than a prominent eye, and therefore is stronger. The third reason is because the visual force is more concentrated and unified, while it is more covered within the concavity of the bone, so that it thus takes a narrower and straighter path to the object seen, and is less scattered and dilated, so that it thus falls into the position of the visual pyramid. For this reason when a man wishes to view something at a distance with care, he applies the hollow of his hand to the bone of the eye, in order that the visual force may be more concentrated and less dispersed. A proof of this is the fact that the eyes of animals, as in the case of fish, which do not have the protection of eyelashes, do not see well at a distance, since the visual force is dispersed and dilated too much, since the eye does not have anything surrounding and confining it. Aristotle takes this view in the nineteenth book on Animals, and it is supported by experience. For a man in a well or in some other deep place will be able to see stars by day which he will not see at the surface of the well, as Pliny states in the second book of the Natural Philosophy, and experience teaches this. One cause of this is owing to the narrowness of the path, because the visual force is compressed for a long time, so that it proceeds more directly to the place of the stars that are above the well, and this is because of the covering that it has in the depth of the well. Thus the eye deep-set in the head in the same way is able to see better for this reason, which will be touched upon below. Ptolemy expressly states this in the second book of the Optics in these words, "Those who have deep-set eyes see further"; the reason for this is the visual force which is produced because of the compression, that is, the concentration and unification, and because of the narrowness of the place. For when it issues from narrow places, the vision is extended and elongated, because the force of the eye of necessity is more concentrated and compressed into a smaller place, and for this reason is stronger; because every united force is of stronger action. Its compression is evident in the figure. If *bc* is the opening of the uvea, and *a* is the pupil of some eye deeper-set, it is evident that its force falls between the lines *e* and *f;* but if it were at *d,* the force would pass through the *gh* lines, which are more divergent. Some with considerable grasp of this science of perspective reject the reason which Ptolemy and Aristotle assign and which experience teaches, alleging certain conclusions which are rightly shattered by so weighty an authority and by the force of experience. I make this statement, because it is not necessary in this place to dissolve all doubts that can be urged against the truth, for an infinite number can be adduced against any truth whatever.

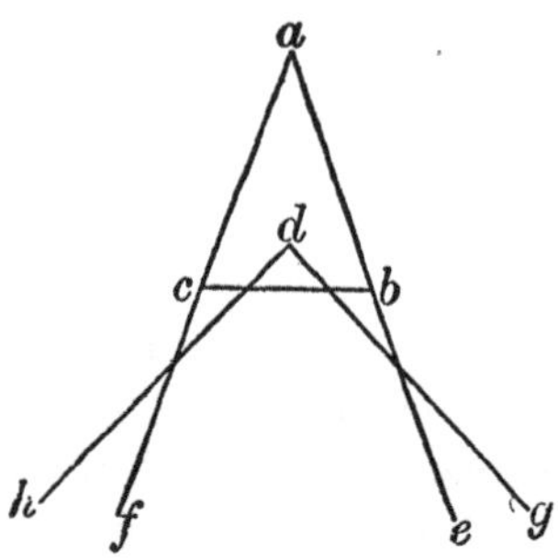

FIG. 37.

But besides this explanation from concavity, another can be given regarding the humors of the eye and depends on the small amount of the albugineous humor; for when this is plentiful it has the same effect on the glacialis as upon one who looks into a large, deep body of water, the large amount of which so shortens vision that it cannot see farther. Avicenna is of this opinion in the nineteenth book on Animals; and Averroës in his book on Sense and the Sensible cites as a reason for far vision the protection of the eyelashes, saying that the eyelashes preserve and guard the eyes from heat and cold and from many impediments, and that for this reason there is greater strength in eyes possessing eyelashes, which have far vision. He cites the case of many animals that see farther than man, because they have thicker eyelashes.

We must next consider why many old people, when they wish to see things distinctly, as in reading letters, see better at a distance than when the object is near. For they hold objects that they wish to examine further from them. The reason for this Ptolemy explains in the second book of the Perspective, for their eyes abound in moisture; for old people have much superfluous accidental moisture. When therefore the moisture is scanty the vision passing through it is immediately set free from it, and is thus able quickly to see distinctly an object appearing close at hand; but when there is an abundance of accidental moisture the eye is confused, and is not so quickly freed on account of the quantity of the moisture. For this reason an object must be further removed from the eyes before it is seen distinctly; for this moisture, which flows in the eye and on the surface of the cornea and the eyelashes, not only occupies the body of the eye, but renders moist the air near the eye; and therefore from a point close at hand such eyes cannot see, but the vision must be freed from that moisture and must pass through a longer distance. The first explanation touched upon the cause of vision at a distance through the natural moisture of the eye; here we are talking about the accidental superfluous moisture. But we are speaking here of vision at a distance in a different sense from that in the former case; for what we here term near and far is contained there under one of those terms, namely, under the term near, because remoteness is there taken as great distance, and here we are speaking of distance and a discriminated perception which always takes place near at hand. But there can be greater proximity or remoteness, which are degrees in proximity opposed to the remoteness of which we were speaking in what precedes.

Chapter II

In this chapter the causes are given for clear or distinct vision in darkness and in light.

SINCE mention has been made of the discrimination of vision, because it is one thing to see a long distance, another thing to discern, we must consider what are required for discrimination and distinctness of vision as regards the structure of the eye. Without doubt the humor glacialis must be of good size, so that on it the parts of the object can be clearly and sensibly distinguished. For if it is small, the species of the parts of the object seen will occupy positions too close and greater confusion will be caused. Moreover, it must be pure, and likewise the humor albugineus; for impressions and spots are quickly detected in a clean thing, as is evident in the case of spots on a clean rag, and therefore the species of things impressed on the eye will appear better and more clearly if the humors are pure and cleansed. But the humor albugineus must be of proper size; for otherwise it will be like deep water with respect to the humor glacialis, as has been said, and the vision will be obscured. Moreover, the weblike structure and the cornea are thin and fine and quite pervious, in order that they may not cause a shadow on the glacialis. Moreover, they must not be wrinkled, but they must have a smooth and expanded surface, since if they have wrinkles they cause a shadow; and this is the reason why old people are deprived of a distinct and discriminated vision, since the weblike structure and the cornea have wrinkles in aged people, just like the skin of their whole body.

Not only must we consider keenness of vision in seeing for a greater distance, and discrimination in vision through the nature of the structure of the eyes; but we must take up this question why some eyes see in the dark and in the twilight and in little light better than in the full light, and other eyes the reverse. Without doubt those that have a great deal of the humor albugineus need much light before it becomes clear; particularly if it is thick, and especially impure; and therefore such eyes do not see well at night with insufficient light. There is the same result if the humor glacialis is large and thick and impure, and if the weblike structure and the cornea are thick and wrinkled: for the action of light must be full and vigorous in order that these may be rendered clear, and in the day and in full light such eyes see well, and not otherwise. But those that have a scanty but clear and pure moisture in the humors, and have the web and the cornea quite transparent and without wrinkle, are not able to see in the strong light; because a strong light confuses such an eye from its excessive clarity, and to such a degree takes possession of the vision with the glare of light, that as regards external visible things the eye is unable to make distinction in them. As, for example, a man that has stood in a strong light, on turning to dark places is unable to discern objects in these places owing to the strong species of light that acts to excess in his eyes; for a more vigorous excitation of the mind always obscures weaker ones. Therefore men with eyes of this kind see well in the twilight and by the light of a candle at night and by the light of the stars. For all that has been said in regard to distinctness of vision and vision in the dark and in the light we have the authority of Aristotle and of Avicenna in their books noted above.

Concerning vision, however, in the dark and in the light another reason can be given, namely, when the eye has a great deal of light of its own; for the eye in man, in the horse, in the cat, and in many other animals shines of itself, as is apparent when a man in the dark moves the eye from its position by the finger. But some people have more light, some less; and those having much are able to see in dim light and in the dark, if the other causes affecting the humors and the web and the cornea are present which I have mentioned. But those having a small amount of light do not possess this power; especially if the other causes affecting the humors and the web and the cornea are not present.

Chapter III

Concerning various errors of vision owing to the structure and complexion of the eye.

The fact that the object seen appears single results from excellence in the structure of the eye; when, namely, the pupils have the same position with respect to the common nerve; since in that case the species come from the two eyes to the same place in the common nerve, and a single judgment is formed. When, however, nature is at fault, as happens in the case of one who squints, since the glacialis of one eye does not have a position similar to the glacialis in the other, the species naturally come to different places in the common nerve, and for this reason the man who squints must spend much effort in reducing the eyes to a similar position, so as to avoid error and grasp the unity of the object. Moreover, although the eye is correct in structure, nevertheless manifold difficulties arise from excessive heat or cold, as men frequently stand in hot places, for example, smiths, bakers, cooks, and the like; for the eye is resolved and destroyed by intense heat and loses its natural arrangement, and the judgment of sight is hindered. Likewise, as Averroës says in his book on Sense and the Sensible, hindrance to vision is caused when the organ is cooled by things in which intense cooling is inherent. For the eye is weakened and obscured in places where there is much snow or much water, and for this reason the shores of the sea appear confused and lacking in light, and so likewise the regions of snow.

So likewise the eye loses its natural formation in those suffering from strong drink, infirmity, and anger. For this reason an object appears double to the intoxicated and the infirm, and the thing seen seems to vacillate and be moved, because excessive moisture of the vapors ascending to the eyes of those enervated by the power of wine or of the warm humor dominant in the condition of disease, disturbs the eyes and causes them to be displaced from their natural position, as in the case of an angered man. For the heat ascends around the heart, and inflames the blood, and resolves the vapors, which easily rise to the eyes and penetrate owing to their channels, and displace the eyes from their natural position, so that the species cannot be formed in one place on the common nerve, nor in one place on the glacialis of the same eye, according to Averroës in the passage mentioned. For because of the movement and ebullition of the vapors the species is formed in different parts, like the species of the sun and of the moon in agitated water, and while it is being formed in one part it is not yet obliterated from another, and for this reason two species appear, just as the species of the sun and of the moon in agitated water appears double to us.

Likewise from suffering in the head a resolution of vapors to the eye sometimes happens, by which the vision is deranged, as in dimness of vision and in vertigo, and then the object seen is thought to be displaced, as Ptolemy explains in the second book of his Optics. This aids in understanding the cause of scintillation, and in this way one thing appears two and the object seen appears to move. Avicenna says in the third book on the Soul that, when some one of the causes is present that have been described in the books on Physics, which deranges circularly the spirit that is in the anterior ventricle of the brain; and when the visual force has given to it the impression received, the species does not remain fixed in one place, but is displaced circularly in accordance with the motion of the resolved spirit, and in this way the object will seem to be moved circularly. Moreover, Avicenna in the third book on the Soul gives three causes affecting the structure of the common nerve and of the eye that result in an object appearing double. For the vitreous humor, which extends from the eye to the common nerve, and the visual force, which flows from the common sense to the common nerve and to the eye, are thin bodies and very mobile, and for this reason easily changed from their position, and this displacement can take place either to the right and left, or before and behind. But if they waver to the right and left, in that case the two species coming from the two eyes cannot be formed in the same place on the common nerve, but one falls to the right, and the other to the left, and thus two appear, and of necessity one thing appears two. Likewise if they waver forward and backward, one species will be formed more in the front and the other more behind, and therefore two separate species will appear. When both motions occur, there will be a movement of the species, like a vertigo, and the object will appear to waver in accordance with the two positions.

The third cause he assigns to the uvea, which easily receives motion. For by the movement of the nervous spirits and by much intense heat it can be disturbed within, and by the effort of the eye is dilated, and by its own nerve of which it is composed it can be dilated or contracted as that nerve is contracted or distended. It can also be disturbed externally from many causes, as, for example, when the finger compresses the uvea or the same thing happens in other ways, and by these disturbances the opening is sometimes changed in a straight line, sometimes obliquely, and this variation can take place in one eye and will not happen in the other, or will take place in both eyes at the same time, but in different ways. Therefore the species itself will be diversified in both eyes and consequently in the common nerve, so that the object and the species appears larger or smaller, and thus one species will have a larger place in the common nerve than the other, and therefore an object that is single will appear double.

Besides all that has been said regarding the fact that one object appears as two, there are still several natural cases with their causes. It sometimes happens, then, that extraneous humor coagulates below the uvea between the glacialis and the opening; and sometimes this happens from above downward as regards length, sometimes crosswise as regards width, sometimes circularly; and in this way a division of the glacialis is made, so that it views the same object through its two different parts, and therefore one object appears to it as two. Physicians touch upon this cause in their books and teach the cure, as is shown by Avicenna in the care of the eye. But if this cause did not exist it is thought that deception in this regard can take place. It is believed, in fact, that the appearance of two objects can be produced otherwise from the fact that vision is produced by external transmission, and because species can be seen in some manner. For there are many men to whom a single object when seen appears double, and this is not owing to a difference in their eyes, since it occurs if they look at the same thing with one eye. Therefore with one eye they see two objects when all the causes already mentioned have been excluded. Wherefore this cannot take place except for the reason that the species of the object, while it is near the object, is stronger and has more of the nature of the object, and therefore is more sensible than when multiplied at a greater distance. When, therefore, a man has weak eyes, the faculty of vision not only is weak as regards itself, but more weakened from distance will stop at the species of the object near the object, and the species will take the place of the object, so that the weakened faculty is terminated at the species, just as the vision weakened from distance is terminated at vapor. But a strong eye penetrates the air in which the species is near the object, and the species is hidden because of the strength of vision, just as vapor in the air near the eye is hidden and penetrated because of the vigor of the eye. This will frequently happen to people with weak eyes, particularly when they look carelessly. Sometimes also people with good eyes in some way or other have this experience, namely, when they are very careless in looking and half close the eye, and both they themselves and the object seen are in dark places and at times of mists, as at twilight.

This must be the reason, or it is because one eye has two pupils. Although this is possible, I have not met with such a case. For Solinus in his book on the Wonders of the World makes mention regarding a certain region in which women usually had two pupils. Pliny, also, in the sixth book shows that throughout different parts of the world there are people with double pupils in their eyes. When this happens one object can appear double.

SECOND DISTINCTION

Concerning the method of direct vision as regards the species of sight and of the object seen; in four chapters.

Chapter I

THE causes of vision as regards the species of vision and of the objects visible must be considered next. It is now easy to give the reason why the eye sees from the outside air charged with vapor and cloud, and when the eye is in such air does not see or perceive the vapors and clouds. For this point has been touched upon incidentally above, but must now be considered in relation to its principal cause. For the strength and weakness of the visual ray must be the cause of this; for when it is near such air and in it the visual force is strong from its proximity to the eye, because of its strength it can penetrate the denser air, and nothing is seen or perceived by the vision, except that which can terminate the vision, as has been stated before. Therefore the air charged with vapor is not perceived in this case. But when the eye is at a distance from air so charged, the species of the eye weakened finds resistance and is terminated, and therefore is able to see a cloud at a distance. From this experiment it is evident that vision is the result of external transmission. For if the species of the object seen were alone responsible, no explanation of this vision could be given.

We must consider species not only as regards itself, but as regards position, without which the species cannot be formed or formed distinctly, unless position be carefully noted. We must state, then, that one eye produces its species to the object seen, and the object seen produces its species to the eye in the same place, whence they have a common axis, which is perpendicular to all parts of the eye, because it passes through the center of all the parts. Therefore since the perpendicular approach is strong and potent, the eye judges most vigorously by means of the species of the part that comes along this line, and grasps this part with final certitude, as far as one eye can grasp it. Therefore it is said in the book on Vision that no visible object is seen at the same moment in its entirety, and this must be understood of final certitude, but one part only on which the axis mentioned falls. Nevertheless with the same look the whole object is seen which is the base of the visual pyramid, and more as has been stated, and therefore several things are seen at the same time. But except for complete certitude both the parts on both sides nearer the end of the axis are seen more clearly, and those more remote are seen less clearly. When, however, the eye is moved, so that it may make sure of the separate parts seen, the axis passes over these parts, and certifies them successively one after another.

Since naturally the two eyes have a similar position with respect to the common nerve, the axes of the eyes will have a uniform position for every point on which they fall, and they will fall necessarily on the same point of the object, and then that point will be seen better and with more certainty by the two eyes than by one. We are to understand, moreover, that from the common nerve an imaginary straight line is directed between the two eyes to the object seen, meeting the axes of the eyes in the same part of the object seen, and this line is the common axis, and that point on which these three axes fall is seen with final certitude, as is shown in the figure, and other parts of the object are seen with more or less distinctness according to the different positions they occupy with respect to this axis. For point *a* is seen most distinctly, because the three axes are concurrent at it, and *b* and *c* are seen more distinctly than *e* and *d*.

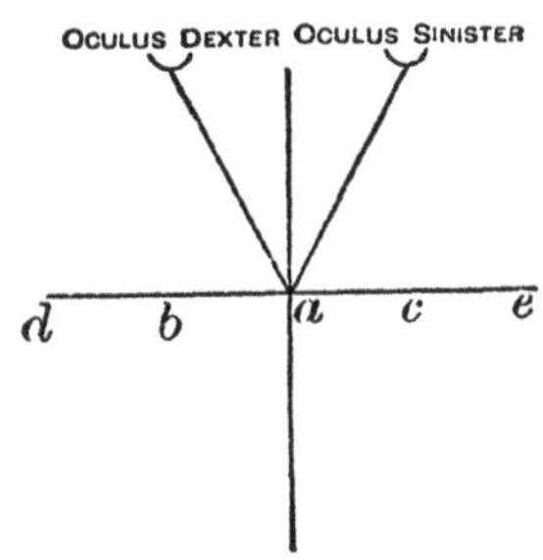

FIG. 38.

But that part is most distinctly seen to which, when the axes terminate, its own species is perpendicular, for then the strength of the impression is double. For this reason *a* is seen more distinctly by means of the axes than *c* and *b*, for the species of vision of *a* itself is perpendicular to it, as is clear. The species on *b* and *c* is not a perpendicular species in this way, for *b* and *c* are perpendicular to the eye but not perpendicular at the points *b* and *c*. But the species of *a* itself is perpendicular to itself and to the eye.

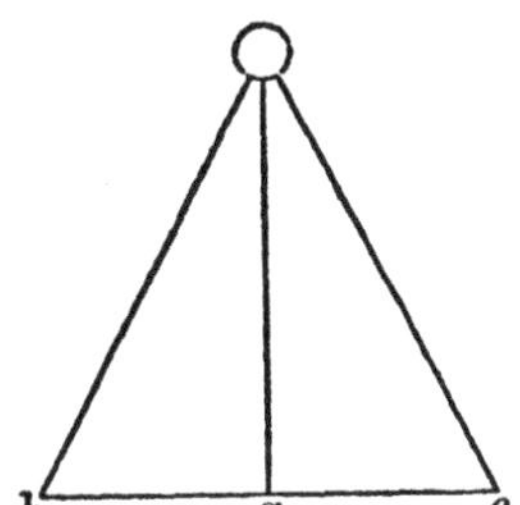

FIG. 39.

Chapter II

In which is shown by two different experiments and by different figures how one object is seen as two.

But not only for the reasons mentioned in the preceding distinction is one object seen as two, but for many other reasons. This experience happens more frequently and clearly when one or both eyes are moved from their position, as, for example, when a man without definite purpose revolves one of his eyes from its natural position, or when he places his finger on the eye and displaces it. For then the two species come to different places on the common nerve and one object appears to be two. It happens also that an object appears double at a great distance from the axes because of the sensible difference of angle, which the species forms in the eyes, as is shown in the figure. For let *mc* be the object seen, and *a* and *b* the eyes, and *ac* and *bc* the axes fixed on the point *c*, which is seen by them with final certitude; then some part can be so far removed from the axes, as the point *m*, that it will be seen double because of the sensible difference in the angle that it forms in the eyes, since *mac* is far larger in the eye *a* than *mbc* in the eye *b;* and therefore both species, because of this difference, cannot come to the same part of the common nerve, but the larger angle must occupy a larger part of the common nerve, and for this reason *m* appears double to the sight. Not only does this happen when the point seen falls to the right and to the left of the axes, as represented in the figure, but when it is at some distance to the right with respect to one of the axes and to the left with respect to the other, namely, when it falls between the intersection of the axes or beyond it, as is shown in the figure. For if the axes of the eyes *a* and *b* are fixed attentively on the part *o* of the visible object *mon,* then the visible object *k* below the intersection of the axes will appear double, and *h* visible beyond the intersection will likewise appear double necessarily. For an experimenter can prove this by taking a stick the width of a

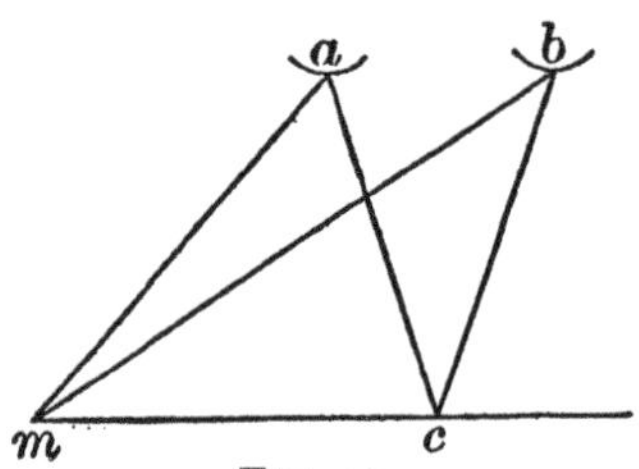

Fig. 40.

palm, and the length of four, five, or six, with a smooth surface. Let him also take three different visible objects, formed out of either wax or wood as large as the last joint of the little finger in the shape of a pyramid, and let them be of different colors for greater distinctness, and be arranged in the order *hok,* so that there be a sensible distance between them. The middle one of these should be in the middle of the board and one of the others at the remoter end of the board, and the third one midway between the middle object to be viewed and the eye. The board at the end toward the eye should be somewhat concave, in order that it may be applied easily to the eyes, and that the board may be applied easily at the end of the nose toward the eyes. The eyes must then fix their axes on the middle object, which will be seen as one, and both of the other two will appear double, for *k* and *h* will appear as four and *o* will be seen as one.

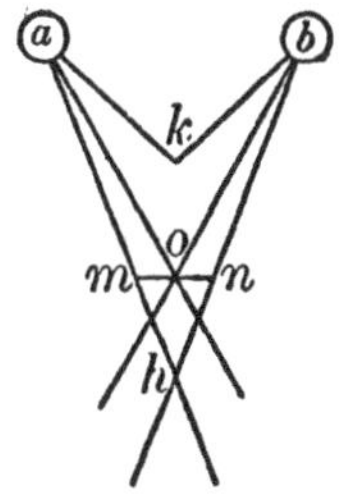

FIG. 41.

CHAPTER III

In which a figure is arranged for these two errors by means of one experiment together with a consideration of several disappearing images.

BY means of the same board, if it has sufficient width, an experiment can be made in regard to the first of these errors. Let, then, one figure suffice for both errors and let it be similar to the figure here given. For *a, c,* and *b* will be seen as they are, provided the eyes are focused rightly, so that the extremities of the visual pyramids may fall on *c* and *b,* and the axes on *a.* But if other objects, or those same objects, namely, *a, b, c,* are arranged on the longer diameter *hd,* so that *c* is placed in the position *m,* and *b* in the position *n, m* and *n* will be seen as four; and just as *c* will be seen as one, when it is on the shorter and transverse diameter, so also will *l,* which is not too far distant from it and from the axes; but *f,* which is quite a distance out of line, will be seen as two, as was noted in the first error. For it will be seen by the eye *p* under the angle *fpk,* which is less than a right angle, and it is seen by the eye *k* under the angle *fkp,* which is a right angle, and therefore it must appear as two owing to the difference in angle. An experimenter can make many experiments in this matter without the board. For he can at night raise a finger between himself and a candle. If then he fixes the axes on the candle, the one finger will appear as two. In these cases we must bear in mind that if the right eye be closed, the left image will disappear, and when the left eye is closed the right image will vanish. The blessed Augustine expressed wonder at this, and writes in the eleventh book on the Trinity, second chapter, that it would be a lengthy matter to give the reason for this phenomenon. It is, in fact, a lengthy matter for one ignorant of perspective, since he must first learn this science. Likewise it is a lengthy matter to reach the reason for this thing, since the teacher must give careful consideration to those things that have been written in this fifth part of this plea, and especially to those things that are said and written in this chapter on the causes of vision as regards the species. It is certain, then, that when the axes of the eye are fixed on the visible object *a,* the visual force nevertheless tends to the visible object *m,* but the species coming from the right eye tends to the left, if it proceeds further, and likewise the species of the left eye tends to the right, for these species intersect at the point *m,* and separate in such a way that the right species passes to the left side, and the left one to the right side, as is clear to the sense. Since, then, *m* appears double, the image corresponding to the right eye must be beyond *m* to the left, because the species of the right eye tends in that direction. Therefore when the right eye is closed, the left image must vanish; and the reasoning is similar with regard to the left eye and the right image, as the figure shows.

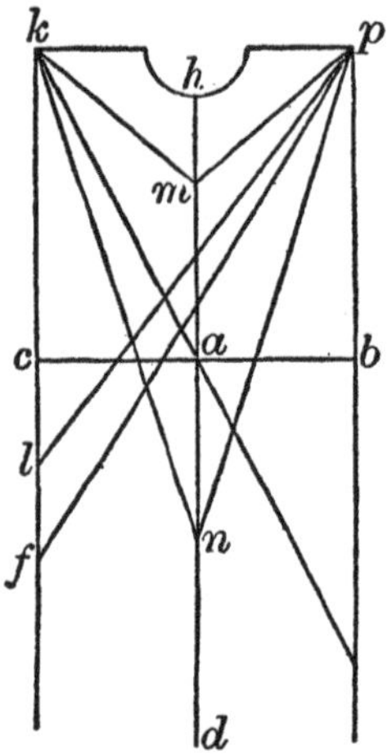

FIG. 42.

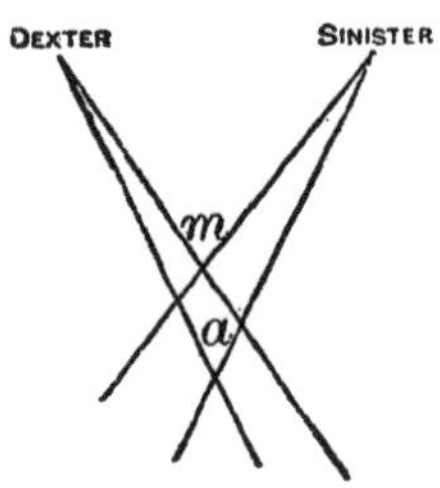

FIG. 43.

And yet the right image will not always disappear when the left eye is closed, nor the left image on closing the right eye; but it easily happens that when the right eye is closed the right image disappears, and when the left eye is closed the left image disappears, as any one can prove in the case of the fixed stars in summer at twilight before the darkness of night, and the same experiment could be made on far distant fires.

Chapter IV

Concerning the fact that an object appears double owing to the elongation of all particular axes from the common axis, with an exposition of all the ways in which an object can be placed and a statement of the cause of good or bad vision in all those positions.

Not only is there a difference in false vision, because there is a difference of position of the object seen with respect to the axes, as has been said, but the axes themselves can diverge too far from the common axis, as is shown in the figure. For the angle *abd* in eye *b* is far greater than angle *acd* in eye *c,* and for this reason there will be an error, and *a* will appear double, at which the axes *ba* and *ca* terminate, and this happens because they diverge from the common axis *ed.* This takes place when the eyes turn their axes from the common axis in the direction of one eye, and the other eye follows as far as it can turn, and therefore the axes are not terminated at *d,* nor are *bd* and *cd* the axes, but *ba* and *ca,* for the axes are thus turned and bent from their natural position. They can terminate also at the same point of the object of vision, but in a different position with respect to the common nerve; and in this case that point appears double, as is evident by elevating one eye from its right position or both eyes. From these oblique positions it is clearly evident that, whether the axes deviate too far from the common axis, or the object of vision or part of the object of vision is too

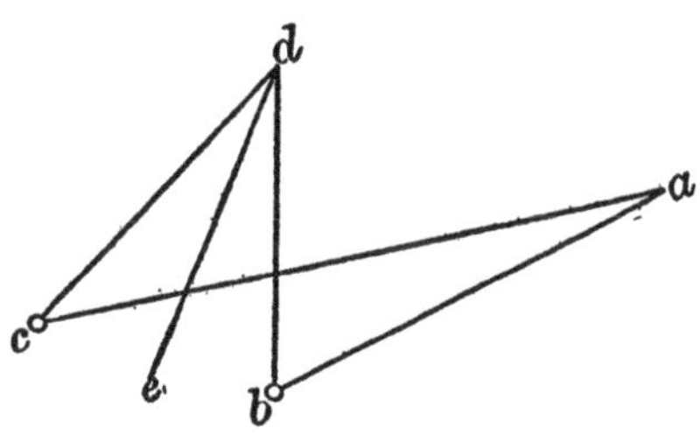

Fig. 44.

far from the axes of the eyes, the object will be seen imperfectly, and this happens either in one eye or in both, or there may also be a displacement of the ends of the object from the position in front of the face. The object directly opposite the face, on which object the special axis and the common axis are concurrent, will be seen with final certainty. If, in fact, the object is near the axes, it will be sufficiently visible in the same direction; and in this way the vision of the same thing will be lessened up to a position of such great obliquity that one object will appear as two, just as we have stated before. The reason for all these things, according to Alhazen in the third book, and according to Ptolemy likewise in the third book, is that when the object is seen in front of the face, or near it, the species of the object comes to the surface of the glacialis under a large angle and occupies a large part of the pupil, in which the parts can sensibly be distinguished; but when the obliquity is great the species comes under a small angle to the organ of sense and occupies a small part of it, so that the species of the parts of the object of vision are crowded together without affecting the sense, and are collected in a confused manner, and therefore no definite judgment is formed concerning them. All these matters are evident, which must be known in regard to position with respect to the eye. For any one can easily distinguish now the differences of position. For there is the position in front of the face when the object is placed before the eye or the eyes, and one end is not raised nor does the other deviate. There is also the oblique position through the raising of one end and the depression of the other; and this is called the oblique position. Moreover, in the position in front of the face there is the position of one part with respect to the axes, and the other parts deviate more; and likewise in the oblique position certain parts are more remote from the axis, certain parts nearer to it. We have stated how comprehension and error occur in all similar cases. A position, however, that is outside not only the visual, but the whole pyramid, excludes vision entirely.

THIRD DISTINCTION

In which consideration is given to the triple mode of vision in reference to the eight conditions of vision mentioned previously. The distinction contains eight chapters, in the first of which examples are given in regard to perception through sense alone.

CHAPTER I

HAVING determined these matters in regard to the eye and the position of the species, we must now consider further concerning the triple mode of vision in reference to the eight necessary conditions without which vision is not possible. Light, then, and color are perceived through sense only, and no error is made in regard to them when these eight conditions are fulfilled without excess or defect; under these conditions light is seen as it is and color is seen as it is. When, however, these conditions are in excess or defect, error or defect in vision occurs. For the light of the stars is not seen by day, when the solar light in full strength comes to the eye, because that light is in excess with respect to the eye and the light of the stars; as, for example, when a man is at the surface of a well the principal solar light comes to his eyes excelling the principal light of the stars. But if he descended to the bottom of the well he would see the stars that were over his head, not only for the reason touched upon previously, but because the solar light would be tempered, since only the accidental light of the sun would come to the eye and no principal ray, and therefore the stars would be seen by means of their principal rays and would not be hidden.

But in regard to the Galaxy it is strange that it cannot be visible in the celestial sphere, or in the sphere of air, but only in the sphere of fire. Now the Galaxy is in one sense a circle in the celestial sphere, which is called the Milky Way, consisting of many small stars clustered, and this part of the heavens produces, according to Aristotle in the first book of the Meteorologics, the impression of continuous light through the meeting of the solar light with the lights of small stars of this kind, and this impression is called likewise galaxy. Hence the term is used in two senses, for cause and effect: and the light appears as continuous and lengthened out, although the stars are distinct. But distance causes this, as would be the case if one looked from a distance at a pot perforated in many adjacent parts, in which fire is contained. For owing to the distance the fire would appear continuous to him because of the proximity of the holes, a proximity that vision would not discern because of the great distance. This impression, however, the eye would not perceive in the celestial sphere, although this light passes through this sphere, nor in the air, although it passes through the air likewise. But in fire it is perceptible, because the celestial sphere is of such excessive rarity that light cannot be incorporated in it so as to be visible. But air is of greater density than is here required. For since this impression is weak owing to the weakness of the lights of the small stars, it can quickly be lost in the darkness of the place, nor does it suffice to clear away the shadows of the denser air. But in the highest part of the sphere of fire the Galaxy can be visible to the eye, because it is properly proportioned for visibility because of its mean between the excessive rarity of the transparent heaven and the greater density in the air than visibility of this kind requires. Therefore in fire it can be visible, as Aristotle states, although he does offer too little explanation regarding the cause. But the light of the other larger fixed stars and likewise of the planets has so much strength that it is not dimmed in the air, nor in the sphere of fire, nor in the celestial sphere, and therefore uniformly lights up the whole medium, namely, the spheres of the heavens and of fire and of air; and therefore vision does not perceive in some part a difference in the incorporation of their light. For this reason there is not some distinct impression of visibility by means of the lights falling from other stars than that received from the stars of the Galaxy, wherefore philosophers make no mention of some other impression, which is pure light, differing from that which is called Galaxy. Thus it is manifest that Galaxy cannot be visible in the sphere of air nor of the heaven, owing to the medium's lack of the proper density. In the air because of the lack of sufficient rarity, in the heaven because of its excess. We must note, however, that a consideration of refractions in the sphere of fire is important for our conception of the Galaxy; for all rays that fall at oblique angles on the surface of fire are refracted. But of this the explanation is to be sought elsewhere.

It is likewise strange that the light of dawn appears to us at that time and not earlier; for the whole heaven is illumined except in the shadow of the earth. I am here calling the whole world heaven from which outside the shadow of the earth the air is illumined, and the sphere of fire and the celestial spheres. We see clearly the stars at night in every part of the horizon; therefore it seems that we should likewise see the spheres of the heavens and of fire and of air illumined outside the shadow. For between us and the stars the air is of necessity illumined, also the sphere of fire and the spheres of the heavens, not only by the light of the stars, but also by that of the sun, which crosses in the air through the sides of the shadow, and therefore just as at dawn we see the air illumined, so is it probable that the illumined air should be seen earlier. But excessive distance together with excessive rarity solves this problem. For the stars are dense bodies and have much fixed light of their own, and from both these causes can be seen by means of their species diffused in the shadow to the eye. But the illumined parts of the spheres, and the sphere of fire and the air outside the shadow are bodies of too great rarity, and do not have any fixed light, but merely light passing through them, and for this reason the light is weak and they are far distant. For these reasons they cannot be seen by night, because the species coming from them into the shadow is too weak and cannot affect the eye. But when the sun approaches its rising, and this is when it is in the eighteenth degree of the circle of its depression below the horizon, as Ptolemy shows in the Almagest, its rays fall in the air nearer to us and enter the top of the shadow and its sides, and in this way the nearer illumined air can cause a stronger species to reach the eye, and it begins to be visible and becomes more so as the nearer air is illumined. There is the added reason, moreover, that the nearer air is denser, and especially so when the light comes to air full of vapor, the greatest altitude of which is fifty-one miles and some fraction of a mile, as is shown in the book on Twilights. Owing to this density more light is retained in such dense air, just as in the case of the stars, and therefore a stronger species reaches the eye. These facts show us that the dawn is not caused by the refraction or reflection of light in the clouds, as those who are informed know, but it is caused by the accidental species of the sun, which comes to us from its rays passing through air filled with vapor and cloud, just as the accidental species from a ray of the sun falling through a window comes through the whole house.

It happens also that a small light or a luminous body of small size appears sometimes to be of great extent in the case of certain things because of their sensible distance and because of the rapidity of motion in shooting stars, as Ptolemy states in the second book of the Perspective. For fiery impressions in the air from ignited vapors like stars, which are called in Arabic *Assub,* ascending and descending, are bodies of small size. But on account of distance and rapidity of motion they appear to possess a long light. A similar phenomenon is found in the case of sparks ascending rapidly from a fire; for the rapidity of their motion frequently causes them to appear extended at great length, when they are, however, of small size.

Color, like light, is perceived by the sense only: and if those nine conditions, concerning which we have stated that they are necessary for vision, are fulfilled without excess or defect, vision will perceive color correctly. If, however, these conditions are in excess or defect, vision will not perceive it correctly. As, for example, if a colored body is sufficiently dense to terminate vision, the color seen will be judged to belong to that body. But if it is not sufficiently dense, as in the case of a crystal, behind which a colored body when placed will appear as the color of the crystal, although it is not. Moreover, when the eye has looked at strong colors and then turns to light places, the species of the strong color will remain in the eye for a time, and at first the color will appear to it red, then purple, and afterwards black, and so vanishes, as Aristotle states in the second book on Sleep and Waking. The error is due to an excess of light changing in this way the judgment of color. Ptolemy, moreover, in the second book of the Perspective shows that different colors appear as one for two reasons. The first reason is because in an object of different colors one appears, owing to excessive distance, so that the angle containing the whole object has not the suitable size. When the separate angles, which contain the different colors, are not perceived, from the compression of the parts that are not discerned, it results that the color of the whole object is one differing from those of the separate parts. The second reason is found in the motion of a swift body, as in that of a hoop with several colors, in which one and the same ray does not pause on one and the same color, since the color recedes from it because of the speed of its motion; and so the same ray falling on all the colors is not able to distinguish between the first and the last, nor between those that are in the whole hoop, but they appear as one, and it seems that that color is as it were a mixture of them all. Moreover, concerning these matters there are an infinite number of examples, and some have been placed at the beginning and in different places; and therefore I pass them by.

CHAPTER II

Concerning perception through knowledge.

THE things that are perceived through knowledge are distinctions of one universal from another, of one particular from another, or of a universal from a particular, as we have stated; whence the difference and discrimination of visible objects are considered in this perception, and this perception in the first place and *per se* applies to light and color, since they are especially objects of sense. When, therefore, the moon is outside the shadow of the earth it has a clear and white light; when it is in the upper part of the shadow it has a reddish light; but when it is in the lower part the light is not visible; when, moreover, it is in conjunction with the sun at new moon its light does not appear. Therefore vision perceives these differences, but the sense alone is not able to make certain of this matter, but after it has seen these differences many times and has viewed the moon, it can recognize that which appears at such a time or under such and such circumstances and what is not visible under others. The explanation of this matter is, however, difficult, yet it is certain that the light of the moon is caused by the sun, in the same way as the light of all the stars, as has been stated above. But as Ptolemy says in the second book of the Optics, when the moon is in conjunction with the sun that which prevents the illumination of the moon toward us is its proximity, since almost half of the lunar body is opposite to us, because the other part is illumined. Therefore the solar rays cannot reach the part toward us; nor can the principal rays reach this part, as is evident; but they are far divergent and separated. Therefore the accidental rays are not sufficient to light up the moon. But when it is in the shadow, the earth, which hides the moon from the sun, is far removed from the moon, and for this reason the principal solar rays are much more convergent owing to the narrowness of the shadow in its upper part; and for this reason the accidental rays are able to come from near at hand in great force to the body of the moon, and therefore light it up. But since accidental rays are weaker in their action than principal rays they do not clearly nor fully illumine the moon, and therefore it is somewhat reddish or pale, shading to red, and it is redder when it falls less into the thick part of the shadow, and paler the more it falls into the shadow. And yet the moon can enter the thick part of the shadow and descend in its own eccentric toward the earth, because the accidental rays are not able to light it up, owing to the great divergence of the principal rays of the sun itself, and the whole moon remains in shadow, just as at conjunction, because the moon does not have light of itself, but is a dense body.

Similarly an example can be cited in regard to color. For if the eye views a body of a single color through the medium of a very thin parti-colored cloth, if the apertures of the cloth are large it will see the color of the body as it is; but if they are small and fine, the color of the object placed behind it will appear mixed, and therefore from the differences in the rarity of the medium through which it is viewed the color will appear different, and in this way there will be an error in knowledge. The reason for this is in doubt. For it is evident that the colors of the parts of the body do not come to the vision except through the apertures of the cloth between the threads, and the colors of the threads come according to their lines and peculiar twists to points different from those to which the species of the colors of the parts of the body placed behind the cloth come. Wherefore the colors of the parts of this body are formed at points on the eye different from those at which the colors of the threads are formed. For this reason it does not seem that the color will be mixed. But we must state that such is the case. For when the apertures are minute, the colors passing through them are very near to the threads, and for this reason the colors of the threads and the colors passing through them are very close and touch at points at an insensible distance apart. Since this is the case, the difference of these colors will not be perceived, and hence a mixed color will appear to the eye.

CHAPTER III

OF those things that are perceived through reasoning likewise notable examples can be given. First among these is distance or remoteness. In regard to this the first consideration is that excessive distance hinders vision, because the same object at a distance makes a small angle, which when near would make a large one, as is evident in the figure. Since from the small size of the angle it results that a small part of the pupil is occupied, the object cannot sensibly be distinguished as it should in the eye, and for this reason the object is imperfectly perceived because of excessive distance. Distance is perceived and determined, if it is moderate by a continuous series of sensible objects lying between the eye and the remote object. As, for example, when a person is near a wall and sees another wall beyond the first more elevated than it and quite a distance from it, he cannot judge of the distance between them because there are no bodies in a continuous series between them, or he cannot perceive these bodies because of the interposition of the first wall beneath which he is standing.

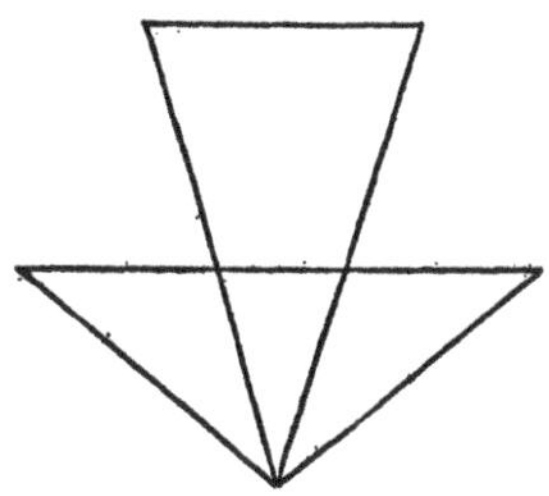

FIG. 45.

Hence it happens that when we are not near high mountains, but are in level country, we cannot judge of the height of clouds; nay, we reckon that they are more remote from us than they are, because there are not sensible bodies lying between us and them, but the air only, which is not sensible to us, and for this reason we cannot judge of the height of the clouds under these circumstances. But when we are near high mountains, although mountains are not very high with a maximum height of about eight miles, we shall see clouds on the tops of the mountains and the summits higher than the clouds. From this fact Alhazen concludes in the second book that the clouds are not of great height, although vapors easily rise fifty-one miles, as is shown in the book on Twilights. But not all vapors form clouds, for only aqueous vapors do so; and not all vapors are the substance of clouds, but terrestrial vapors rise higher although they are heavier, because dryness brings out the full effect of heat and a dry substance retains heat well after it has been received, as, for example, stone retains heat better than wood. Hence it happens that a stronger and more abundant heat of the sun is retained in terrestrial vapors than in aqueous, and for this reason they rise higher. Now, moreover, that no objection might be raised in regard to the height of clouds by means of the contrary statement in the book on Twilights, I have made mention of this book.

Distance, then, is judged when there is a series of objects between the eye and the object of vision, and this is true in the case of moderate distance, and when the eye has viewed those objects and has made certain of their measurements. But if any one of these conditions is not fulfilled, distance will not be determined. Moreover, a moderate distance with respect to vision, that is, one whose extent is determined by vision, is a distance at the end of which no part with an extent sensible with respect to the whole distance is hidden from vision. A moderate distance with respect to the object of vision, in which vision perceives one part of the object of vision, is a distance at the end of which no part of that object of vision bearing a sensible proportional relationship to the size of the object of vision is hidden from vision. But a distance which is excessive with respect to vision is that at the end of which an extent bearing a sensible proportional relationship to that whole distance is hidden from vision. And a distance excessive with reference to the object of vision is one that hides from vision parts in proportional relationship sensible with reference to the whole. Therefore an error is made in the determination of distance owing to excessive remoteness of objects from the eye. For in this way trees at a great distance, although they are quite far apart, appear continuous or to be near one another. For this reason also the planets will be judged to be in the same surface as fixed stars because of the immense distance of the stars from the eye, although the planets are a very great distance from the fixed stars. Moreover, a figure of many equal sides placed directly opposite the eye looks like a circular figure, and the circle looks like a straight line, and a sphere will be judged to be a plane figure. For owing to excessive distance the angles of the figure, although they are sensible with respect to the whole at the required distance, yet when they are at an excessive distance will be hidden from vision with respect to the whole, and therefore an angular object will be judged to be round, because when such an angle is hidden, the figure appears round. Likewise when the gibbosity of the arc of a circle is presented to the sight, although the middle of the gibbosity of the circle is nearer the sight than the ends of the arc at the diameter, yet this nearness is not apparent to the sight because of the excessive distance; and for this reason the approach of the part nearer the sight is hidden, and the gibbosity itself is removed in the judgment of vision; wherefore the curved line will appear straight. This is the reason why, when the moon is in its first or third quarter, the line of the circular base of the luminous solar pyramid enveloping the body of the moon appears to be a straight line, when at other times it appears to be curved. A sphere will appear as a plane surface for the same reason. The nearness of its convex surface exceeds imperceptibly the nearness of the limits of that convex surface because of the excessive distance. Therefore both sun and moon appear to be plane surfaces although they are spheres.

Chapter IV

Concerning the shape of the moon as regards its different phases; also some additional matters at the end.

By what has already been said the reason can be shown for the difference in the form of the lunar light as it fills and wanes. For sometimes the base of the pyramid of solar light enveloping the body of the moon appears to be a curved line, as, for example, before the seventh day and after it, but on the seventh it appears to be a straight line. The reason for this Aristotle in the second book of the Heavens and the World and Averroës hide by imperfect and cursory statements; and the book of the Problems of Aristotle in Latin by the obscurity of its translation leaves us in doubt. For it is quite strange why the same curved line appears sometimes straight, sometimes curved, from the same distance. For if in accordance with the reason given in regard to excessive distance the curved line should appear straight, how then can it many times, nay, through almost the whole lunar month, appear curved, and only twice be judged straight?

Let us then withdraw from the realm of sense and deal with the real fact that the moon is a spherical body, and in the first place let us suppose that it is situated near us at a moderate distance, and the sun likewise, yet so situated that they have the same relation to each other, and that they have their motions as in the heavens, and that they bear the same relation to our vision as when they are in the heavens, except that the distance is not excessive. Then here we shall see two pyramids; one is the visual, whose base is the portion of the moon's surface exposed to us, and the other is the luminous solar pyramid, whose principal base is the surface of the sun. Since this pyramid is cut off, because the surface of the moon meets it before it can reach its own vertex, it must have there a secondary base, although much smaller than the first, and the greater portion of the moon is presented to the sun. But before the part of the moon presented to our sight begins to be illuminated at any point, namely, while the sun and moon are in conjunction, the base of the pyramid of solar light does not include any part of the base of the visual pyramid. But as soon as the moon is free from its conjunction, so that some part of its illuminated portion is exposed to us, the base of the solar pyramid occupies that part of the base of the visual pyramid, and therefore their circles at once intersect at two points, and include within themselves the illuminated part. And as the portion of the base of the solar pyramid cuts more of the base of the visual pyramid, so does the light increase in the portion exposed to us. The circle, therefore, of the base of the solar pyramid at a moderate distance will always appear as a curved line and never as a straight one; because excessive distance alone causes it sometimes to appear as a straight line. But at an excessive distance, just as is now the case, it must frequently appear curved. For the circles on a sphere are to be understood in two senses, namely, parallel circles like the equator on the celestial sphere and the tropics and other circles parallel to the equator on both sides of it; and there are other circles intersecting one another and equidistant circles of this kind, as, for example, those passing through the poles of the sphere like the colures on the celestial sphere, and all are great circles on the sphere. Of the former one alone is a great circle, namely, the largest of the parallel circles, as, for example, the equator in the heavens; for those circles alone are great circles on a sphere that pass through the center of the sphere and divide the sphere into two equal parts.

Let us, then, consider the sphere of the body of the moon, restricting ourselves to that portion exposed to us in such a way that we have a clear notion of the largest of the parallel circles, which divides the body of the moon into two equal parts, also after it the other parallels, as many as we desire, whose pole is the lunar pole toward us. Let us also have a clear notion of the great circles intersecting one another at the poles of the lunar sphere. This one of these circles, however, is sufficient for our purpose, whose gibbosity is directly exposed to our sight, and which cuts the moon into two parts east and west and passes through the pole nearer us and through the other pole on the opposite side. Let us call this the colure. It is certain, then, that excessive distance hides the gibbosity of this colure, as stated above. Hence of necessity this curved line appears to be straight. But no distance can hide the gibbosities of the parallels, since the whole circumference of each of these has the same position with reference to the eye, and one part of the circumference is not nearer the sight than any other part, so that it is visible to the sight at every distance, and therefore no part of these circles can be hidden. For this reason they are always visible in their circumference and revolution, just as we see in the case of the circular line which is the circumference of the base of the visual pyramid. For that circle is one of these parallels and the largest of those that can be exposed to our vision, and it is evident that the vision of each one judges this to be a perfect circle; and it would do so in regard to all the parallels that might be marked in the portion exposed to us, but no one of these has been marked so far as the moon in itself is considered. The visual pyramid, however, marks one parallel, which is the circumference of its base, and the solar pyramid marks another parallel, which similarly is the circumference of its base. When, moreover, this circumference of the base of the solar pyramid touches the portion of the moon exposed to us, its gibbosity of necessity is directed toward the sun, as we see by the eye. The reason for this is found in the fact that it is a portion of a circle which is almost a great circle on the sphere; for it is a part of one of the parallels described around the portion of the lunar sphere exposed to us, and this parallel is greater than any other that can be described there. But it is not a large circle cutting the lunar sphere into equal parts, because half of the lunar sphere is not visible to us, but a portion slightly less than half, as we are now assuming, and as what follows will explain in the proper place. Since, however, the arc of the solar pyramid is almost a portion of a great circle, its concavity in the portion of the moon visible to us cannot lie toward the sun. For if it were the arc of a very small circle, its concavity could lie toward the sun, as is evident. The gibbosity of that arc will lie toward the sun at new moon; and when it first touches the circumference of the base of the visual pyramid, the arc of the visual pyramid and the arc of the solar pyramid are equal, and lie together and terminate at the ends of the chord of the visible portion of the moon itself. But when the solar pyramid occupies some part of the portion of the moon exposed to us, the arc of the solar pyramid leaves the arc of the base of the visual pyramid, always, however, conterminal with it, and they separate, each containing the illuminated portion, and they intersect at the ends of the chord of the portion of the moon exposed to us; and then the arc of the solar pyramid is like the arc of a parallel almost the size of a great circle. Moreover, we must imagine that it was at first on the circumference of the base of the visual pyramid, and afterwards was continuously moved from it toward the colure of which mention was made before, which passes through the poles of the moon, namely, through that pole exposed to us and through the one opposite to it, dividing the moon into two halves, as if it were moved through a quarter of the moon from the circumference of the base of the visual pyramid, until it becomes the colure. When, therefore, it becomes the colure, its gibbosity is of necessity concealed from us because of the excessive distance, and the circle will appear as a straight line. But not before; since it was always previously the arc of a parallel, and parts of the circumference of parallels have the same position with respect to the eye. For no part comes nearer to the eye than any other part, as has been noted before.

Wherefore, from the first illumination of the moon until the portion of the base of the pyramid comes to the position of the colure, it will always be in the form of an arc, both actually and according to the judgment of the sight. But when on the seventh day it reaches the position of the colure, it appears to the sight of necessity in a straight line, the diameter, as it were, of the surface of the moon which appears to us plane. Then when it is in the other half of the moon, it again becomes curved, because it is now assuming the nature of a parallel. Since, moreover, it is the arc of a parallel almost a great circle, its concavity must be toward the sun. For in that half of the moon so large a parallel could not be drawn, whose gibbosity is toward the sun, for it might easily be possible in the case of a small parallel, as is evident to one considering the matter. If, then, that arc of the pyramid be moved until it lies on the arc of the base of the visual pyramid, and the whole circumference of that parallel lie on the whole base of the visual pyramid, the whole surface of the moon exposed to us will of necessity be illuminated and the same will be the base of both pyramids, and the moon will then be full. Then, owing to the varied position of the sun with respect to the moon, the arc of the base of the solar pyramid begins to recede from the base of the visual pyramid, and the moon begins to wane in light, just as in the period of new moon it increased, so that the base of the solar pyramid always appears as a curved line because of the explanation in regard to the parallel, until the base of the solar pyramid arrives at the position of the colure on the twenty-first day, and then it again appears as a straight line, just as it did on the seventh day. Thus, therefore, can the form of the lunar light through its varied phases be understood by means of an excessive distance of this kind and by means of the explanations of the circles.

Therefore it is evidently impossible that the heaven is a plane figure, although the vision so judges it, or at least judges that it approaches closely to a plane figure. For just as we have here said, the angles of a polygonal figure of equal sides appear in consequence of excessive distance to have rotundity. For little by little the angles disappear, because they have no relationship to excessive distance, nor does the access of the angle to the eye exceed sensibly the part of a circle inscribed in such a figure, or of an inscribed sphere, because the size of the angle is small or not sensible at all in respect to such great remoteness. Not only is this statement made by Alhazen in his third book, but the same author maintains this in his book on Vision, when he says, in the tenth proposition, "Rectangular magnitudes seen from a distance appear circumferences." But since rectangular figures of this kind must also be equilateral, another translation substitutes for the above, "Squares because of distance appear round," and not only squares, but every equilateral figure, because the same explanation applies to them all. But in regard to other polygons, in which circles and spheres cannot be inscribed, the statement is not true, because there can be a sensible access of the angle to the eye although the distance be great. I have added this for the understanding of this occultation. If, then, the angles of an equilateral body little by little disappear, until the body appears round, so that such a body does not disappear at one and the same time owing to the flattening of its angles, a round body, because it does not have angles, will at once wholly disappear, and at a great distance will appear to be a plane figure. Therefore if the heaven were extended to the east and west in a plane figure, the stars would disappear all together at the same moment at their setting, since they are of spherical figure. But we see the reverse, because part after part sets and disappears, and in like manner rises. Wherefore this cannot happen in a plane distance and therefore the heaven is not a plane extension but a curved one.

Chapter V

Concerning the comprehension of magnitudes.

As the comprehension of distance has been explained as a reasoning process, so too with magnitude. Both the author of the book on Vision and many others thought that magnitude was appreciated by the size of the angle at the eye, whence at the beginning of that book it is assumed that objects seen under a larger angle appear larger, and under a smaller angle smaller, and things seen under equal angles appear equal. But this does not suffice, as Alhazen shows by examples; since if in a circle different diameters are marked, as *ab, cd,* it is evident to the sense that *ab* is seen under a much smaller angle, and yet the diameters are equal. Similarly in regard to the sides of a square, for the side *ab* makes a much smaller angle at the eye than *dc,* and yet the sides are equal, and the vision perceives that sides of this kind are equal; and in like measure the vision judges that the diameters of the same circle or of equal circles are equal at a moderate distance. Therefore comprehension of the size of the angle does not suffice. Moreover, if an object is seen at a lesser distance and at a greater, provided that the excess is moderate, the angle will vary sensibly. But the vision will not judge that the object seen is greater and smaller; so that if the eye be placed at the vertex of a shorter pyramid, the angle in it will be greater than at the vertex of a longer pyramid, by the twenty-first proposition of the first book of Euclid. And if the distance be trebled, namely, so that the first distance is one cubit, the second two, and the third three, there will be an extraneous difference of angle, and yet it will be judged to be of the same size. Since, then, this is a fact, there can be no determination of the magnitude of an object in accordance with the size of the angle. But it is necessary that the angle be considered and the length of the pyramid, and that the base of the pyramid, which is the object of vision, be compared with these. The length of the pyramid is apprehended by the perception of the size of intermediate bodies, as was stated previously in regard to distance. This is done in the case of objects familiar to the eye and at a moderate distance by comparison and sud-

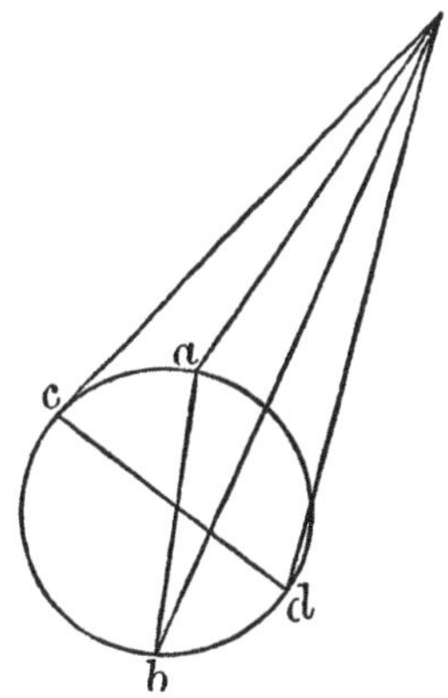

FIG. 46.

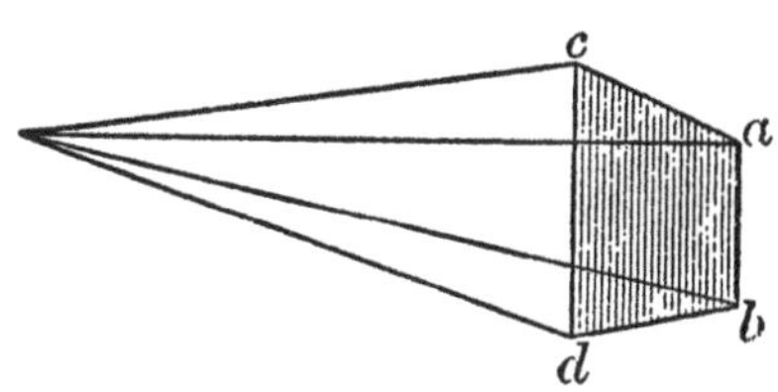

FIG. 47.

den reference to some definite measure ready to the memory, like the size of the man measuring as regards himself as a whole or a part, and all these things are determined by the visual axis passing over visible objects.

But error occurs in apprehension of magnitude when the distance is excessive, since it will appear much less than it really is. The reason for this is found in the excessive distance, which hides from the vision parts proportional to the whole in a sensible relationship; and when the occultation of parts perceptible to the sense shall have taken place, the angles in which they fall are not perceived, although they are proportional to the total angle. Hence when the axis passes over the object of vision, the lines from it and many parts are hidden, whence the total appearance is rendered smaller. And also owing to the smallness of the angle, perceived with difficulty, the species will not be well arranged on the sensible part of the perceptive member, and therefore the size will not be clearly judged.

Chapter VI

In regard to spherical bodies, like stars and other objects, it is not possible for the eye to see one half, but the eye of necessity sees a smaller portion. For the rays of the extremities of the object meet at the eye; but rays that would come from the perimeter of half of a sphere would touch the ends of a diameter, by the nineteenth proposition of the third book of the Elements, and would make right angles with the ends of the diameter, as is shown in that proposition. Therefore it remains that they will not meet by being produced.

That stars for a constant reason seem larger at rising and setting than when at the meridian, is stated by Ptolemy in the third and in the fifth book, and by Alhazen in the seventh. It can be explained owing to the fact that the vision judges the sky as though it were a plane figure extended above the head to the east and to the west, when the eye looks toward either of these. But what is seen near the head seems nearer, and therefore a star when it is at the meridian seems to be nearer, and therefore when on the horizon it seems to be more distant. But what seems to be more distant seems to be larger, since it is seen under the same angle, just as a matter of fact what is more distant is larger since it is seen under the same angle with a smaller object. As, for example, *ab* is more distant from the eye, and is larger than *cd*, and *cd* than *ef*. Therefore it remains that stars appear of larger size on the horizon than at the meridian. And this is manifest in another way. The remoteness of these stars when they are on the horizon is apprehended by the interposition of the earth; but they cannot be so apprehended when they are at the meridian, because of the imperceptibility of the air. Therefore since the remoteness of these is more perceptible when they are on the horizon than when at the meridian, it follows that they seem at that time more distant than when they are at the meridian. Therefore as before they will appear larger.

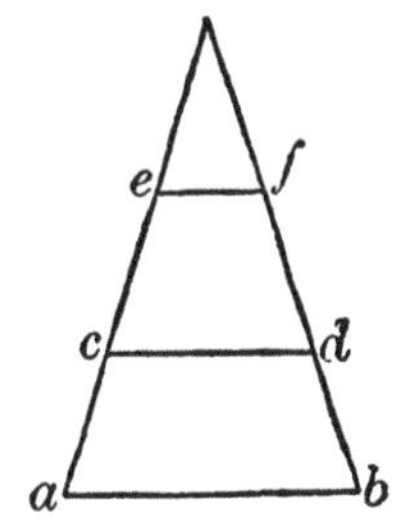

FIG. 48.

The perception of motion and rest, which is said to be the result of reasoning, can also be explained. Motion is apprehended by a comparison of the object moved with another object in respect to which it changes its position. Hence vision does not apprehend motion except when it perceives a thing in two places and in different positions; and the position of the object of vision is not changed without involving time. Wherefore motion is not apprehended except in a perceptible lapse of time. Rest likewise is apprehended by vision from the perception of the object of vision in the same place and in the same position in a perceptible lapse of time. Vision errs frequently in the perception of motion and rest. For sometimes when the clouds are in motion the motion is thought to be that of the moon, and this error results from excessive distance; for it does not occur when the distance is moderate. Hence we see that a pole fastened in water is continuously at rest, and we perceive the motion of the water passing by. The error mentioned above in the motion of the moon occurs when the clouds are numerous and continuous, since motion is not perceived except by the approach of one thing to another or the recession. When, however, the clouds are few and divided, we can discern their motion by their successive approach to a star and their recession from it. Therefore, on the other hand, when the heaven is covered by clouds, because of their continuousness and multitude we do not discern motion in them; and yet we see that the moon is moving very rapidly, because it is passing swiftly through the different parts of the cloud, not only owing to its own proper motion, but also owing to the motion of the clouds. When an observer walks in the direction in which the moon lies or other stars, it seems to him that the stars move before him in the direction in which he is going; since at the end of his motion he judges the star to be as far off and to hold the same relative position as at the beginning of his walk. Owing to these impressions the faculty, distinctive through the medium of vision, concludes that the star moves before the face of the observer in the direction in which the observer is walking; and likewise if he flees in the opposite direction it will seem to him that the star is following him. Just as if he should see a man ahead of him, who always keeps the same relative position, they of necessity must be advancing with an equal motion in the same direction, and therefore the observer reasons that the stars are moving in his direction, either preceding or following. For his position with respect to the star is the same as far as the judgment of vision is concerned.

The reason for this is the excessive distance between him and the star, because of which he does not perceive that by his motion he is departing from the star or approaching it, and therefore he judges that he always keeps the same relative position. In the same way it happens that when the sun is in the south and a man is going to the east, it always seems that the sun is in line with him; and if many men stand in the same east and west line, although they be far apart, yet the sun appears directly in front of each one of them. This effect is caused by the excessive distance and the magnitude of the sun. Moreover, the shadows of these men seem to be parallel; and likewise the rays of the sun coming to them seem to be parallel, although they meet at the center of the sun; but owing to the excessive distance their meeting is not perceived, and therefore the rays seem to be parallel and the shadows likewise. When, moreover, one views the planets, although they are moving swiftly in the heavens westward, he judges that they are at rest and does not perceive their motion. The cause of this is found in the fact that, owing to the excessive distance, the vision does not perceive their approach or recession with respect to some fixed object, nor does it perceive a difference in position. Therefore the visual ray is judged to be unmoved, retaining the same position with respect to the object, and therefore the vision judges that the object is at rest.

Moreover, when one turns frequently in a circle, on coming to rest he thinks that vision and other objects are moving in a circle, since after the observer has been set in motion the visual faculty is moved within. And although the observer comes to rest, the visual faculty will not immediately do so, but its motion will continue in the observer after he has come to rest; because the humors of the eye are easily moved, and when motion has been imparted to them they readily retain it, just like water and the visual rays also, which are bodies of fine texture, easily moved and capable of continuing an imparted motion. For they belong to the class of evaporations and resolutions, and therefore the vision continues its motion within after the observer himself has come to rest. Since his vision is in motion, he thinks that the objects of vision are in motion, as Aristotle states in the second book of the Heavens and the World, and Averroës and Ptolemy in the second book and Alhazen in the third book of the Perspective. Moreover, when a man is on a moving ship, trees and other objects on the bank seem to him to be moving, and this is owing to the motion of the visual faculty. Those objects in particular seem to move that are far from the axes; for if the axes are carefully directed to an object near the water, it does not seem to move.

Chapter VII

Concerning scintillation.

Connected with these matters is a difficulty very philosophic in character, but more insoluble than any other except the rainbow. Every night we can view objects in which a doubt of this kind is present. Hence we see nothing so frequently of whose cause we know less: it is the scintillation of the stars. The fixed stars sometimes manifestly scintillate, and sometimes they do not. The planets, however, according to Aristotle do not scintillate. Scintillation is a certain tremor of the star and an apparent motion, and this occurs particularly when the stars are in the east or in the west, as Aristotle maintains in his second book on the Heavens and the World, because they then seem most remote, as we have stated, and therefore the vision is less certain in regard to them. For Aristotle in the first book of the Posteriores and in the second of the Heavens and the World says that the reason for this is owing to the distance of the fixed stars. Hence, since the planets are near, they do not scintillate, as he states. He says, moreover, that this is not a phenomenon in the stars, but is seen only because of the revolution and tremor of the vision, because, as he states, the planets are near, and for this reason the vision has power over them. But the vision, when directed to the fixed stars, quivers because of the distance: because the tremor of vision causes an apparent motion of the star itself, for it makes no difference whether the vision is moved or the object of vision. This is the opinion of Aristotle, as we clearly infer from several translations and especially from that purer one that has been made directly from the Greek.

But although he says without reservation that the planets do not scintillate, he, however, maintains in the second book of the Heavens and the World that the sun in a manner has a tremor, and especially so when it is at rising and near it. The reason for this is found in the fact that this tremor is not the motion of scintillation but a weaker motion than that. For a stronger tremor is necessary for scintillation, nor does it occur except on the horizon or near it; but scintillating stars have a tremor everywhere. Venus and Mercury near the point of their rising and setting seem to have a slight tremor, as experience shows, but it is not reckoned a scintillation, because it is slight and does not occur except in those places. Hence this tremor in the sun and in these planets contains no proof with regard to the scintillation in the fixed stars, nor does it occur everywhere in the heavens as it does in the case of the fixed stars; and therefore those tremors are not reckoned among scintillations. But since the longer orbit of Saturn is of equal remoteness with the fixed stars, Saturn, when he is at his longer distance, will scintillate, owing to the distance, just like the fixed stars: this, however, we do not see. It can be stated, moreover, that since the diameter of Saturn is 29,250 miles, as is evident from the previous statements in those matters touched upon regarding arithmetic, he is nearer to us by this number of miles than the fixed stars, and therefore his distance is not so great.

Moreover, a light intense and quite perceptible is required in the object of vision to the end that it may scintillate, because it seems to project darts of light. Hence Averroës says that the sun has that motion of which mention was made above, more than other planets, owing to the brilliancy of its light. If it be said that not all the fixed stars scintillate, although they are more distant than the planets, we must answer that those stars that are larger and have more light scintillate more than other stars. Hence for scintillation a sufficient brilliancy is required, because a manifest and sensible image is present in scintillation. If, then, the objection be raised that the light of the sun and of the stars is stronger and more brilliant when they are at the meridian than when they are on the horizon, because then the rays of the stars fall more at right angles, why at that time is the tremor more apparent in them, if brilliancy of this kind is required and is conducive to scintillation? We must reply that as a lessened brilliancy is not sufficient for scintillation, so an excessive brilliancy confuses and absorbs the eye to such an extent that the tremor is not perceived.

The fact that stars appear to be more distant on the horizon than at the meridian causes the vision to err in its perception of the visible object; for since it erroneously judges the star to be more distant when it is on the horizon, it makes a greater effort to see the star, and from the effort there results a stronger movement and straining of the eye. For in consequence of the great effort the eye is compressed, and the coats and humors recede from their natural position, and yet they immediately return to the same position, and therefore straining and tremor occur. Nature, moreover, always returns to the place that is injured or in need, and sends more abundantly spirits and supplies of heat, and therefore in an effort of this kind on the part of the eye spirit and natural heat are released from the common nerve, and an excess of this kind stirs up the humors and causes them to recede from their natural position, to which they immediately return again. Thus a tremor of the eye occurs and consequently a tremor apparently in the object of vision. Ptolemy assigns this cause in the second book of the Perspective. Also since the vision estimates the sun and stars to be more distant when they are on the horizon, it therefore makes more effort and is strained. And according to this the precise reason can be found why the brilliant planets, especially the sun and moon, Venus, Mercury, and Jupiter, do not scintillate; because the vision easily determines them and reckons consequently that they are nearer, according to the statement of Ptolemy in the second book of the Optics, because brilliant objects, like the sun and moon, are reckoned near; those, however, that are less luminous are not reckoned so. Alhazen, moreover, says in the third book that a strong color, although it be more remote, yet appears nearer, because it easily affects the sight; and since this is a fact, the vision does not make an effort in regard to planets of this kind, and therefore is not strained. Moreover, as a matter of fact they are nearer than the fixed stars, and therefore a weaker species comes to the sight from the fixed stars, and the species of vision is weakened in consequence of the distance, and for this reason makes more effort in regard to the fixed stars.

But if it be said that the eye puts forth more effort in beholding small stars than larger ones, and that therefore strain will occur more quickly; we must reply that another cause, which relates to brilliancy, is lacking in them. And if it be said that this effort depends upon a man's choice, and therefore depends upon his will, for which reason this effort is under his control; that he can therefore make an equal effort on a large star as on a small one, on a planet as on a fixed star, and on a star on the horizon as on one at the meridian, wherefore an effort of this kind does not have any bearing in this matter, we must reply that it does. For it is the habit of vision to put forth more or less effort according to the distance of the visible object, and according to the weakness of vision itself, and according to the lack of visibility in the object. Moreover, in this matter vision is trained and practiced, and habit is a second nature, as Aristotle says, and especially so in vision, as the authors on Perspective teach. Hence one does not perceive that he sees by reflection and refraction, but thinks that he sees by means of straight lines only, because he has been accustomed to see by straight lines, as will be stated below. Therefore the effort of vision is not under the control of the rational soul, just as the other judgments of habit are not so controlled, although here some strive by a sophistry of this kind to disprove this effort of vision and certain other natural phenomena habitual and similar to vision. If it be objected that vision does not judge concerning the distance of the planets and the stars, for by reasoning alone do we comprehend the distances of the stars, and that therefore vision will judge equally concerning the fixed stars and the planets as far as concerns the phenomena of visible things that rest upon distance; we must reply that such is not the case. For although vision cannot determine the distance of the stars, because bodies do not exist in continued succession between the stars, yet owing to the greater impression made by the more brilliant planets, as has been said, they appear nearer, although vision cannot state how much nearer they are. And again because it receives a weaker species from the fixed stars than from the planets, since species is weakened by distance, vision puts forth more effort in regard to the species of the fixed stars than in regard to the species of the planets, and similarly the species of vision is weakened by distance, and therefore it is not so active when it comes to a very distant object of vision, and for this reason puts forth more effort and a strain takes place in it and greater movement.

If it be objected that just as vision is weakened by distance, and fails and strains and relaxes again, so it fails in consequence of the density of the medium to secure a true comprehension of the object of vision. Therefore if we assume a density of medium proportional to the distance of scintillating stars they would scintillate. This is not believed to be true. Averroës replies to this in the second book of the Heavens and the World;* for he well maintains that density of medium would tend to produce scintillation and tremor, saying that if the sun quivers only at setting and rising, it is because of the density of the medium. If, however, the sun does so everywhere, then it is because of the strength of the light, which is stronger than in the other planets. He maintains, therefore, that density of the medium tends to produce tremor. Therefore although this is generally denied, yet this is not improbable; and therefore stars scintillate more at rising and setting, not only for the reason stated above, but because of the thickness of the medium

* Bacon evidently intends this to be the conclusion of the condition stated at the beginning of this paragraph.

charged with vapor, since a multitude of vapors abounds in the medium around those hours, because the sun raises at that time the vapors, and does not consume them as at midday; because the solar rays are weak at sunrise and sunset, since the rays fall at oblique angles to the horizon, as has been explained in what precedes.

But if the thickness of the medium does not do this, yet the motion of the medium can have this effect, as Averroës shows in the same passage. For the medium is in continuous motion, and therefore the species of an object while it falls in one part remains in this part after it has moved from its position, until a remaining part succeeds to the place of the first part, and therefore each has a species of the visible object, and they produce species similar to themselves in the same eye, and in different parts of it, and in the common nerve; as Averroës likewise shows in his book on Sense and the Sensible by an example in regard to the species of the sun falling on moving water. For the species does not disappear immediately from the part moved, but remains in it for a time, until a remaining part comes into its place, and then both produce a species in parts of the same eye and optic nerve, and in this way the same object seems to have a change of position continuously, and therefore to tremble. Moreover, Ptolemy approves of this view in the second book of the Optics, giving this example in regard to moving water and a moving medium.

Averroës, moreover, offers a third reason because of intermittent vision, giving the example of a man who looks at an object in one place, and if he moves his eye by pressing it with his finger, he will see the object in another place because of the motion of the eye, since the species of the object is not fixed in one definite place, but is in several places. And likewise if a man looks at an object under a condition of frequent change, namely, now with one eye now with the other, or with one eye frequently closed, provided it be opened and closed continuously, the object seems as though it were in different places; and so subject to a tremor, as it were, and to a change from place to place. The eye, moreover, does this when the distance between it and the object is great; for one frequently closes his eyes and opens them. Although the eyelids may be open, the visual faculty in the pupil and nerve is fatigued from the strain, and the act of seeing at all or at least distinctly is interrupted; and therefore the species of a distant object is not fixed on the sight, because the object is thought to move to and fro.

If it be said that when the air is set in vigorous motion by the wind there would be greater and more violent scintillation, our reply must be that, just as the necessary brilliancy, neither too much nor too little, is required for scintillation, as was stated above (and just as a certain distance likewise is required: for stars might be so far distant that they would not scintillate, and just as some have so little light that they do not scintillate) ; so may we speak concerning the motion of the medium. For the motion of the air, which is caused by the motion of the heaven and of the vapors, a motion that is continuously found in it, is sufficient for scintillation; and this is fixed for the scintillation of the stars, so that a greater motion is not required, nor does a lesser one suffice. If it be said also that species multiplies itself in straight lines in the same medium, nor does it depart from a straight direction because of the motion of the medium, because its path does not vary except by refraction and reflection, which do not occur in the motion of the same medium, and therefore the species will not be produced in different places but only in one place, and thus it will not have a different position in the eye; we must reply that it is true that species always keeps the same direction in the same medium, and multiplies itself directly in one path and one place. But, however, since the different parts fall successively in the same place, and the first part having the species retains it after it leaves the direction of the species, and another part comes into the place of propagation of the species before the species disappears from the first part, therefore the two parts having different positions produce species in different parts of the eye and cause the phenomenon that has been mentioned. If, however, it be said that the rays of all stars and planets pass thus through a tremulous medium, we must reply that the other reasons relating to strain of the eye and brilliancy of the object of vision are missing, as is manifest from the solutions of the objections offered.

From all these statements the reason for scintillation can be gathered, although with difficulty, owing to the various objections occurring. For on the part of the eye there must be effort owing to which the eye is strained. For because of this effort an excessive freeing takes place of spirits and forms of heat, which compel the humors of the eye to leave their natural position, and nevertheless from its effort there result a compression and constriction of the eye, in consequence of which the parts of the eye do not retain their position. Again in the third place there arises from this an interruption of vision, because of which the visible object appears to change its position, as has been explained by Averroës. These three things, then, result from eye-strain, which is caused not only by the weakness of the visual power and of the species, owing to the actual distance, but because of the judgment of the apparent distance; and therefore vision makes a greater effort on the fixed stars than on the brilliant planets, which it judges to be near because of their excessive brilliancy. But not only is this effort required, but also sufficient brilliancy on the part of the scintillating body, because a scintillating body seems to scatter darts of light, and for this reason Saturn and small stars do not scintillate. There must likewise be a tremor of the medium, as has been stated, and all these conditions are required for scintillation, owing to the distance of the object of vision, because an object placed at the required distance does not scintillate; and for this reason Aristotle ascribes the phenomenon chiefly to distance, although many attendant conditions must be considered. In this way scintillation in the heavenly bodies is to be understood, of which Aristotle speaks, although there is a special kind of scintillation in these things here below, of which mention will be made later on.

CHAPTER VIII

AND now at the conclusion of this discussion in regard to the way in which we see in straight lines owing to the three methods of vision, by the sense alone, by recollection [*scientia*], and by a syllogistic process, there is rightly a doubt, which faculty of the soul it is, that in recollection and the syllogistic process acts in regard to things made visible through the medium of the visual sense. If we accept the terms recollection and the syllogistic process in the sense in which they are used in matters pertaining to logic, nature, and metaphysics, in accordance with the general practice of philosophers, this faculty must be the rational soul; because the syllogistic process and recollection pertain to it alone, as these terms are accepted in the sciences mentioned. Alhazen calls this part of the soul the distinctive faculty, which according to him reasons and understands; for sometimes these terms and ones like them are found, so that it would seem according to their literal meaning that the faculty is the discerning and rational soul.

But it is well known that a dog recognizes a man seen previously when it sees him again. Moreover, the ape and many other animals act in this way, and distinguish between the objects of vision of which they have a recollection, and they recognize one universal from another, as a man from a dog or a tree, and they distinguish individuals of the same species; and therefore the perception which writers on perspective call perception through recollection belongs to animals as well as to men. Therefore this faculty must be a function of sensitive life. But likewise it is manifest that the same holds true for the perception that is termed perception through the syllogistic process; for motion is recognized through it; as, for example, a dog, when someone raises a stick to strike it, flees; which it would not do if it did not perceive that the stick is changing its position with respect to itself and is approaching it. Similarly when an animal, for example, a dog, cat, wolf, or some other animal, holds an animal on which it feeds, while the prey is still, the plunderer stands motionless; but when the captured animal flees, the plunderer follows until it seizes its victim if it can. It would not do this except for the reason that it perceives that the position of the prey has changed with reference to itself, and therefore it perceives motion and rest and distance. We must grant that animals have some perceptions of this kind by a certain natural purpose and instinct without deliberation, and the faculty that acts is cogitative, which is mistress of the faculties and employs the other faculties of the soul. For there is here required a recollection of this kind of objects of vision for the distinction of the universals and particulars, and this recollection belongs to the imagination itself, as was previously stated, if it is exercised with respect to light and color and the twenty common sensibles, since imagination is the storehouse of the species coming from them. If, however, it deals with the things pertaining to the estimative faculty and the memory, that particular kind of memory serves here in recalling them. For although the lamb flees from the wolf that it has not seen, yet if it has seen the wolf it flees more quickly and carefully when it sees him a second time. And thus a distinction is made through the cogitative faculty by means of memory, which is the storehouse of intentions apart from the senses with regard to sensible matter, as was explained before. Therefore the perception by which things seen previously are distinguished from other things is present in animals. Wherefore there is here a double use of the term; or rather there is a fault in the translation, which does not have a proper word for this kind of perception. Likewise there is ambiguity, because the term syllogism should be limited to its ordinary use. For without doubt in no way can the fact be hidden that animals perceive the distance of objects, and motion, and rest, although in regard to other common sensibles such is not the case.

But as regards formal reasoning we must consider that the arrangement of the reasoning in a logical figure and the distinction of the conclusion from the premises pertain only to the rational soul. But a certain gathering together of several facts into one in consequence of a natural purpose and instinct, the several facts resembling premises and the resultant one a conclusion, because it is gathered from them, can be found in animals. For we see angered apes prepare plots against men and coördinate many things, so that they may secure their revenge, and therefore they assemble into one that which they draw from many sources. We see also spiders arrange their web, not haphazard fashion, but in various geometrical forms, so that flies may be caught easily by them. The wolf eats earth, so that he may be heavier when he seizes a horse or bull or stag by the nose, in order that by reason of the weight of the earth he may weigh down the animal and hold it. I have seen, moreover, a mouser that wanted some fish swimming in a large stone vessel; and when it was unable to catch them because of the water, it drew out the stopper and let the water out until the vessel became dry, in order that it might capture the fish in a dry place. It therefore conceived of several acts, in order that it might secure the end it had in view. The bee makes all its houses hexagonal, selecting one of the figures that fill up space, so that no vacant space may be left between the houses; and it does not wish that there should be this space, lest the honey or the young should fall outside of the hives and perish. With this end in view, then, resembling a conclusion it assembles in its mental process many facts which resemble premises. This is true also of a host of facts, in which animals think of many things in their order with respect to an object that they have in mind, just as if they were drawing for themselves a conclusion from the premises. But they do not arrange the course of their reasoning in form and logical figure, nor from their deliberation do they distinguish the last from the first. Nor are they conscious that they are following a course of reasoning of this kind, because their reasoning takes place thus from intuition alone and natural instinct. This course of reasoning is similar to argument and syllogism, and for this reason authors on Perspective call it argument and syllogism. And assuredly they call this kind of reasoning syllogistic more appropriately than they call the distinction of universals and particulars seen previously the cogitative faculty through knowledge.

Part Three of Perspective

FIRST DISTINCTION

In which the author considers reflected and refracted vision, dividing the subject into three distinctions. The first one is on reflected vision and contains six chapters. The first chapter is on reflection in general.

Chapter I

AFTER our discussion of vision in straight lines we must now speak of vision in other ways, namely, in reflected and refracted lines. Since, moreover, what has been said concerning the parts of the soul and the structure of the eye, and the path of the species in the coats and humors of the eye to the common nerve, and concerning the triple mode of perceiving sensibles by the sense alone, by syllogistic reasoning, and by recollection [*scientia*], is common to vision effected by straight lines, by reflected lines, and by refracted lines, less need be said in regard to these matters. First, then, as regards the reflection of vision we must recollect that as far as concerns the judgment of the human vision a dense body can impede wholly the species, as a wall and dense bodies of this kind; or in part, as water, glass, and crystal. For every dense body in so far as it is dense reflects species, not because violence is done to the species, but because the species takes the opportunity from a dense body impeding its passage to multiply itself by another path open to it. A dense body may be rough or smooth. The parts of a rough body are unsymmetrical, and therefore each part makes its own proper reflection. For this reason the parts dissipate the whole species, nor can it as a whole reach the eye, and therefore there cannot be a sensible reflection or a representation of the object of vision. But owing to the equality and smoothness of the surface of a polished body, as in mirrors, all the parts are symmetrically arranged for a single action, and the species comes back whole and sensible to the eye, and the image is manifest; not, however, such a perfect one as when the eye sees by means of a straight line, because reflection weakens the species, as has been stated in the book on Multiplications.

Since, if the species passed through the medium of the mirror, it would make the angle *a* equal to the angle of incidence, which is *b*, by the fifteenth proposition of the first book of the Elements of Euclid, which states that the opposite angles are equal, the angle of reflection, namely, *d*, must be equal to the angle of incidence, since the ray forms an angle on this side of the mirror like the one formed below it. This, moreover, is proved in a simple manner, as follows: Let *abc* be a plane mirror, *d* the visible object, *e* the eye, and let *ab* and *bc* be equal, *da* and *ec* be perpendicular and equal, let *db* be the radius of incidence, *be* be the radius of reflexion; then since *ce* and *cb* are equal to *ad* and *ab*, and the angles contained within the sides are equal, since they are right angles, it follows by the fourth proposition of the first book of the Elements that the remaining corresponding angles are equal, namely *g* and *f*, which is the proposition. If, therefore, the triangles are equal, the proposition is evident. If, however, one triangle is larger than the other the angles of incidence and reflection are still the same, and are unchanged, and therefore the proposition will always hold. But the author of the book on Mirrors supposes that the triangles are similar, and therefore *ab* will be to *cb* as *da* to *ec*, and therefore the angles *g* and *f* are equal.

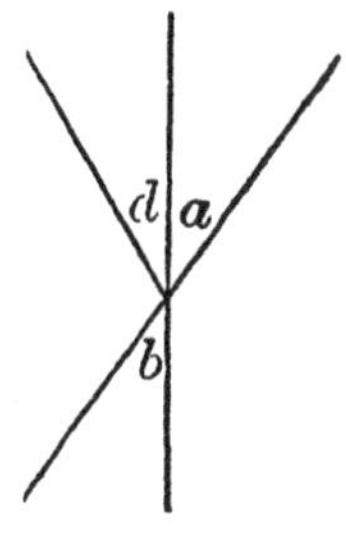

FIG. 49.

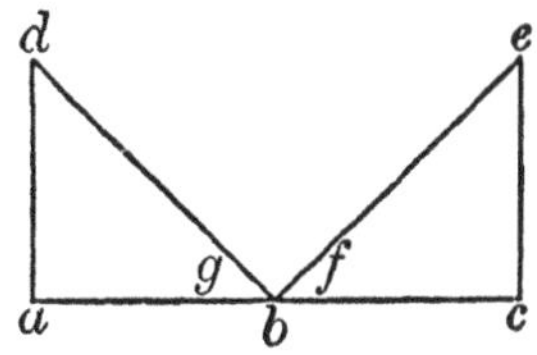

FIG. 50.

By this means also we show that the law holds with regard to convex and concave mirrors. For *a* and *b* make equal angles with a plane mirror, which is *dc*, but the angles of contact [tangency] are equal. Therefore if these angles are taken from those formed with the plane mirror, the angles that are left, namely, *f* and *g*, will be equal, which we wish to prove. In the same way we prove the law in regard to the convex mirror. For *a* and *b* make equal angles with the plane mirror *gf*. Therefore if to these angles are added the angles of contact, which are always equal, the angles *ahd* and *bhc* will be equal, which is the proposition. The proof is given in this way in the book on Mirrors, and in the book of Alkindi on Aspects.

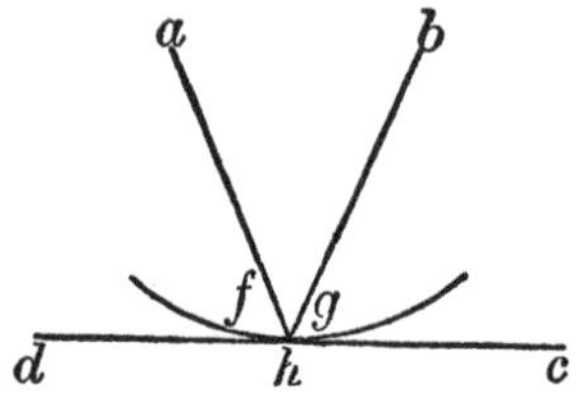

FIG. 51.

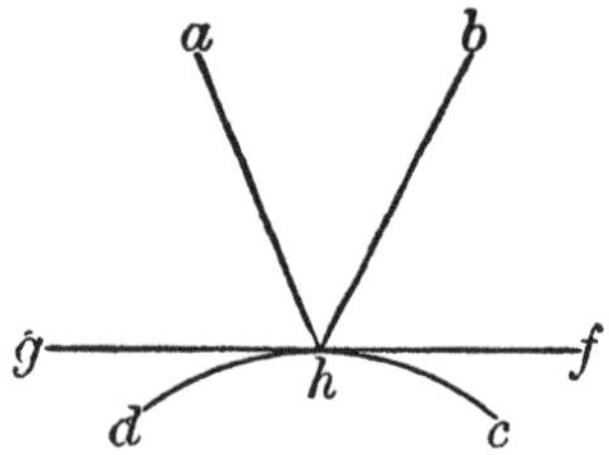

FIG. 52.

CHAPTER II

IN the next place, we must bear in mind carefully that the mirror contains nothing nor is anything seen in it, as is generally thought. But the object opposite it from which the species comes is seen, as Alhazen shows in many ways in the fourth book. For as the end of the straight line *oa,* since vision is effected by means of this line, is a visible object itself, so must the end of the reflected line, namely, *oda,* be *a.* Moreover, the species is not seen except by accident, as has been explained above. We might also assume that the same conditions would hold with regard to the species as with regard to a spot impressed on the mirror or with regard to some part of the mirror that has been marked, on which the species would be impressed, but which would not of necessity have a determined position, in order that one might see the spot on the mirror or some part of the mirror that had been marked. Therefore in vision effected by means of reflection a determined position of the eye is not required. This reasoning, however, is false. For unless the eye is at *o,* it will not see anything by reflection at *o.* For if this reflected species were anywhere else, it would not reach the eye, owing to the equality of the angles of incidence and reflection. This, then, can be shown in many ways, but since all having a knowledge of perspective are in agreement on this point, we need not dwell upon it fur-

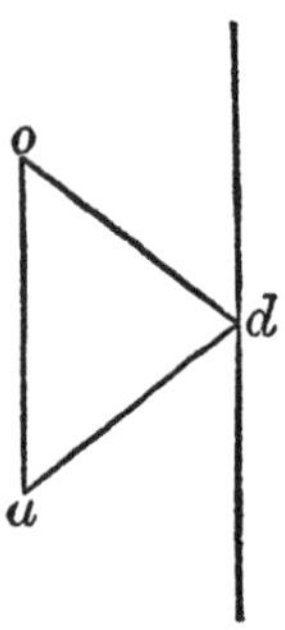

FIG. 53.

ther. Many truths become manifest to the student in consequence of this fact; for, in the first place, it follows that the species is not fixed in the substance of the mirror, nor is it impressed upon the mirror, as is generally supposed, but merely passes by means of the surface of the mirror in the opposite direction in accordance with the equality of the angles of incidence and reflection.

Moreover, since this is the case, if the light coming from the moon and the stars were the light of the sun reflected from their surfaces, as the rank and file of philosophers think, in seeing the moon and a star we should see the sun. This, therefore, is evidently false not only because of the equality of the angles of incidence and reflection, as has been shown in what precedes, according to Averroës in the second book of the Heavens and the World, but for the reason now given. In the same way it follows that a comet is not the reflection of the solar light from the surface of a star, as many have maintained that the tail following the comet is nothing else than the light of the sun reflected from the surface of a star to us, or that the light coming to us from a comet is the light of the sun reflected from the surface of some star. But this is a false notion. For in that case in seeing a comet we should see the sun, which is not true. Moreover, since the rainbow is merely the image of the sun reflected from a moist cloud, as all agree and as the fact is established, in accordance with the proof given below in the section on Experimental Science, then in seeing the cloud we should see the sun and nothing else; but this, however, seems absurd, since our vision judges that it beholds colors and a colored bow. But the sun does not possess such a figure nor such colors. But we shall make sure of this matter later on in a broader discussion.

Moreover, in addition to these facts we must note that the object does not appear to the vision by reflection in its own position, because vision is accustomed to see by means of straight lines at their extremities, and for this reason vision does not perceive the bending of reflection; and therefore it judges that the object is always in the visual ray, and the place of the image, which we call the appearance of the object, is at some point of this ray. This result is due to the fact that vision is caused by external propagation, and for this reason the vision judges that the object is in the direction of the species of the eye. But it is not always, however, in the same place, but as in many instances is at the intersection of the visual ray with the cathetus, which is the perpendicular drawn from the object to the mirror. Sometimes it is not at that intersection, but in the visual ray alone, because it can be equidistant from the cathetus, as in the case of a concave mirror, as we shall explain. When it intersects the cathetus, the intersection takes place in many ways. For sometimes they meet behind the head, sometimes in the eye, sometimes on the surface of the mirror, sometimes beyond the mirror, and this last in several ways. For it happens sometimes that the object appears as much beyond the mirror as it is distant from the mirror, and this diversity happens owing to the difference in mirrors. Therefore we must now take up this question.

Chapter III

Mirrors, then, are of seven kinds, in which, according to authors on Perspective, vision differs; namely, spherical, conical, cylindrical, polished on the outside and on the inside. These make six; the seventh is the plane mirror. For each of the first three classes can be concave, that is, polished on the inside, or convex, that is, polished on the outside, making six in number; the plane mirror has only one structure. In the case of these mirrors, then, I wish, following the views of authors on Perspective and particularly of Ptolemy and Alhazen, to consider as briefly as I can the modes of vision according to the diversity of mirrors.

In plane mirrors, then, very little error occurs, because the objects appear in their proper form and size. For the position alone is altered, since the right appears on the left and the reverse and the top at the bottom, whence towers appear inverted in the water, since the reflection takes place from the plane surface of the water. There is, however, an error in plane mirrors that is common in all mirrors, namely, the object does not appear in its own place, nor is the place of the image there. When, moreover, we speak of the place of the image, we mean the appearance of the object, for this is all that we mean by the term.

Neither the object, then, in these mirrors nor the place of the image appears in the place of the object, but always in the visual ray, for the two reasons given. But in plane mirrors the image is determined at the intersection of the visual ray with the cathetus, namely, as far beyond the mirror as the object of vision is distant from the mirror. This law does not hold in the case of other mirrors. It can be shown by proof. For let *a* be the object of vision, *o* the eye, *ad* the cathetus, *od* the visual ray. I say that *db* is equal to *ab* itself. But *bd* is the distance of the image of the object from the surface of the mirror, and *ab* is the distance of the object from the same surface of the mirror; it appears, however, as far beyond the mirror as it is distant from the mirror. For the angles *e* and *f*, since they are right angles, are equal, and *g* and *h* are equal, by the fifteenth proposition of the first book of the Elements of Euclid, and *h* and *c* are equal, because they are the angles of incidence and reflection. Therefore it is evident that *c* and *g* will be equal. Since, then, the angles *e* and *g* of the triangle *egd* are equal to the angles *f* and *c* of the triangle *acf*, and the included side is common to both triangles, it is evident by the twenty-sixth proposition of the first book of Euclid that these triangles are equal in all their parts. Therefore the sides *ab* and *bd* will be equal. Wherefore vision judges that the object is as far beyond the mirror continuously and in a straight line as it is in front of it in the case of plane mirrors. By this and previous statements the error of many is removed, who believed that the species of the object was there in reality and that it diffused itself through the medium of the mirror and appeared there. But no visual species exists in the mirror, as has been stated, nor does it enter the mirror, so that vision should be effected by means of an entry of this kind, but it merely passes by means of the surface of the mirror to the end of the reflection in the opposite direction, according to the equality of the angles of incidence and reflection. Hence the place of the image does not seem to be at the intersection of the visual ray with the cathetus because of the reality of its existence in that place, but by reason merely of its appearance.

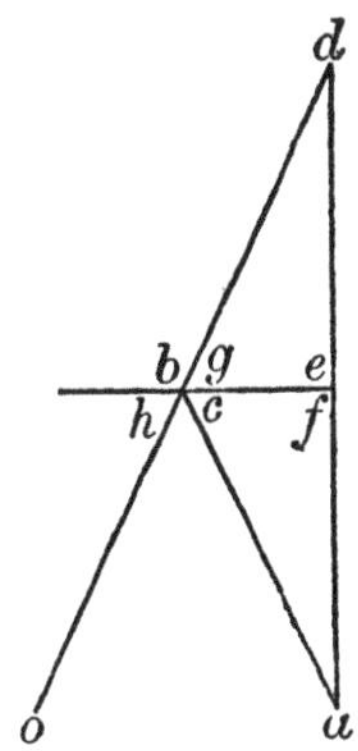

FIG. 54.

In spherical mirrors polished on the outside [convex] also according to the judgment of the vision the object appears at the intersection of the visual ray with a line drawn from the object to the center of the sphere. This intersection can be beyond the mirror, or within, or on the surface of the mirror. The same holds true in the cylindrical and conical mirrors. All the errors, moreover, that are in plane mirrors, occur in convex mirrors also, and more, since in these the object of vision frequently appears less than it is; sometimes, however, equal or greater, but very rarely. But the image appears less, because the width of the surface of the mirror is less, from which the rays intersecting at the eye are reflected, than it is in plane mirrors; for the rays reflected from a convex mirror are more dispersed than those from a plane mirror. Therefore that the rays may pass to the sight just as from a plane mirror, the reflection must be from a shorter surface than from a plane mirror. The representation, moreover, of the image follows the condition of the reflecting surface. In these mirrors, therefore, hardly anything appears as it is, the arrangement of the parts excepted, which is the same in the mirror as in the object. In these mirrors that which is straight appears curved; for since the reflection of the outer rays is on a convex surface, the termini are more distant from the center of the eye than the extremities of the median ray, and that which happens in the turning is judged to be present in the object. Very rarely, however, does it happen that what is rectilinear appears so. It does so when the sight is in the surface in which the line seen and the center of the sphere lie. The mathematical proof of this fact is too lengthy to be given here. Again we must note that in convex mirrors the distance of the image from the mirror is less than that of the object of vision, while in plane mirrors it is equal. The reason for this is found in the fact that in convex mirrors the ray intersects more quickly with the cathetus than in plane mirrors, as is evident to the inquirer.

In cylindrical mirrors polished on the outside the same error is present as in convex spherical mirrors, and more in addition. For in these an object seen at a distance appears smaller than in convex spherical mirrors. The reason for this is apparent if one considers the difference between the surfaces of the former and the latter. In these mirrors, then, very large objects appear very small, and rectilinear objects appear much more curved than in convex spherical mirrors. But we must note that in these mirrors the reflection sometimes takes place from the length of the cylinder, as when a line seen is equidistant from the axis of the cylinder, and in this case the reflection is the same as in plane mirrors, with this exception, that, since the line from which the reflection takes place has breadth, it appears somewhat curved. Sometimes the reflection is from the cross direction of the cylinder, and in this case the image is very much distorted and very much shortened. Sometimes the reflection takes place from a midway position, or from one approaching more nearly the length or breadth. Images will be formed in accordance with these several positions.

In conical mirrors polished on the outside, likewise, the same errors occur as in convex spherical mirrors, because the image is smaller than the object of vision, and what is rectilinear appears curved, and the reflection differs in these as in cylindrical mirrors, because the reflection takes place from the length of the cone, or from its width, or from a position between. Moreover, in these the shape appears conical. For in general it is true that the species comprehended by reflection is assimilated to the shape of the surface of the mirror. In these, also, the further an object is distant from the mirror the smaller it looks, and the nearer it approaches the larger it appears.

Chapter IV

Among all mirrors the greatest deception is in the concave spherical ones; for the deception in regard to size occurs in these as in the others, since sometimes it is greater, sometimes less, sometimes equal; and besides this there is a deception in regard to number, since sometimes one object appears to be two, sometimes three, sometimes four, according to different positions, so that this number it is impossible to exceed. Moreover, in these a disarrangement of parts is apparent, since the object appears sometimes erect, sometimes inverted; and thus it is manifest that in these mirrors nothing appears except with deception. In concave mirrors, however, straight lines appear sometimes straight, sometimes convex, sometimes concave; and convex lines appear sometimes convex, sometimes concave; and sometimes concave lines are apprehended as convex, as is proved in the sixth book of Alhazen, seventh chapter; and this happens in accordance with a difference of position with respect to the mirror. In these mirrors, then, sometimes the cathetus is equidistant from the visual ray, and in this case the place of the image is the same as the point of reflection, and this is true because it is a point of divisible reflection, and if we consider one part, the image should appear beyond the mirror, but if we consider the other, it should appear this side of the mirror, as will be evident. But since the figure is one and continuous, the whole appears at a distance midway between the two, namely, at the point itself of reflection. But when the cathetus and the visual ray intersect, the object appears at their intersection, and this occurs in several ways, according to difference of position. For sometimes the place of the image is in the mirror, sometimes beyond it, sometimes this side of it; and this is either between the sight and the mirror, or in the center itself of vision, sometimes behind the eye.

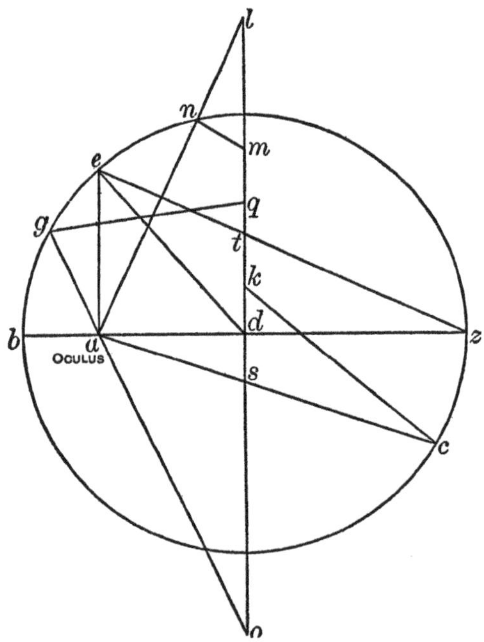

FIG. 55.

All these facts are made evident in the figure given below. For the form *t* is reflected from *e* to *a* by the ray *ea,* equidistant from the perpendicular *td,* and appears at *e;* and *m* is reflected from *n* to *a,* and intersects the perpendicular *ml;* and *k* is reflected from the point *c* to the eye *a,* and appears at *s;* and *q* falls to *g,* and is reflected to *a;* it intersects, moreover, the cathetus behind the eye, namely, at *o;* and *z* falls to *e,* and is reflected to the eye. But the ray *ae* nowhere intersects the cathetus drawn from the point *z* through *d,* except in the center itself of the eye, whence *z* appears there. In all these different images nowhere is any one of them really perceived unless its position is beyond the mirror, or between the sight and the mirror. Hence those that appear in the center of the eye or behind the head appear as things not determined by vision. For vision is not constructed by nature to perceive the position of forms unless they are in front of the eye. When, moreover, the eye is at the center of a concave mirror, it is the only thing visible to itself; for no form is reflected to the center except that which comes from the center. The perpendicular, in fact, returns upon itself. If, however, the eye is placed on the periphery or outside it, the eye is not visible to itself, but the reflection is in the opposite direction. If, however, the eye is placed below the periphery, no one of the objects that are in the semidiameter in which it lies, is visible. If, however, some visible object is placed at the center, it cannot be seen by reflection, for its species is not reflected except upon itself. In regard to the number of images we must note that when the eye is so placed that from four parts of the mirror the reflection of the form of the same object is made, and there is an intersection in different places of each of the rays with the cathetus, there will be four images. When the reflection is from three parts of the mirror there will be three images; when from two parts, there will be two images; when from one part, there will be one image, as is very accurately stated in the fifth book, part second. And note that all proofs dealing with places of reflection seek to answer the question, where the angle of incidence can be equal to the angle of reflection, and as many images appear at the same time as there are points of this kind under the same position and with respect to the same eye; if, however, the rays in different places intersect the perpendicular at a sensible distance. For when the distance of the point seen is greater from one eye than from the other, the positions of the images will be different with respect to either eye, but the difference is imperceptible, and for this reason the two appear as one.

We must note that distance and proximity are reflected in different ways from concave mirrors; as is evident from the eleventh proposition on Mirrors. For the visible object, *ed,* falls to the mirror by the rays intersecting at *z,* for although reflection is made from every point, yet only those intersecting each other meet at such a distance in the eye. For the visible object *kn,* which is within the confluence of the rays, appears otherwise than it is; because, in general, height and depth that are within the confluence of the rays appear inverted; but objects outside of the confluence appear erect, as they are, according to the statement of that proposition. This is evident, because the ray *ba,* which is more elevated, is reflected to *e,* which is higher in the object seen and intersects the cathetus at a higher point, because at *l;* and *bg,* which is a lower ray, is reflected to the point *d* in the visible object *ed,* and meets the cathetus at a lower point, namely, at *m,* whence the object appears as it is. But the lower ray *bg* is reflected to *k,* which is higher in the visible object *kn,* and the ray *ba* to *n,* whence *k* necessarily appears at *f,* and *n* at *c,* and thus the object is inverted. This proof follows closely the first statement of the first book on vision, "Under more elevated rays objects of vision appear more elevated, under lower rays they appear lower." But, however, we must bear in mind that the demonstration in the book on Mirrors has an inaccurate figure, since the catheti

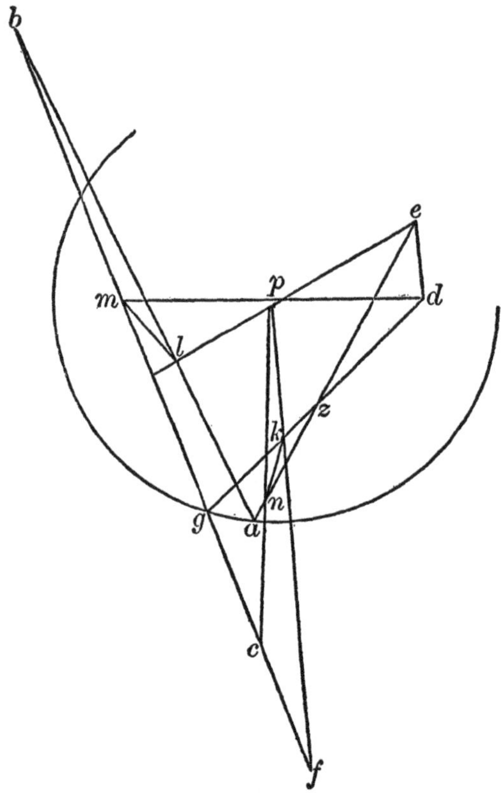

FIG. 56.

should fall to the center of the sphere, which rule is not there observed, and here therefore I give the correct figure.

In cylindrical mirrors polished within the same phenomena occur as in concave mirrors, both in the size of the object of vision and in the number of images, since in the inversion of what is seen there is also a reflection in different ways in these mirrors, as in exterior cylindrical mirrors, from their length, width, and midway position; and moreover, the images are diversified and the positions of the images are varied, according to difference of position with respect to the cylinder, just as in concave spherical mirrors. In concave conical mirrors similar phenomena occur as in concave cylindrical and spherical mirrors; the reflection varies also in these, as in conical mirrors polished on the outside, from the length, breadth, and middle position; and similarly are the images diversified in figure and size.

Chapter V

In addition to the reflections already mentioned we can give some fine examples in natural objects. For although it has often been said that in consequence of the difference in the angle at which light and reflection fall from colors and objects upon the eye is there a difference in color and in the shining back of the light in bright objects, we must understand that this is owing to the incidence and reflection at different angles. For the action of light is stronger when it falls and is reflected at right angles, and it is weaker when it falls at angles less than right angles, and it is much weaker when it falls at very acute angles. Thus light falling in different ways can either make clear, or hide, or soften the intensity of color, or augment it in various ways, as is evident in the pigeon's neck and the peacock's tail, and in many other things. I do not say, however, that there are not real colors in the peacock's tail and in the pigeon's neck, although some deny it, especially in the case of the pigeon's neck; for the figure formed by colors in the peacock's tail manifestly contains a certain color, but because of the tenuity of the feathers in the pigeon's neck, since they have no thickness, and because of their great proximity, we do not perceive the real colors that are in them. I do not, however, deny that owing to the different incidence of the light at various angles those colors are now more manifest, now more hidden, now brighter and livelier, now obscured and weakened. This is the case also in the colors of the rainbow; for in reality they exist only in appearance, as the incidence of light at fixed angles produces them. The rainbow is caused by reflection not by refraction, since it varies according to the view of the observer, as Experimental Science will show.

Drunken men and those weak in health, according to Aristotle in the third book of the Meteorologics, and according to Seneca in the book on the Rainbow, see themselves, and it seems to them that they see themselves walking in advance. In explanation of this Seneca says that the species coming from them, that is, their visual powers, are weak, and therefore the air, although not dense, is able to resist the species and reflect it back to the sight, and for this reason the species is formed in the air in front of them and returns to the eyes and to the whole body. Hence they see themselves just as they would in a mirror. Here alone is vision effected by the species of the eye, and not by the species of the object of vision, except that the eye together with the whole body is seen, and the species of the eye is reflected to all the anterior parts of the body, and the man is seen in front of himself, because the position of the image is in front of the man at the intersection of the visual ray with the cathetus. But that this vision is effected only by the species of the eye is evident, because the species of other parts of the body and of the clothes are strong, so that they penetrate the air, which reflects the weak species of the eye. It is thus evident that the sight produces from itself its own species. This vision is weak because it is effected through the species of the eye alone; and the species of vision is weaker and grows so more quickly than that of the other part of the eye, because the eye possesses a more tenuous and weaker substance. If it be said, then, that the visual rays could be reflected when they are directed toward the heavens and fail in the profundity of the air or of the transparent sky, as was said before, and similarly in deep water, and that a man would then see himself when he views that deep water from a distance, which is not true; we must reply that this does not happen unless the air near these persons be somewhat dense, which can happen in the case of drunken men, owing to the moisture from the vapors of the wine that are set free, and in the case of the infirm, owing to the vapors always released near them from disease. Hence evil and foul vapors are always near these people in the air, by which the air near them is tainted and rendered dense, so that in their vicinity it acts like a mirror, which would not happen in another more remote part of the air. Likewise the air near the ground is made dense by vapors arising from the earth and from the water; and therefore from both causes of density the air near at hand can have the quality of a mirror. Therefore owing to this difformity in parts of the air it happens that one part has the quality of a mirror and another part does not have it with respect to the eyes of drunken men and of those in weak health. But the strong eye looking in water finds it of uniform density. The same is true in the sphere of the heavens at a distance; and vision does not find there a denser part that might have the quality of a mirror, nay, it always finds it rarer, and therefore the species is multiplied until it fails without reflection. But if it be said that a strong eye finds vapors in the air and parts of clouds, so that reflection might take place, we must reply that it passes through vapors because of its strength and also through the light and thin clouds; but it does not pass through thick clouds. One does not, however, see himself because of the distance of the mirror and because of the lack of a perfect polish; for they do not possess polished and regular surfaces at all. Distance, however, is the chief obstacle, just as if a large mirror were placed at a distance of one league or two or three, however large it might be, a man would not on this account see himself; and therefore he could not see himself by means of the clouds, since they are distant from us about fifty miles, as is shown in the book on Twilights.

The vision of drunken men and of those in weak health can be explained in another way, namely, by direct vision. For their species are in the air in front of them, and because the visual faculty is weak it therefore ends at once and the species in the air is like an object, just as we stated before in regard to the weak eye which sees sometimes the species along with the object so that one object appears to be two, errors which do not happen in the case of a strong eye for the reason given there.

When the eye views a candle and lowers the eyelashes, it sees the candle project rays in the form of a radiant pyramid whose vertex is in the candle, and the dispersion of the rays is quite sensible to the eye. The reason for this is found in the fact that the rays of the candle fall on the lids and the lashes are polished, possessing the property of a mirror, and a reflection for this reason takes place from them to the eye, when they are so inclined that the eye is able to receive the rays reflected at angles equal to those of incidence. Therefore this phenomenon does not occur in any position at all of the lashes, but only in one determined.

When a man views something bright and polished, like a metal cross on a campanile or lofty tower, he will see a body of this kind scintillate strongly when the rays of the sun or moon fall on it and are reflected to the sight. The reason of this is the sensible variation of the angle owing to the motion of the star. For although on account of the distance of the star from us we do not perceive its motion, yet owing to the moderateness of the distance of such a scintillating body we are able to judge in regard to the fall of the light at different angles owing to the motion of the star, and therefore it seems to project rays according to the different positions and to scintillate. For if we should imagine a wheel whose center is the center of the world, and the circumference is the revolving heaven, then although the radii extended from the center to the circumference, that is, branches or rods fastened at the center and extended to the circumference, would not seem to move near the circumference owing to distance from the sight, they would, however, be seen to move perceptibly about the center by an eye placed near it. Therefore similarly the rays of the sun falling on an object here below will make a sensible variation in the angle, although the motion in the sun would not be apparent to us. This is the new kind of scintillation, which I promised in what precedes that I would mention.

Optical Science

Chapter VI

It is generally known among students of perspective that when a mirror is placed in a vessel containing water and a double image appears in it, one will be that of the sun, and the other that of some star situated near the sun. But it cannot be a fixed star because the sun hides these, nor is it one of the planets, since the planets are distant sometimes more, sometimes less. But the images always have a uniform distance. Moreover, the phenomenon happens in the light of the moon as well as in that of the sun; likewise in the light of a candle; which they neglect to try. Wherefore it is not a star which appears, but it is the double image of the sun or moon or candle reflected from the double mirror. For the surface of the water is like that of a mirror, and by it one image is formed and the other is formed by the mirror. It is thought that the image formed by the water is larger and more perceptible, since the ray that forms the other image is much weakened because it is first refracted at the surface of the water, then it is reflected from the mirror, and thirdly is refracted at the surface of the air. But reflection and refraction greatly weaken the species, so that it is not able sufficiently to represent the object; and for this reason the former image is weaker and smaller and less perceptible. But this reasoning yields to mine that the greater image is made by reflection from the mirror; because the mirror is dense and has lead as one of its parts, which impedes the passage of the species; and for this reason the mirror has the means of receiving the image and returning it. For water owing to its rarity has less of the quality of a mirror, and therefore gives back a weak image. Now as to the objections raised in regard to the refractions, we must state that the weakening which happens through them does not make the image less than that formed by the water, but less than it would be if the mirror were in a dry place outside the water.

It is generally believed that it is absolutely true that in a broken mirror as many images appear as there are broken parts. But this is not the case except when the broken parts do not keep the same position but have a different one. For if they retain the same position that they had in the unbroken mirror, only one image will appear, because the species coming is made one and remains one whether the mirror be whole or broken, provided the parts retain their same position; because there is one point of reflection, and there is one place on which it falls. When, however, the parts of a broken mirror are placed differently, the species necessarily are counted, because in its changed position the points of reflection are different and in different places, and therefore different images appear.

SECOND DISTINCTION

Of the third part, which is on refracted vision, in four chapters. The first chapter treats of vision by refraction in general.

Chapter I

HAVING showed how vision is effected by straight and reflected lines, we must now show, in the third place, how it is effected by refracted lines. Although this is more difficult than the aforesaid subjects, yet we now are in a position to understand by means of what has been said, because this topic has many points of agreement with the others, as was stated in that part which deals with direct vision. We have seen how all rays must be refracted in the vitreous humor except the axis of the pyramid of rays, which passes through the centers of the coats and humors, and that no ray of the visual pyramid is refracted at the cornea, or at the albugineous humor, or at the anterior of the glacialis, since the whole pyramid falls perpendicularly to those three bodies. And the rays would pass to their center if the vitreous humor were not before that point, and therefore the vertex of the pyramid is of necessity cut off and the pyramid is shortened and truncated. But many objects can be seen besides those from which this pyramid comes; but not by means of rays reflected on the eye, since in that case they would recede from it; and therefore they must be seen by means of refracted rays. For let *ac* be the anterior glacialis and *bd* the cornea, and *feo* the pyramid of rays, then the ray *pn* comes from a visible object outside of the visual pyramid. This ray does not fall perpendicularly on the cornea, nor does it enter the opening of the uvea, or, if it were to enter, it would not pass to the glacialis, but would cross beyond to the side of the eye, to the point *l,* for example. Therefore, since the visual force exists only in the glacialis, *p* will not be seen by means of the ray *pl;* but since the cornea is

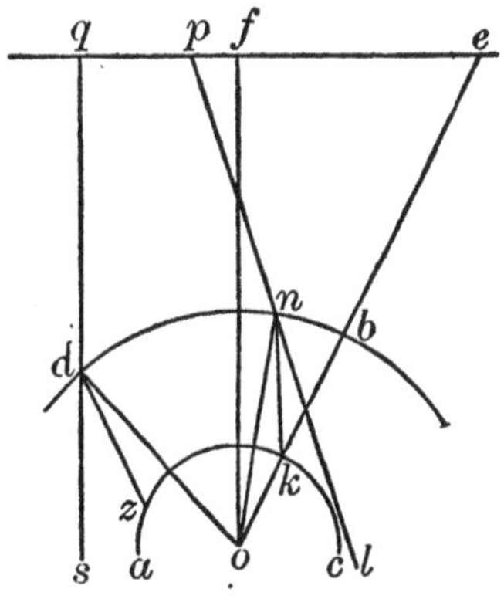

Fig. 57.

denser than the air, and the ray *pn* does not fall perpendicularly on the cornea, although it falls inside of the pyramid before it comes to the cornea, it must be refracted in its progress. Similarly if from a point outside of the pyramid of rays a ray falls on the cornea, and outside of the visual pyramid, as *qd,* it will not pass to *s,* but will be refracted at the point *d* in the surface of the cornea, between the straight path *ds* and the perpendicular *do,* to the point *z* in the glacialis. And so *p* will be seen between the straight path, which is *nl* and the perpendicular *no* to be drawn from the point of refraction, and the refracted ray will go to the point *k* in the glacialis, and in this way *p* will be seen by the refracted ray, namely, *pk.* And for this reason it is less clearly seen than objects that are in the base of the pyramid, since they are seen by straight and perpendicular rays. The same holds true in regard to the visible point *q,* as is evident from the figure.

Similarly whatever is seen by straight and reflected rays, is of necessity seen at the same time by means of refracted rays, and thus it is seen with greater certainty, because it is seen in two ways; and in this manner are the excellence and certainty of vision rendered complete. For point *p* is seen by the perpendicular ray *pg,* which goes to the center *o,* and nevertheless it is seen by means of *pe.* For *pa* does not go to *d,* but is refracted at point *a* of the cornea, between the straight path *ad,* and the perpendicular *ao,* to the point *e* in the surface of the glacialis.

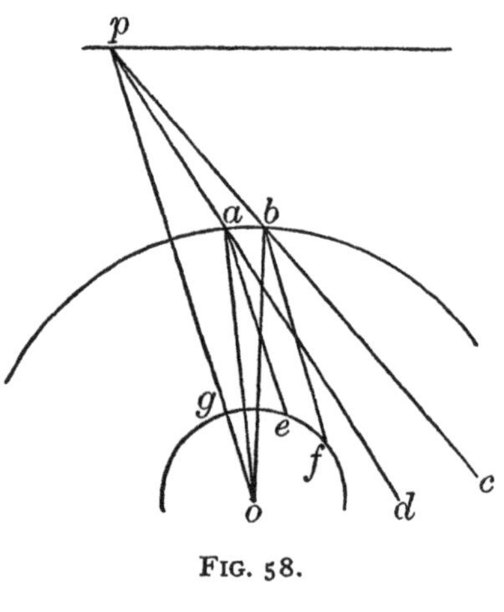

FIG. 58.

Moreover, *p* is seen not only by one refracted ray, but by an infinite number. For from *p* itself an infinite number of oblique rays can be drawn to the surface of the cornea, and each of these will be refracted and will fall into the opening, so that it will come to some point of the glacialis, as is evident in the case of the ray *pb;* for it does not pass to *c,* but is refracted at the point *b* of the cornea between the straight path, which is *bc,* and the perpendicular, which is *bo,* so that it passes to the point *f* of the glacialis; and the same is true of an infinite number of rays. Therefore vision is much improved and perfected by countless refracted rays of this kind, in which every visible object is seen, except that which is seen by the perpendicular ray. Moreover, we understand that anything that is opposite the opening can be seen by refraction and will not be seen by the straight rays, namely, when some obstacle of small breadth is interposed; as, for instance, a small straw, standing opposite the eye between it and some visible object, hinders the passage of the species of some part of it in a straight line. Then oblique rays will fall on the cornea from that object; since besides the one perpendicular which would fall if the obstacle were not there, an infinite number of oblique rays fall, as we have now seen. Therefore the object will be seen by refracted rays only and not by straight rays, as is evident by experience if one holds between his eye and some object a straw or a needle; and particularly can one try this with a candle.

Chapter II

Concerning the differences in the appearance of the position of the image by refraction in planes.

We must understand that vision by refraction is at the intersection of the visual ray with the cathetus, as has been stated in regard to reflection. But this can take place in various wonderful ways. To the end that we may understand all the different appearances of this kind, we must consider how differences of this kind occur in plane, concave, or convex bodies. The variation is also dependent on the position of the eye, whether it be in a rarer or denser medium and the object of vision conversely. If the eye is in a transparent rarer medium and between the eye and the object of vision there is a denser medium, like water of a plane

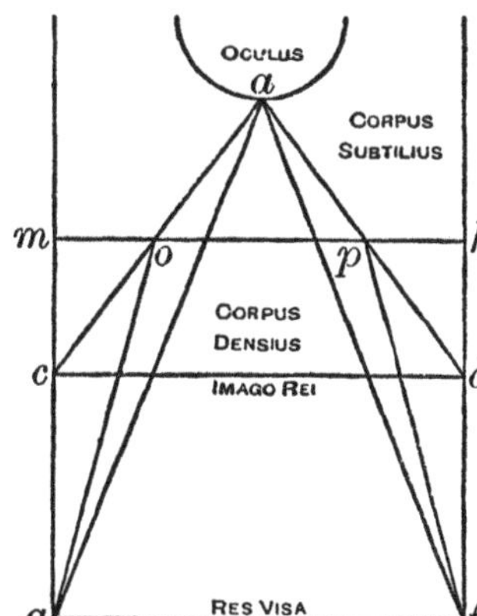

Fig. 59.

surface, or crystal, or glass, and other transparent substances of this kind, then the object appears far larger than it is; for it is seen under a larger angle and appears much nearer than if the medium were uniform. The proof of this fact is evident in this figure. For the visible object *f* will be seen at *d,* where the visual ray *ad* intersects the cathetus *fh;* and similarly *g* will appear at *c,* where the visual ray *ac* intersects the cathetus *gm,* and therefore the whole object *gf* will appear in the position *cd* nearer the eye, and will be seen under a larger angle, than if the body were one. For under the angle *oap* it will be seen through these two bodies, but under the angle *gaf* it would be seen through one medium without refraction. If, however, the eye is in a denser medium, and the object of vision is a rarer one, then the converse is true. For the object will appear smaller, not only because it will be seen under a smaller angle, but also because it will appear more remote. For *o* will be seen at *h,* and *f* at *k* beyond the object of vision, so that *of* will appear at *kh;* for the visual ray *ab* intersects at *h* the cathetus *hc,* and the visual ray *ad* intersects at *k* the cathetus *pfk;* and it is seen under a smaller angle than if it were seen through only one medium. For now the whole object is seen under the angle *dab* because of the refraction; and without refraction it is seen under the larger angle *fao.*

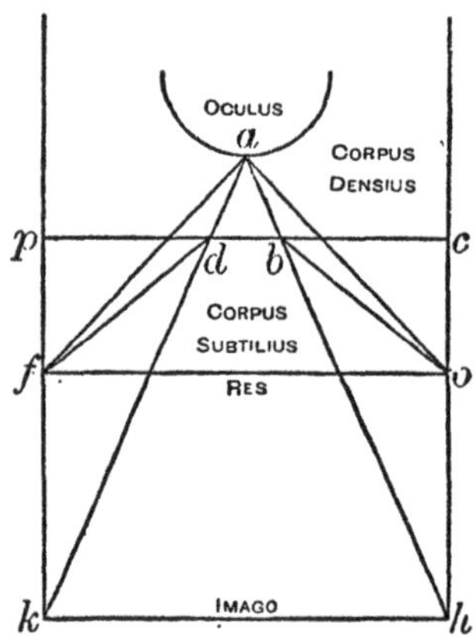

Fig. 60.

Chapter III

Concerning the diversity in the position of the image in the case of spherical surfaces.

If, however, the bodies are not plane by means of which the eye sees but spherical, there is a great diversity. For either the concavity or convexity of the body is toward the eye; if the concavity, then there are four cases. For there are two if the eye is in the rarer medium, and two if it is in the denser medium. If, therefore, the eye is in the rarer medium, and the concavity of the medium is toward the eye, the eye can be between the center of the medium and the object of vision, or the center be-

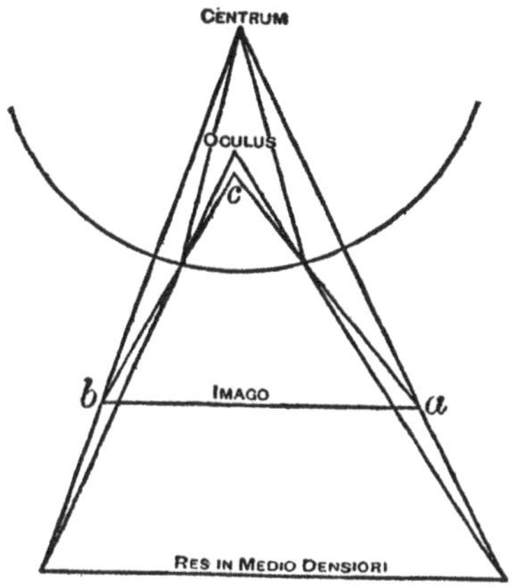

FIG. 61.

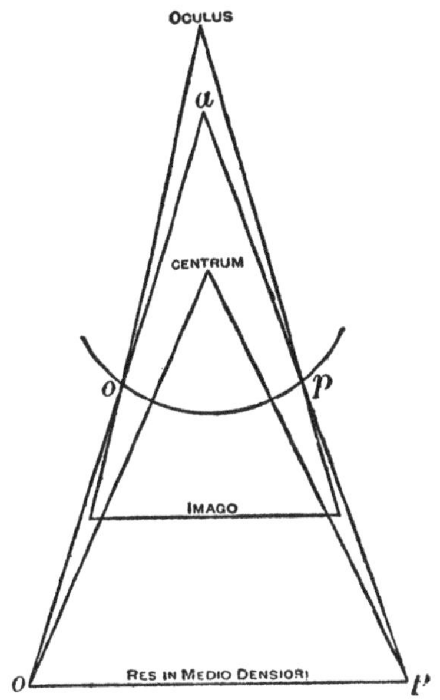

FIG. 62.

tween the eye and the object of vision. The force, moreover, would not proceed here from the center of a denser or rarer medium, because the center of both is the same, and the concavity of both is toward the eye, because the center of the containing and of the contained spheres is the same. I shall, therefore, state all these cases first, and then illustrate them in the figures; for it is necessary to do so owing to the brevity of the rules in each case and the large size of the figures. All these facts are made evident in these figures, which are here given in the order of the cases mentioned above.

If, then, the eye is in the rarer medium and the concavity is toward the eye, and the eye is between the center and the object of vision, the object will appear nearer than it is. For the visual angle will thus be greater than if straight lines should be drawn from the eye without refraction to the extremities of the object and under a larger angle, and yet the image is less than the object. If, however, the eye is in the rarer medium and the concavity is toward the eye and the center of the denser body is between the eye and the object, the object will still appear nearer. But the angle will be less and the image smaller. If, however, the eye is in the denser medium, and the concavity toward the

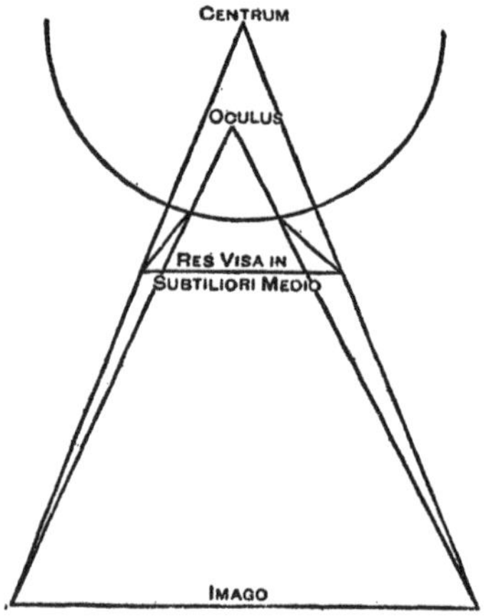

Fig. 63.

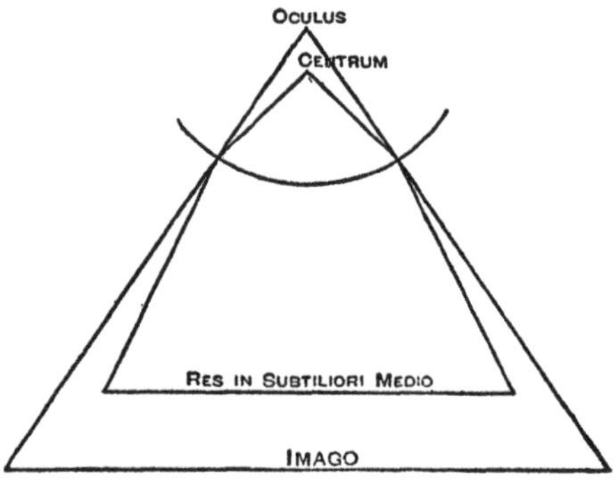

Fig. 64.

eye and the eye is between the center of the concave body and the object, then the object will appear more remote beyond its actual position, and under a smaller angle, and the image will be larger. But if the center of the concave body is between the

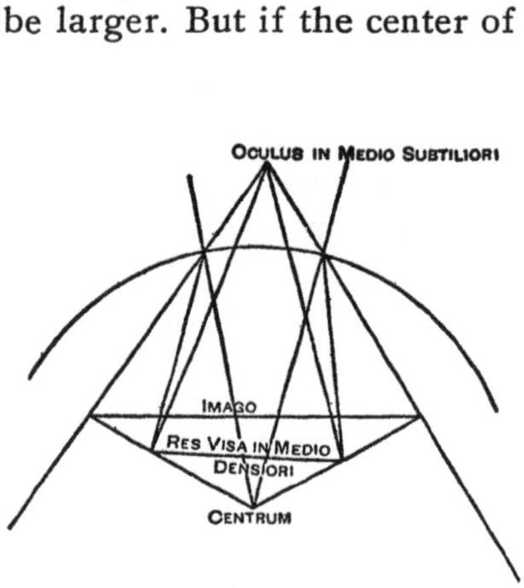

Fig. 65.

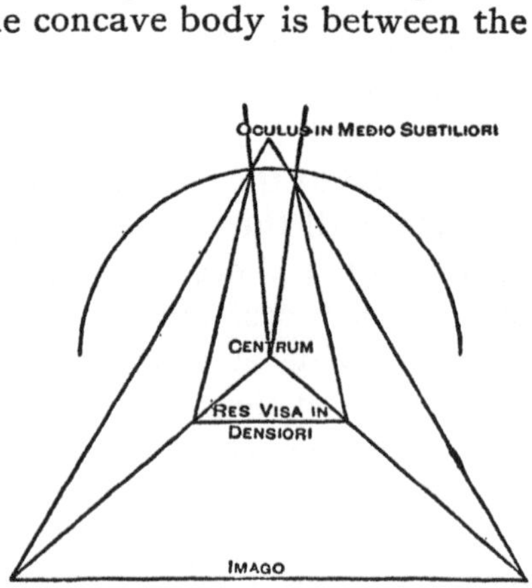

Fig. 66.

eye and the object of vision, the other conditions remaining the same, the object of vision will still appear more remote and under a larger angle, and the image will be greater.

If, however, the convexity of the body is toward the eye, there will similarly be four cases; for there are two if the eye is in the rarer medium, and two if the eye is in the denser medium. If, then, the eye is in the rarer medium and the convexity of the medium in which the object of vision is situated is toward the eye, then the object of vision can be between the center and the eye, or the center between the eye and the object of vision. If the object is between the eye and the center, the image will be nearer and larger and the angle greater. But the

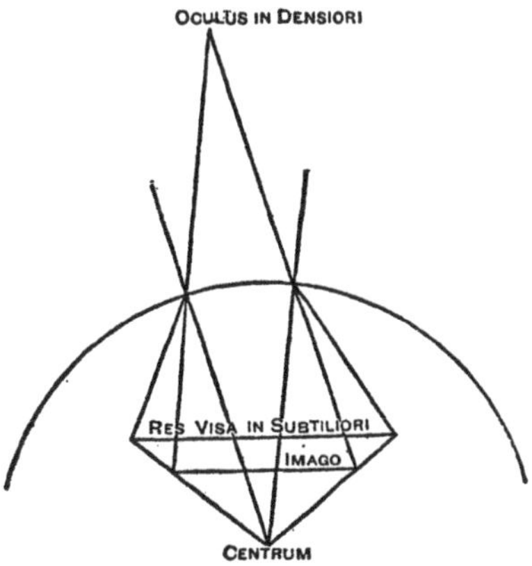

FIG. 67.

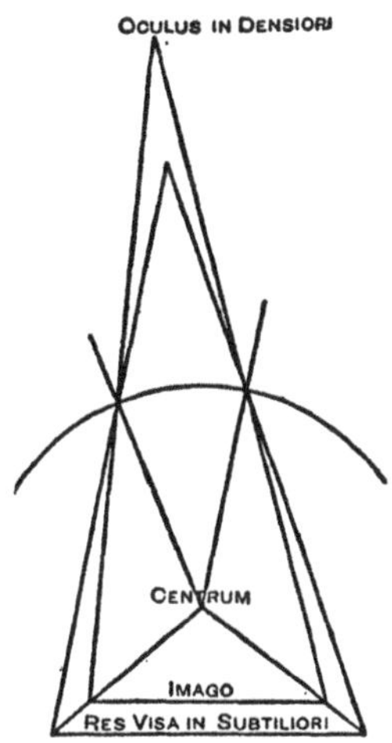

FIG. 68.

position of the image will be more remote. If, however, the eye is in the denser medium, and the object of vision is between the eye and the center, the image will be remoter and smaller, and will be seen under a smaller angle. But if the eye is in the denser medium, and the center is between the eye and the object, the image will be nearer and smaller and will be seen under a smaller angle, and the size of the angle under which the object is seen is perceived to be smaller than it would be if there were only one medium. And this is also the case when it contains that angle which the lines of the straight path make, and which end outside of it at another point; but it is shown to be greater, as is evident in the first figure, when the angle converges to another point within it, but then the angle which the lines of the straight path make is less than the angle under which the object is seen, and therefore the angle under which the object is seen is greater than if there were only one medium. For then is it seen under the angle *bca,* included by straight lines. And in the following figure it would be seen under the angle *oap,* included by straight lines, if there were only one medium, and thus under a greater angle than is the angle included by the broken lines under which the object is seen through the two media. We are to understand the matter in this way in all the other following figures.

Chapter IV

Concerning examples regarding refractions of this kind.

Since these figures in regard to the mode of vision by refraction have been described, examples can be given in objects of vision. First, in regard to a rod, which seems broken when one part is in the air and the other part in the water and the eye is in the air. For on this point contention is common among philosophers, when they dispute on any matter, nor is the question ever solved by them, since they are ignorant of this third part of Perspective. Now since the eye is in the same medium as the upper part of the rod, the eye will see it by direct vision, just as it is. But since the eye is in a rarer medium with respect to the lower part of the rod, which is in the water, the first rule given above in regard to a plane medium, or the fifth in regard to the denser medium in which the object is situated, whose convex surface is toward the eye, is applicable here. Nor is the force of which we may speak present in the waters to which we are accustomed, namely, of rivers and ditches, since although water has naturally a convex surface, wherever it is, owing to its constant tendency to flow to a lower position, as has been stated above, waters familiar to us in streams and fountains and other hollows have as far as regards our sense the upper surface plane. And however we may speak, the object of vision in the water must appear nearer the eye than is its true position, and larger, as is shown in both figures. Therefore the part of the rod which is in the water will not appear to the sight as continuous and in a straight line with the other part, but nearer the eye, and therefore the rod of necessity must appear of a curved and angular shape, as if it had been broken on entering the water, as shown in the figure. For let *fb* be the rod, *a* the eye, and *hm* the surface of the water; *b* will send its species to *c*, but will not pass to *o* by a straight path, but will be refracted in the rarer medium to *a*, so that the straight path is between the refracted ray and the perpendicular drawn from the point of refraction, namely, *gc*. But the object appears at the intersection of the visual ray with the cathetus. The cathetus is *bdp*, and the visual ray *ac* intersects at the point *d* of the cathetus. Therefore *b*, the extremity of the rod will be seen at *d*, and in the same way each particle of the section that is in the water will be seen in the straight line in which *d* itself lies. Therefore the whole portion that is in the water will appear in the line *nd*. Wherefore the whole rod will be seen in the line *fnd*, and therefore in the curved line with an angle at *n*, and thus the rod will appear to be broken. Moreover, since a man can see in the water, if he had the required skill to stay under the water, he could see the rod broken at the surface of the air, just as he now sees it in the water because of the second rule with its figure regarding a plane body, or because of the third rule with its figure, where the eye is in the denser medium whose concavity is toward the eye.

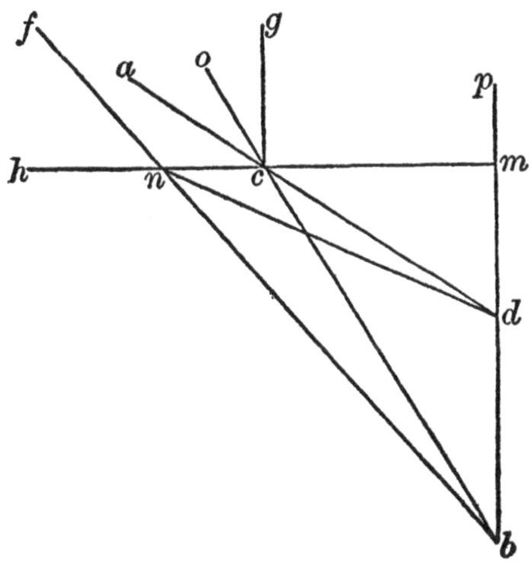

FIG. 69.

There is a similar occurrence if an object is placed in a vessel and the observer steps backward until it is no longer seen; the immersed object will be visible again at the same distance between it and the observer, if water is poured in, as is stated at the beginning of the book on Mirrors. Any one can try this, although it seems wonderful or rather not true to those unacquainted with the fact. The reason for this is evident from the rules given, namely, the first on plane surfaces and the fifth on concave surfaces. For since the eye is in the rarer medium and the object in the denser, the object must appear nearer and more elevated toward the eye where the intersection of the visual ray with the cathetus occurs, and the image will appear larger. Wherefore it will seem to the sight that the object placed in the vessel is raised from the bottom of the vessel to the surface of the water. Nor is another drawing required here than the one made in the places mentioned above, and therefore that one suffices.

If, then, we view the sun or moon and stars at their rising and setting through aqueous vapors, as we often do in summer and in autumn, we see those luminaries of unusual size, as any one experiences. But the reason of such a phenomenon is found in the first rule with its figure, where the eye is in the rarer medium and the object is in the denser, whose concavity is toward the eye, and the eye is between the center and the object of vision. For vapors of this kind are spherical and concentric with the world, because they recede equally from the center; and therefore their concavity will be toward the eye, and the eye will be between them and their center, which is the center of the world. Therefore, as we have shown from that drawing, the image of the object is nearer and is seen under a larger angle, and therefore the object appears larger and nearer.

If the objection is raised that the image is smaller than the object, for which reason it should appear smaller; we must reply that the larger size of the angle together with the apparent nearness is the prevailing factor in this matter. For a nearer object, other things being equal, appears more clearly in respect to its size, and particularly so when it is seen under a larger angle. If again it be said that the rays of the stars find vapors and clouds not only at the horizon and near it, but also toward the meridian and at the meridian, but the sun when it is near the meridian or at the meridian does not appear of unusual size; therefore it should not near its rising and setting.* Probably some trained in the science of perspective have judged that

* This sentence is incomplete in the text.

vapors are not the cause of this phenomenon, owing to this objection. But they have been deceived, because they cannot give another reason; for that reason which was given above in regard to the size of the stars at the horizon is applicable at all times, but this appearance of size is temporary, not constant, and therefore has a temporary cause. Moreover, we see, when the air is serene and dry in the east or west lacking vapors, that the stars have their accustomed magnitudes; but when the air is charged with vapors the appearance of unusual size occurs at this time. It is therefore evident that vapors are the cause. But the objection is removed by the fact that the rays of the stars near their rising and setting fall entirely at oblique angles and therefore are refracted at the surface of the air according to the tenor of the rule mentioned. But when the star approaches the meridian the rays approach right angles in their direction, for which reason they are not so much refracted as when the star is at its rising. If it be objected that all rays of planets are refracted about the tropic of Cancer, as has been previously explained, because they do not fall to the center of the world, but toward the horizon, the fact must be conceded. But, however, they are far less refracted and they approach nearer to the perpendicular when the star is near the meridian; and therefore, although at that time it appears of larger size owing to vapors, it is not, however, of the unusual size of which we are here speaking. But the larger size of the angle of refraction and the wider divergence from the direct path cause the object to appear larger and nearer.

If, however, we consider the stars and the media according to their natural arrangement, apart from vapors and the remaining reason of constant application, concerning which mention was made above, the third rule in regard to spherical bodies applies, whose convexity is toward the eye and the eye is in the denser medium, since it is in an elemental medium, and the object is in a rarer medium, namely, in a celestial medium, and the eye is between the center and the object of vision. The stars will appear smaller than they are and than they would appear if there were only one medium, since they are seen under a smaller angle, and thus there will be an error of vision in regard to the stars. If it be said that the image is far larger than the object, and therefore it will appear larger: again if it be said that the position of the image is far beyond the object, and therefore it will appear to be more distant, and therefore objects will appear to be larger, for we have shown above that things which seem to be more distant appear larger; we must reply to the first objection that the size of the angle is the prevailing factor in these appearances. Therefore since the star is seen under a smaller angle, the magnitude of the image offers no difficulty. To the second objection we reply that because of the transparency of the clear media, intervening bodies are not perceived, and therefore the distance of the image is not perceived, because from too remote a point, as has been explained previously, distance is not judged by the sight unless intervening objects are perceived. Therefore although the position of the image is more remote, and would appear here to the vision through an error, yet as a matter of fact the vision does not perceive this remoteness, and therefore the object should not appear large on this account.

If a man looks at letters or other small objects through the medium of a crystal or of glass or of some other transparent body placed above the letters, and it is the smaller part of a sphere whose convexity is toward the eye, and the eye is in the air, he will see the letters much better and they will appear larger to him. For in accordance with the truth of the fifth rule regarding a spherical medium beneath which is the object or on this side is its center, and whose convexity is toward the eye, all conditions are favorable for magnification, since the angle is larger under which the object is seen, and the image is larger, and the position of the image is nearer, since the object is between the eye and the center. Therefore this instrument is useful to the aged and to those with weak eyes. For they can see a letter, no matter how small, sufficiently enlarged. If, however, the larger portion of a sphere or the half of it is used, then according to the sixth rule the angle is larger, and the image is larger, but it is more remote, because the position of the image is beyond the object, since the center of the sphere is between the eye and the object of vision. Therefore this instrument is not so powerful as it would be if it were the smaller portion of a sphere. Moreover, instruments of crystalline bodies of plane surface, according to the first rule on plane surfaces, and instruments of concave spherical bodies, according to the first and second rules on spherical surfaces, can produce this same effect. But among all these instruments the smaller portion of a sphere whose convexity is toward the eye is the best magnifier, for the three reasons effective at the same time, as I have stated.

That a candle appears larger at a distance than near at hand, provided that the distance is not too great, is due to the fact that it is seen not only by direct, but also by refracted rays, and the vision does not perceive the refraction, and for this reason it judges that it sees by means of straight lines where the visual ray intersects the cathetus drawn from the object. Hence the object of vision seems for this reason to be dilated to *gr*, because the points at its extremities are seen not only by means of the ray *au*, and by another ray *cp*, but also by the ray *ao* refracted at the point *b* on the surface of the eye and by the ray *cd* refracted at the point *m*, and *ob*, the visual ray, intersects at the point *g* the cathetus *ca*, and the refracted ray *dm* intersects at the point *r* the cathetus *ac*, and therefore the diameter of the object of vision appears to be *rg*, and considerably larger than *ac*. Another reason, particularly in the case of weak and careless vision, might be the fact that the species near the object is strong, and therefore naturally fitted to terminate weak and careless vision, and for this reason the size of the object seems greater in accordance with the extent of space in which the species appears sensible. For this frequently happens to weak eyes, to infirm people, to those intoxicated, and to those who look at the object of vision in a negligent and indifferent manner. Other examples can be given, in which much scientific knowledge is displayed just as in these; but since the present discussion is more in the nature of a plea than of a formal treatise, those now given must suffice.

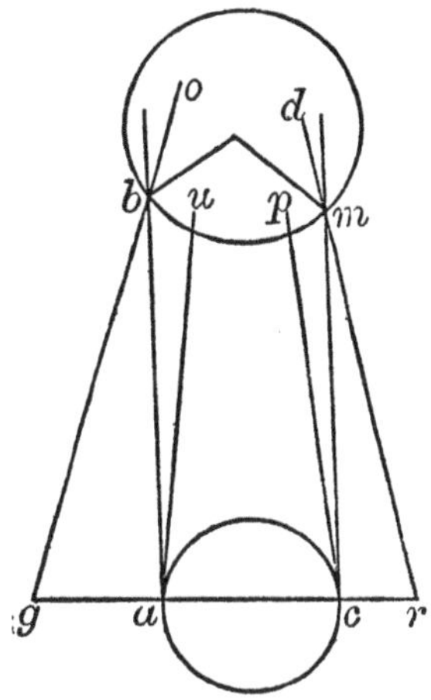

FIG. 70.

LAST DISTINCTION

Concerning the relation of perspective to theology and to practical applications, in four chapters.

CHAPTER I

WE have now spoken of the principles of perspective as far as they are necessary for scientific knowledge and for the understanding of their practical applications. I wish now in conclusion to intimate how this science possesses an ineffable usefulness in regard to divine truth. First we must bear in mind that, since this science attests natural phenomena, as is evident by what has been said, and consequently it is clear that it elucidates and explains the other sciences, this science must be serviceable to divine truth, because it requires a knowledge of the sciences and of temporal matters. Moreover, since divine truth unrestrictedly considered must be understood and expounded, and since it is ordained for the direction of this world, this science of perspective is necessary in both ways. For in God's Scripture nothing is so much enlarged upon as those things that pertain to the eye and to vision, as is evident to one who reads the Scripture through; and therefore nothing is more necessary for the natural and spiritual meaning than definite knowledge of this science. This fact I wish merely to intimate in passing, since it is not necessary to delay at this point on many examples of this kind, because when the truth of the matters set forth in Scripture is known it is very easy for every theologian to draw from them profitably their spiritual meaning. For when it is said, "Guard us, Lord, as the pupil of thine eye," it is impossible to know God's meaning in this prayer unless one first considers how the guarding of the eye is effected, so that God deems it right to guard us according to this similitude. For when something is stated as an example and similitude, that which is exemplified cannot be understood unless the meaning of the example is also understood. As, for example, when the Lord says, "Be wise as serpents," the Lord intended that his disciples should consider the qualities of the serpent in which its wisdom consists, and the nature of the dove in which the sweetness of simplicity is found.

But we shall not understand the guarding of the pupil except through the science of perspective. For the pupil is the anterior glacialis, which is supported by two humors in front of it and behind it; and is contained in one web and three coats, and receives moreover a constant and unrestricted influx of spirits and forces from the plenteous source of supply in the optic commissure; and thus it requires seven things for its protection. This, then, is the literal exposition, to which the Psalmist wishes the spiritual to be likened when he prays for the guarding of the spiritual pupil, that is, of the soul; for the perfect guardianship of which seven things are necessary, namely, virtue, gift, beatitude, spiritual sense, fruits, and revelation according to the states of rapture, and in addition the continual influx of the gifts of grace from the plenitude of the Crucified. But the principal virtues are seven, namely, three theological, charity, faith, and hope, and four cardinal, justice, fortitude, temperance, and wisdom; by means of which our spiritual pupil must be guarded. Moreover, the gifts of the Holy Spirit are seven, and the petitions of the Lord's prayer are confined to seven. But the beatitudes are eight, as is evident from the fifth chapter of Matthew, and therefore we shall add to the seven guardians of the bodily pupil as an eighth that of the eyelashes, so that as many bodily defenders may correspond to the eight spiritual ones. Now the spiritual senses are five, and those things that are directly allotted to the protection of the pupil are five, namely, the web, the albugineous humor, and the three coats. For the vitreous humor can in this view of the subject be understood with the pupil, because it is continuous with the anterior glacialis in the weblike structure. And without question the whole is called the pupil, although more particularly the anterior glacialis. Or the five may be designated thus, cross, humor, web, coat, and eyelash. For these five are the essential guardians, although they are divided in some manner into branches, as are the humor and the coat. But there are twelve fruits, as the Apostle enumerates in the fifth chapter of the Epistle to the Galatians, and therefore if we take the pupil strictly as the anterior glacialis, and consider all things remote as well as near it that can be found for its protection, we shall find twelve, namely, the eight mentioned above, and the eyelids and the eyebrows, which have special function in the protection of the eye, as was noted above. And since the visual nerves descend from the anterior part of the brain, in which are located the common sense and the imagination, from which the forces and spirit flow to the vision, nor is the act of vision completed until the visual species reaches these two, as was explained above, there will be in general twelve things allotted to the protection of the bodily pupil, just as there are twelve fruits in the protection of the spiritual eye. And we have said that there are required for vision not only internal reception, but external transmission, and coöperation by its own force and species: similarly the spiritual vision requires not only that the soul should receive from without, namely, from God, graces and virtues, but that it should coöperate through its own virtue. For the exercise and agreement of our free will are required together with the grace of God to the end that we may see and secure the state of salvation.

Chapter II

Eight conditions are also required for vision; namely, light, distance, and the others already noted, and this is paralleled exactly in the spiritual vision by the eight beatitudes. But this is also evident in another way; for conditions similar to these eight are required for spiritual vision. For just as we see nothing corporeally without corporeal light, so is it impossible for us to see anything spiritually without the spiritual light of the divine grace. And just as a moderate distance is required for vision, so that a body is not visible if it is too far off or too near, the same condition is required for spiritual vision. For remoteness from God through unbelief and a multitude of sins takes away spiritual sight, and so also do an excessive presumption of intimacy with God and scrutiny of his majesty. But those who with moderation approach his feet exclaiming with the apostle, "O depth of the riches of God's wisdom and knowledge, how incomprehensible are his judgments, and how unsearchable are his ways," will learn from his teaching according to the prophet, and "will go gradually from strength to strength, until the God of Gods appears in Sion." And so in regard to the other six conditions it is easily evident to one considering the subject how their similitudes coöperate in the spiritual vision; and therefore I need not dwell upon them individually.

Moreover, since vision is threefold, namely, by sense alone, by memory [*scientia*], and by reasoning, similarly it is necessary for man to have threefold vision. For by sense alone we perceive few things and imperfectly, as, for example, light and color, and we have this perception weakly, namely, whether these objects of vision exist or which they are; but by memory we perceive of what kind and quality they are, whether the light is that of the sun or of the moon, whether the color is white or black. But by reasoning we perceive all that pertains to light and color in accordance with all of the twenty common sensibles. Therefore the first kind of perception is weak, the second is more perfect, and the third is the most perfect. The conditions are similar in spiritual vision; for what a man knows by his own sense alone is little, because he lacks the double perception in addition to this, namely, that acquired through teachers from youth to old age. For we can always learn through those wiser than we are. And therefore we lack the third kind of perception, which comes through divine illumination.

But in another fashion is vision threefold; namely, direct, refracted, and reflected. The first is more perfect than the others, the second is less certain, and the third the least certain. For this reason we have shown above that action along the straight line is the most potent, and that refraction is less weak than reflection. These laws hold in vision as well as in other things; and they must be effective in spiritual vision as well as in our physical vision. A comparison can be drawn here in many ways; for directness of vision belongs to God; deviation from the straight line through refraction, which is weaker, befits the angelic nature; reflected vision, which is weaker, may be assigned to man. For just as a mirror because of its suitability aids vision, and gives to the species an opportunity of multiplying itself to the eye, so that vision may be effected, so does the body animated by the sensitive soul by its own nature and fitness aid the intellectual soul in its perception, and give it perception in those matters in which the intellect is dependent on the corporeal sense. Therefore the perception of man, however perfect it may be, is weaker than the angelic perception for this reason, and rightly may be called a mirrored perception because of the similitude stated. I am speaking of the natural man with the exception of the Blessed Virgin, and in accordance with the common state of man and of angel. Man has a threefold vision; one perfect, which will come in a state of glory after the resurrection; the second in the soul separated from the body in heaven until the resurrection, which is weaker; the third in this life, which is the weakest, and this is correctly said to be by reflection. As the apostle says, "We now see by means of a glass darkly, but in glory face to face," and after the resurrection in perfect directness, and before it in a deviation from that directness of vision in our soul. For this reason the soul will not enjoy perfect vision until it is united to its own body, just as it will not possess other blessings in full measure before that time, as theologians are well aware, and because there is a certain natural desire in the soul for its own body, which cannot be realized except by the resurrection. Moreover, in our present state is vision threefold, namely, direct in those that are perfect; refracted in those that are imperfect; and in the evil and in those who neglect the commandments of God by reflection, according to the Apostle James; for it is compared to a man beholding his natural face in a glass.

Chapter III

As the wisdom of God is ordained for the direction of the universe, so is this science of vision evidently and beneficially ordained for its beauty. I shall give some examples both of refraction and reflection. For by reflection one thing appears to be many and infinite in number; for in this way sometimes many suns and moons are seen in the sky, as Pliny states in the Natural History. This does not happen except when the vapor has been arranged like a mirror and is complex and in a different position. And what nature can do, art perfecting nature can accomplish in greater measure. Hence mirrors can be so made and placed in such a position and arranged, that a single object of vision can be multiplied at will. Therefore one man will appear many, and a single army will appear to be several. The principles relating to this phenomenon have already been touched upon, one of which concerns the broken mirror, whose parts are placed in different positions and produce different images according to their number. The other principle concerns the case of the water and the mirror, from both of which a different image is reflected. If, therefore, as many mirrors as we wish should be arranged in both these ways it is evident that a single object will appear in as many images as we wish, and thus to the advantage of the state images of this kind might profitably be produced. They might also be used against unbelievers to inspire terror. Moreover, if one knew that the air is dense, so that reflection could be obtained from it, he might produce many unusual appearances of this kind. In this way we believe that demons show camps and armies and many wonders to men; and by reflected vision all things hidden in secret places in cities, armies, and the like can be brought to light. For the dragon, which poisons both animals and men and corrupts them with its breath, was detected in its lair by the philosopher Socrates, as the histories bear witness.

Similarly mirrors might be erected on an elevation opposite hostile cities and armies, so that all that was being done by the enemy might be visible. This can be done at any distance we desire, since, according to the book on Mirrors, one and the same object can be seen by means of as many mirrors as we wish, if they are placed in the manner required. Therefore they can be placed more closely and more remotely, so that we might see an object as far off as we pleased. For in this way Julius Caesar, when he wished to subdue England, is said to have erected very large mirrors, in order that he might see in advance from the shore of Gaul the arrangement of the cities and camps of England. Mirrors, moreover, can be so arranged that as many objects as we desire may be visible and all that is in the house or in the street; and any one looking at these objects will see them as they really are, and when he hastens to the places where they appear, he will find nothing. For the mirrors will be concealed in such a manner with respect to the objects, that the positions of the images are in view, and the images appear in the air at the intersection of the visual rays with the catheti, and therefore those looking at them would run up to the places where they appear, and would judge that the objects are there when there is in reality nothing except an image. And thus in accordance with such principles as have now been touched upon in regard to reflection and in accordance with similar ones, not only could results be attained advantageous to our friends and terrible for our enemies, but very great comforts can come to us from philosophy, so that every deceit of jesters may be dimmed by the beauty of the wonders of science, and men may rejoice in the truth, banishing far from them the tricks of the magicians.

Chapter IV

THE wonders of refracted vision are still greater; for it is easily shown by the rules stated above that very large objects can be made to appear very small, and the reverse, and very distant objects will seem very close at hand, and conversely. For we can so shape transparent bodies, and arrange them in such a way with respect to our sight and objects of vision, that the rays will be refracted and bent in any direction we desire, and under any angle we wish we shall see the object near or at a distance. Thus from an incredible distance we might read the smallest letters and number grains of dust and sand owing to the magnitude of the angle under which we viewed them, and very large bodies close to us we might scarcely see because of the smallness of the angle under which we saw them, for distance in such vision is not a factor except by accident, but the size of the angle is. In this way a child might appear a giant, and a man a mountain. He might appear of any size whatever, as we might see a man under as large an angle as we see a mountain, and close as we wish. Thus a small army might appear very large, and situated at a distance might appear close at hand, and the reverse. So also we might cause the sun, moon, and stars in appearance to descend here below, and similarly to appear above the heads of our enemies, and we might cause many similar phenomena, so that the mind of a man ignorant of the truth could not endure them.

PART SIX OF THIS PLEA

It is also the sixth part of the Opus Majus, on Experimental Science.

CHAPTER I

HAVING laid down fundamental principles of the wisdom of the Latins so far as they are found in language, mathematics, and optics, I now wish to unfold the principles of experimental science, since without experience nothing can be sufficiently known. For there are two modes of acquiring knowledge, namely, by reasoning and experience. Reasoning draws a conclusion and makes us grant the conclusion, but does not make the conclusion certain, nor does it remove doubt so that the mind may rest on the intuition of truth, unless the mind discovers it by the path of experience; since many have the arguments relating to what can be known, but because they lack experience they neglect the arguments, and neither avoid what is harmful nor follow what is good. For if a man who has never seen fire should prove by adequate reasoning that fire burns and injures things and destroys them, his mind would not be satisfied thereby, nor would he avoid fire, until he placed his hand or some combustible substance in the fire, so that he might prove by experience that which reasoning taught. But when he has had actual experience of combustion his mind is made certain and rests in the full light of truth. Therefore reasoning does not suffice, but experience does.

This is also evident in mathematics, where proof is most convincing. But the mind of one who has the most convincing proof in regard to the equilateral triangle will never cleave to the conclusion without experience, nor will he heed it, but will disregard it until experience is offered him by the intersection of two circles, from either intersection of which two lines may be drawn to the extremities of the given line; but then the man accepts the conclusion without any question. Aristotle's statement, then, that proof is reasoning that causes us to know is to be understood with the proviso that the proof is accompanied by its appropriate experience, and is not to be understood of the bare proof. His statement also in the first book of the Metaphysics that those who understand the reason and the cause are wiser than those who have empiric knowledge of a fact, is spoken of such as know only the bare truth without the cause. But I am here speaking of the man who knows the reason and the cause through experience. These men are perfect in their wisdom, as Aristotle maintains in the sixth book of the Ethics, whose simple statements must be accepted as if they offered proof, as he states in the same place.

He therefore who wishes to rejoice without doubt in regard to the truths underlying phenomena must know how to devote himself to experiment. For authors write many statements, and people believe them through reasoning which they formulate without experience. Their reasoning is wholly false. For it is generally believed that the diamond cannot be broken except by goat's blood, and philosophers and theologians misuse this idea. But fracture by means of blood of this kind has never been verified, although the effort has been made; and without that blood it can be broken easily. For I have seen this with my own eyes, and this is necessary, because gems cannot be carved except by fragments of this stone. Similarly it is generally believed that the castors employed by physicians are the testicles of the male animal. But this is not true, because the beaver has these under its breast, and both the male and female produce testicles of this kind. Besides these castors the male beaver has its testicles in their natural place; and therefore what is subjoined is a dreadful lie, namely, that when the hunters pursue the beaver, he himself knowing what they are seeking cuts out with his teeth these glands. Moreover, it is generally believed that hot water freezes more quickly than cold water in vessels, and the argument in support of this is advanced that contrary is excited by contrary, just like enemies meeting each other. But it is certain that cold water freezes more quickly for any one who makes the experiment. People attribute this to Aristotle in the second book of the Meteorologics; but he certainly does not make this statement, but he does make one like it, by which they have been deceived, namely, that if cold water and hot water are poured on a cold place, as upon ice, the hot water freezes more quickly, and this is true. But if hot water and cold are placed in two vessels, the cold will freeze more quickly. Therefore all things must be verified by experience.

But experience is of two kinds; one is gained through our external senses, and in this way we gain our experience of those things that are in the heavens by instruments made for this purpose, and of those things here below by means attested by our vision. Things that do not belong in our part of the world we know through other scientists who have had experience of them. As, for example, Aristotle on the authority of Alexander sent two thousand men through different parts of the world to gain experimental knowledge of all things that are on the surface of the earth, as Pliny bears witness in his Natural History. This experience is both human and philosophical, as far as man can act in accordance with the grace given him; but this experience does not suffice him, because it does not give full attestation in regard to things corporeal owing to its difficulty, and does not touch at all on things spiritual. It is necessary, therefore, that the intellect of man should be otherwise aided, and for this reason the holy patriarchs and prophets, who first gave sciences to the world, received illumination within and were not dependent on sense alone. The same is true of many believers since the time of Christ. For the grace of faith illuminates greatly, as also do divine inspirations, not only in things spiritual, but in things corporeal and in the sciences of philosophy; as Ptolemy states in the Centilogium, namely, that there are two roads by which we arrive at the knowledge of facts, one through the experience of philosophy, the other through divine inspiration, which is far the better way, as he says.

Moreover, there are seven stages of this internal knowledge, the first of which is reached through illuminations relating purely to the sciences. The second consists in the virtues. For the evil man is ignorant, as Aristotle says in the second book of the Ethics. Moreover, Algazel says in his Logic that the soul disfigured by sins is like a rusty mirror, in which the species of objects cannot be seen clearly; but the soul adorned with virtues is like a well-polished mirror, in which the forms of objects are clearly seen. For this reason true philosophers have labored more in morals for the honor of virtue, concluding in their own case that they cannot perceive the causes of things unless they have souls free from sins. Such is the statement of Augustine in regard to Socrates in the eighth book of the City of God, chapter III. Wherefore the Scripture says, "in a malevolent soul, etc." For it is not possible that the soul should rest in the light of truth while it is stained with sins, but like a parrot or magpie it will repeat the words of another which it has learned by long practice. The proof of this is that the beauty of truth known in its splendor attracts men to the love of it, but the proof of love is the display of a work of love. Therefore he who acts contrary to the truth must necessarily be ignorant of it, although he may know how to compose very elegant phrases, and quote the opinions of other people, like an animal that imitates the words of human beings, and like an ape that relies on the aid of men to perform its part, although it does not understand their reason. Virtue, therefore, clarifies the mind, so that a man comprehends more easily not only moral but scientific truths. I have proved this carefully in the case of many pure young men, who because of innocency of soul have attained greater proficiency than can be stated, when they have had sane advice in regard to their study. Of this number is the bearer of this present treatise, whose fundamental knowledge very few of the Latins have acquired. For since he is quite young, about twenty years of age, and very poor, nor has he been able to have teachers, nor has he spent one year in learning his great store of knowledge, nor is he a man of great genius nor of a very retentive memory, there can be no other cause except the grace of God, which owing to the purity of his soul has granted to him those things that it has as a rule refused to show to all other students. For as a spotless virgin he has departed from me, nor have I found in him any kind of mortal sin, although I have examined him carefully, and he has, therefore, a soul so bright and clear that with very little instruction he has learned more than can be estimated. And I have striven to aid in bringing it about that these two young men should be useful vessels in God's Church, to the end that they may reform by the grace of God the whole course of study of the Latins.

The third stage consists in the seven gifts of the Holy Spirit, which Isaiah enumerates. The fourth consists in the beatitudes, which the Lord defines in the Gospels. The fifth consists in the spiritual senses. The sixth consists in fruits, of which is the peace of God which passes all understanding. The seventh consists in raptures and their states according to the different ways in which people are caught up to see many things of which it is not lawful for a man to speak. And he who has had diligent training in these experiences or in several of them is able to assure himself and others not only in regard to things spiritual, but also in regard to all human sciences. Therefore since all the divisions of speculative philosophy proceed by arguments, which are either based on a point from authority or on the other points of argumentation except this division which I am now examining, we find necessary the science that is called experimental. I wish to explain it, as it is useful not only to philosophy, but to the knowledge of God, and for the direction of the whole world; just as in the preceding divisions I showed the relationship of the languages and sciences to their end, which is the divine wisdom by which all things are disposed.

Chapter II

Since this Experimental Science is wholly unknown to the rank and file of students, I am therefore unable to convince people of its utility unless at the same time I disclose its excellence and its proper signification. This science alone, therefore, knows how to test perfectly what can be done by nature, what by the effort of art, what by trickery, what the incantations, conjurations, invocations, deprecations, sacrifices, that belong to magic, mean and dream of, and what is in them, so that all falsity may be removed and the truth alone of art and nature may be retained. This science alone teaches us how to view the mad acts of magicians, that they may be not ratified but shunned, just as logic considers sophistical reasoning.

This science has three leading characteristics with respect to other sciences. The first is that it investigates by experiment the notable conclusions of all those sciences. For the other sciences know how to discover their principles by experiments, but their conclusions are reached by reasoning drawn from the principles discovered. But if they should have a particular and complete experience of their own conclusions, they must have it with the aid of this noble science. For it is true that mathematics has general experiments as regards its conclusions in its figures and calculations, which also are applied to all sciences and to this kind of experiment, because no science can be known without mathematics. But if we give our attention to particular and complete experiments and such as are attested wholly by the proper method, we must employ the principles of this science which is called experimental. I give as an example the rainbow and phenomena connected with it, of which nature are the circle around the sun and the stars, the streak [*virga*] also lying at the side of the sun or of a star, which is apparent to the eye in a straight line, and is called by Aristotle in the third book of the Meteorologics a perpendicular, but by Seneca a streak, and the circle is called a corona, phenomena which frequently have the colors of the rainbow. The natural philosopher discusses these phenomena, and the writer on Perspective has much to add pertaining to the mode of vision that is necessary in this case. But neither Aristotle nor Avicenna in their Natural Histories has given us a knowledge of phenomena of this kind, nor has Seneca, who composed a special book on them. But Experimental Science attests them.

Let the experimenter first, then, examine visible objects, in order that he may find colors arranged as in the phenomena mentioned above and also the same figure. For let him take hexagonal stones from Ireland or from India, which are called rainbows in Solinus on the Wonders of the World, and let him hold these in a solar ray falling through the window, so that he may find all the colors of the rainbow, arranged as in it, in the shadow near the ray. And further let the same experimenter turn to a somewhat dark place and apply the stone to one of his eyes which is almost closed, and he will see the colors of the rainbow clearly arranged just as in the bow. And since many employing these stones think that the phenomenon is due to the special virtue of those stones and to their hexagonal shape, therefore let the experimenter proceed further, and he will find this same peculiarity in crystalline stones correctly shaped, and in other transparent stones. Moreover, he will find this not only in white stones like the Irish crystals, but also in black ones, as is evident in the dark crystal and in all stones of similar transparency. He will find it besides in crystals of a shape differing from the hexagonal, provided they have a roughened surface, like the Irish crystals, neither altogether smooth, nor rougher than they are. Nature produces some that have surfaces like the

ROGER BACON

From a Fifteenth Century Manuscript in the Bodleian Library

Irish crystals. For a difference in the corrugations causes a difference in the colors. And further let him observe rowers, and in the drops falling from the raised oars he finds the same colors when the solar rays penetrate drops of this kind. The same phenomenon is seen in water falling from the wheels of a mill; and likewise when one sees on a summer's morning the drops of dew on the grass in meadow or field, he will observe the colors. Likewise when it is raining, if he stands in a dark place and the rays beyond it pass through the falling rain, the colors will appear in the shadow near by; and frequently at night colors appear around a candle. Moreover, if a man in summer, when he rises from sleep and has his eyes only partly open, suddenly looks at a hole through which a ray of the sun enters, he will see colors. Moreover, if seated beyond the sun he draws his cap beyond his eyes, he will see colors; and similarly if he closes an eye the same thing happens under the shade of the eyebrows; and again the same phenomenon appears through a glass vessel filled with water and placed in the sun's rays. Or similarly if one having water in his mouth sprinkles it vigorously into the rays and stands at the side of the rays. So, too, if rays in the required position pass through an oil lamp hanging in the air so that the light falls on the surface of the oil, colors will be produced. Thus in an infinite number of ways colors of this kind appear, which the diligent experimenter knows how to discover.

Chapter III

In a similar way also he will be able to test the shape in which the colors are disposed. For by means of the crystalline stone and substances of this kind he will find the shape straight. By means of the eyelids and eyebrows and by many other means, and also by means of holes in rags, he will discover whole circles colored. Similarly in a place where the dewfall is plentiful and sufficient to take the whole circle, and if the place where the circle of the rainbow should be is dark proportionally, because the bow does not appear in the light part, then the circle will be complete. Similarly whole circles appear frequently around candles, as Aristotle states and we ourselves experience.

Opus Majus

Chapter IV

Since, moreover, we find colors and various figures similar to the phenomena in the air, namely, of the iris, corona, and streak, we are encouraged and greatly stimulated to grasp the truth in those phenomena that occur in the heavens. Further, let the experimenter take the required instrument and look through the openings of the instrument and find the altitude of the sun above the horizon, and keeping the instrument immovable let him turn in the opposite direction and look through the openings of the instrument until he sees the summit of the bow, and let him note the altitude of the rainbow above the horizon; and he will find that the higher the sun's altitude is, the lower is that of the bow, and conversely. By this means he knows that the rainbow is always opposite the sun, and that one line passes through the center of the sun, and through the center of the observer's eye, and the center of the circle of which the bow is an arc to the sun's nadir, which is the point in the heavens opposite the center of the sun. As the extremity of this line toward the sun is elevated above the horizon, the other is depressed, which passes through the center of the bow, and conversely; just like the rule at the back of the astrolabe, one of whose ends is depressed as the other is elevated. The trained experimenter can test until he discovers the opposed altitude of the rainbow and of the sun, namely, that beyond which there can be no appearance of the rainbow. He must then give his attention to the calculation of the altitude of the circles.

We must bear in mind that the horizon is a circle, and that at its center an axis is raised in the starry heavens to the zenith of the observer's head, and let a circle pass through that point and through two parts of the horizon, as, for example, if we desire, through east and west, and let it pass beneath the earth to the point opposite the zenith overhead. This circle is the circle of the altitude of a fixed star, and it passes through its body, for when the star rises above the horizon, we say that it ascends according to the degrees of that circle until it reaches the line of the meridian. Then the star is at its greatest altitude, and this circle intersects the horizon at right angles; and they divide each other into equal parts, and each is a great circle of the sphere. The altitude, then, of a fixed star above the horizon is the arc of this circle intercepted between the star and the limit of the horizon. But the altitude of Saturn and of the other planets must not be reckoned on a circle intersecting the horizon, but on a smaller and concentric circle, because the circle passing through the body of Saturn or of another lower planet does not pass through the extremity of the axis of the horizon, but through another point lower in the axis, which is a point beneath the zenith overhead, and therefore this circle does not pass through the limits of the horizon, which is the circle *ab,* but through the circle concentric with it, namely, through the circle *cd,* and this circle of the altitude of Saturn passes through the body of Saturn, which in this circle is elevated above the horizon and depressed beneath it. Moreover, the circle of Jupiter passes through a smaller concentric circle and through a point lower in the axis beneath the zenith overhead, and similarly with regard to the other planets; so that the lower a body is, so much smaller a circle does it have passing through a lower point in the axis of the horizon and intersecting a smaller circle concentric with the horizon. Although the statement made above is in accordance with the actual fact, yet in ordinary speech we do not distinguish those points in the axis from the zenith overhead, nor the circles equidistant from the horizon do we distinguish from the horizon itself, but we call them all horizons. We also reckon the circles of altitude all equal and regard them as passing through the zenith overhead, although they are unequal. And although for clearness I thus placed them all in the same surface, and because this can happen,* yet in many cases it happens that they are in different surfaces and intersect one another in many ways.

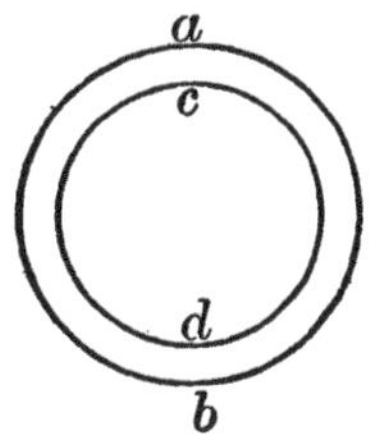

FIG. 71.

Further, we must also consider that the visual ray is equidistant from the horizon and the circles concentric with it, and that therefore the arc of the circle of the altitude which is between the horizon and the visual ray must be noted, which is in accordance with the amount of elevation of the body of the ob-

* The words, "and because this can happen," seem superfluous, as Bridges remarks.

server. But the altitude commonly taken of an object is said to be the arc of the circle of the altitude intercepted between the elevated object and the horizon, yet, strictly speaking, if the altitude of an object is taken, it is the arc of the circle of the altitude between the object and the visual ray; because the eye is not at the center of the horizon, but is above the center in the axis of the horizon; and therefore objects that are in the air and in the moon are considered according to an altitude of this kind. For they have a difference in aspect because they are near the earth, but the sun and the other remoter bodies are not affected in this way, owing to their great distance. For the amount of elevation of the observer has no sensible difference in regard to the remoteness of those bodies, but it bears a considerable relationship to those that are visible in the air, like comets and rainbows. From these statements, then, it is evident that the altitude of the rainbow, taken strictly, is the arc of its circle of altitude intercepted between the summit of the bow and the visual ray parallel to the horizon, which must be known because of what follows.

The experimenter, therefore, taking the altitude of the sun and of the rainbow above the horizon will find that the final altitude at which the rainbow can appear above the horizon is 42 degrees, and this is the maximum elevation of the rainbow. This elevation contains the arc between the summit of the bow and the visual ray, and this is its altitude, strictly speaking, although besides this there is the arc, which is between the ray and the limit of the horizon, or the limits of the circle concentric with the horizon, through which the circle of the altitude of the bow passes. And the rainbow reaches this maximum elevation when the sun is on the horizon, namely, at sunrise or sunset; and also when it is near sunrise and sunset below the horizon, not up to the end of the evening twilight, nor up to the beginning of the morning twilight, but near sunrise and sunset, as has been said. Then if moisture is ready in advance higher in the clouds, the gibbosity of the rainbow will appear, although the sun is too far below the horizon, since his rays can reach vapors high in the air, although they cannot reach vapors near the horizon. The experimenter, moreover, knows how to make the test, because when the sun is at an altitude of 42 degrees the rainbow does not appear in the sky, except that a small portion of its blue gibbosity can appear near the horizon, if moisture is already present there. When the sun rises higher, the rainbow can nowhere appear. Therefore Aristotle and Seneca state that in summer in the heat of the day the rainbow does not appear. The reason for this is because at the latitude of Paris the altitude of the sun at noon of the equinox is 41 degrees and 12 minutes, the sun at that time being at almost such an altitude that the rainbow cannot appear, and therefore a little later it must be elevated so much higher that at noon it is more than 42 degrees on the circle of altitude above the horizon, and therefore in the heat of summer the rainbow does not occur at noon until the sun descends to an altitude less than 42 degrees.

Chapter V

After, therefore, the experimenter discovers this, he must then by his experiments learn the size and shape of the rainbow, and for this purpose he conceives a cone of which the apex is at the eye and the base is the circle of the bow, the portion of which appears colored. The axis of the cone is the line mentioned above, which passes through the center of the eye and the center of the sun and the center of the bow to the nadir of the sun. The base of this cone is elevated and depressed according to the elevation and depression of the sun, as we have stated in regard to that line, and as it is depressed it sinks sometimes to the earth, cutting it and cut by it. This cone can be conceived of as not cutting the earth but as wholly elevated above the horizon and its whole base above the horizon, and this diversity occurs in accordance with the shortness and length of the cones. For it can be of such short length that its base is above the horizon, and then the whole circle can appear colored, as is the case in sprays of the required kind if they are near, just as I have explained in falling rain and dew, when water is sprinkled by anybody in the required manner, and when the water falls from a high place. When, however, the cones are extended so far that they touch the earth, the limit is reached for the appearance of the circle as a whole.

If, however, they are lengthened to such an extent that the base is cut by the earth, then according to the size of the section of the base a portion of the colored circle will appear. And in the first place a larger portion can appear when less of the base of the cone is cut; and then if half of the circle is cut, the other half will appear; and a smaller portion, if more is cut. Thus as more or less is cut, and the cone lengthened, a larger or smaller portion will appear. Moreover, the cone can be so extended that nothing appears; and thus according to the length and shortness of the cones are the bases larger or smaller, and the colored circles larger or smaller, and larger or smaller portions cut off, and the portions belong to larger or smaller circles, as is now evident to any one considering the subject. This appearance of circles and of portions of them is considered with reference to the shortness and the length of the cone, as I have said, although we do not consider the elevation and depression of the sun. But, however, the whole circle of the rainbow never appears in the clouds, but as in many cases a smaller portion, and sometimes a semicircle and a larger portion. For the base of the cone always cuts the earth in the formation of the rainbow in the clouds because of its distance from the eye, and therefore the complete circle is never visible. But when the sun is at its rising and the rainbow is elevated above the horizon about 42 degrees, and the moisture in an infinite number of falling drops is already present, the larger portion of the circle appears through the space of the arc between the visual ray and the horizon; and when the sun is at the altitude of that line* the rainbow is then a semicircle; and when the sun rises above that line the bow is a smaller portion of the circle. Thus as the sun rises higher, the rainbow becomes less and less, because more is cut off from its circle by the earth.

We must consider here, however, that according to Aristotle, particularly in the new translation, the smaller portion of the circle apparent is the portion of a larger circle, and the larger portion is the portion of a smaller circle. Hence when the sun is rising, although more than a semicircle appears colored, yet it is the portion of a smaller circle than when the sun is at a great elevation, nay, at its maximum elevation above the visual ray, at which least of the rainbow can appear. Hence when its gibbosity alone is visible near the earth, while the sun is at the maximum altitude fixed for the formation of the rainbow, that

* *I.e.,* the visual ray.

small portion is part of the base of a larger circle. For we must not think that the cone remains the same and the base the same at the elevation and depression of the sun, but we are to understand that they must be constantly renewed in the conception we should have regarding the rainbow, so that the circle is judged to be larger when the bow is lower, and smaller when the bow is higher. But since a larger portion of the circle of the rainbow appears when the sun is rising, according to the amount of elevation of the observer, if he should be on a high mountain or on a lofty tower he would see a far larger portion according to the amount of his total elevation.

We must also consider that when the sun is just below the horizon the rainbow can appear. But it is the smaller portion, not the larger nor the semicircular. This result is caused solely by the lack of material, for at that time the solar rays do not touch the vapors near the earth, but do reach those situated on high, and therefore the colored semicircle cannot be completed nor a greater portion, but only the crown of the arch, although the greater portion of the circle is above the horizon.

We must understand also that the visual cone and the cone of the iris are not the same, although they are equal, and lie frequently in the same place. For they say, when the eye is unmoved, these cones coincide. When, however, the eye turns to the right or left, turns upward or downward, the head remaining fixed, the visual cone falls in part or in whole outside the position of the cone of the rainbow. This is possible because the center of the eye during this motion is immovable, through which the axis of the cone of the rainbow passes, and therefore this cone of the rainbow remains fixed in its position, although the visual cone moves and changes as the eye looks lower or higher or to the side. And then nothing of the rainbow will be visible except as much as contains the base of the visual cone, since nothing is seen except that which falls below the visual cone; I mean nothing is seen by means of the species coming at right angles. The object thus seen is the principal object of vision, as in this case is the rainbow (or a part of it as contained by the visual cone from its position). In accordance with this will the arc of the bow appear larger or smaller. This point must be carefully considered, in order that the true explanation of the rainbow may be known, and of this I shall speak below.

Since we have now explained that, when the altitude of the sun is 42 degrees or greater, the rainbow cannot appear, it can easily be shown at what time of year and in what localities the rainbow can appear at noon.

Chapter VI

Since we have now explained the altitude, magnitude, and diversity in shape of the bow, we can easily show the time and place of its formation. For the reason why the bow cannot appear in summer in the heat of the day has been given by calculating the altitude reckoned suitable for its formation. But then before noon or after it, while the sun is below 42 degrees, the bow can appear. But this can occur in the known climates that are called the seven; but it does not necessarily occur in all regions from the equator to the pole. We must consider, therefore, that the greater altitude of the sun above the horizon is at noon, and the lesser altitude of the sun in all climates in which the sun shines above the horizon, both in winter and in summer, is in Capricorn, namely, at the winter solstice; and for this reason the days are then shorter. But the rainbow cannot be formed in all of these localities when the sun is at the meridian. For in the region whose latitude, that is, the distance of the zenith overhead from the equator or middle of the world, is 24 degrees and 25 minutes or less, and this is for dwellers near the boundary of the second climate, approximately under the tropic of Cancer beyond Jerusalem, a rainbow can never occur when the sun is at the meridian. For if it could occur, it would take place when the sun is at a less altitude above the horizon, and this would be when the sun is at the winter solstice, namely, in the first degree of Capricorn. But this cannot be, since at that time the altitude of the sun is 42 degrees above the horizon, and it has been stated that the rainbow cannot appear when the sun is 42 degrees above the horizon. Moreover, that the altitude of the sun in that region near the winter solstice is 42 degrees is evident, because it is a fourth part of the heavens from the horizon to the zenith overhead, and the zenith overhead is distant from the equator 24 degrees and 25 minutes, as has been stated; for let this be subtracted from a quarter of the heavens, namely, from 90 degrees, and again let the total declination of the sun be subtracted, which is 23 degrees and 35 minutes, which added together make 48 degrees, and this being subtracted from 90 there will remain 42 degrees, the altitude of the head of Capricorn. Therefore when the sun is at this point it will be at noon 42 degrees above the horizon; wherefore the rainbow cannot be formed on that noon. Therefore no rainbow can be formed at any noon of the entire year, because both before and after the solstice the noon will be of greater altitude.

But if we travel from that locality northward to a point where the latitude of the region is 66 degrees and 25 minutes, that is, beyond Scotland, where at the winter solstice there is no day, except that suddenly half of the solar body appears above the earth, the rainbow can occur in all these regions when the sun is at the meridian, and in Capricorn, because the altitude of the head of Capricorn is less, and therefore the altitude of the sun at noon is less. The bow can appear larger and larger, not only on that day at the hour of noon in all regions, but on the days near the solstice and as we travel farther and farther northward and the latitude of the region becomes greater. But the bow can appear while the sun is at the meridian, not only while the sun is near Capricorn, but at many other times. Accordingly the bow can appear in the seventh climate, where the latitude is 48 degrees and 42 minutes, from the autumnal to the vernal equinox when the sun is at the meridian. For the sun at that time on the meridian does not have in the seventh climate an altitude of 42 degrees, not even on the days of the equinoxes, since the altitude of the head of Aries and of Libra above the horizon is only 42 degrees and 10 minutes, as before stated, and this is made evident by subtracting the latitude of the region from 90. For if 48 degrees and 48 minutes, which is the latitude of the region, be subtracted from 90, there remain only 41 degrees and 12 minutes, and this is less than 42 degrees, which is the altitude of the sun at which the bow cannot appear.

If we travel beyond this climate to the north, the bow will still be able to appear in the heat of summer, when the sun is at the meridian. For the altitude of the head of Aries is small, and similarly the altitude of the head of Cancer when the latitude of the region is 54 degrees or 60, and especially when the latitude is 66 degrees and 25 minutes. For then Capricorn has no altitude, because the horizon passes through it, and at the solstices for a moment the horizon lies beneath the zodiac, without declination from it. And since the zodiac moves with the heavens, it is suddenly separated from the horizon, and the head of Capricorn revolves beneath the horizon through the natural day, except that half of the sun appears above the earth. Hence it has no altitude above the horizon at the winter solstice. And at that time the bow can appear when the sun is at the meridian, while the half appears above the earth, and the bow can be larger than a semicircle. And since the half of the sun suddenly sets, the bow for this reason disappears at once. Similarly on other days of almost the entire year in that region the bow can appear, because the greatest elevation of the sun occurs when the sun is in Cancer, but the sun is not elevated at that time above the horizon except through its two greatest declinations, one of which is 23 degrees and 35 minutes, and the other the same, the sum of which is 47 degrees and 10 minutes. This does not exceed the maximum altitude of the sun at which the bow can appear except by 5 degrees and 10 minutes. Therefore on a few days as compared with the whole year the formation of the bow is prevented when the sun is at the meridian.

But if we advance beyond these localities, the sun will never be in the south,* but in the north and east and west only, because to dwellers beyond these localities the sun appears in front of them toward the pole of the world. This happens sometimes for a month without night, sometimes for two months, sometimes for more, until the sun is visible for six months to dwellers near the pole. Therefore that region whose latitude is 66 degrees and 25 minutes is the farthest point at which the bow can appear when the sun is at the meridian; just as the first point at which it can appear when the sun is at the meridian was found to be the locality whose latitude begins to be in excess of 24 degrees and 25 minutes.

Thus, then, we must consider at what time, and particularly when the sun is at the meridian, and in what climates the bow can appear. In these regions the phenomenon can occur in the true north, namely, when the sun is at the meridian, and in

* See Bridges' note on this passage, Vol. II, page 184.

other parts between the east and the north, and between the west and the north. In summer, moreover, it can occur in the evening between the east and the south, and in the morning in the southwest, because at the summer solstice the sun in the morning has a considerable declination to the north, and then in the locality opposite in the southwest the bow can appear; and in the evening the sun has a declination to the north, and therefore the bow can appear in the southeast. But the bow can never appear in the true south in these climates from the equator as far as latitude 66 degrees and 25 minutes. It can, however, very easily occur at noon in localities beyond this latitude, and this is true up to the pole, since the sun in those regions is always in the north or in the east or in the west; and the bow appears opposite the sun, and it can then appear at every hour of the day. For the sun is never elevated above the horizon except by its maximum declination of 23 degrees and 35 minutes. For this reason the bow can always be visible to these people during the day, when the material is prepared.

From these statements, therefore, it is evident that the bow follows the motion of the sun in two ways; in one according to the elevation and depression of the sun in its circle of altitude, as has been already stated; and in the other according to the motion of the sun above the horizon in its daily motion, because the bow always follows the nadir of the sun, and the nadir moves continuously in an opposite direction to the sun.

Chapter VII

The experimenter further tries to find out whether the bow is caused by incident rays or by reflection or by refraction, and whether it is an image of the sun, as was assumed in the statements made on Perspective, and whether there are real colors in the cloud itself. He must also inquire regarding the variety and cause of its shape; since a statement was made above merely in regard to the size of its form, namely, that sometimes it is a complete circle, and sometimes the larger portion, and sometimes the smaller. But to understand these matters we must employ definite experiments. These are of the following kind, namely, if the observer of the bow moves in a direction parallel to the bow, it follows him at the side; if, however, he moves toward the bow, it recedes; but if he withdraws, the bow follows him, and not only with a slow motion, but with the same speed with which the observer is moving. Hence if a man runs or rides, the bow will seem to travel with an equal speed, as one can clearly perceive if houses or woods or other objects are in front of him in the vicinity of the bow. For he will see that the bow approaches these quite rapidly if they are behind it, or withdraws if they are in front of it.

Likewise by experiment we know that the sun, because of its very great distance from us, of which we cannot judge, seems always to be at the same distance when one recedes before it; and for this reason the sun seems to follow him who recedes, because otherwise the distance would not be judged to be the same. Similarly when one approaches the sun the sun seems to flee before him, because it always appears at the same distance. This appearance cannot be preserved unless the eye judges that the sun is moving in front of it. Moreover, we see clearly an apparent motion of this kind when the sun passes over woods or other elevated objects in the air, from which it recedes or to which it approaches. Moreover, we know that the sun seems to be in the direction of the observer and appears to be moving parallel to him because of its excessive distance. For this reason the sun always appears opposite the observer of the bow, wherever he may move in a direction parallel to the bow. As, for example, if the sun is in the south and the bow in the north, the observer noting their motions eastward sees the sun on one side opposite him and the bow on the other, so that a line passes from the center of the sun through the center of his eye and of the bow; and the motion of the bow varies because the sun seems to move in a direction parallel to the observer and to the bow. For if the sun were stationary according to the judgment of the sight, the bow would move in accordance with the motion of the observer, but would pass through more space than the observer and would pass the observer.

For thus in general it is true that, the visible object being at rest, if the observer moves, the image will move if the mirror is continuous; but if the object of vision moves at the same time in the direction of the observer, the motion of the image will pass through a space equal only to the motion of the observer.

If, however, the visible object is at rest, the image will pass through a greater space. For let *a* be the immovable center of the sun, *c* the center of the bow in the line *abc,* and let *b* and *c* move, so that *b* passes to *d,* and *c* to *f;* it is then evident that the space *cf* is greater than *bd,* and the line *ac* and the line *af* are not parallel but meet in the direction of *a,* and diverge in the other direction, as is apparent in the figure. But if *a* moves toward *g* in the direction of *d,* then *c* will be at *h,* and the lines *bd, ch,* and *ag* will be equal, and the lines *ac* and *gh* will be parallel. But the center of the bow is always in a straight line with the center of the eye in front of it. Wherefore we must assume that the sun seems to move in the direction of the observer, so that in this way the bow may seem to move in a direction parallel to the beholder. Moreover, we know that the sun is visible to the eye of every one. If a thousand men are arranged in a line from east to west their shadows will be sensibly parallel, and also the solar rays falling on them. This is because we do not perceive the convergence of the solar rays to the center of the sun because of its excessive distance. Moreover, owing to this same distance we do not perceive the daily motion of the sun in a short time, but the sun seems to us to be stationary in the sky.

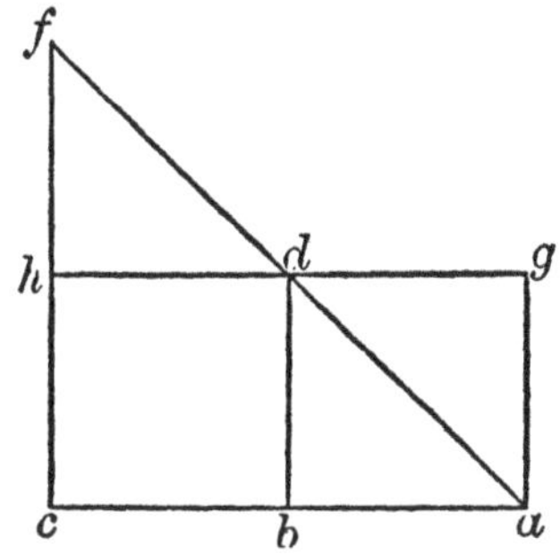

FIG. 72.

From these facts we learn by experiment that there are as many bows as there are observers. For if two observers at the same time stand and look at the bow in the north, and one of them goes westward, the bow will move in a direction parallel to him, and if the other goes to the east, the bow will move in a direction parallel to him, or if he stands still in the original position, the bow will be stationary. It is evident, therefore, that there are as many bows as there are observers; and for this reason it is impossible for two observers to see one and the same bow, although an ignorant person does not grasp this fact. For the shadow of each observer divides the arc of the rainbow into two equal parts, and therefore, although the shadows are sensibly parallel, they do not meet at the middle of the same bow, and therefore each observer must see his own bow. This is evident, because if the two bows move in different and opposite directions, they will move in accordance with the motions of the observers, and therefore there are as many bows as there are observers.

From these arguments it follows, then, that the bow is seen only by reflected rays of the sun, because if it were seen by incident rays, the bow would be an object fixed in one place in the cloud, which would not vary according to the motion of the observers nor according to the number of the observers. For when the solar rays pass through a rare cloud, the cloud appears to be colored white, and if they pass through a very thick cloud, the cloud appears black, and if they pass through one of moderate density, it appears to be of some intermediate color. But the color of the cloud appears the same to different observers, nor does it follow the motion of the observer, because it is not seen by reflected rays, but by incident or refracted rays. Similarly when a color is produced by incident rays through a crystalline stone, refraction takes place in it, but the same color in the same position is seen by different observers. It is evident, therefore, that the bow is not caused by rays falling directly or by refracted rays. Therefore it must be formed by reflected rays, because there are only the three kinds of principal rays. Nor will the bow be seen by means of accidental rays, since they do not change position unless caused by reflection. Hence from the property of reflection the same thing is evident; for the place of reflection and of the image changes according to a change in the observer. But this is the case here; therefore the bow is caused by reflection.

Nor can it be said that the substance of the bow remains and the position of the image varies in the refraction of the rays, since the change of position in the bow will be in accordance with a very sensible distance, namely, with a speed equal to that of a man running or riding as fast as possible. But the position of the image of an object at rest, although it may vary, does not vary with the same speed as the bow, because the whole change of image is in one common place, and it is therefore called a passage of space. Moreover, the image of an object seen by refraction does not follow the observer if he recedes, nor does it recede if he approaches, nor does it move in a direction parallel to him; which is evident when we look at a fish at rest in water, or a stick fixed in it, or the sun or moon through the medium of vapors, or letters through a crystal or glass. But the objection may be raised that only the place of the image is moved, and the bow can therefore be caused by refraction as well as by reflection. But that the only change is in the place of the image is evident, because the mirror, namely, the matter suitable for the bow, is immovable. But since such is the case, although the observer approaches the mirror there is no change on this account except in the place of the image, so that the visual ray more quickly intercepts the cathetus, and the place of reflection will not escape change, especially since the mirror is not on the surface, but in the depth of the cloud, since all the raindrops have the nature of a mirror, and therefore the place of reflection changes position sensibly in accordance with the judgment of the sight.

But on the other hand, since the sun as a matter of fact is at the same distance with respect to the mirror, there is no difference in the hour of the appearance of the rainbow which we need notice, whether the sun be moving from Cancer toward the south or from Capricorn toward the north, because in a natural day it does not move through one degree. And therefore in the hour of the appearance of the bow there is not a difference of distance of the sun from the cloud that we should take into account. Wherefore the sun's ray will reach the same moisture or part of the moisture, namely, the first part or the second; and therefore some definite part, or the whole moisture itself, becomes an immovable mirror. And it is not sufficient to say that the bow flees before the observer as fast as the sun seems to pursue him, although this is true, so that in this way the place of the reflection and of the bow would seem to vary, because the rays of the sun penetrate the whole of the moist matter, and reflection takes place from every raindrop. And on this account, unless there were another reason, one bow would appear immovable and fixed through the whole of the moisture, having a thickness in accordance with the whole of the moisture.

Wherefore we must state that the bow does not appear except in a definite arrangement of the air as regards a greater and lesser lucidity, so that there may be the required illumination in the direction of the eye, and the determined obscurity in the locality of the bow, because the object is weak and does not appear except in a dark place. When, therefore, the air between the eye and the bow is arranged in the required manner, the cone extends from the eye, whose base is the circle of the rainbow, and terminates in the moist cloud at the place of the required darkness, and the bow appears in that place. But when the eye* shifts in the direction of the bow, it now reaches a remoter portion of the air which is brighter, and it approaches an earlier position of the bow, so that owing to the nearness of the observer the same place with the same arrangement of the air is no longer found, but it is brighter than is required for the appearance of the bow. For it cannot appear in every arrangement of position mentioned, since although the moisture reaches to the eye, the bow, however, will not be seen except in a place of relative darkness. Therefore the base of the cone must fall in the cloud farther on in a place of the required obscurity. And thus the whole cone changes its position according to the motion of the observer, just as if the cone should be carried before the face of a man as he walked, whose vertex is fixed in the eye and whose body is extended in the air to the place of the bow, where, because of its distance from the eye, darkness is formed owing to the thickness of the vapors and of the overshadowing cloud. It is like the case of a man beholding a cloudy vapor at a distance which he cannot see near at hand, for owing to the distance the species of the vision is weakened, and can be terminated by the thickness of the vapors, which it can penetrate near at hand. For this reason it is evident that although reflection takes place everywhere in moisture, yet the bow does not appear everywhere, but only in the place determined. Thus that first objection is answered.

The second is cleared away by the same reasoning; for the mirror in this case does not remain the same, nor in the same place, because of the darkness that is required for the appearance of the colors, and for this reason the bow appears now in one part of the moisture, now in another.

If it is said that when the observer moves to the east the sun moves to the west, therefore the point of reflection will be changed according to the motion of the sun, and therefore the

* See Bridges' note on this passage, Vol. II, page 189.

bow will approach the sun more than it will the observer, it is evident by what has already been said that it will not do so; because the motion of the sun to the west in the hour of the appearance of the bow is insensible, and therefore it is considered stationary, that is, not to be moving westward.

CHAPTER VIII

BUT then it is evident that there will be nothing in the locality of the bow except the appearance of colors, and the appearance will not occur except when the bow is visible. For we have stated that the bow changes with a change in the observers. But sight does not produce colors. For the vision cannot cause colors in a cloud, as is evident; wherefore there will be nothing except in appearance. Moreover, vision by reflection has a visible object at the end of the reflected visual ray, as explained above, and therefore not at the point of reflection or elsewhere; because the object of vision is the terminus of reflection and of the reflected line, nothing therefore is seen by reflection in the cloud.

But if it be said that vision is sensibly and actively excited by the bow, and quite lively colors appear in the bow, and for this reason there cannot be merely the appearance of colors; we must reply that just as there is only the appearance of colors, so is there only the appearance of sensation and of lively change of vision; since a change of this kind has no other cause than defect in vision, as we see in the examples. For when in summer one lowers his head to the ground in the morning, in order that he may see the drops of dew on the ends of the blades of grass, if he views them carelessly or languidly and half closes his eyes, he will see in appearance all the colors of the rainbow. And similarly in the colored circle around the candle; and similarly also when one rises from sleep in summer with his eyes scarcely open and full of moisture, and looks at some small hole through which a solar ray is passing, the colors will be visible to him; and especially to those with weak eyes and to those who are nearsighted. Therefore the phenomenon must be owing to defect in vision, and for this reason there is an appearance only and not a reality. Moreover, this phenomenon is intensified, just as though it existed in the locality of the bow, because the visual ray quickly intersects the cathetus, owing to the small size of the raindrops; and therefore near the point of reflection a phenomenon of this kind occurs, and for this reason in the matter itself and in the locality of the bow.

But if it be said that solar rays passing through a crystal produce real and fixed colors, which produce a species and have objective reality, we must reply that the phenomena are different. The observer alone produces the bow, nor is there anything present except reflection. In the case of the crystal, however, there is a natural cause, namely, the ray and the corrugated stone, which has great diversity of surface, so that according to the angle at which the light falls a diversity of colors results. And viewing them does not* cause the colors to be present here, for there is color before it is seen here, and it is seen by different people in the same place. But in the case of the bow the phenomenon is the result of vision, and therefore can have no reality but merely appearance. This is evident, moreover, by the statements above in regard to the visual cone. When this falls as a whole in the position of the cone of the bow, the whole bow appears in the size possible for it with reference to the altitude and depression of the sun and the arrangement of its own matter; so that if the sun is on the horizon and the matter is prepared, a semicircular bow appears or a larger portion of the sphere, provided the whole visual cone falls in the position of the cone of the bow. But if the sun is on the horizon and the matter is prepared, but the visual cone does not fall as a whole in the position of the cone of the bow, only so much of the arc of the rainbow will appear as the base of the cone of vision cuts off from the place of the bow; therefore it is evident that the bow has no existence except in the vision.

Chapter IX

We must next consider that the bow is the image of the sun, and to grasp this idea we must consider the opinions of Seneca in his book on the Rainbow, since we must imagine a cloud suspended in the air, and rain descending from it. The vertex

* The sense of the passage seems to require the insertion of the negative.

of the cone falls upon the earth, and the circular base touches the concave surface of the cloud. There are, then, raindrops of small size in an infinite number, and reflection takes place in every direction as from a spherical mirror, and since they descend without an interval they seem from a distance to be continuous, and therefore the image of the sun seems continuous and not multiplex in accordance with the multitude of raindrops. There is a similar phenomenon in the circles colored by spray. And thus the objection is removed that might be made in regard to the multitude of images. But in regard to the shape, size, and color there is doubt; and we must state that spherical mirrors, and especially small ones, change the size and shape of the object of vision in many ways, and render shapes monstrous, and destroy all relationship of the parts of an object to one another, as was shown in the section on Perspective. And after destroying shape and size, and consequently the proper arrangement of qualities, which require shape and the requisite size, they are also able to change the color and to cause a color to appear from that which is without color, and conversely. And because light incorporated in matter resembles color, although in reality it is not color, as rare clouds appear white and thick ones appear black when the rays penetrate them, so in this case can the light of the sun falling on mirrors of this kind show what resembles color, especially since there is nothing present except an appearance, as we have stated.

Chapter X

In addition to these matters we must consider the diversity of colors. All authorities say it is caused by varieties in the texture of the moist cloud, and this is stated in the common text of Aristotle. Hence they say that when the substance of the cloud is somewhat dense, it seems to be darker; and when it is less dense, it seems to be blue; and if it is still less dense, it seems to be green; and if less still, it appears wine-colored and red; if it is still less dense, it appears grayish and pink; and if it is rather rare, it appears to be white. Accordingly the example of Aristotle is given in regard to flame, which, owing to the rarity and density of terrestrial smoke, appears bright and reddish and dark, and it happens so in different ways. But that example is not applicable here, as is evident from the experiment with crystalline stone. For in it colors are produced, and there is no difference in the density and rarity of the substance, and yet real colors are produced in it. And again, since there is no real color in the rainbow, but merely the appearance, we need only give the cause of the appearance or the appearance of the cause. Moreover, in the experiment on sprinkling drops of water nothing is caused by these differences in density, nor in the drops of dew on the ends of blades of grass; therefore it is evident that this reason of difference in density is not the true one. And in this we are not opposing Aristotle, since many other false statements are contained in the chapter on the rainbow and elsewhere in the current translations, as is evident throughout the whole philosophy of the Latins, if any one examines the different translations and the Greek original from which the Latin version was taken. But the opinion of Aristotle has not in all cases been translated, but there has been error in the copies of the Greeks and Arabs, or rather fault in the translators, no one of whom has perfect knowledge of the languages or sciences, as was stated before.

It is necessary, moreover, that another false statement that is found in this chapter on the rainbow should be corrected; and therefore we are to have this understanding in regard to the false statement already corrected. And that the mind may not be in suspense, I say that the statement in the current text is false, namely, that the rainbow is caused by the moon only twice in fifty years. For whenever the moon is full and is not veiled by cloud, and there is moist substance opposite it, the bow can appear just as it does through the agency of the sun. The experience of any one can prove this, and it has been proved by definite experiments.

Chapter XI

But with regard to the shape of the bow there is very great difficulty. For some think that it should be formed in the same way as the base of the cone of dropping dew. But this is impossible. For in drops of spray it is evident by experiment that in a mass of spray of irregular shape the form of the rainbow appears. Similarly those are in error who say that the bow is caused by refraction, so that it has the shape of a pyramid conformed to the curve of the surface of a cone expanded opposite the sun, for which reason, according to these thinkers, it is thought to be bow-shaped. And because the vertex of the pyramid mentioned, as they state, is near the earth, and the pyramid expands in a direction opposite the sun, half or more of that figure must fall on the surface of the earth and the remaining half or less must fall opposite the sun on the cloud. But this explanation is not correct, because it has been proved that the bow is not the result of refraction, since in the case of spray we cannot give a reason of this kind for the shape. For they say that the rays are refracted first at the point of contact of the air and the cloud, and afterwards at the point of contact of the cloud and the higher part of the moisture, so that by these refractions rays converge in the lower and denser part of the moisture; for the denser part is lower because it is heavier. For there the rays refracted as from the vertex of a cone spread out not into a cone, as they say, but into a figure conformed to the curve of the surface of a cone.* It is evident, therefore, that by three refractions the rays take this form. But there cannot be three refractions in spray formed in the rays, but only one; and yet the same shape is found as in the sky; wherefore the cause is not a refraction of this kind.

Lastly, why will the rays not form a cone but a figure conformed to the curve of the surface of a cone? For this will not be in accordance with the law of refraction, because refraction must form a regular cone, since the refraction of all rays falling on one circle of a spherical body is at right angles, and therefore the figure will not be curved more in one part than in another, but will be wholly regular and uniform. Nor can this curvature be caused by the moisture, because the moisture is not of such a form, but according to them is a rounded mass of conical form composed of an infinite number of drops, and therefore there is no explanation of that form in this way.

Another explanation therefore must be sought; and it can be stated that the bow must be in the form of a circular arc. For in accordance with differences in the position of an object

* See Bridges' note on this passage, Vol. II, page 194.

with respect to light different colors appear, as on the dove's neck. Since, therefore, the same color in one circle of the bow appears from one end to the other, all parts must have the same position with respect to the solar ray and the eye. But such a position of identity cannot exist except in a circular figure because of the equal declination of the parts. Moreover, on a vapor of spherical shape two kinds of circles can be marked. One kind passes through its poles like colures on a sphere; the other kind consists of parallel circles intersecting these colures transversely. The circular path of the colure is always vanishing owing to distance, because it does not keep the same position at its extremities and in the middle with respect to the eye, but the parallels are quite apparent in their circular path because of the equality of their position with respect to the eye, as has been stated concerning the moon in the section on Perspective. Since therefore the circle of the rainbow is one of the parallel circles it can be seen in its circular path.

But if the objection be raised against this view that then not only the arc of the cone in its gibbosity would be colored but also elsewhere throughout its whole surface, just like the full moon, which is fully illuminated on the side turned to us, and therefore the whole surface of the base, which is above the earth, would be colored. In reply to this objection it can be said that drops are not present everywhere from which rays are reflected to the eyes at angles equal to the angles of incidence. For this occurs only in the position of a circle, as is apparent in the reflections and refractions of the concave mirror and of other mirrors; and therefore since there can be four or five circles of the rainbow in the drops from which reflections to the eye are made at angles equal to the angles of incidence, color can be produced in them, and the phenomenon of the rainbow appear.

Chapter XII

Some used to think that the moist cloud was colored throughout, because there is reflection in all the drops, and the bow is caused by reflection; moreover, because in accordance with the number of observers and diversity in the position of the same observer the bow appears everywhere. But to the first statement we must reply that everywhere there are conditions suitable for the appearance of the bow, but as an actual fact the bow appears only in raindrops from which there is reflection to the eye; because there is merely the appearance of colors arising from the fantasy and deception of the vision, as has been stated. Therefore real color is nowhere present in the moist cloud, nor is the appearance of color present except in the drops from which reflection comes to the vision. But a reflection comes from every drop at the same time, while the eye is in one position, because of the equality of the angles of incidence and reflection. To the second statement we must reply that we can imagine colors appearing everywhere from two causes, in one way because the color was fixed, just as in other things; and it would produce a species from itself to the eye, and it would then be everywhere, and thus wherever there were sprays one would see color. But such is not the case here, because there is no real color, but only a fantasy of vision; and therefore the cause of the appearance is an erroneous judgment of the vision alone. Therefore the bow appears only in the place from which reflection reaches the eye; and from this cause, not from the first one, it changes according to a change in the observer.

There are five principal colors, namely, white, blue, red, green, black. For red is equidistant from the extremes, and is midway with respect to white and black. For Aristotle's statement in his work De Sensu et Sensato that there are seven colors is true if we divide blue into several shades, as dark blue and purple, and similarly by dividing green into different shades; but the five principal colors are distinct by nature. For the number five is better than all other numbers, as Aristotle says in the book of Secrets, and this is true as far as distinguishing definitely the number of things is concerned, although the number three is better as regards the numeral quality found in anything. For we draw this from nature, as Aristotle says in the first book on the Heavens and the World, because in everything a trinity exists, as Aristotle there shows, both in the Creator and in the creature. Because the number five distinguishes things more definitely and better, nature for this reason rather intends that there shall be five colors. Therefore these five colors are in the rainbow, rather than other colors, in accordance with the general arrangement of nature, which carries into effect and purposes that which is better. The color of blue termed celestial is black with a certain sweetness of luster, and is therefore reckoned under black.

It is thought by scientists that these colors are caused by the humors and colors of the eye; for these colors exist merely in appearance, and the parts of the eye have some colors naturally, although they are weak, in accordance with the existence of which the colors appear in the bow. Moreover, there are five bodies in the eye, namely, three humors and two coats, the uvea and the cornea; the consolidativa takes no part in vision, and therefore the five colors appear according to the properties of these five parts. Sometimes those bodies do not have distinct colors. For they vary in the eyes of different people, and different bows appear to vary, one in one way, another in another in the same hour, also at different times according to the variation of the air between the eye and the bow in its arrangement of light and dark portions. Moreover, a proof that the colors of the bow can appear through the eye is the fact that, when a hexagonal stone is placed before the eye in a dark place, colors of this kind are formed, and therefore in a similar manner in suitable matter in a dark place in the cloud they can at all events appear.

After what has now been said the cause of the corona is more easily apparent, and likewise the cause of the phenomenon appearing in a rectangular figure, which Seneca calls a rod, and Aristotle a perpendicular; for these phenomena are colored and others not. Therefore Seneca says, "If you form the vapor erect, it is a rod [streak in the heavens]; if you bend it into the arc of a circle, it is the rainbow; if you bend it into a complete circle, it is a corona"; which Aristotle calls a circle around the sun and the moon, and a halo. They state another difference also, namely, that the rainbow is formed opposite the sun, the streak to the side of the sun, and the corona beneath it. But the statements in regard to the circle and corona can be made without misrepresentation. Concerning the perpendicular, however, and the rod [streak] it is different. For it is merely a small portion of the corona, which by reason of the excessive distance seems to be a straight line, when the matter to complete the corona is lacking. A mass, moreover, of dense vapor sometimes forms in the direction of the sun beneath it. This vapor is denser than the air, and yet is not of excessive density but of such density as is required for a phenomenon of this kind. I state, then, that when that vapor is denser than the air, and is of a nearly spherical form, equidistant from the earth as nearly as it can be owing to the nature of a heavy body, all the rays proceeding from the point *o** beyond the perpendicular *od* are refracted at the surface of the vapor between the straight path and the perpendicular drawn from the point of refraction. All the rays falling on one circular path around that axis are refracted at equal angles, because all the angles of incidence are equal, as *a* and *f*, and the angles of refractions within the body are equal, and this is true in regard to all angles of rays falling on the circumference, which passes through the points *agf*, and in the same way the rays falling on the circumference passing through *bg* are all refracted at equal angles, because they are incident at equal angles, as is evident by geometrical demonstration. The rays, therefore, of a circular path, as *oa* and *of*, and all rays that in the circumference pass through *a* and *f*, fall and travel after refraction in the body of the equidistant vapor. Then when the air, which is a rarer body, meets the concave surface of the vapor, the rays are refracted and deflected from the perpendicular drawn from the point of refraction, and meet necessarily at a point, as at *o*. The same is true of those rays that fall on another circular path, as *ob* and *cg*, and they are conterminous on the same circumference, and they intersect by a second refraction at a remoter point, as at *d*, because rays falling on a larger circular path intersect at a remoter point. And thus it is always true of other rays falling on other circumferences that all the rays of one circular path fall at a point different from the point at which the rays of the other circular paths meet. And just as the rays of one circular path at their entrance into the vapor form a circle on the vapor, so when they leave the vapor they pass from a circular base and meet in a point of the cone. For all this light coming from a point in the sun and passing through the vapor is broken up into an infinite number of rays and is like a cylinder in shape, and has a circular form at both ends while it is in the vapor. But after it has passed out of the vapors the rays are deflected into the form of a cone, whose base is at the end of the cylinder of light, and its

* No figure is given in the MSS.

vertex falls on the earth. If, therefore, the eye receives the vertex of a cone, it will see the base of this cone, and because that denser vapor is of such a density that it is visible, especially since it has light in itself, it can be seen beneath the bases of the cone, and therefore in a circular form.

But just as many experiments are needed to determine the nature of the rainbow both in regard to its color and its shape, so too are they required in this investigation, where we must note that rays can be considered in two ways, namely, either as coming from a point and to a circle in the vapor which intersect at the vertex of a cone; and other rays that fall from the same point and to another circle form another cone, and so on. But the vertices of all these cones fall on the axis and the perpendicular coming from that point from which the refracted rays come, and incomplete cones are formed, nor do the rays pass to the points of intersection because of the interposition of the observer, and therefore no one will see by means of these incomplete cones. Wherefore if several people are looking at a corona, we must choose one of the two explanations, either that from other points of the sun rays come to be refracted into a cone, so that to the individual observers rays come from separate points in the sun, or that from the base of a cone whose vertex falls on the eye of an observer, other cones formed from accidental rays come to the other observers; so that the base of the principal cone which comes to the eye of one person is seen by others by means of an accidental cone, just as those in a corner of a house see by means of accidental rays the light falling through a window. But this second explanation cannot be the correct one, because a judgment by means of principal rays and one by means of accidental rays cannot be equal, but those viewing the corona at the same time form equal judgments. Moreover, each one judges that the sun is the center of the corona; therefore each one sees the sun, but accidental rays do not lead the sight to an object but to a species, just as a person in a corner of a house cannot see the sun, although he may see a ray falling through an opening. It is evident, therefore, that to the eye of an observer one cone comes whose rays come from a definite point in the sun, so that from one point rays come to one eye, and to a second eye from a second point, and to a third eye from a third point. But experiment explains all these things, just as in the case of the rainbow. Hence reasoning does not attest these matters, but experiments on a large scale made with instruments and by various necessary means are required. Therefore no discussion can give an adequate explanation in these matters, for the whole subject is dependent on experiment. For this reason I do not think that in this matter I have grasped the whole truth, because I have not yet made all the experiments that are necessary, and because in this work I am proceeding by the method of persuasion and of demonstration of what is required in the study of science, and not by the method of compiling what has been written on this subject. Therefore it does not devolve on me to give at this time an attestation possible for me, but to treat the subject in the form of a plea for the study of science.

In regard to the circle around a candle it is thought that it is formed by the vapors set free from the foods and drinks on the table; or that it can appear when the air is moist; or that it is produced at a distance from the candle in the vapors of the moist eye. Thus, therefore, any scientist will readily admit that experiment, not reasoning, determines the conclusions in regard to these matters. But I now pass to the second prerogative of this Experimental Science.

Chapter on the Second Prerogative of Experimental Science

This mistress of the speculative sciences alone is able to give us important truths within the confines of the other sciences, which those sciences can learn in no other way. Hence these truths are not connected with the discussion of principles but are wholly outside of these, although they are within the confines of these sciences, since they are neither conclusions nor principles. Clear examples in regard to these matters can be given; but in what follows the man without experience must not seek a reason in order that he may first understand, for he will never have this reason except after experiment. Hence in the first place there should be readiness to believe, until in the second place experiment follows, so that in the third reasoning may function. For if a man is without experience that a magnet attracts iron, and has not heard from others that it attracts, he will never discover this fact before an experiment. Therefore in the beginning he must believe those who have made the experiment, or who have reliable information from experimenters, nor should he reject the truth, because he is ignorant of it, and because he does not arrive at it by reasoning. I shall state, then, those things that I hold to have been proved by experiment.

Example I

MATHEMATICAL science can easily produce the spherical astrolabe, on which all astronomical phenomena necessary for man may be described, according to precise longitudes and latitudes. The device of Ptolemy in the eighth book of the Almagest is used with reference to circles as well as stars, as I have stated, by means of a certain similitude, but the subject is not fully explained by that device, for more work is necessary. But that this body so made should move naturally with the daily motion is not within the power of mathematical science. But the trained experimenter can consider the ways of this motion, aroused to consider them by many things which follow the celestial motion, as, for example, the three elements which rotate circularly through the celestial influence, as Alpetragius states in his book on Celestial Motions, and Averroës in the first book on the Heavens and the World; so also comets, the seas and flowing streams, marrows and brains and the substances composing diseases. Plants also in their parts open and close in accordance with the motion of the sun. And many like things are found which, according to a local motion of the whole or of parts, are moved by the motion of the sun. The scientist, therefore, is aroused by the consideration of things of this kind, a consideration similar in import to that in which he is interested, in order that at length he may arrive at his goal. This instrument would be worth the treasure of a king, and would supersede all other astronomical instruments and clocks, and would be a most wonderful instrument of science. But few would know how in a clear and useful manner to conceive of such a miracle and of similar ones within the confines of mathematical science.

Experimental Science

Example II

Another example can be given in the field of medicine in regard to the prolongation of human life, for which the medical art has nothing to offer except the regimen of health. But a far longer extension of life is possible. At the beginning of the world there was a great prolongation of life, but now it has been shortened unduly. Many have thought that the reason for this prolongation and shortening of life is found in the influence of the heavens. For they considered that the arrangement of the heavens was best at the beginning, and that as the world grows old all things decay. They think that the stars were created in more advantageous positions, in which stars have their dignities, which are called house, exaltation, triplicity, face, and boundary; and in a better relationship of these to one another in accordance with the diversity of aspects or the invisible projection of rays. They also think that they have gradually receded from this position, and that in accordance with this recession they impose a shortened span of life up to some fixed boundary, at which there will be a state of rest. But this idea has many contradictions and difficulties, of which I must now speak.

Whether this shall prove to be true or not, another reason must be given, which is ready at hand for us and is plain, which cannot be contradicted, and which we know by experience. Therefore in regard to this we must strive, that the wonderful and ineffable utility and splendor of experimental science may appear and the pathway may be opened to the greatest secret of secrets, which Aristotle has hidden in his book on the Regimen of Life. For although the regimen of health should be observed in food and drink, in sleep and in wakefulness, in motion and in rest, in evacuation and retention, in the nature of the air and in the passions of the mind, so that these matters should be properly cared for from infancy, no one wishes to take thought in regard to them, not even physicians, since we see that scarcely one physician in a thousand will give this matter even slight attention. Very rarely does it happen that any one pays sufficient heed to the rules of health. No one does so in his youth, but sometimes one in three thousand thinks of these matters when he is old and approaching death, for at that time he fears for himself and thinks of his health. But he cannot then apply a remedy because of his weakened powers and senses and his lack of experience. Therefore fathers are weakened and beget weak sons with a liability to premature death. Then by neglect of the rules of health the sons weaken themselves, and thus the son's son has a doubly weakened constitution, and in his turn weakens himself by a disregard of these rules. Thus a weakened constitution passes from father to sons, until a final shortening of life has been reached, as is the case in these days.

Not only is there this accidental cause, but there is also another, consisting in the disregard of morals. For sins weaken the powers of the soul, so that it is incompetent for the natural control of the body; and therefore the powers of the body are weakened and life is shortened. This weakening passes from father to son, and so on. Therefore owing to these two natural causes the longevity of man of necessity has not retained its natural course from the beginning; but for these two reasons the longevity of man has been shortened contrary to nature. Moreover, it has been proved that this excessive shortening of the span of life has been retarded in many cases, and longevity prolonged for many years by secret experiments. Many authors write on this topic. Wherefore this excessive shortening of life must be accidental with a possible remedy.

Since I have shown that the cause of a shortening of life of this kind is accidental, and therefore that a remedy is possible, I now return to this example which I have decided to give in the field of medicine, in which the power of medical art fails. But the experimental art supplies the defect of medicine in this particular. For the art of medicine can give only the proper rules of health for all ages. For although noted authors have spoken inadequately concerning the proper regimen of the aged, it has been possible, however, for medicine to give such a regimen. This regimen consists in the proper use of food and drink, of motion and rest, of sleep and wakefulness, of elimination and retention, of the air, and in the control of the passions of the mind. But if from birth a man followed a proper regimen to the end of his life, he would reach the limit of life set by God and nature, in accordance with the possibility of a proper regimen. But since it is impossible for this regimen to be followed by any one, and since few, nay, scarcely any one at all, from youth pay any heed to this regimen, and very few old people observe it as it is possible, therefore the accidents of old age of necessity come before old age and senility, namely, in the period of the prime of life, which is the age of human beauty and strength. In these times this period of life does not continue beyond forty-five or fifty years.

All these accidents of old age and senility are white hair, pallor, wrinkling of the skin, excess of mucus, foul phlegm, inflammation of the eyes, and general injury to the organs of sense, diminution of blood and of the spirits, weakness in motion and breathing and in the whole body, failure in both the animal and natural powers of the soul, sleeplessness, anger and disquietude of mind, and forgetfulness, of which the royal Hali says that old age is the home of forgetfulness; and Plato that it is the mother of lethargy. Because of the lack of a proper regimen of health all these accidents and many more come to men in the prime of life, that is, either in greater or lesser degree, in accordance with the better or worse control they have exercised over themselves in the matter of their health, and in accordance with a better and stronger constitution and a better or worse control exercised over their morals. But the medical art does not furnish remedies against this corruption that comes from lack of control and failure in regimen, just as all physicians expert in their own art know, although medical authors confess that remedies are possible, but they do not teach them. For these remedies have always been hidden not only from physicians, but from the whole rank and file of scientists, and have been revealed only to the most noted, whom Aristotle mentions in the first book of the Topics in the division of the probable. Not only are remedies possible against the conditions of old age coming at the time of one's prime and before the time of old age, but also if the regimen of old age should be completed, the conditions of old age and senility can still be retarded, so that they do not arrive at their ordinary time, and when they do come they can be mitigated and moderated, so that both by retarding and mitigating them life may be prolonged beyond the limit, which according to the full regimen of health depends on the six articles mentioned. And there is another farther limit, which has been set by God and nature, in accordance with the property of the remedies retarding the accidents of old age and senility and mitigating their evil. The first limit can be passed but the second cannot be.

Because of these two limits the Scripture says more than once, "Thou hast set its bounds which cannot be passed"; for it is impossible for the ultimate limit to be passed, but the first limit can be passed, although it is rarely passed; but the second limit cannot be passed. The proper regimen of health, therefore, as far as a man can possess it, would prolong life beyond its common accidental limit, which man because of his folly does not protect for his own interest; and thus some have lived for many years beyond the common limit of life. But a special regimen by means of remedies retarding the common limit mentioned, which the art of directing the health does not exceed, can prolong life much further. What is possible is shown by Dioscorides, who says that there may be some medicine to protect man from the swiftness of old age and from cold and from the drying up of his members, so that by its means the life of man may be prolonged. In the Tegni near the end Hali maintains this. Again he says, "Those who have lived a long time have used medicines by which their life has been prolonged." Avicenna, moreover, speaks in regard to such matters as follows in the second book of the Canon, "There is a medicine that settles and divides every constitution as it should be." But medical authorities have not given those medicines, nor have they been stated in their books, since these writers pay attention only to the art of caring for the health, and that, too, insufficiently as regards the elderly and the aged, as has been stated. But learned men devoted to experimental science have given thought to these matters, influenced thereto not only by their utility but by the action of animals which in many ways avoid a premature death, as, for example, the stag, the eagle, the snake, and many other animals that prolong their life by natural action, as authors state and experience has shown. Influenced by these examples they believed that God himself granted this power to brutes for the instruction of mortal man. Therefore they lay in wait for animals, in order that they might learn the powers of herbs, stones, metals, and other things, with which they improved their bodies in many apparently miraculous ways, just as we gather with the utmost certainty from the books of Pliny, Solinus, Avicenna on Animals, Tullius on the Divine Nature, from the philosophy of Artephius, and from other books and various authors, and many people have had experience in this matter. For in Paris lately there was a scientist who sought for snakes and took one and cut it into small portions, except that the skin of the belly on which it crept remained intact. This snake crept as it was able to a certain herb by the touch of which it was immediately cured. The experimenter collected the herb of wondrous virtue. Since human reason is superior to all the wisdom of animals, scientists thus encouraged by the examples of animals have thought out better and greater means.

And especially was this wisdom granted to the world through the first men, namely, through Adam and his sons, who received from God himself special knowledge on this subject, in order that they might prolong their life. We can learn the same through Aristotle in the book of Secrets, where he says that God most high and glorious has prepared a means and a remedy for tempering the humors and preserving health, and for acquiring many things with which to combat the ills of old age and to retard them, and to mitigate such evils; and has revealed these things to his saints and prophets and to certain others, as the patriarchs, whom he chose and enlightened with the spirit of divine wisdom, etc. And below he says that there is a medicine called the ineffable glory and treasure of philosophers, which completely rectifies the whole human body. This medicine is said to have been discovered by Adam or by Enoch and secured through a vision, as he himself states, although it has not been fully attested which of these first produced this medicine. But these matters and the most secret of secrets of this kind have always been hidden from the rank and file of philosophers, and particularly so after men began to abuse science, turning to evil what God granted in full measure for the safety and advantage of men.

Many examples, moreover, of these facts are written. Artephius, who traveled over all the regions of the East in his search for knowledge, found Tantalus, teacher of the king of India, seated on a golden throne and discoursing on nature and on the motions of the heavens. This same Tantalus humbled himself and became a pupil of Artephius, who is said in the book of his own philosophy to have lived actually for many centuries by means of secret experiments. Pliny, moreover, in the twenty-second book of the Natural History states that a man stood in the presence of Augustus, who had prolonged his life beyond a hundred years. To the astonishment of the bystanders he was strong, robust, and active to such a degree that the emperor in wonder asked him what he did so as to live in this way. The man replied in a riddle, as Pliny says, that he had applied oil on the outside and mead on the inside. Moreover, as stated in the book on the Accidents of Old Age, in the time of King William of Sicily a man was found who renewed the period of his youth in strength and sense and sagacity beyond all human calculation for about sixty years, and from a rustic ploughman became a messenger of the king. While ploughing he found a golden vessel in the fields hidden in the earth, which contained an excellent liquor. Thinking the liquor was dew from the sky he drank it and washed his face, and was renewed in mind and body beyond measure. And in the book just mentioned it is recorded that a man anointed with an excellent unguent the whole surface of his body with the exception of the soles of his feet, and lived for several centuries without decay except in his soles, which he had neglected to anoint, and for this reason he nearly always rode. Moreover, the author of this book bore witness that he had seen a man, and had talked with him, who had lived for several centuries, because he took a medicine prepared by scientists for a great king, who lost hope for himself and wished the medicine to be tried on an ignorant person. Thus the man's life was prolonged, and he had official letters from the Pope of that time and from others in regard to this fact.

Therefore the excellent experimenter in the book on the Regimen of the Aged says that if what is tempered in the fourth degree, and what swims in the sea, and what grows in the air, and what is cast up by the sea, and a plant of India, and what is found in the vitals of a long-lived animal, and the two snakes which are the food of Tyrians and Aethiopians, be prepared and used in the proper way, and the *minera** of the noble animal be present, the life of man could be greatly prolonged and the conditions of old age and senility could be retarded and mitigated. But that which is tempered in the fourth degree is gold, as is

* Old French word meaning mine used here probably in the sense of blood.

stated in the book on Spirits and Bodies,* which among all things is most friendly to nature. And if by a certain experiment gold should be made the best possible, or at any rate far better than nature and the art of alchemy can make it, as was the vessel found by the rustic, and it should be dissolved in such water as the ploughman drank, it would then produce a wonderful action on the body of man. And if there is added that which swims in the sea, namely, the pearl, which is a thing most efficacious for preserving life, and there is added also the thing that grows in the air.† This last is an *anthos* [flower] and is the flower of seadew, which possesses an ineffable virtue against the condition of old age. But the *dianthos* that is put in an electuary is not a flower, but is a mixture of leaves and fragments of wood and a small portion of flower. For the pure flower should be gathered in its proper season, and in many ways it is used in foods and drinks and electuaries. To these must be added what is cast up by the sea. This last is ambergris, which is spermaceti, a thing of wondrous virtue in this matter. The plant of India is similar to these, and is the excellent wood of the aloe, fresh and not seasoned. To these ingredients there is added that which is in the heart of a long-lived animal, namely, the stag. This is a bone growing in the stag's heart, which possesses great power against premature old age. The snake which is the food of the Tyrians is the Tyrian snake from which Tyriaca‡ is made, and whose flesh is properly prepared and eaten with spices. This is an excellent remedy for the condition of old age and for all the corruptions of the constitution, if it is taken with things suitable to one's constitution and condition, as we are taught in the book on the Regimen of the Aged. Aristotle, moreover, in the book of Secrets recommends strongly the flesh of the Tyrian snake for our ills. The snake that is the food of the Aethiopians is the dragon, as David says in the psalm, "Thou hast given it as food to the tribes of the Aethiopians." For it is certain that wise men of Aethiopia have come to Italy, Spain, France, England, and those lands of the Christians in which there are good flying dragons, and by the secret art they possess lure the dragons from their caverns. They have saddles

* *I.e.,* gases and solids.

† Bacon does not complete these sentences, but the sense is clear enough.

‡ Antidote against poisonous bites.

and bridles in readiness, and they ride on these dragons and drive them in the air at high speed, so that the rigidity of their flesh may be overcome and its hardness tempered, just as in the case of boars and bears and bulls that are driven about by dogs and beaten in various ways before they are killed for food. After they have domesticated them in this way they have the art of preparing their flesh, similar to the art of preparing the flesh of the Tyrian snake, and they use the flesh against the accidents of old age, and they prolong life and sharpen their intellect beyond all conception. For no instruction that can be given by man can produce such wisdom as the eating of this flesh, as we have learned through men of proved reliability on whose word no doubt can be cast.

If the elements should be prepared and purified in some mixture, so that there would be no action of one element on another, but so that they would be reduced to pure simplicity, the wisest have judged that they would have the most perfect medicine. For in this way the elements would be equal. Averroës, moreover, asserts in opposition to Galienus in the tenth book of the Metaphysics that if the mixture was made with an equality of the miscibles, the elements would not act or be acted on, nor would they be corrupted. Aristotle also maintains this view in the fifth book of the Metaphysics, where he has stated definitely that no corruption occurs when the active potencies are equal; and this is an assured fact.

For this condition will exist in our bodies after the resurrection. For an equality of elements in those bodies excludes corruption for ever. For this equality is the ultimate end of the natural matter in mixed bodies, because it is the noblest state, and therefore in it the appetite of matter would cease, and would desire nothing beyond. The body of Adam did not possess elements in full equality, and therefore the contrary elements in him acted and were acted on, and consequently there was waste, and he required nourishment. For this reason he was commanded not to eat of the fruit of life. But since the elements in him approached equality, there was very little waste in him; and hence he was fit for immortality, which he could have secured if he had eaten always of the fruit of the tree of life. For this fruit is thought to have elements approaching equality; and therefore it was able to continue incorruption in Adam, which would have happened if he had not sinned. Scientists, therefore, have striven to reduce the elements in some form of food or drink to an equality or nearly so, and have taught the means to this end. But owing to the difficulty of this very great experiment, and because few take an interest in experiments, since the labor involved is complicated and the expense very great, and because men pay no heed to the secrets of nature and the possibilities of art, it happens that very few have labored on this very great secret of science, and still fewer have reached a laudable end.

Those men, however, of whom mention has been made, who prolonged their life for centuries, had a medicine of this kind prepared with more or less skill. For Artephius, who, it is stated, lived a thousand and twenty-five years, had a better medicine than the aged ploughman, in whom his youth was renewed for sixty years. That liquor which the rustic drank is thought to have approached an equality of elements far beyond ordinary foods and drinks; but yet it was far from possessing full equality. For there are many degrees in the approach to final equality, which the medicine of Artephius failed to secure, as well as that which caused the man to live for five hundred years who had the papal letter in attestation of so great a miracle, of whom mention was made above. Nor is it strange if Aristotle did not live so long, nor Plato, nor many other famous philosophers; since in the Categories Aristotle says he was ignorant of the quadrature of the circle, which bears no comparison to a secret of this kind. Avicenna, moreover, says in the third book of the Physics that he did not yet know the category of habit; and now I judge that this is easily learned, and we wonder that these men were ignorant of matters so evident. For all wisdom is from the Lord God; and therefore sometimes to the simple are granted things that the most learned and famous cannot know. But medicine cannot give us these things nor does it mention them; but the greatness of the secret belonging to an experimental science of this kind has proved them. What, then, the remedies are and what things they contain, are discovered especially in the book of the Secrets of Aristotle and in the philosophy of Artephius, and in the book on the Conditions of Old Age, and in the treatise on the Regimen of the Elderly and the Aged, and in the books of Pliny, and elsewhere in many ways.

Example III

In the third place, the dignity of this science can be exemplified in alchemy. For that whole art is scarcely so perfected that the greater metals may be produced from the lighter ones, as gold from lead, and silver from copper. But that art never suffices to show the natural and artificial grades of gold and the modes of its grades. For experimental science has brought both to light, since it has discovered both the four natural grades and their seventeen modes and the artificial ones. By experiment it can be produced beyond twenty-four. Thus the vessel in which the liquor was contained, by means of which the ploughman became the messenger of the king, had a purity of gold far beyond the twenty-four, as its test and worth showed. But when those twenty-four degrees are found in a mass of gold, the gold is the best that can be produced by nature. But when there are twenty-four degrees of gold and one part of silver or one degree, the gold is inferior to the former, and thus the diminution of the degrees of the gold goes as far as sixteen, so that there are eight grades of gold with an admixture of silver. But the mineral power in the belly of the earth is not able sometimes to digest matter into the nature of gold, and does what it can by digesting it into the form of silver. And that I am not led astray in this matter by my imagination is proved by the fact that men are found in several parts of the world who are clever at producing those sixteen modes, and have discovered pieces and masses of gold in those seventeen modes. They then made a mixture of silver and air with gold in the aforesaid modes, so that they might have seventeen lumps of gold made artificially, by means of which they might learn the natural modes of gold. Since this art is not known to the majority of those who are eager for gold, many frauds consequently are perpetrated in this world. The art of alchemy not only omits these modes, but this gold of twenty-four degrees is very rarely found, and with the greatest difficulty. There have always been a few who during their life have known this secret of alchemy; and this science does not go beyond that. But experimental science by means of Aristotle's Secrets of Secrets knows how to produce gold not only of twenty-four degrees but of thirty and forty degrees and of as many degrees as we desire. For this reason Aristotle said to Alexander, "I wish to disclose the greatest secret"; and it really is the greatest secret, for not only would it procure an advantage for the state and for every one his desire because of the sufficiency of gold, but what is infinitely more, it would prolong life. For that medicine which would remove all the impurities and corruptions of a baser metal, so that it should become silver and purest gold, is thought by scientists to be able to remove the corruptions of the human body to such an extent that it would prolong life for many ages. This is the tempered body of elements, of which I spoke above.

Chapter on the Third Prerogative or the Dignity of the Experimental Art

But there is a third dignity of this science. It arises from those properties through which it has no connection with the other sciences, but by its own power investigates the secrets of nature. This consists in two things; namely, in the knowledge of the future, the past, and the present, and in wonderful works by which it excels in the power of forming judgments the ordinary astronomy dealing with judgments. For Ptolemy in the book introductory to the Almagest says that there is a more certain road than that through the ordinary astronomy, and this is the pathway of experiment, which follows the course of nature, to which many of the philosophers who are believers are turning, just like Aristotle and a host of the authors of judgments formed from the stars, as he himself says, and as we know by proper practice, which cannot be gainsaid. This science was discovered as a complete remedy for human ignorance and inadvertence; for it is difficult to get accurate astronomical instruments, and it is more difficult to get verified tables, especially those in which the motion of the planets is equalized. The use, moreover, of these tables is difficult, but still more difficult is the use of the instruments. But this science has discovered the definitions and the means by which it can answer easily every question, as far as the power of a single branch of philosophy can do so, and by which it can show us the forms of the celestial forces, and the influences of the heavenly bodies on this world without the difficulty of the ordinary astronomy. This part of the science relating to judgments has four principal divisions or secret sciences.

Moreover, certain bear witness that activities of this science which display philosophy consist in changing the character of a region, so that the habits of its people are changed. One of such witnesses was Aristotle himself, the most learned of philosophers. When Alexander asked him in regard to the nations which he had discovered, whether he should exterminate them because of the ferocity of their character, or should permit them to live, he replied in the book of Secrets, "If you can alter the air of those nations, permit them to live; if you cannot, then kill them." For he maintained that the air of these nations could be changed advantageously, so that the complexions of their bodies would be changed, and then their minds influenced by their complexions would choose good morals in accordance with the freedom of the will. This is one of the secrets.

Moreover, certain assert that change is effected by the sun. There is, as an illustration, the example of Aristotle when he said to Alexander, "Give a hot drink from the seed of a plant to whomsoever you wish, and he will obey you for the rest of your life." Some maintain that an army may be stupefied and put to flight. Of this number is Aristotle, who says to Alexander, "Take such a stone, and every army will flee from you." They bear witness that these statements and innumerable others of this kind are true, not meaning that violence is done to the freedom of the will, since Aristotle, who maintains this view, says in the Ethics that the will cannot be coerced. The body, moreover, can be changed by the influence of things, and the minds of people are then aroused and influenced to desire voluntarily that to which they are directed; just as we see in the book of Medicine that through potions and many medicines people can be changed in body and in the passions of the soul and in the inclination of the will.

There are, moreover, other inventions belonging more to nature which do not have as their object a marvelous change in the will, and they are diversified in character. Some of these possess an excellence of wisdom with other advantages, as, for example, perpetual baths most suitable for human use that do not require any artificial renewal; and ever-burning lamps. For we see many things that cannot be impaired by fire, nay, that are purified by fire, like the skin of the salamander and many other things of this kind, which also can be so prepared that they are externally luminous of themselves, and retain the power of fire, and give forth flame and light. Moreover, against foes of the state they have discovered important arts, so that without a sword or any weapon requiring physical contact they could destroy all who offer resistance. There are many kinds of these inventions. Some of these are perceived by no one of the senses, or by smell alone, and of these inventions Aristotle's book explains that of altering the air, but not those of which I spoke above. These last are of a different character, since they act by means of an infection. There are others also that change some one of the senses, and they are diversified in accordance with all the senses.

Certain of these work a change by contact only and thus destroy life. For malta, which is a kind of bitumen and is plentiful in this world, when cast upon an armed man burns him up. The Romans suffered severe loss of life from this in their conquests, as Pliny states in the second book of the Natural History, and as the histories attest. Similarly yellow petroleum, that is, oil springing from the rock, burns up whatever it meets if it is properly prepared. For a consuming fire is produced by this which can be extinguished with difficulty; for water cannot put it out. Certain inventions disturb the hearing to such a degree that, if they are set off suddenly at night with sufficient skill, neither city nor army can endure them. No clap of thunder could compare with such noises. Certain of these strike such terror to the sight that the coruscations of the clouds disturb it incomparably less. Gideon is thought to have employed inventions similar to these in the camp of the Midianites. We have an example of this in that toy of children which is made in many parts of the world, namely, an instrument as large as the human thumb. From the force of the salt called saltpeter so horrible a sound is produced at the bursting of so small a thing, namely, a small piece of parchment, that we perceive it exceeds the roar of sharp thunder, and the flash exceeds the greatest brilliancy of the lightning accompanying the thunder.

There are also very many things that slay every poisonous animal by the gentlest touch, and if a circle is made around these animals with things of this kind the animals cannot get out, but die, although they are not touched. But if a man is bitten by a poisonous animal, by the application of the powder of such things he can be healed, as Bede states in his Ecclesiastical History and as we know by experience. And thus there are innumerable things that have strange virtues, whose potencies we are ignorant of solely from our neglect of experiment.

But there are other inventions which do not possess such advantage for the state, but are to be looked upon as miracles of nature, such as experiments with the magnet, not only on iron, but on gold and other metals. Moreover, if the experiment on iron were not known, it would be viewed as a great miracle. And surely in respect to the action of the magnet on iron there are phenomena unknown to those who use the magnet which show in a wonderful way the dissolutions of nature. Just as also from these the faithful experimenter knows how to experiment on the mutual attraction of other things, as, for example, the stone that passes to the acid, and bitumen that ignites from fire placed at a distance from it, as Pliny states in the second book of the Natural History; and certain other things that are mutually attracted although locally separated. This is truly wonderful beyond all that I have seen and heard. For after I saw this, there has been nothing difficult for my intellect to believe, provided it had a trustworthy authority. And that this fact may not be hidden from your Reverence, this phenomenon occurs in the parts of plants divided and locally separated. For if a sapling of one year's growth is taken, which springs forth beside the roots of the hazel, and is divided longitudinally, and the divided parts separated by the space of a palm or four fingers, and one person holds on one side the extremities of the two parts, and another similarly on the other side, always with an equal and gentle grasp, so that the parts are kept opposite each other in the same position they had before the division, within the space of half a mile* the parts of the twig begin to approach each other gradually, but with greater force at the end of the experiment, so that at length they meet and are united. The ends, however, remain apart, because they are prevented from meeting owing to the force exerted by those holding the parts. This is a very wonderful thing. For this reason magicians perform this experiment, repeating different incantations, and they believe that the phenomenon is caused by virtue of the incantations. I have disregarded the incantations and have discovered the wonderful action of nature, which is similar to that of the magnet on iron. For just as the one attracts the other because of the similar nature of the iron and the magnet, so do the parts in this case. Hence the natural force, which is similar in both parts of the twig, causes them to unite. If they were arranged in the required way, they would meet at the extremities just as in the middle and more quickly, as, for example, if the ends were minutely pierced and threads passed through the ends, so that they could be suspended in the air without hindrance. This is true not only of hazel saplings but of many others, as in the case of willows and perhaps in that of all saplings if they were arranged in the required manner. But since in such matters the mind thinks more aptly than the pen writes, I forbear for the present. I am here merely writing down the statements of scientists and noting their achievements. The genius of these men I admire more than I understand.

Concluding thus the subject of this science experimental without restriction, I shall now show its advantage to theology, as I have done similarly in the case of the other sciences. Since I have now shown the intrinsic nature of this science, it is evident to all that next to moral philosophy this science is the most useful, and it is so in the first place to theology itself in its unrestricted sense because of the literal and spiritual meaning in which it consists. For I showed above that the literal meaning consists in expressing the truth in regard to created things by means of their definitions and descriptions, and I likewise showed that reasoning does not arrive at this truth, but that experiment does. Wherefore this science next to moral philosophy will present the literal truth of Scripture most effectively, so that through suitable adaptations and similitudes the

* Time taken to walk half a mile.

spiritual sense may be derived, owing to the peculiar nature of the sacred Scripture and in accordance with the methods employed by the sacred writers and by all sages.

Then this science as regards the commonwealth of believers is useful, as we saw in its special knowledge of the future, present, and past, and in its display of wonderful works on behalf of Church and state, so that all useful activities are promoted and the opposite are hindered both in the few and in the multitude, as was explained. And if we proceed to the conversion of unbelievers, it is evidently of service in two main ways with numerous subdivisions, since a plea for the faith can be effectively made through this science, not by arguments but by works, which is the more effective way. For to the man who denies the truth of the faith because he cannot understand it I shall state the mutual attraction of things in nature, just as I described it. Likewise I shall tell him that a jar may be broken without human force, and the wine contained in it remain motionless and without flow for three days; and that gold and silver in a pouch, and a sword in its scabbard may be consumed without injury to their containers, as Seneca states in the book of Natural Questions. I shall tell him, moreover, that the birds called kingfisher in the depth of winter compel the stormy sea to be calm and restrain itself until they have laid their eggs and brought forth their young, as Basil and Ambrose in the Hexaemeron and philosophers and poets state. For these facts and similar ones ought to influence a man and urge him to accept the divine verities. Since if in the vilest creatures verities are found by which the pride of the human intellect ought to be subdued so that it may believe them although it does not understand them, conviction should follow, or injury will be done to infallible truth, since a man ought rather to humble his mind to the glorious truths of God. Surely there is no comparison.

But there is still another very useful way; since the formation of judgments, as I have said, is a function of this science, in regard to what can happen by nature or be effected in art, and what not. This science, moreover, knows how to separate the illusions of magic and to detect all their errors in incantations, invocations, conjurations, sacrifices, and cults. But unbelievers busy themselves in these mad acts and trust in them, and have believed that the Christians used such means in working their miracles. Wherefore this science is of the greatest advantage in persuading men to accept the faith, since this branch alone of philosophy happens to proceed in this way, because this is the only branch that considers matters of this kind, and is able to overcome all falsehood and superstition and error of unbelievers in regard to magic, such as incantations and the like already mentioned. How far, moreover, it may serve to reprobate obstinate unbelievers is already shown by the violent means that have just been touched upon, and therefore I pass on.

We must consider, however, that although other sciences do many wonders, as in the case of practical geometry, which produces mirrors that burn up every opposing object, and so too in the other sciences, yet all things of such wonderful utility in the state belong chiefly to this science. For this science has the same relation to the other sciences as the science of navigation to the carpenter's art and the military art to that of the engineer. For this science teaches how wonderful instruments may be made, and uses them when made, and also considers all secret things owing to the advantages they may possess for the state and for individuals; and it directs other sciences as its handmaids, and therefore the whole power of speculative science is attributed especially to this science. And now the wonderful advantage derived from these three sciences in this world on behalf of the Church of God against the enemies of the faith is manifest, who should be destroyed rather by the discoveries of science than by the warlike arms of combatants. Antichrist will use these means freely and effectively, in order that he may crush and confound the power of this world; and by these means tyrants in times past brought the world under their sway. This has been shown by examples without end.

But I now cite the one example of Alexander the Great in place of all other examples that might be cited, who when he set out from Greece to conquer the world had only 32,000 foot soldiers and 4500 horsemen, as Orosius states to Augustine in his book Ormesta Mundi, bringing war with so small a force upon the whole world. It is uncertain which is the more wonderful, that he conquered or that he ventured the attack. In his first battle with King Darius he overcame 600,000 Persians with a loss in his army of 120 horsemen and nine foot soldiers.

In the second battle he conquered 400,000 men, and of his own army 130 foot soldiers and 150 horsemen fell. After this he easily subdued the rest of the world, which had become terrified. But Orosius says that he conquered not less by skill than by the valor of the Macedonians. Nor is it to be wondered at, since Aristotle was with him in these wars, as we read in the life of Aristotle. Seneca, moreover, states in the Natural Questions that Alexander conquered the world under the guidance of Aristotle and Callisthenes, who were his teachers in all knowledge. But Aristotle was his chief teacher; and it is easily apparent from what has been said how by the paths of knowledge Aristotle was able to hand over the world to Alexander. Moreover, the Church should consider the employment of these inventions against unbelievers and rebels, in order that it may spare Christian blood, and especially should it do so because of future perils in the times of Antichrist, which with the grace of God it would be easy to meet, if prelates and princes promoted study and investigated the secrets of nature and of art.

PART SEVEN OF THIS PLEA

MORAL PHILOSOPHY: FIRST PART

I HAVE shown in what precedes that the knowledge of languages, mathematics, perspective, and experimental science are most useful and necessary in the pursuit of wisdom. Without these branches no one can advance as he should in wisdom, taken not only in an unrestricted sense, but also in relation to the Church of God and to the other three activities already described. I now purpose to disclose the principles of a fourth science that is better than all those previously mentioned and nobler. This science is preëminently active, that is, formative, and deals with our actions in this life and in the other. For all other sciences are called speculative. For although certain are active and formative, they are concerned with the actions of art and nature, not with morals, and they investigate the verities of things and of scientific activities which have reference to the speculative intellect and are not concerned with things pertaining to the active intellect, which is called active because it directs action, that is, good or evil action. Hence the term active is here taken in a restricted sense as applying to the actions of conduct, in accordance with which we are good or evil; although if we take the term active in a broad sense for all formative science, many other sciences are active. But this science is called in a self-explanatory way active because of the chief actions of man, which relate to virtues and vices, and to the felicity and misery of the other life.

This active science is called the moral science and the civil science, which instructs man as to his relations to God, and to his neighbor, and to himself, and proves these relations, and invites us to them and powerfully influences us thereto. For this science is concerned with the salvation of man to be perfected through virtue and felicity; and this science aspires to that salvation as far as philosophy can. From these general statements it is evident that this science is nobler than all the other branches of philosophy. For since it is the final inner purpose of human wisdom, and since the purpose is the noblest part in anything, this science is of necessity the noblest. Similarly this science alone or in chief measure is concerned with the same questions as theology; because theology considers only the five aforesaid, although in a different way, namely, in the faith of Christ. This science, moreover, contains a great deal of excellent testimony in regard to this same faith; and from afar surmises its principal articles to the great aid of the Christian faith, as what follows will make clear. But theology is the noblest of the sciences; therefore that science which is most closely related to it is nobler than the others. But that the very great utility of this science may be apparent, we must investigate its parts, to the end that we may draw what we wish from the parts and the whole.

Since, moreover, moral philosophy is the end of all branches of philosophy, the conclusions of the other sciences are the principles in it in accordance with the relationship of preceding sciences to those that follow; because the conclusions of preceding sciences are naturally assumed in those that follow. For this reason it is fitting that these conclusions in preceding sciences should be carefully proved and attested, so that they may be worthy of acceptance in the sciences that follow, as is evident from metaphysics. Therefore the principles of moral philosophy are attested in preceding sciences; and for this reason these principles should be drawn from the other sciences, not because they belong to these sciences, but because these sciences have prepared them for their mistress. Hence wherever they are found they are to be ascribed to moral philosophy, since in substance they relate to morals. And although they may be stated in other sciences, this is by grace of moral philosophy. Wherefore all such are to be considered in regard to moral philosophy and ascribed to it. Therefore if we wish to use them as they were intended to be used, they must be collected in moral science from all the other sciences. Nor is it strange if philosophers through the whole of speculative philosophy have diffused ethical principles, because they knew that these related to man's salvation; and therefore in all the sciences they have diffused beautiful thoughts, so that men might always be directed to the blessing of salvation, and that all might know that the other sciences are sought after only for the sake of that science which is the mistress of human wisdom. Therefore if I adduce from other places authorities other than those contained in books on morals, we should consider that these ought properly to be placed in this science. Nor can we deny that they have been written in books of this science, since in Latin we do not possess except in parts the philosophy of Aristotle, Avicenna, and Averroës, who are the principal authors in this subject. For just as theology perceives that truths bringing salvation belong to it, wherever it finds them, as I stated at the beginning and touched upon later, so also does moral philosophy claim as its right whatever it finds written elsewhere pertaining to it. This science is called moral science by Aristotle, and by others civil science, because it shows the rights of citizens and states. And since cities used to bear sway over countries when Rome ruled the world, this science is called civil from city [*civitas*], although it formulates the rights of kingdom and empire.

This science, moreover, in the first place teaches us to lay down the laws and obligations of life; in the second place it teaches that these are to be believed and approved, and that men are to be urged to act and live according to those laws. The first part falls under three heads; for first comes naturally man's duty to God and in respect to angelic beings; secondly, his duty to his neighbor; and thirdly, his duty to himself, just as the Scripture states. For in the first place in the books of Moses are the commands and laws in regard to God and divine worship. In the second place are those regarding a man's relationship to his neighbor in the same books and in those that follow. In the third place instruction is given in morals, as in the books of Solomon. Similarly in the New Testament these three alone are contained. For a man cannot assume other duties.

Not only on account of the first but of all that follows is it necessary that the principles of this science, by which the others are verified, should be set forth at the beginning. Certain of these principles are purely principles and are capable of being expressed only metaphysically. Others, although they are principles with respect to what follows, either are first conclusions of this science, or although they rejoice in some privilege of a principle, yet because of their very great difficulty, and because they meet with less contradiction,* and because of their very great utility with regard to what follows, ought to be sufficiently demonstrated. Just as Aristotle in the beginning of his

* See Bridges' note on the meaning, Vol. II, page 226.

natural philosophy proves the first principle of that science, namely, that there is motion in opposition to those who maintained that there is only the one immovable. But we must note that metaphysics and moral philosophy are very closely allied; for each is concerned with God, the angels, life eternal, and many questions of this kind, although in different ways. For metaphysics through the common principles of all sciences investigates qualities metaphysically, and through the corporeal investigates the spiritual, and through the created discovers the Creator, and through the present life deals with the future life, and furnishes much introductory matter to moral philosophy. Metaphysics inquires into these subjects because of civil science, so that we have accordingly the right to unite this science with metaphysics, in order that the principles that must be explained in metaphysics may be assumed here, lest I should confound sciences differing from each other if I try to prove in this science what properly belongs to metaphysics.

I state, therefore, that God must exist just as this fact must be proved in metaphysics: second, that God's existence is naturally known to every man; and third, that God is of infinite power and of infinite goodness, and coupled with this that he is of infinite substance and essence, so that it follows that he is best, wisest, and most powerful. Fourth, that God is one in essence and not more than one. Fifth, that not only is God one in essence but triune in another way, which must in general be explained by the metaphysician, but here must be unfolded in special doctrinal form. Sixth, that he has created all things and rules in the realm of nature. Seventh, that besides corporeal things he has created spiritual substances which we call intelligences and angels; because intelligence is a term denoting a nature, but angel is a term denoting an office. This science inquires also how many there are, and what are the activities of these intelligences, in accordance with their relation to metaphysics, as far as it is possible for them to be known by human intelligence. Eighth, that besides angels he has made other spiritual substances which are the rational souls in men. Ninth, that he has made these immortal. Tenth, that the felicity of the other life is the highest good. Eleventh, that man is capable of this felicity. Twelfth, that God has the moral direction of the human race, just as he directs all else in the realm of nature.

Thirteenth, that to those who live aright in accordance with God's direction God promises future felicity, just as Avicenna teaches in the tenth book of the Metaphysics, and that to those who live an evil life a horrible future unhappiness is due. Fourteenth, that worship is due to God with all reverence and devotion. Fifteenth, that just as a man's conduct toward God is regulated by the reverence required, so is his conduct toward his neighbor regulated by justice and peace, and his duty to himself by integrity of life. Sixteenth, that a man cannot by his own effort know how to please God with the worship required, nor how he should conduct himself as regards his neighbor and as regards himself, but he requires that the truth be revealed to him in these things. Seventeenth, that revelation must be made to one only; that this one must be the mediator of God and mankind, and the vicar of God on earth, to whom the whole human race is subject, and who must be believed without contradiction when it has been proved with certainty that he is such as has just been described; and he is the lawgiver and the high priest who in things temporal and spiritual has full power, as it were, a human God, as Avicenna says in the tenth book of the Metaphysics, whom it is lawful to adore after God.

By these principles metaphysics is united with moral philosophy, and approaches it as its final purpose. Avicenna thus beautifully joins them at the end of his Metaphysics. But the remaining principles belong to this science and are not to be explained in metaphysics, although Avicenna adds a number of them. But in the beginning of his book he gives as a reason for this the fact that he had not finished his Moral Philosophy and did not know whether he would complete it; and therefore he mingled with these principles many which properly belong to moral philosophy, as is evident to the inquirer. After these have been considered, the lawgiver at the beginning must take up the properties of God in particular, and of angels, and the felicity and misery of the other life, and the immortality of our bodies after the resurrection, and similar subjects to which the metaphysician could not aspire. For he is concerned in all these matters chiefly with the question as to existence; because it is his special function to make this question clear in regard to all matters, because he considers entity and being in their common properties. But other sciences take up the other questions involved in things; namely, what each one is, of what kind and size, and the like, in accordance with the ten categories. The moral philosopher, however, is not to explain all the secrets of God and of angels and of other questions; but he should explain those things that are needful for the multitude, and in which all men should agree, lest they fall into doubts and heresies, as Avicenna teaches in the Principles of Moral Philosophy.

I say, therefore, that moral philosophy in the first place explains the triune nature of God, a truth which the lawgiver holds through revelation more than through reason. But reason, the source from which philosophers have made many statements in regard to divine things in particular which transcend human reason and fall under revelation, has been touched upon already in the section on mathematics. For it was shown there how they could possess many noble truths regarding God, which were disclosed by a revelation made to them, as the apostle says, "For God revealed these things." But he revealed them rather to the patriarchs and prophets, who, it is agreed, had a revelation, and from whom the philosophers learned all things, as was clearly proved above. For the patriarchs and prophets handled divine things by the methods of theology or prophecy, but also by those of philosophy, because they discovered the whole of philosophy, as has been proved in the second part of this work. Moreover, the metaphysician was able to teach acceptably that God exists, and that he is naturally known, and that he is of infinite power, and that he is one, and that he is triune. But the mode in which the trinity exists he was not able to explain fully; and therefore this must be attested here.

There is, then, the blessed Trinity, the Father and the Son and the Holy Spirit. For Claudius, one of the expositors of Holy Scripture, in that book in which he combats this heresy, God feels nothing with a sense of passion but with a feeling of compassion, says, "Plato with praiseworthy daring, wonderful genius, and immutable purpose sought, found, and proclaimed three persons in the Godhead; God the Father, also the paternal mind, art or wisdom, and the mutual love of both." He not only taught that we must believe in one supreme indivisible Godhead in three equal Persons, but he proved that he exists in this way. These truths are evident from his book on Divine Things. Porphyrius, moreover, as Augustine says in the tenth book on the City of God, chapter XXIX, predicated a Father and his Son, which he called the paternal intellect and mind, and the medium of these, whom, as Augustine says, we think he himself calls the Holy Spirit, and following his custom he calls them three Gods, where, although he employs his terms loosely, he yet perceives the truth to be maintained. Augustine, moreover, in the same book, in chapter XXXII,* states that a certain Platonic philosopher, whose name he does not mention, recited the beginning of the Gospel according to John as far as the Incarnation of Christ, in which the distinction in the divine Persons is clearly stated. Augustine also in the tenth book on the City of God, chapter XXXVI and chapter XXXVII, asserts that Porphyrius states in the first book on the Return of the Soul that sins cannot be purged except by the Son of God. Aristotle states in the beginning of the Heavens and the World that in divine worship our object is to magnify one God through the number three, excellent by reason of the properties of the things that have been created. And therefore since every creature, as is evident from the metaphysics, is a vestige of the Trinity, there must be a Trinity in the Creator. And since Aristotle completed the philosophy of his predecessors as far as his times permitted, he must have had a far more definite perception concerning the blessed Trinity of Persons, so as to confess the Father and the Son and the Holy Spirit. For this reason in the law of Aristotle there were three sacrifices and three prayers, as Averroës says in his commentary on the beginning of the Heavens and the World. This is also made clear by the politics of Aristotle, which is the Book of Laws. Avicenna, moreover, the chief expositor of Aristotle, maintains in the principles of moral philosophy that there is a Holy Spirit.

But he was able to perceive the truth far more clearly concerning the Father and the Son, because it is more difficult to understand the procession of the Holy Spirit from two distinct persons than the generation of one of these from the other. For this reason philosophers were more deficient in their comprehension of the Holy Spirit than in their knowledge of the

* See Bridges' note, Vol. II, page 230.

Father and the Son. And therefore those who were able to gain a knowledge of the Holy Spirit knew much more about the other Persons. Ethicus the philosopher in his book on Things Divine, Human, and Natural, maintains that in God are the Father, and the Word of the Father, and the Holy Spirit; and that there are three Persons, namely, the Father and the Son and the Holy Spirit. This dogma must also be held by reason. This reasoning, however, should not have been given before those things that have to be expressed concerning God in particular, nor before the authorities of the great philosophers, which are introduced for this same purpose in this science as in a place appropriate for them.

I say, therefore, that God is of infinite power, and infinite power is capable of infinite action; therefore something infinite can be made by God, but not anything by substance, because in that case there could be a plurality of gods. The contrary to this has been demonstrated in the section on mathematics. Therefore that which is begotten by God must be God, since it has the essential nature of the begetter; yet it is different in person. And since that which is begotten has infinite power, since it is infinite goodness, it is able to bring forth infinite goodness; therefore it is able to bring it forth in another person. Then, therefore, either the Father brings forth the same person; and in that case the Holy Spirit will proceed from both, or he will be brought forth from the Son alone; and then he will not appertain to the Father, nor will there be full relationship, and in that case there will not be complete accord in the divine Persons, which is contrary to reason. Moreover, there cannot be parity of love according to this view, because the Father would love the Son more than he loves the Holy Spirit, because he begets the Son, and does not bring forth the Holy Spirit. But since the Holy Spirit is God, because he has the divine nature, an infinite love must be due him; and therefore the Father will love him just as he does the Son with an infinite love. And also since the love of the Father cannot be less than an infinite love, because his love is in accordance with his power, it remains, then, that the love of the Father for the Holy Spirit will be as great as that of the Son for the Holy Spirit. Wherefore both the Holy Spirit and the Son must be brought forth from the Father. That, moreover, there are not nor can there be more persons, cannot and ought not to be explained here, but should be assumed, until it is proved in the fourth part of this science, on which the full measure of persuasive argument will be bestowed. But it was necessary that the trinity of Persons, namely, the Father and the Son and the Holy Spirit, should be proved and expounded here, because it is the fundamental principle in this science for establishing divine worship and for many other things. Nor should it be stated, on the other hand, that no science must prove its principles. For how this is to be understood has been explained above. But the other questions that can be asked in regard to God and in which there should be probable doubt, are the conclusions of the fourth part, and therefore they will be determined there.

Not only have the philosophers spoken of God unrestrictedly, but of God incarnate, who is the Lord Jesus Christ, and of those things that pertain to him. For truths of this nature are necessary for the human race, and there is no salvation for man except through a knowledge of these truths. And therefore it was necessary for the salvation of all from the beginning of the world that truths of this kind should be known as far as suffices for salvation. I state this fact because some men have more knowledge, some less, of truths of this kind. It was fitting also that philosophers devoted to wisdom should know something of this truth, whether they were to be saved or not; so that the world might be prepared and made ready for this perfect truth, in order that it might be more easily received when the time came. And the plea in regard to it was stated in greater fullness in a preceding section, and for this reason a general plea suffices here, in order that by experience we may learn that the philosophers had knowledge of many remarkable facts in regard to Christ and the glorious Virgin. Moreover in the section on astronomy the formal judgment of Albumazar was cited from the sixth book of the larger Introduction, where he asserts on the authority of all from the beginning of philosophy that a Virgin should bring forth a son who would be called Jesus Christ. Moreover, in the book of Conjunctions he made a similar statement. But this judgment, although it is stated in the section on Astronomy and there verified like a conclusion, is properly a principle in this science. Hence this science receives this noble principle that has been proved in astronomy; and in this way this science is served, as a mistress by her maidservant. The same is true in certain other sciences, as we note from what precedes, and as we shall state below. Porphyrius, moreover, said that sins cannot be taken away except by the Son of God, as stated above. Moreover, on that statement, "His divinity proved from the instruction of Plato's scholars," the expositor says that in the tomb of Plato a statement written in golden letters was found on his breast containing these words, "I believe in Christ who will be born of a virgin, will suffer for the human race and will rise again on the third day." But the philosopher Ethicus also in the book already mentioned says, "The just shall deserve to see their Lord Christ and their King and the marks and prints of the nails, and the Word of the Father and the Beginning, creating all things with him." The human mind can be influenced to accept the truth of the virgin birth, because certain animals remaining in a state of virginity conceive and bear young, as, for example, vultures and apes, as Ambrose states in the Hexaemeron. Moreover, mares in many regions conceive by virtue of the winds alone, when they desire the male, as Pliny* states in the fifth book of the Natural History, and Solinus tells us in his book on the Wonders of the World. Aristotle maintains in the second book on Vegetation that the fruits of the female palms mature from an odor coming from the male trees.

Since a knowledge of Antichrist pertains to the Christian faith, because this faith holds that Antichrist will come, whom Christ will destroy, therefore the statement has been added to the articles of the faith that we believe Antichrist will come. Therefore one principle of this science is concerned with the coming of Antichrist in confirmation of the things pertaining to the Christian faith. Ethicus the philosopher therefore states that about the time of Antichrist there will be a race of the stock of Gog and Magog opposite the fertile districts of the North around the Black Sea, the worst of all the nations, which in company with the worst seed of those shut up behind the Caspian gates of Alexander will cause a great devastation of this world, and will meet Antichrist and will call him God of gods. Albumazar also in the book of Conjunctions verifies similarly this principle, stating and showing that a leader shall

* Pliny, IV, 22; VIII, 42.

come with a foul and magical law after the law of Mahomet, who shall destroy the other laws for a time. But he shall not last long, owing to the magnitude of the evil. This has been explained above. Without doubt careful consideration must be given to this principle of this science. For the race of the Tartars has come forth from those places, as we know, since they dwelt behind those gates to the northeast, shut up in the mountains of the Caucasus and of the Caspian, and are leading with them peoples who already are masters from the mountains mentioned as far as the confines of Poland, Bohemia, and Hungary, countries far this side of the fertile lands of the North. For it is true that other races have emigrated from those places and have invaded the world to the south as far as the Holy Land, just as the Tartars are now doing, as Jerome writes in his letters and the histories tell us. Moreover, the race of the Goths and of the Vandals which later invaded the South are from the regions of the North. Therefore the invasion of the Tartars is not sufficient to fix the time of the coming of Antichrist, but other facts are required, as what follows will explain.

Another principle concerns the judgment to come. For an article of the Christian faith refers to this, in regard to which the philosopher Ethicus has said, "The devil, who had been established in the beginning and had been the first to fall, will be punished before all wicked men and will be shut up in hell. Since he shone forth first in the order of creation, and was illustrious as a miracle of God's ways, he will be the first to suffer all the penalties in the last terrible judgment to come, such as have been appointed for him from the beginning in the hollow of the lake, and have been reserved for the judgment before the tribunal of the king, in order that the wicked may see the very truculent author of their death."

Another principle relates to the creation, which, since it requires proof in metaphysics, is known; therefore it is to be touched upon here only as regards morals. Aristotle in his book on the Regimen of Kingdoms expressly assumes it and names Adam and Enoch, and therefore knew of the first man and the beginning of the world. Albumazar, moreover, in his book on Conjunctions, doing an excellent service for moral philosophy, assumes a first man, namely, Adam, and shows how long was the period from him to the deluge, and how long from the deluge to Christ, and how long from Christ to Mahomet, and how long from him to the foul law. Avicenna also explicitly assumes the creation. Moreover, Ethicus the philosopher says, "First of all God made all creatures, and with a supreme effort formed a massive whole, and those things which he made out of nothing he amplified in many ways." Trismegistus also in his book on Deity says to Asclepius, "In the Creator are all things before he had created them," so that he is in agreement with John the Evangelist who says, "In himself was the life of that which was made." Yet this Trismegistus lived near the time of Moses and Joshua, as Augustine states in his book on the City of God.

In regard to the beings first created we find another principle of this science; of this kind are the angels, good and bad. In the first place, therefore, because of the motions of the heavenly bodies, which they have discovered to be about sixty, philosophers have assumed that there are that many good angels, because those motions are voluntary and therefore are caused by angels. This, moreover, is evident from the Metaphysics of Aristotle and of Avicenna. In the second place, they have passed on to a further consideration and have found an almost infinite number corresponding to us, just as individuals in this world below are multiplied under one species, and are distinguished from one another in number, as sensible individuals, as we find stated in the book on Causes; with a difference, however, because the angels are separated from one another in a way that they do not suffer corruption, but remain in a stable being. These individuals, however, that are known to us are separated from one another in a way that they finally suffer corruption. Moreover, if we wish to wonder further at the words of Ethicus, we can state, as he himself does in his book, that there are twenty orders of angels; who he also assumed stand in the celestial glory.

But far more is Apuleius of Mandara to be wondered at in his book on the God of Socrates, in which in his exposition of many wonderful things regarding the nature of angels he transcribes the opinion of Plato in the Symposium, namely, that to each man there is an angel deputed to guard him against all evils, and to urge and arouse him to what is good. And after the soul has been separated, the angel becomes a witness before

God as the judge of all the good and evil things which that soul did in the body. He asserts, moreover, that angels carry the petitions of men to the heavenly beings and bring back gifts from them to men; and that one angel is charged with one duty, another with another; and he makes many statements of this kind in the following words: "There are certain divine powers in an intermediate state between men and gods, through whom our desires and merits reach the gods. They are bearers hence of prayers and thence of gifts, who quickly carry our petitions hence to the region beyond, and thence bring back the answers; or they are certain agents acting for both parties, and each of them, according to his special function, provides that salvation be brought us by these messengers, either by confirming dreams, or by directing those set over us, or by clearing up what is too obscure for us, or by inspiring prophets, or by other means which we recognize as from God. We must judge that all these things are done by the divine will, power, and authority, but by the ministry and activity of the angels. From this host of angels Plato then states one is assigned to each man during his life as a witness and a guardian, who, manifest to the mind, is always present as a judge not only of his acts but also of his thoughts. And when our life is over that same angel who was assigned to us must hasten thither and carry as it were his guardianship to judgment; and he must assist there in pleading the case and sentence must be given in accordance with his testimony." "I then admonish you all who hear this divine opinion of Plato according to my version, so form your minds with regard to all your acts and thoughts, that you may know that a man has nothing secret in his soul nor beyond it because of those guardians. Nay, your angel carefully perceives all things, sees all things and understands them, this guardian, individual overseer, household observer, personal defender, intimate advocate, constant watcher, personal judge, inseparable witness, condemner of what is evil, approver of what is good, provider in your uncertainties, forewarner in your doubts, defender in your perils, aid in your needs, who is able in dreams, and in signs, and in person when necessity requires, to curtail your ills, prosper your good, control prosperity, ease adversity." For this is a wonderful statement and wholly favorable to the Christian, and contains nothing unworthy either in the letter or the spirit. May we clearly see that it contains well-known articles belonging to the faith; nor should the philosopher be here interpreted in a worse sense, since his statement evidently contains nothing except that which is consistent with the truth in a wonderful way. I make this statement because others sometimes try to obscure Catholic sentiments found in books of the philosophers. But we should receive gladly such statement in testimony of our faith; and since it is certain that these men learned them through a revelation made to them and to the holy patriarchs and prophets, as we have already shown.

Porphyry, moreover, as Augustine states in the tenth book of the City of God, said that there are angels who descending from above proclaim divine messages to men, and that there are others who on earth declare those things that belong to the Father and state his majesty and profundity. Concerning the devil and his angels philosophers have said many things. For the philosopher Ethicus has expressed a trustworthy opinion about him, as stated above, in regard to his creation, his sin, and his primordial and final damnation after the judgment. He has also ventured to say that the tenth order fell after sin into the punishment of hell. Apuleius and Plato and others distinguish two kinds of daemons, since *daemon* in Greek is the same as *sciens* [versed] in Latin. Therefore also there are *calo-daemones* and *caco-daemones,* that is, good and bad daemons. For *calon* means good and *cacon* means bad. Concerning the good we are to understand what has been said with reference to the guardianship of men. But the evil ones are devils by name and yield to the irrational passions of the mind, according to Apuleius; whence they are subject to displeasure and torment of mind and pleasure from evil and rage and the other malign impulses. These are they of whom the poets speak, as Apuleius states, and whom they represent to be gods and the lovers of some people and hateful to others; and certain of them are incubuses and lead men to sins and vices, and afterwards to the punishment of hell, of which Hermes Trismegistus and the philosopher Ethicus speak.

The immortality of the soul is touched upon in metaphysics. But we must mention it here in reference to morals and particularly in reference to the resurrection of the body, concerning which the metaphysician could not express an opinion, either general or particular. Moreover, not only Aristotle and Avicenna have given general conditions for the immortality of souls of which mention was made previously, but philosophers in treating of morals have spoken of it. For in the first place Cicero in the Tusculan Disputations states as a philosophical proposition the immortality of the soul, and through that whole book investigates it, and considers various arguments for it which are manifest from that book, nor can they be given here because of their great length. And similarly in his book on Old Age this same question of immortality is settled by Marcus Tullius. Moreover, in his book on the Divine Nature, Hermes Mercurius beautifully says, "The God and father of all and lord, and he who alone is all in all, gladly reveals himself; not, however, as regards his locality nor quality nor quantity; but the intelligence alone of the mind illuminates the man who, with the darkness of error dispersed and the brightness of truth perceived, identifies himself with a full sense of the divine intelligence. Freed by his love of this from the share of nature in which he is mortal he conceives a confident belief in a future immortality." Moreover, Avicenna says in the Morals that Mahomet spoke of the glory of the body; but we know, as he says, that the glory of our souls is greater, since we are not asses, reckoning only the delights of the body; and therefore he finds fault with his own lawgiver and wishes another to investigate who promises not only the glory of our bodies, but rather that of our souls. In this Seneca agrees throughout, and Socrates and Plato, as is evident in the Phaedo. Avicenna, moreover, says in the Morals that the resurrection of our bodies must be assumed, as the entire man will be glorified in soul and body if he obeys the commands of God. Not only Avicenna and others of the school of Aristotle have perceived this truth, but Democritus also, who belonged to an earlier age and was a philosopher of great repute, as Pliny states in his Natural History, book VII. Moreover, Plato himself has also said that souls cannot exist forever without bodies, but return to them, just as Augustine teaches in the twenty-second book of the City of God. Varro also in his book on the Race of the Roman People states that many philosophers have said that the same soul will return at length to the same identical body. If, therefore, Plato maintained that souls return to bodies, and Varro that the soul returns to the same body, and if Porphyry, the greatest of philosophers according to Augustine, maintains that a purged soul will never go to evil nor to this world but to God the Father; the resurrection must then follow from the statements of the philosophers.

This conclusion, moreover, is necessary, since from the source of philosophy they have elicited the fact that virtue belongs to the entire union formed of soul and body, that is, man, not to soul only nor to the soul in man, but to man through the soul, just like understanding and building, as Aristotle says in the first book on the Soul.* And they have therefore assumed that felicity belongs to the entire union, whence they have not assumed that man is a soul in a body but in reality is composed of soul and body, so that the essence of man is formed of soul and body, and that his essence is not soul alone in a body. For that which is nobler as regards man they have assumed to be the precise subject of virtue and felicity, but this is a union as far as possible of this kind, since he himself who is composed of soul and body is a noble substance. And although spiritual felicity and virtue are in man by reason of his soul, yet they do not belong to the soul, so that they have their place there, but they are in man because of the united man himself; and therefore they have maintained that felicity which is man's goal, completes the whole man, both as respects the body, as due to it, and as respects the soul. And therefore they have maintained that the body is at length united with the soul, so that both may be perfected in accordance with its special character. For they knew by reason that form is appropriated by its own matter, and contrariwise. And therefore incorruptible form appropriates incorruptible matter, and contrariwise. For they knew that the desire of form is not satisfied except in its own matter. And they have assumed that the desire of the soul is wholly satisfied by felicity. Wherefore they have assumed that it would be in a body.

The reasonings and persuasive arguments of philosophers on this point are of this nature. For they knew that the power of God is infinite, and therefore it can cause the same body to return. And the agent of finite power is able to make the same

* *De Anima,* I, 4, 12.

thing in species, as, for example, nature from a corrupt seed makes other seeds of the same species. Wherefore much more so will the agent of infinite power be able to make the same thing in number [*i.e.*, the same individual]. For infinite power infinitely exceeds finite power. But the production of the same thing according to number infinitely exceeds the production of the same thing according to species. But Aristotle says in the ninth book of the Metaphysics that the living is formed from the dead if a resolving takes place to the first matter. Since, therefore, God is able to cause this resolving, as is evident, a resurrection can take place. Likewise we have a strong supporting argument in the phoenix, which after being reduced to its own ashes comes to life again and becomes a phoenix. But a stronger one is drawn from the worm, which after birth quickly dies, and again coming to life remains immortal, as the philosophers and sacred writers state, and as we learn from the books of the Hexaemeron which determine the works of the six days.

In regard to the felicity of the other life and the misery prepared for the evil, the moral philosopher must assume principles in addition to that which has been touched upon in metaphysics. For these questions must be considered in metaphysics in general, but here in particular. For excellently do the philosophers determine the causes by which we are hindered from a knowledge of eternal life. These hindrances are four in number: sin, care of the body, trammels of the sensible world, and lack of revelation. For revelation is not a matter within our power. But on the question whether there be an eternal promise, they have been able to secure some knowledge, as I have said, and also in general on the question of the reality and nature of this future life. They have not, however, been able to give us particular and definite instruction on this subject, chiefly because of the four reasons first stated. Hence Avicenna in the Fundamentals of Moral Philosophy after many statements concludes, "Our condition as regards these matters is like that of a deaf man who has never heard in his privation of the power of imagining the delight of harmony, although he may be certain of the reality and nature of its sweetness." And not only is the intellect in this condition in knowing, but also the desire and the will in longing for, loving, tasting, or enjoying the sweetness of eternal life, to quote the words of Avicenna. For he compares us to a paralytic to whom delicious food is brought, but he does not perceive its sweetness until he is cured of his paralysis and the disease removed. Such is our condition, he says, with respect to the sweetness of life eternal, both because of our sins and our union with a mortal body; for sins affect the desire of the rational soul, and the weight of the body burdens it. Hence Avicenna gracefully says that we in our age and in this body have been plunged in many vices, and therefore do not perceive that delight, although there are in us some of its causes. And therefore we do not seek after it nor are we attracted to it, until we have removed from our necks the yoke of pleasure and anger, and their sister vices, and thus shall we taste something of that delight; and thus perhaps we shall gain too slight a perception of it, as through something interposed, because revelation is still necessary. Therefore he says that then shall we have a clearer perception, especially when the questions concerning God, and the felicity and immortality of our souls, and the resurrection of our bodies have been settled; and similar truths of high import have been revealed to us. Then, as Avicenna says, "The comparison of our perfect delight with our imperfect one will be like a comparison of the delight of our senses in smelling the odors of delicious foods with the delight we experience in actually eating them."

But the attention paid by the soul to the body, as he says, causes the soul to forget itself and that which it ought to love; just as a weak man is forgetful of that which it is necessary to restore in place of that which is taken from him. For, as he himself states, the body seizes upon the substance of the soul and makes it foolish, and causes it to forget its natural longing and search for the perfection that befits it, and its perception of the delight of its own perfection. Not that the soul has been impressed on the body and immersed in it; but because there is a bond between these two, which is the natural desire to direct the body and to excite its affections.

The third hindrance is the attention one bestows on this world of sense, although a man may not sin nor be occupied with bodily interests. For since we are given over to the world of sense, we neglect the insensible and spiritual world, as Avicenna shows. In these statements he clearly and cogently touches upon the causes that hinder us in our consideration and love of felicity.

On the other hand, his words point out to us what our aids are in knowing and loving and tasting the delight of future felicity. The first of these is the purification of the soul from sins. And the second is the withdrawal of the soul from its natural desire to direct the body. And the third is the elevation of the mind from this world of sense, that it may cleave to the world of intelligence. And the fourth is the attestation through revelation and prophecy in regard to those things about which the human mind cannot form judgment, such as the questions of high import of which he is speaking. For in such matters, as he says, "We believe the testimony of the prophet and the legislator who receives the law from God." He who possesses these four would not place felicity in this world but misery and death, as will be explained clearly below; and with Aristotle, Theophrastus, and other true philosophers would have leisure for the contemplation of future felicity, as far as a man can, in order that a good and merciful God might reveal the truth in fuller measure, just as we have proved that he revealed it to others besides those who were born and reared under the old dispensation or the new one, as must be stated in metaphysics. Since, moreover, they perceived that for a knowledge of felicity it is necessary to separate themselves from sins and from excessive love of the body and from the world, in order that they might be able to receive from God the fourth aid, namely, enlightenment within, so that they might understand the articles of the truths of the faith, casting all else aside they gave themselves up to a wise contemplation of future felicity.

For wisdom, as Aristotle states in the fourth book of the Ethics, is almost the same as felicity, since wisdom is not mere knowledge but is an intellectual force, as he himself determines, producing rather an affection for future felicity than an understanding and a beginning of it, since both are the knowledge and love of God. But wisdom possesses both of these as far as it is possible in this life, and felicity comprehends them imperfectly. For this same fact is evident in regard to felicity and is made manifest in regard to wisdom by the first book of the Metaphysics and by the sixth and the tenth books of the Ethics. Therefore Aristotle himself, the most excellent of all philosophers, renounced all things, that he might have leisure for contemplation suited to a sage, because this life resembles most the divine life; and therefore it was seen by the sage to be the most worthy, as Cicero writes in reference to Aristotle in the fifth* book of the Academics; and similarly in reference to Theophrastus, his successor in philosophy.

After they were prepared for divine enlightenment and had received it they maintained that this felicity is a beatitude of the whole man both in body and soul, which eye has not seen nor ear heard, as Avicenna says. This felicity is a state perfect in the aggregation of all good things, as the philosophy of Boetius teaches in the third book of the Consolation. Moreover, in the same book he proves that this can only be through participation in the highest good, which is God, because complete participation in good does not exist except in a participation in God who is the perfect good. Therefore men cannot be blessed and happy except by enjoying the goodness of God. Therefore his philosophy contains a noble corollary, namely, that the blessed are gods; but there is only one God of nature with many, namely, all the blessed, sharing his deity. Aristotle, moreover, in the first book of the Moral Philosophy teaches that human desire cannot have its limit in any good except in the highest good by which it is terminated; because the desire of the rational soul transcends all finite good and passes on to the infinite good. Therefore of necessity it participates in the highest and infinite good, which is God, if its desire must be satisfied. But it must evidently be satisfied through felicity, wherefore the soul must enjoy God for ever. And then as regards the speculative intellect, the soul will become, according to Avicenna, an intelligible world, and there will be described in it the form of the whole universe and the order of all things from the beginning, namely, God, both through all spiritual substances and the heavens, etc., until there is perfected in it the arrangement of the universe, so that it may thus pass into the world of intellect, perceiving that which is perfect beauty and true grace. And as regards the active intellect he says that it will be perfected in pure goodness, and its delight will not be of the kind of delight pertaining to the senses, which is

* Bridges suggests that the reference may be to *De Finibus,* lib. V, cap. 25.

caused alone by bringing together the surfaces of sensible bodies affecting our senses; nay, it enters the soul, and is infused into its substance, and is a delight befitting the natural disposition that is in pure and spiritual living substances. It is, moreover, more excellent and more noble than any other delight; and this is the delight of felicity, as he affirms.

Not only have they spoken of the felicity, but also of the misery of the other life which is reserved for the wicked. Hence they have maintained that God has prepared for those who obey him a promised felicity which eye has not seen nor the heart of man apprehended; and for the disobedient a dreadful punishment, as Avicenna states. Moreover, Cicero, Trismegistus, Socrates, and many others have spoken expressly of these promises. Hence Cicero says in the first book of the Tusculan Disputations* that two paths and two courses are open to men. Those who have kept themselves upright and pure and have had the least possible contact with the body and in their human bodies have imitated the divine life find the return easy to him from whom they came forth, namely, to God. But the path of those who have defiled themselves with human vices is at length far removed from the counsel of God. Hermes Mercurius in his book on the Divine Nature speaks as follows: "After the departure of the soul from the body has taken place, the judgment and test of its merit pass to the supreme power, which on perceiving the soul to be good and just permits it to remain in places suitable for it. But if this supreme power perceives the soul to be smeared with the stains of faults and marred with vices, it thrusts the soul down to the depths and delivers it up to eternal punishments, so that in this case eternity is a disadvantage to the soul that has been brought under the yoke of eternal punishment by an unalterable sentence." Therefore we learn that we must stand in awe and fear and be on our guard lest we be ensnared by these things. For unbelievers after their transgressions will be compelled to believe not by words but by examples, not by threats but by the actual suffering of punishments. Ethicus the philosopher and Alchimus in their books teach that the wicked are to suffer in hell with the devil, in

* *Tusculan Disputations,* lib. I, cap. 30. Here as in many other passages Bacon has substituted the singular of the word God for the plural.

order that the impious may see the very savage and furious author of death, whom they followed into many useless and harmful desires. The just will deserve to see their Lord God, as I explained above.

The principles having been stated with respect to what follows, although they are conclusions with respect to the statements given previously, and with respect to others of this kind, we must now pass to the laws of divine worship before we consider the other rights, public or private, in human society. Moreover, it is evident that because of our reverence for God a worship without end is his due. This is due him also because of the blessing of our creation, which is the effect of his infinite power, and because of future felicity. Because of the first, Avicenna says in his Moral Principles that obedience to God's precepts is his right; because of the second, he says that obedience must be given to the commands of him to whom creation belongs. Because of the third, he says that God has prepared a promised felicity for those obedient to him and for the disobedient a terrible future punishment. Worship is also due him because of the cleansing of the human race from sins by the Son of God, a cleansing of which Porphyry has spoken, since this is more than the mere act of creating. Also because of the acceptance of our humanity in the unity of the Divine Person, of which Albumazar, Plato, and Ethicus have spoken, for this must be infinite joy to us; and because of the prints of the nails and the passion and the redemption, which Plato and Ethicus have affirmed. And not only this but all that has been stated previously influences men to worship God and leads them to conclude that this worship should be rendered.

For this reason Marcus Tullius in the first book of the Tusculan Disputations says, "What else in fact is philosophy, the mother of all arts, if not, as Plato says, the gift, as I say, the invention of God? She taught us in the first place his worship, and then the rights of mankind on which human society is based." The same writer also says in the second book on the Divine Nature, "The best and likewise the most sacred, the most pure and pious worship of God is for us to reverence him always with a pure, blameless, unpolluted mind and voice. For not only the philosophers but our ancestors as well have separated superstition from religion. For those who were wont to pray whole days and to sacrifice, that* they themselves and their children might survive, were called superstitious, a word which afterwards became more general. But those who paid diligent attention to all that pertains to the worship of God and, as we may say, read over again these matters were called *religiosi,* religious, from *relegendo,* reading over again; as *eligentes,* those making a choice, from *eligendo,* making a choice; *legentes,* readers, from *legendo,* reading; and *intelligentes,* intelligent, from *intelligendo,* understanding. The one is a term of reproach, the other of praise." For Augustine in the fourth book on the City of God accepts and explains this opinion of Cicero in regard to superstition and true religion, and in the eighth book of the same work, his opinion regarding religion and the religious; maintaining that they were called religious because they choose God and revere him, again and again making the choice through a true and continuous worship.

Moreover, in this worship, according to Avicenna and others, temples, prayers, oblations and sacrifices, and fasts must be appointed, and also very great peregrinations to the seat of the legislator, in order that he may be kept in memory and held in veneration. Aristotle also in his considerations of this worship says that never should we be more modest than when we are engaged in divine worship; if we enter temples let us be composed, if we are about to approach the sacrifice let us lower the countenance, if in our prayer we are to give an indication of modesty. Therefore Avicenna says, "The teacher ought to instruct one who prays in the proper means by which he may be prepared for prayer, just as a man is wont to prepare himself to meet a human king in cleanliness and with a becoming appearance. Such a teacher should cause his pupil to form the habit of cleanliness and of presenting a becoming appearance, and should train him in the manner of a man preparing himself to meet a king with humility and downcast look, with limbs restrained, and without needless turning of the body and disturbance." Philosophers such as Avicenna and others decide that great festivals should be appointed, for the reason that they cause nations to assemble and give them boldness in the law and inspire an emulation of it. Such festivals also give an opportunity

* It is to be noted that Bacon has substituted *ipsi et sui liberi* for *sibi sui liberi.*

for the prayers of the multitude to be heard; and because of these festivals they discover blessings from God. And similarly for the same reasons general fraternities should be formed.

How, moreover, in particular prayer can fitly be offered Hermes Mercurius teaches in his book on Divine Matters, beginning thus, "God himself lacks nothing excellent, but let us adore him by giving thanks. For thanks when rendered by mortals are God's highest incense." He also says, "Thanks be to thee, thou supreme and most excellent Being, by thy grace alone have we obtained the light of thy knowledge. O sacred name and the one name to be honored, by which God alone must be blessed because of a paternal religion, since thou deignest to give to all paternal care and religion and love and all things that tend to make life sweeter, and grantest us rational sense and intelligence. Sense that we may come to know thee; reason that we may search thee out more clearly; intelligence that we may rejoice in knowing thee, and also that after being saved by thy divine power we may rejoice, because thou hast revealed thyself wholly to us. Let us rejoice because thou hast seen fit to consecrate us to eternity while we are still in our bodies. For the knowledge of thy majesty is the only source of human rejoicing. O true life of life! O universal productiveness of nature! we know thee, eternal in thy duration. Therefore in this whole prayer adoring thee we ask this only that thou wouldst keep us ever in the love of the knowledge of thee and never suffer us to be parted from this kind of life. With this prayer we turn to a pure supper free from animal food."

On the sacrifices, oblations, and ceremonies of these people we need not dwell, because they were superstitious and useless for the most part, except in so far as they took some things from the priests of the Jewish law. Hence also the philosophers themselves practiced such customs because of civil statutes and because of popular prejudice, not because of their truth, as Seneca states in the book which he composed against superstitions. For although as a senator he was obliged to dissemble in regard to the public statutes, yet he says that ceremonies of this kind do not pertain to the advantage but the custom of the people at large; and that no person of discernment should give them consideration in his private life. Also the observances of the Jews he considered absurd and unbecoming a wise man, just as God also allowed them to have these ceremonies rather lest they should become idolaters than because of their truth, as the sacred writers bear witness. And thus the fundamentals of the first part of Moral Philosophy are brought to a close.

MORAL PHILOSOPHY: SECOND PART

Chapter I

Concerning the observance of the laws of marriage and of the state.

THE second part treats of the laws and statutes regulating human relations. In the first place, the preservation of the human species is considered in the line of propagation with a view to bind the people by laws in their increase. Therefore legislators give the laws of marriage and determine how they must be made and how obstacles may be removed; and especially do they decree that fornicators and sodomites be excluded from states, who are inimical to the fabric of the state, since they draw men away from that which is a better thing in states, namely, marriage, as Avicenna and others maintain.

In the next place, laws are given in accordance with which the mutual relationship existing between subjects and prelates and princes is arranged, and that between slaves and their masters as regards every class of servitude and ownership. Laws are also given in accordance with which the fathers of families must live in directing their offspring and family, and the master as regards his pupils. Next, teachers are appointed and artificers in the different sciences and arts; and from the youths to be instructed those best fitted to pursue studies and perform duties of this kind according to the advice of wise men are selected; and the rest are assigned to military duty to execute justice and check malefactors. And it is necessary, as Avicenna says, that in instituting law this should be the first intention, namely, to arrange the state in three parts, those who administer, those who serve, and those skilled in the law, and that in each of these groups a dignitary be appointed. After him other dignitaries inferior to him should be appointed, and after these still others until nearly all are included, to the end that no one should be useless in the state, but that each one may have an honorable status, and that from each profit may be derived by the state. Hence in Plato that state is said to be most justly organized in which each citizen comes to know his proper function. Therefore, as Avicenna says, the prince of a state ought to prohibit idleness and lack of occupation. Moreover, those who cannot be restrained should be expelled from the state except in the case of infirmity or old age; and then a place should be assigned in which people of this kind should remain and have a keeper appointed over them. It is necessary, moreover, that in the state there should be a means of collecting a public fund, which in part should come in accordance with the law governing contracts, partly from fines inflicted for malicious prosecution, partly from the estates and spoils of rebels, partly from other sources; and that this public fund should be secured in part for those who are not able to make a living owing to infirmity and old age, and in part for teachers of law and medicine, and in part for public uses.

And then the legislator teaches men to frame laws relating to patrimonies, inheritances, and wills; because Avicenna says that the substance necessary for life is partly branch, partly root. But the root is patrimony and anything bequeathed and given by will, of which three roots patrimony is the most secure. But the branch of substance comes from gains derived from the various kinds of business. Then laws should be published regarding contracts in all kinds of business, buying, selling, giving out and taking contracts, borrowing, lending, expending, saving, and the like, so that in contracts whatever can do harm may be excluded, as Avicenna says.

Then laws must be framed in accordance with which in all causes and cases justice may be shown and the cases closed, so that peace and justice may be fostered among citizens. Afterwards, as Avicenna says, pursuits must be prohibited by which inheritances and property are lost, and the peace and concord of citizens are disturbed; and those practicing these pursuits are people who wish to get the upper hand in the case of some gain, as, for example, the wrestler, the gambler, and the like. Similarly, pursuits should be prohibited which are in opposition to what is advantageous, as he shows in his instruction in regard to thieving, pillaging, and other acts of this kind.

And further ordinances should be passed, as Avicenna states, that men should aid and defend one another, and be united against the foes of law in the effort to subdue them. If, moreover, another state or government has good regulations and laws, therefore no attack is made on it, except that the time will come when there must be no other law than that one, the establishment of which, since it is the best, must then be extended throughout the whole world. In this statement the Christian law is hinted at, as will be explained below. If, moreover, there are some among these who are at variance with the law, let them first be corrected, in order that they may come to their senses; but if they are unwilling to do so, let them be put to death.

Chapter II

THE last point required here is that the legislator should arrange for his successor. This is done, according to Avicenna, in this way. For he must do this with the consent of the nobles and of the people; and he should choose a man who can rule well and is prudent and of good morals. He should be courageous, kind, skilled in management and in the law, than whom, in fact, no one is more skilled, and this fact should be evident to all. But if after this men should so disagree that they desire to choose another, they have actually denied God, and for this reason the legislator should place in his law enactments that the whole state should with one accord attack and slay any one who wishes to obtrude himself by power or money. But if they shall have the power to do this and shall not have done it, they have actually contradicted God, nor is he blood-guilty who kills such men, provided that the facts are already known to the people. But if the man who should receive the office is not worthy and this fact has been proved, let another be appointed.

Thus terminates the statement of the application of the principles of the second part with what follows as consequences to the principles in general. In this part the civil law is included, which is now in force among the Latins, as is evident from the principles of this part. Moreover, it is certain that the Latins received their rights and laws from the Greeks; namely, from the books of Aristotle and Theophrastus his successor, with the exception of the laws of the twelve tables which in the first place came from the laws of Solon the Athenian.

MORAL PHILOSOPHY: THIRD PART

Chapter I

Concerning the guidance of a man in reference to himself.

THE third part of moral and civil science is concerned with personal conduct, to the end that every one should keep his life pure and free from the foulness of vices because of future felicity and the horror of eternal punishment. It is clearly evident that this should be the third part, since it is plain that that part which contains the worship of God is the first, as I have stated. But public good takes precedence of private good, as Aristotle says in the first book of the Metaphysics. But the part preceding contains the public good; this part urges upon men private good. For love is the greatest virtue, and is ordained for the common good, and peace and justice are its companions, virtues which transcend the morals of individuals. For man is a social animal and it is in accordance with his own nature, as Avicenna says in the fifth book on the Soul, that he should not live alone like a beast which in its life suffices itself alone. Therefore the laws regulating men with regard to the last topic are more important.

According to Aristotle and Averroës, in the tenth book of the Metaphysics, a hermit who is not a part of the state, but lives for himself alone, is neither good nor bad. And Cicero in his book on Duties, quoting the words of Plato, highly commends his statement that we were not born for ourselves alone. Our country claims part of our origin, and friends a part, and as it is a tenet of the Stoics that all things are created for the use of men, they taught that men were generated for the sake of men, so that they might be able to benefit one another. Since Cicero himself says in the fifth book of the Academics, "Nothing is so noble as the imparting of benefits." For it is innate in man to have some organized society, a relationship expressed by the Greeks in the word *politicon*. Hence in his book on the Happy Life Seneca says, "This statement is required of a man, that he benefit all if it can be done, or many; if not, a few; if not, his nearest relatives; if not, himself." Wherefore the second principal part of moral philosophy must be concerned with public laws, as I have stated; and the third part will deal with the life and integrity each one should strive after. This, moreover, is the proper arrangement according to Nature's dignity and for simplicity of statement, although Aristotle does not adopt this method in his books; because he proceeds by the path of inquiry, and therefore from what is better known to us, not to Nature. But since we have been informed through him and others what the power of this science requires, we can arrange its divisions in accordance with the order that Nature's dignity demands.

Moreover, under this heading philosophers have said many admirable things in regard to the virtues and vices, so that every Christian may be confounded when we perceive that unbelievers had such lofty conceptions of the virtues, and we seem basely to fall short of the glory of the virtues. But we should be much encouraged to aspire to the acme of virtue, and inspired by noble examples we should produce still nobler fruits of the virtues, since we have a greater aid in life than the philosophers themselves, and we are assured that we receive incomparably greater assistance through the grace of God. I shall first quote certain passages relating to the virtues and vices in general, and shall then pass to special branches.

CHAPTER II

ARISTOTLE shows in the first book of the Ethics that virtue is twofold; one, according to him, is in the sensitive part of the soul, in so far as it is obedient to reason, or in reason dominating the parts of the sensitive soul, and so regulating them that they obey its rule. Virtue of this kind is called moral and habitual, by which men form good moral habits. He teaches, moreover, in the second book of the Ethics that there are twelve moral virtues, and he treats of these in the fourth book and calls them means, because each of these is a mean between two vices opposite to each other and to the virtue. For one extreme falls short of the virtue and the other passes beyond it; as is the case with avarice, which falls short of liberality, and with prodigality, which exceeds it. For the liberal man gives only what he should give, the miser gives nothing or very little; the prodigal scatters abroad his all. Such is the case also with the vices that form the extremes for the other means.

These twelve according to Aristotle are as follows. First is fortitude in attacking what is terrible and in bearing what is adverse. Second is chastity both in the things pertaining to taste and in those pertaining to touch. Third is liberality as relating to moderate expenditures. For the fourth virtue is munificence, which is concerned with important matters, like temples, divine worship, hospitals, and other public blessings. And just as munificence and liberality differ in respect to money, so do magnanimity and a certain other virtue in respect to the merit and honor of virtue. Hence the magnanimous man is one who is not cast down by adversity nor puffed up by prosperity, but glories only in virtue. Hence magnanimity is the ornament of all virtues, as Aristotle says, by which a man raises his mind above all things human, and this virtue is of merited worth in all, nor does the magnanimous man meditate on injury or enmities. For he does not consider the avenging of injuries, nor does he preserve the memory of evils, nor does he deign to consider anything worthy of his anger, nor does he trouble himself about the folly and negligence of men, but overlooks all things by the greatness of his mind; nor does he rejoice in regard to honor done him, or only moderately, since for perfect virtue, as Aristotle says, a reverence equal to it is not found, nor as much as it deserves; nor is he disturbed by anything, since there is nothing that is extraordinary in his sight, prepared as he is for retributions and benefits, like God as much as a man can be.

The sixth mean, which bears the same relation to liberality as liberality does to munificence, has no name in Latin, but is concerned with matters of less merit than is magnanimity, and becomes any man according to his station. The seventh virtue is gentleness, which restrains angry passions. The eighth is friendliness in social intercourse, to the end that a man be pleasing to his fellow man without adulation, and that he be not envious or disagreeable but forbearing. The ninth is sincerity, which removes from a man pretense and deceit, so that he is such in his heart as he appears to the outside world. The tenth consists in cheering others in the proper way, to the end that scurrility be excluded, and that a man be not boorish and rude, disquieted and glum toward the friendly advances of others. For quiet relaxation and friendly cheer with a vein of humor are necessary in human intercourse, as the philosopher says. The eleventh is a sense of shame at sins, which is especially needful for young people, so that they may not easily incline to vices. Therefore Aristotle says, "We praise youths with a sense of shame." For Seneca in his book of the first Letters relates the story of a youth who sinned, and when a sage asked whether the youth had a sense of shame, and received the reply that he had, he said, "Well, then the situation is saved." The twelfth is justice.

And besides these virtues there are other nobler ones which do not depend on sense but belong absolutely to pure reason, and therefore are called intellectual virtues, which are intellect, knowledge, art, prudence, and wisdom. But these may be considered purely in the region of speculation, when they are directed to truth alone. And in this view they are not virtues but intellectual and cogitative attitudes of mind, or they may be directed to matters of practice, and accordingly they have as their object the good. And thus they are virtues, namely, when they aim at the salvation of the soul; and this is the case when they are concerned with divine worship and public good among citizens, and integrity of life and morals, and the consideration of the life eternal and of like matters. Intellect is an acquired mental condition concerned with the principles underlying practice, and knowledge is an acquired mental condition concerned with conclusions. Art is the knowledge of good works in their outcome, and prudence is the acquired mental condition directing such works. But wisdom is the perfect knowledge of spiritual blessings together with the sweetness of love, a sweetness in which there is peace of the human mind as far as it is possible for the mind to enjoy such in this life; and therefore this is the beginning of future felicity, and is almost the same as felicity, as Aristotle says in the sixth book of the Ethics.

In the same book he states that in every kind of virtue there is one that is natural and another that is acquired, just as we find a man chaste by nature and courageous, and the same is true in regard to the other virtues. He also makes a similar distinction at the end of the tenth book; and he there says that the natural virtues are not derived from ourselves, but are given us by divine grace. This is a wonderful statement. Therefore still more are the acquired virtues derived from God, which are far more noble, although the habit of virtuous action preserves and strengthens and moulds them, for which reason they are called the virtues of habit and are said to depend on our actions.

The ancient philosophers uphold virtue as the whole good and the only good of man in this life, as Seneca teaches throughout in his book on the Happy Life, and Cicero everywhere in the fifth book on the Tusculan Disputations. He also solemnly assures us in his book on the Paradoxes that he will never consider anything good except virtue. He confirms the view by the judgment of one of the Seven Wise Men. When all his fellow citizens were fleeing from their homes owing to a hostile attack, and they asked him why he was not carrying his possessions with him as the rest were doing, he replied, "I am carrying my all with me"; meaning that nothing was his except virtue. This man was Bias of Priene, as Valerius Maximus informs us in the fourth book. Seneca also in his book to Serenus on the topic, "How neither injury nor contumely happens to a wise man," states that the philosopher Stilbon, on being asked by the tyrant who had invaded his state and had seized all his temporal possessions whether he had lost anything, replied, "Nothing, for my all is with me"; and he bore witness that he was not only unconquered but had suffered no loss. For he had with him his real blessings, on which no hand is laid. But those things which were scattered and carried off as plunder he did not consider his, but the mere accidents of fortune following its nod.

Moreover, virtue is the life of man: since Seneca says in the First Letters that men die before they begin to live, speaking of those who busy themselves in sins up to the time of their natural death. For this reason the very noble Apuleius in his book on the God of Socrates expresses astonishment that men do not care for their souls so that they may live. He says, "I am surprised at nothing so much as at the fact that, although men wish to live in the best way and although they know that they live in the mind alone, and that they cannot live in the best way without cultivating the mind, they do not, however, cultivate their mind. Thus, if a man wishes to see clearly, he must take care of his eyes; if you wish to run swiftly, you must take care of your feet; similarly each of your members must receive special care. When I see all men acting on this principle, I do not know how adequately to explain to myself and to express my wonder why they do not also cultivate their mind by means of reason. The art of living is equally necessary for all, not so the art of painting and of harp-playing; and the same is true of the other arts, not to go through them all, which you may be ignorant of without shame. But when you say, I am not ashamed of my ignorance of the correct way to live—but you will never venture to say this. And chiefly must we wonder that those things which they are least desirous of seeming to be ignorant of, they never the less neglect to learn, and they fail to learn this same art and at the same time hold an ignorance of it reprehensible."*

This statement is directed chiefly against those who pose as models and teachers of morals to others; of whom Aristotle says in the Elenchi, "They make very fine speeches, but their wishes are at variance with their words." Cicero in the second book of the Tusculan Disputations asks in regard to such people, "How few of these philosophers will you find whose life and manners are conformable to the dictates of reason! who regard their profession, not as a means of displaying their learning, but as a rule for their own life! who conform to their own precepts and obey their own decrees! You may see some carried away by such levity and vanity, that it would have been better for them to have remained ignorant; some covetous of money, others eager for glory, many slaves of their lusts, so that their discourse is strangely at variance with their mode of life, which indeed seems to me a most disgraceful thing. For just as if a professed teacher of grammar should speak ungrammatically; or if he who wishes to be considered a musician sings out of tune, such inconsistency being all the worse since each of them fails in the very subject in which he professes to be expert; so a philosopher who fails in the conduct of his life is the more disgraced, because he errs in the very branch which he claims to teach, and does not regulate his own life by the rules he professes to teach." When Zaleuchus, the lawgiver of the Locrians, with remarkable justice and zeal for virtue wished that the law which he had made inflicting the loss of both eyes as a penalty

* "This quotation like many others made by Bacon," as Bridges says, "is nearly but not quite accurate."

for adultery should be complied with in the case of his own son, the state refused because of the integrity of the father. But the righteous judge put out one of his own eyes and an eye of his son to avoid breaking the law which he had made. Valerius Maximus records these facts in the sixth book.

But all philosophers maintain that the grace of virtue by its own wondrous beauty should attract each man, since it is becoming, as Cicero says in his book on Duties, that by its own power it draws us and allures us by its own worth. Apuleius also, in the third book on the Teaching of Plato, says that virtue is beauty of mind. Cicero in the fourth book on the Tusculan Disputations says, "As in the body there is a certain proper formation of limbs, together with a certain sweetness of complexion, which is called beauty, so in the mind beauty consists in an impartiality, and constancy of opinions and judgments, joined to a certain firmness and stability, following virtue or preceding virtue itself." Apuleius says that virtue is not only beauty of mind but health and strength also. Cicero also holds the same opinions, as is evident from many passages in the second, third, fourth, and fifth books on the Tusculan Disputations.

Although it is impossible or difficult for the man who has become hardened in vicious habits to experience a change of heart, as Aristotle says at the end of the Ethics, yet Seneca says in the second book on Anger, "There is nothing so hard and so difficult, that the human mind cannot master it and render it familiar by constant practice. For none are so iron-hearted and of such an independent frame of mind that they may not be brought under control by discipline. Whatsoever the mind has imposed upon itself it has obtained. Some have gained such control of themselves as never to laugh. Others have forbidden their bodies the use of wine or fluid of any kind. Still others satisfied with a short sleep prolong the period of wakefulness tirelessly. Some have learned to run on very slender and steep ropes, and to carry great weights beyond the ordinary strength of man, and to dive into the sea to a great depth without any change of breath. There are a thousand other things in which perseverance overcomes every obstacle." And he shows that nothing is difficult if the mind has enjoined patience upon itself. For as Averroës says in the second book of the Physics, many have become accustomed to nourish themselves on poison, as Aristotle writes in the book of Secrets. For he himself discovered that a beautiful girl sent to Alexander with other gifts had been fed on poison. Therefore persistent labor conquers all things, as Vergil says.

There is, then, no difficulty in virtue for him who loves it. For virtue is eager for danger. Virtue considers the end it has in view, not what it will be called upon to suffer, as Seneca writes in his work De Copia Verborum, and makes the same statement in another book in which he inquires why many misfortunes happen to the good. And he makes there the additional statement, "Since that which it is destined to suffer is part of the glory. Soldiers glory in their wounds and rejoice to show their blood flowing from wounds that are not mortal. He who returns wounded attracts more attention than those who return unharmed from the battle-line, although they have been in the same engagement." He states, moreover, in the seventh book on Benefits that it is easy to know and to possess the virtues, as far as they themselves and nature are concerned; and he adds, "Whatever will make us better and happy, nature has placed in plain view or within easy reach, if a man has consecrated his mind to virtue. And whithersoever virtue summons he considers the path easy, if he is a social animal and views the world created for the common good as one household, and always lives as though in the public eye, more fearful of himself than of others. Such a man withdrawn from the storms of tribulations has set his foot on a firm and peaceful soil. He has made useful knowledge his own. For of no thing is the discovery difficult except of that whose only reward when discovered is in the discovery itself." And this lies in the domain of the purely speculative, not in matters pertaining directly to the happy life. Seneca, moreover, in the second book on Anger, refuting those who seek the solace of vice instead of the difficult path of virtue, speaks as follows, "There is nothing in the vices which you may say cannot be removed; we are ill with curable maladies, and nature aids us, created as we were to be whole, if we desire to be cured. Nor is the road to the virtues hard and rough, as it has seemed to some; we advance along an easy path since God himself is our aid. Much harder is it to do what you are doing. What is more restful than peace of mind, what more laborious than anger? what milder than forbearance, what more vexatious than cruelty? Chastity enjoys leisure, lust is very much preoccupied. Finally the guardianship of all the virtues is easy, the vices are cultivated at a great cost." Cicero, moreover, in the third book on the Tusculan Disputations says, "The knowledge of the virtues is innate and if permitted to increase, Nature herself would conduct us to the happy life." Seneca expresses the same sentiment in the book of the First Letters. That virtue, however, as he says in the first book on Anger, is surest which has greatly distrusted itself for a long time and has controlled itself and has advanced slowly and intentionally.

CHAPTER III

THEY discourse about sin in common just as they do about virtue. For Cicero in the fifth book on the Disputations says, "No one is permitted to sin"; because sins hinder complete perfection of mind and acquisition of eternal felicity, since Algazel says in the Logic this felicity comes from perfection of soul. Now perfection of the soul consists of two things, cleanness and adornment. Cleanness is the purification of the soul from filthy habits and sanctification from foul impressions; adornment consists in depicting on the soul the certitude of divine truth and the being of the whole universe in its sequence, by means of a revelation in which there is no error nor concealment; like a mirror, for example, which does not possess perfection unless a beautiful form is apparent in it; but this is not possible unless it has been cleansed completely from dirt and rust and beautiful forms afterwards placed before it. The soul, then, is like a mirror; for there are depicted in it the forms of the whole universe when it is clean and has been freed from foul habits. These are the words of Algazel. Sins, therefore, render a man blind, because every evil man is ignorant, as Aristotle says in the second book of the Ethics. Socrates says that it is not possible for any one to do a base act except through ignorance; since when he reaches the state of a desire for sin, he loses his knowledge and his intellect is in abeyance. So great is the depravity and turpitude of sin that the Philosopher has said, "If

I were to know that the Gods would pardon and that men would pardon, I would refuse to sin." For this reason Cicero says in the second book of the Academics that men select darkness and solitude for committing sins, because the very baseness of sin deters by its foulness. Seneca, moreover, in the fifth book on Natural Questions says that because of their shamefulness crimes shun the sight of themselves, for concealing which no night is sufficiently dark. Apuleius also, in the third book on the Beatitude of Plato, says, "Evil or sin is filthiness of soul, and not only this but weakness and disease." Cicero expresses the same thought in many ways in the Tusculan Disputations.

Sin not only blinds, defiles, and weakens the rational soul, but lowers it to the level of the brutes, as the philosophers show in many passages. Hence Seneca in his book on the Happy Life says, "I place men in the same class whom a dull nature and ignorance of themselves have reduced to the level of brutes. There is no difference between them, since brutes do not have any reason, and such men have a depraved reason and one that is clever to their own hurt and in the wrong direction." Moreover, philosophy proves in the fourth book of the Consolations that evil men lose their identity, because the identity of a thing consists in retaining its order and preserving its nature. But sin is contrary to the order of nature, therefore evil men cease to retain their identity. It is necessary, therefore, to thrust down below the rank of mankind those whom their wickedness excludes from human status. The result is, then, that you cannot regard as a human being the man whom you see transformed by vices. And the philosopher infers that you would compare to a wolf the robber who is inflamed with greed and employs violence in possessing himself of the property of other people. You will compare to a dog him who is insolent and wags a quarrelsome tongue. He who lurks by the way and finds delight in pilfering is on an equality with foxes. If a man roars in uncontrolled anger, he may be regarded as having the intelligence of a lion. He who is timid and ready to flee, fearing what need not be feared, may be considered as similar to deer. The lazy and stupid dolt lives the life of an ass. The fickle and inconstant man who changes his pursuits differs in nothing from the birds. He who is plunged in defiling and unclean lusts is foully possessed by the pleasure of a hog. Therefore it comes to pass that he who ceases to be a man by the loss of his goodness is turned into a beast.

Since, moreover, we gladly excuse as far as possible our sins, whether they be small or great, many or few, Seneca in the second book on Anger asks, "Which vice, pray, has lacked an advocate?" but we do not thus excuse the vices of others. For in the same book he says, "We keep the vices of others in view, but our own in the background." And since with the holy thou shalt be holy and with the innocent man thou shalt be innocent, Seneca says in the book of the Second Letters, "Nothing else so clothes our minds with the garments of respectability and recalls to rectitude those who are wavering and inclining to evil as association with good men." And in the third book on Anger, "We form our habits from our associates, and as certain diseases of the body pass from one person to another by contact, so the mind gives its evils to those with whom it comes in contact. The drunkard inspires his messmates with a love of wine. Association with the impure weakens the character even of a strong man, if it be permitted. Avarice passes on its poison to those near it. The virtues conversely have the same action in softening everything with which they have to do. A mild region and more salubrious climate do not confer so great a benefit on bodily health as that derived by the morally weak from association with better people. You will understand how strong such influence is if you see that wild animals grow tame by living with us, and that no savage beast retains its ferocity if it has dwelt for a long time in the habitation of man. There is the added fact that not only does a man who lives with peaceful men become better by their example, but he does not find occasions for his vice and does not practice it."

Moreover, in the same book for the correction of all vices he says that all the senses must be brought to a condition of stability. For they are patient if the mind, which must be called daily to render an account, has ceased to corrupt them. Sextius followed the practice, when the day was over and he had retired to rest at night, of questioning his own mind, "What one of your faults have you corrected today? What vice have you withstood? In what way are you better? Can anything be better or finer than this habit of examining the whole day? How wonderful is that sleep which follows self-examination; how tranquil, deep, and free it is after the mind has been praised or admonished; and as spy and secret censor of itself finds out about its own behavior." And in the same book he says about himself, "I take advantage of this opportunity and daily before myself plead my case when the lamp has been removed. I go over with myself the whole day, and I consider what I have said and done, I hide nothing from myself, I pass over nothing. For why should I have any fear of my mistakes, when I can say, See that you do not do that again, I pardon you this time?"

Moreover, since that man is happy who dashes his little ones against the stones, Seneca says to Marcia, "All our vices will become fixed within us, unless they are crushed as they spring up. The violence of a disease should be checked at its outbreak by a gentle remedy, but we must take more vigorous measures against chronic disorders. For wounds are easily cured while they are fresh with blood." And since this matter should receive particular attention in the period of our youth, Aristotle says in the second book of the Ethics that it makes no little difference that we are trained in this way or in that from our youth. For he compares the innocence of youth to a green twig, which can be bent in every direction and its curvature easily removed. And Seneca says in the second book on Anger, "It is easy to mould minds while they are still tender; for with difficulty are we freed from vices that have grown along with us." Seneca likewise says to Helvia, "In this case mould both the manners and nature of the young girl. Precepts that are impressed at a tender age sink more deeply. You will bestow much upon her even if you give her nothing but your example." And therefore in the second book on Anger he says, "Everything that is tender clings to the things nearest and grows to resemble them. The manners of youth soon repeat those of their nurses and tutors. A boy brought up in the home of Plato, on returning and seeing his father vociferating and shouting, said, I never saw this in the home of Plato. I do not doubt that he would have imitated more readily his father than he did Plato, had he been brought up in his father's house."

Moral Philosophy

Chapter IV

If we consider what the philosophers have said and done in particular in regard to habits of this kind, we find much that is most pleasing. All their teachings are reduced to a contempt of riches in opposition to avarice, or to a despising of honors as opposed to pride, or to the shunning of pleasures as opposed to luxury and gluttony, or they have reference to the violent feelings and passions of the mind and are directed against anger, envy, and sloth; to the end that thus the seven mortal sins may be avoided. All these sins except anger relate to prosperity. For the mind grows feeble in the midst of riches and pleasures while it delights itself in these sins. For as regards avarice, pride, luxury, and gluttony it is evident that they are aroused by a strong desire for enjoyment. Envy also springs from the good and the prosperity of another. Sloth, however, owes its being to the fact that the mind is absorbed in pleasures and in the other interests incident to prosperity. For in such circumstances a man loses interest in the true good which consists in virtue, and finds every virtuous effort disagreeable. Inert and lazy becomes the man for whom death of the soul is at the portals, to bring quickly to a close his spiritual life. Anger alone has no regard for prosperity and battles in adversity. For this reason I shall touch first upon their statements pertaining to the contempt for prosperity; then I shall consider how adversity should not be dreaded.

Aristotle, the greatest of philosophers, accordingly is the most conspicuous example, who in contempt of the world with all its riches and honors and pleasures left his native land and ended his life in exile. Theophrastus, his noted successor in philosophy, did the same thing, as Cicero states in the fifth book on the Tusculan Disputations. But these were not the only instances, but there were many others among the most noted philosophers and fathers of philosophers, as Xenocrates, who was head of the old Academy, according to Censorinus in his book on the Natal Day, and Carneades, who founded the third Academy called the new one. For the disciples of Plato were called the Academics from the place in which Plato studied, and they were divided into many schools after the death of their master. But these were not the only ones; there were many other famous men, whom Cicero names to the number of sixteen: and still others in endless numbers, as he himself says, passed their years in exile and in constant travel abroad, who once having set out never returned home. Socrates, on being asked of what country he was a citizen replied, "I am a citizen of the world"; for he considered that he was an inhabitant and a citizen of the whole world, as Cicero states in the fifth book on the Tusculan Disputations. Seneca accordingly says to Marcellus in exile, "It is not a misfortune if you are without a country. You were trained to know that every place is a philosopher's native land. The lowly hut to be sure contains the virtues. It will be more beautiful than any temple since we shall see in it justice, continence, prudence, piety, a plan of duties properly coördinated, a knowledge of things human and divine. No place is narrow which contains this number of such great virtues." Ptolemy also, in the maxims given in the Almagest, says, "He is superior among men who does not care under whose power the world is." Nor is it strange if this philosopher speaks thus who spent most of his labor in learning about the heavenly bodies and the principal divisions of the world. For he proves in the first book of the Almagest that the whole earth has no sensible size in comparison with the heavens, a fact established more fully above. Seneca says to Helvia, "The mind is narrow which earthly things delight. For the higher men rear their towers, and the larger they build the roofs of their dining halls, the more do they shut out the heavens." Xenophon, the pupil of Socrates, said, according to Censorinus, that to be in need of nothing is a characteristic of God, but to be in need of as little as possible is the next thing to God.

Seneca in the fifth book of the Natural Questions, rightly judging the earth but a point in respect to the heavens, a fact proved by astronomy, considers of no moment and reduces to naught all things which are distributed in this world among men, as compared with heaven for which man was made. He accordingly says,* "He who now broadens and prepares his mind and makes it worthy of fellowship with God, possesses the complete and perfect good of the human lot; when treading under foot every evil he seeks what is above and enters into

* This quotation is from the prologue to the first book of *Naturales Quaestiones*.

the bosom of nature, he delights wandering among the very stars to view the pavements of the rich and the whole earth with its gold, not that only which it has produced, but that which it reserves in secret for the avarice of posterity." "This is a mere point which is divided by sword and fire among so many nations. Oh how ridiculous are the boundaries of mortals! The Dacian may not pass beyond the Ister. The Euphrates must restrain the Parthians, and the Danube divide the Sarmatian and the Roman territories, the Rhine is to set a bound for Germany. The Pyrenees are to rear their crest between Gaul and Spain. Between Egypt and Ethiopia a vast uninhabited stretch of sand must lie. If any one should give human understanding to ants, will they not divide one anthill into many provinces? When you see armies marching with flags flying and the cavalry now scouting ahead, now spread out from the flanks, that is merely the running to and fro of ants busying themselves in their narrow path. What difference is there between them and us except the extent of a small body? This is only a point on which you sail, on which you reign, on which you make war. Above are the vast spaces into the possession of which the mind is admitted; and as though freed from chains returns to its origin. And this is a proof of its divine origin that divine things delight and interest it, not as things alien to it but as its own possessions. It will then despise the narrow confines of its former abode. For the space from the remotest parts of Spain to India is traversed in a very few days if the ship have a favorable wind. But that celestial region requires a journey of thirty-six thousand years for the fastest star."

Moreover, Apuleius in his book on the God of Socrates fitly says that all good things, external and belonging to the body, are to be despised and not to be taken into account in praising a good man. His words are, "In considering men do not set any value on things that do not really belong to them, but consider the man himself within, view him as my poor Socrates. Now I call those things not really his which his parents procured, and which fortune bestowed, no one of which I join to the merits of my Socrates, neither high birth, nor lengthy lineage, nor riches that cause others to envy. Therefore you may number all gifts together; is he high born? You are praising his parents. Is he rich? I do not trust fortune. Nor do I enumerate such gifts further. Is he in good health? He will be worn out by sickness. Is he active? He will pass into old age. Is he handsome? Wait awhile and he will not be. But is he trained in the liberal arts and is he especially learned, and wise, as far as a man is permitted to be, and of good counsel? At length now you are praising the man himself. For this is neither a heritage from his father, nor is it dependent on chance, nor is it assistance dependent on the will of others, nor is it a transitory thing from the body, nor is it changed by age. All these things my Socrates had, and for this reason he despised the possession of all else." Therefore Cicero in his book on the Paradoxes glories in the fact that he never counted gold or silver or honor or anything belonging to this world among blessings. Sallust, moreover, in his book on Catilina says, "Fortune makes all things famous according to its caprice rather than in accordance with fact. First there was the desire for money, then for empire. These became as it were the sources of all evils. For avarice has destroyed faith, probity, and the other good qualities; in place of these she has taught us pride, cruelty, neglect of God, and to consider that all things are to be bought and sold. Avarice has a fondness for money, which no philosopher has desired. This vice steeped as it were in noxious poisons weakens the body and mind of man; always boundless, insatiable, it is lessened neither by plenty nor want." And Seneca in the third book on Anger says, "In regard to money there is the utmost vociferation; this wearies out the forums, mingles poisons, places swords in men's hands, this is gathered from our blood, because of it the throng presses upon the tribunals of the magistrates, kings rage and overthrow states: the law courts echo with the roar of the cases, judges summoned from far distant regions sit to determine which one of the two litigants has the juster avarice." And in his work De Copia Verborum he says, "The avaricious man's only correct action is performed when he dies." Sallust in the book mentioned above says, "The ambition for power has forced many men to become false; to have one thing locked in their heart and another ready on their tongue. And when men passed their lives in the land without greed, all were satisfied with what they had. But when insolence of power appeared, fortune changes along with character." Therefore Seneca addressing those who were proud because of the blessing of fortune says, "All things which force you, swollen with pride and superior to the common lot, to forget your own frailty, things which you guard with iron bars, which plundered from the blood of others you defend with your own, because of which after the rupture so often of the ties of marriage, friendship, and colleagueship the world was crushed between the two contenders, all these things I say are not yours. Already they are looking to another owner; either an enemy, or an unfriendly successor, or fire or some other misfortune will assail them. Do you ask how you are to make them yours? By bestowing them as gifts. Take thought then for your possessions and make your ownership of them sure and impregnable; you will render them not only more respectable but more secure. That in which you reckon yourself rich and powerful, as long as you are in possession, rests under a sordid name; it is house, or slave, or money. When you have donated it, it becomes a benefaction." And therefore he asks in the book of the First Letters, "Who is worthy of God except the man who has despised riches?"

Chapter V

Concerning the sensual pleasures philosophers make many excellent statements. Cicero in his book on Old Age speaks as follows: "Since God has given to man nothing more excellent than mind, nothing is more hostile to this divine gift than sensual pleasure. For such pleasure hinders deliberation, is hostile to reason, and blindfolds the eyes of the mind. For the great Archytas of Tarentum, a Pythagorean and a teacher of Plato, was of the opinion that no one can doubt that a man so long as he is enjoying sensual pleasure is incapable of handling any matter requiring thought, reason, and reflection, since such pleasure, if indulged in too much or too long, extinguishes completely the light of the soul."* Hence Cicero in the third book on the Tusculan Disputations says,† "But these definers say

* Bacon has changed the order of sentences in this quotation, as he frequently does in quotations from classical authors. See *De Senectute,* XII, 40, 41.

† *Tusculan Disputations,* IV, 9, 21.

that the source of these disturbances is lack of moderation, which is a defection from the entire intellect and from correct reasoning, a state so averse to the dictates of reason that the appetites of the mind can in no way be governed and restrained." Moreover, Pliny for this reason says in the fourteenth book of the Natural History, "After pleasure has begun to live life fails." Aristotle also in his book on the regimen of Life said to Alexander, "Avoid the inclinations to bestial pleasures, for the carnal appetites incline the mind to the corruptible pleasure of the bestial soul if no discretion be used. Therefore the corruptible body will rejoice, and the incorruptible intellect will be saddened. The inclination to carnal pleasure therefore generates carnal love. But carnal love generates avarice; avarice generates the desire of riches; the desire of riches generates shamelessness; shamelessness generates presumption, and presumption generates infidelity. And therefore pleasure brings with it stains on the soul of man, by which the soul is darkened." Aristotle for this reason also says in the book of Problems that we are called incontinent and blameworthy more on account of taste and touch than because of the other senses, since we are overcome by the basest pleasures and by those in which we participate with the brutes. For the latter take no delight in seeing and hearing and smelling, as does man. For they do not seek an odor for itself, but because of the food to which they are led by the odor. But a man not suffering from hunger finds delight in a perfume, as in the case of roses and other odoriferous flowers. He also takes great delight in hearing and in seeing; but brutes take pleasure only in what is known to pertain to taste or to touch. For this reason regulations and directions for correct living are drawn up in regard to the pleasure of both taste and touch; and we are praised more when we abstain from these pleasures, and we are blamed because of them since they are low and bestial. Moreover, Seneca says in the second book on Anger,* "When pleasures have corrupted both mind and body, no form of adversity seems tolerable, not because it is hard, but because a soft man is enduring it." For this reason since there are two principal modes of pleasure, one in lust, the other in intoxication and drunkenness,

* *De Ira,* II, 25.

Cicero asks in the fifth book on the Tusculan Disputations,* "Why should you not call that man most miserable whom we see raging and inflamed with lustful passions, madly craving all things with an insatiable desire, and thirsting the more keenly and ardently the more deeply he drinks everywhere from the cup of pleasure?" Seneca, moreover, says in the seventh book on Benefits† that "pleasure is fragile, brief, exposed to disgust, the more eagerly quaffed the more quickly subsiding into the opposite, a thing of which we must either repent or be ashamed, in which there is nothing splendid nor befitting the nature of man next to God, a thing that is low belonging to our base members, foul in its end." And Seneca in the second book of the Declamations says, "The young man sins, the voluptuous old man is mad, for he brings dishonor on his old age, and is rendering the age of youth more imprudent." And Cicero in the fourth book on the Tusculan Disputations, teaching how to cure this lustful person, shows the worthlessness of such pleasure in these words,‡ "Now the cure for one who is affected in this manner, is to show how light, how contemptible, how very trifling he is in what he desires; how he may turn his affections to another object, or accomplish his desires by some other means; or else to persuade him that he may entirely disregard it: sometimes he is to be led away to objects of another kind, to study, business or other different engagements and concerns: very often the cure is effected by change of place, as sick people that have not recovered their strength are benefited by change of air. But above all things the man thus afflicted should be advised what madness love is: for of all the perturbations of the mind there is no one which is more vehement; for, (without charging it with debaucheries, adultery, or even incest, the baseness of any of these being very blameable; not I say, to mention these,) the very perturbation of the mind in love is base of itself, for, to pass over all its acts of downright madness, what weakness do not those very things which are looked upon as indifferent argue, injuries, suspicions, unfriendliness, war, then peace again" renewed, all of which follow upon men inspired by lust? Hence it follows that

* *Tusculan Disputations*, V, 6.

† *De Beneficiis*, VII, 2.

‡ *Tusculan Disputations*, IV, 35. Yonge's translation.

"The man who seeks to fix
These restless feelings, and to subjugate
Them to some regular law, is just as wise
As one who'd try to lay down rules by which
Men should go mad.

Now is not this inconstancy and mutability of mind enough to deter any one by its own deformity? We are to demonstrate, as was said of every perturbation, that there are no such feelings which do not consist entirely of opinion and judgment, and are not owing to ourselves. For if love were natural, all would be in love, and always so; nor would one be deterred by shame, another by reflection, another by satiety." For this reason Aristotle in the Secrets of Secrets says to Alexander, "Kind Emperor, do not yield to the desire for sexual intercourse with women, since such intercourse is a characteristic of swine. What glory is it for you if you practice a vice that is characteristic of unreasoning animals and follow the actions of brutes? Believe me without question that sexual intercourse involves the destruction of our bodies, the shortening of life, the corruption of the virtues, the transgression of the law, and the adoption of effeminate manners." Therefore, that every flaw of character may be avoided, Seneca says to Helvia,* "If you consider that sexual desire was given man not for his enjoyment but for the purpose of propagating the race, every other desire will pass by that man unscathed whom this secret and ruinous passion deep-seated in our very vitals has not harmed."

Concerning the pleasure taken in gluttony Seneca speaks fittingly to Helvia,† "Oh, how wretched are those whose appetite is not excited except at the sight of expensive foods! But not their fine flavor or the pleasure they give the palate but their rarity and the difficulty in procuring them render such foods dear. Otherwise, if they care to return to a sane state of mind, what need is there of so many arts ministering to the belly? What need to ravage forests, what need to search the deep? Everywhere are the different kinds of food, which nature has distributed in all places. But men pass by these as though they were blind, . . . and although they can appease their hunger

* *Ad Helviam,* XIII.
† *Ibid.,* X.

at a small cost, they whet it at a great one. From far and wide they bring together all that is known to pampered gluttony; and they convey from a remote ocean that which is admitted with difficulty into a stomach weakened by dainties. May God confound those, the luxury of whose gluttony thus passes beyond the boundaries of their own land." Moreover, Seneca says in the book of the First Letters regarding intoxication and drunkenness,* "Oh, how many things men do when intoxicated for which they blush when sober! Consider drunkenness to be nothing else but voluntary insanity. Prolong the drunkard's state to several days, will you have any doubt of his madness? Drunkenness undoubtedly elicits, kindles, and discloses every vice; it removes the sense of shame that hinders our evil purposes; . . . when the power of wine overpowers the mind, every hidden evil comes forth to view. It is then that the lustful man brooks no delay in gratifying the demands of his passions; it is then that the unchaste man proclaims his disease; it is then that the ill-natured man does not restrain his tongue or his hand." And since such is the case Aristotle says in the second book of the Ethics that a twofold malediction is due the drunkard. And after many statements on this subject Seneca says in the place quoted above, "Cruelty generally follows wine drinking, for a man's soundness of mind is corrupted and he is made savage. As prolonged illnesses render it difficult for the eyes to bear even a very weak ray of the sun, so continued spells of drunkenness bestialize the mind. For when people are frequently beside themselves, the habit of insanity becomes fixed, and the vices produced by wine retain their ascendancy without it." Seneca, moreover, in the first book on Natural Questions writing against those with a luxurious and perverted taste in drinking says,† "Good God, how easy it is to slake a natural thirst! do you consider such the thirst of voluptuaries?" As though he were to say their thirst is not a thirst but a fever. "And indeed it is the more severe in this way because it is detected neither by touching the veins nor by heat diffused over the skin; but luxury dries up the heart itself, an unconquerable evil."

I wish now at the end to add to the authors cited a letter of

* *Epist.*, lib. XII, 1.

† *Naturales Quaestiones,* IV, 13.

the philosopher Anacharsis assailing pleasure, which he wrote to the rich Hammo in these words:* "Anacharsis to Hammo greeting. My clothing is the same as that with which the Scythians cover themselves; the hardness of my feet supplies the want of shoes; the ground is my bed, hunger my sauce, my food milk, cheese, and flesh. So you may come to me as to a man in want of nothing. But as to those presents you take so much pleasure in, you may dispose of them to your own citizens, or to the immortal gods." Cicero in the fifth book of the Tusculan Disputations quotes this letter with approval. Seneca in the book of the First Letters says, "He who comes forth into the light of this world, should be content to live on bread and water." And in his book on the topic, Why many evils happen to good men, he exhorts us to shun luxury, "Shun luxury, shun an enervated prosperity, by which the mind is cloyed, and unless something happen to admonish of our human lot, it falls into a state of torpor like continued drunkenness." Moreover, in the Phaedo of Plato we read that he who frees as much as possible his soul from association with the body and is least under the influence of carnal pleasures is manifestly a philosopher. The meditation of philosophers is the freedom and separation of the soul from the body. He teaches us that the man has confidence in regard to his own soul who permits carnal pleasures to depart as though they were strangers, and adorns his mind not with an embellishment that is foreign but with one belonging to it, temperance, justice, fortitude, liberality, and virtue. Wherefore Cicero in his work on the Immortality of the Soul concludes for many reasons that the "whole life of philosophers is a meditation on death"; to separate the soul from the body, what else is it than to learn how to die? and therefore he continues, "Let† us separate therefore ourselves from our bodies, that is, let us accustom ourselves to die, and thus while we remain on earth we shall be enjoying a life like that of heaven, and when we are borne thither from these chains . . . then finally shall we live." For this "life in fact is death," as he himself elegantly puts it. Moreover, Seneca says to Marcia,‡ "Tarrying in the body is never dear to great souls; they long

* *Tusculan Disputations,* V, 32. C. D. Yonge's translation.

† *Ibid.,* I, 31.

‡ *Ad Marciam,* XXIII, XXIV.

ardently to go forth and break loose, they endure with difficulty these narrow bounds, wandering as they do through the whole universe and accustomed from their lofty pinnacle to look down upon human affairs. Hence it is that Plato exclaims that the whole soul of the philosopher reaches out to death, desires it, meditates on it, always is carried away with a longing for it." "These things that you see about us, bones, tendons and skin stretched over them, face and ministering hands, and all the rest with which we have been enveloped, are chains of our souls and shadows; by these the soul is overwhelmed, routed, stained, kept from the truth and from what is its own, and driven to what is false; its whole struggle is with this flesh that weighs it down; it presses forward to that place whence it was sent forth; there eternal rest awaits it." Therefore Avicenna says in his Morals, "A man will not be set free from this world and its allurements, unless after becoming wholly dependent on that celestial world he desires that which is there, and the love of those things that are there withdraws him completely from the consideration of that which is behind him."

Chapter VI

I HAVE spoken of prosperity and how the six mortal sins are nourished by it. Now I must speak of adversity and how anger contends with it, although anger suffers a base defeat. In the first place, I shall state the principles regarding those matters that are to be considered relating to the remedy for anger, and in the second I shall take up the main points in regard to the glorious endurance of adversity. The first step in resisting anger is the comprehension on our part of its horrible consequences, so that each of us bearing these in mind may make the avoidance of this vice his special study. Anger is a very serious sin. For the angry man blasphemes God, loses his neighbor, confounds himself, scatters abroad his temporal blessings, and is not deterred in venting his rage by the fear that he is neglecting his eternal good and making himself liable to the punishments of hell. First, then, we must consider that anger is contrary to all human nature. For the description of man marking him as such is that he is an animal kindly by nature. Therefore Seneca in the first book on Anger proves that anger is contrary to human nature in this way:* "If we consider man, what do we find milder than he is while he is in the right frame of mind? But what is more cruel than anger?" He continues, "Anger is eager to inflict punishment, but the presence of such a desire in the very peaceful breast of man is by no means in accordance with his nature." But elsewhere he reasons thus: "Man is born to help his fellow man, but anger is instrumental in his ruin." He, that is, man, "wishes to associate with his fellow man, anger tries to separate them: he wishes to benefit, anger wishes to harm: he wishes to aid even those who are unknown to him, anger desires to destroy those who are dearest: he is ready to expend himself in relieving the misfortunes of others, anger is prepared to run into danger, provided it can involve others. Who then is more ignorant of nature than he who assigns this savage and destructive vice to nature's best and most faultless work," namely man? "For human life consists of favors, and is bound together into common fellowship and assistance by concord, and not by fear but by mutual love." And in the first book he compares men in anger to beasts, which are different from men in their nature. He says,† "Anger displays itself and appears in the face; and the greater it is the more manifest is its expression. Do you not see that in the case of all animals, as soon as they rise to attack, signs precede, and their whole bodies cease to retain their usual quiet habit? The jaws of the boar are covered with foam, his teeth are sharpened by rubbing, the horns of the bull are tossed in the air and the dust is scattered by the stamping of his feet, the lion roars, and the neck of the angered snake swells, the look of the mad dog is forbidding." Moreover, he expresses the opinion in his third book that angry men are further removed from human kindness than are other beasts. "The‡ face, believe me, of a wild beast, whether goaded by hunger or by the dart piercing its vitals, even when half dead it makes a final attack with its jaws on the hunter, is less repulsive than the face of a man inflamed with anger." Moreover anger is contrary to man's nature, because it first destroys and confounds his own substance. For in the first place it alters a

* *De Ira*, I, 5.

† *Ibid.*, I, 1.

‡ *Ibid.*, III, 4.

man externally as regards his whole body into a foul and horrible appearance. Hence Seneca in the opening part of his book writes as follows :* "You have asked me, Novatus, to write how anger can be controlled, and not without reason do you seem to me to have feared this especially foul ungoverned state of mind. In other passions there is an element of rest and peace, but this state of mind is marked by violent agitation and passion of grief." And in the second book on Anger he says,† "Nothing, however, will profit so much as first of all to view the hideousness of the thing. Nor is the appearance of any other passion more repulsive. It defiles the most beautiful face, renders savage the most tranquil brow. All beauty departs from those angered; and if they are neatly dressed, at the dictate of anger they will pull their garments awry and pay no heed to their persons; and their hair smooth either by nature or art becomes rough along with their disordered mind. Their veins swell, their breast heaves with their rapid breathing, angry ejaculation of the voice distends the throat. Then the limbs become restless, the hands shake, the whole body is agitated. In what state do you suppose the mind is whose external appearance is so repulsive? How much more terrible within the breast is its look, how much keener its spirit, more eager its passion, destined to burst itself unless it find outward expression? We should picture anger to ourselves as similar to the fire-breathing monsters which the poets have imagined in the lower world —anger which raises its voice loudly with hissing, bellowing, and raucous sounds, which labors under universal hatred of itself, being especially anxious, if it cannot do harm otherwise, to bring ruin on lands, seas, and sky." And in the third book he says,‡ "It is necessary, therefore, to expose its foulness and ferocity and place before our eyes how monstrous is man raging at man"; and "how we have described the angry man's countenance as fierce and now pallid with the blood withdrawn and banished from the surface, and now red with all the color and energy brought into the face as though covered with blood, the veins swollen, and the eyes now restless and darting from their sockets, now fixed and set in a stare. Add the sound of his teeth

* *Ibid.*, I, 1.
† *Ibid.*, II, 35.
‡ *Ibid.*, III, 3, 4.

as they clash not differing from the noise made by boars sharpening their weapons by grinding them together. Add the cracking of his joints as his hands crush each other, and his oft smitten breast, his rapid breathing and deep-drawn groans, his swaying body, uncertain words, his lips quivering with sudden exclamations and sometimes compressed, and hissing forth some dreadful imprecation." Moreover, anger is contrary to man's nature because it injures the body with many afflictions and scourges it with deformities. For Seneca says at the end of the second book, "We must see how many anger has injured by its own action. Some have ruptured their veins by excess of passion, and a shout raised beyond one's strength has caused a hemorrhage, and humor discharged into the eyes with too much force has destroyed the sight, and the sick have suffered a relapse." And we have seen frequently healthy men under the influence of violent anger attacked by various maladies.

But similarly anger is unnatural in man as regards his soul. For the intellectual soul has two divisions; one is speculative and employs reason; the other is active and fitted for the love of virtue. Anger disturbs the first and is especially inimical to it, and frequently reduces it to madness and insanity. For in the second book Seneca says,* "Reason gives time to both sides of a question and seeks for time to investigate the truth. Anger is precipitate. Reason wishes that the decision be rendered in accordance with justice. Anger cares nothing for the just decision rendered by reason. Reason considers nothing except the case in hand. Anger is influenced by foolish and irrelevant matters. Rage does not brook control; it is angered at truth itself if truth appears to oppose its wish. With shouting and uproar and tossings of the body it pursues with insults and curses the objects of its hate. Reason does not act in this way, but, if it be necessary, Reason silently quietly destroys families including wives and children that are dangerous to the state, pulls down houses and levels them to the ground, and roots out names inimical to liberty. All these things Reason does without gnashing its teeth and shaking its head or any other indecorous act."

Moreover anger reduces reason to insanity and madness. For

* *De Ira*, I, 18, 19. (Not from the second book.)

Seneca at the end of the second book on Anger says,* "There is no quicker road to insanity. Many accordingly have continued in their state of passion and have never recovered their banished reason. Madness drove Ajax to death, anger drove him to madness." And in the third book he says,† "What then? Does any one call that man sane who caught up as it were by a tempest does not walk but is driven along; and enslaved by an evil passion does not entrust to another his vengeance, but exacts it himself, mad alike in mind and deed?" And in the first book he says,‡ "That you may know that those are not sane whom anger possesses, look at their appearance. For as there are definite indications of madness, the bold and threatening look, the gloomy brow, the savage face, the hasty step, the restless hands, the changing color, the frequent and labored sighs, so are the same signs visible indicative of anger: the eyes blaze and flash, a deep flush covers the whole face as the blood surges up from the depths of the breast; the lips quiver, the teeth are pressed together, the hair is rough and erect, the breathing labored and strident, the joints emit a sound as they are bent, with groaning and bellowing and abrupt speech in words offering little explanation and hands frequently clapped together, and stamping on the ground with the feet and the whole body agitated and making dire threats, the face foul and horrible to look upon. . . . You are in doubt whether the vice is more detestable or hideous." Finally anger is altogether unnatural in man, since the angry man does not shrink from exposing himself to every kind of danger and has no fear of death, but freely runs the risk of death in order that he may take vengeance. Hence at the end of the second book he says, "Angry men are prepared both to fight with swords and to fall upon them." And at the beginning of the third book he says,§ "The angry man is destructive not without his own destruction and sinks those things that cannot be sunk except with him who sinks them." Not only does he destroy himself but he loses his neighbor; and not only does he drive his enemies to death, but he is the cause of the death of his friends. Seneca in the second book gives

* *Ibid.*, II, 36.
† *Ibid.*, III, 3.
‡ *Ibid.*, I, 1.
§ *Ibid.*, III, 3.

us an excellent example illustrating this matter. A tyrannicide on being captured by the tyrant deceived him with such caution, that the tyrant himself slew all his own friends;* and the tyrant at length dies. Seneca says, "The incident of the tyrannicide is well known who on being thwarted in his attempt and seized was tortured by Hippias, so that he might disclose his associates. The tyrannicide thereupon named the friends of the tyrant surrounding him, to whom he knew the tyrant's safety was especially dear. After ordering all those who had been named to be killed, the tyrant asked if there was any other survivor. You alone, said the tyrannicide. For I have left no one else to whom you are dear. His anger, therefore, caused the tyrant to aid the tyrannicide and slay with his own sword his guard."

Just as anger confuses the speculative part of the soul and is unnatural to it, so also does it affect the active part and far more so. For this part of the soul has virtue as its dower, and in it virtue has its place; and anger is a vice. Since, therefore, according to Seneca in his book of the First Letters, and according to Cicero in his book on Duties, and according to the same author in the second book on the Tusculan Disputations, all the virtues are interconnected; so that he who has one has them all owing to the general conditions of their association, although they differ in their special conditions, of necessity anger does not permit any virtue to accompany it, since anger excludes kindness which is joined chiefly to other virtues. This is what Seneca says in the second book on Anger, "You must first remove virtue from your mind, before you entertain anger, since vices do not associate with virtues; and a man can no more be at the same time both angry and good, than he can be sick and well." Finally anger destroys the noblest virtues, namely, kindness, clemency, magnanimity, piety, joy, and peace of mind. For Aristotle in the fourth book of the Ethics teaches that kindness is a virtue the opposite of anger. And Seneca says in the first book on Clemency,† that no one of all the virtues is more befitting man than clemency, since there is none more human. But anger excludes clemency, because it degenerates into cruelty;

* Bridges' text reads *inimicos*. The sense of the passage requires *amicos*.

† *De Clementia*, I, 3.

and cruelty is rightly the opposite of clemency, as Seneca says in the second book.

Although clemency is fitting in all, yet, as Seneca says, clemency befits no one more than a king and a prince. For the kings of Israel are clement, says the Scripture. For as Seneca says,* "To have power to injure is a pernicious force; the greatness of that man is firm and secure whom people know to be as truly for them as he is above them: from whom as he proceeds they do not flee, as though he were some dangerous animal, but they emulously run to him as to a bright and beneficent star"; and he adds, since "savage and inexorable anger does not befit a king." "It is proper for a prince to hold the same position with reference to his subjects as God does to him." "If† God forgiving and just does not immediately follow up the faults of sinners, how much juster it is for a man placed over men to exercise his power with clemency, and to consider whether the aspect of the world is more pleasing to the eye and more beautiful on a calm and clear day or when all things are shaken by frequent peals of thunder; . . . and yet the appearance of a quiet and well-ordered government is not different from that of a calm and bright sky!" "This is true clemency," as he infers,‡ "to have no stain—a clemency that did not have its origin in regret for cruelty. This is in the midst of the greatest power the truest moderation of mind. Clemency renders princes not only more honorable but safer. Tyrants rage at will; kings not unless with reason and from necessity. Clemency creates a great distinction between a king and a tyrant. No one can have faithful and willing servants, whom he uses not otherwise than as beasts." For severity long continued loses authority. "On the other hand there is the man in whose mind there is nothing inimical, nothing savage; who exercises his power peacefully and beneficially; affable in his speech, easy to approach, amiable of countenance which especially wins popular favor, . . . of whom men speak the same in private as in public."

Moreover, Seneca cites a fine example. He says,§ "Nature invented the idea of a king, as we may learn both from other

* *Ibid.*, I, 3.

† *Ibid.*, I, 7.

‡ *Ibid.*, I, 7.

§ *Ibid.*, I, 19.

animals and from the bees, whose king has a spacious couch in their midst and in the safest place. As the superintendent of the work of others he does none himself; and on the loss of their king the whole swarm separates; nor do they permit more than one king at a time, and by means of combat they discover the better candidate. The king is distinguished in appearance and unlike the others both in size and beauty. Bees are very subject to anger and for their size are most pugnacious, and they leave their stings in the wound: the king himself has no sting. Nature did not wish him to be savage nor seek a revenge that would be so costly, and took away his weapon, and left his anger unarmed: this is an important example for great kings . . . since the mind of men ought to be as much more under control as it is capable of doing more serious harm. Would that the same law applied to man, so that his anger would be broken with his weapon, and that he could not do harm more than once." And he infers, "He is in error who thinks that a king is safe when nothing is safe from the king. Security must be obtained by mutual security. There is no need for a king to hedge himself about with walls and towers . . . clemency will render a king safe in the open. The love of his subjects is an impregnable fortification. For what is more beautiful than to live with the good will of all?" "We* exhort, therefore, that if clearly injured he retain his self-control, and remit the penalty, if he can do so with safety, and that he be far more willing to pardon injuries to himself than those to others. For as he is not generous who is liberal with what belongs to another, but he who takes away from himself what he bestows on another, so I shall call clement, not the man who is easy going in the matter of another man's grief, but him who when he is stirred by goads applied to him does not spring forward." And again Seneca says,† "The mere act of clemency on the part of a ruler causes a feeling of shame at wrong-doing." "And not less disgraceful to a prince are many acts of punishment than many deaths to a physician." For "by nature the human mind is stubborn and resists what is unpleasant and difficult, and follows more readily than it is led. And as spirited horses of noble stock are controlled with a light rein, so integrity of its own free will

* *De Clementia,* I, 20.

† *Ibid.,* I, 22, 24.

and motion follows clemency." And at length directing his speech to the Roman prince he says,* "The kindness of your disposition will be handed down and gradually diffused throughout the whole of your empire, and all things will be formed in your likeness, as health derived from an excellent source spreads among all men."

Thus, then, I have introduced certain facts to show the beauty and nobility of clemency, in order that the principal proposition may be made clear, namely, the madness of anger, which destroys this glorious virtue, and turns it into cruelty. Now I wish to make some statements in regard to the other virtues enumerated above with clemency, all of which frenzied anger destroys. That one especially united to clemency is magnanimity, which, as Aristotle teaches in the fourth book of the Ethics, is the ornament of all virtues. It is characteristic of this virtue that he who possesses it is forgetful of injuries and disregards them, as stated above, and does not deign to give way to anger nor consider any one worthy of his anger. Hence Seneca says in the second book on Anger,† "It is the mark of a great mind to disdain injuries; the most insulting way to exact vengeance is not to let him who has injured us appear worthy of our vengeance." And in the book on the Four Cardinal Virtues Seneca says;‡ "If you are magnanimous, you will never consider that an insult has been offered to you. You will say of your enemy, He did not harm me, but he intended to do so: and when you see him in your power, you will think that you have been able to take your revenge. For you should know that to pardon is the most noble and greatest kind of vengeance." And in the second book on Anger he says,§ "It is characteristic of the small and pitiable person to attack the one who pains him. Mice and ants, if you move your hand towards them, turn their mouths; weak creatures think that they are hurt if they are touched." Accordingly in the first book he says,|| "The most irritable are infants, old people and the sick; and everything that is impotent is querulous by nature." And

* *Ibid.*, II, 2.

† *De Ira*, II, 32.

‡ *De formula honestae vitae.*

§ *De Ira*, II, 34.

|| *Ibid.*, I, 13.

in the third book he says,* "Sick people are never touched without a complaint. Anger is small and narrow. For no man is not inferior to him by whom he considers himself despised. But the great mind and the true appraiser of itself does not take vengeance for an injury because it does not feel any. As missiles fly back from a hard object, and solids are struck with pain to him who strikes them, so no injury causes a great mind to feel it, since it is weaker than that which it attacks. How much nobler as though impenetrable to every weapon to spurn all injuries and insults! For taking vengeance is a confession that one is hurt; it is not a great mind that is bowed by injury. Either a stronger man or a weaker injures you; if a weaker, spare him; if a stronger, spare yourself. There is no surer proof of greatness than to be unmoved by anything that can happen." For who abandoning himself to personal grievance and madness does not cast aside his sense of shame first? Who in his blind rush and assault on another has not cast away whatever feeling of shame he possessed?

Since, therefore, nothing so befits a man as magnanimity, as Seneca says in the second book on Clemency, and since anger confuses and brings to naught this most glorious virtue, man should root out completely anger from his mind, and this is especially true of a ruler. Since in the first book Seneca says† that "it is the mark of a great mind to put up with injuries although possessed of the greatest power, nor is anything more glorious than a prince who refuses to punish personal offences." And in the second book on Anger he cites an example pertinent to this point,‡ "That prince is magnanimous and steadfast who like some mighty wild beast listens unmoved to the barking of his subjects." He adds, moreover, in the third book,§ "There is no question but that he who disdains his assailants will free himself from the throng and will stand upon higher ground; for it is the part of true greatness not to feel that one has been assailed. Thus the huge wild beast regards indifferently the barking of dogs." For the lion does not deign to hasten at the shouts and din of the hunters. "Thus in vain the wave assaults

* *De Ira,* III, 5, 6.

† *De Clementia,* I, 20.

‡ *De Ira,* II, 32.

§ *Ibid.,* III, 25.

the huge rock. He who does not become angry, remains unshaken by injury; but he who yields to anger has been moved." Hence this is the advice he gives against anger, "Beware lest your magnanimity and strength relied upon by most men fail"; most particularly has he in mind those possessed of great authority. And in the third book he says,* "The higher part of the world and that which is the more orderly and is nearest to the stars is never gathered into a cloud nor driven into a storm nor whirled into confusion; it is free from disturbance of any kind. It is the parts below that are blasted with lightning. In the same way the lofty mind, always serene and unmoved in its tranquillity, checking every disposition within itself to anger, is modest, commands respect and is properly disposed. You will find no one of these qualities in an angry man." Hence anger has no greatness in it, as he says in the first book. "We† must not think that anger makes any contribution to magnanimity. For anger is not greatness but a mere swelling of the mind. Nor to bodies already strained by a supply of harmful liquid is disease a mere augmentation, but a destructive abundance. All whom an insane mind lifts above human thoughts believe that they are breathing something lofty and sublime: but there is nothing solid beneath; those things are prone to fall that have increased without foundations. Anger has nothing on which to stand: it does not spring from what is firm and destined to be permanent; but is full of wind and folly." "Like ulcered and diseased bodies groaning at the lightest touch, anger is especially the vice of women and children. . . . There is no reason why you should believe the words of men in anger whose din is great and threatening; within their mind is very timid, because they are disturbed and hurt at trifles. Hence an angry man resembles closely a crocodile, an animal that is very bold when one is timid and very timid when one is bold. For it flees from one who pursues it and pursues him who flees from it: exactly what happens in the case of those who are subject to fits of anger. For they tower over those who are submissive to them, but shrink away when they are resisted with spirit. Hence they would not be disquieted so frequently at trifles unless their minds were weak. Since, therefore, anger is incompatible with

* *Ibid.*, III, 6.
† *Ibid.*, I, 20.

magnanimity, a virtue most worthy of honor, anger must be banished completely from our hearts."

And not only is this true because of this virtue of magnanimity and the remaining virtues mentioned above, but also because of other virtues, namely, mercy, piety, patience, joy, and peace of heart. Cicero in his plea to Caesar that he would spare Marcellus says, "No one of your many virtues is more admirable or pleasing than your quality of mercy; men in no other way approach more nearly to God than by granting deliverance to their fellow men. Your fortune holds nothing greater than that you should be able, your nature nothing better than that you should wish, to save many." Therefore in the case of each one of us Seneca says in the third book,* "Who am I whose ears it is a crime against heaven to wound? Many have pardoned their enemies; am I not to pardon the lazy, the negligent, the talkative? Youth excuses the child, sex the woman, freedom the stranger, familiarity one of the household." "He is a friend: he did what he did not wish to do: he is not a friend, he did what he should have done. Let us yield to one who is more prudent: let us pardon one who is more foolish. On behalf of each offender let us make this reply to ourselves that the very wisest men commit many faults; that no one is so circumspect that his carefulness is not sometimes forgotten . . . no one so fearful of offences as not to fall into them while he is seeking to avoid them. But if the most prudent sin, who is without a good excuse for his error? He is unjust who reproaches individuals with a vice that is general. The color of the Aethiopian is not conspicuous among those of his own race, nor is red hair gathered in a knot unseemly in a German. You will judge nothing noteworthy or unbecoming in an individual which is customary in his race. These peculiarities which I have mentioned are excused owing to the custom of some particular district and corner of the world; consider now how much more reasonable is pardon in the case of faults that are common throughout the whole human race. We are all inconsiderate and careless; we are all unreliable, querulous, solicitous of favor. Why under milder words do I hide a public ulcer? We are all evil; therefore each one of us will find in his own bosom that which he complains of in another." "Let us be

* *De Ira*, III, 24, 25, 26, 27.

more considerate of one another: we are evil men living among evil men. One thing can make us peaceful, an agreement to observe mutual affability. That man has already harmed me, I have not yet harmed him. But already you have perhaps injured some one; but you will do so. Do not consider only this hour, this day. Look into your whole habit of thought; even if you have done no evil, you have the ability to do it. How much better it is for a wrong to be healed than avenged! Vengeance consumes much time, exposes itself to many injuries, while it is smarting under one. We are angry for a longer time than we are injured; how much better it is to go off in the opposite direction and not oppose vices to vices! Would a man seem quite consistent if he sought satisfaction from a mule with his heels and from a dog with his teeth? Those creatures, you say, are ignorant that they are offending. In the first place how unjust is he in whose eyes the fact that the offender is a man is an obstacle in the way of his obtaining pardon. In the second if the fact that other animals lack sense protects them from your anger, let any one who lacks sense be equally privileged. For what difference does it make whether he has other qualities unlike those of brutes, if he has like the brutes that which defends them in every fault, namely darkness of understanding? He has offended; well, is this the first time? Is this the last? He will offend, and another will commit an offence against him, and all life will wallow in the midst of errors."

Therefore he says in the second book,* "Do not become angry at individuals, we must extend a general pardon: indulgence must be shown to the human race." "Does any one become angry at children whose youthful years do not yet know the distinctions in things? The excuse is a more weighty one and a juster one that the offender is a man than that he is a child." "What removes the anger of a wise man? The throng of offenders. He perceives both how unjust and perilous it is to be angry at a vice that is general." "The wise man will not be angry at sinners, because he knows that no one is born wise. He also knows that very few in any age turn out to be wise, since he has the condition of human life under observation. No sane man becomes angry at nature. For what if he should take a notion to be surprised that apples do not grow in uncultivated

* *Ibid.*, II, 10.

thickets? What if he should be surprised that thorns and briars do not swell with some useful fruit? No one becomes angry when nature defends a vice. The wise man, therefore, is calm and just to errors, he is not the foe but the corrector of sinners, he looks upon all men in just as kindly a way as a physician does sick people." It is evident, therefore, that mercy and anger cannot be in accord.

Similarly also the other noble virtues, piety, patience, joy, and peace of heart, are incompatible with anger. And since there is no question about these, and our discussion of the other virtues that are related to these has been an extended one, there is no need to dwell upon them. Piety which is effective in all things, according to Aristotle, is destroyed by anger, because the mind of an angry person is impious and cruel owing to an excessive desire for vengeance. That it destroys patience and peace of mind is evident; and therefore it removes and quenches all good, and breaks every protection against adversity. For as long as a man retains patience and happiness of mind and peace, he fears nothing, he is disturbed in no manner, he does not feel an injury, but master of his own mind he will despise whatever is termed adverse. Therefore the angry man by his anger destroys himself both in soul and body in this way that anger of itself acts on its own subject by destroying whatever is laudable in him. But not only do angry people thus incur danger, they even when goaded by passion and frenzied by the desire for vengeance expose their life to every danger, and with no fear of death impetuously oppose themselves to the assaults of their enemies, and are ready to fight with swords and to fall upon them, as Seneca says in the second book. And not only do they lose themselves but those nearest to them; and not only those whom they reckon hostile about whom there is no question, but also those who are friendly. For, as Seneca says in the second book, "Men in anger invoke death upon their children, and are hostile to their closest friends and must be avoided by those dearest to them." He says, "Anger has brought grief to a father, divorce to a husband, and hatred to a magistrate." And he aptly cites instances where men in anger have attacked their own friends.

Moreover, in the third book he states that Praexaspes,* one

* *De Ira,* III, 14, 15, 16, 17.

of the dearest friends of Cambyses, advised the monarch, who was addicted to over-indulgence in wine, to drink more sparingly, saying that drunkenness is disgraceful in a king, whom the eyes and ears of all make note of. But the king ordered the son of his monitor to proceed beyond the threshold, bent his bow and pierced the heart of the youth. And he infers, "I do not doubt that Harpagus* gave some such advice to the Persian King, at which he took offence, and served up to Harpagus his own children as a banquet." "And Darius, who was first to obtain the Persian throne after the Kingdom had been wrested from the Magian, on being asked by the aged noble Oeobazus to leave as a solace to his father one of his three children and to employ the services of the other two, said that he would send them all to him, and cast them down slain in the sight of their parent." And he continues, "I shall cite to you the case of King Alexander, favorite pupil of Aristotle, who slew at a banquet Clitus, a very dear friend who had been brought up with him; Alexander also threw to a lion Lysimachus, a friend equally dear to him. Lysimachus, moreover, severely mutilated his friend Telesphorus by cutting off his ears and nose, and kept him for a long time in a cage as though he were some strange and very rare animal." When Alexander traveled around the world, under the guidance of Aristotle and Callisthenes, he put Callisthenes to death, one of his great men, as Seneca states in his book on Natural Questions. Moreover, he provoked Aristotle to such an extent, that he was forced to free himself and the world by poison which he had sent Alexander to drink, as Pliny narrates in the thirtieth book of the Natural History. But these few examples are sufficient, because they are well known and have been written in regard to those matters under discussion.

Chapter VII

Not only does anger destroy its own subject and loses for him his neighbor and his friends, but dissipates his wealth, makes him negligent in regard to his good name, and careless of honor. For the love of vengeance surpasses every other affec-

* *Ibid.*, III, 15, 16, 17.

tion of the mind and overpowers every other vice. Since Seneca says at the end of the second book on Anger,* "Anger has spurned avarice, the hardest and least pliant of evils, forcing it to scatter its wealth and set fire to home and possessions collected into a single heap. The ambitious man in fact has cast away insignia valued at a great price and has rejected an honor granted. There is no passion of the mind over which anger will not bear sway." And not only does the angry man destroy himself, his neighbor, the blessings of fortune, but he offends and loses God. Not indeed so much because anger is a sin, but because in particular men in anger blaspheme God, of which there are numberless examples, and there have been and will be. And in contrast to such people, Seneca in the second book on Anger says,† "God does not wish to harm nor is he able. For his nature is mild and peaceful, as far removed from doing injury to another as to himself. Therefore madmen and those ignorant of the truth impute to him the raging of the sea, violent storms, and the persistence of winter. . . . No one of these happens for the purpose of injuring us, nay, there is nothing that is not for our preservation." Hence he states in the first book‡ that "Caius Caesar because his revelry was disturbed by thunderbolts challenged God to combat. How great was such madness! He thought either that he could not be harmed even by God, or that he could harm God."

Although many evils have been stated which arise from anger, since a man loses his body, his reason, his virtue, his neighbor, his possessions, his honor, and his God, yet its exceptionally bad quality can still further be shown in which it surpasses all other sins. Hence Seneca says in the third book,§ "Anger must be compared with the worst vices, in order that its nature may become apparent. Avarice acquires and hoards that which a better man might use. The angry man, however, by his anger loses more than the mere object of his wrath. Anger is worse than extravagance since the latter enjoys its own pleasures, the former the pain of another. Anger surpasses malignity and envy; for they wish that a man become unhappy, anger

* *De Ira,* II, 36.
† *Ibid.,* II, 27.
‡ *Ibid.,* I, 27.
§ *Ibid.,* III, 5.

wishes actually to make him so. They are pleased with fortuitous ills; anger cannot wait for fortune. Anger wishes to harm the object of its hatred not merely that he be harmed by some one else or by fortune." And at the beginning of the book he says,* "Anger does not disturb men's minds in the manner of other vices," "its impetuous and headlong violence does not proceed gradually: anger is hasty and does not make its attack on the objects only at which it aims, but vents its fury on everything that crosses its path. Other vices urge on our minds, anger drives them headlong. Even if a man may not oppose his low impulses, yet the impulses may cease of themselves. This vice just like thunderbolts and hurricanes and all other irrevocable things, because they do not proceed but fall, exerts its force more and more. Other vices deviate from reason, but this vice from sanity. Other vices have gentle means of approach and grow without attracting attention; in anger there is an overthrow of our minds. Nothing, therefore, drives on a man more than does anger, frenzied, as it is, and subject to its own violence, insolent when it succeeds, crazed when it is foiled. Reduced to a state of weariness not even by a repulse, when fortune saves its adversary, anger turns its teeth against itself. It matters not what motive has aroused anger, for it passes from trifles to the greatest excesses. It passes by no period of life, it excepts no class of men. For certain races, thanks to their poverty have escaped luxurious living. Certain races because of their active and wandering life have escaped sloth. . . . There is no race that anger does not goad. . . . Finally other vices seize upon individuals. This is the one passion that sometimes lays hold of a people. Nowhere has a whole people burned with passion for the female sex, nor has a whole state built its hope on money or gain. Ambition seizes upon individuals. Frequently anger seizes upon men in a body. Men, women, old men, children, princes and commons have had but one feeling, and the whole multitude aroused by a very few words has outstripped the very man who incited them. They have rushed forthwith for arms and fire-brands." And in the first book he says,† "If you care now to view the effects of this vice, you will find that no bane has cost the human race more dearly. You

* *Ibid.*, III, 1, 2.

† *Ibid.*, I, 2.

will behold murders, and poisons, and the mutual baseness of parties to an action and the ruin of cities, and the destruction of whole races . . . fires not confined within the walls of cities, but vast stretches of country lit up with hostile flames. Behold the foundations of the noblest cities scarcely discernible; anger has cast them down. Behold tracts solitary and deserted for miles without an inhabitant: anger has depopulated them." And since Seneca says in the third book,* "We shall protect ourselves against anger, if we represent to ourselves from time to time all its evils . . . it must be accused before us and condemned. Therefore its evils must be examined and brought to light"; for this reason in what precedes I have recounted all the fundamental evils of anger to the end that it may be completely rooted out, and that no vestiges of it may remain in our hearts.

CHAPTER VIII

NOT only does a consideration of these evils warn us to renounce anger, but I shall cite some noteworthy examples of philosophers and of illustrious princes which properly have the effect of quieting down any man who is angry and banishing wrath from his mind. Solinus in his book on the Wonders of the World states among other marvels that Socrates, the parent of the great philosophers, never changed countenance, but always remained in the same habit of mind and face. And Seneca in the second book on Anger states the same fact with regard to Socrates. And Jerome writing against Jovinian touches on the subject of Socrates' bad wife. When Socrates was asked why he did not drive away so cross a wife, he said, "I am being trained at home so that I may be able to bear more easily injury and abuse abroad." And in the same place Jerome relates that on a certain occasion Socrates was drenched with dirty water from above after prolonged abuse from his wife, but merely remarked: "I knew that rain would follow that thunder." Cassianus tells us in his book of Dialogues of one who reproached Socrates because he was a corrupter of boys; but when his pupils wished to assault this detractor Socrates restrained

* *De Ira,* III, 5.

them with the words, "Well I am, but I keep myself under control." Seneca in the second book on Anger says, Socrates said to his slave, "I would beat you were I not angry." And in the third book he says,* "Men say that Socrates on receiving a cuff on the ear merely remarked that it was annoying that men did not know when they should go forth with a helmet." "Plato, the pupil of Socrates, when about to beat a slave with his own hand realized that he himself was angry and kept his hand suspended and stood in the attitude of one about to strike. On being asked by a friend what he was doing Plato replied, 'I am exacting punishment from an angry man.' " He had already forgotten the slave because he had found some one more deserving of punishment. Seneca tells this story in the third book on Anger, and he introduces after this another example. "Because of a certain offence Plato was considerably exasperated, 'Do you then, Speusippus, correct this young slave with blows; for I am angry.' For this reason he did not flog him. . . . 'I am angry,' he said, 'I shall do more than I ought, I shall go too far; this slave should not be in the power of a man who is not master of himself.' " And in the second book he gives an example that was quoted above, in regard to a boy brought up in Plato's home, who on returning to his own home and finding his father loudmouthed, said that he did not see such behavior at Plato's. And Archytas of Tarentum, the other teacher of Plato, being somewhat angered at his steward said, "How would I have treated you if I were not angry!" These words are quoted by Cicero† in the fourth book of the Tusculan Disputations. Moreover, Eusebius in the Chronicles says that Xenophon, the pupil of Socrates next in importance to Plato, said to one who reviled him on a certain matter, "You revile me: I on the witness of my conscience have learned to despise abuse." Diogenes the philosopher, as Seneca relates in the third book on Anger, when a forward youth spat on him as he was discussing anger, bore the insult meekly and like a philosopher. "I am not indeed angry," said he, "but I am in doubt, however, whether I should be." And Seneca continues, "How much better seems the reply of that sage who as he was pleading a case had discharged into the middle of his brow a mouthful of thick spittle by Lentulus, a

* *Ibid.*, III, 11, 12.

† *Tusculan Disputations*, IV, 36.

turbulent individual. He wiped off his face and said, 'I am in a position, Lentulus, to state to all men that those are in error who say that you do not possess mouth' " [*i.e.*, effrontery]. Democritus the philosopher resembled Socrates in the serenity of his countenance. For as Seneca tells us in the second book on Anger,* "Never did he appear in public without laughing: nothing in fact seemed serious to him of all that was being done with serious intent." But Heraclitus the philosopher, as Seneca states in the same passage, whenever he went out and saw so great a throng of people around him living wretchedly, nay, rather, perishing miserably, would weep and pity all.

Now these examples and the like refer to the philosophers. But there are deeds of others also that should be imitated. When Cambyses, a very evil king, had shot an arrow into the heart of the son and asked the boy's father whether he had a sure hand, the father replied that Apollo did not shoot an arrow with truer aim; he did not revile the king nor did he utter a word of sorrow, although he perceived that his own heart had been pierced as well as that of his son. For if he had said anything as an angered man, he could have done nothing as a father. Seneca relates this story in the third book on Anger. And he continues, "When the Persian king served up to his friend Arpagus as a banquet the latter's children and asked the father whether he was pleased with the food, Arpagus replied, 'At a king's table every dinner pleases' . . . from this I infer that anger arising from great wrongs can be concealed and forced to utter words at variance with one's thought."

He adds, moreover,† "But although for those who serve control of their passions and especially of this savage and unbridled one is advantageous, it is still more so for kings. For all is lost when fortune permits a man to go as far as anger dictates." "For what was easier for King Antigonus than to order two soldiers to be led forth, who while leaning against the tent of the king were doing that which men do with the greatest risk and also with the greatest satisfaction who are discontented with their king? Antigonus had heard all since between the speaker and the listener there was only a curtain. Gently pushing the curtain aside he said, 'Move further away

* *De Ira*, II, 10.

† *Ibid.*, III, 16, 22, 23.

DOCTORS OF THE UNIVERSITY OF PARIS

From *Chants Royaux* in the Bibliothèque Nationale, Paris

lest the king hear you.' The same king on hearing certain of his soldiers calling down curses on their king for bringing them into that particular road and into mud from which they could not extricate themselves, approached those who were in the most serious difficulty, and said to them, 'Now curse Antigonus, through whose fault you have fallen into these difficulties, but bless him who has rescued you from this morass.' Antigonus bore insults as tranquilly from his enemies as from his subjects. Accordingly when the Greeks were besieged in a certain small fortress, and relying on the strength of their position held the enemy in contempt and passed many jokes on the ugliness of Antigonus, now deriding his small stature and now his flat nose, he said, 'I rejoice and I hope that it is a good omen if I have a Silenus in my camp.' After conquering them by hunger, he enrolled in his cohorts those useful for war, the rest he sold; and he said he would not have done this save for the fact that it was expedient for these men who had such a bad tongue to have a master."

"If there was," as Seneca says in this passage, "any virtue in King Philip, father of Alexander, it was his patience under insults, a great aid in protecting a kingdom. For when Athenian envoys came to him and were asked what the Athenians desired, one of them replied, 'That you hang yourself.' And when the indignation of those present flared up at such a rude answer, Philip bade them be quiet, and that the envoy be dismissed safe and unharmed. 'But you other envoys,' said he, 'are to report to the Athenians that those exhibit far more haughtiness who make such statements than those who hear them without exacting vengeance.' " Alexander the Macedonian, grandson of Antigonus and son of Philip, although he was haughty and evil in his ways, yet, as Seneca says in the second book,* "after reading a letter from his mother, in which she warned him to beware of poison at the hands of Philip his physician, received the potion and drank it without fear. He relied on his own judgment rather in regard to his friend. I praise this in Alexander, because no man was more subject to anger. The greater the moderation in kings the more is it to be praised." Seneca in the third book says,† "It is recorded of

* *Ibid.*, II, 23.
† *Ibid.*, III, 11.

Pisistratus the Athenian tyrant that when a drunken table companion had declaimed vigorously against his cruelty, and there were not lacking those desirous of offering their hand to Pisistratus, and first one and then another tried to inflame his resentment, he bore the affront calmly and replied to those urging him on that he was no more incensed than he would have been had his detracter run into him blindfolded." Seneca in the second book relates the following anecdote in regard to Marcus Cato, who according to the histories was second to none in virtue and in wisdom was foremost among the Romans:* "When a man ignorant of his identity unintentionally struck Cato in the bath (for who knowingly would have done him an injury?) and offered an apology, Cato replied, 'I do not recall that I was struck': he thought it better not to acknowledge the injury than to avenge it." And in the third book he says,† "The deified Augustus did and said many things worthy to be remembered, from which we see that anger had no control over him. Timagenes, a writer of histories, had made certain statements against Augustus, against his wife and his whole house, and his pleasantries were not lost. For pleasantry of a daring kind is freely passed from mouth to mouth. Caesar frequently warned him to be more guarded in his speech. He put up patiently with the abuse; he never complained to Pollio, the host of his enemy. He merely said this to Pollio, 'Enjoy him, my dear Pollio, enjoy him.'" And Seneca concludes, "Therefore each one should say to himself, Am I more powerful than Philip? Yet an insult offered him was not punished. Have I more power in my home than the deified Augustus had in the whole world? Yet he was content to withdraw from his traducer."

Chapter IX

We have, however, remedies against anger drawn from a consideration of the evils inherent in it: moreover, we are influenced by the examples of the wise and powerful. In addition to these Seneca teaches us more direct remedies against this

* *De Ira,* II, 32.

† *Ibid.,* III, 23, 24.

vice. One of these is a knowledge of the facts before we give way to anger. For, as he says in the second book,* "Of those things that offend us some are told us, others we ourselves hear or see. We should not give ready credence to all that is told us; for many lie to deceive, many because they have been deceived. One tries to win favor by an accusation and feigns an injury in order that he may seem to have deplored its commission. There is the malignant man and he who wishes to separate firmly united friendships. . . . The question of a trifling sum would not be settled if you should act as judge, without a witness. The witness would have no standing except under oath. You will give both parties a chance to state the case, you will give them time, you will hear the case more than once. For the truth will be more evident, the oftener the case is reviewed. Do you condemn a friend at once? Before you hear him, before you question him, before he is permitted to learn who his accuser is or the charge, are you going to give way to anger? . . . The very man who told you will cease making the statement if he is required to prove it. 'There is no reason,' he says, 'to drag me into it. If I am brought into it I shall deny all; I shall never tell you anything in the future.' At the same time he both urges you on and himself withdraws from the contest and strife. He who is unwilling to speak to you except in private, is almost in the position of one who does not speak at all. What is more unjust than to believe in private and to be angry in public?" And again he says,† "Credulity causes a great deal of harm. In many cases we should not even listen, since in some things it is better to be deceived than to distrust." Of some things we are witnesses, but "suspicion and conjecture, two of the most deceiving provocatives, must be removed from the mind. That man greeted me with scant courtesy. This one did not embrace me. Still another broke off abruptly the conversation we had begun. This man did not invite me to dinner. I noticed aversion on the face of that man." "Suspicion will never lack a reason: we are in need of frankness and a kindly judgment in all things; we should believe nothing except that which meets our eyes and is manifest; and as often as our suspicion shall have appeared groundless, let us blame our credulity. For such cor-

* *Ibid.*, II, 29.

† *Ibid.*, II, 24.

rection will form in us the habit of not believing a story on insufficient evidence." And for this reason he says in the third book,* "We shall find many innocent if we begin to investigate before we yield to anger. But now we follow our first impulse; then although our suspicion is groundless, we persist, in order that we may not seem to have begun without cause; and unfairest of all is the fact that the injustice of our anger makes us the more obstinate. For we retain and augment our feeling of anger, just as though the excess of our passion were a proof of its justice. . . . You will find that very same thing in man which you see happen in the brutes. . . . The color red excites the bull, the adder rises at a shadow, a napkin angers bears and lions. All animals that are wild and savage are disturbed at trifles." And similarly he says in the second book,† "Trifles are a terror to such as have no stability in themselves. The motion of a chariot and the change in the appearance of the wheels drive lions back into their den. The squeal of a pig terrifies elephants." And in the third book he says,‡ "The same thing happens in the case of restless and stupid natures. They are struck by suspicion to such a degree that they call moderate favors injuries. In favors is found the most frequent, at any rate the most bitter, source of anger. For we become angry with our dearest friends because they have given us less than we expected, less than they have bestowed on another. . . . 'I have less than I hoped for . . . than I deserved.' This source of anger is especially to be dreaded; from it spring most destructive enmities and such as are likely to attack all that is most sacred. More friends than enemies slew the deified Julius, friends whose insatiable hopes he had failed to satisfy. . . . For no one has made a more generous use of victory from which he claimed nothing for himself except the power to regulate. But how could he have satisfied such insatiable desires, since all gauged their wishes to the full extent of the power of a single individual? . . . This has turned the arms of their own subjects against kings, and has driven the most faithful adherents to plot the death of those for whom and before whom they had vowed to die. No man is contented with his own lot when he

* *De Ira*, III, 29, 30.

† *Ibid.*, II, 11.

‡ *Ibid.*, III, 30, 31.

considers that of his neighbor. . . . So great is the injustice of men, that although they have received much, the fact that they could have accepted more they regard as an injury. He gave me a praetorship: but I had hoped for a consulship. He has given me that which he had to give some one: he has contributed nothing of his own. Come, be thankful rather for what you have received. Wait for the remainder and be glad that you are not yet filled. We should reckon it among our pleasures that there still remains something to hope for. . . . Consider how many more you precede than you follow."

The second remedy is delay in exacting punishment. Hence Seneca says in the second book,* "The greatest remedy against anger is delay. Beg of anger at the beginning, not that it pardon, but that it judge; it will cool down if it waits." And in the third book he says,† "Give yourself no license while you are angry. Why? Because you wish for free license in all things. Fight with yourself: if you cannot conquer your anger, it is beginning to conquer you. If it is hidden, if an outlet is not given to it, let us destroy its signs, and let us keep it as far as possible hidden and secret. This will prove a great trial to us. For anger is eager to leap forth and inflame the eyes and change the face. But if we have allowed it to become manifest on our exterior, it triumphs over us. Let it be hidden in the lowest depths of our breast: let us carry it, let not it carry us. Nay, let us turn all signs into their opposites. Let our look be mild, our voice gentler than usual, our step slower. Gradually the inner man is conformed to the outer. In Socrates a lowered voice and restrained speech were indicative of anger. It was then apparent that he was holding himself in check. . . . Let us beg all our most intimate friends to treat us with the utmost candor when we are least in a condition to brook it, and not to lend themselves to our anger. Against this potent evil and one that is pleasing to us . . . it is best to be on the look-out for obstacles with which to oppose our known vices, and before all else so to order our mind, that even when shaken by sudden and very grievous blows it may either have no feeling of anger, or bury such a feeling if excited by some great and unexpected

* *Ibid.*, II, 29.

† *Ibid.*, III, 12, 13.

injury in the depths of the heart and give no manifestation of its pain."

In the second book he likewise combines both the remedies just mentioned in a fitting and useful way. He says,* "Therefore against the first causes we must fight: for the cause of anger is the thought that we have been injured. To this thought, however, we must not give ready acceptance. Not even to what is open and manifest must we accede at once. For some falsehoods bear the appearance of truth. Time should always be given; time discloses the truth. Let us not lend ready ears to those who prefer charges: let this vice of human nature be noted by us and suspected, because we willingly believe that which we are unwilling to hear, and we yield to anger before we sit in judgment on the matter. What of the fact that we are influenced not only by charges but by mere suspicions? And become angry at innocent people by misinterpreting a look or a laugh? We ought then to plead the cause of the absent against ourselves and to hold our anger in suspense. For a delayed punishment can be exacted; an exacted punishment cannot be recalled." And he, moreover, says in the third book,† "It is not expedient to see all things, to hear all things; let many injuries pass unnoticed, most of which he does not receive who is ignorant of them. You don't wish to be angry? Then don't be inquisitive. The man who asks what has been said about him, he who digs up malicious remarks even though uttered in secret, disquiets himself. Our interpretation leads us to view certain things as injuries; therefore some of them should be postponed, others laughed at, and still others pardoned. Anger must be restrained in many ways: many things may be turned off into a laugh or a jest." And in addition to all these he sets forth a threefold argument against anger in the second book: for he argues as follows:‡ "It is wrong to injure your country; therefore it is also wrong to injure a fellow citizen, for he is part of your country. The parts are sacred if the whole is worthy of veneration. Therefore it is wrong to harm any man, for he is your fellow citizen in the larger city, that is, in the world." The second argument is drawn from a simile in reference to the

* *De Ira,* II, 22.

† *Ibid.,* III, 11.

‡ *Ibid.,* II, 31, 34.

members of the same body, and is as follows: "What if the hands should wish to harm the feet, and the eyes the hands? As all the parts are in accord, because the conservation of the whole depends on the preservation of the individual members, so men should spare their fellows, because they were born to form a society." The third argument is a most fitting and excellent one, the conclusion of which he places first. He says, "Therefore we must abstain from anger; either the man is an equal you are to attack, or a superior, or an inferior. To contend with an equal is a doubtful matter, with a superior is madness, with an inferior is paltriness." And he says,* "Is he a good man who has inflicted injury? Do not believe it. Is he a bad man? Do not imitate him. He will pay the penalty to another which he owes to you: and he who has sinned has paid the penalty to himself."

Many other matters should be considered with reference to anger, all of which are taken up mainly in the books of Seneca on Anger and on Clemency. But since the discussion in the present instance is in the nature of a plea and not of a complete presentation, I am for this reason concluding it at this point. I have spoken more fully on this vice, since it will always injure and confound the whole human race so long as man retains his mortal state. For the vice is a very bad one and particularly unnatural in man, flaring out to his peril. For this reason I have written at greater length and in more detail in regard to it.

CHAPTER X

SINCE the subject of anger has just been considered, which fights with adversity and is conquered, some virtues will be touched upon which conquer adversity and take no thought for this world's prosperity. Seneca composed a book to Lucilius whose title is, Since the world is ruled by providence, why do many evils happen to good men? in which he says, "When you see good men and such as are pleasing to God laboring, sweating, mounting the steep road, but evil men wanton and welter in pleasures, consider that we are pleased with the modest be-

* *Ibid.*, II, 30. Bacon substitutes *imitari* for *mirari* in this quotation.

havior of our sons, but with pertness of house-born slaves: that the former are restrained by a more rigorous training, and the latter are encouraged in their pertness. God treats a good man as a favorite, he tries him, hardens him, prepares him for himself. No evil can happen to a good man; opposites do not mingle. As so many streams, so great a quantity of rain water, and such an abundance of medicinal springs do not change the taste of the sea, do not even alter it; so the shock of adversity does not unsettle the spirit of a good man. He remains unmoved and gives to every event the coloring of his own character: for he is superior to all external happenings. I do not say that he is insensible to these things; but he conquers them, and his quiet and placid nature is supported against all adversity. He considers all adverse things as part of his training. . . . Virtue grows feeble without an adversary. Then are its greatness and its ability apparent when it shows its power of endurance. Whatever happens men should take in good part and should turn into good. It is not what you bear but how you bear it that matters. Do you not see how differently fathers and mothers show their affection? The former bid their children rise early and begin their studies, do not permit them to be idle on holidays, and draw the sweat from them, sometimes tears. But mothers desire to fondle their children and keep them in the shade: they are eager to save them from tears, labor, sadness. God views good men as a father, and loves them intensely, and says, Let them be trained by hardships, by pains, by losses to develop true strength. Animals fattened by inactivity grow weak: they succumb not only from labor, but from movement and their own weight. A happiness that is unimpaired can endure no blow. But after waging unceasing conflict with his own misfortunes, a man grows callous from the injuries he receives, he yields to no evil, but even if he falls, he still fights on his knees. . . . Behold two champions worthy the attention of a god, a brave man matched with an evil fortune."*

"Nothing seems to me more unfortunate than the man who has never been touched by adversity. For he has never had an opportunity to make trial of himself. . . . He is not considered fit to win a victory over fortune; as though fortune were to say, 'Am I to take that man as my opponent? He will lower his arms

* *De Providentia*, I, II.

at once. There is no need of exerting my full power against him . . . he cannot sustain my look . . . I am ashamed to fight with a man who is ready to be conquered.' A gladiator considers it a disgrace to be pitted against an inferior, and he knows that an opponent is conquered without glory who is conquered without danger. Fortune reasons in the same way: she seeks the bravest as her antagonists, she passes by certain with disgust, she attacks those that are most steadfast and upright, that against such she may direct her force. It is only adversity that discovers the shining example. . . . The greater the suffering the greater will be the glory: . . . prosperity falls to the lot of the common herd and to men of mean abilities."*

"But it is characteristic of a great man to overcome misfortunes and terrors common to mortals. But he who is always fortunate and passes his life without vexation of mind is ignorant of the other side of nature. You are a great man; but how do I know it if fortune does not give you an opportunity of showing your excellence? . . . I consider you an unhappy man because you have never been unhappy. You have passed your life without an opponent. No one will know what you could do; not even you yourself. For you cannot know yourself without trial. No man has learned his own ability without effort. Therefore some have voluntarily exposed themselves to adversity, which seemed to have forgotten them, and have sought for their virtue, destined to pass without notice, an opportunity to display itself. Great men," he says, "rejoice sometimes in adversity like brave soldiers in wars. . . . Courage is eager for danger; it thinks of its goal, not of what it is to suffer on the way: since that also which it is called upon to suffer is part of the glory. Soldiers take pride in their wounds; and joyfully display their blood flowing from wounds that are not mortal. Although those who return unscathed from battle have met the same danger, yet he who returns wounded, receives greater admiration. . . . You perceive the quality of the pilot in the storm, of the soldier in the battle. How am I to know what fortitude you possess against poverty if you are wallowing in riches? How am I to know your steadfastness when you are the object of ignominy, calumny and popular hatred if you are growing old amid plaudits? . . . Do not be afraid of those things that God applies as goads

* *Ibid.*, III.

to our characters: misfortune is virtue's opportunity. We may rightly call those wretched who grow torpid from excessive good fortune, whom as on an unruffled sea a lifeless calm detains . . . adversity presses harder on such as have never experienced it; the yoke is heavy on tender necks. The recruit pales at the thought of a wound; the veteran views his own blood undismayed; he knows that he has often been victorious after his blood was shed."

"God therefore hardens . . . trains those whom he esteems, whom he loves. But those to whom he seems indulgent he is keeping soft for evils yet to come. You are in error if you believe any one exempt: his own turn will come to the man who has long enjoyed prosperity. He who seems exempted is merely deferred. Why does God afflict all good people with ill health or grief or other misfortunes? Because in camp also the bravest are commanded to incur the greatest dangers. The general sends his finest troops to waylay the enemy by night. . . . No one of those who march forth says 'The general has done me an injustice'; but each one, 'He has a good opinion of me.' Those should say the same thing who are bidden to suffer afflictions that bring tears to timid and mean-spirited souls. Shun luxury, flee from good fortune weakening to our moral fibre, in which our souls become steeped and, unless something happens to warn us of our human lot, we remain in a state of torpor resembling a lasting intoxication. . . . Since all things that have passed beyond a due measure are harmful, excessive good fortune is most dangerous. . . . God follows the method in the case of good men that teachers employ with their pupils, exacting more work from such as are more promising. . . . Never is virtue's apprenticeship a mild one. Fortune lashes and lacerates us. Let us be patient: it is not mere brutality, it is a contest: the oftener we enter it the stronger we become. That part of our body is most robust which we have exercised most. We must offer ourselves to fortune, that we may be hardened against her by her own training. Gradually she will make us a match for her. Constant jeopardy will give us a contempt for dangers. Thus are sailors' bodies hardened from enduring the buffeting of the sea; thus do the hands of farmers become worn, the arms of soldiers acquire strength to hurl their weapons, the limbs of runners become agile; that member in each is strongest which he has exercised. The mind by endurance arrives at a point where it can despise the power of misfortunes. Why wonder that good men are shaken, that they may become steadfast. The tree is not firm nor strong unless it is subject to frequent assaults of the wind. For by the mere shaking it becomes firmly placed and more surely rooted. Fragile are the trees that have grown in the sunny valley. It is therefore for the advantage of good men themselves, in order that they may be undaunted, to be trained amid adversity and to bear with equanimity those things which are not ills except to the man bearing them ill."*

"This is God's purpose which he would show to the sage. . . . Good men labor, they sacrifice and are sacrificed and that too of their own free will; they are not dragged along by fortune, they follow her and keep pace with her. If they had known, they would have preceded her. . . . I can utter only one compaint in regard to thee, O God; thou didst not make thy will known to me before; for I would have taken up those tasks earlier which I am assuming after being summoned. . . . Dost thou wish some part of my body? Take it. I am offering no great thing. I shall soon leave it all behind. Dost thou wish my spirit? I shall not hinder thee in receiving again what thou hast given. Thou shalt receive from a willing hand whatsoever thou askest. What then? I should have preferred to offer it rather than to hand it over. We have received that which is as perishable as we are. Always cheerful and steadfast let us believe that nothing which we lose is ours. What is a good man called upon to do? To offer himself to fate. It is a great consolation to be borne away with the universe. . . . Fire tests gold; misfortune good men."†

"Yet why does God permit any evil to befall good men? He removes all evils from them, crimes, disgraces, wicked thoughts, covetous designs, secret lust, and avarice threatening the property of another. God protects and sets them free. . . . Democritus cast away his riches, considering them as a burden on a virtuous mind. What then? Are you surprised if God permits that to happen to a good man which a good man at times wishes to happen to himself? . . . Imagine, therefore, that God says to you, what complaint have you to make of me, you who have

* *De Providentia,* IV.

† *Ibid.,* V.

chosen virtue? I have surrounded others with false blessings, and I have mocked their empty minds with what I may call a long and deceptive dream; I have adorned them with gold and silver and ivory; within they are destitute of good. Those men whom you look upon as fortunate, if you see them on the side they hide and not on that which they show, are wretched, sordid, base, after the similitude of their own walls adorned on the outside. Such felicity has no firm basis and exists in appearance only. It is only a shell and a thin one at that. Therefore while they are permitted to stand and appear to men as they themselves desire, they are brilliant figures and impose on their fellows. When something happens which disturbs and discloses them, it then becomes apparent how much deep and real deformity is hidden by a splendor not their own. The blessings I have given you are sure and lasting, the more you consider them and investigate them in every way, the better and greater will they appear. I have permitted you to despise what other men fear and to disdain what they wish. You do not shine on the outside, your excellence is within. Thus the universe despises all outside of itself, and rejoices in its own spectacle. Every blessing has been placed within. Your happiness consists in not having need of it. But you will say many things happen of a sad and dreadful nature and hard to bear. Because I was not able to withdraw you from them, I have armed your minds against them all. Bravely endure them; in this you are superior to God; he is exempted from the endurance of evils; you endure and triumph over them. While life is prayed for, learn of death."*

Chapter XI

Seneca, moreover, in his book to Serenus on the topic, Why a philosopher is affected by neither injury nor insult, maintains that a philosopher cannot receive injury or insult. And he proves this by the statement, "I have not meant to attribute to the sage an imaginary honor, but to place him in such a position that he can receive no injury. What then? Will there be no one to attack him, no one to disturb him? There is nothing in nature

* *De Providentia,* VI.

so sacred as not to be subject to sacrilege: but divine things on this account are not on a less elevated plane. . . . That is not invulnerable which is not struck, but that which is not injured: this I tell you is the distinguishing mark of the sage. Is it not certain that you can depend more on the strength that is not conquered than on that which is not assailed, since untried powers are doubtful, but you rightly regard that strength of character as most reliable which successfully resists all assaults? Thus you may rest assured that the sage is of a finer nature if he is harmed by no injury, than if no injury happens to him. I shall call that man courageous whom wars do not subdue nor hostile force terrify when directed against him; I do not call the man courageous who enjoys a luxurious ease among indolent people. This therefore I maintain, that the sage is subject to no injury. Accordingly it matters not how many shafts are hurled at him since he is invulnerable to them all. He resembles the hardness of certain stones which does not yield to iron. The diamond cannot be cut or marked or worn away, but actually dulls every tool applied to it. He may be compared to certain substances that cannot be consumed by fire, but even when enveloped in flame retain their rigidity and form; or to rocks projecting out into the deep which break the waves nor do they show any traces of the sea's violence although they have been lashed for so many ages."*

"Thus is the mind of the sage firm and gathers such strength, that it is as safe from injury as the objects to which I have compared him. . . . For the sage is too far removed from contact with baser things for any hostile violence to expend its force upon him. All the attacks of shamelessness, ill will, and pride directed against the sage are made in vain."†

"Let us distinguish, if you approve, Serenus, injury from insult. The former is more severe in its nature, the latter is milder and grievous only to weaklings. Men are not harmed by insult, only offended. Such are the weakness and vanity of some men's minds, that they consider nothing more bitter: thus you will find a slave who prefers to be beaten with a whip rather than slapped in the face, and who considers death and stripes more tolerable than words of insult. We go so far in our folly

* *De Constantia,* III.

† *Ibid.,* IV.

as to afflict ourselves not only with pain but even with the thought of it. We are like children who are afraid of a shadow or the ugliness of masks or an ill-favored face. . . . Injury has the intention of harming some one. But wisdom leaves no place for harm. In the sight of wisdom there is only one thing that harms, namely, the baseness of sin, which cannot enter where virtue and integrity are already established. Every injury is a loss to him who is subject to it, nor can any one receive an injury without some loss in prestige, in person, or in external interests. But the sage can lose nothing. For he has stored away his all in himself, he trusts nothing to fortune, he has his possessions in safety, content with virtue, which has no need of fortuitous things and therefore can be neither increased nor diminished. For virtue in possession of the highest good can increase no more, and fortune snatches away nothing except that which it has given. But fortune does not bestow virtue, and therefore it does not take virtue away. For virtue is free, inviolable, immovable, unshaken. Virtue so hardens a man against misfortunes, that he cannot even be bent, much less be conquered. . . . Virtue does not lower its eyes nor change its expression at the sight of either adversity or prosperity. Therefore the sage will never experience a loss of which he will be sensible. For he is in possession of virtue alone, of which he can never be deprived. He uses all else as though merely borrowed: but who is disturbed at the loss of what is not his? But if injury can harm no one of the sage's possessions, because they are guarded by his virtue, injury cannot befall the sage. King Demetrius, surnamed Poliorcetes, had captured Megara. When Stilbon the philosopher was asked by Demetrius whether he had lost anything, he replied, 'Nothing; I have all my possessions with me.' And yet his patrimony had been plundered, his daughters carried off by the enemy, his country reduced to a foreign yoke, and he himself questioned by a king enthroned on high and surrounded by the arms of a victorious army. But the philosopher robbed the king of his victory, and bore witness that, although his city had been captured, he himself was not only unconquered, but had suffered no loss: for he had in his possession the true riches, on which no man could set hand. But those things that had been pillaged and destroyed, he did not consider his, but extraneous and subject to the caprice of fortune; therefore he had not set his heart on these things as though they were his. For the possession of all that comes to us from without is hazardous and uncertain. The sage looks upon adversity . . . and bears prosperity with moderation, neither yielding to the former nor relying on the latter. He is one and the same under all circumstances, nor does he reckon anything his except the better part of himself. . . . Those defences which guard the sage are protected from flames and from every assault; they offer no entrance, they are lofty, impregnable, equal to God!"*

"You may not make your usual statement that this sage of ours is nowhere found. Such a beautiful human character is no mere fiction; nor are we imagining some exalted being who has no existence, but one whose existence is an assured fact and has been proved by us . . . seldom found perhaps, and then only after a lapse of many years. For great and exceptional phenomena do not frequently occur. For the rest I am afraid that Marcus Cato is even superior to our model sage. Finally, that which injures must be stronger than that which is injured. But wickedness is not stronger than virtue: therefore the sage cannot be injured. It is only evil people that try to injure the good. Good men live at peace with one another. Evil men are not so harmful to the good as they are to one another. But if only the weaker man can be injured, but the evil man is weaker than the good . . . injury cannot befall the sage. For you need not be reminded that no one is good except the sage. If Socrates, you say, was condemned unjustly, he received an injury. It is necessary to observe here that it may happen that some one does me an injury and I do not receive it; as, for example, if a man place in my city house an article he has stolen from my house in the country, he has committed a theft, but I have suffered no loss. . . . If a man has intercourse with his own wife under the impression that she is the wife of another, he will be an adulterer, although she will not be an adulteress. Some one has given me poison, but it has lost its efficacy owing to the fact that it has been mixed with food; by administering the poison he rendered himself guilty of a crime, even though he did no harm. . . . All crimes are accomplished facts, so far as their guilt is concerned, even before they have produced their

* *De Constantia*, V, VI.

effect. . . . If I have received an injury, one must of necessity have been done me. If an injury has been done, it does not necessarily follow that I received it; for many things can happen to ward off injury."*

"Moreover justice can suffer nothing that is unjust, because contraries cannot unite; . . . therefore injury cannot befall the sage. Nor is there any reason why you should be surprised at this: if no one can do him an injury, no one can confer even a benefit upon him. The sage lacks nothing that can be conferred upon him as a gift. . . . No one therefore can either harm or benefit the sage; since what is divine neither craves aid nor can suffer injury. Moreover, the sage closely approaches God, is like God with the exception that he is mortal. Striving to mount to those beings on high, well ordered, inaccessible to fear, ever moving in an equal and harmonious course, tranquil, beneficent, created for the general good, salutary to himself and to others, the sage will set his heart on nothing lowly; he will suffer no affliction who supported by reason passes through human vicissitudes with a divine courage. There is no source from which he may receive an injury. Do you think that I refer only to an injury received from man? I mean not even one received from fortune which after a contest with virtue has always been worsted. . . . And if the sage bears with equanimity the injuries of fortune, how much more will he endure those of men in power whom he knows to be the instruments of fortune?"†

"Now there is the added fact that no one receives an injury with an untroubled mind, but is perturbed when he experiences it. But the upright man is free from perturbation. He is master of himself‡ and of his deep and calm repose. For if an injury touches him, he is disturbed and hindered thereby. But the sage is free from anger which the appearance of an injury excites, nor would he be free from anger unless he was also free from injury, which he knows cannot happen to him. For this reason he is so firm and so happy, for this reason constantly elated with joy. So far is he from being saddened at the attacks

* *De Constantia,* VII.

† *Ibid.,* VIII.

‡ The reading *moderator sui* omitting *erroribus* seems to give better sense than the reading followed by Bacon.

made on him by men and things, that he views an injury as a means of experimenting on himself and testing his virtue."*

"After having passed through the first part, let us pass to the second," namely, insult. "An insult is less serious than an injury: we can complain of it rather than exact punishment for it, and the laws have not considered it worthy of a penalty. The resentment which it excites is the effect of an abject mind, which an action or a word is capable of dishonoring. A certain man refused to receive me at his home today, although he admitted others: he listened with a disdainful air to that which I had to say, or he openly laughed at it . . . and other affronts of this kind, which I can call only the complaints of a squeamish mind. Those affected by such insults are as a rule men who are pampered by luxury and prosperity. Characters weak and effeminate in nature owing to too much ease, and wanton from the lack of real provocation, are aroused by affronts, the greater parts of which are taken as such because of a faulty interpretation of their intent. Thus he who is affected by an insult shows that he possesses neither prudence nor self-confidence. For he does not doubt that he has been scorned, and this sentiment cannot fail to lower and depress him in his own estimation. But the sage is despised by no one. He knows his own greatness and is convinced that no one is in a position to despise him, and does not overcome all these—I shall not call them miseries of our minds but rather plagues, nay, he does not even feel them. There are other things, however, which do assail the sage, even if they do not overthrow him, as, for example, physical pain and debility, or the loss of friends and children, and the ruin of his country in the flames of war. I do not deny that the sage feels these misfortunes. For we are not attributing to him the hardness of stone or iron. There is no virtue in supporting misfortunes one does not feel. In what then does the virtue of the sage consist? He receives certain blows, but overcomes those received and heals and checks them. But these minor ills, namely insults, he does not even feel, nor does he employ against them that virtue with which he is accustomed to endure hardships, but he either does not notice them, or judges them worthy of a laugh merely."†

* *Ibid.*, IX.

† *Ibid.*, X.

"Moreover since most of our affronts are offered by proud and insolent men, who are unable to stand prosperity, the sage can repel such pride with that noblest of virtues, magnanimity. He passes over such trifles, regarding them merely as idle dreams and visions of the night without reality. . . . *Contumelia* [insult] comes from *contemptus* [contempt], since men insult only those for whom they have contempt. But no one really despises a greater and a better man, even if he comports himself in the same way as those who do despise. For children slap a parent's face, and the child disarranges and pulls the hair of the mother, and spits on her, or exposes in the sight of its parents what should remain covered, and uses obscene expressions. We do not call any one of these acts an insult. Why? Because he who has committed them is not in a position to despise. . . . But what folly it is at one moment to be pleased, and at another to be vexed by the same actions, and to call a remark made by a friend an affront, but if made by a young slave a droll bit of repartee?"*

"The same feeling that we have toward children the sage has toward all who even after youth and gray hairs still remain like children. Have those men secured any advantage, whose minds are evil and whose errors have grown greater, who differ from children merely in the size and shape of their bodies, but are not less inconstant and unreliable, men who are eager for pleasures without discrimination, restless beings and kept quiet not by their nature but by fear? No one would say therefore that there is any difference between such men and children, since the latter display avarice for knuckle-bones, nuts, and small change, while the former are greedy for gold and silver and cities. . . . Therefore children and those more advanced in years are equally in error, but with the difference that the latter fall into other and greater evils. Rightly therefore does the sage regard their affronts as mere pleasantries. And sometimes he admonishes and punishes them as though they were children, not because he has suffered an injury, but he corrects them because they have inflicted one, and with the idea that they will cease to act in this way. For so also are cattle mastered with the lash, nor do we become angry at them when

* *De Constantia,* XI.

they refuse to be ridden, but we curb them, in order that the pain may conquer their obstinacy. Therefore you will perceive that the objection has been removed . . . why, if the sage has not suffered an injury or insult, does he punish those who have inflicted them? For he is not avenging himself, but he is correcting them."*

"Does a physician become angry at an insane patient? . . . The sage has the same attitude of mind toward all men as the physician has toward the sick under his care. The sage knows that all these grave personages who pass before him clad in purple and whose visage proclaims good health are of unsound mind. He regards them in no other light than as sick men lacking self-control. Consequently he does not even become vexed if in their sickness they have ventured to illtreat him who is trying to cure them, and he shows the same good sense in reckoning as equally unimportant both their dignities and their depravities. As he will not be flattered at the attentions of a beggar, nor will he regard it as an insult if a man from the dregs of the people does not return his salutation, so will he feel no pride if many rich men pay him honor. For he knows that they differ in no respect from beggars, nay, that they are more wretched. For the former need little, while the needs of the latter are great. And again the sage will not be affected if the King of the Medes or Attalus King of Asia passes him in silence and with a haughty look as he bows. Am I to be vexed if one of those . . . whose shops are filled with a horde of the worst kind of slaves, has failed to greet me by name? No, without doubt. . . . Therefore the sage will not be affected by the insult of any one. For they all differ from one another, but the sage considers them on an equality owing to the equality of their folly. For if he should once lower himself to the point where he is affected by injury or insult, he can never be secure. But security is the sage's particular blessing, nor will he consent to pay honor to the man who has insulted him by feeling that an insult has been inflicted upon him. For one must necessarily value the esteem of the man whose contempt hurts."†

"But what is the sage to do if he is slapped in the face? He

* *Ibid.*, XII.

† *Ibid.*, XIII.

is to do what Cato did when he was insulted in this way; he did not fly into a passion, he did not seek vengeance, he did not even pardon the injury, but said that it had not been done. He displayed a greater elevation of soul in disavowing the injury than in pardoning it. For who is ignorant that the sage does not hold the ordinary opinion with respect to what is believed to be evil or good? . . . He pays no heed to what men in general regard as disgraceful or wretched. But as the stars hold their way in a direction opposite to that of our world, so the sage moves forward in opposition to popular opinion."*

"After you have calculated what you think you can endure, you assign a slightly further limit to the patience of the sage, but his virtue has placed him in another world where he has nothing in common with you. This being granted, the throng of hardships and difficulties hard to bear and of things revolting to the ear and the eye will not overwhelm the sage; he will offer the same resistance to all as to each in particular. He who pretends to determine that which the sage can or cannot endure, he who limits the grandeur of the sage's soul within certain bounds, is doing wrong: fortune vanquishes us unless she is completely vanquished. . . . The home of the sage is small, without luxury, without uproar, without sumptuous equipment, is not guarded by janitors; but over this threshold tended by no doorkeepers fortune does not pass. She knows that she has no place there where nothing is hers."†

"Epicurus said that injuries are supportable by the sage; we say that he suffers no injuries. Nor may you say that this is repugnant to nature. We do not deny that it is a disadvantage to be lashed and to be struck, and to lose one of our members; but we refuse to consider all those things injuries. We do not take from them a sense of pain, but we do remove the name of injury, which cannot be accepted without impairing virtue. . . . We are taught by both these examples to despise injuries and insults which I may call the semblances and suspicions of injuries. To disdain such one need not be a sage but merely a man of discernment. . . . If I have deserved that these things should befall me, I am justly punished. If I am not guilty, he

* *De Constantia,* XIV.

† *Ibid.,* XV.

should blush who commits an injustice. . . . The fruit of an insult consists in the sentiment of indignation which it excites in him who is subjected to it."*

"Let us take as our models those whose patience we praise. Let us imitate Socrates who took good-naturedly the sarcasms launched at him in the comedies, and laughed as merrily as he did on the occasion when he was doused with dirty water by his wife Xanthippe. When Antisthenes was reproached with the fact that his mother was a barbarian and a Thracian, he replied that the mother of the Gods was born on Mount Ida."†

"We must not engage in quarreling and contention. . . . Whatever is done by foolish people must be disregarded, and honors and injuries having their origin in the vulgar throng are to be viewed in the same class; we are not to grieve over the latter nor be elated at the former. . . . Moreover liberty does not consist in exemption from suffering. Liberty is a superiority of the mind to injuries."‡

Chapter XII

Seneca in his book addressed to his mother Helvia speaks about the remedies not only for grief and suffering, but also those for poverty, exile, insult, and many other misfortunes of this kind. Therefore he proposes the following remedy to her on the occasion of an affliction which she found hard to bear: "Accordingly I shall exhibit to that heart all its own griefs and bereavements. This will be effecting a cure not by gentle means, but by burning and cutting. What end have I in view? That the mind victorious over so many sorrows may be ashamed not to bear bravely a single wound on a body marked by the scars of so many. Let us leave therefore tears and lamentations to those weak minds, so enervated by a long period of prosperity, that the least shock of adversity overthrows them; but let those whose whole life has been spent in adversity, endure griefs even the most severe with a firm and unalterable con-

* *Ibid.*, XVII.
† *Ibid.*, XVIII.
‡ *Ibid.*, XIX.

stancy. Constant misfortune has the advantage of finally hardening those whom it is ever assailing."*

"As recruits cry out at trifling wounds and are more afraid of the hand of a physician than of a sword, but veterans although seriously wounded permit their wounds to be treated with patience and without a groan, as though their bodies belonged to other people." . . . and therefore he recounts to his mother all the misfortunes she has patiently endured, in order that he may remove her grief for her son.†

Then Seneca wishes to console her by his own example as follows: "But I shall overcome my grief, I believe, first if I shall show that I am suffering nothing that can render me an object of pity . . . I say that I am not unhappy. I shall add to relieve your anxiety that I cannot even be made unhappy." And he proves his statement as follows: "We are born under conditions which would be favorable if we do not abandon them. Nature's intention was that there should be no need of great magnificence for life to be happy. Every one can make himself happy. The power to turn the balance is slight in external things and it exerts little influence one way or the other; the sage is not carried away by prosperity nor cast down by adversity. For he has always striven to rely entirely upon himself, and to seek all his joy within himself. What then? Do I claim to be a sage? By no means. For if I were able to make such a declaration, I should not only deny that I am unhappy, but I should declare myself to be the most fortunate of men and almost the equal of God: as it is, I have done that which suffices to soften all asperities, I have put myself in the hands of wise men, and, too feeble for my own defense, I have fled to the camp of those who defend with ease themselves and their friends. They have bidden me stand constantly as though on guard and to foresee all attempts of fortune and all her assaults long before they happen. She is dangerous to those who do not expect her: her attack is easily sustained by him who is always on the lookout for it. For the arrival of the enemy dismays those only whom it has taken off their guard; but those who have made due preparation beforehand for war, calmly and easily sustain the first assault which is the most violent. I

* *Ad Helviam Matrem,* II.

† *Ibid.,* III.

have never trusted fortune even when she seemed to bring peace. All those things that her favor conferred upon me, money, honors, influence, I have placed where she could recall them without moving me. I have left a great interval between them and me. Therefore she has taken them from me, she has not torn them away. Adverse fortune never crushes a man unless good fortune beguiles him first. Those who have set their heart on her gifts as though they were their own permanent possessions, who have wished to be looked up to because of them are humbled and distressed, when their empty and childish minds, ignorant of all stable happiness, are deprived of deceitful and transient pleasures. But the man who has not inflated himself with prosperity also does not collapse when fortune changes; toward both conditions he possesses a mind of tested firmness: for in the midst of prosperity he has tested his own strength against adversity. For myself I have never believed that true happiness is to be found in those things which all men desire: besides I have found them empty . . . and smeared with a deceptive rouge, with nothing on the inside corresponding to their exterior."*

Later on he takes up the consideration of particular ills, namely, exile, poverty, insult, as follows: "Now in these things which are termed evils I find nothing so terrible and so hard to bear as the popular idea led me to apprehend. . . . But the decisions of the people in a great measure are repealed by the sages."

"Disregarding then the judgment of the majority whom the outward appearance in such matters without further examination leads astray, let us see what exile really is: it is to be sure a change of place. Let me add, that I may not seem to be narrowing its effect, and to be relieving it of its worst features, many disadvantages are attached to this change of place: poverty, ignominy, contempt. . . . I wish to consider in the first place what hardship mere change of place imposes."

"Consider, if you please, this throng for which the dwellings of an immense city (he means Rome) scarcely suffice: the greater part of this throng is deprived of their country. . . . Bid all these people be summoned and ask each where his home is: you will find that most of them have left their own

* *Ibid.*, IV, V.

abodes and have come to the greatest city in the world, but a city that is not theirs. Then leave this city which can be said in a sense to belong to all. Visit all other cities; there is not one of them that does not contain a large foreign population."*

His second reason is drawn from the nature of the soul. He says: "A restless and an unquiet mind has been given to man: it never remains at rest, it is ubiquitous, and directs its thoughts into all things known and unknown. The mind of man is always on the move, objects to inaction, and takes the utmost delight in constant novelty. You need not wonder at this, if you consider its origin; it is not formed from an earthly and sluggish body; its descent is from the pure celestial spirit. But perpetual motion is characteristic of the heavenly bodies which are borne along in their rapid course. Contemplate the stars that give light to the world; not one of them is stationary. . . . How idle then to suppose that the mind of man suffers with regret a transition and change of place, although nature either delights or preserves herself by a continual and very rapid change!"†

He draws his third reason from the experience of all nations. He says: "Turn now your thoughts from the heavenly bodies to the affairs of men: you will see that whole nations have changed their abodes. How can we account for Greek cities in the midst of barbarous lands? What of the speech of Macedonia among the Indians and the Persians? Scythia and that whole stretch of lands inhabited by savage and unconquerable races show Grecian cities planted on the shores of the Black Sea. The whole side of Italy that is washed by the Tyrrhene sea was greater Greece. Asia claims the Tuscans. The Tyrians dwell in Africa. The Carthaginians emigrated to Spain, the Greeks to Gaul, the Gauls to Greece. The Pyrenees offered no obstacle to the passage of the Germans. The wandering spirit of man has cut for itself roads through the unknown. Some after the trials of a long journey have selected an abode, yielding to their fatigue rather than making a deliberate choice. . . . This at least is manifest that nothing has remained in the same place in which it had its origin." He then returns to his

* *Ad Helviam Matrem,* VI.

† *Ibid.,* VI.

argument in the words, "But what else are all these migrations of nations but exiles on the part of peoples?"*

"Due allowance being made for all the other disadvantages that are inseparable from exile, Varro, the most learned of Romans, remarks as a sufficient consolation for change of place, that wherever we go we are bound to find the same nature. Marcus Brutus considers it sufficient that his own virtue accompanies the exile wherever he goes. If each of these methods of consolation, taken separately, does not suffice the exile, we can have no doubt of their efficacy when they are united. For how trifling is our loss? Two things that are most excellent will always accompany us, universal nature and personal virtue. But if he says that universal nature consists in whatever is best, this lies outside of the powers of man: it can neither be given nor taken away. I speak of this world, the grandest and most magnificent of the works of nature; of this mind made to contemplate and admire the world, of which it is the noblest part, ours as a lasting possession destined to remain with us as long as we ourselves shall last. Let us hasten then with alacrity and resolution wheresoever circumstances may direct. . . . From all places equally we lift our eyes to the heavens, all things divine are separated by equal intervals from all things human."†

"Narrow is the mind that earthly things delight," as quoted above from this book. "Brutus in his work on virtue says that he saw Marcellus in exile in Mitylene living as happily as human nature permits, and never more interested in the liberal arts than at that time. . . . Brutus adds that he felt when on the point of returning without Marcellus that it was he who was going into exile, and not Marcellus who was being left in exile. How much more fortunate was Marcellus when he merited the praise of Brutus because of his exile, than when he won the praise of the republic because of his consulship! . . . You so trained your mind, Marcellus, by methodical studies, that you knew that every place for the sage was his native land."‡

"Marcellus then supported exile well, nor did change of place produce any change in his mind, although poverty followed

* *Ibid.*, VII.

† *Ibid.*, VIII.

‡ *Ibid.*, IX.

upon it. No man considers poverty an evil, unless he has reached that point of madness which permits avarice and luxury to subvert all things. . . . The desires of the body are few: relief from cold, hunger, and thirst. He who desires more than these toils to serve, not his needs, but his vices. It is not necessary to ransack every sea, nor to burden the stomach by the slaughter of animals, nor to dig up shellfish from the unknown shore of a remote sea: may the gods and goddesses destroy those whose extravagance passes beyond the limits of an empire exposed to hatred. . . . From every land all things known to a pampered palate are brought to Rome. That which is admitted with difficulty by a stomach enfeebled by delicacies is transported from a remote ocean. They vomit that they may eat, and they eat that they may vomit, and do not deign even to digest feasts that have required a search of the whole world for their viands. What harm does poverty do the man who holds such luxuries in contempt? . . . Unhappy people"—as was noted above; and he continues; "I am disposed to say: Why do you launch ships? Why arm bands against wild beasts and against men? Why run hither and thither in such a commotion? Why pile treasures on treasures? Do you never stop to think how small your bodies are? Is it not the height of folly and madness, when you can hold so little, to desire a great deal? Therefore although you may increase your rating and extend your boundaries, you will never give your bodies a greater capacity. . . . Such is the lot of men when they measure wealth not by the standard of reason, which has fixed bounds, but by that of a debased habit of life, whose caprices are insatiable and incomprehensible. For cupidity nothing is sufficient, for nature even a little suffices."*

"Everything that nature has rendered necessary for man she has made easy for him to obtain. . . . That man is poor not through the fault of fortune but through his own. . . . He who accordingly keeps himself within the limits prescribed by nature, will not have a sense of poverty; he who disregards these limits will be pursued by poverty even in the midst of the greatest wealth." For "exile is" always "able to provide necessaries, but not even a kingdom suffices to supply things that are needless. It is the mind that makes men rich. This accompanies

* *Ad Helviam Matrem,* X.

the exile, and in the midst of the most cruel solitudes, after it has found enough to sustain the body, overflows with its own blessings and enjoys them; money has no more to do with the mind than with the immortal God. All these things . . . stones, gold, silver, and great circular tables are earthly burdens, which the unpolluted mind cannot love, remembering as it does its own nature, itself light, free from contamination, and ready to spring on high whenever released: meanwhile to the full extent permitted by hampering limbs and the burden that rests heavily upon us, the mind on the wings of thought traverses the realms of the gods. Therefore the mind can never be exiled, the mind that is free and related to God and equal to all space and all time. . . .This feeble body, the prison and the chain of the soul, is tossed hither and thither. It is against this that punishments, robberies, and maladies do their worst; the mind itself is sacred and eternal, and the hand of violence cannot be laid upon it."*

"Consider that the poor, who form the larger part of humanity, have not a more gloomy or anxious air than the rich; nay, I am inclined to think they are more joyous owing to the fact that their mind is distracted by fewer cares. . . . As for me, when I look back at instances in the past, I am ashamed to draw any consolation from poverty. Luxury has increased to such an extent in our day, that the provision made for exiles exceeds the patrimony of the chief men of the state in days gone by. We know that Homer had but one slave, Plato three, Zeno, founder of the unbending and virile philosophy of the Stoics, had none. Will any one say therefore that these philosophers lived in a wretched way, without seeming himself to be most wretched to all men on this account?"†

"The reply may be made: Why do you unnaturally separate these evils, which can be borne singly, but when united cannot be endured? Change of place is endurable if one change merely his dwelling-place. Poverty is endurable, if disgrace be not added, which even alone is wont to crush men's spirits. To him who seeks to terrify me with an aggregation of evils I must make this reply: The mind that is superior to any one phase of fortune will be superior to them all. The mind once hardened

* *Ibid.*, XI.

† *Ibid.*, XII.

by virtue is invulnerable to every assault. If avarice, a very deadly bane of the human race, has relaxed its grip on you, ambition will not cause you delay. If you do not regard your last day as a punishment, but as a law of nature, the fear of no other thing will venture to enter that breast from which you have driven the dread of death. If you consider that sexual desire was given man not for the sake of enjoyment but for the purpose of propagating the race, every other desire will pass by untouched him whom this ruinous passion deep seated in our very vitals has not injured. Reason does not triumph over the vices singly but over all of them together; her victory is general." And he continues in regard to disgrace: "Do you suppose the sage is disturbed by disgrace, who relies wholly on himself, and who is free from the opinions of the common herd? An ignominious death is more even than disgrace. Yet Socrates with the same look with which he had treated the thirty tyrants with disrespect on a certain occasion, entered the prison, destined to remove disgrace even from such a place. For that abode could not seem a prison in which Socrates was. . . . No man is scorned by another unless he has first been scorned by himself. A groveling and an abject mind is exposed to insult; but he who raises himself above the most cruel misfortunes and triumphs over evils by which other men are crushed, regards the very misfortunes as a badge of distinction, since," no doubt, "our disposition is such, that nothing challenges our admiration more than a man courageous in adversity. As Aristeides was being conducted to punishment in Athens, all who met him lowered their eyes and groaned at the fate, not of a just man, but of justice itself. One man, however, as he passed, spat in Aristeides' face . . . but Aristeides wiped his face off and with a laugh said to the magistrate who accompanied him: Warn that fellow not to yawn so offensively. . . . If a great man has fallen, he retains his greatness; he is no more an object of scorn than are the ruins of temples when trodden under foot. To such ruins the devout pay the same respect as to temples still standing."*

After giving his mother these remedies against adversity, he takes up again all the misfortunes that have befallen her, in order that he may teach her how one is victorious who has al-

* *Ad Helviam Matrem,* XIII.

ready vanquished so many ills by strength of character. He says accordingly: "But the harder these trials are, the greater the courage you must summon to your aid; it is an enemy known and often vanquished, against whom you must struggle with fresh ardor. Blood in this instance has flowed from a body that has already suffered wounds, through whose very scars you have been stabbed."*

Chapter XIII

Writing to Gallio on the remedies for misfortunes Seneca says: "You are going to die, death is natural to man and is not to be regarded as a punishment. I entered the world on the condition that I should leave it. It is foolish to fear what you cannot avoid. . . . You are going to die; I am not the first nor the last: many have preceded me, all will follow me. . . . Am I ignorant that I am a rational and a mortal animal?"†

"Men think evil of you. But they are evil men. Now to displease evil men is worthy of praise. An opinion can have no weight when he who condemns is himself worthy of condemnation. . . . Men are speaking ill of you. I should be disturbed, if they were acting with sound judgment: as the matter stands, their action is prompted by a diseased state of mind. They do not know how to speak well of a man. They act as they do, not because I deserve it but because such is their habit. . . ."‡

"You will suffer exile. You are mistaken: there is only one country for us all. I am not forbidden to dwell in my native land but in a certain locality. . . . No foreign land is a place of exile but simply another country. . . . A man's country is where he does well. But well-being is in the man himself, not in his location. . . . If he is a sage, he is simply dwelling abroad; if he is a fool, he is suffering exile."§

"Grief threatens. If it is slight, let us bear it: patience is a small matter. If our grief is severe, let us bear it, glory is not

* *Ibid.*, XV.
† *De Remediis Fortuitorum*, II.
‡ *Ibid.*, VII.
§ *Ibid.*, VIII.

a small matter. Grief is a hard thing. Nay, you are soft. Few have been able to endure grief. Let us be among the few."*

"Poverty is grievous to me. Nay, you are grievous to poverty. The fault is not in poverty but in the man suffering from it. Poverty is free from encumbrance, blithe, safe. . . . You are poor because you think you are. The birds lack nothing. The beasts of the field live for the day. . . . That man has received a great sum of money. Therefore pride also. . . . Do you consider him a man? He is a money-chest. Who is envious of a treasury or of full coffers? He is merely the coffer whom you consider the owner of the money. If he is prodigal, he will not have it; if he is a miser, he does not have it. He whom you reckon fortunate often grieves, often sighs. Many accompany him. Flies follow honey, wolves carrion, ants grain. That throng is after the booty not the man."† Death, exile, grief, pain, and the like are not the punishments but the tributes of life.

Chapter XIV

Consoling Marcia for the death of her son, Seneca writes as follows: "Grief renews and strengthens itself daily . . . and sorrow becomes the pleasure of an unhappy mind. Do not imagine it a great achievement to preserve a courageous bearing in the midst of prosperity when life moves forward propitiously. A calm sea and favoring wind do not display the skill of the pilot; something adverse must occur to test his courage. . . . The envy of fortune is aroused by nothing else to such a degree as by tranquillity of mind. The steersman of the ship is disgraced who permits the helm to be wrested from his grasp by the waves, who abandons the sails to the gale, and hands over the ship to the tempest. But we must praise him even in shipwreck whom the sea has overwhelmed while still retaining his ship and struggling to save it."‡

"Longing for one's own is natural. Who denies this, as long as it is within moderation? . . . But our feeling passes far

* *De Remediis Fortuitorum*, IX.

† *Ibid.*, X.

‡ *Ad Marciam*, I, V, VI.

beyond the requirements of nature. Consider how violent is the grief of animals and yet how brief. For one or two days the lowing of the cow is heard, and the frenzied wandering of the mare does not last longer. Wild beasts after following the tracks of their young, on their return to their plundered lairs allow their rage to subside within a short time. Birds with loud cries mourn over their empty nests, yet within a brief space of time they resume quietly their flight. Man is the only animal that suffers a prolonged grief for the loss of its young. . . . But to convince you that it is not natural to succumb to our griefs, consider that the same loss inflicts a deeper wound on women than on men, on barbarians than on cultured nations, on the ignorant than on the learned. But those things that have derived their force from nature retain the same. It is apparent that that which varies is not natural. Fire burns all ages, the citizens of all cities, men and women alike. Iron will display its power to cut in every body. Why? Because it has derived its powers from nature. . . . Each one in his own way feels poverty, grief, ambition, according to the impress of habit upon him: and according to a dreadful preconceived idea in regard to things that should not be feared, an idea that has rendered him feeble and incapable of endurance."*

"In the second place what is natural does not decrease with time: grief, however, is consumed by time. But you will say, 'I did not think that it would happen.' Do you think that something will not happen which you know can happen to many and which you see has actually happened to many? An excellent maxim and worthy of no common origin is the line,

That which can happen to anyone can happen to each.

That man has lost his children; you can also lose yours. This man has been condemned; and your innocent life is in imminent danger. This error deceives. . . . He robs present evils of their power who has foreseen that they were coming."†

"All this accidental splendor that surrounds us, Marcia, children, honors, riches, spacious halls, throng of clients, crowded vestibules . . . and all the rest depending on an uncertain and capricious fortune are the property of another and are merely

* *Ibid.*, VII.

† *Ibid.*, VIII, IX.

loaned to us. No one of these things is given us. They are like those borrowed decorations that embellish our theaters, which are to be returned to their owners, some on the first day, others on the second, while a few remain to the end of the play. Therefore there is no reason why we should admire these things as though we were located among our own possessions. The use and enjoyment of them is ours. It is God who gives these games, who determines the time of restitution. We must hold in readiness what has been given us for an indefinite time only, and when called upon must make restitution without complaint. It is only a very bad debtor who reviles his creditor. . . . The mind must often be admonished to love such objects as destined to leave us, nay, as now doing so. You are to possess the blessings of fortune without a permanent claim. . . . You have no promise for the coming night, none for the present hour. You must hasten; you are threatened from the rear. . . . All your possessions are subject to plunder; you are ignorant, poor wretches, that your life is only a flight."*

"Why should we weep over the details of life? It is deplorable as a whole. New misfortunes will press upon you, before you have finished with the old. . . . What is man? A fragile vase shattered by any jar. There is no need of a great tempest to shatter you. You will decompose under any shock. What is man? A weak and fragile body, naked, without natural defense, dependent on the aid of others, exposed to all the outrages of fortune . . . the prey and the victim of any wild beast . . . unable to endure cold, heat, toil. . . . An odor, a savor, insomnia . . . and food and the things without which man cannot live, prove deadly to him. . . . He is not capable of enduring every climate, change of water and the blast of an air to which he is not accustomed; in a word, the most trifling causes make him ill . . . meanwhile how great a commotion is stirred up by this animal so contemptible! In what great thoughts does he indulge, forgetful of his own condition! His projects are limitless, eternal, his plans include grandsons and great-grandsons. But in the midst of his great designs death overwhelms him. That which we call old age is merely the circuit of a very few years."†

* *Ad Marciam*, X.
† *Ibid.*, XI.

"As a rule great and lasting blessings do not fall to the lot of anyone. Good fortune does not last nor continue to the end unless it be moderate."*

And he introduces examples of patience in bearing the loss of sons. "Greece is not the only nation to be proud of a father who in the midst of a sacrifice on learning of the death of a son merely bade the flute-player be silent, removed the chaplet from his head, and duly completed the rest of the ceremony. Pulvillus the pontiff acted in a similar manner when the death of his son was reported to him while he was consecrating the Capitol. He pretended that he did not hear what had been told him, and repeated the solemn words of the ritual, without suffering a groan to interrupt his prayer."†

"Cornelia, daughter of Scipio, mother of the Gracchi, recalled her twelve children by as many deaths. . . . The Gracchi, to whom one refuses the title of good citizens but not that of great men, the mother saw slain and deprived of sepulcher. To those who offered her consolation and called her unhappy she said, 'I shall never call myself unhappy, I who have borne the Gracchi.' Octavia and Livia,‡ one the sister of Augustus, the other his wife, lost sons both in the flower of their youth, and both regarded as a future emperor. Octavia refused all consolation. . . . Through the rest of her life she gave way to her grief as she had done at the funeral. She would have considered it a second bereavement had she ceased to weep; she wished for no portrait of her beloved son, she did not permit his name to be mentioned in her presence. . . . Livia displayed the sensibility of a mother and the firmness of the wife of an emperor. She did not cease to celebrate the name of her son, to picture him in private and in public, to speak of him and to hear him mentioned with the utmost pleasure."

"Human life is assailed by evils of every kind; they never accord us a long peace, scarcely a truce. . . . It is in fact the nature of mortals that nothing is more pleasing than that which they have lost."§

"It is then opinion that torments us; it is we ourselves who

* *Ibid.*, XII.

† *Ibid.*, XIII.

‡ See *Ibid.*, II and III, where this passage belongs.

§ *Ibid.*, XVI.

place the estimate on our misfortune. The remedy is in our hands."*

"All the works of man are perishable and of short duration; they do not occupy any place in the immensity of time . . . whose increment however great will be negligible. There is only one way to live for a long time, that is to live sufficiently. . . . You will find no difference between the longest life and the shortest if you compare the time one has lived with that which he has not lived."†

"Do you complain, Marcia, that your son did not live so long as he might have lived? How do you know if a longer life would have been to his advantage? If his own interest did not require his death? What man can you find today whose interests are on so sure a foundation, that he has nothing to fear as time goes on? Human affairs are always subject to change, are never stable . . . and for this reason death is to be desired by those who are most fortunate, because in such a state of instability and in such a disordered condition of affairs the past alone is certain. Nothing is so deceptive as the life of man, nothing so insidious; no one would have accepted it, had he known the nature of the gift."‡

"Besides this since everything in the future is uncertain and threatening, the pathway to the gods is easiest for the souls that have been set free betimes from human intercourse. For they have contracted very little defilement and sluggishness, since they are set free before an earthly character is too deeply impressed. They fly back on lighter wing to their origin . . . a lengthy sojourn in the body is never pleasing to great minds."§ There are other passages also as noted above.

Chapter XV

I HAVE treated this third part of Moral Philosophy at greater length than I purposed at the beginning. But the beauty of the moral sentiments delights us; and especially are these princi-

* *Ad Marciam,* XIX.
† *Ibid.,* XXI.
‡ *Ibid.,* XXII.
§ *Ibid.,* XXIII.

ples brought out with great clearness of thought by the efforts of philosophers. And we should accept them with greater eagerness since we Christian philosophers do not know how to reason on the subject of morals, nor can we make so eloquent an appeal. Would that we might approve by our conduct those principles which the philosophers so wisely propound. For although in regard to the virtues bestowing grace, namely, faith, hope, love, and the like, we are able to perceive truths of which even the philosophers are ignorant, yet in the virtues that in general are requisite for integrity of life and for the intercourse of human society, both in word are we no match for them, and in deed are we less efficient, as is clearly shown by a consideration of the wisdom taught by them. This is quite blameworthy in us and worthy of the scorn of all men. It should therefore be necessary for Christian philosophers to consider the unclouded glory exhibited by the philosophers. Moreover, we are encouraged thereto by the examples of the sacred writers. For they diligently applied to divine things the teaching of the philosophers, and especially those principles pertaining to morals and to the rules of correct living.

But there is a particular reason for my lingering on these works of Seneca, since although I have searched for books of this kind from my childhood, I have never succeeded in finding them until the present time. I refer to the De Ira, Ad Helviam, Cur bonis mala accidunt, An in sapientem cadunt contumelia et injuria, Ad Marciam, and the three following works. Since I do not know whether these books have been seen by your Excellency, I have attempted to quote them here somewhat freely.

Chapter XVI

Seneca begins his consolation addressed to Paulinus on the shortness of human life as follows:

"Most men complain of the injustice of nature, because our life is short. . . . We do not have too little time, we waste too much of it. Life would be long enough and would suffice to complete the greatest enterprises, if it were properly arranged as a whole. But when life has been dissipated in extravagance and in neglect of opportunity, when it is devoted to no useful occupation, at length under the prodding of the last grim necessity, we perceive that life has passed without our perception of its passage. The fact is, we have not been granted a short life, we have made it so. As riches falling into bad hands are dissipated in a moment; while a modest fortune if entrusted to a good guardian is increased by his care; even so our life if well employed extends over a long period.*

"Life is long if you know how to use it. But one is possessed by an insatiable avarice, another devotes himself constantly to superfluous efforts. Still another is steeped in wine. And yet another stagnates from sloth. . . . The whole period is not life but is time. Vices surround and press upon us from every side nor do they suffer us to rise or to lift our eyes to examine truth, but they press us down plunged and held fast in our lust. We are never permitted to recover ourselves; when perchance we have some moments of respite, our passions never permit us to be perfectly tranquil; we are like the sea whose waves continue in motion even after the wind has ceased. That man is attached to such a one, while this man is attached to someone else. No man belongs to himself."†

"Though all the geniuses that have enlightened the world by common consent express their wonder at this one thing, they will never express adequately their astonishment at this blindness of the human spirit. No man is found who is willing to distribute his money; among how many does each of us distribute his life! They are careful in guarding their patrimony; they are prodigal when it comes to wasting their time, the only thing in which avarice is laudable. . . . This is your hundredth year or more. Come, reckon up your life. Tell me how much of this time the creditor, how much the mistress, the patron, the client, have consumed; how much in the coercion of slaves, how much in running about to oblige your friends? Add the diseases we have brought upon ourselves. Add the time also of which we have made no use: you will see that you have fewer years than those you count. Recall how many times when you were firm in your resolution . . . when you made use of your own self, when your face remained serene and your spirit undaunted, . . . how many have pillaged your life without your

* *De Brevitate Vitae,* I.

† *Ibid.,* II.

perceiving what you were losing . . . you will perceive that your death is premature. What then is the reason? You live as though you expected to live always. Your frailty never occurs to you. Like mortals you fear all things, like immortals you desire all things. You will hear many say, 'At fifty I shall retire! My sixtieth year will free me from business.' . . . Who will permit those plans of yours to mature as you now arrange them? Are you not ashamed to reserve for yourself only the remnant of your life, and to destine for the good of your soul only that time which you can employ in no other way? How late it is to begin to live at a time when we must cease our existence . . . to postpone sane counsels, and to desire to begin life at a limit few have attained."*

"You will hear the most powerful and exalted personages say that they desire repose; you will find that they praise it and prefer it to all the blessings they enjoy. They wish meanwhile to descend from their lofty position if they can do so with safety. For fortune topples down upon herself although neither assailed nor shaken by an external force. The deified Augustus, on whom God bestowed more than upon any other prince, prayed without ceasing for repose and sought for relief from the affairs of state. All his conversation always returned to this one theme, his hope for rest. With this sweet yet deceptive solace he was wont to console his labors. . . . But those things can be accomplished with more glory than promised. . . . Repose seemed so desirable that, not being able to enjoy it in fact, he pictured it in imagination. He who saw all things dependent on himself alone, who regulated the destinies of men and of nations, regarded the day on which he should lay aside his greatness as the happiest in his life. For experience had taught him how much effort those benefactions cost which spread such luster throughout the world, how much disquietude they covered,"† but did not remove.

"Marcus Cicero . . . while driven hither and thither with the Republic and while endeavoring to keep it from going to ruin . . . how many times had he reason to curse that consulship of his, which he had praised not without cause but without end? How tearful the words he utters. . . . It is useless to recall

* *Ibid.*, III.

† *Ibid.*, IV.

here the examples of many who, reputed to be very happy, have borne true witness against themselves in their utter hatred of every act of their lives. . . . Though your life extend beyond a thousand years, it will be reduced to a very short span."*

"In the first place then I number those who have leisure for nothing except wine and lust. For none are occupied more disgracefully. As for the rest although they are possessed by a vain idea of glory . . . they err in a more manly way; it is a disgraceful wasting sickness that afflicts men given up to wantonness and lust. . . . Finally all are agreed that nothing can be handled properly by a man preoccupied with other matters; one cannot cultivate eloquence or the liberal arts, when his mind is distracted by other interests and is incapable of receiving any deeper impression, and rejects all that is forced upon it. Nothing is less in the possession of the preoccupied man than his own life; there is no science more difficult than that of life. All the other arts have many prepared to teach them. Mere boys seem to have mastered some of them to such a degree that they could even teach them: we require the whole of life to learn how to live; and what will perhaps surprise you still more, we require the whole of life to learn how to die. So many very great men, after freeing themselves from all hindrances, after renouncing riches, occupations, and pleasures, up to the very end of their lives have tried to learn how to live. Yet many of them departed this life confessing that they were still ignorant of this art. . . . It is the mark, believe me, of a great man who is raised far above human errors to permit none of his time to be wasted. His life therefore is a very long one, since whatever its span, it has been wholly at his service. No part of that time was unemployed or wasted in idleness, no part was under the control of another. A rigid economist of his time, he finds nothing worthy to be received in exchange for it. Therefore his allotted time sufficed him: but of necessity time failed those, from whose life the public took a large portion," and their vices robbed them of still more. . . . "Count up, I say, and examine closely the days of your life; you will see that very few of them remained at your own disposal. . . . But he who employs some of his time to his own advantage, who arranges all his days with a view to life itself, neither wishes for the mor-

* *De Brevitate Vitae*, V, VI.

row nor fears it: he actually directs fortune as he wills; his life is in safety. Nothing can be added to him, nothing taken away. Such an addition is like a portion of food given to a man already sated and filled; he receives that which he does not desire. You should not accordingly think that a man has lived a long time because of his gray hairs and wrinkles; he has not lived a long time, but he has existed a long time. What then! Are you to think that he has made a long voyage whom the savage gale caught on leaving the port and bore hither and thither? Or likewise he whom shifting and raging winds have driven around in a circle over the same spot? Such a man has not made a long voyage, but has been much buffeted."*

"I am constantly surprised to see people asking for the time of others; I am also surprised at those who grant requests of this kind with such readiness. Both regard the motive of the request, neither pays heed to the thing requested: as if nothing is asked for, as if nothing is given, the most precious thing in the world is made a jest of. It deceives them because it is incorporeal, because it does not come under their eyes." . . . But "no one puts a value on time. They employ it as a thing of no value. But behold these same men, when they are sick and believe themselves in danger of death, embracing the knees of their physicians; if they fear capital punishment, behold them ready to expend their all, in order that they may live; so great is the discord of their passions. But if each of us could know the number of his years in the future as he does the number in the past, how would those of us tremble who saw that they had few left, how sparing they would be of them? And yet it is an easy matter to regulate what is certain even though it be small: we should guard with greater care that which may fail us at any moment."†

"They arrange their life at the expense of life. They direct their thoughts far afield. Postponement is responsible for life's greatest loss . . . it robs us of our present while promising what is more remote. Life's greatest hindrance is the expectation which depends on tomorrow and loses today. You make disposition for that which has been placed in the hand of fortune; you lose that which has been placed in your own. Whither

* *Ibid.*, VII.

† *Ibid.*, VIII.

are you directing your gaze, to what point are you straining? All that is to come is uncertain: begin to live now. Behold, the greatest of bards warns us . . . *'The best day in the life of unhappy mortals is that which flees first.'* "*

"Why do you delay? Unless you seize it, it is gone. And when you have seized it, it will still be in flight. Therefore you must combat the swiftness of time with promptitude in making use of it. The poet seems to reproach us admirably for our endless planning for the future by saying not that it is the best age that flees first, but only that it is the best day. It is of the day the poet speaks to you, of this very day which is in flight. There is no doubt, then, that the best day is the first that takes flight for unhappy mortals, that is, for mortals burdened with the cares of life, whose minds, still those of mere children, are crushed by old age which they have reached unprepared and defenseless. For they have made no provision; old age assails them suddenly, at a moment when they did not expect it."†

"Fabianus, not one of your pedantic philosophers, but a true philosopher of the old type, used to say, 'We must combat the passions with force, not with subtlety; we must repulse their evil band not with light blows but with a vigorous assault.' . . . When we reproach men for their errors, we should teach them and not merely deplore their blindness. Time is divided into three parts, past, present, future. Of these what we are doing is short: what we are going to do, doubtful: what we have done, certain. Fortune has lost her rights over the last: it is no longer at the disposal of any one. It is the past that men lose who are wrapped up in their own affairs. For they have no leisure to review the past; and if they should have the leisure, the recalling of a thing to be repented of is not pleasant. It is therefore in spite of themselves that they recall times that have been misspent, nor do they venture to return to those subjects whose faults are evident on review. The man whose conscience, always an infallible judge, has censored all his actions, is the only one who will look back to the past with pleasure. But he whose ambition has been inordinate, whose pride has rendered him contemptuous, who has been insolent in his success, who has insidiously deceived, greedily stolen, prodigally squandered, of

* Vergil, *Georgics,* III, 66, 67.

† *De Brevitate Vitae,* IX.

necessity fears his own memory. Yet this is the portion of our time that is sacred and consecrated, beyond the reach of all human accidents, withdrawn from the realm of fortune, which neither want, nor fear, nor disease assails. This can be neither disturbed nor snatched away; the possession of it is lasting and unshaken. Each of our days is present only moment by moment, but the days that have passed will present themselves at our bidding, and will submit to inspection and detention in accordance with our will. . . . It is the untroubled and tranquil mind that can survey all the portions of its life; the minds of the preoccupied, as though under a yoke, are unable to turn and look back. The life of such men therefore passes away into obscurity . . . thus it matters not how much time is granted them . . . it passes through minds shattered and riddled with holes. The present is very brief; so much so, in fact, that some do not think that it has any existence. For time is always on the wing, it flows along and speeds ahead, it is over before it arrives. . . . The present, then, is the only time that concerns the preoccupied, but this so brief that it cannot be seized, and is lost so far as it affects them."*

"Do you wish then a proof of the shortness of their life? Consider how they wish to prolong their life. Decrepit old men beg in their prayers for a few additional years. They pretend that they are younger than they are. They delude themselves with the lie, and willingly deceive themselves, as though they were deceiving the fates at the same time. When at length some infirmity has warned them of their mortality, in what great fear do they die, not as if they are departing from life, but as though they are being dragged out of it. They cry out that they have been foolish not to have lived, and that if they may only recover from that illness they will live in repose. It is then that they realize how vain has been their preparation of things they were not to enjoy, how profitless all their labor has proved to be. On the other hand how ample is the life that is passed far from all distraction . . . no part of it is dissipated . . . nothing is subject to fortune . . . nothing is lost through negligence . . . nothing is superfluous. . . . Therefore no matter how short life may be it is amply sufficient, and when the last day

* *Ibid.*, X.

comes, the sage will not hesitate to advance to death with a firm step."*

And since Seneca has conceived of the sage as being in a state of repose, and the foolish and the vicious as preoccupied and not enjoying repose, he has divided them into three classes. For some men have minds busied with the cares that sins and vanities impose, although they do not commit such or at least only a few, possibly because they are not able. Of these men Seneca says: "Certain have an occupied leisure; in their country house, in their bed, in the midst of solitude. For although they have withdrawn from men, they are a burden to themselves, whose life must not be called one of leisure, but a slothful occupation." Those do not enjoy leisure whose desires embrace many projects. Of the man who is continually given over to sins and follies and plunged in vices Seneca says: "This man does not enjoy leisure. You should employ another term in his case. He is sick, nay, he is dead." Of those who are not addicted either in thought or deed to such sins of the flesh, yet find delight in curious trivial details of learning, Seneca says: "No one doubts that" when "you devote yourself to scholarship of a useless kind, your laborious efforts have no real value. . . . If you keep these results to yourself, your inner man is no better off; if you publish them, you appear not more learned but more of a bore. A useless zeal for such information has appealed to Romans also. . . . I heard recently a certain savant recount what each of our Roman generals had been the first to inaugurate: Curius Dentatus was the first to lead elephants in a triumph. . . . Who was the first to persuade the Romans to go aboard a ship? It was Claudius, surnamed Caudex for this reason, because by this word the ancients designated a framework of many planks. For the same reason the public registers are called codices, and boats which, following ancient usage, carry provisions on the Tiber are called *codicariae*. Such knowledge is not likely to profit us. . . . Will you suffer any one to give himself concern over the fact that Lucius Sulla was the first to put unbound lions in the circus to be slain by javelin-throwers sent by King Bocchus, while in general they were presented bound? . . . What benefit is it to know that Pompey was the first who showed in the circus a combat of

* *De Brevitate Vitae,* XI.

eighteen elephants with criminals? . . . He considered it a notable kind of spectacle to destroy men in a novel way. . . . It were better that such things as these be forgotten, lest some man in power should learn about them and be tempted to imitate their inhuman cruelty. . . . Then there are a host of other things which are either fictions or resemble lies. . . . For our Fabianus used to say he was in doubt whether it is better to apply oneself to no studies at all than to become entangled in these."

Seneca has cited these things to illustrate his meaning, and includes them as things to be avoided. But I have introduced them rather because of their value in explaining ancient words, in which we all make mistakes regarding their meaning not only in profane but also in sacred writings. For, omitting any consideration of the other words, we learn from this passage the meaning of the word *Circus,* from which is derived *ludi circenses* [contests in the Circus Maximus], of which Jerome makes mention in his Epistle to Paulinus which serves as a preface to the Bible. The contests are not called *circenses* because they are *circa enses* [having to do with swords], although Isidore gives this explanation in the eighteenth book of the Etymologies, nor is the true explanation found in any of the other ones suggested. But the contests are called *circenses* from *circus* [circle], which was a place circular in form, in which gladiators and javelin-throwers used to fight with wild beasts as a public spectacle, as is still the custom in many provinces where men gather in a circle around those who fight with wild beasts.

But we must return with Seneca to our topic. He says: "Alone of all men those enjoy repose who have leisure for wisdom; they alone live; for they not only conserve carefully their own time," but "they judge all the ages to be theirs. All the years that have preceded them are their possession. Unless we are most ungrateful, we shall recognize the fact that it is for us those very illustrious authors of thoughts regarded with reverence have been born and have prepared life. We are conducted through the labor of others to very beautiful things which have been drawn by them from darkness into the light. No age is forbidden us, we are admitted to all ages. And if we have sufficient greatness of spirit to advance out of the narrow confines of human weakness, we see open to us a vast stretch of time. We may, if we desire, dispute with Socrates, doubt with Carneades, . . . conquer human nature with the Stoics, escape from it with the Cynics, since nature permits us to enter the fellowship of all time. Why not whole-heartedly transfer ourselves from this narrow and fleeting passage of time to those boundless and eternal spaces which offer us fellowship with nobler men? . . . Those who wish to have daily Zeno, Pythagoras, Democritus, and the remaining masters of the liberal arts, and Aristotle and Theophrastus as their most intimate friends . . . will receive from them whatever they wish. Men will have their full permission to drink their fill. What happiness, what a noble old age awaits the man who has made himself their client! He will have counsellors with whom he may consult on matters of the least and of the greatest moment, whose advice he may ask daily concerning himself, from whom he may hear the truth without offense . . . in whose likeness he may mould himself. . . . These men will show you the road that leads to eternity, and will raise you to that place from which no one is cast down, this is the true way of extending your mortal life. . . . Honors, monuments, all that ambition has ordered by her decrees or reared by her efforts are quickly overthrown. . . . What wisdom has consecrated no age will destroy. . . . The life of the sage therefore has a wide range. His life is not closed by that same limit that bounds the life of other men. He alone is set free from human limitations. All ages render him service as though he were a god. The time that has passed he grasps with memory, that which is present he makes use of, that which is to come he anticipates"* and arranges. . . .

"The life of those is very short who forget the past, neglect the present, fear for the future. . . . Their joys also lack tranquillity; for they do not rest on solid foundations, but are troubled by that same insecurity from which they spring. . . . All their greatest blessings are sources of anxiety, and the highest good fortune is always least to be trusted. To safeguard good fortune additional good fortune is required, and prayers must be offered for the prayers that have just been answered. For all that comes as the gift of fortune is unstable. The higher

* *De Brevitate Vitae,* XIV, XV.

fortune rises the more liable it is to fall. That which threatens ruin cannot be a source of pleasure to any one. Most wretched, therefore, as well as very brief must be the life of those who secure with great labor what they retain with still greater effort. They laboriously acquire what they want, they retain with anxiety what they have acquired. . . . They substitute new occupations for old ones. Hope excites hope, ambition arouses still further ambition. No end of their woes is sought for, only the cause for them is changed. . . . There will never be lacking either unfortunate or harrowing causes of care, all repose will be gnawed away, so to speak, by the occupations of life. Repose will never be enjoyed. Life will always be desired."*

"Separate, then, yourself from the common throng, dear friend, and withdraw at length into a more peaceful haven, after your stormy voyage which has lasted longer than one of your years should reasonably expect. Consider how many waves you have breasted, how many storms you have endured, some of which you have encountered as a private citizen, while others in your public capacity you diverted from the state to yourself. You have given a sufficient proof of the quality of your virtue in labors and difficulties: now give it a trial under peaceful conditions. The greater part of your life, certainly the better part, has been devoted to the state: devote some of your time also to yourself. . . . You administer the affairs of the whole world as disinterestedly as those of others, as diligently as your own, as faithfully as those of the republic. In your public position in which it is difficult to avoid hatred you win affection. But, believe me, it is better to know the value of one's own life than that of the public grain supply."†

"We must press on to better things while our blood is warm and we are still in our prime. In this new kind of life you experience the enjoyment of all the liberal arts, the love and the exercise of virtue, forgetfulness of the passions, the art of living and of dying, a profound repose. The state of all men preoccupied with worldly affairs is indeed an unhappy one, but most wretched is the lot of those who toil without profit for themselves, whose sleep is subject to the sleep of others, whose step

* *Ibid.*, XVI, XVII.

† *Ibid.*, XVIII.

must be accommodated to the step of another. . . . If these men desire to know how short their life is, they should consider how small a part of it is their own."*

"At fifty years of age the law exempts the soldier from service, at sixty it does not summon the senator. Men obtain more easily from the law than from themselves an end to their labors."†

Chapter XVII

As a solace for a man's death and that of his loved ones Seneca eloquently says: "Who is so vain, so presumptuous, as to wish that he and his should be the only ones exempt from the laws of a necessity by which nature summons all things back to the same end? Who will pretend to shelter his family from a destruction that menaces the world itself? It is therefore the greatest consolation to think that nothing has befallen us except that which all before us have suffered and all are destined to suffer. And nature accordingly seems to have dealt her severest blow to all men alike, so that the equality of our fate might console us for its severity."‡

"We can accuse destiny over and over again, we cannot alter it. . . . But what is so abject and effeminate as to allow oneself to be consumed with grief?" And since the man was a great one to whom he is speaking, he says, "Many things are not permitted you which are permitted men in an humble and obscure position. High rank involves great slavery. He is happier, believe me, for whom fortune is unnecessary, than the man who has her at his command. All those blessings which delight us with their specious but deceptive pleasure, money, rank, power, and many other things which excite the admiration and cupidity of blind mortals, are retained with effort and viewed with envy, and crush those very same men whom they adorn. They threaten more evil than they bring good. They are hazardous and uncertain. We never are in secure possession, for supposing that we need not fear for the future, yet the mere

* *De Brevitate Vitae,* XIX.

† *Ibid.,* XXX.

‡ *Ad Polybium,* I.

guardianship of great wealth is filled with anxiety. If you wish to take the view of those with a deeper insight into truth, all life is a punishment: cast upon this deep and restless sea, with its change of ebb and flow, at one moment we are raised on the crest of the wave and at the next are precipitated into its trough; we are unceasingly buffeted and have no secure resting place. We are in suspense and float hither and thither, one colliding with another, and at some time or other we suffer the shipwreck of which we are in constant fear. For those embarked on this troubled sea, exposed as it is to every storm, there is no haven except death. . . . There is, believe me, great happiness in the mere happiness of dying. Nothing is certain for an entire day. Where the truth is so involved and hard to see, who can divine whether death has envied the dying one or has dealt kindly with him?"*

"He is unjust who does not leave to the donor the power to deal with his gift; miserly is the man who treats as gain that which he has received, and puts down as loss that which he has returned. Ungrateful is the man who calls the termination of a pleasure an injury, foolish he who finds profit only in present blessings. . . . Is not he to be regarded as an unjust man who is vexed at having paid back borrowed money, especially when it was borrowed without interest? Nature has given life and then, exercising her own right, takes it away. Nor is nature at fault whose terms were well known, but the greedy hope of the mortal mind, a hope that constantly loses sight of what nature is, and never remembers its own lot except when it is admonished. How much juster was he who, when his son's death was announced to him, expressed a sentiment worthy of a great man: 'When I brought him into the world I knew that he must die.' . . . Let us rejoice over that which shall be given us, and let us return it on demand. . . . I do not know whether it is more foolish to be ignorant of the law of mortality or more shameless to refuse to submit to it."†

"The deified Augustus lost his very dear sister Octavia . . . sons-in-law, children, grandsons; and during his whole life no one realized more than he that he was a mere mortal. Yet his heart, capable of confronting all things, mastered these

* *Ibid.*, IV, VI, IX.

† *Ibid.*, X, XI.

many bitter afflictions, and the deified Augustus became victor not only over foreign nations but over his own griefs also." "But to pass over all other instances . . . twice has fortune assailed me with grief for a brother, twice she perceived that I could be wounded, but that I could not be vanquished. . . . Thus, however, have I controlled my feelings, that I neither left undone anything that should be required of a loving brother, nor did I do anything that could be censured in a prince." "For not to preceive one's misfortunes is unworthy of a human being, and not to bear them is unworthy of a man."*

Chapter XVIII

We must now consider Seneca's views on the blessed life and on peace of mind. For the statements in those books strengthen and fortify the mind against adversity, and render it careless of outward prosperity. Seneca, a master of moral precepts, begins his essay addressed to Gallio on the Blessed Life as follows:

"All men, brother Gallio, wish to live happily, but have no clear vision when they are called upon to determine what makes life happy. . . . So that if one has lost his way, his speed merely increases his distance from his destination, since he is on a road leading in the opposite direction. . . . But as long as we wander about with no other guide but the uproar and discordant shouting of those summoning us in different directions, life will pass away amid errors—life which is short, even if we labor night and day with excellent intentions . . . since the conditions in this case are not the same as those in our other travels. For in the latter there is some well-defined road, and natives of the locality on being asked do not suffer us to go astray. But in life all the most beaten and frequented paths lead us farthest astray. We must therefore exercise the utmost care not to follow like sheep the flock ahead of us, which is moving forward not in the required direction but along the trodden path. And yet there is nothing that involves us in greater ills than lending an ear to mere rumor, considering as we do those things best which have received general assent."

* *Ad Polybium,* XV, XVI.

The other causes of error I have collected and noted in their proper place in the first part of this work. "For we are harmed by attaching ourselves to those advancing ahead of us, and as long as we prefer to trust others rather than to judge for ourselves, we form no judgment in regard to life. . . . Our cure depends on our separation from the crowd."*

"There is no occasion for your answering in the conventional way; 'This side seems to have the majority'; since for this very reason it is in the wrong. Human affairs are not so well ordered, that better counsels appeal to the majority;" and Seneca makes other statements as above. "Let us seek to discover what conduct is best, not what is most usual; and what will place us in possession of happiness, not what has won the approval of the multitude, the very worst judge of the truth. . . . I have a better and more certain light by which to distinguish the true from the false; let the soul discover the good qualities of the soul. If the soul ever has time to breathe and to withdraw into itself, how it will confess the truth to itself after self-examination and will say, 'All that I have done up to the present time I should prefer undone. . . . I am not yet a friend to myself. . . . Those things that attract attention, at which men pause, which one man points out to another with admiration, are bright and fair externally, but within are hideous.' "†

"Let us seek for some good possessing not mere appearance but solid and abiding worth, one that is more beautiful in that which is less visible to the eye . . . nor is such a boon remote from us; it will be found. It is only necessary to know in which direction to extend your hand; now we pass by as though in darkness what is near at hand. . . . But not to take you by too long a circuit . . . I assent to nature. Wisdom consists in keeping close to nature and forming oneself in accordance with her law and model. Happy then is the life that is in harmony with its own nature. Such a life cannot be realized unless the mind is sound and in constant possession of its sanity; it must also possess courage, strength, magnanimity, patience, resignation to circumstances. The care of the body and of what concerns it is not excluded, but must not be a source of anxiety. We can procure the commodities of life without being disturbed by

* *De Vita Beata,* I.

† *Ibid.,* II.

any of them; we can use the gifts of fortune but we must not be enslaved by them. You perceive, even if I do not add, that perpetual peace and liberty are the result when those things have been removed that irritate and terrify us. For in place of pleasures and in place of what is trivial and perishable . . . a great joy, constant and unalterable, succeeds, also peace and harmony of mind and greatness united with kindness."*

"We can also give other definitions of our good in different terms but with the same meaning. . . . The greatest good is a mind that despises all that depends on fortune, a mind that rejoices in virtue; or, it is an invincible force of mind, fortified by experience, calm in action, endowed with much benevolence and regard for one's fellow men. We may also define the happy man as one for whom there is no good and there is no evil except a good mind and an evil one. Such a man fosters what is honorable, is contented with virtue, and is neither exalted by fortune's favors nor broken by her rebuffs. He knows no higher good than that which he can bestow upon himself; his greatest pleasure is the contempt of pleasures. . . . For why should we not define the happy life as the possession of a mind free, upright, undismayed, firm, placed beyond fear, beyond desire, whose only good is integrity, whose only evil is shame. Such a mind regards as of no moment all other things, which neither take from the happy life nor add to it, which come and go without augmenting and decreasing the highest good. Constant happiness and profound joy must attend the man with such principles, since he finds his happiness in what he has, and does not crave more than he possesses in his own store. . . . Contempt of fortune alone can bestow this liberty; then will come into being a peace of mind based on security, an elevation of thought, a great and unalterable joy from the knowledge of the truth after fears have been banished, and an affability and serenity of mind, which will delight the sage not as mere blessings but as blessings arising from his own goodness."†

"The happy life then is based on a correct and certain judgment and is immutable. For then is the mind pure and free from all evils when it has escaped not only from lacerations but even from stings, resolved always to stand her ground and to

* *De Vita Beata,* III.

† *Ibid.,* IV.

maintain her position even against an angered and hostile fortune."*

And he shows that carnal pleasure is not the supreme good in the following statement: "Pleasure enters the worst life, but virtue does not admit an evil life. . . . Pleasure itself is a source of man's unhappiness, which could not be if virtue and pleasure could mingle, because virtue often lacks pleasure, but never needs it. Why put together dissimilars, nay, opposites? Virtue is something high, lofty, royal, invincible, indefatigable; pleasure is low, servile, weak, perishable, whose post and abode are brothels and cook-shops. You will find virtue in the temple, in the forum, in the senate house with hands hardened by labor; you will discover pleasure more frequently in hiding and eager for dark retreats around the public baths and sudatories . . . steeped in wine and perfume." Likewise he says, "The highest good is immortal . . . contains neither satiety nor regret. An upright mind never varies, nor does it become hateful to itself. . . . But pleasure vanishes at the very moment when it delights most. Pleasure has narrow borders and quickly fills them, then wearies, and languishes after the first impression is passed. Nor is that ever certain whose essential nature is movement. Thus it is not even possible that there should be any real substance to a thing that comes or passes with the utmost speed, destined to perish in its very use. For at the point where pleasure arrives it ceases, and even at its commencement is near its end."†

Likewise, "What of the fact that the evil enjoy pleasure as well as the good? The vicious find as much pleasure in their shame as the virtuous in their moral rectitude. It is for this reason that the sages of old have taught that men should lead the best life, not the most agreeable one, to the end that pleasure should not be the guide but the companion of good and honorable intentions. For it is nature that we must take as our guide: it is she that reason follows and consults. It is the same thing, then, to live happily and to live according to nature. I shall now explain my meaning. If we conserve with care but without fear our bodily gifts and such things as are fitting to nature, regarding them as given but for a day and likely to take flight,

* *Ibid.*, V.
† *Ibid.*, VII.

if we do not enslave ourselves to them, nor permit ourselves to covet the possessions of others, if we regard all that is pleasing to the body and all that comes to us from without as, so to speak, merely auxiliary and light-armed troops in the camp, we shall be living according to nature. Those things should serve, not master us; thus then they become useful to the mind. A man should be incorruptible and unconquerable as regards external things, and should admire himself alone, confident in mind and prepared for either of the two contingencies, the artificer of his own life. His confidence should not be without knowledge, his knowledge not without constancy; his decisions should stand, nor should there be any alteration in his decrees. It is evident, even if I do not state the fact, that such a man will be well constituted and properly regulated, and in all his conduct displays greatness combined with cheerfulness . . . let it return into itself. For this world also which embraces all things, this God who rules the universe extends outwardly, but yet wholly returns into himself from all directions. Our mind should do the same: When the mind following its senses has with their aid extended itself to external objects, it should be master of them and of itself. In this way a single force will be produced and a power in harmony with itself, and pure reason will come into being, certain in its conclusions and without inconsistency or vacillation in its ideas. . . . The mind that has arranged itself properly, that has coördinated its own parts, and brought them into tune, so to speak, has attained the highest good. For nothing base, nothing slippery remains on which it may stumble or fall. Such a mind will do all things at its own bidding, it will meet nothing unexpectedly, but easily, readily, without question will turn its every act into good. For inaction and irresolution indicate lack of harmony and inconstancy within. Wherefore you may declare with confidence that the highest good consists in peace of mind. For the virtues must be present where harmony and unity prevail: the vices are always at variance with one another."*

"But you yourself, they say, cultivate virtue for no other reason than your desire to derive some pleasure from it. . . . If virtue shall guarantee us pleasure, virtue is not sought after for this reason . . . but the effort of virtue, although directed

* *De Vita Beata,* VIII.

to another object, will secure for us pleasure also. As in a field that has been ploughed for a crop of grain, flowers spring up amid the grain, yet so much labor has not been expended on their behalf although they do please the eye. The sower had something else in view, the flowers came merely as an accessory. So also pleasure is not the recompense nor the cause of virtue, but the accessory. . . . The supreme good consists in the judgment alone and habit of a virtuous mind. When such a mind has rounded out its own goodness and has enclosed itself within its own boundaries, it has attained the supreme good and desires nothing more. For nothing exceeds the whole, nor extends beyond the end. Therefore you make a mistake when you inquire what ulterior motive I have in seeking virtue; for you are asking for something higher than the summit. Do you ask what I am seeking from virtue? I am seeking virtue itself. Virtue has nothing better to give me; virtue is its own recompense. . . . For when I tell you that the supreme good consists in the firmness, foresight, keenness, sanity, freedom, concord, and beauty of a strong mind; why speak to me of pleasure? I am seeking man's good."*

And since the objector would at once say that he is not speaking of carnal but of spiritual and mental pleasure, Seneca for this reason introduces a statement of the supposed objector excluding carnal pleasure in order to assign another pleasure to virtue. Because of this other pleasure, then, the objector says, "I am not seeking the pleasure of the belly, an organ which is larger in domestic animals and in wild beasts. You profess ignorance of my meaning, he says. For my assertion is that no one can live pleasantly unless at the same time he also lives virtuously. Such a life is not possible for dumb animals nor for men who measure their good in the terms of food. I aver clearly and openly, says he, that this life which I call pleasant is inseparable from virtue." But Seneca meets the objection with the statement that pleasure is found not only in gluttony and debauchery, but in other spiritual vices also, since men take pleasure in these other vices as well as those mentioned above. And therefore virtue cannot exist for the sake of mental pleasure, since pleasure accompanies sins as well as virtue, being in fact common to both virtue and sin, because

* *Ibid.*, IX.

men rejoice and find pleasure in evil as well as in good. This is his further statement: "Who does not know that all men with the least sense are the most sated with pleasures? And that wickedness is overflowing with pleasurable things? that the mind itself supplies many low kinds of pleasure? Chief among these are arrogance, too high an opinion of oneself, a fancied superiority to others, and a blind and heedless love for one's own possessions." Since then the reply might be made that the good man at least experiences pleasure and joy and therefore has pleasure of mind, he says, "Virtue judges pleasures before she admits them, nor does she place a high value on such as she approves, for she merely admits them, and does not rejoice in the use she makes of them but in the limits she prescribes." And he says to the objector, "You embrace pleasure, I restrain it. You enjoy it, I use it. You consider it the greatest good, I not even a good. You do all things for the sake of pleasure, I do nothing."*

"I do not call that man a sage to whom something is superior, and still less if this something be pleasure. For, dominated by pleasure, how shall he withstand labor, danger, and want . . . how shall he endure the sight of death, of pain? How shall he endure the convulsions of nature and such a host of bitter foes, he who has been vanquished by so feeble an adversary?" And because the objector might still press his point, Seneca continues his opposition, "Do you not see how many counsels pleasure is likely to give?" "She will be able to give no bad advice since she is joined to virtue," is the objector's reply. But Seneca thus refutes the statement: "Do you not see what would be the nature of a supreme good that requires a keeper, that it may be a good? But how shall virtue rule pleasure which it follows, since it is incumbent on one who obeys to follow, and on one who commands to rule. Do you place your general in the rear? The eminent function of virtue in your system is to test in advance the pleasures . . . and as happens to many they are mad with a jovial madness and laugh in their delirium. But the pleasures of the sage are mild and restrained and almost listless. They are kept under subjection and are scarcely noticeable, since they come unsought, and when they present themselves, they are not received with honor nor welcomed with

* *De Vita Beata,* X.

any joy on the part of those experiencing them. For they mingle them and intersperse them through life like play and jest in the midst of serious matters. Let them cease, then, to try to ally two things that are incompatible, namely, to combine pleasure with virtue, a fault by which they pay court to all the most corrupt. The corrupt man, absorbed in his pleasures, always overfed and drunken, believes that he is living virtuously since he is living pleasurably. For he is told that pleasure cannot be separated from virtue; he then gives to his vices the name of wisdom and makes a display of what he should conceal. Therefore steeped in vices they seek for some defense and cloak for their lusts, and they lose the only good thing they possessed among their evils, namely, their feeling of shame at wrongdoing. . . ."*

"That which suffices nature is insufficient for debauchery. . . . Whosoever has given his assent to virtue has given the appearance of noble qualities. He who follows pleasure seems soft, without vigor, degenerate, a man destined to fall into evil ways, unless someone distinguish for him among the pleasures, in order that he may know which of them confine themselves within the limits of natural desire, which are uncontrolled and unrestrained and the more they are sated the more insatiable they become. . . . No good is burdened by its own greatness."†

"And if that union of virtue and pleasure pleases you, let virtue precede, let pleasure attend upon it, and be like the shadow to the body . . . we shall have pleasure, but we shall be its masters and controllers. Whatever pleasure secures from us will be won by entreaty not by compulsion. But those who have entrusted pleasure with the leadership have deprived themselves of both; they lose virtue, and they do not possess pleasure but pleasure possesses them, victims equally of its absence which torments them, and of its excess which suffocates them; wretched if they are deserted by it, still more wretched if they are overwhelmed. . . . The greater and the more numerous they are, the smaller is he whom the world calls happy, and the slave of many masters. . . . He who follows pleasure, renounces liberty first, and expends himself for the sake of his belly; he does not purchase pleasure for himself, but he sells

* *Ibid.,* XI, XII.

† *Ibid.,* XIII.

himself to his pleasures. Yet what prevents, he asks, virtue and pleasure being united into one, and the supreme good being formed in such a way that it becomes at the same time both honorable and pleasing? Because no part of the honorable can be dishonorable. Nor does the supreme good retain in itself its own purity, if it beholds in itself something inferior. . . . But he who associates virtue and pleasure weakens whatever vigor is found in the one good by the frailty in the second, and sends under the yoke that invincible liberty than which nothing is known to be more precious. For he begins to have need of fortune, which is the very worst slavery. . . . You do not, therefore, grant to virtue a solid and immovable base, but you bid virtue stand in a changing and uncertain position. For what are so uncertain as the expectation of what is fortuitous and the diversity of bodies and of things affecting the body? How can this man obey God and accept with a tranquil mind whatever happens? . . . How can he put a favorable construction on his own ills if he is sensitive to the least incitements of pleasure and pain? But not even of his country is such a man a good guardian and avenger, nor is he a defender of his friends if he is given over to pleasures. Let, then, the supreme good mount to a point from which no force can drag it down, to a point inaccessible to pain, fear, hope, and to all else that may impair the full rights of the supreme good. But virtue alone can mount to this height; that ascent can be mastered by her alone. She will stand firm and will endure whatever betides not only with patience but also with pleasure, and will realize that all difficulties are in accordance with nature's law. She is like the courageous soldier who endures his wounds, counts his scars, and pierced by darts dies still loving the commander for whom he falls. He will bear in mind that old precept, Follow God. Whosoever complains and laments and groans is none the less forced to obey, and, in spite of his resistance, to execute the orders imposed upon him. But what folly it is to be dragged along rather than to follow! It is both ignorance of one's condition and folly to grieve . . . to be astonished and indignant at those things that happen to the good as well as to the evil. Let us meet with high courage all the universal order requires us to endure. We have been caused to take this oath to bear the lot of mortals and not to be perturbed by those things which it is not in our power to avoid. . . . Obedience to God is liberty."*

"True happiness, then, rests in virtue. What advice will this virtue give you? Do not consider anything as either good or evil that does not proceed from virtue or vice. She will further counsel you to be firm and in opposition to evil to form God from the good, as far as you may do so in accordance with the divine will. What does she promise you for such an enterprise? Great blessings and equal to those possessed by the gods; you will suffer no compulsion, you will lack nothing, you will be free, safe protected from loss. You will never attempt anything in vain, you will never be arrested by any obstacle, all things will happen to you in accordance with your wish. Nothing adverse will happen, nothing contrary to your expectation. . . . What then? Virtue suffices for a happy life. Why should it not suffice? Nay, it is more than sufficient. For what can a man lack who is without desires? What need of external things can he have who has gathered all things into himself?"†

But since no men or at least very few attain such moral excellence, and since Seneca might be subjected to criticism for certain imperfections of his own, in not conforming as a teacher of virtue all his actions to the requirements of virtue, Seneca replies when criticism of this kind is directed against him as follows:

"Now I make this reply to you; I am not a sage, and, to add more fuel to your fiery criticism, I shall not become one. Exact of me, therefore, not that I equal the best, but that I be better than evil men: this suffices me daily to lessen somewhat my vices and to correct my errors. . . . I prepare palliatives rather than cures for my gout, content if it returns less frequently and is less painful: I am a weak runner compared with your powers."‡

"Your speech and your life do not correspond, you say. Plato and Zeno were subjected to this same reproach. All those teachers used to tell not how they themselves were living, but how they ought to live. I am speaking of virtue, not of myself, when I abuse the vices. . . . Nor shall that malevolence of

* *De Vita Beata,* XIV, XV.

† *Ibid.,* XVI.

‡ *Ibid.,* XVII.

yours, imbued as it is with much venom, deter me from what is best. Not even that poison with which you attack others, with which you kill yourselves, shall hinder me from continuing to praise the life, not that I am leading, but the one I know that I ought to be leading. Nor shall it hinder me from adoring virtue and from creeping after her at a great distance."*

"The philosophers are useful in their statements regarding virtue. . . . For if they acted in accordance with their teachings, what life could be happier than theirs? Meanwhile there is no reason why you should despise good words and hearts full of good thoughts. The pursuit of useful studies is to be commended apart from their result. Is it surprising that by routes so difficult men do not reach the summit? But recognize powers that are attempting great things even though they fail. It is a noble thing to attempt lofty things and to conceive in the mind projects too great even for a great soul to accomplish. Such a man has proposed to himself this line of conduct: I shall view death on its arrival with the same look that I have when I hear it mentioned; I shall resign myself to labors, however great they shall prove to be, supporting my body with my mind; I shall be equally contemptuous of riches whether I have them or lack them, neither sadder if I know that they are in the possession of others, nor prouder if I find them casting their luster about me: I shall look with indifference upon fortune whether she comes or goes: I shall view all hands as belonging to me, and mine as belonging to all other people: I shall so live persuaded of the fact that I have been born for others. . . . She has given me to everybody and everybody to me; whatever I possess I shall not guard in a niggardly way nor scatter abroad lavishly: I shall believe that I possess nothing more than what I have bestowed with discernment: I shall not count my benefits, I shall not weigh them, nor shall I value them except in accordance with the merit of him who receives them. If he is worthy of them, I shall not think that I have done much; I shall never take opinion, but my conscience, as the motive in all that I do; I shall believe that whatever I do, when I am the sole witness, is done in the sight of the whole world; my purpose in eating and drinking will be to satisfy the desires of nature, not to fill and empty my stomach; I shall be agreeable

* *De Vita Beata,* XVIII.

to my friends, mild and affable to my enemies; I shall be appeased before I am asked, I shall meet more than half-way reasonable requests; I shall know that the world is my country and God is its ruler; that he is above me and about me as the censor of my deeds and my words; and when either nature shall demand back my spirit or reason release it, I shall go hence after bearing witness that I loved a good conscience, honorable pursuits, that I impaired no man's liberty. . . . He who shall propose, wish, try to do these things, will be making his journey toward God; verily, such a man even if he does not keep to his course 'falls in a daring and a great enterprise.'* You indeed who hate virtue and him who practises it do nothing new. For weak eyes fear the sun and . . . nocturnal animals . . . are dazed at his rising and everywhere seek their lurking-places . . . groan, employ your wretched tongue in abusing the good . . . you will sooner break your teeth than sink them in."†

"Why does he who is devoted to philosophy live in opulence? Why does he say that riches should be despised and yet still hold on to them? Why does he think that life should be despised and yet still live? Why does he think that health should be despised and yet guard it with the greatest care? . . . And thinks and says that exile is only an empty name: what harm is there in changing countries: and yet . . . he is growing old in his native land?"

He answers this question by showing how the sage can be good who wishes to have riches and does not wish to be poor: by drawing a distinction between the use made of riches by the sage and by the foolish, he shows how the sage should regard riches. The sage as well as the man of evil habits requires the necessities of life: and so out of the abundance owing to the world's resources he wishes to have what is necessary for himself and for others, that he may aid the indigent, and handle the business of the state. And thus since he possesses his riches for an honorable and useful purpose, he will not lose his virtue, which the evil man will throw away because of his abuse of riches. He says therefore that "these things should be despised by the sage, not that he should not possess them, but that he should not possess them with uneasiness of mind. He does

* Ovid, *Metamorphoses,* II, 328.

† *De Vita Beata,* XX.

not drive them away from himself, but attends them on their departure without regret. Where indeed shall fortune deposit more safely her riches than with him, from whom she will receive them back again without complaint? . . . The sage does not love riches but prefers them. He receives them into his home, not into his heart. . . ."*

And he says to the foolish man "If riches forsake me, they will carry away nothing but themselves; you will be amazed and you will think you have been left without yourself, if they should depart from you. With me riches have a place, with you they hold the highest place. . . . Riches belong to me, you belong to riches."†

"The philosopher will have ample riches, not stolen from any one nor stained with another's blood, secured without injury to any one, free from sordid gain, riches which pass out as honorably as they enter, and are complained of by no one unless he be envious. . . . Such a man will neither boast of his patrimony gained by honest means nor will he blush because of it. He will have . . . reason for pride if, on opening his house and admitting the public to view his possessions, he shall be able to say, Let each one take what he recognizes as his own. O great man, rich in the most excellent way! . . . Let riches come, let them enjoy his hospitality. He will neither be prodigal with them nor will he hide them; the former is the act of a silly mind, the latter that of a timid and petty one. . . . He will possess his wealth but will regard it as fickle and likely to vanish: nor will he suffer his riches to weigh upon either himself or another. . . . He will give to the good . . . or to those whom he shall be able to render good. He will give with the greatest discretion choosing the most worthy. . . . He will give from upright and laudable motives. For a gift unworthily bestowed is to be classed among disgraceful losses. His purse will open easily, but will not have a hole in it, a purse from which much comes out and nothing falls out."‡

"And note this first; there is a difference between him who is interested in wisdom and the man who has actually acquired it. The former will say to you, 'I speak excellently, but I am

* *De Vita Beata,* XXI.

† *Ibid.,* XXII.

‡ *Ibid.,* XXIII.

still involved in very many evils. There is no reason why you should judge me in accordance with the rule I have laid down for myself. . . . I am still working on myself and fashioning myself and trying to attain to a high standard. If I advance as far as I propose, then require that my acts correspond to my words.' But the second man, having attained the height of human wisdom, will talk to you in a different way and will say to the foolish man: 'First, you should not permit yourself to pass judgment on your betters; I already possess a proof of rectitude in being displeasing to evil men. . . . I deny that riches are a good: for if they were, they would make men good. Now since they are found in the possession of evil men, riches cannot be called good: I refuse them this name. . . .' "*

"Place me in the most luxurious home with gold and silver at my command. . . . I shall not respect myself because of those things which, although they are in my possession, are yet not part of me. . . . Drive me away among the beggars: I shall not, however, despise myself on this account, because I am found among those who hold out the hand for alms. For what matters the lack of a morsel of bread when one has still the opportunity to die? . . . I shall not think myself a whit more happy because I have a soft cloak, and because my guests recline on purple. . . . I shall not be more unhappy if my wearied head finds repose on a bundle of straw. . . . No hour is without some complaint; I shall not for this reason call myself wretched in the midst of these adversities. . . . For I have provided that no day should be a sinister one for me. . . . Therefore," declares the sage, "my life is not at variance with my teaching. But you do not hear aright. The sound merely of my words reaches your ears, you do not seek their meaning."†

"What then is the difference between me with my folly and you with your wisdom, if we both desire to have riches? A very great difference. For riches in possession of the sage are in servitude, in the possession of the fool are in command. The sage intrusts nothing to riches; riches mean everything to you. You accustom yourselves to them and cling to them as though some one had promised to you their eternal possession: the sage never meditates so much on poverty as when he is in the midst

* *Ibid.*, XXIV.
† *Ibid.*, XXV.

of affluence. . . . You waste away in the midst of your own possessions nor do you think how many chances threaten every moment to carry off your precious spoils. He who deprives the sage of his wealth leaves him all things that are really his. For the sage lives pleased with the present and without anxiety for the future. Of nothing has Socrates so firmly convinced me as that I should not regulate the conduct of my life in accordance with your opinions. Marshal from all sides your usual array of words: I shall consider that you are not uttering insults but wailings like very unhappy children. . . . Your hallucinations I put up with as does Jupiter the absurdities of the poets, one of whom has attached wings to him, another horns, and still another has introduced him as an adulterer. . . ."*

"Lo! Socrates, from that prison which he cleansed by his entrance and rendered more august than any senate-house, proclaims, 'What madness is this of yours, what nature hostile to God and man that prompts you to decry the virtues and to violate all that is sacred by your wicked statements?' Praise the good if you can; if not go your way. But if you find pleasure in exercising that foul freedom of yours, direct your attacks on one another. For when you rave against heaven, I do not say you commit sacrilege, but you lose your pains. . . . It is an advantage for virtue to be brought forth and put to the test, nor do any perceive her greatness better than those who have felt her power from attacking her. It is only by striking a piece of flint that one realizes its hardness. I offer myself to their attacks like a solitary rock in a restless sea, which the waves lash unceasingly on every side, and yet they do not move it from its place or wear it away through so many ages by their incessant attack. Spring at me, make your assault. By my endurance I shall overcome you. All that rush upon what is firm and invincible expend their force to their own hurt. Therefore seek some soft yielding substance in which to drive your spears. But have you the time to examine the faults of others and pass judgments on any man? Why, you ask, does this philosopher live too well housed, why does that one dine too sumptuously? Are you observing the slight blemishes on others when you are yourself covered with many ulcers? Such conduct is like that of him who laughs at the moles or warts on the persons of

* *De Vita Beata,* XXVI.

the most beautiful when he himself is devoured by a terrible skin disease. Reproach Plato because he demanded money, Aristotle because he accepted it, Democritus because he neglected it, Epicurus because he expended it. Cast in my teeth my inheritance. How very happy you will be as soon as you are in a position to imitate our faults. Nay, regard rather your own vices, which riddle you everywhere, some raging externally while others burn in your vitals. What shall I say more? Is not a kind of whirlwind, even though you scarcely discern it, rotating and enveloping your minds as they flee from and pursue the same objects?"*

Chapter XIX

"For among our other ills the very worst is that we change our vices . . . and this also plagues us that our judgments are not only bad but also fickle. We move to and fro, we grasp at one thing after another, we abandon what we have been seeking, we seek again what we have abandoned. Our life is merely an alternative between our desires and our regrets. For we depend wholly on the judgments of others, and we consider that best which is praised and sought after by the multitude, not that which is praiseworthy and desirable in itself. We judge a road to be good or bad only by the number of footprints among which there are none of those returning."†

"Man is too mortal for the comprehension of things immortal." And since the pursuit of wisdom and virtue is still disapproved of by many on the ground that it involves the abandonment of civic duties, he therefore cites this charge against the leisure of the sage: he says, "Nature has destined me for two things, action and contemplation."‡

"Virtue banished to a leisure without action is an imperfect and languid good and never displays what it has learned. . . . He ought not only to think of what he should do but also take a hand sometimes in realizing his meditations. . . . The sage withdraws into seclusion, that he may be assured of the op-

* *Ibid.*, XXVII, XXVIII.
† *De Otio*, I.
‡ *Ibid.*, V.

portunity to do what will benefit posterity. We surely are ready to admit that Zeno and Chrysippus accomplished greater results than if they had led armies, held offices, promulgated laws; in effect they did promulgate laws, not for a single state, but for the whole human race. Do you think, then, that such a leisure is unworthy of a good man, a leisure in which he forms the ages to come, and addresses not a small number of his fellows but the men of all nations present and future? . . . They have found the means of rendering their leisure more useful to men than the activity and sweat of others. Therefore they accomplished much although they took no part in public affairs. . . ."*

"To what state shall the sage withdraw? To Athens in which a Socrates is condemned, from which an Aristotle flees to escape condemnation? in which envy overwhelms the virtues? Shall he go to Carthage where sedition is always rife and liberty is deadly to all good people? . . . If I should care to pass in review all governments, I shall not find one which can endure the sage or one which he can endure."†

Chapter XX

Now finally I shall introduce some quotations from Seneca's work addressed to Serenus on peace of mind, because this accompanies the happy life; and it is certain that happiness cannot be possible in this life without peace of mind. Nor will the life of true happiness in the future be complete without it. All adversities can of necessity be easily borne if we possess this peace of mind. But since we are just as prone to flatter ourselves as others, and by palliating our vices measure our happiness of mind by a false standard, Seneca says with excellent point, "We look familiarly on the imperfections in our own household, and our self-love blinds our judgment. I think that many could have attained to wisdom if they had not believed that they were already sages, if they had not concealed certain vices in themselves, if they had not skipped over others with their eyes closed. For there is no reason why you should think that we are

* *De Otio,* VI.

† *Ibid.,* VIII.

ruined more by flattery of others than by our own. Who has ventured to tell himself the truth? Who set in the midst of herds of flatterers and courtiers has not been his own most eager flatterer?"*

"What you wish for, Serenus, is to remain unshaken, a perfect state of mind resembling that of God himself. The Greeks call this stability of mind εὐθυμία, on which there is an excellent treatise by Democritus: I call it tranquillity. . . . We are seeking then to discover how the mind, enjoying always a perfect equality, may follow a uniform course, live at peace with itself, contemplate its lot with satisfaction, experience this joy without interruption, and maintain itself in a state of calm never suffering undue elevation or depression: this will be tranquillity." And first he touches upon the vices that destroy our peace of mind, in order that we may know how to avoid them and return to our former state of mind. He says accordingly: "Some men are plagued with a fickleness of mind and a feeling of irksomeness and habitual inconstancy of purpose, men who always prefer the plan they have rejected, . . . they are weak and indolent. Add those who resemble the victims of insomnia, tossing about and trying this position and that until fatigue finally brings them repose. Such men, constantly changing their manner of life, finally remain in that state in which not a dislike of change but old age too sluggish to make one finds them. Add also those whom the vice of sloth rather than reason preserves from inconstancy; they live not as they wish but as they have begun. Vice has innumerable forms but one general effect, namely, displeasure with oneself;" for which reason peace of mind departs. "This arises from lack of mental balance and from desires that are timid or from such as are crowned with little success, where either their daring is not commensurate with their wishes, or they do not obtain what they wish, and live entirely upon hope. Such men are always in a state of agitation and suspense, the inevitable result of being dependent on their own wishes. . . . Those men are possessed by regret for what they have begun and by fear of making a new start; and that agitation of a mind which finds no egress creeps upon them, since they are able neither to control their wishes nor to gratify them. Thus their life arrested so

* *De Tranquillitate Animi,* I.

to speak in its course drags along slowly and painfully, and the mind languishes in the midst of its disappointed hopes. All these ills are aggravated, when in their resentment at an ill fortune that is so burdensome they have taken refuge in leisure and private studies, which cannot be endured by the mind that is interested in public affairs, eager for action, restless by nature, and that possesses in itself few sources of solace. For this reason, when the delights are withdrawn which their occupations furnish, they cannot endure home and solitude and confining walls. The mind left to itself does not care to view its own image. This is the cause of weariness and self-disgust, of the agitation of a mind never at rest, and of the pain and bitterness with which it endures its own inaction. . . . Desires pent up without egress stifle themselves: then come grief and langor and the thousand fluctuations of an inconstant mind held in a gloomy and deplorable state of suspense by the hopes it has entertained. Then follows the condition of mind of the man who detests his own leisure and complains that he has nothing to do, and who views with an envious eye the success of another. For sloth causing unhappiness breeds envy, and men desire the fall of others, because they have not been able to advance themselves. . . . For the mind of man is naturally active and prone to movement. Every occasion of exciting and diverting itself is pleasing to the mind; still more pleasing to the worst characters who waste away voluntarily in their pursuits. As certain ulcers crave the harmful touch of the hand and find pleasure therein, and whatever irritates gives pleasure to a foul itch of the body, just so I may say that these minds in which desires like malignant ulcers have broken out find their pleasure in labor and vexation of spirit. For there are certain things that give pleasure to our bodies also, as for instance turning about in bed and changing a side that is not yet tired, and exposing ourselves to the air in first one position and then in another. . . . It is characteristic of a sick man not to endure anything for a long time and to employ changes as a means of relief. . . . We are weak in our endurance, we cannot stand either labor or pleasure for any length of time, and grow weary of our own and of all other interests. Some have been driven to suicide when they realized that by frequent change they had arrived again at the point of departure and had deprived themselves of the possibility of further change. Life and the world itself became a source of disgust to them."*

Seneca then introduces remedies for these vices. And he first touches upon them, quoting the opinion of another writer. He says, "You inquire what remedy in my judgment should be used against this feeling of disgust. The best would be, as Athenodorus says, to keep oneself always on the alert in the management of affairs, in the administration of the state, and in civic duties. For as some spend the day in the sunshine in the exercise and care of the body, and as athletes find it most serviceable to spend most of their time in exercising their arms and nursing their strength, to which alone they have devoted themselves; so is it most fitting for us in preparing our mind for the arena of public life to be always at work. For since he has determined to render himself useful to his fellow citizens and to mankind, he is occupied and of service at the same time when he has placed himself in the midst of affairs and administers public and private matters to the best of his ability."

Seneca, however, does not agree with him in this, but recommends the pursuit of wisdom. He says therefore: "But since in this mad race of ambition, with so many detractors putting the wrong construction on upright conduct, integrity has insufficient protection, and since there will always be more to hinder than to promote success, we should withdraw from the forum and from public affairs. But a great mind has an opportunity to develop itself freely in private life. The power of lions and of other animals is held in check by the cages that restrain them; not so with man, whose greatest actions are performed in retirement. Let the sage conceal himself indeed, but with the intention of serving in his solitude both public and private interests with his talents, his voice, and his counsel. For he who defends the accused and votes on questions of peace and war is not the only one that benefits his country. The man who instructs the youth, who in so great a dearth of good teachers instills virtue in the youthful mind, who draws back those who are eagerly pursuing money and pleasure, and retards at least for a time their ruin, such a man, I say, even in his private position is working for the public. Is the magistrate who judges between citizens and strangers, or the city praetor who pro-

* *De Tranquillitate Animi,* II.

nounces on litigants the sentences dictated to him by his assessor, more useful than he who teaches what justice is, what loyalty is, patience, fortitude, contempt of death, knowledge of God, and the great benefit of a good conscience? Therefore if you should devote to study the time you take from active employments, you will not have deserted your post nor failed in your duty. For he is not the only one performing military service who stands in the line of battle and defends the right or the left wing, but he also serves who guards the gates, and fills a less perilous post, but not, however, an idle one, by standing guard and supervising the armory. These duties, although they are performed without risk to life, are classed as military services. By devoting yourself to study you will escape all the tedium of life, nor will you desire the return of night because of the irksomeness of the day. You will be neither burdensome to yourself nor useless to others. You will win the friendship of many, and all good men will be attracted to you. For virtue although obscured never remains hidden, but sends forth indications of its existence. Whosoever is worthy will find her by her footsteps. For if we break off all relationship and renounce the human race, and live occupied with ourselves alone, this solitary life deprived of every kind of liberal study will be followed by a total lack of occupation. We shall begin to erect buildings in one place and tear them down in another . . . and to make a bad use of our time, . . . a very serious fault. Often an aged man has no other proof to offer in establishing the fact that he has lived for a long time except his age."*

"Let him be in private, in public, at table, a good companion, faithful friend, temperate table companion. If he no longer performs the duties of a citizen, let him fulfill those of a man. We have not shut up our minds within the walls of a single city, but have sent them forth to have intercourse with all lands and we have claimed the world as our country, so that we may give our virtue a wider field. . . . The work of a good citizen is never useless. He has been heard, he has been seen; his look, his nod . . . even his step are of use. As certain medicines are efficacious by their odor apart from their taste and contact, so virtue though at a distance and unperceived spreads abroad an atmosphere of helpfulness; whether virtue walks abroad at

* *De Tranquillitate Animi,* III.

will . . . or enjoys only a precarious freedom in this respect . . . or inactive, silent, narrowly circumscribed . . . in whatever guise she makes her appearance, she is beneficial. What of this? Do you consider of little moment the example of a man who knows how to rest? It is a very excellent thing, then, to mingle leisure with action whenever a life of action is prevented either by adverse fortune or by the state of the republic. For never have all avenues of approach been so closed, that honorable effort found its passage completely barred."*

"Can you find a city more wretched than was that of the Athenians when it was distracted by the thirty tyrants? The thirty had slain thirteen hundred citizens, nor had they reached by any means the end of their atrocities, but their barbarity was increasing under its own stimulus. . . . Socrates, however, continued to live in their midst, consoling bereaved parents, encouraging those who were despairing of the state, reproaching the rich who were trembling for their wealth . . . and exhibiting a lofty example to those who were willing to imitate him by boldly entering the presence of the thirty. Yet Athens herself executed this man in prison, and even liberty did not endure the man who as a free citizen had defied a throng of tyrants."†

"But we must now consider in the first place ourselves; then the matters on which we are engaged; finally the persons for whom or with whom we must act. Before all else it is necessary to form an estimate of ourselves, because frequently we overestimate our powers. One man will fail because of overconfidence in his own eloquence, another has drained too heavily his patrimony, still another has taxed his weak body with too strenuous work. Some have a timidity rendering them unfit for civic affairs which require resolution. The inflexibility of others does not meet with success at court. Some do not keep their anger under control, and under any provocation are indiscreet in their language. Others do not know how to restrain their raillery," in their reprimands, "and do not refrain from dangerous witticisms. For all these classes of people a quiet life is better than one of action. . . . We must then estimate the matters themselves which we are undertaking, and we must com-

* *Ibid.*, IV.

† *Ibid.*, V.

pare our powers with the projects we propose to attempt. For the power must always be greater than the resistance. Of necessity he must be crushed who carries a burden too great for his strength. There are, moreover, certain matters which are not so important in themselves, but which carry the seed of many others. We must shun activities from which spring still other occupations of a new and complex nature. We must not travel to a point from which we may not return at will; let us charge ourselves with matters which we can terminate or at least hope to do so. Let us abandon projects that go too far afield and do not terminate at the point we have set."*

"In our choice, then, of individuals we must examine whether those are worthy on whom we are bestowing a portion of our life; whether the expenditure of our time has any good effect on them. For some people go so far as to charge against us the favors we render them. Athenodorus says that he would not even dine with a man who would not feel under obligation to him for this favor. . . . You must consider whether your nature is better fitted for a life of action or for a leisure devoted to study and contemplation; and you should follow your natural bent. . . . For natures that are forced make a poor response. . . . But nothing so delights the heart as a sweet and faithful friendship. How great a blessing it is when there are ready hearts to which every secret can be intrusted with safety, whose conscience you fear less than your own, whose conversation relieves your solicitude, whose opinion aids your judgment, whose gaiety dispels your sadness, the mere sight of whom is a pleasure! We shall choose, to be sure, those who are free as far as possible from passions. For vices creep along and pass to all that are exposed to them and harm by contact. Therefore as we are careful in times of pestilence not to sit near persons already infected and inflamed with the disease, because we shall incur danger and shall suffer even from their breath, so in our choice of friends we shall make every effort to select such as are least corrupt. Association of healthy people with sick folk is the beginning of disease. Nor should I give you this advice not to follow or attract any one but the sage. For where will you find him for whom we have been searching so many ages? The best is the least bad. . . . Now indeed in such a dearth of

* *De Tranquillitate Animi,* VI.

good men our choice should be less exacting. Avoid, however, in particular those gloomy and mournful individuals whose delight is in complaining of all things in general. Although he may be of a kind and loyal heart, a gloomy companion with a perpetual sigh on his lips is an enemy to your peace of mind."*

"Let us pass to riches, the greatest source of human anxieties. For if you compare all our other worries, such as death, sickness, fear, desire, endurance of pain and labor, with those ills caused by our money, you will find that the latter press far more heavily upon us. Therefore we must consider how much less painful it is not to have than to lose what we have had; and we shall perceive that the less we have to lose the less is poverty able to torment us. For you are in error if you think that the rich support losses with greater fortitude: the largest and the smallest bodies suffer equal pain from a wound. Bion cleverly says, 'Plucking out hair is just as painful to heads that are bald as to those covered with hair.' . . . It is more endurable . . . not to acquire than to lose; and for this reason you will see those more happy on whom fortune has never smiled than those whom she has deserted. Diogenes, a man of lofty spirit, perceived this truth, and was careful to have nothing which could be taken away from him. Call this condition poverty, need, indigence, give to this state of security any opprobrious name you wish; I shall cease to believe in his happiness only when you find me some one else guaranteed against loss. Either I am deceived, or it is a mark of kingship, that there is one man who cannot be harmed among the covetous, robbers. . . . He who is doubtful regarding the happiness of Diogenes may well be in doubt whether the condition of the immortal God is not deficient in happiness, since he has neither estates, nor gardens, nor farms enriched by the labor of slaves, nor a large income from money in bank. Are you not ashamed, you who are enraptured with riches? Just look at the world; will you not see God giving all, retaining nothing? Do you consider the man who has divested himself of all fortuitous things poor or like the immortal God? Is it your idea that a rich man is happier? Diogenes' only slave ran away from him, nor did Diogenes consider it worth while to have him brought back when he was informed of his whereabouts. 'It is disgraceful,' said he, 'that

* *Ibid.*, VII.

Manes can live without Diogenes, but Diogenes cannot live without Manes.' He seems to me to have said, 'Attend to your own business, Fortune; you have no longer any claim on Diogenes. Has my slave run off? Nay, it is a free man that has departed.' . . . But since we do not possess such strength of character, we should at least curtail our wealth, so that we may be less exposed to injury at the hands of fortune. Those are better fitted for warfare whose bodies can be contained within their armor, than those of too great bulk, whose size exposes them to wounds. The best measure of wealth is to be neither too near to poverty nor too far removed from it."*

"We shall be pleased with this measure, if we begin by taking pleasure in frugality, without which no riches suffice, . . . especially since the remedy is at hand, and poverty itself can be converted into riches by means of frugality. Let us accustom ourselves to do without pomp and to estimate things in accordance with their usefulness and not in accordance with their embellishment. Our food and drink should suffice to remove hunger and thirst; . . . let us learn to go about on our own legs, and not regulate our dress and mode of living after new fashions; . . . let us learn to increase our continence, restrain our luxury, temper our pride, soften our anger, look upon poverty in the proper light, cultivate frugality, even if the majority is ashamed to apply to natural longings relief secured at a small cost, let us learn to restrain as it were in fetters unbridled hopes and thoughts soaring into the future, and to exert ourselves in seeking riches from ourselves rather than from fortune. . . . Let us accustom ourselves, then, to dine without a number of guests, to require the service of fewer slaves, to get our garments for the purpose for which they were intended, and to dwell in narrower confines."

Seneca speaks of the aid we derive from study as follows: "Our expenditure on studies is most praiseworthy so long as it is kept within reasonable limits. To what end such numberless books and libraries, whose owner in the course of his whole life scarcely reads through the titles? A multitude of books burdens, does not instruct, the learner; and it is much better to devote yourself to a few authors than to lose your way in a host of them. Forty thousand of the books in Alexandria were

* *De Tranquillitate Animi,* VIII.

destroyed by fire. The vaunting of this superb monument of royal magnificence I leave to another, as Livy, for example, who says that it was a masterpiece of taste and care on the part of sovereign power. As a matter of fact it was not an example of taste and care, but of the extravagance of study, nay, not even of study, since they did not have study in view but merely ostentation. . . . In all excess there is vice. Why should you pardon the man who eagerly acquires bookcases laden with the works of either unknown or discredited authors and yawns in the midst of so many thousands of books, a man whose chief pleasure is derived from their titles and frontispieces? . . . I could certainly pardon this madness if it had its origin in an excessive zeal for study. As a matter of fact, they search with such care for the works and busts of revered men of genius in order that they may use them as mural decorations."*

"You have fallen into a difficult sort of life, and either public fortune or your own particular fortune has fastened a noose upon you without your knowledge, which you can neither loosen nor break. Consider how prisoners at first endure with difficulty the chains and shackles on their legs; then when they have made up their minds not to worry over but to endure them, they are taught by necessity to bear their shackles with fortitude and by familiarity with ease. You will find in every kind of life delights, relaxations, and pleasures, if you wish. . . . On no other count has nature deserved more gratitude from us than for her discovery of habit as an alleviation of woes, familiarizing us quickly with the severest evils, fully aware as she is of our heritage of cares. No one could hold out if adversity oft repeated had the same impact as its first blow. We have all been linked with fortune: of some the chain is golden, of others lax, of still others tight and foul. But what difference does it make? We are all prisoners together, and those who have bound others are bound themselves also . . . one is bound by honors, another by wealth. Some are the slaves of their high birth, others of their obscurity. Some are subject to a strange yoke, others are subject to a yoke of their own. These are kept in one place by exile, those by their priestly office. Every life is servitude. One must accustom himself, therefore, to his lot and complain of it as little as possible. . . . There is nothing so

* *Ibid.*, IX.

bitter that a tranquil mind does not find some comfort in it. . . . Apply reason to difficulties; hard things can be softened and narrow places broadened. . . . Moreover, we should not permit our desires to go far afield, but should limit their excursions to the immediate neighborhood. . . . Renouncing objects that we cannot obtain, at least without difficulty, let us seek for those things that are at hand and offer encouragement to our hopes; and let us realize that all things are equally worthless, outwardly differing in appearance, but inwardly alike vain. And let us not envy those above us. This seeming elevation is but the edge of a precipice. . . . There are many . . . on a lofty pinnacle from which they can descend only by falling; but this very fact is evidence that their burden is a very great one, because they are compelled to be burdensome to others, and that they have not merely been raised to their high position but have been pilloried there. Let justice, human kindness, liberality, prepare resources for the lot which awaits them, in the hope of which they may hang on the edge of the abyss with greater security. Nothing, however, will serve better to emancipate these men from these mental upheavals than always to fix some limit to the increase of their greatness, and not to leave the decision of their retirement to fortune, but to halt far this side of the boundary set. In this way the mind will feel the goad of desires, and will not be carried on into a limitless uncertainty, since its desires are limited."*

"This discourse of mine is not addressed to the sage, but to those who are imperfect and whose health is not assured. The sage must advance neither timidly nor step by step. For so great is his confidence in himself that he never hesitates to meet Fortune and will never give ground to her, . . . since not only his slaves and possessions and rank but his body also and eyes and hands, and whatever renders life dearer to him, nay, his very self he reckons among things committed merely to his care, and he lives as though lent to himself and is prepared to return them on demand without regret. Nor is he cheap in his own eyes because he knows that he does not belong to himself; on the contrary, he will do all things with all the care and caution with which a scrupulously honest man guards all that is intrusted to his keeping. But whenever he is bidden return the

* *De Tranquillitate Animi,* X.

loan, he utters no complaint to Fortune, but says, 'I thank you for what I have possessed and had. It is true that I have cared for your property at great expense, but at your demand I give, I yield it up gratefully and willingly. . . . My treasures, my home, my family I give back, I restore.' If it is Nature who has summoned us to restore what she previously intrusted to us, we shall say to her also, 'Take back a better soul than you gave. I do not refuse nor do I seek to flee. I restore to you voluntarily that which you gave me without my knowledge.' . . . He will have a bad life of it who does not know how to die properly. Therefore the first value set upon life must be revised and the spirit must be reckoned among things of little value. . . . The fear of dying is often the cause of death. Fortune, who stages shows for her own amusement, says, 'Why should I reserve you, evil and cowardly creature? You shall be wounded and stabbed the more, because you do not know how to present your throat; but you shall live longer and die more easily who do not draw back your neck from the iron nor resist with your hands but receive it with courage.' . . . But if we know that from the instant even of our conception the verdict has been passed upon us, we shall live following the order of nature; and the same strength of mind will prevent our regarding any event as unexpected. For by foreseeing the possibilities the future has in store for us we shall lighten the force of all evils which bring no surprise to those who are prepared and ready for them. Misfortunes come to those who believe themselves secure and are concerned with their own happiness alone. For sickness, captivity, ruin, fire, are by no means unexpected. I knew that Nature had shut me in a troubled lodging. I have heard so often the funeral lamentations in my neighborhood, I have seen pass so often before my door the torches and wax tapers attending the funeral processions of those who died young. Often the crash of a falling building has met my ear. A single night has carried off many of those with whom the forum, the senate-house, conversation had brought me in contact, and severed hands joined in friendship; ought I to be surprised that the dangers have at last reached me which have always been roving about in my vicinity? . . . Publius, a more forceful writer than the authors of tragedy and comedy, has said:

'That which can happen to an individual man can happen to all.'

If any one has grasped the full meaning of this maxim, and has viewed all the numberless and daily mishaps of others with a realization that he himself is equally exposed to their assaults, he will arm himself long before he is attacked. Too late is the mind trained to endure dangers after it has been exposed to them. I did not think that this would happen. . . . What riches are there which are not followed by indigence, hunger, and beggary? . . . What kingdom is there for which overthrow and ruin do not lie in wait? . . . Nor are such changes separated by long intervals, but the short space of an hour reduces a monarch to a suppliant. Know then that every condition is subject to change, and whatever can happen to another can happen to you also. You are rich; not richer than Ptolemaeus, I suppose? . . . He begged for the rain drops: he perished of hunger and thirst. . . . On the same day on which the senate escorted Sejanus, the populace tore his body in pieces. . . . Therefore in such constant vicissitudes of rising and falling fortune, if you do not regard the possibilities the future holds in store for you, you are lending force to your own adversity, a force which he has already broken who has foreseen it."*

"We must curtail restless activity, such as engages most men as they wander about their homes and the theaters and the forums. They busy themselves in the affairs of others always with an assiduous air. If you ask one of these men as he leaves his house, 'Where are you going? What have you in mind?' He will answer you, 'I do not know; but I shall see some people, I shall do something or other.' Without any definite aim they go to and fro seeking for things to do; . . . their inconsiderate and useless course is like that of ants crawling on trees which mount to the top and then descend without carrying anything. . . . Therefore every labor should have some definite purpose, some definite end in view. . . . Not real diligence but false views of life actuate these restless madmen. . . . In the same way each one of those whom vain and trifling motives lead hither and thither through the city and who sally forth to swell the crowd . . . some run as to a fire or to salute some one who will not return their greeting, or to escort the dead body of a man unknown to them. . . .† On this evil hangs that heedless

* *De Tranquillitate Animi,* XI.

† Bacon has altered this sentence greatly.

vice of prying into secrets of state and of learning many things which are neither told nor heard with safety."*

"It is this consideration, I suppose, that led Democritus to begin as follows, 'To live in tranquillity one must abstain from many public and private matters.' He had reference to superfluous things. . . . For he who engages in many enterprises frequently gives Fortune a hold upon him, with whom it is the safest course not to have any dealings, but always to keep her in mind and to put no confidence in her loyalty. . . . This is why we say that nothing can happen unexpectedly to the sage. We are not exempting him from the accidents to humanity, but from humanity's errors: nor do all things turn out for him as he wished, but as he thought. But first of all he perceived that his projects might meet obstacles. But lack of success in a cherished plan is necessarily less painful when one has not promised himself its realization."†

"We ought to cultivate a complaisant disposition, that we may not indulge too far our fixed ideas and projects, and may accommodate ourselves to changed conditions, accepting without fear such as are dictated by our position or by wisdom, without, however, becoming the victims of inconstancy, a vice most hostile to tranquillity of mind. For obstinacy, of necessity, is an anxious and wretched condition of mind; . . . inconstancy is much worse, since it in no way sets a curb on itself. There are two excesses equally deadly to tranquillity, the impossibility of changing and the impossibility of enduring anything. Undoubtedly the mind must retire into itself from all external interests; the mind should trust itself, be pleasing to itself, have regard for its own interests, withdraw as far as possible from the concerns of others, and cling to itself; it should not feel losses, and should put a kindly interpretation even upon adversity. Zeno the philosopher on learning of a shipwreck in which he lost everything merely said, 'Fortune bids me take up philosophy with fewer encumbrances.' A tyrant was holding out the threat of death and loss of sepulcher to the philosopher Theodorus. 'You can please yourself,' the philosopher said, 'I have a measure of blood at your disposal; as regards my burial you are indeed a fool if you think that it mat-

* *Ibid.*, XII.

† *Ibid.*, XIII.

ters to me whether I rot above ground or beneath it.' Canus Julius, a man . . . of true greatness, . . . on being condemned to death after a long dispute with Caius Caesar said, 'I thank thee, excellent prince.' . . . He was playing chess when the centurion who was leading to execution a band of other victims bade him be summoned also. At this order he counted his pieces and said to his companion, 'See that you do not after my death state falsely that you won.' Then nodding to the centurion he said, 'You will be my witness that I am one point ahead of him.' . . . His friends were sad at the thought of losing such a man. 'Why are you down-hearted?' he said. 'You ask whether the soul is immortal: I shall soon know.' He did not cease to search for truth even at the very end, and to make his own death a subject of investigation. . . . And at length the eminence was close at hand on which a daily sacrifice was offered to Caesar. A friend said to him, 'Of what are you now thinking, Canus? Or what idea is in your mind?' 'I have determined,' said Canus, to observe in that fleeting moment of death whether the soul is conscious of its own departure.' He promised that, if he discovered anything, he would go about and make known the state of the departed. Lo! a calm in the midst of a storm. Lo! a soul worthy of eternity, a soul that summons its own condition to bear witness to the truth, and at the very end of life makes an inquiry regarding its own departure, continuing to learn not only up to the moment of death, but even from death itself gaining information."*

"But it is vain to have rid ourselves of the causes of personal unhappiness" unless we overcome those that are general. "For sometimes a hatred of the human race takes possession of us, and the throng of so many wrongs that prosper passes in review; . . . Honesty is rare . . . and innocence unknown. . . . We must accustom ourselves, therefore, to look upon the vices of the multitude not as hateful but as ridiculous, and to imitate Democritus rather than Heraclitus. For the latter used to weep, the former to laugh, whenever they appeared in public. All our actions seemed tragic to the one and comic to the other. . . . There is the added fact that he is more serviceable to the human race who laughs at it than he who mourns over it; the former leaves something at least to kindly hope, while the latter fool-

* *De Tranquillitate Animi,* XIV.

ishly weeps over what he regards as without hope of correction. And after a general survey he is possessed of a greater mind who does not restrain his laughter than he who does not refrain from tears. . . . Let each one of us represent to himself all those things that render him sad or joyous, and he will perceive the truth of what Bion has said, 'All the actions of men are like mere beginnings, and their life is neither a more revered nor a more serious matter than their conception.' But it is better to view without emotion public morals and human vices and to refrain from both laughter and tears. For to torture oneself with the evils of others is a never-ending misery, and to take delight in their ills is an inhuman pleasure."*

"We shall pass to the case that habitually and rightly depresses us; . . . I mean the unhappy end of some good men. Socrates, for example, is forced to die in prison, . . . Cicero to offer his neck to his own clients, Cato, . . . the living image of the virtues, by falling on his sword to put an end to his own life and the republic as well. . . . Consider how each of them bore his fate, and wish for the courage of those who showed themselves brave. If they perished like weaklings and poltroons, no loss was suffered. . . . I shall weep over no man who rejoices" in meeting dangers, "I shall weep over no man who weeps when confronted with peril: the former has himself wiped away my tears, the latter by his tears has rendered himself unworthy of any."†

Chapter XXI

Since according to Scripture the body which is corrupted burdens the soul, and our earthly habitation is depressing to the sense that is cognizant of many things, for peace of mind, therefore, human frailty must necessarily relax the mind at times through the comforts and recreations necessary for the body. For otherwise the spirit becomes anxious, dull, fitful, and gloomier than it should be, languid with the tedium of goodness, complaining, and prone to frequent fits of impatience and anger. For this reason the most godly men would seek relief

* *Ibid.,* XV.

† *Ibid.,* XVI.

sometimes from their spiritual cares, soften at times the rigor of their abstinence, and moderate their excessive vigils. The most blessed John the Evangelist as a relief to his human frailty used to amuse himself with a partridge, as we learn from the Collations of Cassianus; and the blessed Benedict, when he was rebuked by a passing traveler for diverting his monks with physical pleasure, replied by asking the stranger to stretch a bow more and more. On the stranger's replying, "I will not do so, because I should break the cord," Benedict drew the parallel that he could force his monks to the rigor of penitence to such a degree that they could not endure it, but would break under the strain. And for this reason the Patriarch Jacob excused himself from accompanying Esau, fearing lest, if he tried to keep pace with him, his sons and his wives and his cattle would die.

Hence it is that Seneca wisest of men, who was granted a revelation accorded by God to few, as the Apostle says, realizing the weight of human frailty, took up the question of relaxation and recreation for the body at the end of his essay on Peace of Mind, in order that by introducing remedies of this kind our vigor may be more strongly restored. For even Cato says, "Intersperse sometimes joys with your cares," so that we may say to God with the Psalmist, "My strength will I ascribe unto thee." Seneca accordingly touching upon solaces of this kind speaks as follows: "The world and the retreat are two things that should intermingle and one succeed the other. The latter will cause us to long for our fellow men, the former for ourselves, and the one will be a remedy for the other. Solitude will relieve the irksomeness of the throng, and the throng that of solitude. The mind should not be kept at the same tension, but should be relieved by witticisms" and recreations for the body besides. "Socrates did not think it beneath his dignity to play with children. Cato used to refresh his mind with wine when he was wearied with affairs of state. Scipio, after so many triumphs, did not disdain to move his martial body to the rhythm of the dance, not affecting, as is the custom today, those soft attitudes and wanton movements which give to our dancing an effeminate air, but dancing in that manly fashion which characterized the heroes of old on festive days, and which would not provoke unfavorable criticism even if they were seen by the enemies of their country. We must give our minds relaxation; they are better and more active after a rest. Just as we must not constantly force fertile fields, since an uninterrupted productiveness will soon exhaust them, so unremitting toil will break our mental vigor. Repose for a short period and relaxation will restore our powers of mind. A certain dullness and languidness are produced by incessant mental effort. If play and jest did not possess some natural attraction, men would not be so strongly drawn to them. Frequent indulgence in them, however, will rob the mind of its force and gravity. For sleep is necessary to refresh the body, but if it continues day and night it will become death. There is a great difference between relaxing a thing and decomposing it. Our legislators appointed feast days, in order that men, assembled for public amusements, might find some intervals of relaxation necessary for their work; and illustrious men were in the habit of observing each month certain holidays. Some of them in fact would divide each day between leisure and business, whom no matter of business detained beyond the tenth hour. They would not even read a letter after this hour for fear that it might prove to be the source of fresh worry, but during the two remaining hours they would refresh themselves from the fatigue of the whole day. . . . Our ancestors forbade any new motion in the senate after the tenth hour. The mind requires indulgence and rest from time to time to serve in the nature of nourishment and renewal of its powers. It is also essential to take walks in the open country; a free and abundant air gives the spirit a new tone. Sometimes a ride, a journey, a change of scene, a banquet and a somewhat great indulgence in drink will give us renewed vigor; . . . for wine drowns cares, lifts the soul out of its depression, and is a cure for sadness as well as for certain ills of the body. Liber, who is Bacchus, has been called the inventor of wine, not owing to the freedom of speech he imparts, but because he frees the mind from the servitude to care and makes it more daring in all its efforts. . . . It is generally believed that Solon, one of the seven wise men, and the philosopher Arcesilaus were fond of wine. Cato has been charged with drunkenness. He who casts this reproach will find it easier to prove the vice of drunkenness honorable than to prove Cato base. But moderation in both liberty and wine is beneficial; we must not repeat the remedy too often, lest the mind acquire a bad habit. And yet there are times when the mind should be aroused to a state of exultation and freedom and permitted to throw off for a season its gravity and sobriety. For if we are to believe a Greek poet, 'It is sometimes pleasant even to lose one's reason,' or Plato, 'In vain has a man in his senses knocked at the door of the Muses,' or Aristotle, 'There has never been a great genius without a strain of madness.' The mind unless stirred cannot utter anything sublime, surpassing the speech of others. It is only when it has spurned all that is common and ordinary and has mounted aloft with divine inspiration that it has uttered something too sublime for human lips. The mind cannot reach the realms of the sublime while it remains sane. It must leave the beaten track, dash forth, take the bit in its teeth, and carry its rider where he would have feared to mount himself."*

Seneca† ends his discourse at this point, maintaining that he strengthened the vigor of his mind by proper relaxation at times, and that he would make greater efforts than if he devoted himself constantly to spiritual and mental activities. The prophet Elijah bade the lute be brought, in order that his soul, stirred by the delight of physical harmony, might be caught up to contemplate divine things.

* *De Tranquillitate Animi,* XVII.

† Bridges has the following note: "The foregoing extracts from Seneca contain many variations from the text now usually adopted. Some of these are evidently due to Bacon himself, occurring naturally as the result of condensation. Some may be due to the imperfection of the text used by him. Other errors are doubtless due to the incompetence of Bacon's transcribers. Some of the more obvious mistakes have been corrected. But no attempt has been made to elevate this series of extracts to the standard of a critical edition."

FOURTH PART OF MORAL PHILOSOPHY

Chapter I

I HAVE dwelt upon this third part of Moral Philosophy at greater length than I was required to do, owing to the beauty and helpfulness of moral sentiments, and because the books are rare ones from which I have gathered these roots, as it were, flowers, and fruits of morals. I now wish to take up the fourth part of this science, which although not so rich and abounding in material as the third part, is, however, more wonderful and more exalted than not only this third part, but also than all the other divisions of the subject, since it consists of a plea for belief in the religion which the human race should accept, for loyalty to it, and for confirmation of it by our deeds. There is nothing else within the range of philosophy more necessary for man or of such great utility and worth. For because of this part of the subject is it especially true that all the other sciences are subordinate to Moral Philosophy. For all wisdom is constituted with a view to the discovery of salvation for the human race; and this salvation consists in the perception of those things that guide man to the felicity of the other life. Avicenna says that this felicity is such as eye hath not seen nor ear heard, as I stated above. And since this fourth part of philosophy has as its object the investigation of this salvation and the attracting of mankind to it, all the sciences, arts, and activities are attached to this most noble part of civil science; and this is the end of all human consideration.

For this reason it is most profitable to consider the purpose of this part of the subject; and it is fitting that every Christian should do so for the confirmation of his own profession, and to the end that he may have the means of correcting those in error. Verily, God can never deny to the human race a knowledge of the way of salvation, since he wishes that all men should be saved, according to the Apostle. And his goodness is infinite, wherefore he has always left men the means by which they might be enlightened to gain a knowledge of the ways of truth. Aristotle in his Politics takes up the different kinds of sects, and says that he wishes to consider out of the sects and laws of states four or five of the simple ones, and see which laws corrupt states and kingdoms and which do not. He says also that the four or five simple ones are corrupt sects, meaning that a sect or law is called simple because it has only one end in view, and is called complex when the end in view is complex, since every sect varies according to the end it has in view, as Alpharabius teaches us in his book On the Sciences, expounding the view of Aristotle regarding the sects. According to Alpharabius, and still more clearly according to Boetius in the third book of the Consolation of Philosophy, these simple aims are pleasure, riches, honor, power, fame, or glory of name.

I shall now state the principal nations in which the various sects are found that are now existing throughout the world, namely, Saracens, Tartars, Pagans, Idolaters [Buddhists], Jews, Christians. For the principal sects do not exceed this number nor can they do so until the sect of Antichrist appears. There are, moreover, sects formed from all of these, or from four, three, or two of them, according to various combinations.

But besides these aims there is another, namely, the felicity of the other life, which is sought and striven after in different ways. Because some believe that it consists in the delights of the body, some in the delights of the soul, and still others in the delights of both. In addition there are sects that include this felicity and all the others aims, or at least many of them, and that too in various ways. For although they all have in view a future felicity, yet many men abandon themselves to pleasures, and others are eager for riches, and certain aspire after honors, while others desire sovereign power, and still others crave glory and fame. I shall touch first upon the three divisions of the sects, that the end in view may be made clear. Then I shall deal with the choice of the sect of the faithful, which is the only one that should be spread throughout the world.

Some desire to confine those aims to the present life, without considering that they are losing future felicity, abusing temporal blessing, and yielding themselves up to the allurements of pleasures, as, for example, the Saracens, who, in accordance with their law, take as many wives as they wish.

Some are inflamed with the lust for dominion, as the Tartars, whose emperor says that there should be only one ruler on earth just as there is only one God in heaven, and that he himself ought to be constituted that ruler, as we see in the letter he sent to the Lord Louis, King of France, in which he demands tribute from him. This is recorded in the book of Friar William on the Manners of the Tartars, which he wrote to the King of France just mentioned. From their actions it is clear in what way they have already gained possession of the kingdoms of the East. They care nothing about the refinements of life but are quite barbarous in this respect. They use the milk of mares for drink, and make a practice of eating unclean foods, as we learn from the book just mentioned and from that of Friar John on the Life of the Tartars, and also from the Cosmography of the philosopher Ethicus. For that philosopher and those books on the manners of the Tartars describe this very foul and wicked race, as we stated in the section on Mathematics describing the races and localities of this world.

Real Pagans, like the Praceni and the nations bordering on them, live according to custom and not according to laws based on reason. Their interests are centered in the delights, riches, and honor of this life, and they believe that the life to come is similar in every way to this. Hence at death they have themselves burned publicly, together with their precious stones, gold, silver, equipment, family, friends, and all their wealth and goods, hoping to enjoy all these things after death.

Similarly the Idolaters expect to possess a future in the enjoyment of all the blessings they had in this world. Their priests, however, take a vow of chastity, and rejoice in abstinence from the delights of luxury, as we learned in regard to the regions of the East lying toward the north in the section of this work dealing with the localities of the world. All these expect corporeal blessings in the life to come, and have no discernment regarding spiritual blessings. It is not contrary to their laws to seek for this world's goods in any way possible, nor do they believe that in any event are they deprived of the life to come.

But the Jews hoped for blessings both temporal and eternal; in a different way, however, since with spiritual discernment by virtue of their law they aspired after blessings not only of the body, but also of the soul. But interpreting their law in its literal sense they believed the blessings of the law only corporeal. Similarly according to their law they seek temporal possessions not as a matter of right and wrong, but on the authority of God and in accordance with his laws. For although they plundered many nations and subjugated them, they had a just right to do this. For the Promised Land was due them by hereditary right, because they were descended from a son of Noah; and the sons of Cham invaded those lands without any right, since they did not fall to their lot at the beginning. For Egypt, Africa, and Ethiopia were given to the sons of Cham, as we learn from Scripture, the sacred writers, and the histories. We have touched upon this topic also in a previous section.

The Christians, securing spiritual blessings by spiritual means, according to their law are able to have temporal blessings, owing to their human frailty, in order that they may practice the things of the spirit in this life, to the end that they may at length reach the life eternal both in body and in spirit. And yet in that life they will live without the extrinsic things men employ in this present life. For the animal body will become spiritual, and the whole man will be glorified, and will live with God and the angels.

These then are the principal sects. The first is that of the Pagans, who have less knowledge about God and are without a priesthood, but each man fashions a god to his own liking and worships whatever he pleases and sacrifices at will. Then come the Idolaters, who have priests and synagogues and huge bells like those of the Christians, with which they are summoned to their office and to certain prayers and regular sacrifices; and they maintain that there are many gods, but that no one of them is omnipotent. The Tartars are third, who adore and worship one God as omnipotent. But they nevertheless venerate fire and the threshold of the home. For they cause all things to pass through fire; for this reason they cause the effects of the dead and gifts and messengers to pass through flames and other things, that they may be purified. For their law declares that all things are purified by fire. Moreover, he who treads on the threshold of the house is condemned to death. Both in these two matters and in certain others they are quite barbarous. In the fourth class are the Jews, who should have had a better knowledge of God in accordance with their law, and should have truly desired the Messiah, who is Christ. And those have actually done so who knew the law spiritually, as the holy Patriarchs and Prophets. In the fifth place are the Christians, who accept the law of the Jews spiritually and complete it by adding the faith of Christ. Last of all will come the law of Antichrist, who shall subvert the other laws for a time, except that the elect in the Christian faith shall remain firm, although with difficulty, owing to the fury of the persecution. There are, then, six laws according to this division, and six according to the former one, each with its special aim, pleasure, wealth, honor, power, fame, and felicity of the life to come apart from these temporal blessings.

In the relationship of Mathematics to the Church discussed in a preceding section the sects were considered with reference to planetary influences and were found to be six, namely, the law of Saturn, the law of Mars, the law of the Sun, the law of Venus, the law of Mercury, the law of the Moon. In whatever way, then, we may distinguish the sects, we shall find that they are always six in number. For the first and principal distinction is this one in regard to the planets; upon that the other distinctions follow, since the celestial influence inclines a man to receive the laws, either as a controlling influence, or as an important one, or as one that renders the reception easier. For although the rational soul is not forced to anything, yet, as we proved above, a man's habit of thought is much altered with regard to the sciences, morals, and laws. By these changes the soul is stimulated as far as the action of the body is concerned, and is influenced to acts both public and private, with complete freedom of the will, however, in all matters. And hence it comes to pass that, in accordance with the conjunctions of Jupiter with the other planets, six changes appear in the laws and habits in men's hearts, as we noted above. There cannot be more than the six sects; and at times the followers of one sect incline to another, owing to the strong influence of the constellation, and sometimes they change their own sect entirely, or in great part, or modify it with the tenets of another sect in accordance with the occurrence of the several constellations. It is in this way that sects become composed of the parts of several others. For the Saracens, although they follow in the main the law of Venus, yet derive much from the Jewish law and from the Christian law, since they employ various baptisms and sacrifices quite similar in part. They say, moreover, that Christ is the Son of the Virgin and the greatest of the prophets, and they retain many words of the Evangelists in their law, and they all wish to die the death of Christians. They made this declaration in the disputation they held with Christians and Idolaters in the presence of the great emperor of the Tartars, as we learn in the book on the Manners of the Tartars. But the Tartars profess chiefly the law of Mars. For they venerate fire, are eager for war, and devote their leisure to the principles of philosophy, like the ancient Chaldeans, to whom the law of Mars is attributed. They, however, are modified by the law of Mercury. For they permit their sons to be instructed in the Gospel and in the lives of the Fathers. When they become sick they ask for Christian priests and the cross and holy water, as we learn in the book of Friar William, and as the experience of men worthy of our belief renders us certain in this matter. The law of the Jews is that of Saturn; and the law of the Christians is called that of Mercury by the astronomers for certain reasons noted above. The sects of the Pagans and Idolaters are reduced to the sect of Mars and that of the Egyptians, a sect worshiping the sun, which is the leader of the heavenly host. For these two sects worship the creature instead of the creator, since the Idolaters worship images made by hand and the heavenly bodies. And therefore in as far as they worship the celestial natures they agree with the Egyptians. In as far as they worship these objects here below they are in accord with the law of Mars. But the Pagans worship both terrestrial and celestial things. For whatever they meet with of a useful character, as the sun, moon, an animal, a grove, water, fire, or anything else they worship from affection. But whatever they find terrible they worship from fear.

After the statement of these principal sects as regards the following they have among the nations, the influence of the planets upon them, and the diversity of their aims, the consideration follows of the necessary means to be employed in persuading men of the truth of a sect. I said above, in the section on Mathematics, in regard to the conversion of unbelievers, that persuasion of the truth as alone contained in the Christian religion is a twofold matter, since we may appeal to miracles which are beyond us and beyond unbelievers, a method in regard to which no man can presume; or we may employ a method familiar to them and to us, which lies within our power and which they cannot deny, because the approach is along the paths of human reason and along those of philosophy. Philosophy belongs in an especial sense to unbelievers, since we have derived all of our philosophy from them, and not without the most weighty reason, to the end that we may have on our own behalf a confirmation of our faith, and that we may be able to speak convincingly on behalf of the salvation of unbelievers. Nor should the statement of Gregory be urged as an objection that faith has no place where reason furnishes the proof. For this statement is to be understood of the Christian who relies only or mainly on human reason. But this should not be: nay, we must put our faith chiefly in Church and Scripture and the Sacred Writers and Catholic doctors.

But as a solace for human frailty, to the end that it may avoid the attacks of error, it is useful for the Christian to have effective reasons for those things which he believes, and he should have a reason for his faith for every case requiring it, as the blessed Peter in his first Epistle teaches us, saying, "But sanctify in your hearts Christ as Lord; and be ready always to give an answer to every man that asketh you a reason of the faith and hope that are in you." But we are not able to argue in this matter by quoting our law nor the authorities of the Sacred Writers, because unbelievers deny Christ the Lord and his law and the Sacred Writers. Wherefore we must seek for reasons in another way which is common to us and to unbelievers, namely, philosophy. But the power of philosophy in this particular is in perfect accord with the divine wisdom, nay, is a vestige of the divine wisdom given by God to man, that by this vestige man might be stimulated to comprehend divine truths. Nor do these things belong exclusively to philosophy, but are common to theology and philosophy, to believers and unbelievers, given by God and revealed to philosophers, to the end that the human race might be prepared for particular divine truths. And the reasons of which I speak are not alien to the faith or outside of its principles, but are drawn from its roots, as will appear from what is to be said.

I might indeed be able to present these methods in a simple

and crude manner within the grasp of the unlettered multitude of unbelievers, but this is not expedient. For the multitude is too imperfect, and for this reason a plea for the faith that is within its grasp is crude, undigested, and unworthy of the learned. I wish, therefore, to proceed on higher ground and to present a plea concerning which the learned can judge. For in every nation there are some active individuals who are fitted to receive wisdom who can be persuaded by force of reason; so that when these men become enlightened, the persuasion through them of the multitude is much easier.

I assume in the first place that there are three kinds of knowledge; one comes from study of our own devising along the pathway of experience. The second is that which is learned from others. The third precedes these and is the pathway to them, and is called natural knowledge, being so named because common to all. For that is natural which is shared by all members of the same species, as, for example, burning is natural to fire, as Aristotle explains in the fifth book of the Ethics; and Cicero makes this same statement in the first book of the Tusculan Disputations, and we see it exemplified in numberless ways. For we say the cries of animals have a natural signification because they are common to the individuals of their species; and things of this kind are naturally known by us in which we all agree, as, for example, that every whole is greater than its part, and the like, both simple and complex. For we know that the rational soul is formed to learn the truth and to love it, and the proof of this love is shown in our deeds, according to Gregory and all the sacred writers and philosophers. Certain, however, think that there are two distinct divisions in the rational soul, or two faculties; that by one of these the soul learns the truth, and by the other desires to hear the truth when learned. But certain believe that there is one substance of the soul that performs both functions, because these acts are related to each other, since the knowledge of the truth is owing to the love of it; for it is one and the same faculty. According to these thinkers the mind first perceives truth, and then loves it when known and carries it into practice. Hence Aristotle maintains in the third book on the Soul* that the speculative intellect through the extension of truth to the love of it becomes

* *De Anima,* III, 7.

the active intellect. Nor does he ever make a specific difference between the speculative and the active intellects, as he does between the intellect and the sense and the vegetative soul. For he proves in the second book on the Soul that these three are different in species, because their actions are different in species, namely, intelligence, feeling, and growth; nor are they related to one another. But the knowledge of the truth is related to the love of it and is because of it; and for this reason there is one faculty, or nature, or substance of the rational soul that learns the truth and loves it. Hence in the third book on the Soul Aristotle thus begins: "Concerning the part of the soul with which it learns and understands I must now speak;"* maintaining that it is the same part that has both functions; just as it is in the sensitive soul; because it is the same faculty that perceives and desires, as is evident in every sense. For touch perceives warmth and seeks for it, and so taste with savor, and the same is true of the other senses.

But as regards these matters it is not of great moment how we define the faculty. For we know that the rational soul is formed to learn the truth and to love it. But the truth of a religion is perceived only so far as the knowledge of God abounds in the individual, because every religion is referred to God; and therefore he who wishes to arrive at a definite knowledge of a religion must begin with God. But the knowledge of God, so far as the question of his existence is concerned, is known to all naturally, as Cicero shows in his book on the Immortality of the Soul. And he proves it with this statement,† "No nation is so savage and barbarous as not to have some conception of God, nor is there any nation without some form of divine worship." But should Avicenna say in the first book of the Metaphysics‡ that this science seeks to prove the existence of God, we must reply that this is true as regards full certitude. For the natural knowledge possessed by the individual in regard to God is feeble, and is weakened by sins which are numerous in every one. For sin darkens the soul, and especially is this true with regard to divine things.

Therefore this knowledge needs the support of argument and

* *Ibid.*, III, 4.

† *De Natura Deorum*, I, 16, 23.

‡ *Metaphysics*, I, 1.

faith. But the knowledge of God's unity and of his other attributes is not known by nature. For in these particulars men were never in agreement, some maintaining that there are many gods, others believing that the stars are gods, and still others that things here below are deities, as, for example, the pagans and the idolaters. And therefore they must err in their religion. All others who say that there is only one God do not understand other truths regarding him. And therefore the advocate of a religion in the first place must know how to answer the questions that are asked about God in general. He must not, however, enter into a discussion of all the particular truths at once, but should proceed gradually and begin with the easier topics in this way. For as the geometer gives his definitions in order, that the symbols and terms he employs may be known, so should we proceed in religious instruction; for unless we know the meaning of the terms used, we shall arrive at no proof.

God, then, is the first cause antecedent to which there is no other, a cause that did not emerge into being nor will it be possible for it to cease to be, of infinite power, wisdom, and goodness: the Creator of everything and the director of every man according to the susceptibility of individuals to such guidance. And in this definition Tartars, Saracens, Jews, and Christians agree. The learned also among the Idolaters and Pagans cannot contradict when the reason for this has been grasped, nor can the multitude as a consequence whom the learned direct as rulers and guides. For there will be a double method of proof on this point: one is the consent of all other nations and religions and of the rest of the human race. But the minority must conform itself to the majority; and the part is disgraced that is not in accord with the whole. It is generally agreed that the wiser men are found among the other sects, and the Pagans and Idolaters are not ignorant of this fact. For when we meet them they are easily convinced and clearly perceive their own ignorance; as was shown by the emperor of the Tartars, who summoned before him Christians, Saracens, and Idolaters to confer on the truth of their religions; and forthwith the Idolaters were confounded and convinced. This fact is stated in the book on the Manners of the Tartars addressed to the present Lord King of France. And when Christians confer with Pagans, like the Praceni and the other adjoining nations, the latter are easily convinced and perceive that they are in error. The proof of this is found in the fact that they would become Christians very gladly if the Church were willing to permit them to retain their liberty and enjoy their possessions in peace. But the Christian princes who labor for their conversion, and especially the brothers of the Teutonic order, desire to reduce them to slavery, as the Dominicans and Franciscans and other good men throughout all Germany and Poland are aware. For this reason they offer opposition: hence they are resisting oppression, not the arguments of a superior religion.

Then the advocate of the religion of the faithful has on the part of Metaphysics and this Moral Science another method of procedure in giving his proofs, which I merely wish to hint at, until the treatise is finished which your Highness has requested. And in fact we can present to the assiduous man who is amenable to the force of reason the following as an acceptable proposition. Causes do not go back endlessly, since they cannot be infinite in number nor can they be so conceived of. For all things that exist and that are perceived are included within some number, as Aristotle says in the third book of the Physics. There is not therefore a cause preceding a cause endlessly. Therefore we must stop at some first cause, which does not have a cause antecedent to it, and the whole number is reduced to a single one. And in every class one must be first to which all the rest are reduced. But if this should be the first cause without any other cause antecedent to it, it is evident that it did not come forth in being through some cause; nor is anything else the cause of its being, nor does it cause itself to exist after non-existence, since in that case, while it was non-existent, it would have existence, so as to cause itself to be. For everything that causes something else to be after non-existence, has being while it is performing this act; therefore nothing is the cause of its own being. Wherefore this first cause never had non-existence, and therefore has always existed. But if this is granted, then it will always exist, since there are many things that will always exist, and yet have not always existed, as angels, souls, the heavens, earth, and the like; and for this reason that which never had non-existence will far more easily preserve its existence eternally. Moreover, that which never had non-existence is infinitely removed from non-existence, and therefore it is impossible that it should pass into non-existence. For some things that have come forth into being can cease to exist, because they are not infinitely removed from non-existence. For at one time they did not exist: therefore since non-existence is infinitely removed from that which has always existed, there is no comparative relationship between them. And therefore such a thing will not be able to cease to be; and this statement is more readily accepted than any of the other statements here made, and for this reason is rather a matter of perception than one in need of proof.

But it is evident that the thing that has always existed and shall exist is of infinite power. Because if it is of finite power, then is its power imperfect, since in every finite thing something can be added, and every imperfect thing is naturally subject to change; but it is not possible to assume some change unless a first one be assumed. For what is first naturally precedes what comes after. And therefore since the first change is in regard to existence and non-existence, this change must be possible in that which has finite power. But this change does not happen in that which has always been and always shall be: wherefore there is no limit to its power.

Likewise Philosophy argues in the third book on Consolation* in this fashion. In every class where the imperfect is found it is natural that the perfect be found. And therefore in the genus of power we must find a perfect power, since we find an imperfect one. But that is perfect from which nothing is lacking nor can something be added, according to Aristotle in the third book of the Physics and the fifth of the Metaphysics. And that to which nothing can be added is infinite; because to everything finite in so much as it is such, an addition can be made, and something else can be understood beyond it. Therefore perfect power must be infinite. But in things apart from this cause which we are seeking there is not a power that is perfect and infinite, therefore in this cause there will be such power. But if its power is infinite, then is its essence infinite, since power does not exceed essence. For essence is either equal to power or greater. Already the proofs on this point have been stated in the discussion of matter. It is manifest, therefore, that the essence of the first cause is infinite.

* *Consol. Philosophiae,* Bk. III, Prosa X.

And surely if its essence and power are infinite, its goodness must be infinite, since a thing whose essence is finite has finite goodness. Therefore infinite essence will have infinite goodness; and otherwise there is no comparative relation of goodness to essence in this cause; which cannot be in so great a majesty. And if its goodness were finite, it would be imperfect, and something could be added to it, and it could be lessened, and it could thus be subject to change; and for this reason it is natural for it to have non-existence, just as we proved above with regard to power. But that which has infinite majesty in essence and power and goodness cannot possibly lack knowledge, since a thing which is of such a nature has usefulness, nor can it be referred to the infinite majesty like the elements and stones and vegetation.

Then we see that things lacking infinite power, as animals and men and angels, have knowledge, owing to the nobility of their nature: therefore since the nature of the cause now sought for is infinitely nobler than anything of this kind, it will have the power of knowledge. But since all the other things that are in it are found to be infinite, this cause has infinite wisdom. Moreover, if it were finite it would be imperfect and subject naturally to increase and diminution, as we see in the case of other intelligent beings, and as is true in all imperfection. And therefore the first change which is with regard to existence and non-existence could be found here, as we argued above. Therefore the wisdom in this cause must be infinite; but if its power is infinite it can create this world, and its infinite wisdom knows how to make the best possible arrangement in regard to it, because it is characteristic of the best to do the best, and to communicate its goodness to others, as far as it is possible for them to receive it. Therefore this cause has necessarily produced the world.

Unless an objector might say that the world never came forth into being, and did not have a non-existence. But in that case it would be of infinite power, just like this cause, and therefore would be equal to it and would be God, a conclusion that no one deems worthy of a hearing, and which no religion maintains. For if the world has always been, and never had non-existence before existence, then it did not come forth into existence after non-existence. Nor at the same time did it have existence and non-existence, because contradictories are not true either at the same time or by nature. Therefore the world was brought into being, but by no agency except this first cause. Unless you say that there are more causes than one of this kind, which cannot be true, since no one of them would be of infinite power. Because if one is of infinite power it can do whatever it wishes, and therefore can act contrary to the will of another; therefore this other is not of infinite power since its will can be thwarted. And at any rate in one world there will be one God. For one God suffices for one world. And Aristotle in the eighth book of the Physics says that it is better to assume that there is one than more, since one suffices; but more worlds cannot coexist, as we showed in the chapter on Unity of the World. It is manifest, therefore, that there is only one God. Likewise if there were more worlds, one God would still suffice, because he is of infinite power, therefore he would be able to produce and rule all those worlds, since all of them, no matter how many in number, would not make something infinite.

But Aristotle* in the eighth book of the Physics and in the eleventh of the Metaphysics concludes that there is one prime motor, or one first cause, since there is one prime motion, namely, the daily one, and one world. And that there cannot be more worlds he himself proves in the first book of the Heavens and the World,† because the earth of this world would be similar in nature and species to the earth of another world. But things of the same nature have a natural motion to the same place, just as all heavy bodies naturally tend to move to the same point below. Therefore the earth of another world would tend to move to the same place to which the earth of this world naturally moves, since it is of the same specific nature. But the earth of another world could not move to the center of this world unless it passed through the circumference of the other world so as to penetrate it and at length fall to the center of this world. Since therefore this is impossible, it is not possible that there should be two earths or two worlds. There is therefore one God, who is the first cause of all causes, who has always existed and always will exist, possessing infinite majesty, power, wisdom, and goodness, the Creator and Ruler of all

* *Nat. Auscult.*, Bk. VIII.

† *De Coelo*, I, 8, 9.

things. Nor can there be more than the one God blessed for ever and ever, Amen.

After the Christian advocate has sufficiently verified this fundamental principle of religion, he should then argue further that man is bound to do God's will and to serve him with all reverence. For his majesty is infinite, as we have shown. Therefore boundless reverence is due him. But the blessing of creation is a boundless one in this, that it cannot be brought to pass except by infinite power. For no finite power is able to create, because there is an infinite distance between non-existence and existence. Therefore the transition from non-existence to existence must be through the infinite power of an agent. Wherefore the creature owes the Creator boundless reverence. Hence Avicenna says in the Elements of Morals that it is God's right that we should obey his precepts. For the creature must obey the commands of its Creator.

But a third reason is found in the infinite felicity he will bestow on those who obey him and the endless punishment he will inflict on those who are disobedient to him. Hence Avicenna expresses these reasons why we must serve God in the statement that he has prepared a promised felicity which eye has not seen nor ear heard for those who obey him, and for those who are disobedient he has prepared a terrible future punishment. This is also made apparent by Metaphysics and the first part of this Moral Science. For the soul is immortal, as Aristotle, Avicenna, and all the noble philosophers say; and Cicero has composed a book on this immortality. Seneca, moreover, avows this truth everywhere, and all sects aspire to this other life. Even the simple Pagans believe that they will live after death in body and soul, as I explained before; and therefore they hold that there is a resurrection of the dead. Hence the Saracens hold this belief also. But they have in mind only the physical enjoyments of this present life. For Avicenna says in the Elements of Morals that Mahomet offered only a glorification of our bodies, not of our souls, except in so far as the soul shares in the enjoyment of the body. The Jews also hold that there is a resurrection, and the Christians as well. And if the first and the last, that is, Pagans and Christians, agree in this belief, all the intermediate sects must necessarily believe in it; and for this reason the Idolaters like the others must avow it.

And the philosophers in particular teach this tenet, as I showed above in the first part of this Moral Science.

But it cannot be that a man pleasing to God in this life will have after the resurrection a life such as we have now, because this life is full of all miseries, and is not in accordance with the goodness of human nature. And therefore when the resurrection shall grant us an immortal state, we must of necessity be free from these miseries. Nor shall those who serve God meet with any misery, because it is just that they be given a reward according to the bounty of the divine goodness. And in the second place, the good as a rule have more misfortunes here than the evil. Therefore if God is a just judge he does not assign in this present life his rewards to the good, nor consequently will they be similar in the other life to the blessings in this. Therefore of necessity the just must be free from all misery; and since that state is perfect, there must be a glory of both body and soul. But the desire of the rational soul transcends all finite good. And therefore its desire cannot be satisfied except by an infinite good, and this is God; and therefore our human glory will participate in the goodness of the Deity; and from this the noblest corollary of philosophy follows, found in the third book of the Consolations,* namely, that by participation in the Deity men will become gods, although God is one by nature.

Similarly since the evil here offend the divine Goodness, which is infinite, and fall into the crime of treason against God, their punishment of necessity in the other life must be inexpressible and of infinite duration. Since these are the facts, a man should desire to please God by doing his will, to the end that he may so serve God that he shall escape intolerable punishment, and obtain infinite beatitude.

But if it behooves a man to do the will of God for the reasons given, namely, the dignity of an infinite majesty, and because of the benefit of creation into the existence of nature and of preservation therein, and because of the retribution to come, he should learn the divine will, and the things that belong to God, and the state of future felicity and misery. But man cannot of himself know these things, as is evident because of the errors and heresies and differences, nay, the contradictions of the principal sects, and not only of these, but because of the

* *Consol. Philosophiae,* Bk. III, Prosa X.

articles of the same principal sect. For among Christians we see so many differences: since some are heretics, some schismatics, some true Christians. The difference among heretics has no limits. Similarly among the Jews; for some were Pharisees, some Sadducees, and others differing from both, as the Gospels teach us, and Josephus makes clear in his books of Antiquities. Similarly the Saracens, for Avicenna and other philosophers contradict the rank and file and the priests. For they show that not only does glory belong to our bodies, but also, and in greater measure, to our souls; and they decide that that sect will soon be destroyed. So too the Idolaters have different sects. For those called Ingures, whose literature is possessed by the Tartars, hold that there is but one God, a belief not shared by many other Idolaters, as we learn in the book on the Manners of the Tartars. But the Tartars likewise are in serious disagreement. For some of them lean to the Christian rite, some to that of the Saracens, and others to that of the Idolaters, although they are members of the same principal sect. Among the Pagans there are still greater diversity and error, because the individual fashions a god according to his liking and worships what is pleasing to him.

But the power of knowledge in man does not suffice for the material things of this world and objects of sense, as is obvious to all. For no man knows with certainty and sufficiency the nature of the smallest thing, as, for example, a small plant, or fly, or anything else. And we see the utmost disagreement on the part of the learned in regard to the natures of material things. Much more then will man err in regard to things immaterial, as spiritual substances, and the state of the other life, and especially as regards those things which pertain to God and to God's will. Wherefore man must have revelation.

This is in accordance with Aristotle's statement in the second book of the Metaphysics, that the human intellect is so related to the most manifest things of this kind in its own nature as the eye of the bat and of the owl to the light of the sun. And Avicenna says in the first book of the Metaphysics that man's relation to those things that belong to the Lord of the ages in his kingdom is like that of a man deaf from his birth to the delight of musical harmony or of a paralytic to the delights of food. Wherefore man will not be able to attain certainty in these matters to a greater degree than the bat, owl, deaf man, or paralytic in the subjects mentioned. Moreover, Alpharabius says on this topic in his Morals that the untaught child bears the same relation to the man most learned in human wisdom as such a learned man to the divine verities: and therefore he will not be able to advance except through instruction and revelation. And he adds that if man could attain to the verities of sacred religion the world would not require revelation and prophecy. But both these, as he says, have been granted to the world and are necessary. Therefore man cannot unaided know verities of this kind. Avicenna makes this clear in his Elements of Moral Philosophy in his statement that religion requires a revelation from God; and that mortal man cannot pass to these immortal things. This accords with Seneca's statement in his book on Peace of Mind, that man is too mortal for immortal things.

Then, as man cannot unaided attain to these things, so should he not presume of himself to decide on these questions without revelation and instruction, for two reasons. One reason is found in the fact that those things that pertain to the other world, and in particular the will of God and matters pertaining to God, are of infinite greatness. Therefore they transcend our wretched humanity; nor is man worthy to try to discover and understand these matters of his own powers; wherefore it suffices for him to believe the one instructing him. Nor is he also worthy to believe those verities because of his sins and his wretched state. Let him rejoice, then, that he is taught by another, but not unless he be taught by God or by angels on God's authority. For it is obvious, therefore, that man should not attempt to inquire regarding these divine verities before he is taught and believes. Moreover, God himself has infinitely better and more certain knowledge of all that concerns him than the creature can have. The authority and wisdom of God are infinite, with respect to which no human authority has comparison or wisdom, and especially with respect to those things that are divine.

It remains, then, that the authority of God should alone be sought, since his goodness is infinite which wishes to reveal to the human race what is necessary for salvation. And without doubt we clearly see in the sects that all believe that their religions possess a revelation. For this is evident in the case of Christians and Jews. The Saracens also believe that Mahomet had a revelation, and he himself makes such a claim; otherwise he would not have been believed. Moreover, if he did not have a revelation from God, he had one from demons. The Tartars similarly say that God revealed their religion, as we find written in the book mentioned and as we are assured. The Idolaters and Pagans in like manner believe that God reveals all pertaining to religions of this kind: because man would not believe man otherwise in this matter. For every one who establishes a sect ascribes the authority to God that he may gain greater credence.

Since then these sects are so constituted we should be more fully persuaded that revelation should be made to only one perfect lawgiver, and that one perfect law should be given by God. And this is apparent because of divisions and heresies. For if there were more heads, the human race could not be united, since each man strives to defend his own view. Moreover, since there is one God and there is one human race to be regulated in accordance with God's wisdom, of necessity this wisdom must choose unity from the twofold unity just mentioned, or otherwise it would not conform to either one God or one human race. For if there were more gods and more worlds, and, so to speak, more human races, then there could be more than one divine wisdom. But the existence of many gods and many worlds is not possible. Therefore God's wisdom cannot be plural. Moreover, since the perfect religion should show man what he ought to know about God and all that is useful for the individual, another religion will be either superfluous if it treats of the same subjects, or erroneous if it promulgates opposing doctrines. Wherefore there can be only one perfect sect of the faithful, and likewise only one lawgiver who receives this revelation from God; because if there is only one religion, there is also only one lawgiver; and the contrary would also be true. This is also the teaching of Avicenna in the Elements of Morals, and of Alpharabius in the Morals. For, as Avicenna says, there must be one mediator of God and men, and a vicar of God on earth to receive the law from God and to promulgate it. Of this lawgiver Alpharabius says, "When it has been proved that he comes to us by inspiration from God he cannot possibly be false. And when we have established a truth of this kind, his word must remain unquestioned. Nor is further scrutiny of his statements permissible, nor must there be further consideration if we are to put implicit faith in him."

This point settled, we must then inquire who should be proclaimed as lawgiver and which sect should be spread throughout the world. The principal rites are first that of the Pagans, who live according to custom alone, and have no priests, but each man is his own teacher. The second rite is that of the Idolaters, who have a priesthood, agree in certain regulations, and assemble in one place at required hours to celebrate their rites. For they have large bells like those of the Christians, as I stated above. The Idolaters also differ from the Pagans. For the Idolaters worship images, while the Pagans worship natural objects like woods, waters, and the like in endless number. The third rite is that of the Tartars, who pursue eagerly philosophy and the arts of magic. The fourth is that of the Saracens. The fifth is that of the Jews. The sixth is that of the Christians. For there are no more principal sects in this world, nor will there be until the sect of Antichrist appear.

Since the Pagans and the Idolaters maintain that the creature is God and avow that there is a multitude of gods, and since both of these tenets are impossible for the reasons given, it is evident that their rites are erroneous. Hence the Emperor of the Tartars, who is named Mangu Khan, of whom mention was made above in what was said about Places in the World, assembled together Friar William with his Christians, and Saracens and Idolaters, in order that they might dispute in regard to their religions. Both the Christians and the Saracens at once proved the Idolaters in the wrong and the latter ceased to defend their religion, as Friar William states in his book mentioned above, addressed to the Lord Louis, illustrious King of France. And since the foul state of the Pagans and the Idolaters is evident, we shall not consider them further at present.

Although the Tartars worship the one true God, yet they nevertheless incline to idolatry. For they adore fire, believing that all things are purged and expiated through it. Hence they cause their sons and messengers and gifts to pass through fire. Similarly they honor the threshold of the house; for he who steps on the threshold is condemned to death. Hence likewise the Azotii did not step on the threshold of the temple of Dagon, as recorded in the fifth chapter of the first book of Kings. Moreover, they do not have any priests except philosophers who also lean to magic arts and employ the responses of demons, as we learn in the book on their manners and as we know for a fact. Hence they do not pass beyond philosophy either true or magical. But I showed above that philosophy is related to religion. For he who is perfect in the wisdom of philosophy bears the same relation to revealed religion as the untaught child to the aforesaid sage, as I explained above. It is evident, then, that the religion of the Tartars is not that which we are seeking, and the Tartars themselves acknowledge this. For the Emperor of the Tartars mentioned above acknowledges that the religion of the Christians is given by God to man and is the best. Moreover, the Tartars have their sons instructed in the Gospel and in the lives of the Fathers, and they accept holy water and the cross and Christian priests when they are sick and have recourse to them as to a last refuge, as we learn in the book on their manners, and as we ascertain through experience. For the empress wife of that emperor summoned Friar William and Christian priests, along with their holy water and cross, seeking their counsel, when she had become ill. It is an easy matter, therefore, for one who considers the facts to perceive that these religions should be eliminated. But the three others are more rational, namely, the religion of the Jews, of the Saracens, and of the Christians.

But that the Christian religion alone should be preferred we can decide in the following ways: first, from the views of the philosophers noted above in the relationship of mathematics to the Church, and in the first part of the Moral Philosophy. For in that section are presented their noble testimonies regarding the articles of the Christian faith, namely, regarding the blessed Trinity, Christ, and the Blessed Virgin, creation of the world, angels, souls, judgment to come, life eternal, resurrection of the body, punishment in purgatory, punishment in hell, and the like, which are contained in the religion of the Christians.

Philosophy is not, however, in such conformity with the religion of the Jews and with that of the Saracens; nor do the philosophers bear witness in their favor. It is evident then that, since philosophy is the forerunner of religion and disposes men to it, the Christian religion is the only one that should be adhered to. Moreover, the philosophers not only give way to the Christian religion, but they destroy the other two. Since Seneca, in the book which he composed against the religion of the Jews,* shows in many ways that it is most irrational and erroneous, so far as the carnal Jews are bound to the letter alone, according to their belief that it suffices for salvation.

The Saracen philosophers, moreover, find fault with their own law and decide that it will soon come to an end. For Avicenna in the ninth book of his Metaphysics proves Mahomet in error because he has set forth only physical delights and not spiritual ones. Albumazar, also, in the first book of the Conjunctions teaches that that sect will not last for more than 693 years, and already 665 have passed, and he holds that it can cease in less time, as I noted above in the section on Mathematics. And it is evident that the Tartars have destroyed nearly the whole dominion of the Saracens on the north, east, and south as far as Egypt and Africa. Thus it happens that their Caliph, who holds the position of a pope among them, was destroyed thirteen years ago, and Baldach, a city belonging to that Caliph, was captured with a great number of Saracens. But we can see this same thing in another way through the testimony of the Sibyls. For, as shown above, they proclaimed that Christ is God, and made known all the principal articles of the Christian faith. But they do not bear witness to the other religions: nay, they have said that this religion alone possesses salvation.

But there is also another way of considering these matters by reviewing the peculiar characteristics of the lawgiver and of the religions. The religion of the Jews does not in truth end with Moses, but looks forward to a Messiah who is Christ; although the Jews do not expect the Christ who is the head of the Christian Church, but another, who, they imagine, is still to come. Therefore it is evident that their law does not suffice, since they expect a more perfect lawgiver than Moses. But that this is Christ the Lord of the Christian can be proved by their law and by their authorities. For the prophecy of Daniel by a com-

* Quoted by St. Augustine, *De Civitate Dei,* VI, 11.

putation of years evidently extends up to Christ; for he came after that time. Moreover, at the advent of Christ the priesthood of the Jews ceased and also royal power among them, as they themselves cannot deny. But this was fulfilled at the time of Christ. For the kingdom of the Jews passed first to Herod and then to the Roman empire, as the histories of the Jews teach us. This information is contained in the books of Josephus, a Jew who narrated the destruction of the Jews by Titus and Vespasian. But Josephus says in the same work that in his time "Jesus Christ, a most holy man, appeared, if it is right to call him a man, concerning whom all things were fulfilled which our prophets said regarding him, as he himself gloriously bears witness." Likewise he himself says that, when the Lord was crucified, the voice of the heavenly powers was heard to say in Jerusalem, "Let us abandon our abode." Since Josephus is the chief authority on history among Jews, Greeks, and Latins, as all learned men and the sacred writers acknowledge, the fact is established by him that the Messiah promised by the Jewish law is the Lord Jesus Christ whom the Christians worship. We see this clearly from the fourth book of Esdras. For he states that God the Father thus spoke, "For my son Jesus shall be revealed with those that shall rejoice with him and that remain four hundred years: and after four hundred years my son Christ shall die, and the world shall be changed."* But so many years passed from Esdras to Christ. Likewise in the book of the twelve Patriarchs there is teaching of the clearest nature regarding Christ. For each Patriarch gave his tribe definite proof in regard to Christ, and there was corresponding fulfillment.

And if it be said that these books belong to the Apocrypha, that is, of unknown authorship, this fact does not destroy their veracity, since these books are accepted by Greeks, Latins, and Jews. For the blessed Ambrose in his homily on the Gospel of Luke quotes the authority of Esdras. And the Church in its office employs many other passages from that book. For there are many books current among Latins, Hebrews, and Greeks, the authorship of which is uncertain: nay, we Latins are certain of the authorship of only a few books, and in many cases we are in error. For when we think that Avicenna composed the book on the Heavens and the World which is commonly

* *2 Esdras,* VII, 28, 29.

attributed to him, we are in error. And many commentaries are thought to be the work of Averroës, but are to be considered rather those of Alpharabius, as, for example, the commentary on the Physics of Aristotle. And in theology the authors of Ecclesiasticus and of Wisdom are unknown, some regarding them as works of Solomon, others of Philo, while still others believe them the works of some one else. Hence we cannot urge as an objection that a book is of unknown authorship, provided that the book has the sanction of the majority of learned men.

Moreover, if we consider the Jewish religion literally as it is presented, it is abominable, irrational, and intolerable. It is abominable because the slaughter of those rams, calves, bulls, and goats by the priests in the temple is horrible and most unclean; and it is wholly irrational because it contains abuses without end, as is apparent to every one; and it is intolerable because the burden of it is without limit; nor were the Jews able to carry out fully the ceremonials of that law; as the blessed Peter says, "Neither we nor our fathers." And this is evident from the series of Scripture bearing on the law; since owing to the intolerable burdens they turned to idolatry and to strange gods from age to age. And God himself bears witness that ceremonies of this kind did not please him, as we learn from many Psalms, where the Psalmist says, "Sacrifice and oblation thou hast not desired, but mine ears hast thou opened." And in another Psalm, "I shall not accept from your house calves nor he-goats from your flocks." And in the first chapter of Isaiah and elsewhere the same fact is made clear. Hence lest they should become wholly Idolaters, God kept them busy in ceremonies of this kind. In the second place, the law promises only temporal and physical blessings; but the perfect religion promises eternal and spiritual blessings, as is evident from what precedes.

In the same way we may consider the religion of the Saracens and their lawgiver. In fact, Mahomet himself in the Koran, which is the book of his law, says that Christ was born of the Virgin Mary by the influence of the Holy Spirit without a human father, and that he is the greatest prophet of God. Hence Christ takes precedence of Mahomet, and for this reason the law of Christ must take precedence. Since, therefore, there should be only one law in the world, as we have proved, it will be that one which is better and more worthy; and this is the religion of Christ. Moreover, the philosophers of the Saracens find fault with their own religion. But the religion of truth, although it is above philosophy, is not, however, contrary to it, nor is it called in question by philosophy but is approved by it, as I have shown in regard to the religion of Christ in the section on Mathematics and in the first part of the Moral Philosophy. Likewise they point out that this sect will be destroyed and the time for its destruction is made known. But the religion of life eternal will always continue, even up to life eternal, because it leads to eternal life; therefore this religion is not that of Mahomet. Moreover, the lawgiver of this sect was most depraved in his life; for he was a very wicked adulterer, as we find recorded in the Koran. For every beautiful woman he took forcibly from her husband and violated her. But adultery is contrary to every law and religion, for this it is that introduces discord among citizens; and likewise causes false heirs to appear, and therefore destroys the laws of the state.

But if we examine more closely these three sects, we can see more clearly which of them should be preserved. We must assume as a fundamental principle in this consideration that the histories of all nations are to be accepted on an equal footing when we take up the form in which the disputation is to proceed. For if Christians should refuse to accept the histories of the Saracens and of the Jews, they with equal right will reject the histories of the Christians. Let us uphold for the sake of argument the histories of these sects, in order that we may see which sect should prevail in this matter. It is said then in Gospel history that among those born of women none has arisen greater than John the Baptist; but this same John says that he is not worthy to unloose the latchet of the shoe of Jesus Christ: and therefore neither Mahomet nor Moses is worthy, since there is no comparison of them with Christ, nor are their laws comparable with his. Moreover, Alpharabius in his book on Sciences shows the ways in which religions are proved, and he presents two of these as more noteworthy than the others. One test is that there should be a perfect lawgiver to whom prophets preceding and following bear witness. But prophets preceding and following do not bear witness to Moses nor to Mahomet. Therefore Christ alone is the legislator whom we are seeking.

That the earlier prophets bear witness to Christ, he himself testifies, saying in the last chapter of Luke, "As is written in the law of Moses and in the Prophets and in the Psalms concerning me." Also in the second chapter of John, "For Moses wrote of me"; and we have further evidence in the prophecy of Daniel mentioned previously; also in Isaiah who says, "Lo a virgin shall conceive and bring forth a son"; and Mahomet is in accord with Isaiah when he says that Christ is the son of the Virgin. Josephus also testifies that all things were fulfilled in Christ that were said of him by the prophets. But the fact is made apparent not only in this way, but also through the prophets that followed him, namely, the numerous Sacred Writers, who not only gave testimony to Christ, but also testify that the earlier prophets spoke all things concerning him.

But these later holy prophets are to be believed for six reasons. The first and weightiest reason for believing them is their perfect sanctity, since Cicero in the Topics, in which he considers the question of authority, rightly assigns this first and *per se* to virtue; because a good and holy man does not wish to lie. The second condition is an ineffable wisdom by which they have known how to avoid all error. For they have knowledge not only of things present in their sight, but also of things that are absent; and not only of things physical, but also of things spiritual as affecting the consciences of men. Filled with the prophetic spirit these men had knowledge of the present, past, and future, as the histories of the Christians inform us in numberless instances. The third condition was the indescribable power of miracles with which these sacred writers were endowed. But Alpharabius says that men should believe because of miracles. For this is the second method offered by Alpharabius of proving a religion, namely, by means of miracles. The fourth condition found in these sacred writers was their steadfastness even unto death in defense of their doctrine, a quality they would not have exhibited had they not known that there was supreme truth in their teaching. The fifth condition is found in the fact that, while living in different parts of the world, they promulgated the same sentiment, although no one of them received the doctrine from another. Therefore they must have spoken by the inspiration of God, and hence without error. The sixth ground is found in the fact that they were private citizens and laymen, such as Peter and others, and poor, humble, downtrodden men: and yet their view prevailed over their emperors, philosophers, and pontiffs. It is evident, therefore, for these reasons, that those men are worthy of belief. But those men say that the earlier prophets bear witness to Christ: therefore this is true. Moreover, they themselves bear witness to Christ in numberless instances: nay, their whole life, not only their doctrine, was a confirmation of the Christian law. Since, therefore, the earlier and the later prophets in countless numbers have borne witness to Christ, as have also Moses and Mahomet, while lacking such witness themselves, as is evident from the histories of the Jews and the Saracens, Christ must, therefore, be the perfect lawgiver.

This same fact is proved by the second method of Alpharabius in regard to proving religions, namely, by miracles. For although Moses did great wonders in fact, and Mahomet in appearance only and by fraud, yet they are not to be compared with the miracles of Christ. For conceding all their miracles and all the histories, we shall find in the Gospel history of John in the last chapter this statement: "But there are many other things which Jesus did, the which, if they should be written every one, I suppose that even the world itself could not contain the books that should be written;" owing to the number and greatness of the miracles, a statement that cannot be accepted with regard to the histories of Moses and Mahomet. He likewise says, "But that ye may know that the Son of man hath power to forgive sins (he saith to the sick of the palsy), I say unto thee, arise." But to forgive sins and heal the soul is a miracle boundless in its nature, and greater than the performance of an infinite number of physical miracles. And on this fact rests a very strong argument in favor of Christ. For no one can forgive sin unless he be God; therefore Christ was God. Moreover, all histories confess that he is God, and these histories were given to the world by sacred writers whose sixfold qualification was stated above, who could not lie. Therefore the fact stated must be true. The earlier prophets moreover confess that he is God. Hence Isaiah says, "He shall be called Emanuel, that is, God with us." And David says, "Thy seat, O God, is for ever and ever." And there are very many other passages of similar import according to the exposition of the prophecies made by the later Sacred Writers who cannot err. Since then Christ is God, which is not true of Mahomet and Moses according to the testimony of even Jews and Saracens, it is evident that he and he alone is the perfect lawgiver, and that there should be no comparison of Moses and Mahomet or of any one else with him.

Moreover, this same fact is obvious from a comparison of the laws. For all sanctity and perfection of life are taught in the Christian law, and the impurity of sin is not permitted. But with Mahomet many sins are allowed, as is evident in the Koran, and no perfection of life is observed, since they are absorbed in sensual pleasures because of their polygamy. Similarly the Jews above all else were eager for offspring. For the barren woman lay under the curse of the law, and virginity was not commended by them, nor voluntary poverty, nor the submission of one's will to that of another, three things indicative of the utmost perfection. For no one can doubt that virginity is a most pure and holy state, and voluntary poverty is approved by all philosophers, as we learn above quite clearly from the third part of the Moral Philosophy. But to submit oneself entirely to the will of another is the supreme and the most difficult test of all, as every one knows. And therefore since these virtues are taught in the law of Christ, and not in the law of Mahomet nor in that of Moses, it is evident that the law of Christ is altogether superior. Moreover, by the law of Christ we are assured with reference to God and the future life of the good and of the evil, and likewise with reference to the blessed Trinity and other spiritual truths. But these verities are not presented in the law of Moses nor in that of Mahomet, as is evident. Therefore the law of Christ should be preferred to the others. But it is such a law that should be published throughout the world, as we have shown above. Therefore this law must be the Christian law.

Since we have proved that the Christian faith is to be accepted, all its articles must evidently be accepted also. For if the whole is approved, each part of it also must be conceded. Since, however, a certain article seems burdensome to human frailty, for this reason some deny it, others are doubtful in regard to it, still others accept it with difficulty, some look upon it as hard, others imperfectly understand it, a few grasp it easily with full peace and sweetness of mind. I refer to the Sacrament of the Altar, as regards which, according to the Apostle, many are weak and sickly among you and many sleep. Therefore in regard to it I have thought it proper to undertake to show that this sacrament is a most true and certain one, which ought to be most ardently desired and most earnestly sought for, which we should most fervently await, with all reverence worship, with joy and devotion retain, with most certain faith contemplate. If this sacrament did not exist, we should seek to obtain it from God with the utmost diligence, and to purchase it not with corruptible gold and silver but with life and death. If it should exist and be unknown, it should be received as soon as it is offered, nay, it should be snatched from the mouth of him who proffers it, nay, without a teacher it ought to be known and loved above all things, so that he that is faithful should rejoice in knowing and possessing nothing else; since this being known, all things are known that pertain to salvation; this being unknown, all things are unknown; this being possessed, all things are possessed, and he who lacks this is proved to possess nothing. Since, therefore, it is necessary for every one to understand this, although notwithstanding many are ignorant of this sacrament, while others on being converted to the faith are more disturbed by this particular doctrine than by any other, and many Christians, who strive to judge with their human sense in regard to divine things, waver or are imperfect; it is best to consider how both unbelievers and Christians can be most surely convinced in regard to this verity. I shall therefore give the main points of this persuasive reasoning according to the grace given me.

Therefore that this fact is most certain, namely, that the Lord Jesus is present in this sacrament as true God and true man, is shown first by the assent of the whole Christian Church; and since that Chuch has been proved to be the true one, and since this sacrament is one that this Church professes, it is evident that it will contain the truth. Moreover, in addition to the general proofs applicable to the whole religion, this blessed truth possesses its own particular modes of proof just like the whole religion. For this sacrament is clearly taught by holy Scripture in the Gospels and by the Apostle; "Your fathers did eat manna and are dead; he that eateth me liveth because of me; and, I am the living bread which came down from heaven;

if any man eat of this bread, he shall live for ever." And in the sixteenth chapter of the book of Wisdom, "Thou feedest thy people with angel's food"; and below, "For thy substance and the love thou hast for thy sons thou showest." This is the bread on which the angels desire to look, according to the Apostle Peter. And therefore it is not that manna of the ancient people, since that is not the food of angels; when he says, "Thy substance, etc.," we are evidently not to understand this of the ancient manna. And the Apostle who first persecuted the truth of Christ later confesses that he received this sacrifice from the Lord. Moreover, this truth is made manifest not only by Holy Scripture, but by all the Sacred Writers, who bear witness in regard to this article, who could not lie because of the conditions I have enumerated above regarding them. But this is evident by common agreement and definition of all Catholic doctors, masters, and readers in the law of God. For all agree in their testimony that Christ is present in this sacrament.

In the next place, there are an innumerable number of miracles which are found in the Sacred Writers and in the histories. But I shall here record two, which have not yet been chronicled, of undoubted authenticity and subject to no doubt. For a certain devout matron desiring to have offspring had her hopes excited by a heretical bishop, who promised her a son. Summoning a necromancer who knew how to evoke demons, the woman, the bishop, and the necromancer met at a secret place. After the necromancer had gone through all the rites to his satisfaction in the matter of circles and incantations, a demon appeared in the guise of a boy wearing a crown, who asked them what they wanted. Both the necromancer and the bishop asked the woman what she desired. But while she terrified and seized with the fear of God made no reply, the demon retired to a certain wall near by, and taking the crown from his head worshiped and bent the knee. After doing this he returned to them and again asked what they wanted. And when the evil men kept asking the woman, and she did not venture to speak from fear, the devil returned to the place where he had first adored, and bent his head, but did not lay aside his crown nor genuflect. On his return to them they inquired why he had acted as he had on the first and on the second occasion. He replied that a man lay sick in a house beyond the wall, and a priest brought the body of Christ to him; and "I was compelled," said the demon, "to lay aside the crown from my head and worship and genuflect, because it is written, At the name of Jesus every knee should bow of things in heaven, and things in earth, and things under the earth. And when I returned the second time, the priest had given the body of Christ to the sick man, and was carrying the empty vessel, and therefore I did not lay aside my crown nor genuflect, but I merely bent my head and adored out of respect for the priest who carries the body of Christ and for the sacred vessel." Then the heretical bishop became a true Christian, and immediately began to preach the faith of Christ, and to confound heretical evil. Oh, how true a proof and how noble a one this is, adequate praise of which cannot be expressed! Moreover, many are alive who know for a certainty the truth of this miracle.

It came to pass, moreover, that in the Franciscan order there was a brother who for many years was unable to partake of this sacrament, nor would he have stood before the altar with his brothers when they received the sacrament, had any one offered him the whole world; and yet he himself was ignorant of the reason of his own inability to receive the sacrament. And as it pleased God, a certain wise and holy brother said to him that he was either in cruel mortal sin, or that he was not baptized. The man himself, since after careful self-examination he had no consciousness of mortal sin, began to have doubts regarding his baptism. Inquiring of his parents in regard to the manner of his baptism, he learned that he and another boy were presented to the priest at the same time, who baptized the other child and innocently neglected to baptize him. After learning this fact he had himself properly baptized, and afterwards received the Sacrament of the Altar like any other man. From this it is evident that the truth of this sacrament is proved by baptism, and the truth of baptism by this sacrament. And therefore, since the Church holds that baptism must first be performed in order that the rest may follow, it is quite evident by this miracle that what the Church believes in regard to this most noble sacrament is true.

But besides these methods of proof there are reasons of the following nature. For just as the Creator is related to the creature and as far as he is related to the being of nature, so is the

Redeemer related to the redeemed or to those who are in the state of grace, and more abundantly so. For it is more to re-create than to create; but it is in accordance with the law of the Creator's majesty that he be omnipresent to the creature, nor is there any creature that does not share his presence. Therefore it will be in accordance with the power of the infinite Re-creator that he be present to every one who has been re-created and is in the state of grace. But his presence is not granted to us except in this sacrament. Therefore of necessity this sacrament must exist.

Likewise the boundless need of the creature requires the presence of the Creator. For the creature otherwise would fail, because it would tend to a state of non-existence, unless it is supported by the presence of majesty; as philosophy and theology agree. But so great, nay, greater, is the need of the re-created for the Redeemer; therefore the Re-creator must be in the re-created and be present to him, if he is to remain in the state of re-creation, that is, in the state of grace. For as the state of nature is related to the Creator, so is the state of grace to the Re-creator. And, therefore, just as the creature would fall into a natural state of non-existence without the presence of the Creator, so would the re-created fall into a state opposed to grace unless he be supported by the presence of the Redeemer.

Moreover, the power of Christ is infinite, since he is God, and his wisdom is infinite, and his goodness likewise. Wherefore if from the infinity of his power he is able to do this, and from the infinity of his wisdom he knows how to do this, therefore from the infinity of his goodness he wishes to do this; because it is characteristic of the best to do what is best. And therefore since he possesses infinite goodness he must do infinite good, and as far as possible this must be received from him. But this good can be received by the re-created, that is, one in the state of grace; therefore it must necessarily happen to him.

And yet again this is evident by a comparison. For any creature in the state of nature receives from the Creator as much as he can receive in his condition: as Cicero says in his book on the Nature of the Gods,* "All parts of the world are so constituted, that they cannot be excelled in their utility, nor can they

* *De Natura Deorum,* II, 34.

be more beautiful in appearance." Therefore similarly whatever is re-created, being in the state of grace, will receive from the goodness of the Re-creator all that it is capable of receiving. But it can receive the blessing of his substance; therefore it will receive this blessing, as the book of Wisdom says, "Thy substance hast thou given them." And therefore every man in the state of grace must receive his Savior as the Church has appointed.

For the rest original sin cannot be removed without offering this sacrifice to God. But mortal sins are multiplied daily. Therefore it is necessary that this sacrifice make satisfaction to God the Father for the sins of the world. Likewise God assumed this humanity so as to permit his blood to be shed by dying for us, and thus he redeemed us. But this is more than feeding us with this flesh and this blood. Therefore if he has wished of his goodness to do what is more, much rather will he wish to do what is less, since it is useful, convenient, and necessary for us.

These arguments and the like can give conclusive proof that this sacrament is true, so that no one can nor should gainsay it. But mere assent is not the sole requirement; we should accept this truth with the greatest ease, nay, we should accept it more easily, more willingly, more devotedly than any other truth that can be submitted to us in this life; nay, we should so accept it that in this life we may be at peace just as in the sweetness of life eternal; that we should wish for nothing except this; and if we were without it, we should consider that we neither knew nor possessed anything with regard to this glorious truth by which we are deified and taken into life eternal.

As regards the ease with which we may accept it there are several modes of approach. The first method is the consideration of what is contained in this sacrament; the second is the mode of existence; the third is the manner in which it is made available; and there are certain other modes, as I shall note. In this sacrament are contained the culmination of majesty and glory and fullness of salvation and perfection of beauty; truths that are evident since God and Man, glorified and one, are here existent in one person. For God in accordance with his own nature possesses infinite majesty and goodness and beauty, as we see from the topics discussed above. For whatever is in him is infinite. Similarly man, received into the glory, goodness, and the beauty of God, shares the threefold infinitude by reason of the union of an indivisible person. Therefore there is full infinitude of glory, salvation, and beauty in the content of this sacrament.

The same is true in the mode of its existence. For since this Person in his divine and human nature of the same majesty is similarly present here and in heaven, so in every church, and in every one in the state of grace is he received similarly and at the same time. This is a mark of infinite power, because he is not limited to this nor to that, and this attests his infinite majesty. Similarly the fact that Christ as a whole is present in every part of the Host is a mark of infinite power; because he is not limited to a part, nor are his parts divided in the parts of the Host. There is therefore a threefold reason for his majesty in the mode of his existence. Moreover there is a fourth reason for his majesty evident in this sacrament. For his humanity, although in itself it is a creature, yet it there transcends the laws governing the creature, nor has it the mode of existence required of the creature, namely, that it confine itself within the limits of space and physical locality, but it is free in these respects like Deity; and therefore so far as its mode of existence is concerned it is raised to the divine mode of existence, so as to transcend every law of existence that trammels the creature. And this can be effected by the infinite power of God, just as the assumption of humanity into a union of persons was effected by this same power.

In like manner there is in this sacrament the infinitude of our salvation proceeding from the infinite goodness. For in conferring upon us the power at will through our ministry of causing the Savior to be present in a miraculous manner in this sacrament, God concedes more to us than if he granted any one of us the power to make a universe of himself, nay, more than if he granted him the power to make an infinite number of universes. For the advantage of an infinite number of universes to us is nothing in comparison with that of this sacrament. We need not say, therefore, with the Prophet, "If he delay, await him." Nor with that other prophet, "Would that thou wert breaking through the clouds and descending"; because God has placed in our hands the power to cause him to be present among us when we so desire, according to the dispensation of his divine goodness by which we say, "Do this in remembrance of me." For he commanded the clouds, and opened the doors of heaven, and rains upon us the manna of eternal life when we wish to utter five words. Wonderful goodness of God! which, since nothing is easier for us than to form a word, has granted us the power by merely a word to cause the Lord our Savior to be with us at our will. For we are not required to ascend into heaven, to pass over seas, nor to plow nor reap for this bread, nor to plant vineyards nor tread out the wine for this drink, but we are required to utter with ease five words, in order that our God and Lord may be with us, who is blessed for ever more. For this reason it is said in the thirtieth chapter of Deuteronomy, "For this commandment which I command thee this day, it is not hidden from thee, neither is it far off. It is not in heaven, that thou shouldest say, Who shall go up for us to heaven, and bring it unto us, that we may hear it, and do it? Neither is it beyond the sea, that thou shouldest say, Who shall go over the sea for us, and bring it unto us, that we may hear it, and do it? But the word is very nigh unto thee, in thy mouth, and in thy heart, that thou mayest do it." Therefore, in order that we may have eternal life, we need only believe in our heart, and utter a brief sentence compressed into the number of five words. For this number in its distinct comprehension of various things is better than all other numbers, according to Aristotle in his book of Secrets.

Moreover, that this is veiled from our sense is due to our inability to comprehend him. For we would not be able to sustain with our senses the majesty of God, but we should fail completely, owing to our reverence and devotion and wonder; just as the Apostles after the resurrection were not able to endure the sensible presence of the Lord. Even the blessed Dionysius the Areopagite was unable to bear the presence of the Virgin Mary after the ascension of the Lord. Nay, when he crossed from Greece to the Holy Land, that he might see the mother of the Lord, on entering the place where the blessed and glorious Virgin prayed, he fell as though dead, crying out that he should be led forth, freely confessing that no believer ought to look upon the glorious Virgin out of reverence for her. Therefore far less able would any one be to bear the sensible presence of the Lord; and this we prove by experience. For those who exercise themselves diligently in faith and in love of this sacrament are unable to sustain the devotion which springs from pure faith, but give way to a flood of tears and the mind is dissolved wholly in the sweetness of devotion, raised above itself, not knowing where it is or its own condition. Wherefore it would be impossible for a believer to sustain the sensible presence, and therefore God has arranged with the utmost goodness for us that the sensible vision should be veiled from us.

Moreover, we could not sustain it from horror and loathing. For the human heart could not endure to masticate and devour raw and living flesh and to drink fresh blood. And therefore the infinite goodness of God is shown in veiling this sacrament. But its advantage in other respects is very great, so that we may learn to aspire to real blessings. For all our principal blessings and evils are invisible; namely, God, the angels and the saints, eternal life, virtue and grace, demons, purgatory, and hell. And this could not belong to the class of our principal blessings unless it were veiled from our senses, for which reason the Lord says, "Unless I go away, the Comforter will not come to you": to the end that we should know that we should cleave to invisible blessings which do not appear; according to the Apostle, "Those things that exist, that is, are apparent, let us reckon as non-existent, and those things that do not appear let us reckon as existent and as good." It was necessary also for us for our infinite advantage, that this infinite good should be veiled from our senses, and that we should know this means of salvation by the intuition of our mind alone.

In the third place, we gain facility in believing by the use of this sacrament. For if the mode of its existence possesses an infinitude of salvation, majesty, and beauty, much more does the use itself of this sacrament in which we are united to God. Three things are here to be considered. For we are nourished by the glorious body and we are permitted to drink of his glorified blood, nay, we are renewed by the whole Christ, God and Man. But this is eternal life and infinite salvation as far as this mortality can sustain it. For God is not satisfied to feed us sensibly with his creatures and to give us drink, but he cherishes us spiritually with himself and with his flesh and blood. Then from participation in God and Christ we become one with him and one with Christ and are gods, as Philosophy in the third book of the Consolations concludes, because from participation in the Deity many become gods, although he is one by nature. And therefore by participation in Christ we become Christ's. And for this reason the Scripture says, "I have said, ye are gods"; and elsewhere, "Do not touch my Christs." And what more can a man seek in this life?

Here the manuscript breaks off abruptly.

INDEX

Index

Index

Index

Index

Index

Printed in Australia
Ingram Content Group Australia Pty Ltd
AUHW021852260224
390945AU00003B/4